Achieving sustainable cultivation of sorghum

Volume 1: Genetics, breeding and production techniques

It is widely recognised that agriculture is a significant contributor to global warming and climate change. Agriculture needs to reduce its environmental impact and adapt to current climate change whilst still feeding a growing population, i.e. become more 'climate-smart'. Burleigh Dodds Science Publishing is playing its part in achieving this by bringing together key research on making the production of the world's most important crops and livestock products more sustainable.

Based on extensive research, our publications specifically target the challenge of climate-smart agriculture. In this way we are using 'smart publishing' to help achieve climate-smart agriculture.

Burleigh Dodds Science Publishing is an independent and innovative publisher delivering high quality customer-focused agricultural science content in both print and online formats for the academic and research communities. Our aim is to build a foundation of knowledge on which researchers can build to meet the challenge of climate-smart agriculture.

For more information about Burleigh Dodds Science Publishing simply call us on +44 (0) 1223 839365, email info@bdspublishing.com or alternatively please visit our website at www.bdspublishing.com.

Related titles:

Achieving sustainable cultivation of sorghum Volume 2: Sorghum utilization around the world
Print (ISBN 978-1-78676-124-8); Online (ISBN 978-1-78676-126-2, 978-1-78676-127-9)

Achieving sustainable cultivation of maize Volume 1: From improved varieties to local applications
Print (ISBN 978-1-78676-008-1); Online (ISBN 978-1-78676-010-4, 978-1-78676-011-1)

Achieving sustainable cultivation of maize Volume 2: Cultivation techniques, pest and disease control
Print (ISBN 978-1-78676-012-8); Online (ISBN 978-1-78676-014-2, 978-1-78676-015-9)

Achieving sustainable cultivation of wheat Volume 1: Breeding, quality traits, pests and diseases
Print (ISBN 978-1-78676-016-6); Online (ISBN 978-1-78676-018-0, 978-1-78676-019-7)

Achieving sustainable cultivation of wheat Volume 2: Cultivation techniques
Print (ISBN 978-1-78676-020-3); Online (ISBN 978-1-78676-022-7, 978-1-78676-023-4)

Chapters are available individually from our online bookshop: https://shop.bdspublishing.com

climate•SMART•publishing
IN AGRICULTURAL SCIENCE

BURLEIGH DODDS SERIES IN AGRICULTURAL SCIENCE

NUMBER 31

Achieving sustainable cultivation of sorghum

Volume 1: Genetics, breeding and production techniques

Edited by Professor William Rooney
Texas A&M University, USA

burleigh dodds
SCIENCE PUBLISHING

Published by Burleigh Dodds Science Publishing Limited
82 High Street, Sawston, Cambridge CB22 3HJ, UK
www.bdspublishing.com

Burleigh Dodds Science Publishing, 1518 Walnut Street, Suite 900, Philadelphia, PA 19102-3406, USA

First published 2018 by Burleigh Dodds Science Publishing Limited

Library of Congress Control Number: 2017956487

British Library Cataloguing in Publication Data
A catalogue record for this book is available from the British Library

ISBN 978-1-78676-120-0 (print)
ISBN 978-1-78676-122-4 (online)
ISBN 978-1-78676-123-1 (online)
ISSN 2059-6936 (print)
ISSN 2059-6944 (online)

Typeset by Deanta Global Publishing Services, Chennai, India
Printed by Lightning Source

Contents

Part 3 Sorghum production techniques

Series list

Title	Series number
Achieving sustainable cultivation of maize - Vol 1 From improved varieties to local applications Edited by: Dr Dave Watson, CGIAR Maize Research Program Manager, CIMMYT, Mexico	001
Achieving sustainable cultivation of maize - Vol 2 Cultivation techniques, pest and disease control Edited by: Dr Dave Watson, CGIAR Maize Research Program Manager, CIMMYT, Mexico	002
Achieving sustainable cultivation of rice - Vol 1 Breeding for higher yield and quality Edited by: Prof. Takuji Sasaki, Tokyo University of Agriculture, Japan	003
Achieving sustainable cultivation of rice - Vol 2 Cultivation, pest and disease management Edited by: Prof. Takuji Sasaki, Tokyo University of Agriculture, Japan	004
Achieving sustainable cultivation of wheat - Vol 1 Breeding, quality traits, pests and diseases Edited by: Prof. Peter Langridge, The University of Adelaide, Australia	005
Achieving sustainable cultivation of wheat - Vol 2 Cultivation techniques Edited by: Prof. Peter Langridge, The University of Adelaide, Australia	006
Achieving sustainable cultivation of tomatoes Edited by: Dr Autar Mattoo, USDA-ARS, USA & Prof. Avtar Handa, Purdue University, USA	007
Achieving sustainable production of milk - Vol 1 Milk composition, genetics and breeding Edited by: Dr Nico van Belzen, International Dairy Federation (IDF), Belgium	008
Achieving sustainable production of milk - Vol 2 Safety, quality and sustainability Edited by: Dr Nico van Belzen, International Dairy Federation (IDF), Belgium	009
Achieving sustainable production of milk - Vol 3 Dairy herd management and welfare Edited by: Prof. John Webster, University of Bristol, UK	010
Ensuring safety and quality in the production of beef - Vol 1 Safety Edited by: Prof. Gary Acuff, Texas A&M University, USA & Prof. James Dickson, Iowa State University, USA	011
Ensuring safety and quality in the production of beef - Vol 2 Quality Edited by: Prof. Michael Dikeman, Kansas State University, USA	012
Achieving sustainable production of poultry meat - Vol 1 Safety, quality and sustainability Edited by: Prof. Steven C. Ricke, University of Arkansas, USA	013
Achieving sustainable production of poultry meat - Vol 2 Breeding and nutrition Edited by: Prof. Todd Applegate, University of Georgia, USA	014
Achieving sustainable production of poultry meat - Vol 3 Health and welfare Edited by: Prof. Todd Applegate, University of Georgia, USA	015

Acknowledgements

We wish to acknowledge the following for their help in reviewing particular chapters:

- Chapter 1: Prof Hari D. Upadhyaya, International Crops Research Institute for the Semi-Arid Tropics (ICRISAT), India; Dr Monique Deu, CIRAD, France; and Dr Moses Muraya, Chuka University, Kenya
- Chapter 2: Dr K. Visarada, Indian Institute of Millets Research, India
- Chapter 3: Dr Bob Klein, ARS-USDA, USA
- Chapter 4: Dr Ramasamy Perumal, Agricultural Research Center – Hays, Kansas State University, USA
- Chapter 5: Prof David Jordan, Queensland Alliance for Agriculture and Food Innovation, University of Queensland, Australia; and Dr Hai-Chun Jing, Institute of Botany, Chinese Academy of Science, China
- Chapter 6: Dr Gilles Trouche, CIRAD, France
- Chapter 7: Prof Gary Peterson, Texas A&M University, USA
- Chapter 8: Dr Vincent Vadez, International Crops Research Institute for the Semi-Arid Tropics (ICRISAT), India
- Chapter 9: Dr Jurandir Magalhaes, Embrapa, Brazil; and Dr Elke Vandamme, AfricaRice, Tanzania
- Chapter 10: Dr Charles Miller, Chromatin Inc, USA
- Chapter 11: Prof Calvin Trostle, Texas A&M University, USA
- Chapter 12: Dr Kraig Roozeboom, Kansas State University, USA
- Chapter 13: Dr William E. Stevens, University of Missouri, USA
- Chapter 14: Dr Michael Brewer, Texas A&M University, USA
- Chapter 15: Dr Louis Prom, USDA-ARS, USA; and Prof. Clint Magill, Texas A&M University, USA
- Chapter 16: Dr Louis Prom, USDA-ARS, USA; and Prof. Clint Magill, Texas A&M University, USA
- Chapter 17: Prof. Phillip Stahlman, Kansas State University, USA
- Chapter 18: Prof. John Leslie, Kansas State University, USA; and Prof. Erick Cheruiyot, Egerton University, Kenya

Introduction

Sorghum is one of the world's major cereals, cultivated in the semi-arid tropics for a growing range of uses, from food and feed to biofuel. Like other crops it faces the need to meet rising demand whilst reducing its environmental impact and adapting to the challenges of climate change. These challenges are addressed in the two volumes of: *Achieving sustainable cultivation of sorghum*:

- Volume 1 Genetics, breeding and production techniques
- Volume 2 Sorghum utilization around the world

The chapters in this volume (Volume 1) review key developments in breeding, from conventional to marker-assisted techniques, as well their application in developing higher-yielding, more stress-resistant varieties. The volume also discusses key research on making cultivation more efficient and sustainable, including ways of optimising nutrition, water, pest and weed management

Part 1 Sorghum the crop species

The first part of the book explores sorghum genetics. The theme of Chapter 1 is the classification and genetic diversity of sorghum, and the chapter presents a revised classification of sorghum. It sets out a generalized classification of cultivated sorghum from which a more precise working group can be defined to explain the variation within the chosen race. The classification is presented in a format that would be useful in breeding programs as breeders look to incorporate exotic germplasm into their breeding lines and hybrids.

Chapter 2 builds on the previous chapter's focus on genetic diversity by examining the comparative and evolutionary genomics of sorghum. Sorghum has a relatively small diploid genome, and retains much of the chromosomal structure and organization of its common ancestor. It also has multiple origins of domestication in different regions of Africa, resulting in a highly diverse germplasm well adapted to drought and other challenging environmental conditions. This makes sorghum ideal for dissecting the architectures of complex yet desirable traits. The chapter discusses what we know about the evolution and structure of the sorghum genome, the impact of domestication and the genetic architecture of key traits such as C_4 photosynthesis, drought and aluminium tolerance.

Maintaining the focus on genetic diversity, Chapter 3 looks at the challenge of ensuring the genetic diversity of sorghum. Sorghum is a staple food crop for millions of the poorest and most food-insecure people in the semi-arid tropics, and ensuring its diversity and conservation are therefore important for global food and nutritional security. The chapter discusses the taxonomy of sorghum, sorghum diversity and conservation, the factors shaping sorghum diversity, genomic prediction of sorghum landraces adaptation, and the use of cultivated and wild genes of sorghum to broaden the genetic base of sorghum cultivars.

Part 2 Sorghum improvement techniques

Part 2 reviews advances in conventional and marker-assisted breeding and their use to develop high-yielding and stress-resistant varieties. The subject of Chapter 4 is conventional breeding techniques for sorghum. The chapter describes a number of standard field operations, such as crossing, emasculation and harvesting, followed by a discussion of various methods of selection and topcrossing. After describing these classical breeding methodologies, the chapter evaluates new directions in sorghum breeding, such as the use of molecular markers and high-throughput means of phenotyping plants. The chapter concludes that, although classical breeding methodologies will continue to be a requirement in future, sorghum breeders will need to make use of evolving technologies if they are to be successful.

Chapter 5 moves beyond conventional breeding techniques to consider advanced technologies to accelerate sorghum breeding. The chapter provides an overview of the different types of molecular markers and describes techniques of marker-assisted selection in sorghum breeding, focusing on genomic selection, wide crosses and mutagenesis breeding. The chapter examines emerging technologies in the field, including doubled haploids, transgenic technology, genome editing and high-throughput phenotyping.

Complementing the preceding chapters, Chapter 6 looks at broader issues in harnessing genetic/genomic resources to improve sorghum yields. In spite of sorghum's inherent yield potential, its remarkable adaptation to marginal environments and its extensive use in various food applications by smallholders in developing countries, sorghum productivity has remained low. The chapter examines the major hurdles undermining sorghum productivity, including the constraints faced by smallholders.The chapter addresses how the gap in yield potential might be bridged in such areas as optimizing plant architecture through unlocking heterosis to produce more vigorous hybrids.

Chapter 7 examines sorghum breeding for biotic stress tolerance. Sorghum breeding is poised to take advantage of remarkable advances in genomics and molecular technologies. While, as Chapter 5 points out, classical breeding methods have been successful in identifying and incorporating resistance to biotic stresses, Chapter 7 focusses on the application of the new tools and technologies that can simplify and accelerate breeding of improved sorghum varieties. The chapter documents current genetic sources of resistance which can be exploited by rapidly developing techniques such as gene editing and gene transfer. The chapter reviews recent developments in dealing with major biotic stresses of sorghum, including diseases (fungal, bacterial and viral), insect pests, *Striga* (witchweed), and nematodes. The chapter also considers the structure and function of resistance or 'R' genes and their deployment.

The subject of Chapter 8 is sorghum breeding for abiotic stress tolerance. Globally, domesticated sorghum is often grown on marginal lands and in environments exposing the plant to temperature extremes and water deficits. Although sorghum exhibits better abiotic stress tolerances than many crops, significant sorghum yield losses are experienced annually in response to unfavorable environments. The chapter focuses on methods for identifying diversity among germplasm collections. It then explores ways of exploiting germplasm diversity to improve soybean responses to temperature and water-deficit stress. The chapter looks at the challenge of identifying quantitative trait loci (QTL) for post-flowering drought tolerance and presents a number of case studies in improving drought tolerance.

Turning to another specific application of breeding technologies, Chapter 9 examines sorghum's tolerance of low phosphorus soil conditions. Many sorghum-growing areas in Africa are affected by phosphorus (P) scarcity. In these areas small-scale farmers depend on sorghum for food security and income, but have difficulty accessing or paying for P fertilizer. With the expected future price increases for P fertilizer, this problem is likely to increase. Enhancing sorghum tolerance to low P soil conditions is therefore of great importance both now and for the future. The chapter first gives an overview of the worldwide P challenge, and then focuses on the challenges of breeding sorghum for low soil P conditions, taking breeding for smallholder farmers in West Africa as a case study. The chapter summarizes the implications for applied sorghum breeding in West Africa, and reviews the future trends and challenges (both breeding and agronomic/socioeconomic) in bringing benefits of sorghum low-P tolerance to West African farmers.

The concluding chapter of the section, Chapter 10, examines advanced testing, multiplication and the release of new sorghum varieties. The chapter focuses especially on hybrid evaluation and release. The release of a new sorghum variety or hybrid is the culmination of many years of breeding and evaluation. The chapter describes the hybrid advancement process, including ways of increasing the parent lines for hybrid production and the guidelines that must be followed to ensure high quality seed. The chapter also examines procedures for registration, certification, plant variety protection and plant patenting.

Part 3 Sorghum production techniques

The final part of the book discusses ways of improving cultivation to make the most of new varieties, from nutrient and water management to better control of diseases, insects and weeds. Chapter 11 provides an overview of improving sorghum crop management. Sorghum is a versatile crop, produced in more than 100 countries. However, the productivity of sorghum in developing countries is much lower than developed countries, due to non-adoption of improved crop management practices. The chapter describes a variety of standard crop management techniques, including discussion of optimum planting conditions and the management of nutrients and water, and considers how practices may differ across major sorghum-producing areas such as India, West Africa, Australia and the USA. The chapter concludes that greatly improved yields can be obtained by matching genotype with environment and using best management practices.

Building on Chapter 11, the subject of Chapter 12 is improving soil and crop nutrition management in sorghum cultivation. The chapter provides an overview of the nutrients needed, quantities used and concepts of nutrient management commonly utilized in commercial grain sorghum production. The chapter describes how these concepts can also be applied to subsistence production as practiced in much of the developing world where sorghum is grown. The chapter's focus is on the management practices and tools utilized for the essential mineral nutrients commonly observed to be deficient in sorghum production, including N, P, K, S, Fe, Mn, Zn, and Cl. The chapter discusses soils and conditions where deficiencies may occur and, tools to access nutrient availability. It also reviews nutrient sources, from chemical fertilizers to common byproducts which can be used as nutrient sources, as well as timing and application methods.

Moving from soil conditions to water use, Chapter 13 looks at the challenge of improving water management in sorghum cultivation. Sorghum's deep, fibrous root system provides greater root volume and therefore potential access to a greater soil water volume enabling sorghum to withstand greater periods of water stress if there is ample stored soil water. The chapter examines dryland production of sorghum and sorghum irrigation methods, including deficit irrigation. The chapter also addresses the relationship between soils and irrigation management.

Chapter 14 moves from management of soil and water to consider the management of insect pests in sorghum cultivation. Integrated pest management (IPM) is a practical approach that uses a combination of methods, including cultural, biological, and chemical controls, to keep pest abundance or damage below economic loss. The chapter describes ways of preventing damaging infestations, as well as evaluating pest abundance and damage to determine when remedial action with insecticide is justified. The chapter describes key insect pests, their life cycles and impact on sorghum crops, with a particular focus on the USA but also considering other countries. The chapter also examines the contribution of beneficial insects.

The theme of Chapter 15 is diseases of sorghum, with a focus on seedling, seed, panicle and foliar diseases. As sorghum continues to grow in significance in large parts of the world, attention is given to the many different types of disease that can affect this crop, and how they might be successfully avoided or managed. All physical structures and tissues of the sorghum plant may become diseased with one pathogen or another, and different diseases can damage the plant at virtually every stage of growth. The chapter provides a comprehensive overview of the main diseases affecting sorghum, describing the different conditions in which they proliferate, their common symptoms, and their impact on crop yield. In each case, the chapter discusses the potential for host resistance and provides information regarding effective disease management.

Continuing the theme of disease management, Chapter 16 considers diseases of sorghum affecting the plant's stalk and root, as well as other diseases. Of all the diseases affecting sorghum, root and stalk diseases are among the most damaging and difficult to control. This is due to their hidden nature, and the fact that their symptoms often resemble abiotic stressors such as drought or freeze damage. The chapter presents an overview of some of the main bacterial, viral and parasitic diseases affecting sorghum root and stalk, describing their common symptoms, environments and typical effects on yield. In each case, the chapter offers recommendations for effective management of the disease, including information regarding tillage practices, fertilizer rates, crop rotation and use of fungicides.

Chapter 17 focuses on weed management in sorghum cultivation. Sorghum remains an important crop particularly in marginal environments due to its low input requirements and high resilience. Weed management is a key production challenge for sorghum, and the herbicide tools available are very limited, particularly for postemergence control of grass weeds. Weedy relatives such as shattercane and johnsongrass present challenges for selective control in sorghum due to genetic similarities. The chapter presents a number of weed management options for sorghum, including identifying critical periods for weed control, with a particular emphasis on preemergence control of grasses and other tough-to-control annual weeds. The chapter concludes that an integrated weed management approach, encompassing a strategic application of pre- and post-emergence herbicides used in combination with non-chemical tactics, will be critical to achieving sustainable weed management in sorghum.

The final chapter in the book, Chapter 18, considers the management of sorghum after harvesting. Sorghum is a versatile crop that has many post-harvest uses. After describing post-production operations in dealing with sorghum, the chapter examines techniques for the transformation of sorghum grain and the variety of resulting sorghum products. Finally, the chapter looks at requirements for sorghum export and quality assurance.

Part 1

Sorghum the crop species

Chapter 1

Classifying the genetic diversity of sorghum: a revised classification of sorghum

J. A. Dahlberg, Kearney Agricultural Research and Extension Center – University of California, USA; and D. T. Rosenow, formerly Texas A&M AgriLife Research and Extension Center, USA

http://dx.doi.org/10.19103/AS.2017.0015.02

1 Introduction

Drs Jeff Dahlberg and Darrell Rosenow have worked to modify the classification schemes of sorghum as proposed by Harlan and DeWet, Doggett, and Murty and Govil. Dahlberg (2000) proposed an Integrated Classification of Sorghum for use by plant breeders throughout the world so that sorghum classification could be more unified in its application in both research and plant breeding. This scheme does not require extensive knowledge of variability found within sorghum, but relies on the classifier's ability to distinguish between the major and intermediate races as a way to move into a more precise classification based on the working groups. Following are excerpts from Dahlberg's 2000 chapter that attempts to clarify the reasoning for the new system.

> This classification and the description of the working groups relies on Murty and Govil's 'Description of 70 Groups in Genus Sorghum based on a Modified Snowden's Classification' (1967), Dogget's (1988) work, and de Wet's 1978 publication. In many cases, the description of the working groups (WG) was taken directly from Murty and Govil. Harlan and de Wet's classification scheme is used to enter into a generalized classification of cultivated sorghum from which a more precise working group can be defined to explain the variation within the chosen race. When possible the numeric values used by Harlan and de Wet (1–15 for races and intermediate races) have been maintained. The new numerical system and descriptors of the working groups follows. Where there are two names for either the race or working group, the first name designates what is to be considered the dominant race or working group in the classification. For example, caudatum-kafir would suggest that caudatum is the most influential of the races showing in the line. Where there is no definition of a working group, the accessions being evaluated is a previously undescribed working group. The system, however, is flexible enough to allow for the addition of new working groups.
>
> The new numeric system for the working groups attempts to simplify and follow the numeric designations outlined by Harlan and de Wet (1972). The digit or digits to the left of the last digit represent the race or intermediate race and their designation. For example, working group 10 indicates that this is a member of the race bicolor (the digit to the left being 1). If the last digit is 0, this indicates the working group that most closely represents the definition for the actual race. For example, working group 30 would be race caudatum, working group caudatum. The definition for the working group caudatum would represent the general description for the race caudatum. The last digit would also provide an indication as to how many working groups might fall within the race. Working group 37 would represent zerazeras and would indicate that there are at least 7 working groups within the race caudatum.

In Sections 2–14 we attempt to outline the classification scheme in a format that would be useful for use as a teaching tool and in breeding programmes as breeders look to incorporate exotic germplasm into their breeding lines and hybrids (see Table 1). Definitions of races and working groups are taken directly from Dahlberg (2000). This is a living document and representative pictures were taken from the Sorghum Conversion Program materials (see Appendix A), while others were maintained from Murty and Govil's publication. The author encourages sorghum researchers to add to this classification and to provide high-quality pictures that may better represent working groups. Table 2 provides the reference descriptors that are currently used within the United States Sorghum Collection, which are identified by PI numbers, to describe a core set of descriptors for accession characterization.

Table 1 Revised working group designations

Race	Race #	Working group	WG #	M&G #[−1]
Bicolor	1	Bicolor	10	10
Bicolor	1	Dochna	11	12–15
Bicolor	1	Nervosum	12	7
Bicolor	1	Nervosum-kaoliang	13	8
Bicolor	1	Nervosum-broomcorn	14	9
Bicolor	1	Sudanense	15	52
Guinea	2	Guineense	20	4
Guinea	2	Conspicuum	21	3
Guinea	2	Margaritiferum	22	5
Guinea	2	Roxburghii	23	1
Caudatum	3	Caudatum	30	33
Caudatum	3	Caudatum-nigricans	31	31
Caudatum	3	Nigricans	32	29
Caudatum	3	Sumac	33	33
Caudatum	3	Nigricans-feterita	34	31
Caudatum	3	Dobbs	35	31(1)
Caudatum	3	Caudatum-kaura	36	34
Caudatum	3	Zerazera	37	39(1)
Kafir	4	Caffrorum	40	22
Durra	5	Durra	50	41
Durra	5	Nandyal	51	46(1)
Durra	5	Cernuum	52	49
Guinea-bicolor	6	Guinea-bicolor	60	20/10
Guinea-bicolor	6	Dochna-honey	61	16
Guinea-bicolor	6	Dochna-roxburghii	62	17
Caudatum-bicolor	7	Caudatum-bicolor	70	36
Caudatum-bicolor	7	Caudatum-dochna	71	37
Caudatum-bicolor	7	Nigricans-bicolor	72	30
Caudatum-bicolor	7	Dochna-nigricans	73	19
Kafir-bicolor	8	Bicolor-kafir	80	11
Kafir-bicolor	8	Caffrorum-bicolor	81	26
Kafir-bicolor	8	Dochna-kafir	82	18
Durra-bicolor	9	Durra-bicolor	90	44
Durra-bicolor	9	Dochna-durra	91	91

(Continued)

Table 1 (*Continued*)

Race	Race #	Working group	WG #	M&G #[1]
Durra-bicolor	9	Durra-dochna	92	45
Durra-bicolor	9	Subglabrescens	93	50
Durra-bicolor	9	Subglabrescens-milo	94	51
Durra-bicolor	9	Milo-kaura	95	51(1)
Guinea-caudatum	10	Caudatum-guineense	100	35
Guinea-caudatum	10	Nigricans-guineense	101	30(2)
Guinea-kafir	11	Caffrorum-roxburghii	110	25
Guinea-kafir	11	Roxburghii-shallu	111	2
Guinea-durra	12	Durra-roxburghii	120	42
Guinea-durra	12	Membranaceum	121	6
Guinea-durra	12	Durra-membranaceum	122	43
Kafir-caudatum	13	Caudatum-kafir	130	38
Kafir-caudatum	13	Caffrorum-birdproof	131	24
Kafir-caudatum	13	Caffrorum-darso	132	23
Kafir-caudatum	13	Caffrorum-feterita	133	27
Durra-caudatum	14	Caudatum-durra	140	40
Durra-caudatum	14	Nigricans-durra	141	32
Durra-caudatum	14	Durra-nigricans	142	47
Durra-caudatum	14	Durra-feterita/Kaura	143	48
Kafir-durra	15	Durra-kafir	150	46
Kafir-durra	15	Caffrorum-durra	151	28
Annual wild	16	*S. bicolor* subsp. *Drummondii*	160	53
S. bicolor subsp. *verticilliflorum*	17	Verticilliflorum	170	61
S. bicolor subsp. *verticilliflorum*	17	Arundinaceum	171	61(1)
S. bicolor subsp. *verticilliflorum*	17	Virgatum	172	62
S. bicolor subsp. *verticilliflorum*	17	Aethiopicum	173	63
Perennial wild	18	*S. halepense*	180	54
Perennial wild	18	*S. propinquum*	181	
Unclassified	19	Unclassified	190	70
Breeding material	20	Unclassified	200	
Mixed	21	Mixed	210	

[1] Numeric classification from Murty & Govil (1967).

Table 2 Field and panicle descriptors (Dahlberg and Spinks, 1995)

Descriptor	Codes
	Field notes
Source	92X0001 = Year, Location (X), Plot #
Pedigree	P.I. Number
Race	Revised (1–21) (Dahlberg 2000)
Working group	10–210 (Dahlberg 2000)
Anthesis	Days to 50% flowering
Plant Ht. to Top of head	Cm
Exsertion	Cm
Panicle length	Cm
Sprouting tendency	0 = na, 1 = Yes, 2 = No
Basal tillering	Number per plant
Nodal tillering	0 = na, 1 = Yes, 2 = No
Midrib colour	1 = Colourless, 2 = Green, 3 = Yellow, 4 = Brown
Juicy/Dry midrib	1 = Juicy, 2 = Intermediate, 3 = Dry
Panicle erectness	1 = Erect, 2 = Recurved
Height uniformity	1 to 5. 1 = very uniform, 5 = not uniform
Plant colour	1 = Purple, 2 = Red, 3 = Tan 4 = PR, 5 = RP, 6 = TR, 7 = TP, 8 = Mixed, 9 = Black
Stalk waxiness	1 = Bloom, 2 = Sparse Bloom, 3 = Bloomless
Rust incidence	1 to 5. 1 = resistant, 5 = susceptible
Anthracnose incidence	1 to 5. 1 = resistant, 5 = susceptible
Yield potential	1 to 5. 1 = high, 5 = low
Desirability rating	1 to 5. 1 = Excellent, 5 = Poor
	Panicle notes
Shattering	1 = non, 2 = mod, 3 = complete
Grain weathering	1 = VG, 2 = G, 3 = Ave, 4 = Below, 5 = Poor
Panicle shape	1 = Lax, 2 = Oval, 3 = Elliptical, 4 = Rectangular, 5 = Broomcorn, 6 = Segregating, 7 = Sprankletop
Branch angle	1 = Erect, 2 = Drooping
Panicle compactness	1 = Compact, 2 = Semi-compact, 3 = Semi-loose, 4 = Loose, 5 = Mixed
Awns	1 = Long, 2 = Short, 3 = Tip, 4 = No, 5 = Seg

(Continued)

Table 2 (*Continued*)

Descriptor	Codes
	Field notes
Glume colour	1 = Purple, 2 = Red, 3 = Tan, 4 = Black, 5 = Sienna, 6 = Mahogany, 7 = Straw, 8 = Mixed, 9 = Segregating
Kernel covering	1 = Involuted, 2 = 25%, 3 = 50%, 4 = 75%, 5 = Fully, 6 = Glumes longer than grain, 7 = Segregating
Glume pubescence	1 = Yes, 2 = Intermediate, 3 = No, 4 = Segregating
Transverse wrinkle	0 = NA, 1 = Yes, 2 = No
Seed form	1 = Single, 2 = Twin
Kernel appearance	1 = White, 2 = Red, 3 = Yellow, 4 = Brown, 5 = Buff, 6 = Mixed, 7 = Purple
Pericarp colour	1 = White, 2 = Red, 3 = Lemon-yellow, 4 = Mixed
Mesocarp	1 = Thick, 2 = Thin, 3 = Segregating
Testa	1 = Yes, 2 = No, 3 = Segregating
Spreader	0 = NA, 1 = Yes, 2 = No
Kernel plumbness	1 = Plump, 2 = Dimple
Kernel shape	1 = Round, 2 = Oval, 3 = Variable, 4 = Other
Endosperm colour	1 = White, 2 = Yellow, 3 = Mixed
Endosperm texture	1 = Completely corneous, 2 = Almost, 3 = Partly, 4 = Almost starchy, 5 = Completely starchy
Endosperm type	1 = Non-waxy, 2 = Waxy, 3 = Sugary, 4 = Possible High Lysine (?)

2 The major races of sorghum: Bicolor, Guinea, Caudatum, Kafir and Durra

2.1 Bicolor

The race bicolor is thought to be the most primitive grain sorghum (Fig. 1). It is believed to have evolved from the subspecies *verticilliflorum* complex; however, Harlan (1975b) is unsure as to its centre of origin (for a more detailed review of the dispersal and development of all the races, see Doggett, 1988). Since its domestication, it has spread throughout many agricultural areas of Africa and Asia from India to Indonesia (de Wet and Price, 1976). The race is characterized by open panicles of medium size with long, slightly stiff rachis branches terminating with long clasping glumes. The glumes are thick and coriaceous with obscure nerves, and the tips of the lower glumes are depressed and hairy. The pedicellate spikelets are persistent. Grains are elliptic to subglabose, but elongated, generally small in size and persistently attached to the panicle. It is enclosed or almost enclosed by tightly clasping glumes. The pedicels are short. Seed, glumes and plant parts generally are highly pigmented. Bicolor plants are fairly low yielding, medium in height and tend to tiller profusely (Mann et al., 1983).

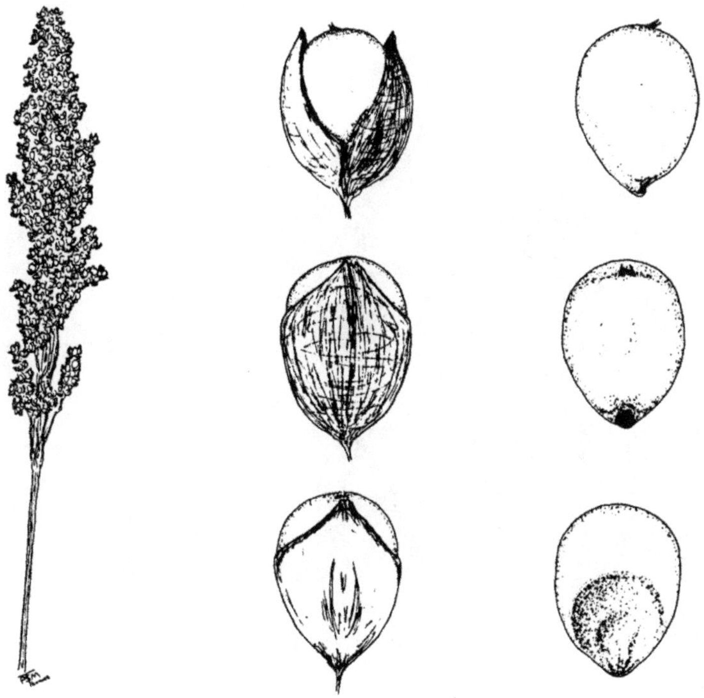

Figure 1 Stylized drawing of the race bicolor (drawing from Mann et. al., 1983).

2.2 Guinea

Harlan and Stemler (1976) proposed that guinea sorghums (Fig. 2) originated some 3000 years ago. Porteres (1962, from de Wet et al., 1972) proposed that the race guinea arose independently from the wild race arundinaceum, which is primarily a forest grass. There were two primary reasons behind his rational proposal: 1. the guineas are grown extensively in the forest regions of West Africa, and 2. it is sympatric with the wild arundinaceum. However, de Wet et al. (1972) could not find 'closer affinities with arundinaceum than the verticilliflorum-aethiopicum complex'. Although guineas are grown in the forest, it is primarily a savanna cereal and probably originated in the West African savanna as an early selection out of the original bicolor race. It extends across the savannas of Africa and is grown widely in South Asia. It is also present in highly humid areas of southern Africa; de Wet et al. (1972) speculated that since the guineas of West and southern Africa are so similar, it is best to assume West Africa as the place of origin for all guineas.

The guinea race is characterized by long, loose, glabrous and pendulous panicles. Sessile spikelets open when mature, thus exposing the grain. The glumes are involuted, open widely and are hairy with conspicuous awns. Pedicellate spikelets are both persistent and deciduous. Grains are small to medium, biconvex and nearly ovate but some are also flattened. Grain tends to be light coloured or slightly pigmented. Plants are medium to tall in stature and tend to be low yielding (Mann et al., 1983).

Figure 2 Stylized drawing of the race guinea (drawing from Mann et. al., 1983).

2.3 Caudatum

The race caudatum (Fig. 3) was associated closely with the speakers of the Chari-Nile languages in Africa and it is widely distributed throughout northeastern Nigeria, Chad, Sudan, Ethiopia and Uganda (Stemler et al., 1975). Caudatums are primarily restricted to the area of early bicolor domestication and because of this, it is thought to have been derived from an introgressed cross with an early bicolor and the wild sorghum complex (Harlan and Stemler, 1976). Doggett (1988) believed that the race must be younger than guinea or durra because it does not occur in India and they do not arise from all the guinea and durra crosses that must have occurred there. Caudatum is one of the most important races agronomically and provides genes for high yield and excellent seed quality. It has become one of the most important sources of germplasm in modern breeding programmes throughout the world.

The race caudatum is characterized by medium to large, oblong, dense to slightly open panicles which can be hairy with stout peduncles and rigid primary branches. Sessile spikelets are obovate to elliptical while pedicellate spikelets are deciduous. Glumes are coriaceous, pubescent and shorter than the large grain. Grains are flat on one side and round or bulging on the other (turtle backed) and usually are chalky white or pigmented. Many have a recessive spreader gene. Plants are generally medium to tall in stature and are usually high yielding (Mann et al., 1983).

2.4 Kafir

The southern region of Africa is dominated by the race kafir (Fig. 4), which also is widely distributed in northern Nigeria. No strong evidence supports an apparent centre of origin; however, it is thought that it was derived from early bicolor of northern Africa and

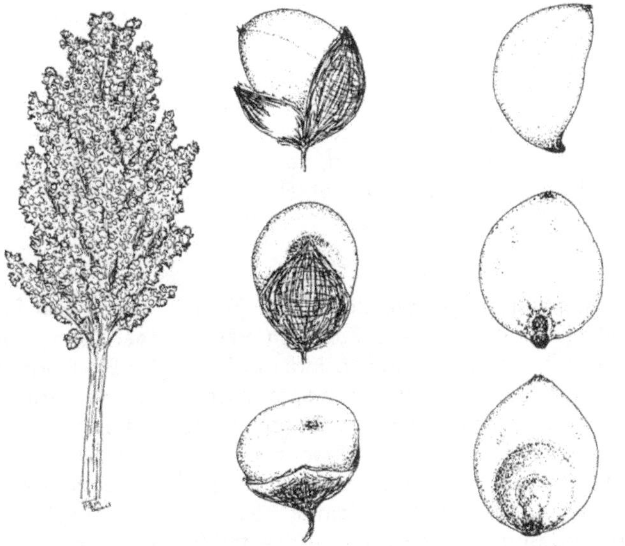

Figure 3 Stylized drawing of the race caudatum (drawing from Mann et. al., 1983).

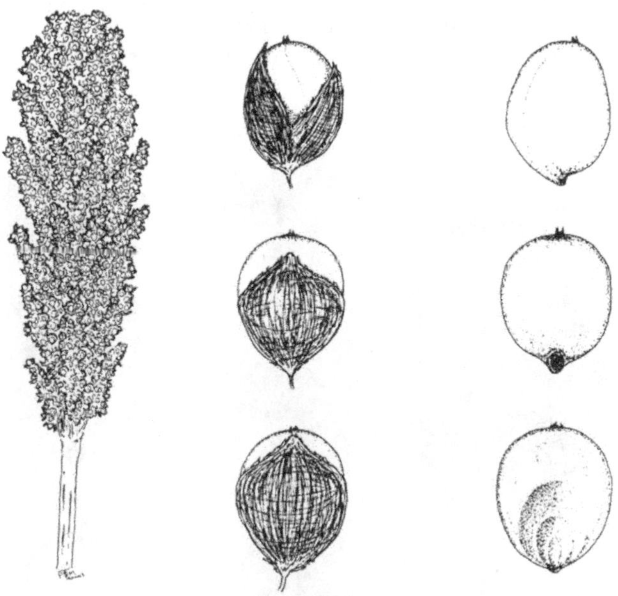

Figure 4 Stylized drawing of the race kafir (drawing from Mann et. al., 1983).

carried south, primarily by Bantu speakers of Africa (Harlan and Stemler, 1976). De Wet and Huckabay (1967) felt that its distribution and morphology suggested a closer affinity with the wild verticilliflorum, and electrophoretic data seems to strengthen that argument (Shechter and de Wet, 1975).

The kafir race is characterized by erect, elongated, mostly semi-compact and cylindrical panicles. The branches and sessile spikelets are hairy but the glumes are almost glossy at maturity. Glumes are moderately coriaceous and much shorter than the grain, which is broadly elliptical, sometimes compressed, flattened or biconvex. Plants are of medium height and generally high yielding (Mann et. al., 1983).

2.5 Durra

Snowden (1936) suggested that the durra (Fig. 5) may have arisen from introgression with the wild aethiopicum. Harlan and Stemler (1976) felt that durras were selected from early bicolors which had been carried to India before 3000 BP. Doggett (1988) argued convincingly that durras originated in Ethiopia since the whole sequence of wild-type bicolor-durra clearly is represented in this country. He suggested that early bicolors introgressed with wild forms and adapted to drier conditions, thus leading to the development of the race durra. The durras were moved westward through Sudan and began to occupy the drier regions below the southern margin of the Sahara. Doggett (1988) then suggested that they were moved through the Horn of Africa and finally into India.

The durras are characterized by stiff, dense and compact panicles that are ovate to oblong in shape, or rhomboidal, and covered with dense velvety pubescence. The peduncle is often recurved, but occasionally erect. Panicle branches are short, semi-erect and hairy, in which the rachis may vary from elongated to hidden. The pedicellate spikelets are large and persistent, while the sessile spikelets are obovate-elliptic to rhomboidal. Glumes are coriaceous on the lower half and slightly to strongly depressed with a central transverse

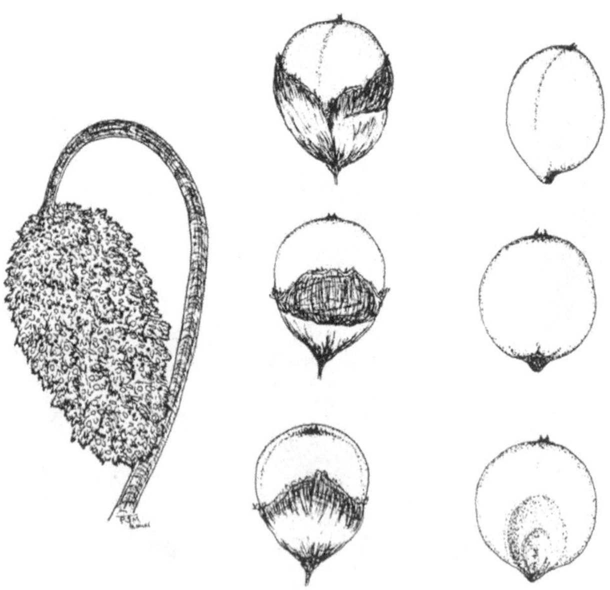

Figure 5 Stylized drawing of the race durra (drawing from Mann et. al., 1983).

wrinkle. The lower half of the glume is strongly nerved with a papery tip. Glumes tend to be lightly pigmented. Grains are generally medium to large, are biconvex, have a broad top and are wedge shaped at the base. Plants are medium to tall in stature and have good quality.

3 Working groups: Bicolor (1)

Panicles are open and medium in size. Glumes are glasping and cover three-fourths or more of the grain, which is elongate, symmetrical and small. Rachis is typically long with long, slightly stiff branches.

WG 10: Bicolor: Formally WG 10 in Murty and Govil's definition. Panicle medium to large with stiff primary branches; sessile spikelets obovate to sub-rotund; awns of upper lemma, if present, deciduous at maturity; glumes thick and coriaceous with obscure nerves; tip of the lower glume depressed and hairy; pedicelled spikelets persistent with short pedicels; grains elliptic to sub-globose and small to medium in size and turtle back shape in several cases; glumes black, purple or red; seed colour light brown to red or purple; glutinous and amber-coloured endosperm; subcoat present. Plants medium to very tall; nodal tillering prominent (Fig. 6).

Figure 6 Working Group (WG) bicolor Sorghum Conversion (SC) 614 (IS 1333C).

WG 11: Dochna: Formally WGs 12–15 in Murty and Govil's definition. Panicle loose, elliptic to oblong; may be sprankletop; semi-open to semi-erect primary branches; sessile spikelets not transversely wrinkle; glumes thin, crustaceous and brittle; lower glume coriaceous with nerves obscure even at the top; upper lemma awned or mucronate; glumes can be hairy at the tip; grains similar in shape to the spikelets and tightly clasped and almost completely covered by the glumes; glume colour black, brown or purple; glumes normally cannot be easily separated from grains; endosperm white and mealy. Plants very tall with 8–10 leaves; some stems are juicy, sweet and/or thin; nodal tillering present (Fig. 7).

WG 12: Nervosum: Formally WG 7 in Murty and Govil's definition. Panicle compressed or contracted at the base and spreading above; rachis elongated and erect with hairy internodes; short and stiff primary and secondary branches; sessile spikelets with prominent nerve on the coriaceous upper glume and a thin papery lower glume; glumes shorter than the grains covering about 2/3 of the grain, glume colour light brown to black; pedicelled

Figure 7 WG dochna SC 19 (IS 12539C).

Figure 8 WG nervosum IS 2339 (Murty & Govil, 1967).

spikelets persistent; seed flat at the base, turtle back on dorsal side with slightly rounded lower side, seed exceeds the gaping appressed glumes in size; colour light to dark brown or purple; endosperm mealy to partly glutinous except at the centre; subcoat present (Fig. 8). Plants medium height (about 2 m); leaves eight to ten; early maturing; nodal tillers present.

WG 13: Nervosum-kaoliang: Formally WG 8 in Murty and Govil's definition. Similar to nervosum (WG 12) except for the following additional features: Panicles wider but shorter, more compact and oval in shape with a thick peduncle; sessile spikelets larger than in nervosum; glumes prominently gaping when seeds mature; grains large and have a prominently convex dorsal side and a flat lower side; endosperm partly glutinous and amber-coloured with limited mealy portion; subcoat present; plants taller and have more leaves than nervosum; nodal tillering very frequent (Fig. 9).

WG 14: Nervosum-broomcorn: Formally WG 9 in Murty and Govil's definition. Panicles large with a weak or inconspicuous rachis and long primary branches with spikelets clustered at the upper one-third of the branch like broomcorn; branches almost equal to the panicle in length and flexuous upwards; sessile spikelets with upper glumes having striate nerve as in nervosum (WG 12) but lower glume wrinkled and depressed across the middle as in bicolor; seeds almost completely enclosed by glumes; endosperm mealy white; plants medium to tall; basal and nodal tillering absent (Fig. 10).

WG 15: Sudanense: Formally WG 52 in Murty and Govil's definition. Panicle ovate to pyramidal, loose, spreading branches; racemes tough; pedicelled spikelets long and persistent and reddish purple; elliptic-oblong, subacute sessile spikelets cream to yellow in colour; sparsely hairy when young, glabrescent when mature; glumes coriaceous, with the nerves visible at the tip; lower glume usually winged on the keels near the tip with the keels sometimes ending in obscure teeth; upper lemma awned; grains small, enclosed by glumes; obovate-elliptic, planoconvex to biconvex; reddish brown; partly glutinous endosperm; subcoat present. Annual plants medium tall with slender culms; nodal and basal tillering prominent with narrow leaves; wild grasses, cultivated for fodder (Fig. 11).

Figure 9 WG nervosum-kaoliang SC 609 (IS 1213C).

Figure 10 WG nervosum-broomcorn IS 34 (Murty & Govil, 1967).

Figure 11 WG sudanense SC 941 (IS 3196C).

4 Working groups: Guinea (2)

Panicles are long, loose and pendulous. Glumes are involuted and cover most of the grain, which is flattened, twisting and ovate. Rachis is long with short branches.

WG 20: Guineense: Formally WG 4 in Murty and Govil's definition. Panicles long and glabrous in appearance; sessile spikelets with keels distinctly winged with minute teeth; primary branches short with closely spaced spikelets as compared to roxburghii (WG 23); panicle internodes shorter and more in number than in WG 23; pedicelled spikelets highly deciduous; sessile spikelets elliptic – oblong or ovate in shape; awns conspicuous; glumes involute, recurved at the tips and widely gaping; grains compressed, large, mucronate; elliptic or ovate sometimes larger than the glumes; grain shattering considerably less than in WG 23; seed colour white to pink; subcoat present; partly glutinous. Plants are tall, nodal tillering prominent; leaves 15–20; relatively free from pests and diseases (Fig. 12).

Figure 12 WG guineense SC 1333.

WG 21: Conspicuum: Formally WG 3 in Murty and Govil's definition. Panicle hairy, rarely glabrous, and very long up to 50 cm in some cases; sessile spikelets lanceolate to elliptic; upper lemma usually awned but awnless types also present or deciduous when seed is mature; grains elliptic, bold, compressed or flattened, turtle back shape in some cases with large embryos; pedicelled spikelets persistent; endosperm partly glutinous, sometimes yellow; seed colour variable; subcoat present, glumes involute but less gaping and clasping the seeds in some cases. Plants tall, with nodal and basal tillering; internodes very long; photosensitive and late in maturity (Fig. 13).

Figure 13 WG conspicuum SC 258 (IS 1201C).

WG 22: Margaritiferum: Formally WG 5 in Murty and Govil's definition. Panicle long, very loose and open with several branches at each node; branches flexuous: pedicelled spikelets highly deciduous when mature; sessile spikelets slightly narrower than in guineense (WG 20) but lanceolate to oblong-lanceolate in shape; awns conspicuous but

Figure 14 WG margaritiferum SC 605 (IS 7979C).

deciduous when mature; glumes involute and gaping; grains shorter than glumes and shatter readily when mature; seed biconvex or slightly compressed; seed colour white to red; subcoat present; endosperm highly corneous and pearly. Plants are tall with occasional nodal tillering (Fig. 14).

WG 23: Roxburghii: Formally WG 1 in Murty and Govil's definition. Panicle and spikelets hairy; sessile spikelets lanceolate to ovate and finely acute and awnless; pedicelled spikelets persistent when mature; panicle shape sub-umbellulate; panicle nodes villous; grains biconvex, flattened or slightly compressed; some cases turtle back in shape;

Figure 15 WG roxburghii SC 195 (IS 1116C).

grains at maturity are freely exposed or partially enclosed, sometimes shattering when disturbed; grains medium in size; predominantly white; subcoat coloured; endosperm highly glutinous and mostly pearly white. Stigma are densely feathery at the base, but not in the upper one-third (Fig. 15).

5 Working groups: Caudatum (3)

Panicles are dense to slightly open with a stout peduncle. Glumes are pubescent or coriaceous and partially cover the grain, which is turtle back in shape. Rachis and branches are rigid.

WG 30: Caudatum: Formally WG 33 in Murty and Govil's definition. Panicle large, oblong, densely villous or hirsute with a strong peduncle and rigid primary branches; sessile spikelets obovate-elliptic when mature; pedicelled spikelets deciduous; glumes coriaceous or thinly coriaceous; much shorter than the large grains which are turtle back in shape, flattened on the face, rounded and bulging on the back; subcoat dark in colour; grains mostly chalky white to dull grey (Fig. 16).

Figure 16 WG caudatum SC 337 (IS 7094C).

WG 31: Caudatum-nigricans: Formally WG 39 in Murty and Govil's definition. Similar to caudatum (WG 30) with panicles similar in size to nigricans (WG 32); sessile spikelet and grain size intermediate to caudatum and nigricans; grains round-topped and with a less pronounced plano-convex shape; clustering of spikelets similar to nigricans (Fig. 17).

WG 32: Nigricans: Formally WG 29 in Murty and Govil's definition. Panicle hairy, including rachis, semi-compact to loose with characteristic clustering of spikelets and stout peduncle; pedicelled spikelets with short pedicel and deciduous; sessile spikelets small, awnless; elliptic-ovate to elliptic-rotund; glumes glabrous and small in size as compared with the grains which are spherical at the top; grains mostly red or brown with dark subcoat and partly glutinous endosperm surrounding a mealy interior; mostly late in maturity (Fig. 18).

Figure 17 WG caudatum-nigricans SC 424 (IS 8100C).

WG 33: Sumac: Formally WG 30(1) in Murty and Govil's definition. Panicle contracted, oblong and semi-compact; rachis hidden; pedicelled spikelets persistent in some cases; sessile spikelets obtuse, hairy at the tip, otherwise glabrescent; glumes strongly coriaceous; grains small to medium, obovate-rotund, red to blackish purple colour; endosperm partly glutinous with a soft centre; subcoat purple (Fig. 19).

WG 34: Nigricans-feterita: Formally WG 31 in Murty and Govil's definition. Similar to nigricans (WG 32) with the following additional features of feterita: sessile spikelets large with grains much larger than the glumes and prominently bulging on the back; nucellar layer mostly coloured; endosperm mostly mealy with a thin glutinous layer on the periphery (Fig. 20).

WG 35: Dobbs: Formally WG 31(1) in Murty and Govil's definition. Similar to WG 34 but with small grains predominantly red in colour and prominent subcoat: endosperm mealy and white (Fig. 21).

Figure 18 WG nigricans SC 557 (IS 1318C).

Figure 19 WG sumac IS 7155 (Murty & Govil, 1967).

Figure 20 WG nigricans-feterita IS 3569 (Murty & Govil, 1967).

Figure 21 WG dobbs SC 690 (IS 2758C).

WG 36: Caudatum-kaura: Formally WG 34 in Murty and Govil's definition. Panicle oblong to elliptic, long, with thick erect peduncle, hairy, branches appressed and rigid; sessile spikelets obovate-elliptic; upper lemma mucronate; grains larger than the glumes and elliptic to rotund or biconvex; seed colour cream to yellow; endosperm glutinous but mealy near the centre; endosperm colour creamy-white to yellow; subcoat colourless; pedicelled spikelets long and deciduous but persistent in some populations. Plants tall with basal and nodal tillering; leaves broad and large (Fig. 22).

WG 37: Zerazera: Formally WG 39(1) in Murty and Govil's definition. Panicle and spikelets as in caudatum (WG 30) but with a shorter rachis; spikelets clustering as in nigricans (WG 32); grains broadly elliptic to rotund, distinctly round-topped; short glumes coloured predominantly ivory yellow to cream or straw; grain size larger than in nigricans but smaller than in caudatum; grain can be involuted within glumes; grain colour yellow, cream or

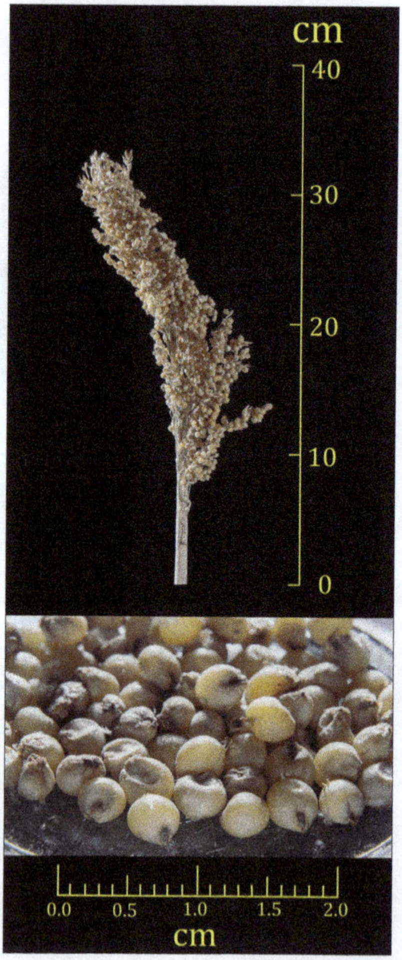

Figure 22 WG caudatum-kaura SC 366 (IS 7248C).

Figure 23 WG zerazera SC 803 (IS 2586C).

dull white; endosperm highly corneous and flinty; white to yellow in colour; subcoat not coloured; grains easily separable from glumes; leaves relatively free from diseases and pests (Fig. 23).

6 Working groups: Kafir (4); Durra (5)

Kafir

Panicles erect, cylindrical; glumes clasping and variable in length which cover symmetrical, spherical grains; rachis long and branches tend to be hairy.

WG 40: Caffrorum: Formally WG 22 in Murty and Govil's definition. Panicle erect, contracted and dense to compact; hairy; branches hairy to villous; sessile spikelets hairy, elliptic-ovate to elliptic-rotund; glumes almost glossy when mature and moderately coriaceous; shorter than the grains; upper lemma mucronate occasionally with a short awn which is deciduous when mature; pedicellate spikelets persistent with a short pedicel; grains broadly elliptic, compressed, flattened or biconvex; grain colour variable from white to brown or red; endosperm mealy, subcoat present. Nodal tillering present (Fig. 24).

Figure 24 WG kafir SC 625 (IS 8003C).

Durra

Panicles tend to be stiff, dense, compact and pubescent; glume tips are of different texture and normally with pronounced transverse wrinkles which cover wedge-shaped obovate grains; rachis stout, hidden and sometimes recurved with short, erect and often hairy branches.

WG 50: Durra: Formally WG 41 in Murty and Govil's definition. Panicle stiff, dense, compact and ovate to oblong in shape covered with short and dense velvety hairs; peduncle thick and often goose-necked but sometimes erect; panicle branches stout, short and semi-erect; rachis elongated and completely hidden; pedicelled spikelets large and persistent; sessile spikelets obovate-elliptic to rhomboid with glumes coriaceous in the lower half and slightly depressed in the middle; lower glume with a strongly nerved, herbaceous tip; glumes yellow, white or straw-coloured; upper lemma usually awned, the awns dropping when disturbed in mature panicle or mucronate; grains much larger than glumes when mature, biconvex in shape with a broad top and wedge-shaped base;

Figure 25 WG durra SC 227 (IS 2478C).

grain white, yellow to red: endosperm white and mealy to partly glutinous; subcoat occasionally coloured; plants usually tall with nodal tillering and rarely basal tillering (Fig. 25).

WG 51: Nandyal: Formally WG 46(1) in Murty and Govil's definition. Panicle contracted and compact, oblong to elliptic-oblong, with semi-erect branches with stout base; peduncle always erect; pedicelled spikelet persistent; obovate-oblong sessile spikelets; glumes thin and papery in the middle; upper lemma awned; grains obovate broadtopped, with a wedge-shaped and constricted base, much exposed on maturity; creamy to yellow in colour; endosperm glutinous and pearly. Plant medium to tall and photosensitive. There are several forms with a wide range of panicle shape mainly due to difference in the degree of compactness. All are highly glutinous with pearly endosperm, although the glume size is comparatively larger than in durra (WG 50), seeds can be easily separated due to their constricted bases (Fig. 26).

Figure 26 WG nandyal SC 199 (IS 1121C).

WG 52: Cernuum: Formally WG 49 in Murty and Govil's definition. Panicle usually larger than durra (WG 50) but less compact and narrower in width; ovate to elliptic or oblong, peduncle mostly recurved; rachis fully hairy, branches short and sub-erect; pedicelled spikelets large, persistent with short pedicel; sessile spikelets awned, broadly ovate to obovate-oblong, pale yellow and fully hairy; awns larger than in durra and persistent; glumes thick and spongy below the middle, thin and papery above; characteristically transversely wrinkled and depressed across the middle; large thin and herbaceous nerved

Figure 27 WG cernuum SC 209 (IS 1151 C).

tip of the lower glume, tips breaking off at maturity; rotundate or orbiculate flattened grain, protruding beyond the glumes; glumes mostly dull white to pale yellow; endosperm pearly white and glutinous; grains medium size as compared to durra. Plants medium tall, photosensitive (Fig. 27).

7 Working groups: Guinea-bicolor/Bicolor-guinea (6); Caudatum-bicolor/Bicolor-caudatum (7)

WG 60: Guinea-bicolor: No formal definition in Murty and Govil. Plant represents intercross between guinea (WG 20) and bicolor (WG 10) (no picture available).

WG 61: Dochna-honey: Formally WG 16 in Murty and Govil's definition. A sweet-stemmed, juicy group; panicle intermediate between dochna (WG 11) and roxburghii (WG 23) but shorter in length (less than 20 cm); primary branches four to five at each whorl; spikelet characteristics similar to dochna; endosperm highly glutinous, subcoat present. Plants intermediate in height and better than other sweet sorghums in basal and nodal tillering (Fig. 28).

WG 62: Dochna-roxburghii: Formally WG 17 in Murty and Govil's definition. Similar to WG 61, except for the following: Panicle more loose and wider with longer primary branches; more nodes of panicle; spikelets closely and uniformly spaced; glumes of

Figure 28 WG dochna-honey SC 15 (IS 12535C).

sessile spikelets partially involute, as compared to dochna (WG 11); panicle shape is closer to dochna than roxburghii (WG 23) and wider at the top. Grains easily separable from glumes; medium to small, obovate to elliptic in shape; dorsal side curved and ventral side almost flat; predominant seed colour white or creamy yellow; endosperm glutinous and pearly white (Fig. 29).

Figure 29 WG dochna-roxburghii SC 315 (IS 5747C).

WG 70: Caudatum-bicolor: Formally WG 36 in Murty and Govil's definition. Panicle, spikelet and grain characteristics similar to caudatum (WG 30) but the glumes of sessile

Figure 30 WG caudatum-bicolor SC 749 (IS 8120C).

spikelets are thick and tightly clasping as in the grain in bicolor (WG 10) and panicle shape intermediate to caudatum and bicolor; endosperm mealy (Fig. 30).

WG 71: Caudatum-dochna: Formally WG 37 in Murty and Govil's definition. Panicle loose like dochna (WG 11) and sessile spikelets smaller than in caudatum (WG 30); pedicelled spikelets persistent as in dochna; grain shape similar to caudatum but threshability poor like dochna; awns present as in dochna. Rest of the features similar to caudatum (Fig. 31).

Figure 31 WG caudatum-dochna SC 411 (IS 6964C).

Figure 32 WG nigricans-bicolor SC 115 (IS 2683C).

WG 72: Nigricans-bicolor: Formally WG 30 in Murty and Govil's definition. Similar to nigricans (WG 32) except the following additional features: Panicle with short primary branches; sessile spikelets with thick coriaceous glumes and grains not necessarily round-topped (Fig. 32).

WG 73: Dochna-nigricans: Formally WG 19 in Murty and Govil's definition. Panicle and spikelet characteristics similar to dochna (WG 11) except that the spikelets are clustered as in nigricans (WG 32). Grains small in size and not exceeding the appressed glume in length (Fig. 33).

Figure 33 WG dochna-nigricans SC 319 (IS 2757C).

8 Working groups: Kafir-bicolor/Bicolor-kafir (8); Durra-bicolor/Bicolor-durra (9)

WG 80: Bicolor-kafir: Formally WG 11 in Murty and Govil's definition. Similar to bicolor (WG 10) with the following additional features: Compact panicle with contracted primary branches; nearly half of the size of those in WG 10; width of the panicle much smaller than in bicolor (4–5 cm); sessile spikelets broadly obovate to elliptic and open at the tip; tip strongly depressed: awns much reduced if present; grains larger than in bicolor, partly enclosed by glumes that are persistent or not readily separated from grain. Endosperm is mostly mealy; subcoat present (Fig. 34).

WG 81: Caffrorum-bicolor: Formally WG 26 in Murty and Govil's definition. Similar to caffrorum (WG 40) in panicle shape, panicle, branches, sessile spikelets and grain characters

Figure 34 WG bicolor-kafir SC 1104 (IS 1104C).

except that the branches of panicle are more flexuous and the glumes coriaceous; and grains with glutinous endosperm, like bicolor (WG 10) (Fig. 35).

WG 82: Dochna-kafir: Formally WG 18 in Murty and Govil's definition. Panicle intermediate in shape between dochna (WG11) and caffrorum (WG 40) but grains mostly enclosed by glumes and spikelet characteristics are similar to dochna (Fig. 36).

WG 90: Durra-bicolor: Formally WG 44 in Murty and Govil's definition. Spikelets and grain characteristics like durra (WG 50) but with the smaller grain size, elongated panicle, and panicle branching as in bicolor (WG 10); the grains are not easily separable from glumes unlike in durra. Grain shape tending to be turtle back, endosperm highly glutinous and pearly white (Fig. 37).

WG 91: Dochna-durra: Formally WG 20 in Murty and Govil's definition. Panicle shape similar to dochna (WG 11); sessile spikelets nearly ovate as compared to obovate to oblong in dochna; glumes of sessile spikelets densely hairy as compared to dochna; grains

Figure 35 WG caffrorum-bicolor IS 8770 (Murty & Govil, 1967).

Figure 36 WG dochna-kafir SC 21 (IS 12541C).

Figure 37 WG durra-bicolor SC 30 (IS 12550C).

more easily separated from glumes due to their larger size; endosperm waxy and subcoat present; grain colour mostly light brown to grey or white (Fig. 38).

WG 92: Durra-dochna: Formally WG 45 in Murty and Govil's definition. Panicle shape is intermediate between dochna (WG 11) and durra (WG 50); rachis is long and erect as in dochna; sessile spikelets are smaller than in durra but shape close to broadly obovate with lower glumes, similar in appearance to durra; glumes of sessile spikelets covering nearly

Figure 38 WG dochna-durra SC 623 (IS 2456C).

3/4 of the seed as in dochna but the seed shape is close to durra with a wedge-type base; glumes are yellow, cream or white; endosperm highly glutinous (Fig. 39).

WG 93: Subglabrescens: Formally WG 50 in Murty and Govil's definition. Panicle medium compact to medium loose, ovate or oblong; peduncle erect or goose-necked with semi-erect branches; pedicelled spikelets persistent on short pedicel; oblong to hexagonal sessile spilkelet, which are always transversely wrinkled and depressed in the middle; glumes thinner than durra (WG 50); sub-coriaceous near the base and at the tip; lower glume with a broad triangular subacute and depressed tip; awned upper lemma; grains at maturity exceeding the glumes, with broad tops, biconvex with a subspherical

Figure 39 WG durra-dochna SC 124 (IS 12615C).

top as compared to the flat top in cernuum (WG 52); wedge-shaped base, white or yellow in colour; endosperm partly glutinous (Fig. 40).

Figure 40 WG subglabrescens SC 1014 (IS 11443C).

WG 94: Subglabrescens-milo: Formally WG 51 in Murty and Govil's definition. Similar to subglabrescens (WG 93) except that the transverse depression and wrinkling of lower glume in the sessile spikelet is pronounced and glumes more papery; seed colour white to red; grains rotundate to orbicular (Fig. 41).

WG 95: Milo-kaura: Formally WG 51(1) in Murty and Govil's definition. Similar to WG 94 except that the grains are yellow with a glutinous yellowish endosperm. Probably hybrid derivatives of milo with caudatum-kaura (WG 36) (no picture available).

Figure 41 WG subglabrescens-milo SC 935 (IS 399C).

9 Working groups: Guinea-caudatum/Caudatum-guinea (10); Guinea-kafir/Kafir-guinea (11)

WG 100: Caudatum-guineense: Formally WG 35 in Murty and Govil's definition. Similar to caudatum (WG 30) in panicle, spikelet and grain characteristics with the following additional features of guineense (WG 20): semi-compact panicle, longer internodes of panicle and slightly involute glumes with minute teeth in mature sessile spikelets; endosperm highly glutinous; subcoat present (Fig. 42).

Figure 42 WG caudatum-guineense SC 1321.

WG 101: Nigricans-guineense: Formally WG 30(2) in Murty and Govil's definition. Clustering of spikelets and short primary branches of panicle similar to nigricans (WG 32) with the following features of guineense (WG 20): Panicle long with well-spaced internodes; rachis covered by the acute angle of the rigid branches; glumes with a slight tendency for involuteness at the time of maturity; endosperm highly glutinous; subcoat coloured as in nigricans (Fig. 43).

WG 110: Caffrorum-roxburghii: Formally WG 25 in Murty and Govil's definition. Panicle and spikelet characters resemble caffrorum (WG 40) except that the primary branches are longer and loose as compared to caffrorum; sessile spikelets more hairy and have partially

Figure 43 WG nigricans-guineense SC 66 (IS 12575C).

involute glumes as in roxburghii (WG 23); endosperm glutinous like roxburghii, grain size and shape similar to caffrorum. Plants usually short with nodal tillers (Fig. 44).

WG 111: Roxburghii-shallu: Formally WG 2 in Murty and Govil's definition. Appearance similar to roxburghii (WG 23) except the following additional features: Panicles usually long and more dense with an angle of the primary branch to main axis more acute than in WG 23; primary branches shorter and seeds held better between the involute glumes than in WG 23; the margins of the glumes are less involute; grains biconvex but nearly ovate; endosperm starchy to partly glutinous; plant height and photosensitivity considerably less than in WG 23 but with same leaf number; nodal tillering present (Fig. 45).

Figure 44 WG caffrorum-roxburghii SC 663 (IS 2232C).

Figure 45 WG roxburghii-shallu IS 2952 (Murty & Govil, 1967).

10 Working groups: Guinea-durra/Durra-guinea (12); Kafir-caudatum/Caudatum-kafir (13)

WG 120: Durra-roxburghii: Formally WG 42 in Murty and Govil's definition. Panicle loose and intermediate in appearance between durra (WG 50) and roxburghii (WG 23); panicle branches larger than in durra but stiff at base like durra; rachis completely covered as in durra; pedicelled spikelets large and persistent as in durra; grain size and grain covering in sessile spikelets similar to roxburghii; glumes tend to be slightly involute when mature; grain shape is biconvex like durra but never flattened like roxburghii; endosperm partly glutinous and white (Fig. 46).

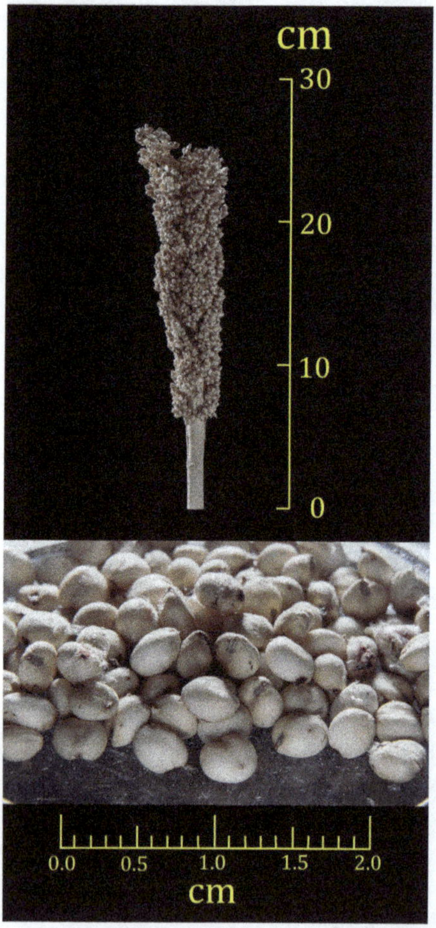

Figure 46 WG durra-roxburghii SC 449 (IS 5763C).

Figure 47 WG membranaceum IS 4671 (Murty & Govil, 1967).

WG 121: Membranaceum: Formally WG 6 in Murty and Govil's definition. Large panicles, oblong to ovate in shape, compact to semi-compact with condensed primary branches; sessile spikelets with large whitish with greenish tinge, papery glumes, sub-coriaceous in the lower region of the glume; glumes striately nerved; pedicelled spikelets persistent; grains much smaller than the glumes and completely enclosed; grains elliptic or oblong and compressed; seed white to pale yellow; endosperm starchy; subcoat present; seeds can be easily extruded by pressure at the base of glumes; frequently sectorial chimeras of small glumes resembling those in zerazera (WG 37) types are found. Plants generally tall and late in maturity (Fig. 47).

WG 122: Durra-membranaceum: Formally WG 43 in Murty and Govil's definition. Panicle, spikelet and grain characteristics like durra (WG 50) except that the glumes of sessile spikelets are papery to sub-coriaceous as in membranceum (WG 121) but with velvety hairs as in durra (Fig. 48).

WG 130: Caudatum-kafir: Formally WG 38 in Murty and Govil's definition. Panicle, spikelet and grain characteristics similar to caudatum (WG 30) except for a few features similar to caffrorum (WG 40), namely long and almost conical panicle; grain size smaller than caudatum and approaching caffrorum; grain shape biconvex as compared to the planoconvex nature of caudatum; glumes dark and not easily separated from grains as in caffrorum; a few pedicelled spikelets persistent (Fig. 49).

Figure 48 WG durra-membranaceum SC 1322.

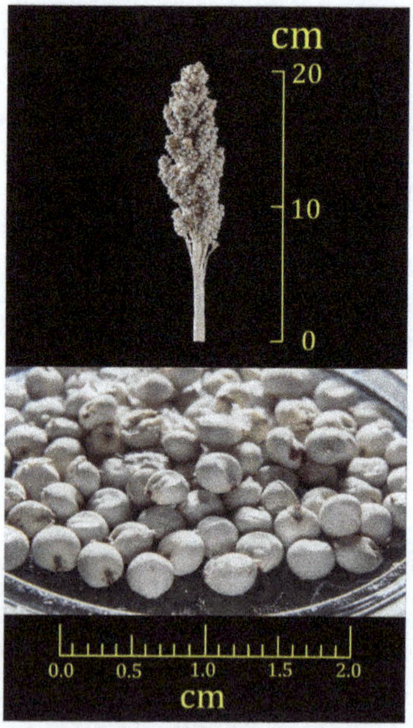

Figure 49 WG caudatum-kafir SC 64 (IS 2573C).

Figure 50 WG caffrorum-birdproof SC 655 (IS 2862C).

WG 131: Caffrorum-birdproof: Formally WG 24 in Murty and Govil's definition. Closely resembles caffrorum (WG 40) but for its invariably purple or reddish-purple grains which are bitter and therefore not liked by birds; endosperm slightly glutinous as compared to the mealy nature in caffrorum (Fig. 50).

Figure 51 WG caffrorum-darso SC 641 (IS 8166C).

WG 132: Caffrorum-darso: Formally WG 23 in Murty and Govil's definition. Panicle and grain characteristics similar to caffrorum (WG 40) except for the following: Panicles densely compact and wider than caffrorum and ovoid to cylindrical; sessile spikelet broadly ovate and subacute; upper lemma awnless; grains distinctly larger and prominently turtle back as compared to caffrorum; glumes slightly gaping when mature; endosperm mealy; subcoat light purple (Fig. 51).

Figure 52 WG caffrorum-feterita SC 184 (IS 12675C).

WG 133: Caffrorum-feterita: Formally WG 27 in Murty and Govil's definition. Panicle and spikelet characteristics essentially the same as in caffrorum (WG 40) except the following features: primary branches larger; grains larger and distinctly planoconvex (turtle back); easily separated from the glumes which are more hairy as compared to caffrorum; pedicelled spikelets partly deciduous; tillering pronounced; height variable (Fig. 52).

11 Working groups: Durra-caudatum/ Caudatum-durra (14)

WG 140: Caudatum-durra: Formally WG 40 in Murty and Govil's definition. Panicle shape, and spikelet characteristics similar to caudatum (WG 30) except that the sessile spikelets are broadly obovate with broad-topped grains as in durra; pedicelled spikelets persistent; grain shape intermediate between caudatum and durra; awns sometimes present (Fig. 53).

Figure 53 WG caudatum-durra SC 808 (IS 2680C).

WG 141: Nigricans-durra: Formally WG 32 in Murty and Govil's definition. Panicle highly condensed; ovoid to sub-cylindric like durra (WG 50); peduncle erect or goose-necked like durra; primary branches stout at the base; sessile spikelets obtuse, sparsely hairy, clustered as in nigricans (WG 32); grains small to medium with spherical top; grain colour dull red to purple. All other characteristics similar to those of nigricans (Fig. 54).

Figure 54 WG nigricans-durra SC 331 (IS 2169C).

Figure 55 WG durra-nigricans SC 53 (IS 12567C).

WG 142: Durra-nigricans: Formally WG 47 in Murty and Govil's definition. Panicle shape, branching and spikelet characteristics similar to durra (WG 50); clustering of spikelets, grain size and colour similar to nigricans (WG 32); grain shape turtle back to biconvex and can be distinguished from WG 141 which has round-topped grains; endosperm waxy white; subcoat coloured as in nigricans (Fig. 55).

WG 143: Durra-feterita/kaura: Formally WG 48 in Murty and Govil's definition. Similar to durra (WG 50) in panicle, and spikelet characters; grains are large and planoconvex as in feterita (WG 30) but wedge-shaped at the base as in durra; endosperm yellow or pale yellow and partly glutinous; grain colour is yellow as in caudatum-kaura (WG 36) (Fig. 56).

Figure 56 WG durra-feterita/kaura SC 144 (IS 12635C).

12 Working groups: Kafir-durra/Durra-kafir (15); Annual wild (16)

WG 150: Durra-kafir: Formally WG 46 in Murty and Govil's definition. Panicle long as in caffrorum (WG 40) and semi-compact but with branches with a stout base as in durra (WG 50); panicle appearance similar to caffrorum but the sessile spikelet and grain features similar to durra; the sessile spikelets are broadly obovate with velvety hairs as in durra; grain size smaller than durra but with a wedge-shaped base and broad top as in durra (Fig. 57).

WG 151: Caffrorum-durra: Formally WG 28 in Murty and Govil's definition. Panicle shorter and much wider than caffrorum (WG 40); nearly globose in shape; peduncle with curved tendency or goose-neck in shape; primary branches at a wider angle with main axis than in caffrorum; sessile spikelets larger, wider and broadly ovate in shape; glumes hairy at maturity; lower glume ending in a tooth; upper lemma usually awnless; grain large and biconvex although the ventral side is less rounded than the dorsal side; glumes mostly white, grey or tawny, grain colour variable; endosperm mealy as in caffrorum (Fig. 58).

WG 160: S. bicolor subsp. *drummondii*: No formal WG in Murty and Govil's definition or as WG 53 Grass-grain. Annual weed found in both cultivated grain sorghums and their wild relatives; plants are annuals with stout culms up to 4 m tall; lanceolate leaf blades

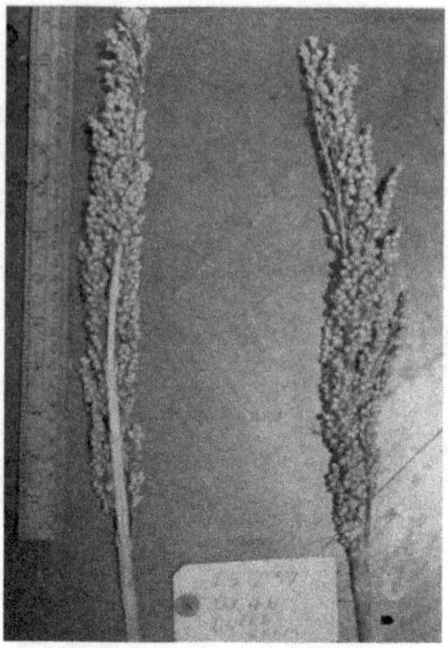

Figure 57 WG durra-kafir IS 2377 (Murty & Govil, 1967).

Figure 58 WG caffrorum-durra IS 8358 (Murty & Govil, 1967).

up to 50 cm long and 6 cm wide; panicles are variable, usually rather contracted, up to 30 cm long and 15 cm wide with somewhat pendulous branches; racemes are more or less crowded, 3–5-noded, tardily disarticulating at maturity; sessile spikelets are lanceolate-elliptic, 5–6 mm long; lemmas are hyaline ciliate, lower around 6 mm, upper around 5 mm in length, shortly lobed; and the lower often lobed; grains are variable in shape and often resemble cultivated races; pedicellate spikelets are male or neuter (Fig. 59).

Figure 59 WG *S. bicolor* subsp. *Drummondii* IS 802 (Murty & Govil, 1967).

13 Working groups: *Sorghum bicolor* subsp. *verticilliflorum* (17); Perennial wild (18)

WG 170: verticilliflorum: Formally WG 61 in Murty and Govil's definition. Plants are tufted annuals or weak biennials, with slender to stout culms up to 4 m in height; leaf blades are linear-lanceolate up to 75 cm long and 7 cm wide; panicles usually large, somewhat contracted to loose, up to 60 cm long and 25 cm wide; branches obliquely ascending, or spreading, and are sometimes pendulous; racemes 1–5 noded and fragile; sessile spikelets are lanceolate to elliptic, 5–8 mm long and usually ciliate; glumes are coriaceous, more or less ciliolate at least in the upper third; lemmas are ciliate, lower up to 7 mm with the upper around 5 mm long, 2-lobed, the upper awned from between the lobes, rarely awnless; grains are lanceolate to obovate-lanceolate; pedicelled spikelets are male or neuter, and often longer than the sessile spikelet (Fig. 60).

WG 171: arundinaceum: Formally WG 61(1) in Murty and Govil's definition. Panicle long with distinct whorls widely spaced; mature, primary branches almost at 90° to the main axis; with the following additional features: Racemes fragile; pedicelled spikelets long and deciduous; lanceolate-elliptic, acuminate sessile spikelets, shatter easily and creamy-white to yellow, sparsely hairy when young, glabrous when mature; glumes coriaceous, sometimes thinner (at least the upper half) with the greenish nerves visible beyond the middle; the lower glume usually winged on the keels near the tip with the keels sometimes ending in obscure teeth; upper lemma awned; small grains enclosed by glume; obovate, brown to dark red, partly glutinous, subcoat variable. Plants medium tall, basal tillering prominent (no picture available).

WG 172: virgatum: Formally WG 62 in Murty and Govil's definition. Panicle narrow-elliptic to oblong; semi-erect primary branches, lower up to about half the length of panicle with upper one remaining very close to the rachis; racemes fragile; pedicelled spikelets long and deciduous; broadly narrow, finely acute sessile spikelets greenish yellow, sparsely hairy when young, glabrescent when mature; glumes coriaceous, at least the upper half, or

Figure 60 WG verticilliflorum IS 3636 (Murty & Govil, 1967).

papery, with the greenish nerves visible beyond the middle; lower glume usually winged on the keels near the tip with the keels sometimes ending in obscure teeth; upper lemma awned; grains small, enclosed by glumes, obovate-elliptic; endosperm glutinous, subcoat present. Annual plants dwarf to medium with slender culms and narrow leaves; wild grasses or cultivated for fodder (Fig. 61).

WG 173: Aethiopicum: Formally WG 63 in Murty and Govil's definition. Panicle narrowly oblong, semi-erect branches; racemes mostly fragile; pedicelled spikelets deciduous; sessile spikelets large, elliptic-lanceolate, acute, tawny to reddish, rarely creamy to yellow; glumes thinly coriaceous especially at the tip, lower glume winged on the keels ending in obscure teeth, upper lemma awned; grains obovate-elliptic, tawny to dark red, small, enclosed by glumes; endosperm variable, subcoat present; culms stout, with long narrow leaves (no picture available).

WG 180: S. halepense: $2n = 40$: Formally WG 54 in Murty and Govil's definition. Perennial with elongated rhizomes; panicle erect, elliptic-oblong, branches compound; pedicels of pedicellate spikelets pilose, half as long as the sterile spikelets, jointed at the base; pedicelled spikelets persistent; sessile spikelets elliptical-oblong, tawny to light brown in colour, glumes coriaceous otherwise papery, glabrous at the base and hairy at the tip, upper lemma awned; grains small, enclosed by the glumes, obovate-elliptic, endosperm partly glutinous, subcoat present. Plants are considered wild grass, medium tall with narrow leaves (Fig. 62).

WG 181: S. propinquum: No formal WG in Murty and Govil's definition. Plants are robust, tufted perennial with stout rhizomes. Culms up to 5 m, erect; leaf blades linear, up to 1 m long and 3–5 cm wide; leaf sheaths glabrous; inflorescences are large, open, up to 60 cm

Figure 61 WG virgatum IS 2746 (Murty & Govil, 1967).

Figure 62 WG *S. halepense* IS 2695 (Murty & Govil, 1967).

long, with slender branches, the lower ones up to 25 cm long, usually bare of spikelets for as much as 5 cm at the base; racemes fragile, 1–5 noded, with slender and ciliated pedicels; sessile spikelets elliptic-lanceolate, 3.5–5.0 mm long, strigose at the base and above the middle, usually awnless; glumes crustaceous below and membranaceous near the tip, lower keeled, often ending in minute teeth, upper glume similarly keeled; lemmas thinly ciliate above the middle, lower 3.5–4.5 mm long, upper one about 3 mm long, usually awnless; anthers three; grains obovate 1.5–2.0 mm long; pedicellate spikelets 4.0–5.5 mm long, male, rarely neuter (no picture available).

14 Working groups: Unclassified (19); Breeding material (20); Mixed (21)

WG 190: Unclassified: Formally WG 70 in Murty and Govil's definition. A majority of the lines are of complex origin and could not therefore be classified. Some of the lines are new accessions to be evaluated (Fig. 63).

WG 200: Unclassified: No formal classification under Murty and Govil's definition. Germplasm or improved lines that have been modified through various breeding techniques to enhance desirable characteristics that will be used in breeding programmes for the development of improved cultivars or new hybrids (no picture available).

Figure 63 WG unclassified IS 4810 (Murty & Govil, 1967).

WG 210: Mixed. No formal classification under Murty and Govil's definition. Much of the collected germplasm comes from farmers' fields and in many cases is a mixture of races and/or working groups. This classification is used when an accession has no clear race or working group (no picture available).

15 Acknowledgements

The author gratefully thanks Marshall W. Johnson, Specialist & Entomologist Emeritus with the University of California Riverside for taking the panicle and seed pictures.

16 Appendix: Sorghum Conversion Program

Appendix Sorghum Conversion Program

Figure	Race	WG	Work. Grp.	PI #	SC #	R-Parent	(PI) or other design.	Designation	Local name	City	Country	Reasoning	B/R Rx
6	1:B	10	10:Bicolor	533940	0614	IS1333C	NSL51157	AS4601	AS 4601 (Pawaga Irungu)	Combatoire	Tanzania	Mod Nur (M)	B
7	1:B	11	11:Dochna	533904	0019	IS12539C	147937	MN635	No.33	Saboru, JinKach	Ethiopia	Div/Elite/IES	BP
9	1:B	13	13:Nervosum-kaoliang	576332	0609	IS1213C	NSL51085		AS 5200		China	Mod Nur (M1)	B
11	1:B	15	15:Sudanense	576347	0941	IS3196C	NSL55899	KS 1044	Purdue No. 81147-1	Sudan (Indiana)	USA	Mod Nur (M)	RP
12	2:G	20	20:Guineense	597975	1333	SC1333C	NSL 365762		CSM-207		Mali	Elite Mali Var/JFS	
13	2:G	21	21:Conspicuum	533749	0238	IS1201C	NSL50453	AS9117	AS9117		Africa	Mod Nur (M)	P
14	2:G	22	22:Margaritiferum	534096	0635	IS7979C	563028	SO85	Malle SO 85		Kenya	Mod Nur (A)	B
15	2:G	23	23:Roxburghii	533748	0195	IS1116C	291218	–	Nauhatta		India	Div/Elite/JCS	R
16	3:C	30	30:Caudatum	533868	0337	IS7094C	562944	–	187 SBI 151		Sudan	Mod Nur (M)	BP
17	3:C	31	31:Caudatum-nigricans	533901	0424	IS8100C	NSL54489	A-96	A-96		Japan	Mod Nur (A)	R
18	3:C	32	32:Nigricans	533939	0557	IS1318C	NSL51735	AS4055	AS 4055 (Mkambwa)		Mozambique	Mod Nur (M1)	BP
21	3:C	35	35:Dobbs	533971	0690	IS2758C	267654	–	Sorghum Soroti, Dobbs	Soroti	Uganda	Mod Nur (A)	R
22	3:C	36	36:Caudatum-kaura	533771	0366	IS7248C	NSL54178	–	Agyou		Nigeria	Mod Nur (M)	R
23	3:C	37	37:Zerazera	533964	0803	IS2586C	152781	MN899	Safara, Kordafan		Sudan	Mod Nur (A)	R
24	4:K	40	40:Caffrorum	534097	0625	IS8003C	NSL51400	HC6028	HG 6028		Japan	Mod Nur (M)	B
25	9:DB	93	93:Subglabrescens	533828	0226	IS2477C	152588	MN727	Aklamoi Red		Ethiopia	Div/Elite/JCS	R
26	5:D	51	51:Nandyal	533810	0199	IS1121C	291222	–	Karad 2-7-11		India	Div/Elite/JCS	R
27	5:D	52	52:Cernuum	533817	0209	IS1151C	291232	–	Aispuri	Maharashtra	India	Div/Elite/JCS	R
28	6:GB	61	61:Dochna-honey	534124	0015	IS12535C	147930	MN630	Lule No. 28	Marawa	Ethiopia	Div/Elite/IES	B
29	6:GB	62	62:Dochna-roxburghii	533851	0315	IS5747C	NSL51358	–	Narkatia Shittajpur	Darbhanga, Biahr	India	Mod Nur (M)	R
30	7:CB	70	70:Caudatum-bicolor	576373	0749	IS8120C	NSL54251		A 112-3		Japan	Mod Nur (M1)	B
31	7:CB	71	71:Caudatum-dochna	533866	0411	IS6964C	NSL54256	–	255 Tirter		Sudan	Mod Nur (M1)	R
32	7:CB	72	72:Nigricans-bicolor	533965	0115	IS2683C	267369	J-73	J-73	Uganda	Uganda	Div/Elite/JCS	BP

(*Continued*)

Appendix (*Continued*)

Figure	Race	WG	Work. Grp.	PI #	SC #	R-Parent	(PI) or other design.	Designation	Local name	City	Country	Reasoning	B/R Rx
33	7:CB	73	73:Dochna-nigricans	533833	0319	IS2757C	267653	–	Sorghum Soroti		Uganda	Mod Nur (M1)	R
34	8:KB	80	80:Bicolor-kafir	576435	1104	IS1104C	NSL 365709		Lula (?)	Sere	Uganda	Div/Elite/ OJW	R
36	8:KB	82	82:Dochna-kafir	534127	0021	IS12541C	147938	MN639	Uki No. 37	Albelti	Ethiopia	Div/Elite/IES	B
37	9:DB	90	90:Durra-bicolor	534131	0030	IS12550C	148097	MN702	Getetie No. 78	Djibwaku	Ethiopia	Div/Elite/IES	BP
38	9:DB	91	91:Dochna-durra	533956	0623	IS2456C	146885	MN586	Sweet Sorghum Jones		Congo	Mod Nur (M1)	RP
39	9:DB	92	92:Durra-dochna	533919	0124	IS12615C	276770	SA2288	Unnamed-R1, 4	Dire Dawa	Ethiopia	Div/Elite/JCS	RP
40	9:DB	93	93/92:Subgla-brescens/ Durra-dochna	576375	1014	IS11443C	329707		Col. No. R-71	Debre Berham Shoa Prov.	Ethiopia	High Altitude	R
41	9:DB	94	94:Subglabrescens-milo	576388	0935	IS399C	NSL84173				USA	Mod Nur	
42	10:GC	100	100:Caudatum-guineense	597968	1321	SC1321C	NSL 365751		Col. El Obeid 8-1, Korky	El Obeid Area, N. Kordofan	Sudan	Drought Res	
43	10:GC	101	101:Nigricans-guineense	533913	0066	IS12575C	152748	MN877	Nyan Dok	Equatoria	Sudan	Div/Elite/IES	R
44	11:GK	110	110:Caffrorum-roxburghii	533948	0663	IS2232C	NSL51585	–	Nebraska 6350	Lincoln, Nebraska	USA	Mod Nur (M2)	B
46	12:GD	120	120:Durra-roxburghii	597950	0449	IS5763C	591377		Karkatia Salimpur	Monghyr, Bihar	India	Mod Nur (M)	R
48	12:GD	122	122:Durra-membranaceum	597969	1322	SC1322C	NSL 365752		Col. El Obeid 8-2, Ferik	El Kharta, N. Kordofan	Sudan	Drought Res	
49	13:KC	130	130:Caudatum-kafir	533757	0064	IS2573C	152736	MN864	Monshal	Equatoria	Sudan	Div/Elite/IES	R
50	13:KC	131	131:Caffrorum-birdproof	533976	0655	IS2862C	267519	–	Framiola DL/59/1539	Pretoria	South Africa	Mod Nur (M2)	R
51	13:KC	132	132:Caffrorum-darso	534104	0641	IS8166C	NSL51499	–	EC 21360, G29		Uganda	Mod Nur (A)	B
52	13:KC	133	133:Caffrorum-feterita	597958	0184	IS12675C	277542		Suki PS295	Pretoria	South Africa	Div/Elite/JCS	BP
53	14:DC	140	140:Caudatum-durra	597944	0808	IS2680C	267366		J-16		Uganda	Mod Nur (M2)	R
54	14:DC	141	141:Nigricans-durra	533824	0331	IS2169C	221662	–	Bonkum	Panskshin	Nigeria	Mod Nur (M1)	P
55	14:DC	142	142:Durra-nigricans	533788	0053	IS12567C	152636	MN775	Feterita Fuyoumi	Anglo-Egyptian	Sudan	Div/Elite/IES	R
56	14:DC	143	143:Durra-feterita/kaura	533921	0144	IS12635C	276799	SA2317	Unnamed-R1, 38	Dire Dawa	Ethiopia	Div/Elite/JCS	R

17 References

Dahlberg, J. A. 2000. Classification and characterization of sorghum, pp. 99–130. In C. W. Smith and R. A. Frederiksen (Eds) *Sorghum: Origin, History, Technology, and Production*. John Wiley & Sons, Inc., New York, NY.

Dahlberg, J. A. and M. Spinks. 1995. Current status of the U.S. sorghum germplasm collection. *International Sorghum and Millets Newsletter* 36: 4–11. ICRISAT, Patancheru, India.

de Wet, J. M. J. 1978. Systematics and evolution of sorghum Sect. Sorghum (Gramineae). *Am. J. Bot.* 65:477–84.

de Wet, J. M. J. and J. R. Harlan. 1971. The origin and domestication of *Sorghum bicolor. Econ. Bot.* 25:128–35.

de Wet, J. M. J. and J. P. Huckabay. 1967. The origin of *Sorghum bicolor*. II. Distribution and domestication. *Evolution* 21:787–802.

de Wet, J. M. J. and E. G. Price. 1976. Plant domestication and indigenous African agriculture, pp. 453–64. In J. R. Harlan, J. M. J. de Wet and A. Stemler (Eds) *Origins of African Plant Domestication*. Mouton Publishers, The Hague.

de Wet, J. M. J., J. R. Harlan and B. Kurmarohita. 1972. Origin and evolution of guinea sorghums. *E. Afr. Agr. Forestry J.* 38:114–19.

Doggett, H. 1988. *Sorghum*. John Wiley & Sons, Inc., New York.

Harlan, J. R. 1975a. Geographic patterns of variation in some cultivated plants. *J. Hered.* 66:184–91. https://doi.org/10.1093/oxfordjournals.jhered.a108610

Harlan, J. R. 1975b. *Crops and Man*. American Society of Agronomy, Madison, Wisconsin.

Harlan, J. R. and J. M. J. de Wet. 1972. A simplified classification of cultivated sorghum. *Crop Sci.* 12:172–6.

Harlan, J. R. and A. Stemler. 1976. Plant domestication and indigenous African agriculture, pp. 465–78. In J. R. Harlan, J. M. J. de Wet and A. Stemler (Eds) *Origins of African Plant Domestication*. Mouton Publishers, The Hague.

Mann, J. A., C. T. Kimber and F. R. Miller. 1983. *The Origin and Early Cultivation of Sorghums in Africa*. Texas Agric. Exp. Stn. Bull. Texas A&M University, College Station, TX, p. 1454.

Murty, B. R. and J. N. Govil. 1967. Description of 70 groups in genus sorghum based on a modified Snowden's classification. *Indian J. Genet.* 27:75–91.

Porteres, R. 1962. Berceaux agricoles primaires sur le continent Africain. *J. Afr. Hist.* 2:195–210.

Shechter, Y. and J. M. J. de Wet. 1975. Comparative electrophoresis and isozyme analysis of seed proteins from cultivated races of sorghum. *Am. J. Bot.* 62:254–61.

Snowden, J. D. 1936. *The Cultivated Races of Sorghum*. Adlard and Son, Ltd., London.

Stemler, A., J. R. Harlan and J. M. J. de Wet. 1975. Caudatum sorghums and speakers of Chari-Nile languages in Africa. *J. Afr. Hist.* 16:161–83.

Chapter 2

The comparative and evolutionary genomics of sorghum

Elizabeth A. Cooper and Stephen Kresovich, Clemson University, USA

1 Introduction

Sequencing of the first complete genome was accomplished over 20 years ago (Fleischmann, 1995), and since then the number of sequenced genomes has grown rapidly to include thousands of species (NCBI Genome Browser, 2016), including many crops of economic importance (Morrell, 2012). With so many sequences available, comparative genomics has emerged as a logical and powerful tool for finding changes at the DNA level that underlie a given species' unique characteristics (Touchman, 2010). With the advent of cost-effective high-throughput sequencing technology, it is now more feasible than ever to generate complete re-sequencing data sets for entire populations, and the future of crop genomics and molecular breeding will soon transition from the use of selected markers to focus on comparisons of entire individual genomes both within and across species (Morrell, 2012).

Many early comparative studies centred on discovering the origins and histories of different taxa, but this approach is not limited to answering questions about evolution; it can be broadly applied to any field in which connecting genotype to phenotype is the goal. At the same time, the field of comparative genomics is rooted in evolutionary theory; correctly interpreting the similarities and differences between two genomes depends on understanding their phylogenetic relationship and the predictions made by theoretical models.

The study of evolution and the advancement of agriculture have an extensively shared history throughout which each field has significantly contributed to and benefited from progress in the other. Experiments conducted by plant and animal breeders in the early nineteenth century greatly influenced the research of both Charles Darwin and Gregor

http://dx.doi.org/10.19103/AS.2017.0015.04

Mendel (Cobb, 2006). Darwin actually devoted the first chapter of *On the Origin of Species* (Darwin, 1859) to drawing a parallel between human-mediated breeding of domesticated species and the selection of better-adapted individuals in nature (Brown, 2010). This connection introduced the idea of natural selection and provided a proof of concept for it as a mechanism of gradual change, which in turn distinguished Darwin's work from earlier versions of evolutionary theory.

Following the Modern Synthesis in the early twentieth century, the laws of Mendelian genetics were incorporated into evolutionary theory to create the field of population genetics and subsequently provide a powerful framework for understanding how selection shapes the variation observed in populations and species (Huxley, 1942). Now, the connection that Darwin originally drew between selective breeding and natural selection has taken on a new importance as modern scientists attempt to use predictions made by evolutionary theory to accelerate breeding and address the coming challenges posed by population growth and climate change.

Even before the integration of Mendel's and Darwin's work, Russian botanist and geneticist Nikolai Vavilov realized early on the implications of evolutionary theory for advancing agricultural practices as he sought ways to end famine in his native country (Harland, 1954). He noted that many related cultivated plant taxa not only shared certain features in common among all individuals, but also exhibited a very similar set of variations among individuals within each taxonomic group (Vavilov, 1922). Based on this observation and his understanding of the descent with modification process put forward by Darwin, he formulated the 'law of homologous series', which predicts that similar sets of morphological variants across species have arisen from shared genetic variants inherited from their common ancestor. This has several important implications for modern crop improvement: 1) closely related species may have the same genetic potential for desirable traits, even if the trait is not currently visible in all of the species, 2) dissecting the architecture of a trait in one species is likely informative for other species, 3) pre-existing diversity in the gene pool of cultivated varieties and their wild ancestors is a valuable resource for continued crop improvement (Tanksley, 1997; McCouch, 2004) and 4) natural diversification is constrained by evolutionary history and the diversity of the ancestor.

In his writing, Vavilov pointed to cereal crops as a remarkable example of homologous variation occurring not only across species, but also across the entire family of grasses as a whole. Cereals have been independently domesticated worldwide (Pringle, 1998; Doebley, 2006), adapted to a wide array of climates and agricultural practices, and are the dominant source of nutrition for much of the world's population. This incredible versatility has stemmed from a single common ancestor and a core set of genes that characterize the grass family (Freeling, 2001).

Although each cereal crop has its own origin, and in some cases multiple origins, the evolutionary path from wild grass to cultivar was strikingly parallel across species due to recurrent selection on the same types of traits during the early stages of domestication (Harlan, 1973; Doebley, 2006; Fuller, 2009). At the molecular level, this has often involved the same underlying genes, but not always (Doebley, 2006; Fuller, 2009; Tang, 2010; Meyer, 2012, 2013). Genomic studies have revealed that, as expected, these domestication events have led to population bottlenecks and reductions in diversity; in some cases this could potentially limit the genetic substrate for further selection (Tanksley, 1997; Wright, 2005; Doebley, 2006). Considering cereal crops and their wild relatives as a single system is one way to find and recover useful variation that may have been lost in one lineage, but not others (McCouch, 2004; Morrell, 2012; Gepts, 2014).

To fully unlock the potential of these economically important crops, it will be necessary to determine the genomic architecture of key traits that underlie their unique adaptability and versatility. Sorghum [*S. bicolor* (L.) Moench] provides an excellent model for dissecting these traits. Unlike its close relatives, including maize, sugarcane and wheat, sorghum has a relatively small and tractable diploid genome, and retains much of the chromosomal structure and organization of its common ancestor (Paterson, 2009; Kellogg, 2013). Sorghum also has multiple origins of domestication in different regions of Africa (Kimber, 2000), and has since spread across most of the African continent, the Middle East, India and East Asia, resulting in a highly diverse germplasm adapted to multiple climates and uses (Morris, 2013). Because of its African origins, sorghum is highly drought tolerant, a characteristic that will become increasingly important for growers in the future as growing populations and climate change pose new challenges.

By comparing sorghum with other crops at the genomic level, scientists can capitalize on their shared history to find hidden genetic potential and create new combinations of desirable traits to accelerate and enhance sorghum breeding. Additionally, because sorghum already has multiple uses as food, fuel and forage material, understanding the unique aspects of the sorghum genome that enable this versatility will also offer insight into the improvement of other cereal crops.

2 Structure and evolutionary origins of the sorghum genome

The family of grasses to which sorghum and other cereal crops belong is called the Poaceae, and with nearly 10 000 species, it is one of the largest and most widely distributed plant families in the world (Clayton, 1986). Evidence from modern grass genomes indicates the entire family arose from a single common ancestor that underwent a whole-genome duplication (WGD) event roughly 70 million years ago (Kellogg, 2004). This initial duplication event has been followed by numerous large-scale rearrangements, deletions, fusions and even additional duplications occurring at different points throughout the diversification of grasses (Kellogg, 2004, 2013; Murat, 2010). The result has been that modern grass species display an astounding amount of diversity in terms of chromosome number, size and organization.

At first glance, it may seem that the presence of dramatic structural changes contradicts the prediction that grasses could be treated as a single system, but further comparison of the available genomes reveals they are largely syntenic, meaning the ancestral gene content and gene order has been preserved in long sections of the genome, even though the sections themselves have been rearranged (Bennetzen, 2000; Freeling, 2001).

The modern-day sorghum genome is relatively small compared to many of its close relatives. It is a diploid and contains ten chromosomes (Kim, 2005; Price, 2005; Murat, 2010; Kellogg, 2013). These chromosomes are identical in number and surprisingly similar in structure to the predicted cereal ancestor (Paterson, 2009; Murat, 2010). The small size and relative stability of the sorghum genome over time make it an appealing model for some of its more complex relatives, such as the highly polyploid sugarcane genome or the extensively rearranged maize genome (Ming, 1998; Bennetzen, 2009; Paterson, 2013).

Despite drastic size differences, sorghum has a very similar euchromatic content to other known crop genomes, and fewer than 10% of genes found in sorghum appear to be lineage specific, with the remainder having known counterparts in at least one or all of the sequenced cereals (Kim, 2005; Paterson, 2009). The majority of large-scale differences are the result of expansion and turnover of repetitive elements in the heterochromatic regions of chromosomes, which collectively comprise over half of the sorghum genome (Kim, 2005; Paterson, 2009, 2010). Rapid expansion of certain repeat families is another signature of evolution that is common across many plant species; these repetitive elements not only drive the genome size changes discussed above, but can also increase the frequency of other structural changes, including inversions, tandem duplications and translocations (Coghlan, 2005).

However, in sorghum, nearly 70% of the genes are found outside the repeat-rich heterochromatin and are located in the euchromatic arms of chromosomes, where a high degree of synteny exists between sorghum, maize and rice (Ahn, 1993; Bennetzen, 2000; Bowers, 2003; Kim, 2005; Paterson, 2010). This not only reinforces the idea that future improvements and discoveries made about gene function in one species are likely applicable to other members of the Poaceae, but also implies that similar selective and adaptive potential should exist across species boundaries.

3 Sorghum genetic development during domestication

3.1 Genes under selection

Predictions made by the law of homologous series with regard to recurrent selection acting on the same characters and therefore the same genetic substrate are most visible when considering the loci selected during domestication. Even though the initial domestication of cereals from their wild ancestors occurred at different times on separate continents and was undertaken by completely distinct cultural groups, there was consistent unconscious selection on the same set of traits that facilitated the harvesting and replanting of seeds (Harlan, 1973; Doebley, 2006; Murphy, 2007; Fuller, 2009; Lin, 2012; Meyer, 2013; Olsen, 2013).

For grasses, one of the most crucial traits in the 'domestication syndrome' is the loss of seed shattering, whereby mature plants will retain seeds until harvest rather than shedding them for natural propagation (Doebley, 2006; Fuller, 2009). This trait has a relatively simple genetic architecture: in sorghum it is caused by mutations in a single locus (*Sh1*) (Paterson, 1995) encoding a transcription factor (Lin, 2012). This gene has orthologs in rice, maize and foxtail millet, and appears to have undergone parallel selection during domestication in all of these species, although in rice and maize there are also additional genes implicated in the non-shattering phenotype (Li, 2006; Konishi, 2006; Fuller, 2009; Tang, 2010; Lin, 2012). Even within sorghum, there is evidence of recurrent selection on this gene, as there are three alternate haplotypes present at this locus, which seemingly correspond to multiple independent domestication events in sorghum (Lin, 2012).

Other phenotypic changes that are common to grass domestication include less stalk branching and more uniform maturation (also called determinate growth), increased seed production and loss of seed dormancy (Harlan, 1973; Fuller, 2009). Not all of the genetic controls for these traits are currently known, but for the ones that are, there is additional evidence for parallel changes in orthologous genes leading to the domesticated phenotype in multiple crops [reviewed in Lenser (2013)].

Following initial domestication to make crops more harvestable and dependent upon humans for propagation, every cereal crop has also undergone diversification and improvement (Meyer, 2013). This second phase includes local adaptation and the spread of crops to new environments, improvements in overall yield, and modifications of grain taste and texture based on cultural preference, among other characteristics. Even though the traits involved in diversification can be more varied for each species than those underlying domestication, there are still several instances of parallel selection on the same gene across species. A well-characterized example of this is the *WAXY* gene, which has been the target of repeated selection to produce 'stickier' grains in sorghum, maize, rice, wheat, barley and millet for a range of end uses (Meyer, 2013; Lenser, 2013).

All of the most clear-cut examples of parallelism occur when traits have a simple genetic basis. Most phenotypes tend to be more complex, and for these there is still evidence of parallel selection, but the patterns are rarely perfect. Often, there is significant overlap in the genes involved, but other lineage-specific genetic changes are also contributing to the trait in each species (Fuller, 2009; Tang, 2010; Lenser, 2013; Gepts, 2014). The non-shattering phenotype exhibits this to a certain degree; a single locus confers loss of shattering in sorghum (Paterson, 1995), but maize and rice each have at least two loci with significant effects (Li, 2006; Konishi, 2006; Lin, 2012).

The more complex a phenotype is, the less likely it is for the exact same set of genes to be recurrently selected across lineages. Short stature was a crucial target of selection during the 'green revolution' as a means of reducing lodging and increasing harvest index (Khush, 2001), and dwarf varieties of several crops were created during this era. However, despite simultaneous selection and similar resultant phenotypes, the genes underlying dwarfism are numerous and varied across species (Lin, 1995; Fernandez, 2009; Lenser, 2013). Multiple pathways can affect plant height; brassinosteroid (BR) and gibberellin (GA) synthesis and signalling are the pathways subject to the most recurrent selection across species (Fernandez, 2009; Lenser, 2013), although often not through the same underlying genes.

Complex biological pathways are typically robust, meaning that there is often more than one way to produce a phenotype, and a single change within the pathway will not always alter the end result (Waddington, 1957). Therefore, it makes logical sense that the same traits could evolve convergently in crops through different genetic changes happening in the same underlying network (Gepts, 2014). For dwarfed varieties of maize, wheat and rice, this appears to be the case, as GA- and BR-related genes are repeatedly implicated in studies of plant height (Peng, 1999; Fernandez, 2009). In sorghum, however, dwarfism has a different architecture. Of the four major loci found to confer dwarfism (Quinby, 1974), only one (*dw3*) is well characterized. This gene, which affects auxin transport, has undergone parallel selection in maize, but is not part of the GA or BR pathways more commonly associated with plant height reduction (Multani, 2003). Another sorghum dwarfing gene (*dw1*) was long hypothesized to be involved in GA metabolism (Brown, 2008), but recent results suggest that this may in fact be a previously unidentified membrane transporter of unknown function (Hilley, 2016).

Flowering time is another complex phenotype that is often related to height (Lin, 1995). Changes in flowering time are important for adapting crops to new environments, and were especially important for converting tropical grasses to temperate environments (Khush, 2001; Olsen, 2013). Like plant growth, flowering time is controlled by interconnected signalling pathways, and to date, many different genes have been identified in various crops as targets of selection. Sometimes the same genes are involved in more than one species and sometimes not (Olsen, 2013), although there is still more consistency across taxa than would be expected by chance (Lin, 1995).

In the case of flowering time, sorghum again displays a simpler genetic architecture than many of its closely related counterparts (Quinby, 1974). The gene with the most significant effect (*ma1*) is a circadian clock-associated transcription factor that has undergone multiple independent mutations throughout sorghum's history (Murphy, 2011; Klein, 2015). In fact, many of the genes under parallel selection during domestication and crop improvement are transcription factors or other regulatory genes that affect conserved pathways (Doebley, 2006; Tang, 2010; Meyer, 2013; Gepts, 2014). As gene expression data becomes more widely available, it will be possible to identify more common networks and their regulators, providing additional insight into the shared adaptive potential across species.

3.2 Whole-genome patterns

Regardless of which locus or trait is selected, an important general genomic signature of the domestication process is the overall reduction in genetic diversity (Wright, 2005; Meyer, 2013; Gepts, 2014). A domestication event is equivalent to a population bottleneck, which increases genome-wide linkage disequilibrium (LD) and decreases effective population size (Ne) (Wright, 2005), both of which are factors that can hinder the efficacy of selection and remove adaptive potential from a population (Meyer, 2013). The crop improvement phase can reintroduce diversity in some cases, but domesticated cultivars will almost always be less genetically diverse than their wild counterparts.

The broad patterns of genomic diversity in sorghum reflect its evolutionary history in several ways. Genome-wide surveys of polymorphism have shown that wild relatives are more genetically diverse than cultivated sorghums (Hamblin, 2004, 2006; Tang, 2010; Muraya, 2011; Mace, 2013; Billot, 2013; Fernandez, 2014), and that regional patterns of diversity among cultivated sorghums corresponded strongly with domestication origins (Mace, 2013; Billot, 2013; Morris, 2013), with populations nearest to the centre of origin being the most diverse.

Sorghum was originally domesticated roughly 10 000 years ago near present-day Sudan (Kimber, 2000), and subsequently spread throughout the African continent, the Middle East, India and parts of Asia. The broad range of climates and cultural practices encompassed by sorghum's range has resulted in distinct racial types that are largely defined by their panicle architecture and seed characteristics (Harlan, 1972), with panicle architecture being another trait that seems to be under parallel selection on orthologous genes in other cereals (Morris, 2013). Density and shape of the panicle is important for preserving the grain under varying wet or dry conditions, but many other mutations have also allowed the spread and local adaptation of sorghum as well. For this reason, sorghum forms genetic clusters based on both botanical race and geography (Perumal, 2007; Casa, 2008; Morris, 2013; Billot, 2013). One exception is the bicolor group, the primitive sorghum race, which shares genetic variants with most other botanical races, reflective of its ancestral status (Morris, 2013).

The guinea margaritiferum types are genetically distinct from other sorghum working groups, although they also display relatively high diversity among themselves, which seems to confirm a separate origin of domestication in West Africa (Mace, 2013; Morris, 2013). The presence of alleles unique to this group could be a particularly valuable resource for improving crop performance on difficult soil types.

Known domestication loci show signatures of strong selective sweeps, which include a sharp reduction in haplotype diversity and extended LD around the selected locus. Genes underlying traits involved in sorghum diversification and improvement show similar patterns of diversity reduction, but typically the LD surrounding these loci is not as extensive as the more severely bottlenecked domestication traits (Meyer, 2013).

Outside the selected regions, LD in sorghum begins to decay significantly within 20 kb, and reaches background levels within approximately 150 kb (Hamblin, 2004). Compared to maize, an outcrossing crop, levels of linkage in sorghum are seven times higher, but they are similar to rice, which, like sorghum, mainly relies on self-pollination. Both linkage and single nucleotide polymorphism (SNP) density vary spatially across the sorghum genome. In heterochromatic regions, SNP density is very low and recombination is essentially non-existent. Gene-rich euchromatic regions display local variation in the recombination and polymorphism rates, which are often correlated (Evans, 2013). Most SNPs (83%) are located in intergenic regions both in wild and cultivated varieties, reflective of a certain level of background selection maintaining gene function (Mace, 2013). However, most genes still contain a few SNPs, and this is especially true for duplicated genes, indicating that standing variation already holds much natural adaptive potential.

Genomic patterns of mutation and recombination, aside from being informative about the evolutionary past of sorghum, are important factors when considering its future improvement as well. Domestication may have been facilitated by linkage among genes controlling sets of related traits, but continued linkage between genes can actually slow the rate of evolution and hinder selection (Hill, 1966). Similarly, reductions in genetic diversity during both domestication and improvement phases restrict the adaptive substrate within groups. In an era when transgenics are feasible, though not yet widely applied in sorghum, these effects may seem trivial and easily overcome by inserting new genes or alleles into the genome. However, standing genetic variation is still the best and most viable source of improved traits. From an evolutionary standpoint, existing mutations have an advantage because they have already persisted through years of selection to remove potentially deleterious alleles that could have unforeseen fitness costs (McCouch, 2004; Barrett, 2008). In sorghum, while variation may have been lost in a certain race or population, comparative genomics has shown that other races, wild ancestors and even more distant relatives in the grass family harbour alternate alleles that can be reintroduced and recombined to create new and improved varieties.

3.3 Targets of improvement

For most of its history, breeding of sorghum has been focused on improving grain yield and quality, and this was a major objective of the green revolution (Khush, 2001). Grain sorghums have been strongly selected for short stature, photoperiod insensitivity and large panicles, but these traits represent only a small subset of diverse phenotypes available in the sorghum germplasm.

Other major types of sorghum include sweet, bioenergy and forage types. All of these types more closely resemble their tropical ancestor than cultivated grain sorghum in that they tend to be tall and photoperiod sensitive, but differ from each other in their patterns of carbon allocation and structural composition. Despite the drastic phenotypic differences between all types, their broader genomic patterns of variation still follow underlying historical racial and geographic structures more than their current classifications (Ritter, 2007; Wang, 2009a; Brenton, 2016). This reflects their polyphyletic origins from a multitude of different racial types, but also shows how much potential for selection and improvement there is in the same set of underlying genes with very few genetic changes. Recent results point to the importance of developmental regulators for increasing both biomass and sugar accumulation, which suggests a mechanism for achieving morphological diversity with minimal genotypic differences (McKinley, 2016).

4 Genomic architecture of key sorghum traits: C_4 photosynthesis, drought and aluminium tolerance

4.1 C_4 photosynthesis

One of the most appealing facets of sorghum's biology is its adaptation to harsh environmental conditions, including heat, drought and marginal soil types. Unravelling the molecular basis of these adaptations will allow for further improvement of sorghum and its cereal relatives to meet the challenges brought on by rising temperatures and a growing human population. Comparative genomics analyses have already offered insights into some of these traits, which range from extremely simple (e.g. soil toxicity) to incredibly complex (e.g. flowering time) in their architecture (Lasky, 2015).

A key feature of sorghum is that it uses C_4 photosynthesis, a derived photosynthetic pathway characterized by less CO_2 loss at higher temperatures and more efficient use of water and nitrogen [reviewed in Sage (2012)]. This pathway represents a remarkable case of evolutionary convergence; it has arisen over 60 different times in angiosperms, and over 20 of these origins have occurred just within the Poaceae. Physiological changes required for C_4 evolution typically include reorganization of the leaf anatomy, changes to the number and distribution of mitochondria, and compartmentalization of photosynthetic enzymes, but the specifics vary widely across plant lineages (Sage, 2012).

Although the shift to C_4 has recurred many times, certain families show a strong tendency towards the evolution of this phenotype. Over two-thirds of modern C_4 plants come from just four families, and the Poaceae contributes by far the largest number of species as well as the first instance of C_4 evolution (Osborne, 2006). Part of this predisposition may stem from the presence of anatomical features that permit C_4 photosynthesis, but a large factor seems to be the genome structure of grasses, and specifically the clustered gene families that have resulted from duplication events (Wang, 2009b; Sage, 2012).

Several gene families produce photosynthetic enzymes, and comparison of the rice, maize and sorghum genomes reveals that several types of duplications have had roles in shaping each of these families (Wang, 2009b). Even though rice is a C_3 plant, it contains at least one member of each enzyme-encoding gene family, since these enzymes are used in traditional photosynthesis as well. Several of these gene families underwent expansion following the ancestral WGD event 70 million years ago; in sorghum and maize, the new gene copies went on to acquire C_4 functionality through adaptive evolution, but the same changes did not happen in rice (Wang, 2009b). In other cases, the gene families only underwent expansion via tandem gene duplication following the divergence of sorghum and rice.

The process of C_4 evolution in sorghum highlights the role that duplication can play in the creation of new genes with acquired functions. It also demonstrates that multiple instances of duplication followed by extended periods of mutation and selection were required before a new pathway could be developed, which may help explain why some grasses, such as rice, possess the same starting genetic material and yet have not developed C_4 photosynthetic capabilities.

4.2 Drought tolerance

Sorghum's drought tolerance is one of its most desirable, but also most complex traits. Plants can use many different strategies throughout their life cycle to tolerate drought, including the development of morphological features to avoid drought stress in arid climates, as well as physiological mechanisms to respond to water loss once drought stress has become unavoidable (Bray, 1997). Many particularly well-adapted species use a combination of strategies, making the underlying genetics especially hard to dissect. C_4 photosynthesis is one mechanism to improve water use efficiency, but sorghum also has other characteristics that make it one of the most drought-tolerant crops in the world.

The stay-green phenotype is the ability of a plant to retain leaf area and continue photosynthesis post-flowering and during grain fill, and it has also been shown that stay-green varieties are more drought tolerant in general (Borrell, 2014). This phenotype has been reported in many species, including other cereals, although the physiological basis is not fully understood. Several QTLs have been identified in sorghum that appear to consistently be associated with stay-green in multiple environments, and evidence suggests that these QTLs play a role in both canopy and root development (Borrell, 2014). Gene expression studies have found thousands of genes differentially expressed in association with stay-green; a gene involved in proline biosynthesis was not only differentially expressed, but also mapped to the region underlying a previously identified QTL in sorghum (Johnson, 2015). Other differentially expressed genes, including a signalling gene that affects stomatal aperture and a transcription factor that affects expression of numerous genes, have known orthologs in other cereals and seem to play similar roles in stress tolerance across the grass family.

Transcription factors and other regulators of gene expression appear especially important in drought-response (vs drought avoidance) phenotypes. Across grass species, a handful of *cis*-acting promoter elements are consistently enriched among genes differentially expressed during drought stress (Dugas, 2011). Certain classes of miRNA that target stress-response networks have undergone gene family expansions in sorghum relative to rice and maize, which may be linked to its enhanced drought tolerance (Paterson, 2009). Therefore, even though complex interactions of thousands of genes are required for drought adaptation, it may be possible to identify common regulatory elements that could be selectively targeted to confer significant improvements in water use efficiency across species.

4.3 Aluminium tolerance

High performance on marginal soil types is another trait essential to future crop cultivation. Adaptations to aluminium toxicity, which is especially prevalent in tropical and subtropical growing regions, have been observed in multiple grass lineages, including sorghum, rice, barley and wheat.

Comparative studies have now demonstrated that there are numerous independent loci that have convergently evolved mutations conferring aluminium tolerance in different lineages (Magalhaes, 2004). In wheat and barley, mutations in *ALMT1*, a malate transporter, are responsible for aluminium-tolerant phenotypes. Although this gene has a sorghum ortholog, it is not responsible for aluminium tolerance in sorghum. Instead, a different transporter, *SbMATE*, allows for the adaptation (Magalhaes, 2007). In rice, which has the highest levels of aluminium tolerance among cereal crops (Foy, 1988), the genetic architecture appears to be much more complex (Famoso, 2011; Tsutsui, 2012), but does not involve the previously identified sorghum or wheat genes. Again, the orthologous genes are present across taxa, demonstrating that the substrate for improvement is present in multiple species, but different evolutionary paths have led to the aluminium tolerance phenotype in each case.

5 Assembling and exploiting a pangenome for maize, rice and soybean

The history of evolution in angiosperms and the impact of gene duplications and family expansions on key traits such as photosynthesis and possibly drought tolerance highlight the importance of structural changes and their effect on phenotypic differences at the species level. There is now emerging evidence for the importance of structural changes at the level of individual variation within species as well (Fu, 2002; Han, 2003; Mace, 2013), which is not surprising given the dynamic nature of grass genomes. Marker-based sequencing approaches will not reveal the extent of structural polymorphism; instead, *de novo* assembly of whole genomes is required. Genome assembly necessitates high coverage and, in a repetitive genome like sorghum, long reads to span heterochromatic regions. Performing a *de novo* assembly of every individual in an association mapping panel would currently be both cost and time prohibitive, but recently there has been increased interest in assembling 'pangenomes' for multiple plant species.

The pangenome concept originated in prokaryotes (Tettelin, 2005), but has now been applied in maize, rice and soybean (Golicz, 2016). The purpose of a pangenome is to create a consensus assembly of multiple individual genomes into a single reference that contains the genomic content for the entire species. Ideally, the pangenome could be constructed from a handful of representative individuals, and then subsequently used as an effective reference for a much larger mapping population without the cost of a complete genome assembly every time.

In maize, which has a highly dynamic genome, it was found that the original reference genome contained roughly only 70% of the genomic content of the pangenome, meaning a lot of the within-species diversity went unobserved in all studies that relied on a single reference. In rice, which seems to be more structurally similar to sorghum and more stable than maize, 8% of the pangenome differed among three cultivars (Schatz, 2014), which is still a significant amount of structural difference. A study of re-sequencing data from a handful of sweet sorghum lines found evidence for over 70 new genes in the sweet lines that were not present in the BTx623 reference, along with thousands of other smaller structural variants (Zheng, 2011). Given that structural variation has been instrumental in shaping the genomes of all grasses over the course of their evolution, it is not surprising that structural differences continue to be an important substrate for selection in present-day domesticated crops.

6 Conclusions and where to look for further information

This chapter has primarily focused on comparative genomics undertaken since the completion of the sorghum reference genome in 2009, but decades of genetic research in sorghum have been done to build the foundation and the resources used in these studies. Over the years, many biparental mapping populations have been developed to exploit the diversity of sorghum in order to dissect the genetics of desirable traits. Prior to sequencing of the sorghum reference genome, many traits had already been mapped using crosses between phenotypically distinct sorghum lines and typically a few hundred to a few thousand molecular markers (Bowers, 2003; Ramu, 2010). Early successes in these QTL studies identified genomic regions affecting crucial traits such as insect and disease resistance (Agrama, 2002; Perumal, 2009), drought tolerance (Haussmann, 2002; Sanchez, 2002), flowering time and plant height (Lin, 1995; Feltus, 2006), and tillering (Hart, 2001). Only in more recent years has the availability of the genome sequence allowed for fine mapping the genes underlying these loci.

Now, hundreds of thousands, and in some cases millions, of SNP markers are publicly available for many diverse sorghum accessions through databases such as Gramene (http://www.gramene.org/) and SorGSD (http://sorghumsnp.big.ac.cn/) (Luo, 2016). These markers can be used to map traits of interest or assist in the selection of lines for breeding new hybrids. Complete reference genomes for sorghum and other grass species can be accessed through the Department of Energy Joint Genome Institute (JGI), a facility focused on using genome sequencing technologies for improving bioenergy and environmental research. JGI offers free access to several databases and tools specifically designed for comparative plant genomics, including Phytozome (https://phytozome,jgi.doe.gov), which hosts the sorghum reference genome along with numerous other reference genomes, and KBase, a software for predicting genes and their functions. They also currently have nearly 100 sorghum genome, transcriptome and methylome projects.

The diversity and trait architectures described in this chapter barely scratch the surface of the complexity of grass genomes, particularly at the intraspecific level. Hundreds of sorghum accessions have been genetically characterized, but the complete germplasm boasts over forty thousand lines (Crop Trust, 2007), undoubtedly harbouring a wealth of useful and unexplored variation. As the available genomic information on not just sorghum, but all cereals continues to grow, we can not only begin to understand how a common set of ancestral genes gave rise to these diverse and successful crops, but we can also harness the natural potential left over from the past in order to prepare for the future.

7 References

Agrama, H. A., G. E. Widle, J. C. Reese, L. R. Campbell and M. R. Tuinstra (2002) Genetic mapping of QTLs associated with greenbug resistance and tolerance in *Sorghum bicolor. Theor. Appl. Genet.* 104: 1373–8.

Ahn, S. and S. D. Tanksley (1993) Comparative linkage maps if the rice and maize genomes. *Proc. Natl. Acad. Sci.* 90: 7980–4.

Barrett, R. D. H. and D. Schluter (2008) Adaptation from standing genetic variation. *Trends in Ecol. and Evol.* 23: 38–44.

Bennetzen, J. L. (2000) Comparative sequence analysis of plant nuclear genomes: microcolinearity and its many exceptions. *Plant Cell* 12: 1021–30.

Bennetzen, J. L. (2009) Maize genome structure and evolution. In: J. L. Bennetzen and Sarah C. Hake (Eds.) *Handbook of Maize: Genetics and Genomics*. Springer, New York (NY), 179–99.

Billot, C., P. Ramu, S. Bouchet, J. Chantereau, M. Deu, L. Gardes, J. L. Noyer, J. F. Rami, R. Rivallan, Y. Li and et al. (2013) Massive sorghum collection genotyped with SSR markers to enhance use of global genetic resources. *PLoS ONE* 8: e59714. https://doi.org/10.1371/journal.pone.0059714

Borrell, A. K., E. J. van Oosterom, J. E. Mullet, B. George-Jaeggli, D. R. Jordan, P. E. Klein and G. L. Hammer (2014) Stay-green alleles individually enhance grain yield in sorghum under drought by modifying canopy development and water uptake patterns. *New Phytol.* 203: 817–30.

Bowers, J. E., C. Abbey, S. Anderson, C. Chang, X. Draye, A. H. Hope, R. Jessup, C. Lemke, J. Lennington, and Z. Li et al. (2003). A high-density genetic recombination map of sequence tagged sites for *Sorghum*, as a framework for comparative structural and evolutionary genomics of tropical grains and grasses. *Sorghum Genetics* 165: 367–86.

Bray, E. A. (1997) Plant response to water deficit. *Trends Plant Sci.* 2: 48–54.

Brenton, Z. W., E. A. Cooper, M. T. Myers, R. E. Boyles, N. Shakoor, K. J. Zielinski, B. Rauh, W. C. Bridges, G. P. Morris and S. Kresovich (2016) A genomic resource for the development, improvement, and exploitation of sorghum for bioenergy. *Genetics* 203. doi:10.1534/genetics .115.183947.*Genetics*

Brown, A. H. D. (2010) Variation under domestication in plants: 1859 and today. *Phil. Trans. Royal Soc. B* 365: 2523–30. https://doi.org/10.1098/rstb.2010.0006

Brown, P. J., Rooney, W. L., Franks, C. and Kresovich, S. (2008) Efficient mapping of plant height quantitative trait loci in a sorghum association population with introgressed dwarfing genes. *Genetics* 180: 629–37.

Casa, A. M., G. Pressoir, P. J. Brown, S. E. Mitchell, W. L. Rooney, M. R. Tuinstra, C. D. Franks and S. Kresovich (2008) Community resources and strategies for association mapping in sorghum. *Crop Sci.* 48: 30–40. https://doi.org/10.2135/cropsci2007.02.0080 *Crop Sci.*

Clayton, W. D. and Renvoize, S. A. (1986) Genera *Graminum*: Grasses of the World Kew Bulletin Additional Series XIII ., Royal Botanical Gardens, Kew, Her Majesty's Stationery Office, London.*Graminum*

Cobb, M. (2006) Heredity before genetics: a history. *Nat. Rev. Genet.* 7: 953–8.

Coghlan, A., E. E. Eichler, S. G. Oliver, A. H. Paterson and L. Stein (2005) Chromosome evolution in eukaryotes: a multi-kingdom perspective. *Trends Genet.* 21: 673–82.

Crop Trust (2007) Strategy for the global *ex situ* conservation of sorghum genetic diversity (https://www.croptrust.org/) (Accessed12 August 2016).*ex situ*

Darwin, C. (1859) *On the Origin of Species*. London: John Murray.*On the Origin of Species*

Doebley, J. F., B. S. Gaut and B. D. Smith (2006) The molecular genetics of crop domestication. *Cell* 127: 1309–21.*Cell*

Dugas, D. V., M. K. Monaco, A. Olsen, R. R. Klein, S. Kumari, D. Ware and P. E. Klein (2011) Functional annotation of the transcriptome of *Sorghum bicolor* in response to osmotic stress and abscisic acid. *BMC Genomics* 12: 514. https://doi.org/10.1186/1471-2164-12-514

Evans, J., R. F. McCormick, D. Morishige, S. N. Olson, B. Weers, J. Hilley, P. E. Klein, W. Rooney and J. Mullet (2013) Extensive variation in the density and distribution of DNA polymorphism in sorghum genomes. *PLoS ONE* 8: e79192. https://doi.org/10.1371/journal.pone.0079192

Famoso, A. N., K. Zhao, R. T. Clark, C. W. Tung, M. H. Wright, C. Bustamante, L. V. Kochian and S. R. McCouch (2011) Genetic architecture of aluminum tolerance in rice (*Oryza sativa*) determined through genome-wide association analysis and QTL mapping. *PLoS Genet.* 7: e1002221. https://doi.org/10.1371/journal.pgen.1002221.

Feltus, F. A., G. E. Hart, K. F. Schertz, A. M. Casa, S. Kresovich, S. Abraham, P. E. Klein, P. J. Brown and A. H. Paterson (2006) Alignment of genetic maps and QTLs between inter- and intra-specific sorghum populations. *Theor. Appl. Genet.* 112: 1295–305. https://doi.org/10.1007/s00122-006-0232-3

Fernandez, M. G. S., P. W. Bercraft, Y. Yin and T. Lübberstedt (2009) From dwarves to giants? Plant height manipulation for biomass yield. *Trends Plant Sci.* 14: 454–61. https://doi.org/10.1016/j.tplants.2009.06.005

Fernandez, M. G. S., J. A. Okeno, E. Mutegi, A. Fessehaie and S. Chalfant (2014) Assessment of genetic diversity among sorghum landraces and their wild/weedy relatives in western Kenya using simple sequence repeat (SSR) markers. *Conser. Genet.* 15: 1269–80. https://doi.org/10.1007/s10592-014-0616-x

Fleischmann, R. D., M. D. Adams, O. White, R. A. Clayton, E. F. Kirkness, A. R. Kerlavage, C. J. Bult, J. F. Tomb, B. A. Dougherty, J. M. Merrick and et al. (1995) Whole-genome random sequencing and assembly of Haemophilus influenza Rd. *Science* 269: 496–512. https://doi.org/10.1126/science.7542800

Foy, C. (1988) Plant adaptation to acid, aluminum-toxic soils. *Soil Sci. Plant Anal.* 19: 959–87.*Soil Sci. Plant Anal.*

Freeling, M. (2001) Grasses as a single genetic system: reassessment 2001. *Plant Physiol.* 125: 1191–7. https://doi.org/10.1104/pp.125.3.1191

Fu, H. and H. K. Dooner (2002) Intraspecific violation of genetic colinearity and its implications in maize. *Proc. Natl. Acad. Sci.* 99: 9573–8. https://doi.org/10.1073/pnas.132259199

Fuller, D. Q. and Allaby, R. (2009) Seed dispersal and crop domestication: shattering, germination, and seasonality in evolution under cultivation. *Annu. Plant Rev.* 38: 238–95 https://doi.org/10.1002/9781119312994.apr0414

Genome Browser (2016) Genome information by organism, http://www.ncbi.nlm.nih.gov/genome/browse/(accessed 12 August 2016).

Gepts, P. (2014) The contribution of genetic and genomic approaches to plant domestication studies. *Curr. Opin. Plant Biol.* 18: 51–9. https://doi.org/10.1016/j.pbi.2014.02.001

Golicz, A. A., J. Batley and D. Edwards (2016) Towards plant pangenomics. *Plant Biotechnol.* 14: 1099–105. https://doi.org/10.1111/pbi.12499

Hamblin, M. T., S. E. Mitchell, G. M. White, J. Gallego, R. Kukatla, R. A. Wing, A. H. Paterson and S. Kresovich (2004) Comparative population genetics of the panicoid grasses: sequence polymorphism, linkage disequilibrium and selection in a diverse sample of *Sorghum bicolor*. *Genetics* 167: 471–83. https://doi.org/10.1534/genetics.167.1.471

Hamblin, M. T., A. M. Casa, H. Sun, S. C. Murray, A. H. Paterson, C. F. Aquadro and S. Kresovich (2006) Challenges of detecting directional selection after a bottleneck: lessons from *Sorghum bicolor*. *Genetics* 173: 953–64. https://doi.org/10.1534/genetics.105.054312

Han, B. and Y. Xue (2003) Genome-wide intraspecific DNA-sequence variations in rice. *Curr. Opin. Plant Biol.* 6: 134–8. https://doi.org/10.1016/s1369-5266(03)00004-9

Harlan, J. R. and J. M. J. de Wet (1972) A simplified classification of cultivated sorghum. *Crop Sci.* 12: 172–6. https://doi.org/10.2135/cropsci1972.0011183x001200020005x

Harlan, J. R., J. M. J. de Wet and E. G. Price (1973) Comparative evolution of cereals. *Evolution* 27: 311–25. https://doi.org/10.1111/j.1558-5646.1973.tb00676.x

Harland, S. C. (1954) Nikolai Vavilov 1885–1942. *Obit. Not. Fell. R. Soc.* 9: 259–64. https://doi.org/10.1098/rsbm.1954.0017

Hart, G. E., K. F. Schertz, Y. Peng, N. H. Syed (2001) Genetic mapping of *Sorghum bicolor* (L.) Moench QTLs that control variation in tillering and other morphological characters. *Theor. Appl. Genet.* 103: 1232–42. https://doi.org/10.1007/s001220100582

Haussmann, B. I. G., V. Mahalakshmi, B. V. S. Reddy, N. Seetharama, C. T. Hash and H. H. Geiger (2002) QTL mapping of stay-green in two sorghum recombinant inbred populations. *Theor. Appl. Genet.* 106: 133–42. https://doi.org/10.1007/s00122-002-1012-3

Hill, W. G. and A. Robertson (1966) The effect of linkage on limits to artificial selection. *Genet. Res.* 8: 269–94. https://doi.org/10.1017/s0016672300010156

Hilley, J., S. Truong, S. Olsen, D. Morishige and J. Mullet (2016) Identification of *Dw1*, a regulator of sorghum stem internode length. *PLoS One* 11: e0151271. https://doi.org/10.1371/journal.pone.0151271

Huxley, J. (1942) *Evolution: The Modern Synthesis*. MIT Press, Cambridge, MA.

Johnson, S. M., I. Cummins, F. L. Lim, A. R. Slabas and M. R. Knight (2015) Transcriptomic analysis comparing stay-green and senescent *Sorghum bicolor* lines identifies a role for proline biosynthesis in the stay-green trait. *J. Exp. Bot.* 66: 7061–73. https://doi.org/10.1093/jxb/erv405

Kellogg, E. and J. L. Bennetzen (2004) The evolution of nuclear genome structure in seed plants. *Am. J. Bot.* 91: 1079–724. https://doi.org/10.3732/ajb.91.10.1709

Kellogg, E. A. (2013) Phylogenetic relationships of Saccharinae and Sorghinae. In: Paterson, A. H. (ed.) *Genomics of the Saccharinae*. Springer, New York (NY), 3–23.

Khush, G. S. (2001) Green revolution: the way forward. *Nat. Rev. Genet.* 2: 815–22. https://doi.org/10.1038/35093585

Kim, J. S., M. N. Islam-Faridi, P. E. Klein, D. M. Stelly, H. J. Price, R. R. Klein and J. E. Mullet (2005) Comprehensive molecular cytogenetic analysis of sorghum genome architecture: distribution of euchromatin, heterochromatin, genes and recombination in comparison to rice. *Genetics* 171: 1963–76. https://doi.org/10.1534/genetics.105.048215

Kimber, C. T. (2000) Chapter 1.1: Origins of Domesticated Sorghum and its early diffusion to India and China. In: Smith, C. W. and Frederiksen, R. A. (Eds.) *Sorghum: Origin, History, Technology, and Production*. John Wiley and Sons, New York (NY), 3–99.

Klein, R. R., F. R. Miller, D. V. Dugas, P. J. Brown, A. M. Burrell and P. E. Klein (2015) Allelic variants in the *PRR37* gene and the human-mediated dispersal and diversification of sorghum. *Theor. Appl. Genet.* 128: 1669–83. https://doi.org/10.1007/s00122-015-2523-z

Konishi, S., T. Izawa, S. Y. Lin, K. Ebana, Y. Fukuta, T. Sasaki and M. Yano (2006) A SNP caused loss of seed shattering during rice domestication. *Science* 312: 1392–6

Lasky, J. R., H. D. Upadhyaya, P. Ramu, S. Desphpande, C. T. Hash, J. Bonnette, T. E. Juenger, K. Hyma, C. Acharya, S. E. Mitchell and et al. (2015) Genome-environment associations in sorghum landraces predict adaptive traits. *Sci. Adv.* 1: e1400218. https://doi.org/10.1126/sciadv.1400218

Lenser, T. and G. Theiβen (2013) Molecular mechanisms involved in convergent crop domestication. *Trends Plant Sci.* 18: 704–14. https://doi.org/10.1016/j.tplants.2013.08.007

Li, C., A. Zhou and T. Sang (2006) Rice domestication by reducing shattering. *Science* 311: 1936–9. https://doi.org/10.1126/science.1123604

Lin, Y. R., K. F. Schertz and A. H. Paterson (1995) Comparative analysis of QTLs affecting plant height and maturity across the Poaceae, in reference to an interspecfic sorghum population. *Genetics* 141: 391–411.*Genetics*

Lin, Z., X. Li, L. M. Shannon, C. T. Yeh, M. L. Wang, G. Bai, Z. Peng, J. Li, H. N. Trick, T. E. Clemente and et al. (2012) Parallel domestication of the Shattering1 genes in cereals. *Nat. Genet.* 44: 720–4. https://doi.org/10.1038/ng.2281

Luo, H., W. Zhao, Y. Wang, Y. Xia, X. Wu, L. Zhang, B. Tang, J. Zhu, L. Fang, Z. Du and et al. (2016) SorGSD: a sorghum genome SNP database. *Biotechnol. Biofuels* 9: 6. https://doi.org/10.1186/s13068-016-0450-0

Mace, E. S., S. Tai, E. K. Gilding, Y. Li, P. J. Prentis, L. Bian, B. C. Campbell, W. Hu, D. J. Innes, X. Han and et al. (2013) Whole-genome sequencing reveals untapped genetic potential in Africa's indigenous cereal crop sorghum. *Nat. Comm.* 4: 2320. https://doi.org/10.1038/ncomms3320

Magalhaes, J. V., D. F. Garvin, Y. Wang, M. E. Sorrells, P. E. Klein, R. E. Schaffert, L. Li and L. V. Kochian (2004) Comparative mapping of a major aluminum tolerance gene in sorghum and other species in the Poaceae. *Genetics* 167: 1905–14. https://doi.org/10.1534/genetics.103.023580

Magalhaes, J. V., J. Liu, C. T. Guimaraes, U. G. P. Lana, V. M. C. Alves, Y. H. Wang, R. E. Schaffert, O. A. Hoekenga, M. A. Pineros, J. E. Shaff and et al. (2007) A gene in the multidrug and toxic compound extrusion (MATE) family confers aluminum tolerance in sorghum. *Nat. Genet.* 39: 1156–61. https://doi.org/10.1038/ng2074

McCouch, S. (2004) Diversifying selection in plant breeding. *PLoS Biol.* 2: 1507–11. https://doi.org/10.1371/journal.pbio.0020347

McKinley, B., W. L. Rooney, C. Wilkerson and J. Mullet (2016) Dynamics of biomass partitioning, stem gene expression, cell wall biosynthesis, and sucrose accumulation during development of Sorghum bicolor. *Plant J.* 88: 662–80.doi: 10.1111/tpj.13269.

Meyer, R. S., A. E. DuVal and H. R. Jensen (2012) Patterns and processes in crop domestication: an historical review and quantitative analysis of 203 global food crops. *New Phytol.* 196: 29–48. https://doi.org/10.1111/j.1469-8137.2012.04253.x

Meyer, R. S. and M. D. Purugganan (2013) Evolution of crop species: genetics of domestication and diversification. *Nat. Rev. Genet.* 14: 840–52. https://doi.org/10.1038/nrg3605

Ming, R., S. C. Liu, Y. R. Lin, J. da Silva, W. Wilson, D. Braga, A. van Deynze, T. F. Wenslaff, K. K. Wu, P. H. Moore and et al. (1998) Detailed alignment of *Saccharum* and *Sorghum* chromosomes: comparative organization of closely related diploid and polyploid genomes. *Genetics* 150: 1663–82.

Morrell, P., E. S. Buckler and J. Ross-Ibarra (2012) Crop genomics: advances and applications. *Nat. Rev. Genet.* 13: 85–96.

Morris, G. P., P. Ramu, S. P. Deshpande, C. T. Hash, T. Shah, H. D. Upadhyaya, O. Riera-Lizarazu, P. J. Brown, C. B. Acharya, S. E. Mitchell et al. (2013) Population genomic and genome-wide association studies of agroclimatic traits in sorghum. *Proc. Natl. Acad. Sci.* 110: 453–8. https://doi.org/10.1073/pnas.1215985110

Multani, D. S., S. P. Briggs, M. A. Chamberlain, J. J. Blakeslee, A. S. Murphy and G. S. Johal (2003) Loss of an MDR transporter in compact stalks of maize *br2* and sorghum *dw3* mutants. *Science* 302: 81–4. https://doi.org/10.1126/science.1086072

Murat, F., J. H. Xu, E. Tannier, M. Abrouk, N. Guilhot, C. Pont, J. Messing and J. Salse (2010) Ancestral grass karyotype reconstruction unravels new mechanisms of genome shuffling as a source of plant evolution. *Genome Res.* 20: 1545–57. https://doi.org/10.1101/gr.109744.110

Muraya, M. M., S. de Villiers, H. K. Parzies, E. Mutegi, F. Sagnard, B. M. Kanyenji, D. Kiambi and H. H. Geiger (2011) Genetic structure and diversity of wild sorghum populations (*Sorghum* spp.) from different eco-geographical regions of Kenya. *Theor. Appl. Genet.* 123: 571–83. https://doi.org/10.1007/s00122-011-1608-6

Murphy, D. J. (2007) *People, Plants, and Genes: The Story of Crops and Humanity*. Oxford University Press, Oxford, NY.

Murphy, R. L., R. R. Klein, D. T. Morishige, J. A. Brady, W. L. Rooney, F. R. Miller, D. V. Dugas, P. E. Klein and J. E. Mullet (2011) Coincident light and clock regulation of *pseudoresponse regulator protein 37* (PRR37) controls photoperiodic flowering in sorghum. *Proc. Natl. Acad. Sci.* 108: 16469–74. https://doi.org/10.1073/pnas.1106212108

Olsen, K. M. and J. F. Wendel (2013) Crop plants as models for understanding plant adaptation and diversification. *Front. Plant Sci.* 4. https://doi.org/10.3389/fpls.2013.00290

Osborne, C. P. and Beerling, D. J. (2006) Nature's green revolution: the remarkable evolutionary rise of C_4 plants. *Phil. Trans. Royal Soc. B* 361: 173–94. https://doi.org/10.1098/rstb.2005.1737

Paterson, A. H., K. F. Schertz, Y. R. Lin, S. C. Liu and Y. L. Chang (1995) The weediness of wild plants: molecular analysis of genes influencing dispersal and persistence of johnsongrass, *Sorghum halepense* (L.) Pers. *Proc. Natl. Acad. Sci.* 92: 6127–31. https://doi.org/10.1073/pnas.92.13.6127

Paterson, A. H., J. E. Bowers, R. Bruggmann, I. Dubchak, J. Grimwood, H. Gundlach, G. Haberer, U. Hellsten, T. Mitros, A. Poliakov et al. (2009) The *Sorghum bicolor* genome and the diversification of grasses. *Nature* 457: 551–6.

Paterson, A. H., M. Freeling, H. Tang and X. Wang (2010) Insights from the comparison of plant genome sequences. *Ann. Rev. Plant Biol.* 61: 349–72. https://doi.org/10.1146/annurev-arplant-042809-112235.

Paterson, A. H. (2013) The Sorghum genome sequence: a core resource for Saccharinae genomics. In: Paterson, A. H. (Ed.) *Genomics of the Saccharinae*. Springer, New York (NY).

Peng, J., D. E. Richards, N. M. Hartley, G. P. Murphy, K. M. Devos, J. E. Flintham, J. Beales, L. J. Fish, A. J. Worland, F. Pelica and et al. (1999) 'Green revolution' genes encode mutant gibberellin response modulators. *Nature* 400: 256–61.

Perumal, R., R. Krishnaramanujam, M. A. Menz, S. Katile, J. Dahlberg, C. W. Magill and W. L. Rooney (2007) Genetic diversity among sorghum races and working groups based on AFLPs and SSRs. *Crop Sci.* 47: 1375–83. https://doi.org/10.2135/cropsci2006.08.0532

Perumal, R., M. A. Menz, P. J. Mehta, S. Katile, L. A. Gutierrez-Rojas, R. R. Klein, P. E. Klein, L. K. Prom, J. A. Schlueter, W. L. Rooney and C. W. Magill (2009) Molecular mapping of Cg1, a gene for resistance to anthracnose (*Colletotrichum sublineolum*) in sorghum. *Euphytica* 165: 597–606. https://doi.org/10.1007/s10681-008-9791-5

Phytozome v11.0 (1997–2015). https://phytozome.jgi.doe.gov/pz/portal.html (accessed 12 August 2016)

Price, H. J., S. L. Dillon, G. Hodnett, W. L. Rooney, L.Ross, J. S. Johnson (2005) Genome evolution in the genus *Sorghum* (Poaceae). *Ann. Bot.* 95: 219–27. https://doi.org/10.1093/aob/mci015.

Pringle, H. (1998) The slow birth of agriculture. *Science* 282: 1446. https://doi.org/10.1126/science.282.5393.1446

Quinby, J. R. (1974) *Sorghum Improvement and the Genetics of Growth*. Texas Agricultural Experiment Station, College Station.

Ramu, P., S. P. Deshpande, S. Senthilvel, B. Jayashree, C. Billot, M. Deu, L. A. Reddy and C. T. Hash (2010) In silico mapping of important genes and markers available in the public domain for efficient sorghum breeding. *Mol. Breed.* 26: 409–18. https://doi.org/10.1007/s11032-009-9365-9

Ritter, K., C. L. McIntyre, I. D. Godwin, D. R. Jordan and S. C. Chapman (2007) An assessment of the genetic relationship between sweet and grain sorghums, with *Sorghum bicolor* ssp. *Bicolor* (L.) Moench, using AFLP markers. *Euphytica* 157: 161–76. https://doi.org/10.1007/s10681-007-9408-4

Sage, R. F., T. L. Sage and F. Kocacinar (2012) Photorespiration and the evolution of C_4 photosynthesis. *Annu. Rev. Plant Biol.* 63: 19–47.

Sanchez, A. C., P. K. Subudhi, D. T. Rosenow and H. T. Nguyen (2002) Mapping QTLs associated with drought resistance in sorghum (*Sorghum bicolor* L. Moench). *Plant Mol. Biol.* 48: 713–26.

Schatz, M., L. Maron, J. Stein, A. Wences, J. Gurtowski, E. Biggers, H. Lee and et al. (2014) Whole genome de novo assemblies of three divergent strains of rice, *Oryza sativa*, document novel gene space of aus and indica. *Genome Biol.* 15: 506. https://doi.org/10.1186/preaccept-2784872521277375

Tang, H., U. Sezen and A. H. Paterson (2010) Domestication and plant genomes. *Curr. Opin. Plant Biol.* 13: 1 6 0–6. https://doi.org/10.1016/j.pbi.2009.10.008

Tanksley, S. D. and S. R. McCouch. (1997) Seed banks and molecular maps: unlocking genetic potential from the wild. *Science* 277: 1063–6. https://doi.org/10.1126/science.277.5329.1063

Tettelin, H., V. Masignani, M. J. Cieslewicz, C. Donati, D. Medini, N. L. Ward, S. V. Angiuoli, J. Crabtree, A. L. Jones, A. S. Durkin et al. (2005) Genome analysis of multiple pathogenic isolates of Streptococcus agalactiae: implications for the microbial 'pan-genome'. *Proc. Natl, Acad. Sci.* 102: 13950–5.

Touchman, J. (2010) Comparative Genomics. *Nat. Educ. Knowl.* 3: 13.

Tsutsui, T., N. Yamaji, C. F. Huang, R. Motoyama, Y. Nagamura and J. F. Ma (2012) Comparative genome-wide transcriptional analysis of Al-responsive genes reveals novel Al tolerance mechanisms in rice. *PLoS One* 7: e48197. https://doi.org/10.1371/journal.pone.0048197

Vavilov, N. I. (1922) The law of homologous series in variation. *J. Genet.* 12: 47–89. https://doi.org/10.1007/bf02983073

Waddington, C. H. (1957) *The Strategy of Genes*. Allen and Unwin, London.

Wang, M., C. Zhu, N. A. Barkley, Z. Chen, J. E. Erpelding, S. C. Murray, M. R. Tuinstra, T. Tesso, G. A. Pederson and J. Yu. (2009a) Genetic diversity and population structure analysis of accessions in the US historic sweet sorghum collection. *Theor. Appl. Genet.* 120: 13–23. https://doi.org/10.1007/s00122-009-1155-6

Wang, X., U. Gowik, H. Tang, J. E. Bowers, P. Westhoff and A. H. Paterson (2009b) Comparative genomic analysis of C4 photosynthetic pathway evolution in grasses. *Genome Biol.* 10: R68. https://doi.org/10.1186/gb-2009-10-6-r68

Wright, S. I., I. V. Bi, S. G. Schroeder, M. Yamasaki, J. F. Doebley, M. D. McMullen and B. S. Gaut (2005) The effects of artificial selection on the maize genome. *Science* 308: 1310–14. https://doi.org/10.1126/science.1107891

Zheng, L. Y., X. S. Guo, B. He, L. J. Sun, Y. Peng, S. S. Dong, T. F. Liu, S. Jiang, S. Ramachandran, C. M. Liu, and H. C. Jing (2011) Genome-wide patterns of genetic variation in sweet and grain sorghum (*Sorghum bicolor*). *Genome Biol.* 12: R114. https://doi.org/10.1186/gb-2011-12-11-r114

Chapter 3

Ensuring the genetic diversity of sorghum

Hari D. Upadhyaya and Mani Vetriventhan, International Crops Research Institute for the Semi-Arid Tropics (ICRISAT), India

1 Introduction

Sorghum [*Sorghum bicolor* (L.) Moench] is the fifth major cereal crop in the world in terms of area under cultivation and level of production, and is a staple food crop for millions of the poorest and most food-insecure people in the semi-arid tropics. Africa (39.8%), the Americas (38.5%) and Asia (16.9%) are the main sorghum-producing regions in the world, and together contribute about 95% of the world's total sorghum production. The major sorghum-producing countries are the United States, Mexico, Nigeria, India and Sudan (averaged over 2010–14). Sorghum production in Asia decreased from 15.5 million tons in 1961 (maximum production being reached during the 1970s and 1980s) to 9.6 million tons in 2014, although this reduction was compensated by increases in the area under cultivation (from 13.2 million hectares in 1961 to 29.0 million hectares in 2014) and in the levels of production (from 10.7 million tons in 1961 to 29.0 million tons in 2014) in Africa (http://www.faostat.fao.org; data accessed in July 2016). The world's sorghum production increased from 40.9 million tons in 1961 to 67.9 million tons in 2014, with maximum production being achieved during the 1980s. The world's sorghum productivity has been increased from 889 kg ha^{-1} in 1961 to 1535 kg ha^{-1} in 2014, which is still low. Adoption of improved sorghum cultivars and management practices contributed to the productivity gains though large differences exist in different parts of the world.

http://dx.doi.org/10.19103/AS.2017.0015.06

Though considerable gains have been achieved in improving productivity levels, sorghum production is constrained by several biotic (diseases: downy mildew, grain mould, anthracnose, rust, leaf blight, etc.; insect pests: stem borer, shoot fly, midge, aphids, etc.) and abiotic (drought, heat, salinity, cold, aluminium toxicity, nutrient deficiency, etc.) stresses. The genetically improved sorghum hybrids and cultivars were reported to be less diverse compared to the wild and weedy relatives and landraces (Mace et al., 2013; Mutegi et al., 2011). An assessment of the genetic diversity of the US sorghum hybrids, which were widely cultivated between 1980 and 2008, indicated that the number of alleles per locus has been constant, with a gain in new alleles being balanced by a loss of other alleles, and that the hybrids released during the 2000s showed the least number of new alleles compared to those in the previous two decades (Smith et al., 2010). The improved sweet sorghum lines from the United States are derived from six landraces. Thus, the majority of the lines have similar genes/alleles (i.e. they are identical by descent) for high Brix (Murray et al., 2009). In Australian sorghum-breeding programmes, the strong selection for resistance to the sorghum midge, stay-green (a drought-tolerant trait) and other agronomic traits has been associated with a decline in genetic diversity (Jordan et al., 1998, 2003). Such a narrow genetic base of cultivars may result in an increased risk of crop vulnerability, such as crop failure due to insect pests and disease epidemics or unpredictable climatic effects. History has been witness to numerous epidemics caused by low levels of nuclear and cytoplasmic genetic diversity. These include the blight that affected temperate potatoes in the northern hemisphere in the 1840s, wheat stem rust that devastated wheat fields in 1917, the southern corn leaf blight epidemic of maize in the United States in 1970 and the Bengal famine in India in 1943 that resulted from brown spot disease in rice (reviewed in Keneni et al. 2012).

The low diversity of crop cultivars is mainly because most crop breeders use their working collection, comprising materials that have been adapted and improved to have the most desirable traits, and avoid using wild and weedy relatives and unadapted landraces in their hybridization programmes. The low use of germplasm accession is mainly due to the lack of reliable information on the traits of economic interest, the linkage load of many undesirable genes and their assumed risks and restricted access to the germplasm collection as a result of regulations governing international exchange, etc. (Dwivedi et al., 2009; Upadhyaya et al., 2014a). However, a large number of sorghum germplasm (236 617 accessions) have been conserved in genebanks globally, which can provide a rich pool of diversity for various traits of economic importance. The greater use of such diversity in sorghum-breeding programmes to develop cultivars with a broad genetic base may result in more sustainable sorghum production. In this chapter, we discuss the origin, domestication and taxonomy, *in situ* and *ex situ* conservation and diversity, factors shaping sorghum diversity, geographical distribution of sorghum germplasm, germplasm gap analysis and ensuring diversity in sorghum using the cultivated and wild gene pool to enhance crop yields and broaden the genetic base of sorghum cultivars.

2 Origin, domestication and taxonomy of sorghum

The sub-Saharan and northeast regions of Africa (particularly Ethiopia and Sudan) are the primary centres of the origin and diversity of sorghum, and India and China are the secondary centres (de Wet, 1978; Doggett, 1988). The genus *Sorghum* Moench has considerable morphological and ecological diversity and is subdivided into five subgenera or sections: *Chaetosorghum*, *Heterosorghum*, *Parasorghum*, *Stiposorghum* and *Sorghum*. The section *Sorghum* has three species: two wild perennials, *S. halepense* (L.) Pers. (2n = 40) and *S. propinquum* (Kunth) Hitchcock (2n = 20), and an annual, *S. bicolor* (L.) Moench (2n = 20), which is cultivated for food, animal feed and bioenergy production. *S. bicolor* has been subdivided into three subspecies: (i) ssp. *bicolor* (including all domesticated grain sorghum), (ii) ssp. *drummondii* (Steud.) de Wet comb. nov, which are derivatives of hybridization among grain sorghums and their closest wild relatives and (iii) ssp. *verticilliflorum* (Steud.) (earlier subsp. *arundinaceum* (Desv.) de Wet et Harlan), which are the wild progenitors of grain sorghums. The three subspecies of *S. bicolor* together form extremely variable crop-weed complex members which are fully inter-fertile.

Based on spikelet/panicle morphology, Harlan and de Wet (1972) distinguished cultivated sorghums into five main races (*bicolor*, *guinea*, *caudatum*, *kafir* and *durra*) and ten intermediate races (*guinea–bicolor*, *caudatum–bicolor*, *kafir–bicolor*, *durra–bicolor*, *guinea–caudatum*, *guinea–kafir*, *guinea–durra*, *kafir–caudatum*, *durra–caudatum* and *kafir–durra*). Wild sorghum *S. bicolor* ssp. *verticilliflorum* (Steud.) Piper includes four botanical races/ecotypes: *aethiopicum*, *virgatum*, *arundinaceum* and *verticilliflorum* (de Wet, 1978). *S. propinquum* is a robust, tufted perennial with stout rhizomes and is closely allied to *S. halepense*, differing from this species primarily in its smaller spikelets, and being a diploid rather than a tetraploid. The *S. propinquum* species crosses extensively with *S. bicolor* wherever they are sympatric to produce fully fertile hybrids.

Some recent classification systems have been proposed by Liu et al. (2014) and Dahlberg (2000). Liu et al. (2014) suggested a new subgeneric classification of *Sorghum* Moench into three distinct subgenera: (i) subg. *Chaetosorghum*, which has two sections (sect. *Chaetosorghum* and sect. *Heterosorghum*), (ii) subg. *Parasorghum* and (iii) subg. *Sorghum*. Dahlberg (2000) proposed an integrated classification of sorghum by combining sorghum working groups and races. Such a classification helps to identify the working groups and associated races and intermediate races.

Harlan and de Wet (1971) proposed three gene pools based on the ease of crossability between species. The primary gene pool of the genus *Sorghum* includes two diploid species, *S. bicolor* and *S. propinquum*. The secondary gene pool includes *S. halepense*, while species belonging to *Chaetosorghum*, *Heterosorghum*, *Parasorghum* and *Stiposorghum* come under the tertiary gene pool.

3 Sorghum germplasm conservation and diversity

In order to conserve the diversity of the sorghum crop, two types of conservation strategy are followed: *in situ* and *ex situ* conservations. As per Article 2 of the Convention on Biological Diversity (UNCED, 1992), *in situ* conservation means 'the conservation of ecosystems and natural habitats and the maintenance and recovery of viable populations of species in their natural surroundings and, in the case of domesticated or cultivated species, in the surroundings where they have developed their distinctive properties', while *ex situ* conservation means 'the conservation of components of biological diversity outside their natural habitats'.

In situ conservation aims to protect, manage and monitor the selected populations in their natural habitats so that the natural evolutionary processes can be maintained, thus allowing new variations to be generated in the gene pool. This will enable the species to adapt to changing environmental conditions (Heywood, 2014). The distinct techniques of the *in situ* conservation strategy include the conservation of crop wild relatives in natural habitat/genetic reserves and on-farm conservation of traditional crop cultivars (landraces). *Ex situ* conservation includes seed storage, *in vitro* storage, DNA storage, pollen storage, field genebanks and botanical gardens (Maxted et al., 1997).

3.1 *In situ* conservation and diversity

Sorghum species are not specifically protected in their natural environment (Bhattacharya et al., 2011), although some wild and weedy species of sorghums were reported to exist in genetic reserves. For example, the Kora National Reserve in Kenya contains *S. arundinaceum* and other species, such as *Vernonia galamensis, Cenchrus ciliaris, Panicum maximum, Acacia senegal, Gossypium somalense* and *Populus ilicifolia* (Kabuye et al., 1986).

On-farm conservation of sorghum landraces is practised by farmers, and several researchers have assessed species diversity occurring in on-farm contexts. Adugna (2014) assessed *in situ* diversity and the genetic structure of eight sorghum landrace populations occurring in three different geographical and agroclimatic regions of Ethiopia (Wello in the regional state of Amhara, the Gibe river valley in the regional state of Oromia and the Metekel zone in the regional state of Benishangul-Gumuz). Adugna used seven phenotypic traits and 12 SSR markers on 20 plants per population. This study showed a considerable variation among the populations for quantitative traits (e.g. 147–470 cm plant height, 0–5 tillers, 11–46 cm head length, 5–40 cm head width, etc.) and divergence in molecular diversity in the populations (F_{ST} = 0.40). Pairwise F_{ST}, a measure of population divergence among populations, indicated a distinct population structure, and the population from Wello in the Amhara regional state was found to have higher diversity (Adugna, 2014).

Rabbi et al. (2010) reported that most farmers in western Kenya cultivate landraces and use farm-saved seeds, and only about 15% of the farmers plant modern cultivars. They also collected on-farm and farmer-saved seeds and assessed diversity in samples from two contrasting agroecosystems in eastern Sudan and western Kenya (23 samples from each site), using 16 SSR markers. The assessment of diversity in landraces and modern varieties indicated a weak genetic differentiation of differently named landraces from western Kenya, while those from eastern Sudan formed distinguishable groups (Rabbi et al., 2010). Okeno et al. (2012) reported a wide variability of wild–weedy sorghum populations with respect to habitats and morpho-types in western Kenya, where true wild sorghum populations in national parks and the sugarcane belt were clearly distinguishable from those of putative hybrids or the intermediate forms found in sorghum fields and their surroundings. Mutegi et al. (2011) collected/assembled 329 cultivated and 110 wild sorghum seed samples from farmers' fields in the four main sorghum-growing areas of Kenya (i.e. Turkana, the western/Nyanza region, the eastern/central region and coastal areas) and from the National Genebank of Kenya. These samples represented agroclimatic and ethnolinguistic diversity. The study reported greater diversity in the wild sorghums (with a gene diversity of 0.69) than in the domesticated sorghums (with a gene diversity of 0.59). Ngugi and Onyango (2012) collected landraces (139 accessions) from various sorghum-growing areas of Kenya and assessed genetic diversity using 11 SSR markers. They reported a mean genetic diversity of 0.484, while, among the populations, North Eastern province had the maximum gene diversity (0.62), followed by Eastern province

(0.60), and Central province had the lowest (0.09). The average genetic distances of accessions in North Eastern province with Central (0.31), Western (0.20), Rift Valley (0.18) and Nyanza (0.15) provinces, and Central province with Eastern (0.18), Western (0.15), Rift Valley (0.12) and Nyanza (0.10) provinces were greater than between the other regions. Muui et al. (2013) collected 49 germplasm samples from four districts, namely Mbeere, Makueni, Kitui and Mutomo, in the Eastern province of Kenya, and reported that the Mbeere district had the most diverse landraces.

An assessment of the genetic diversity of 484 sorghum varieties collected in 79 villages distributed across Niger, using 28 SSR markers, showed a high level of diversity, although diversity varied between eastern and western Niger, and allelic richness was lower in the eastern part of the country (Deu et al., 2008). Genetic differentiation between botanical races, geographical distribution and the ethnic group to which farmers belonged was also significantly associated with genetic diversity partitioning. For instance, Deu et al. (2008) reported a distinct racial distribution between regions of Niger. For example, *guinea* varieties are mainly cultivated in western Niger, *caudatum* and *bicolor* varieties are rare in eastern Niger, *durra* varieties cover all the sorghum-growing areas but are predominant in the eastern region and high racial diversity was observed in central Niger. *Durra* (23.1%) and *caudatum* (21.3%) races are the most prevalent in Niger, followed by *guinea* (10.5%) and *bicolor* (8.1%), whereas intermediate races contributed 25.8%, mostly *durra–bicolor*, *durra–caudatum* and *caudatum–bicolor*.

India and China are the secondary centres of diversity for sorghum. In India, sorghum is mainly grown in Maharashtra, Karnataka and Andhra Pradesh during the season of *rabi* (the post-rainy season, that is, October–November to March–April) and in Madhya Pradesh, Rajasthan, Uttar Pradesh, Uttarakhand, Gujarat and Tamil Nadu during the *kharif* season (the rainy season, that is, June–July to September–October) (Elangovan and Babu, 2015). Collections of landraces in India were reported to have a large variability as the crop has been grown traditionally under varied agroclimatic conditions for centuries and has been adapted to the environmental conditions (Elangovan et al., 2009, 2012; Elangovan and Babu, 2015; Pandravada et al., 2013). In India, Maldandi is a popular sorghum landrace for post-rainy or *rabi* cultivation in the southern and central states, and is predominantly used for food. Over the last 80 yrs, many selections have been made on Maldandi and these are being cultivated as landraces across sorghum-growing regions. A selection from Maldandi, M 35–1, was released in 1969. Genetic diversity among Maldandi landraces using morpho-agronomic and SSR markers indicated the presence of wide variability (Rakshit et al., 2012). In China, sorghum is a minor crop, and is mostly grown in northern and southwestern China and rarely in southeastern China (Diao, 2017). Xu et al. (2014) reported the on-farm conservation of sorghum landraces in the Yunnan province of China. On-farm diversity assessments in China are very few and have reported low and moderate diversity in Chinese landraces. For example, Burow et al. (2012) reported moderate diversity in the 159 sorghum landraces from the colder regions (primarily the northeastern region) of China using 40 SSR markers. Zhang et al. (2011) investigated the genetic variation of 184 Chinese sorghum landraces from a broad geographic area (12 provinces) and 69 representative exotic landraces, using 32 nuclear SSR primer pairs. Their results revealed a lower level of genetic diversity in Chinese sorghum (Nei's allele diversity $H_e = 0.629$) compared with sorghum from across the world ($H_e = 0.745$); however, considerable differences in the levels of genetic variation were found among 12 provinces (H_e from 0.517 to 0.714). Accessions from Jilin Province had the highest level of genetic diversity among all regions in China.

3.2 *Ex situ* conservation and diversity

3.2.1 Global status of sorghum germplasm resources

The genetic erosion of crops occurs mainly due to the replacement of diverse landraces with one or a few modern varieties. Other factors include environmental degradation, urbanization and land clearing through deforestation and bush fires. It is therefore essential to collect and conserve crops' diversity *ex situ*. A total of 236 617 sorghum accessions have been conserved in genebanks globally, of which 98.3% are cultivated and 1.7% are wild and weedy relatives (Fig. 1, http://apps3.fao.org/wiews, accessed on 29 February 2016). Considerably more accessions may exist as duplicates among the genebanks than within a collection of a given crop. The genebanks conserving major collections of sorghum germplasm are listed in Table 1. The ICRISAT genebank in India and the USDA-ARS each holds ~17% and ~15% (respectively) of the total germplasm collections, while ICS-CAAS in China and NBPGR in New Delhi, India, hold about 7–8% each.

3.2.2 Active and base collections

In *ex situ* conservation, the collections are conserved as an active collection (medium-term storage, MTS) and as a base collection (long-term storage, LTS). At the ICRISAT genebank, the active collection is stored under MTS conditions (4°C and 20–30% relative humidity) and remains viable for 10–20 years with ≥85% viability. The collection is used for distribution, utilization and multiplication purposes. Accessions in the base collection are vacuum-sealed in aluminium foil pouches and stored at −20°C with 5–7% moisture content once initial germination has been confirmed (>90%). Seed viability is regularly monitored at 5–10-year intervals in the active MTS collection and at 10–20-year intervals in the base collection (Upadhyaya et al., 2014a).

3.2.3 Field genebanks

Several wild and weedy relatives of sorghum, particularly accessions of perennials and vegetatively propagated species, are maintained in field genebanks. For example, at ICRISAT, India, the sorghum wild and weedy relatives that are of perennial types and vegetatively propagated are being maintained as live samples in the field genebank. These

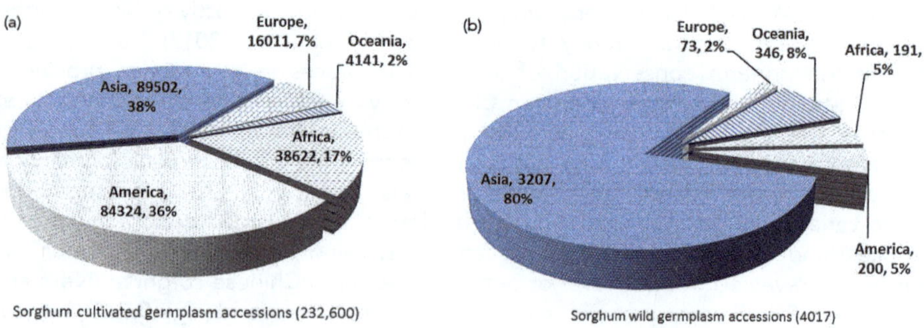

Figure 1 (a) Sorghum-cultivated germplasm accessions (232600). (b) Sorghum wild germplasm accessions (4017).

Table 1 The genebanks conserving major collections of sorghum germplasm in the world

Country	Institution	Wild	Cultivated	Total holding (%)
Global	International Crops Research Institute for the Semi-Arid Tropics (ICRISAT)	461	39 092	39 553 (16.7)
USA	Plant Genetic Resources Conservation Unit, Southern Regional Plant Introduction Station, University of Georgia, USDA-ARS	197	35 976	36 173 (15.3)
	National Center for Genetic Resources Preservation (NCGRP)	2	7 535	7 537 (3.2)
China	Institute of Crop Science, Chinese Academy of Agricultural Sciences (ICS-CAAS)		18 263	18 263 (7.7)
India	National Bureau of Plant Genetic Resources (NBPGR), New Delhi (NBPGR)	2 674	14 792	17 466 (7.4)
Ethiopia	Institute of Biodiversity Conservation (IBC)		9 772	9 772 (4.1)
Brazil	Embrapa Milho e Sorgo (CNPMS)		7 225	7 225 (3.0)
World total	–	**4 017**	**232 600**	**23 6617**

Source: http://apps3.fao.org/wiews accessed on 29 February 2016.

include *S. plumosum*, *S. interjectum*, *S. nitidum*, *S. grande*, *S. timorense*, *S. stipoideum* and *S. intrans*.

3.2.4 Regeneration and maintaining genetic integrity

Regeneration should be carried out in such a manner that the genetic integrity of a given accession is maintained. Accessions should be regenerated when viability drops below 85% of the initial viability or when the remaining seed quantity is less than what is required for three sowings of a representative population of the accession (Upadhyaya and Gowda, 2009; FAO, 2014). At ICRISAT, samples whose seed stock falls below 50 g or whose seed viability is less than 85% are taken out for regeneration. For regeneration, each accession is planted in a row 9 m in length, producing about 90 plants per accession. Since sorghum is an often cross-pollinated crop, the genetic integrity of sorghum accessions is maintained by selfing. Seeds from each 'selfed' panicle are bulked to maintain the accession and dried to 8% moisture content and stored in screw-capped aluminium containers.

3.2.5 Characterization and evaluation

Characterization is the description of plant germplasm. It is essential that the germplasm being conserved is known and described to the greatest extent possible to assure their maximum use by plant breeders. Therefore, characterization should be carried out as soon as possible to add value to the collection (FAO, 2014). Until a collection has been properly evaluated and its useful features become known to breeders, it has little practical use.

After the receipt of new samples by the curator, accessions need to be characterized for morpho-agronomic traits, for abiotic and biotic stresses and for seed quality traits, following the descriptor for sorghum (IBPGR and ICRISAT, 1993).

3.2.6 Distribution and impact

As per the International Treaty on Plant Genetic Resources for Food and Agriculture, germplasm is supplied under the Standard Material Transfer Agreement. The ICRISAT genebank has been the major source of sorghum germplasm accessions, supplying them worldwide for use in crop improvement programmes. Since 1974, the ICRISAT genebank has distributed 268 613 samples of sorghum germplasm accessions to 110 countries (as on June 2017), with the majority of the samples being distributed during the 1980s and 1990s. The global collections held at the ICRISAT genebank have been used to restore germplasm to the source countries when national collections are lost due to natural calamities, civil strife, etc. The ICRISAT genebank has supplied more than 22 000 sorghum samples globally for this purpose. Restorations have included 14 615 accessions to India, 362 to Botswana, 1827 to Cameroon, 1723 to Ethiopia, 838 to Kenya, 1436 to Nigeria, 445 to Somalia and 977 to Sudan. In this way, the national programmes of several countries have regained their precious plant germplasm heritage, which could have been lost if it had not been conserved in the ICRISAT genebank. One of the immediate impacts of germplasm is its direct release as cultivars. Of the germplasm distributed from the ICRISAT genebank, 39 sorghum accessions originating from 14 countries have been released directly as 41 cultivars in 18 countries, with few of them being released in more than one country (Table 2).

3.2.7 Safety backup

Safety duplication guarantees the availability of a genetically identical subsample of the accession. This precautionary measure mitigates the risk of its partial or total loss as a result of natural or human-caused catastrophes. Safety duplication includes both the duplication of the material and its related information, including backup on a database. Safety duplication requires the materials to be deposited in LTS at a different location. The second location is chosen to minimize risks and to provide the best possible storage facilities. The best way to minimize risks that can arise in any individual country is for safety duplication to be undertaken outside that country (FAO, 2014). The ICRISAT genebank has safely duplicated about 91% of its total sorghum germplasm accessions at the Svalbard Global Seed Vault in Norway.

3.2.8 Phenotypic diversity

For breeders to better understand and use the available genetic resources, they need to know about the diversity of conserved germplasm. Such knowledge makes the selection of more diverse adapted parents more efficient in the ongoing endeavour to improve crops. A large portion of US germplasm accessions were evaluated for morpho-agronomic traits, grain nutritional content and for tolerance/resistance to biotic and abiotic stresses and possessed the greatest range of diversity (Wang et al., 2015). The ICRISAT genebank conserves the largest number of sorghum germplasm accessions, and has a wider variation of morpho-agronomic traits (Table 3). In the

Table 2 ICRISAT's sorghum germplasm accessions directly released as cultivars

Accession no.	Country of origin	Country of release	Year release	Release name
IS 6928	Sudan	India	1978	Moti
IS 302	China	Myanmar	1980	Shwe-ni 10
IS 5424	India	Myanmar	1980	Shwe-ni 8
IS 8965	Kenya	Myanmar	1980	Shwe-ni 1
IS 9302	South Africa	Ethiopia	1980	ESIP 11
IS 33892	India	India	1980	NTJ 2
IS 2940	USA	Myanmar	1981	Shwe-ni 2
IS 4776	India	India	1983	U P Chari-1
IS 18758	Ethiopia	Burkina Faso	1983	E-35-1
IS 9323	South Africa	Ethiopia	1984	ESIP 12
IS 18484	India	Honduras	1984	Tortillerio 1
IS 2391	South Africa	Swaziland	1989	MRS 13
IS 3693	USA	Swaziland	1989	MRS 94
IS 8571	Tanzania	Mozambique	1989	Mamonhe
IS 23520	Ethiopia	Zambia	1989	Sima
IS 18758	Ethiopia	Burundi	1990	Gambella 1107
IS 3924	Nigeria	India	1991	Swarna
IS 9830	Sudan	Sudan	1991	Mugawim Buda-2
IS 3541	Sudan	India	1992	CS 3541
IS 3923	Zimbabwe	Botswana	1994	Mahube
IS 23496	Ethiopia	Tanzania	1995	Pato
IS 15845	Cameroon	India	1996	Paiyur 2
IS 9468	South Africa	Mexico	2000	Marvilla No SOFO 430201092
IS 13444	Zimbabwe	Sudan	2000	Arous el Rimal
IS 29415	Lesotho	Eritrea	2000	Shiketi
IS 76	Mexico	Kenya	2001	IS 75
IS 8193	Uganda	Kenya	2001	Kari Mtama 2
IS 8193	Uganda	Rwanda	2001	IS 8193
IS 15401	Cameroon	Mali	2001	Soumalemba
IS 21219	Kenya	Rwanda	2001	IS 21219
IS 25395	Kenya	Rwanda	2001	IS 25395
IS 33844	India	India	2002	Parbhani Moti
IS 21055	Kenya	Kenya	2008	Legio

(Continued)

Table 2 (*Continued*)

Accession no.	Country of origin	Country of release	Year release	Release name
IS 25403	Kenya	Uganda	2011	IS 25403
IS 35	USA	Kenya	2013	Sweetsorg 21
IS 12611	Ethiopia	Mali	2015	Siaroukala
IS 23525	Ethiopia	Mali	2015	Tiokala
IS 23541	Ethiopia	Mali	2015	Pitikala
IS 23555	Ethiopia	Mali	2015	Kala Wassale
IS 23562	Ethiopia	Mali	2015	Jiguikala
IS 23519	Ethiopia	Mali	2015	Zalatimi

ICRISAT genebank collection, the landraces from India were late-flowering, tall and produced stout panicles and larger seeds, while the landraces from Pakistan flowered early in both rainy and post-rainy seasons at ICRISAT Patancheru, and produced stout panicles. Accessions from Sri Lanka were late-flowering and tall in both rainy and post-rainy seasons, and produced more basal tillers and stout panicles (Upadhyaya et al., 2016b). The landraces from Ethiopia were early-flowering and short in plant height, with high panicle exsertion, and greater panicle width and 100-seed weight; the landraces from Kenya had a higher number of basal tillers; those from Sudan were early-flowering and tall in the rainy season and had larger seeds and those from Tanzania had long panicles (Upadhyaya et al., 2017a). The collection from Sierra Leone flowered late in both rainy and post-rainy seasons, produced more basal tillers per plant and had longer panicles. The collection from the Central African Republic grew significantly shorter in the rainy season and taller in the post-rainy season. The landraces that differed significantly from other collections included those from Gambia for panicle exsertion

Table 3 Mean and range of quantitative traits in sorghum germplasm conserved at the ICRISAT genebank, India

Sorghum	Mean	Range
Days to 50% flowering (Rainy)	91	31–199
Days to 50% flowering (Post-rainy)	72	36–154
Plant height (cm) (Rainy)	327	50–655
Plant height (cm) (Post-rainy)	214	50–580
Basal tillers number	2	1–14
Panicle exsertion (cm)	16	0–72
Panicle length (cm)	22	3–90
Panicle width (cm)	9	1–80
Seed Size (mm)	3	0.8–6.0
100 seed weight (g)	3	0.1–9.4

and panicle width, those from Nigeria for seed width and those from Cameroon for 100-seed weight (Upadhyaya et al., 2017c).

Representative sets of germplasm diversity, such as core (Grenier et al., 2001) and mini core (Upadhyaya et al., 2009) collections, have been established to capture the maximum range of diversity and for their utilization in crop improvement. The sorghum mini core collection possesses a large range of variability that allows the identification of trait donor lines that can be used to improve sorghum crops. Variations include for traits such as grain iron and zinc content (Upadhyaya et al., 2016d), stalk sugar content (Upadhyaya et al., 2014b), post-flowering drought tolerance (Upadhyaya et al. 2017b), germination and vigour at low temperatures (Upadhyaya et al., 2016c) and resistance to grain mould, downy mildew, anthracnose, leaf blight and rust diseases (Sharma et al., 2010, 2012).

3.2.9 Molecular diversity

The global composite germplasm collection for sorghum (3367 accessions) includes the collections from ICRISAT-India, CIRAD-France and CAAS-China. It comprises 280 breeding lines and elite cultivars from public sorghum-breeding programmes, 68 wild and weedy accessions and over 3000 landraces from the collections held by CIRAD or ICRISAT and included previously defined core and mini core collections (Grenier et al., 2001; Upadhyaya et al., 2009) and sources for resistance to various biotic stresses and/or for variations in other agronomic and quality traits. The collection was genotyped using 41 SSR markers (Billot et al., 2013). The results revealed a large range of diversity, with an average gene diversity of 0.67, and the breeding lines were less diverse compared to landraces and wild accessions. The wild and weedy accessions were found to have higher private alleles, followed by the landraces. Among races, *kafir* showed the lowest gene diversity (0.41) compared to other races (0.60–0.67). The largest numbers of alleles were concentrated in central and eastern Africa. The cultivated sorghum accessions were structured according to geographic regions and the races within regions (Billot et al., 2013).

Morris et al. (2013) characterized diverse sorghum germplasm (971 accessions) from worldwide collections, including the US sorghum association panel (Casa et al., 2008), the sorghum mini core collection (Upadhyaya et al., 2009) and the sorghum reference set of the generation challenge programme (Billot et al., 2013). They used the genotyping-by-sequencing (GBS) approach and the study showed distinct patterns of population structure: the *kafir* sorghums showed the strongest pattern of population subdivision relative to other races, the *durra* types formed a distinct cluster that was further differentiated according to the geographic origin, the *bicolor* types were not remarkably clustered except for those from China, the *caudatum* types showed modest clustering according to the geographic distribution, the *guinea* types showed five distinct subgroups, four of which were clustered according to their geographic origin and the fifth group included *guinea margaritiferum* types in addition to wild accessions from western Africa. When genotyped using a GBS approach, ICRISAT's sorghum mini core collection (Upadhyaya et al., 2009) showed that it was structured according to both geographic origin and sorghum races (Wang et al., 2013). Accessions of different races from southern Africa tended to be more similar to each other, as were those from East Asia. *Caudatums* from different countries were clustered, while *guineas* from West Africa and *durras* from India were clustered by race and origin. The race *bicolor* was found to be largely distributed among the other four races.

The maximum diversity of wild and weedy relatives with a higher proportion of rare and private alleles is of great importance to the conservation and utilization of sorghum genetic resources. The wild relatives of sorghum are potential sources of important and unique genes for crop improvement programmes. The divergence of wild sorghums from those of the five races of cultivated sorghum was studied by Zhang et al. (2015). The study revealed that the race *bicolor* (Fixation index, F_{ST} = 0.04) had a close genetic relationship with wild sorghums, more so than the other four primary races (F_{ST} between populations being 0.11 for *guinea* and wild sorghum; 0.20 for *durra* and wild sorghum; 0.33 for *kafir* and wild sorghum and 0.14 for *caudatum* and wild sorghum). The races *caudatum*, *durra* and *kafir* showed a relatively high level of population differentiation (F_{ST} between populations being 0.26 for *durra* and *caudatum*; 0.46 for *durra* and *kafir* and 0.33 for *caudatum* and *kafir*), indicating that these are highly differentiated and distinct from each other. Mace et al. (2013) resequenced 44 sorghum lines representing major races of cultivated sorghum, in addition to its progenitors and *S. propinquum*. The study indicated that there was maximum diversity in the wild and weedy relatives, followed by the landraces, when compared to improved inbred varieties. They found more wild-specific alleles (34%) than improved inbred-specific alleles (8%) and landrace-specific alleles (18%). *S. propinquum* differs from other sorghums, with 22% of *S. propinquum* reads unmapped to *S. bicolor* due to the majority of gene gain/loss events.

4 Factors shaping sorghum diversity

In many traditionally managed agroecosystems, landraces present higher levels of genetic diversity than those of improved inbred lines. Thus, landraces form the major component of total germplasm resources conserved in genebanks globally. For example, 86% of the total germplasm conserved at the ICRISAT genebanks is of landraces. The abundant diversity among sorghum landraces is shaped by the mating system, ethnic, linguistic and customary end use and sociological factors, and edaphic, temporal and spatial factors (Upadhyaya et al., 2016a).

4.1 Mating system and gene flow

The mating system is a significant factor in plant evolutionary genetics and genetic diversity, and gene flow plays a major role in structuring the genetic variability within and among populations. Gene flow introduces new alleles into a population or changes the frequency of existing alleles in a population. Gene flow can occur via seed dispersal and colonization of a new population or by cross-pollination from external sources, which can be of the same or a different but related species. For gene flow to occur, the species must be sexually compatible and must flower synchronously.

Sorghum is predominantly a self-pollinated crop (Doggett, 1988) and the outcrossing rate may vary and has been reported to reach up to 40% (Barnaud et al., 2008). In traditional farming systems, biological and human factors interact to shape evolutionary forces. Biological factors comprise both environmental pressures and biological traits of the plant, such as its mating system, while human factors affect the dynamics of diversity in many ways, acting on gene flow, drift and selection. In a study involving Duupa farmers in Cameroon, who grow numerous landraces mixed in a field, a practice that favours extensive pollen flow, Barnaud et al. (2008) evaluated the extent of pollen flow, its links with farming practices and its impact on the dynamics of diversity of sorghum in the fields. The study showed that although the biological traits of sorghum (inflorescence morphology, floral traits, phenology, etc.) and the spatial planting practices of Duupa farmers led to extensive pollen flow among landraces, it is the selection exerted by the farmers that appears to be a key parameter affecting the fate of new genetic combinations from outcrossing events. Adugna (2014) reported a high divergence of populations representing three diverse geographical regions of Ethiopia, indicating a low level of gene flow, while high gene flow was observed in some adjacent populations. The populations from Wello showed a close relationship with remote Gibe and Metekel populations, indicating that the variation followed human migration patterns.

The potential gene flow between a crop and its wild relatives is largely determined by the overlaps in their ecological and geographical distributions. In a survey in Ethiopia and Niger, Tesso et al. (2008) observed an overlap in the occurrence and flowering phenology between cultivated and wild sorghum, an indication of the potential for gene flow between the two congeners. Okeno et al. (2012) reported a wide variability within wild–weedy sorghum populations with respect to habitats and morpho-types and evidence of gene flow between cultivated sorghum and its wild–weedy relatives. Mutegi et al. (2012) examined the magnitude and dynamics of crop-to-wild gene flow and genetic variability in a crop–wild–weedy complex of sorghum under traditional farming methods in Meru South district, Kenya, using 110 cultivated and 373 wild sorghum individuals and genotyped with ten polymorphic SSRs. The results indicated an asymmetric gene flow with higher rates of gene flow from the crop to the wild forms than *vice versa*. The asymmetric crop-to-wild gene flow may reflect (i) differences in the size of the on-farm crop population compared to their wild relatives, (ii) differences in the mating systems between the two and (iii) selection by farmers (Mutegi et al., 2012). Crop population sizes are usually much higher than those of their sympatric wild/weedy relatives, and members of *S. bicolor* are considered to be predominantly autogamous, with occasional cross-pollination occurring at rates that may be different for cultivated and wild sorghums. Pedersen et al. (1998) reported a higher rate of natural outcrossing (up to 61%) in the weedy Sudangrass (wild sorghum forage type) compared to grain-type sorghum (0–13%); Muraya et al. (2011) reported an outcrossing rate of up to 75% in an analysis of 12 populations of wild sorghum from different agroecological zones of Kenya. Extensive introgression, especially within *in situ* conservation areas and/or in the areas of high diversity, would therefore lead to genetic erosion and the possible depletion of important wild sorghum genetic resources.

4.2 Geographical, environmental and social patterns of sorghum diversity

Because landraces are nurtured and cultivated by farmers using traditional methods of selection over decades, there is a strong association between the people of an area and the genetic diversity of the cultivars present in those areas. The key mechanisms in the dynamics of crop genetic diversity are the differences in farmers' ethnic, linguistic, cultural, socioeconomic and family structure, environmental stress leading to animal and human migration, economic factors and, above all, informal seed exchange (Delêtre et al., 2011; Pautasso et al., 2013; Westengen et al., 2014; Labeyrie et al., 2014).

On-farm diversity in Ethiopia is reported to be shaped by field altitude, field size, soil pH, levels of phosphorus, available nitrogen, organic matter, exchangeable potassium and clay content, and varietal mixture is one of the important methods of managing on-farm genetic diversity (Mekbib et al., 2009). Labeyrie et al. (2014) conducted a study to assess the effect of social boundaries in the contact zone among the Chuka, Mbeere and Tharaka ethnolinguistic groups in eastern Kenya. They showed that the spatial distribution of landrace names and the overall genetic spatial patterns were significantly correlated with the ethnolinguistic partition. In a uniform agroecological environment, social boundaries associated with the patterns of ethnolinguistic diversity have impacted the distribution of sorghum varieties and their genetic spatial patterns.

Westengen et al. (2014) examined the genetic structure of sorghum in Africa. On a continent-wide scale, they identified three major sorghum populations (central, southern and northern) that are associated with the distribution of ethnolinguistic groups. The co-distribution of the central sorghum population and the Nilo-Saharan language was observed, suggesting a close and causal relationship between the distribution of sorghum and the languages used within the region between Chari and Nile rivers. The southern sorghum population is associated with the Bantu languages of the Niger–Congo language family, and is consistent with the farming–language co-dispersal hypothesis as it has been related to the Bantu expansion. The northern sorghum population is distributed across the early Niger–Congo and Afro-Asiatic language family areas and areas with dry agroclimatic conditions. At a finer geographic scale, the genetic substructure within the central sorghum population is associated with language-group expansions within the Nilo-Saharan language family. Deu et al. (2008) showed a greater racial diversity and a more even racial distribution of sorghums in Niger and the significant presence of intermediate races. They reported geographical and ethnic patterns of racial distribution, for example, the *guinea* sorghums are mainly cultivated by the Zarma/Songhaï group in western Niger, where rainfall is more abundant, while the Kanuri people, who live in the eastern part of the country around Lake Chad, cultivate almost exclusively *durra* sorghums.

4.3 Trait preference and selection

Farmers' knowledge about the sorghum crop, such as types, names, uses, cropping systems, cultivation methods and so on, is important for the conservation and utilization of the rich diversity of sorghums conserved by farmers *in situ*. Farmers name the landraces based on morphological features, adaptability, locality, uses, etc., and apply selections that shape *in situ* sorghum diversity. In Ethiopia, farmers refer to distinct sorghum types by different names, which reveal several characteristics. For example, the name *Chibite* designates a sorghum with a compact head; *Wotet-Begunche* indicates seeds with a milky taste; *Ahyo* designates tolerance to birds and striga; *Cherekit* indicates white seed colour; *Minchiro* designates loose, drooping panicles; *Wof-Aybelash* means bird-resistant; *Marchuke* and *Mar-Beshenbeko* mean full of honey; *Ayfere* refers to striga resistance; *Shilme* means a fruiting head that contains seeds of different colours and *Gubete* invokes the softness of the roasted seed (Benor and Sisay, 2003). Teshome et al. (1997) indicated that the sorghum landraces of the north Shewa and south Welo regions of Ethiopia are grouped into three clusters with nine characteristics – stem juiciness, grain plumpness, grain shape, grain covering, grain size, grain colour, glume hairiness, glume constriction and lodicule hairs. Farmers in eastern Ethiopia name their varieties in relation to their introduction, use, use-related traits and morphological attributes (Mekbib, 2007). For example, the names *Bisidimo*, *Cherchero*, *Mureta*, *Wobere*, *Weliso* and *Wahelo* are based on place names; *Kassim*, *Manahaile* and *HajiiAli* are named after the introducer; *Alasherif* (sweet stalk), *Chomme* (seed with a lot of fat), *Daddu* (sweet as butter), *Fendisha* (very beautiful), etc., are named according to usage traits; and the names *Aday* (white seeded), (H) *Anchero* (lax-type panicle), *Chamme* (grows fast), *Chelle* (amber-coloured seed), *Firelemi* (twin-seeded), *Keyla* (red-seeded), etc., are based on morphological traits (Mekbib, 2007).

The eastern province of Kenya is a major sorghum-growing zone, and landraces from this region are continuously maintained according to the cultural preferences and traditional practices adopted by the farmers. For example, landraces in the Mbeere region have a diverse seed colour compared with those from the Kitui, Mutomo and Makueni regions, and many of these landraces are unique in their adaptation, food quality, grain yield, quality of harvested products, biotic stress resistance and post-harvest processing (Muui et al., 2013). Dissan farmers from southern Mali named landraces based on days to maturity, yield and taste, for example, *Boboka* for high yield and taste, *Segatono* for striga resistance and *Kalo Saba* for earliness (Lacy et al., 2006).

Sudan is widely recognized as a major centre of diversity of sorghum. Grenier et al. (2004) characterized sorghum germplasm from Sudan in order to understand regional diversity and the distribution of landraces across the five provinces (Gezira-Gedarif, Kassala, Blue Nile, Upper Nile and Equatoria). The study showed a high level of within-region diversity among all the Sudanese sorghums. The landraces from Gezira-Gedarif tended to be shorter in stature, earlier to maturity and less sensitive to changes in the photoperiod, and had long, narrow and compact panicles that may result from adaptation to low rainfall and the early adoption of mechanized farming practices. The taller and later maturing sorghums were from Equatoria, most of which delayed their flowering in response to increased day length. Collections from Kassala showed a higher frequency of landraces with kernels that were more difficult to thresh. The landraces from Blue Nile tended to be agronomically superior, with a higher proportion of landraces having white kernels that were poorly covered and thus were easier to thresh. Sorghums from the Upper Nile tended to have loose panicles with poorly covered kernels that may result from the adaptation to the high rainfall of the southern region.

India is the home of the largest number of traditional sorghum landraces with diverse ecosystems. The landraces have specific traits and traditional utilities, for example, *moli jowar* from Madhya Pradesh fetches higher prices due to its attractive grains, *irungu cholam* from Tamil Nadu yields the best quality porridge and *mathappu cholam* is used for preparing a jelly-like food called *khali*. The *beed maldandi* and *bidri* from Maharashtra produce the best quality jowar roti or unleavened bread. Karnataka's *kodumurugu jola* and *allin jola* are used to make *laddus* and *papads*, respectively, while *allur jola* is used for pops (*allu*). The *pachcha jonna* from Andhra Pradesh, *barmuda local*, *deshi chari* and *gudli local* from Rajasthan and *bendri dagdi* and *khondya* from Maharashtra are excellent fodder types. *Valsangh maldandi local*, *Vadgaon dagdi maldandi*, *tongraligaon maldandi*, *tongraligaon dagdi*, *sultanpur local dagdi*, *sultanpur maldandi*, *harni jogdi* (*dagdi*), *harni jogdi*, *chungi maldandi*, *musti local* (*Maldandi*), *chungi kuch-kachi*, *baddi jowar* and *chakur maldandi* from Maharashtra, *sai jonna* from Andhra Pradesh and *karuppu irungu* from Tamil Nadu are considered by the farmers as drought-tolerant landraces (Elangovan et al., 2009, 2012). In the Adilabad district of Telangana, where sorghum has been cultivated by different tribal communities since ancient times, the *Gond*, *Lambada* and *Kolam* tribal populations are dominant. These farmers named sorghum landraces according to phenotypic traits. The chief phenotypic characteristics used by the farmers in the naming of the sorghum landraces are midrib colour (*Vubiri patti jonna*), grain colour (*Erra jonna*, *Pachcha jonna*), grain size (*Chinna jonna*, *Chinna boda jonna* and *Pedda jonna*), glume colour (*Tekedari jonna*), glume hairiness (*Leha jonna*), panicle shape (*Pandimutte jonna*, *Pelala jonna*), grain shape (*Sevata jonna*), etc. Apart from the use of morphological characteristics, farmers have named landraces by habitat and crop-growing season (*rabi jonna*), and the landrace which never fails to give an assured yield had been named *Sai jonna*, after the holy saint Sai Baba (Pandravada et al., 2013). In Tamil Nadu, the majority of landraces belong to *Irungu cholam*, *Senjolam*, *Vellai cholam*, *Periya manjal cholam*, *Manjal cholam*, *Pei cholam*, *Nattu cholam*, *Senkaton cholam*, *Karareddu cholam*, *Thalaivirchchan cholam* and *Makkatai cholam*. Some of the sorghum landraces are named with the village name prefixed to the landrace (e.g. *Kovilpatti cholam* and *Tenkasi vellai*), and are also named according to the grain colour (e.g. *Vellai cholam* (white), *Manjal cholam* (yellow), *Senjolam* (red) and *Karuncholam* (black glume)). Some of the landraces are named for the shape of the ear head (e.g. *Matthappu cholam*, i.e. the ear heads look like flower fireworks in shape).

China is considered as one of the secondary centres of diversity for sorghum and a large number of germplasm accessions have been conserved (Table 1). The genetic diversity of Chinese sorghum landraces exhibits moderate or less diversity compared to exotic collections, especially those which are found in African sorghum (Li et al., 2010; Burow et al., 2012). Sorghum accessions from China ('kaoliang' in Chinese) have distinct characteristics from those of African and Indian origins. Kaoliang plants have a strong aerial root system, weak tillering and a long internode stalk that is suitable for weaving. The stem of kaoliang is filled with medulla tissue and lacks water when mature, and the main vein of the leaf is white. The panicles of Chinese kaoliang are not prone to shattering and kernels are easily threshed when mature. Chinese sorghum has a strong resistance to cold temperatures and good seedling vigour, but poor disease and pest resistance. The majority of Chinese-produced sorghum is used to produce liquor and vinegar, and a very small proportion is used for human food (Diao, 2017).

5 Geographical distribution of sorghum germplasm

About 42% of the sorghum germplasm collections conserved in genebanks globally have no information about their geographic origin. The remaining accessions mostly originate from Ethiopia (10.0%), India (7.8%), the United States (4.1%), Mali (3.1%) and Yemen (3.0%), while other countries represent less than 3% of the collections. The ICRISAT genebank conserves 39553 accessions originating from 93 countries covering all geographical regions (Fig. 2; Table 4 and 5). In the entire collection, landraces constitute the major portion (86.36%), originating mostly from Africa (69.20%) and Asia (26.19%) (Table 4). Accessions of the races *durra* and *bicolor* and the intermediate race *durra–caudatum* are mostly from Africa and Asia, and *guinea–kafir* is from the Americas, while all other races are mostly from Africa (Table 5).

6 Germplasm gap analysis of sorghum

Gap analysis is a systematic method of analysing the degree of conservation of taxa in order to identify those locations, taxa and particular traits (adaptations) that are not secured or are under-secured in the conservation systems (Maxted et al., 2008). It is an important step to capture and conserve the maximum number of alleles and genes in the crop species. Germplasm gap analysis enhances the collection and conservation of

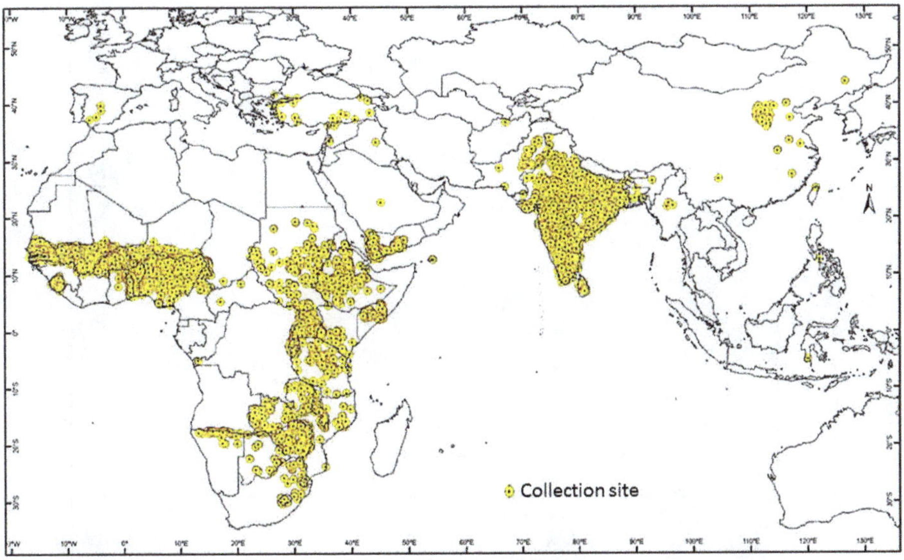

Figure 2 Geographical distribution of sorghum germplasm accessions conserved at the ICRISAT genebank, Patancheru, India.

germplasm diversity through (i) targeting the localities where sets of species are absent from existing collections; (ii) determining which areas are 'under-collected' or 'over-collected' for germplasm relative to the known distribution of a taxon; (iii) locating which regions have the greatest or most dissimilar species richness compared with other regions and (iv) outlining the ecological amplitudes of each species so that a wider representation of the ecotypes or genetically adapted populations of each can be sampled.

Geographic Information System technologies have enabled a better understanding of species distributions and of the representativeness of germplasm collections. Gap analysis of the ICRISAT sorghum germplasm accessions originating from South Asian countries (Afghanistan, Bangladesh, India, the Maldives, Nepal, Pakistan and Sri Lanka) showed geographical gaps of 131 districts located in 27 provinces of the four South Asian

Table 4 Geographical distribution of sorghum germplasm collections conserved at the ICRISAT genebank, India

Region	Africa	Americas	Asia	Europe	Oceania	Unknown	Total
Advanced/ Improved cultivar	19	7	74	–	–	–	100
Breeding/Research material	1 558	1 571	1 530	38	6	133	4 836
Traditional cultivar/ Landrace	23 636	885	8 948	610	39	38	34 156
Wild	330	69	32	7	22	1	461
Total	**25 543**	**2 532**	**10 584**	**655**	**67**	**172**	**39 553**

Table 5 Geographical distribution of basic and intermediate races of sorghum germplasm conserved at the ICRISAT genebank, India

Sorghum germplasm type	Asia	Africa	America	Europe	Oceania	Caribbean	Unknown	Total
Bicolor	512	600	337	114	4	–	4	1 571
Guinea	838	4 081	76	8	–	1	4	5 008
Caudatum	685	6 549	383	94	15	1	27	7 754
Kafir	71	925	305	11	1	–	2	1 315
Durra	4 032	3 652	190	98	2	2	11	7 987
Guinea-bicolor	39	281	21	4	1	–	2	348
Caudatum-bicolor	570	1 145	210	75	5	1	33	2 039
Kafir-bicolor	45	52	46	2	1	1	–	147
Durra-bicolor	843	1 446	62	83	2	1	3	2 440
Guinea-caudatum	713	3 212	227	50	5	1	64	4 272
Guinea-kafir	5	35	65	–	–	–	1	106
Guinea-durra	79	128	18	7	–	–	2	234
Kafir-caudatum	45	250	118	3	–	–	5	421
Durra-caudatum	1 987	2 402	321	85	7	–	10	4 812
Kafir-durra	53	142	76	–	2	–	–	273
Wild	32	330	69	7	22	–	1	461
Unclassified	36	312		14	–	–	3	365
Total	**10 585**	**25 542**	**2 524**	**655**	**67**	**8**	**172**	39 553

countries. Countrywise, these gaps are 110 districts in 20 provinces of India, 13 districts in three provinces of Pakistan, three districts in Bangladesh and five districts in four provinces of Sri Lanka. Uttar Pradesh in India, with relatively low representation (238 accessions) in the collection, had 27 districts where no sorghum germplasm was represented. Other provinces showing gaps of more than five districts include Bihar, Madhya Pradesh, Rajasthan, Odisha, Tamil Nadu and West Bengal. *S. bicolor* subsp. *verticilliflorum*, *S. halepense* and *S. propinquum* were identified as taxonomic gaps in the collection (Upadhyaya et al., 2016b). Geographical gaps in sorghum germplasm conserved in the genebank at ICRISAT were also found in 153 districts located in 50 provinces of ten East African countries (Upadhyaya et al., 2017a). Similarly, 386 districts in 11 west and central African countries (Upadhyaya et al., 2017c) and 108 districts in eight southern African countries were identified as gaps (Upadhyaya, unpublished). Only by identifying the gaps can the genebanks assemble/collect the missing diversity and enrich the germplasm collection.

7 Ensuring diversity in sorghum

7.1 Germplasm enhancement through the use of unadapted landraces and wild germplasm

The loss of genetic diversity has been reported in several crops, including sorghum (Jordan et al., 1998, 2003; Murray et al., 2009; Smith et al., 2010). It results in a reduction of genetic variability for important traits and leaves the crops vulnerable to epidemics of pests or diseases. The introduction of new diversity from exotic germplasm, unadapted and/or wild relatives through breeding is one way of combating this problem, but it is challenging due to the linkage load of many undesirable genes and the assumed risks associated with unadapted and wild germplasm. In Australia, Jordan et al. (2011) have developed 56 backcross-derived populations using an elite line with a local adaptation as the recurrent parent and crossed it with a range of sorghum lines (lines with geographic or racial diversity, phenotypic diversity for key traits, elite lines from breeding programmes in other countries and cross-compatible wild species) that are unadapted to Australian conditions. Selections were performed from BC_1F_1 onwards for key adaptive traits, such as plant height and flowering time, until BC_1F_4. Populations of 30–90 BC_1F_4 lines derived from 56 unadapted parents were then evaluated in hybrid combinations in 21 trials over a four-year period. Despite strong selection for acceptable height and maturity, a considerable genetic variation for the grain yield was retained in the populations and identified lines in each population performed significantly better than the recurrent parent for the grain yield. This method proved to be an effective way to introduce new alleles from unadapted sorghum germplasm into elite breeding material.

7.2 Core and mini core collections to access whole diversity and genome-wide association mapping

Representative subsets of germplasm diversity, such as core (~10% of the entire collection) (Frankel, 1984) and mini core (~10% of the core collection) (Upadhyaya and Ortiz, 2001) collections, enhance the utilization of germplasm, mainly because they comprise a manageable number of accessions that represent the diversity of the entire collection of a given species. Such representative subsets are available in sorghum. The core collection in sorghum consists of 3475 accessions (Prasada Rao and Ramantha Rao, 1995), 2247 accessions (Grenier et al., 2001) or 3011 accessions (Dahlberg et al., 2004), while the mini core collection consists of 242 accessions (Upadhyaya et al., 2009). Under the Generation Challenge Programme, the Global Composite Germplasm Collection (GCGC) of sorghum, which consists of 3384 cultivated and wild accessions, was established (http://www.generationcp.org/issue-59-march-2012/32-research/sorghum/180-sorghum-products). This sorghum GCGC was then genotyped with 41 SSR markers and formed a genotype-based reference set of 383 accessions that captured 78.3% of the SSR alleles detected in the sorghum GCGC (Billot et al., 2013).

These subsets, representing the entire collection of sorghum conserved in the genebank, are sufficiently diverse to serve as a panel for the detailed characterization of those traits that are of economic importance to plant breeding programmes and for the assessment of allelic diversity in the genes associated with the traits of economic interest. The sorghum mini core collection (Upadhyaya et al., 2009) has been extensively used as a panel to map and identify marker trait associations for plant height and maturity (Upadhyaya et al., 2012b, 2013a; Wang et al., 2012), kernel weight, tiller number (Upadhyaya et al., 2012a), resistance to anthracnose, leaf rust and grain mould (Upadhyaya et al., 2013b, 2013c), germinability and seedling vigour at low temperatures (Upadhyaya et al., 2016c).

7.3 Diverse trait-specific sources

The sorghum mini core collection (Upadhyaya et al., 2009) has been extensively evaluated for morpho-agronomic traits and for resistance to grain mold, downy mildew, anthracnose, leaf blight and rust diseases (Sharma et al., 2010, 2012). The sorghum mini core collection was also evaluated for stalk sugar content and for grain iron (Fe) and zinc (Zn) concentrations under drought stress conditions (Upadhyaya et al., 2014b, 2016d). Seven accessions – IS 13294, IS 13549, IS 23216, IS 23684, IS 24139, IS 24939 and IS 24953 – recorded significantly greater mean Brix levels (14.0–15.2%) compared with the best control, IS 33844 (12.4%), across environments. IS 1004, IS 4698, IS 23891 and IS 28141 had almost the same Brix content (~13%), with a significantly higher grain yield (11.7–22.7% over IS 33844) (Upadhyaya et al., 2014b). The sorghum mini core accessions showed a large variation for grain nutrient contents: Fe (25.8–48.9 mg kg^{-1} seed) and Zn (13.5–42.6 mg kg^{-1} seed) (Upadhyaya et al., 2016d). Eleven accessions with high seed Fe – IS 16382, IS 23992, IS 28313, IS 28389, IS 28849, IS 20743, IS 21645, IS 21863, IS 28747, IS 30508 and IS 31681 (Fe 40.3–48.6 mg kg^{-1} seed) – 14 accessions with high Zn – IS 30460, IS 602, IS 17980, IS 19859, IS 28451, IS 30466, IS 30536, IS 5301, IS 8774, IS 4951, IS 25249, IS 24139, IS 24175 and IS 24218 (Zn 32.2–36.4 mg kg^{-1} seed) – and nine accessions with both Fe and Zn – IS 1219, IS 1233, IS 30450, IS 30507, IS 1212, IS 27786, IS 30383, IS 31651 and IS 24503 (Fe 40.8–48.9 mg kg^{-1} seed; Zn 32.8–42.6 mg kg^{-1} seed) – were identified. Six accessions (IS 1004, IS 23514, IS 23579, IS 23590, IS 28141 and IS 31706) and four accessions (IS 1004, IS 27034, IS 28141 and IS 31706), respectively, showed 8–39% and 9–38% greater amounts of Fe and Zn over the control IS 33844, and produced seed yields similar to that of IS 33844. Upadhyaya et al. (2016c) reported six accessions (IS 1212, IS 14779, IS 15170, IS 22986, IS 7305 and IS 7310) and five accessions (IS 602, IS 1233, IS 7305, IS 10302 and IS 20956), respectively, from the mini core collection for a higher percentage of seedling vigour and germination under low temperature conditions. One accession (IS 7305) showed higher germinability and vigour at low temperatures. Upadhyaya et al. (2017b) identified seven accessions (IS 14779, IS 23891, IS 31714, IS 4515, IS 5094, IS 9108 and IS 15466) that were tolerant to post-flowering drought stress. The sorghum mini core collection was also screened against insect pests, such as stem borer, shoot fly and aphids and drought tolerance, and sources for these traits have been identified (Upadhyaya, unpublished). Furthermore, agronomic performance and marker data were used to identify genetically diverse multiple trait-specific accessions with desirable agronomic traits. For example, IS 23684 for high grain nutrients, resistant to diseases (anthracnose, leaf blight and rust) and insect pests (aphids); IS 1212 for earliness, high grain nutrients, tolerant to drought, low temperature stress (high seedling vigour) and resistant to diseases (downy mildew and grain mould); IS 5094 for high grain yield, tolerant to drought and resistant to diseases (downy mildew and charcoal rot) and insect pests (stem borer and shoot fly); IS 473 for earliness and resistant to disease (downy mildew, anthracnose, leaf blight and rust); IS 4698 for high grain yield, high Brix % and resistant to insect pests (stem borer, shoot fly and aphids) and IS 23891 for large seeds, high grain yield, high Brix %, tolerant to drought and resistant to charcoal rot (Upadhyaya, unpublished). These diverse trait-specific sources can be utilized in hybridization programmes for introducing novel diversity into sorghum cultivars.

7.4 Wild relatives for broadening the genetic base of cultivated sorghum

The wild relatives of sorghum continue to play a key role in crop improvement techniques, contributing genes for the adaptation to various climate conditions, including biotic and abiotic stresses. The wild sorghum species represent a diverse source of germplasm that offers a considerable potential in broadening the genetic base and enhancing the genetic potential of cultivars. However, the research on sorghum has been geared towards cultivated sorghums, while studies in wild sorghums are limited. Studies have revealed that the cultivated sorghums harbour lower genetic diversity than their wild progenitors (Casa et al., 2005; Mace et al., 2013; Mutegi et al., 2011; Muraya et al., 2011). This is because during the process of domestication, evolutionary processes such as population bottleneck, founder effect, genetic drift and artificial selection have reduced genetic diversity of the crop in relation to its wild progenitor. The high genetic diversity of the wild relatives can potentially be exploited to broaden the genetic base of cultivated sorghum.

The genetic potential of the wild relatives of sorghum, particularly as sources of resistance to pests and diseases, is well documented. Kamala et al. (2002) reported 45 wild accessions comprising 15 species from four sections – *Parasorghum*, *Heterosorghum* (*Sorghum laxiflorum* Bailey), *Chaetosorghum* (*S. macrospermum* Garber) and *Stiposorghum* (*S. angustum* S.T. Blake, *S. ecarinatum* Lazarides, *S. extans* Lazarides, *S. intrans* F. Muell. ex Benth., *S. interjectum* Lazarides, *S. stipoideum* (Ewart & Jean White) C. Gardener & C.E. Hubb.) – including all accessions from Australia, which exhibited immunity to downy mildew. The cultivated types and wild races of the section *Sorghum* showed the greatest susceptibility, while accessions of *S. halepense* (L.) Pers. were comparatively less susceptible. Two accessions from the primary gene pool, IS 18821 (*aethiopicum*) and IS 18882 (*arundinaceum*), were free from downy mildew and were cross-compatible with cultivated sorghum. These may be directly used to develop cultivars which are resistant to downy mildew. Kamala et al. (2009) reported 32 accessions belonging to *Parasorghum*, *Stiposorghum* and *Heterosorghum* that did not suffer any shoot fly damage under field conditions, while one accession each of *Heterosorghum* (*S. laxiflorum*) and *Chaetosorghum* (*S. macrospermum*) suffered very low shoot fly damage. Under greenhouse conditions, the same accessions either showed absolute non-preference for oviposition under no-choice conditions or were preferred for oviposition, but suffered low deadheart damage. Larvae survival and fecundity were significantly reduced in a few accessions within the section *Sorghum*. This is promising since none of the existing resistant cultivars is known to be completely non-preferred for egg laying (Kamala et al., 2009). Oviposition non-preference is the primary mechanism of resistance to shoot fly in cultivars, and there is strong evidence that antibiosis is an important mechanism of resistance in the wild relatives of sorghum. The combination of antibiosis and ovipositional non-preference would be highly desirable as an operating mechanism for resistance to shoot fly. However, the exact nature of the resistance conferred by these species needs to be unravelled through further studies and biochemical assays for a better understanding of shoot fly behaviour, particularly in relation to its host species (Kamala et al., 2014).

The levels of resistance to spotted stem borer in cultivated sorghum are low to moderate. Kamala et al. (2012) evaluated sorghum's wild relatives to identify accessions with high levels of resistance and studied the mechanisms of resistance. *Heterosorghum* (*S. laxiflorum*), *Parasorghum* (*S. australiense*, *S. purpureo-sericeum*, *S. versicolor*, *S. matarankense*, *S. timorense*, *S. brevicallosum* and *S. nitidum*) and *Stiposorghum* (*S. angustum*, *S. ecarinatum*, *S. extans*, *S. intrans*, *S. interjectum* and *S. stipoideum*) showed very high levels of resistance to stem borer, while *Chaetosorghum* (*S. macrospermum*), four wild races of *S. bicolor* subsp. *verticilliflorum* and *S. halepense* were found to be susceptible. Accessions belonging to *Stiposorghum* and *Parasorghum* (*S. purpureo-sericeum*, *S. versicolor* and *S. timorense*) were significantly less preferred for oviposition; accessions belonging to *Stiposorghum* showed slight leaf feeding, but there was no deadheart formation; very few deadhearts were formed in *Parasorghum*; in *Heterosorghum*, the two accessions of *S. laxiflorum* were highly preferred for oviposition; accessions belonging to Sec. *Sorghum* showed maximum deadhearts; more larvae were recovered and larger numbers of adults emerged (Kamala et al., 2012).

Sorghum's wild relatives have also been reported as sources of genes for resistance to sorghum midge (Sharma and Franzmann, 2001) and green bug (Duncan et al., 1991). *Striga* resistance mechanisms, such as low germination stimulant production, germination inhibition and low haustorial initiation activity, have been reported to occur in wild sorghum (Rich et al., 2004). Some of the wild relatives also produce high biomass and can be utilized in breeding programmes to transfer such traits for breeding high biomass-producing cultivars. Ohimain and Izah (2016) evaluated wild sorghum (*S. arundinaceum*) for productivity and bioethanol potentials and their results showed that the extract from wild sorghum can be distilled and used as ethanol, while the bagasse can be used as fuel to generate electricity via a steam cycle.

Valuable traits, such as resistance/tolerance to biotic and abiotic stresses, are often present but are inaccessible in the wild relatives of cultivated crop species due to strong reproductive barriers that prevent hybridization between them. However, Price et al. (2006) have demonstrated the production of hybrids involving cultivated sorghum (*S. bicolor*) with those of species from the tertiary gene pool (*S. angustum*, *S. nitidum* and *S. macrospermum*) through the use of recessive *iap* allele (dominant allele iap: inhibition of alien pollen) that produces or eliminates the pollen–pistil incompatibilities that prevent hybridization. The production of these interspecific hybrids demonstrated that recessive *iap* allele circumvents pollen–pistil incompatibilities in the genus *Sorghum* and permits hybrids to be made between *S. bicolor* and species of the tertiary gene pool. However, this method of gene transfer can be useful only when the trait of interest is not present in the cross-compatible species, because the agronomic fitness of the derived hybrids is often poor. More research on the hybrid recovery rate involving different species, cytological investigations, etc., is needed in order to utilize this method of gene transfer to achieve introgression between the genomes.

8 Future trends and conclusion

Sorghum landraces and wild and weedy relatives possess abundant diversity *in situ* and considerable amounts of diversity have been conserved *ex situ* in genebanks globally. Germplasm gap analyses have revealed the presence of geographical and taxonomic gaps, necessitating the collection and conservation of landraces and wild and weedy relatives before we lose them forever. Valuable traits are present in the wild and weedy relatives and landraces. However, they have not been fully utilized because most crop breeders use working collections comprising adapted and improved materials that have the most desirable traits. This may lead to a narrower genetic base of sorghum cultivars. Developing representative subsets of germplasm diversity, such as the sorghum mini core collection, has been useful for identifying multiple-trait germplasm for traits of economic importance. The use of such germplasm, and the wild and weedy relatives of sorghum, through deploying appropriate breeding approaches and genomic tools, has the potential to enhance crop yields and broaden the genetic base of cultivars.

9 Where to look for further information

The genebank, ICRISAT, India, conserves the largest collection of sorghum germplasm and passport and characterization details are available on the website, http://genebank.icrisat.org/. Sorghum researchers can obtain seed samples of accessions from the ICRISAT genebank for research purposes through a Standard Material Transfer Agreement (SMTA).

10 References

Adugna, A., 2014. Analysis of in situ diversity and population structure in Ethiopian cultivated *Sorghum bicolor* (L.) landraces using phenotypic traits and SSR markers. *Springerplus* 3, 212.

Barnaud, A., Trigueros, G., McKey, D. and Joly, H. I., 2008. High outcrossing rates in fields with mixed sorghum landraces: How are landraces maintained? *Heredity (Edinb)* 101, 445–52.

Benor, S. and Sisay, L., 2003. Folk classification of sorghum (*Sorghum bicolor* (L.) Moench) landraces and its ethnobotanical implication: A Case study in Northeastern Ethiopia. *Etnobiologia* 3, 29–41.

Bhattacharya, A., Rice, N., Shapter, F. M., Norton, S. L. and Henry, R. J., 2011. Sorghum. In: B. Kilian, K. Mammen and C. Kole (Eds) *Wild Crop Relatives: Genomic and Breeding Resources: Cereals.* Springer-Verlag Berlin Heidelberg, London, New York, 397–406.

Billot, C., Ramu, P., Bouchet, S., Chantereau, J., Deu, M., Gardes, L., Noyer, J.-L., Rami, J.-F., Rivallan, R., Li, Y., Lu, P., Wang, T., Folkertsma, R. T., Arnaud, E., Upadhyaya, H. D., Glaszmann, J.-C. and

Hash, C. T., 2013. Massive sorghum collection genotyped with SSR markers to enhance use of global genetic resources. *PLoS One* 8, e59714.

Burow, G., Franks, C. D., Xin, Z. and Burke, J. J., 2012. Genetic diversity in a collection of Chinese sorghum landraces assessed by microsatellites. *Am. J. Plant Sci.* 03, 1722–9.

Casa, A. M., Mitchell, S. E., Hamblin, M. T., Sun, H., Bowers, J. E., Paterson, A. H., Aquadro, C. F. and Kresovich, S., 2005. Diversity and selection in sorghum: Simultaneous analyses using simple sequence repeats. *Theor. Appl. Genet.* 111, 23–30.

Casa, A. M., Pressoir, G., Brown, P. J., Mitchell, S. E., Rooney, W. L., Tuinstra, M. R., Franks, C. D. and Kresovich, S., 2008. Community resources and strategies for association mapping in sorghum. *Crop Sci.* 48, 30–40.

Dahlberg, J. A. 2000. Classification and characterization of sorghum, pp. 99–130. In C. W. Smith and R. A. Frederiksen (Eds) *Sorghum: Origin, History, Technology, and Production*. John Wiley & Sons, Inc., New York, NY.

Dahlberg, J. A., Burke, J. J. and Rosenow, D. T., 2004. Development of a sorghum core collection: Refinement and evaluation of a subset from Sudan. *Econ. Bot.* 58, 556–67.

de Wet, J. M. J., 1978. Systematics and evolution of sorghum sect sorghum (gramineae). *Am. J. Bot.* 65, 477–84.

Delêtre, M., McKey, D. B. and Hodkinson, T. R., 2011. Marriage exchanges, seed exchanges, and the dynamics of manioc diversity. *Proc. Natl. Acad. Sci. USA* 108(45), 18249–54.

Deu, M., Sagnard, F., Chantereau, J., Calatayud, C., Hérault, D., Mariac, C., Pham, J. L., Vigouroux, Y., Kapran, I., Traore, P. S., Mamadou, A., Gerard, B., Ndjeunga, J. and Bezançon, G., 2008. Niger-wide assessment of *in situ* sorghum genetic diversity with microsatellite markers. *Theor. Appl. Genet.* 116, 903–13.

Diao, X. 2017. Production and genetic improvement of minor cereals in China. *Crop J.* 5(2): 103–14.

Doggett, H., 1988. *Sorghum*, 2nd edn. Longman Scientific and Technical, London, UK

Duncan, R. R., Bramel-Cox, P. J. and Miller, F. R. 1991. Contributions of sorghum germplasm to hybrid development in the USA, pp. 69–102. In H. L. Shands and L. E. Weisner (Eds) *Use of Plant Introductions in Cultivar Development. Part I.* Spec. Pub. 117. CSSA, Madison, WI.

Dwivedi, S. L., Upadhyaya, H. D. and Gowda, C. L. L., 2009. Approaches to enhance the value of genetic resources in crop improvement. In: *Strengthening Information on Plant Genetic Resources in Asia*. Bangkok, Thailand, pp. 68–74, (http://www.fao.org/docrep/014/am265e00.pdf).

Elangovan, M. and Babu, P. K., 2015. Genetic variability and diversity of sorghum landraces collected from uttar pradesh, India. *Indian J. Plant Genet. Resour.* 28, 213–21.

Elangovan, M., Ganesamurthy, K., Rajaram, S., Sankarapandian, K. and Babu, P. K., 2012. Collection and conservation of sorghum landraces from Tamil Nadu. *Electron. J Plant Breed.* 3(2), 753–62.

Elangovan, M., Prabhakar, Tonapi, V. A. and Reddy, D. C. S., 2009. Collection and characterization of Indian sorghum landraces. *Indian J. Plant Genet. Resour.* 22(3), 173–81.

FAO, 2014. *Genebank Standards for Plant Genetic Resources for Food and Agriculture*. Rev. ed. FAO, Rome.

Frankel, O. H., 1984. Genetic perspective of germplasm conservation. In: W. Arber, K. Illmensee, W. J. Peacock and P. Starlinger (Eds) *Genetic Manipulations: Impact of Man and Society*. Cambridge University Press, Cambridge, pp. 161–70.

Grenier, C., Hamon, P. and Bramel-Cox, P. J., 2001. Core collection of sorghum: II. comparison of three random sampling strategies. *Crop Sci.* 41, 241–6.

Grenier, C., Bramel, P. J., Dahlberg, J. A., El-Ahmadi, A., Mahmoud, M., Peterson, G. C., Rosenow, D. T. and Ejeta, G., 2004. Sorghums of the Sudan: Analysis of regional diversity and distribution. *Genet. Resour. Crop Evol.* 51, 489–500.

Harlan, J. R. and de Wet, J. M. J., 1972. A simplified classification of cultivated sorghum. *Crop Sci.* 12, 172–6.

Harlan, J. R. and de Wet. J. M. J., 1971. Toward a rational classification of culti- vated sorghums. *Crop Sci.* 12, 172–6.

Heywood, V. H., 2014. An overview of *in situ* conservation of plant species in the Mediterranean. *Flora Mediterr.* 24, 5–24.
IBPGR and ICRISAT. 1993. *Descriptors for Sorghum [Sorghum bicolor (L.) Moench].* International Board for Plant Genetic Resources, Rome, Italy; and International Crops Research Institute for the Semi-Arid Tropics, Patancheru, India.
Jordan, D. R., Tao, Y. Z., Godwin, I. D., Henzell, R. G., Cooper, M. and McIntyre, C. L., 2003. Prediction of hybrid performance in grain sorghum using RFLP markers. *Theor. Appl. Genet.* 106, 559–67.
Jordan, D. R., Tao, Y. Z., Godwin, I. D., Henzell, R. G., Cooper, M. and McIntyre, C. L., 1998. Loss of genetic diversity associated with selection for resistance to sorghum midge in Australian sorghum. *Euphytica.* 102, 1–7.
Jordan, D. R., Mace E. S., Cruickshank A. W., Hunt C. H., Henzell, R. G., 2011. Exploring and exploiting genetic variation from unadapted sorghum germplasm in a breeding program. *Crop Sci.* 51(4): 1444–57.
Kabuye, C. H. S., Mungai, G. and Mutangah, J. G., 1986. Flora of Korea national reserve, pp. 57–104. In M. Coe and N. Collins (Eds) *Kora: An Ecological Inventory of the Kora National Reserve, Kenya.*. Royal Geographical Society, London.
Kamala, V., Muraya, M., Dwivedi, S. L. and Upadhyaya, H. D., 2014. Wild sorghums – Their potential use in crop improvement. In: Y. H. Wang, H. D. Upadhyaya and C. Kole (Eds) *Genetics, Genomics and Breeding of Sorghum.* CRC Press, New York, pp. 56–89.
Kamala, V., Sharma, H. C., Manohar Rao, D., Varaprasad, K. S., Bramel, P. J. and Chandra, S., 2012. Interactions of spotted stem borer *Chilo partellus* with wild relatives of sorghum. *Plant Breed.* 131, 511–21.
Kamala, V., Sharma, H. C., Rao, D. M., Varaprasad, K. S. and Bramel, P. J., 2009. Wild relatives of sorghum as sources of resistance to sorghum shoot fly, *Atherigona soccata. Plant Breed.* 128, 137–42.
Kamala, V., Singh, S. D., Bramel, P. J. and Rao, D. M., 2002. Sources of resistance to downy mildew in wild and weedy sorghums. *Crop Sci.* 42, 1357–60.
Keneni, G., Bekele. E., Imtiaz, M. and Dagne, K., 2012. Genetic vulnerability of modern crop cultivars: Causes, mechanism and remedies. *Int. J. Plant Res.* 2(3), 69–79.
Labeyrie, V., Deu, M., Barnaud, A., Calatayud, C., Burion, M., Wanbugu, P., Manel, S., Glaszmann, N.-C. and Leclerc, C., 2014. Influence of ethnolinguistic diversity on the sorghum genetic patterns in subsistence farming systems in eastern Kenya. *PLoS One* 9(3), e92178.
Lacy, S. M., Cleveland, D. A. and Soleri D., 2006. Farmer choice of sorghum varieties in southern Mali. *Hum. Ecol.* 34(3), 331–53.
Li, R., Zhang, H., Zhou, X., Guan, Y., Yao, F., Song, G., Wang, J. and Zhang, C., 2010. Genetic diversity in Chinese sorghum landraces revealed by chloroplast simple sequence repeats. *Genet. Resour. Crop Evol.* 57, 1–15.
Liu, Q., Liu, H., Wen, J. and Peterson, P. M., 2014. Infrageneric phylogeny and temporal divergence of sorghum (Andropogoneae, Poaceae) based on low-copy nuclear and plastid sequences. *PLoS One* 9, e104933.
Mace, E. S., Tai, S., Gilding, E. K., Li, Y., Prentis, P. J., Bian, L., Campbell, B. C., Hu, W., Innes, D. J., Han, X., Cruickshank, A., Dai, C., Frère, C., Zhang, H., Hunt, C. H., Wang, X., Shatte, T., Wang, M., Su, Z., Li, J., Lin, X., Godwin, I. D., Jordan, D. R. and Wang, J., 2013. Whole-genome sequencing reveals untapped genetic potential in Africa's indigenous cereal crop sorghum. *Nat. Commun.* 4, 2320.
Maxted, N., Dulloo, E., Ford-Lloyd, B. V., Iriondo, J. M. and Jarvis, A., 2008. Gap analysis: A tool for complementary genetic conservation assessment. Divers. Distrib. 14(6), 1018–30.
Maxted, N., Ford-Lloyd, B. V. and Hawkes, J. G., 1997. Complementary conservation strategies. In N. Maxted, B. V. Ford-Lloyd and J. G. Hawkes (Eds) *Plant Genetic Conservation- The In situ Approach.* Kluwer Academic Publishers, P.O. Box 17, 3300AA Dordrecht, The Netherlands.
Mekbib, F., 2007. Infra-specific folk taxonomy in sorghum (*Sorghum bicolor* (L.) Moench) in Ethiopia: Folk nomenclature, classification, and criteria. *J. Ethnobiol. Ethnomed.* 3(38), 1–18.

Mekbib, F., Bjørnstad, A., Sperling, L. and Synnevåg, G., 2009. Factors shaping on-farm genetic resources of sorghum [*Sorghum bicolor* (L .) Moench] in the centre of diversity, Ethiopia. *Int. J. Biodivers. Consev.* 1, 45–59.

Morris, G. P., Ramu, P., Deshpande, S. P., Hash, C. T., Shah, T., Upadhyaya, H. D., Riera-Lizarazu, O., Brown, P. J., Acharya, C. B., Mitchell, S. E., Harriman, J., Glaubitz, J. C., Buckler, E. S. and Kresovich, S., 2013. Population genomic and genome-wide association studies of agroclimatic traits in sorghum. *Proc. Natl. Acad. Sci. USA* 110, 453–8.

Muraya, M. M., de Villiers, S., Parzies, H. K., Mutegi, E., Sagnard, F., Kanyenji, B. M., Kiambi, D. and Geiger, H. H., 2011. Genetic structure and diversity of wild sorghum populations (*Sorghum* spp.) from different eco-geographical regions of Kenya. *Theor. Appl. Genet.* 123, 571–83.

Murray, S. C., Rooney, W. L., Hamblin, M. T., Mitchell, S. E. and Kresovich, S., 2009. Sweet sorghum genetic diversity and association mapping for Brix and height. *Plant Genome J.* 2, 48–62.

Mutegi, E., Sagnard, F., Labuschagne, M., Herselman, L., Semagn, K., Deu, M., de Villiers S., Kanyenji, B. M., Mwongera, C. N., Traore, P. C. S. and Kiambi D., 2012. Local scale patterns of gene flow and genetic diversity in a crop-wild-weedy complex of sorghum (*Sorghum bicolor* (L.) Moench) under traditional agricultural field conditions in Kenya. *Conserv. Genet.* 13(4), 1059–71.

Mutegi, E., Sagnard, F., Semagn, K., Deu, M., Muraya, M., Kanyenji, B., de Villiers, S., Kiambi, D., Herselman, L. and Labuschagne, M., 2011. Genetic structure and relationships within and between cultivated and wild sorghum (*Sorghum bicolor* (L.) Moench) in Kenya as revealed by microsatellite markers. *Theor. Appl. Genet.* 122, 989–1004.

Muui, C. W., Muasya, R. M. and Kirubi, D. T., 2013. Identification and evaluation of sorghum (*Sorghum bicolor* (L.) Moench.) germplasm from Eastern Kenya. *Afr. J. Agri. Res.* 8(37), 4573–9.

Ngugi, K. and Onyango, C. M., 2012. Analysis of the molecular diversity of Kenya sorghum germplasm using microsatellites. *J. Crop Sci. Biotechnol.* 15, 189–94.

Ohimain, E. I. and Izah, S. C., 2016. Productivity and bioethanol potentials of wild sorghum (*Sorghum arundinaceum*). *Br. J. Renew. Energy* 01, 14–17.

Okeno, J. A., Mutegi, E., de Villiers, S., Wolt, J. D. and Misra M. K., 2012. Morphological variation in the wild-weedy complex of *Sorghum bicolor in situ* in Western Kenya: Preliminary evidence of Crop-To-Wild Gene flow? *Int. J. Plant Sci.* 173(5), 507–15.

Pandravada, S. R., Sivaraj, N., Sunil, N., Jairam, R., Prasanthi, Y., Chakrabarty, S. K., Ramesh, P., Bisht, I. S. and Pareek, S. K., 2013. Sorghum landraces patronized by tribal communities in Adilabad district, Andhra Pradesh. *Indian J. Tradit. Knowl.* 12(3), 465–71.

Pautasso, M., Aistara, G., Barnaud, A., Caillon, S., Clouvel, P., Coomes, O. T., Delêtre, M., Demeulenaere, E., De Santis, P., Döring, T., Eloy, L., Emperaire, L., Garine, E., Goldringer, I., Jarvis, D., Joly, H. I., Leclerc, C., Louafi, S., Martin, P., Massol, F., McGuire, S., McKey, D., Padoch, C., Soler, C., Thomas, M., Tramontini, S., 2013. Seed exchange networks for agrobiodiversity conservation. A review. *Agron. Sustain. Dev.* 33(1), 151–75.

Pedersen, J. F., Toy, J. J. and Johnson, B., 1998. Natural outcrossing of sorghum and *Sudangrass* in the central great plains. *Crop Sci.* 38(4), 937–9.

Prasada Rao, K. E. and Ramanatha Rao. V., 1995. The use of characterization data in developing a core collection of sorghum, pp. 109–15. In: T. Hodgkin, A. H. D. Brown, Th. J. L. van Hintum and E. A. V. Morales (Eds) *Core Collections of Plant Genetic Resources*. John Wiley & Sons, Chichester.

Price, H. J., Hodnett, G. L., Burson, B. L., Dillon, S. L., Stelly, D. M. and Rooney, W. L., 2006. Genotype dependent interspecific hybridization of *Sorghum bicolor*. *Crop Sci.* 46, 2617–22.

Rabbi, I. Y., Geiger, H. H., Haussmann, B. I. G., Kiambi, D., Folkertsma, R. and Parzies, H. K., 2010. Impact of farmers' practices and seed systems on the genetic structure of common sorghum varieties in Kenya and Sudan. *Plant Genet. Resour. Charact. Util.* 8, 116–26.

Rakshit, S., Gomashe, S. S., Ganapathy, K. N., Elangovan, M., Ratnavathi, C. V., Seetharama, N. and Patil, J. V., 2012. Morphological and molecular diversity reveal wide variability among sorghum Maldandi landraces from India. *J. Plant Biochem. Biotechnol.* 21, 145–56.

Rich, P. J., Grenier, C. and Ejeta, G., 2004. Striga resistance in the wild relatives of sorghum. *Crop Sci.* 44, 2221–9.

Sharma, H. C. and Franzmann, B. A., 2001. Host-plant preference and oviposition responses of the sorghum midge, *Stenodiplosis sorghicola* (Coquillett) (Dipt., Cecidomyiidae) towards wild relatives of sorghum. *J. Appl. Entomol.* 125, 109–14.

Sharma, R., Rao, V. P., Upadhyaya, H. D., Reddy, V. G. and Thakur, R. P., 2010. Resistance to grain mold and downy mildew in a mini-core collection of sorghum germplasm. *Plant Dis.* 94, 439–44.

Sharma, R., Upadhyaya, H. D., Manjunatha, S. V., Rao, V. P. and Thakur, R. P., 2012. Resistance to foliar diseases in a mini-core collection of sorghum germplasm. *Plant Dis.* 2009, 1629–33.

Smith, S., Primomo, V., Monk, R., Nelson, B., Jones, E. and Porter, K., 2010. Genetic diversity of widely used U. S. sorghum hybrids 1980–2008. *Crop Sci.* 50:1664–73.

Teshome, A., Baum, B. R., Fahrig, L., Torrance, J. K., Arnason, T. J. and Lambert J. D., 1997. Sorghum [*Sorghum bicolor* (L.) Moench] landrace variation and classification in north Shewa and south Welo, Ethiopia. *Euphytica* 97(3), 255–63.

Tesso, T., Kapran, I., Grenier, C., Snow, A., Sweeney, P., Pedersen, J., Marx, D., Bothma, G. and Ejeta, G., 2008. The potential for crop-to-wild gene flow in sorghum in Ethiopia and Niger: A geographic survey. *Crop Sci.* 48, 1425–31.

UNCED, 1992. *Convention on Biological Diversity*, United Nations Conference on Environment and Development, Geneva.

Upadhyaya, H. D., Ortiz, R., 2001. A mini core subset for capturing diversity and promoting utilization of chickpea genetic resources in crop improvement. *Theor. Appl. Genet.* 102, 1292–8.

Upadhyaya, H. D., Dwivedi, S. L., Singh, S., Sahrawat, K. L., Singh, S. K., 2016d. Genetic variation and postflowering drought effects on seed iron and zinc in ICRISAT sorghum mini core collection. *Crop Sci.* 56, 1–11.

Upadhyaya, H. D., Dwivedi, S. L., Wang, Y., Vetriventhan, M., 2016a. Sorghum genetic resources. In: I. Ciampitti and V. Prasad (Eds), *Sorghum: State of the Art and Future Perspectives*, Agron. Monogr. 58. ASA and CSSA, Madison, WI. doi:10.2134/agronmonogr58.2014.0056.5.

Upadhyaya, H. and Gowda, C. L. L., 2009. Managing and enhancing the use of germplasm – Strategies and Methodologies. Technical Manual no. 10. International Crops Research Institute for the Semi-Arid Tropics, Patancheru 502 324, Andhra Pradesh, India, 236p.

Upadhyaya, H. D., Pundir, R. P. S., Dwivedi, S. L., Gowda, C. L. L., Reddy, V. G. and Singh, S., 2009. Developing a mini core collection of sorghum for diversified utilization of germplasm. *Crop Sci.* 49, 1769–80.

Upadhyaya, H. D., Reddy, K. N., Vetriventhan, M., Krishna, G. M., Ahmed, M. I., Reddy, M. T. and Singh, S. K., 2016b. Status, genetic diversity and gaps in sorghum germplasm from South Asia conserved at ICRISAT genebank. *Plant Genet. Resour.* 13, 247–55.

Upadhyaya, H. D., Sharma, S., Dwivedi, S. L., Singh, S. K., 2014a. Sorghum genetic resources: Conservation and diversity assessment for enhanced utilization in sorghum improvement. In: Y. H. Wang, H. D. Upadhyaya and C. Kole (Eds) *Genetics, Genomics and Breeding of Sorghum*. CRC Press, New York, pp. 28–55.

Upadhyaya, H. D., Wang, Y-H., Dintyala, S. V., Dwivedi, S. L., Prasad, P. V. V., Burrell, A. M., Klein, R., Morris, G. P. and Klein, P. E., 2016c. Association mapping of germinability and seedling vigor in sorghum under controlled low temperature conditions. *Genome* 59, 137–45.

Upadhyaya, H. D., Wang, Y-H., Gowda, C. L. L. and Sharma, S., 2013a. Association mapping of maturity and plant height using SNP markers with the sorghum mini core collection. *Theor. Appl. Genet.*, 126(8), 2003–15.

Upadhyaya, H. D., Wang, Y-H., Sharma, R., Sharma, S., 2013b. Identification of genetic markers linked to anthracnose resistance in sorghum using association analysis. *Theor. Appl. Genet.* 126(6), 1649–57.

Upadhyaya, H. D., Wang, Y-H., Sharma, R., Sharma, S., 2013c. SNP markers linked to leaf rust and grain mold resistance in sorghum. *Mol. Breed.* 32(2), 451–62.

Upadhyaya, H. D., Wang, Y-H., Sharma, S., Singh, S. and Hasenstein, K. H., 2012a. SSR markers linked to kernel weight and tiller number in sorghum identified by association mapping. *Euphytica* 187(3), 401–10.

Upadhyaya, H. D., Wang, Y-H., Sharma, S. and Singh, S., 2012b. Association mapping of height and maturity across five environments using the sorghum mini core collection. *Genome* 55(6),471–9.

Upadhyaya, H. D., Dwivedi, S. L., Ramu, P., Singh, S. K. and Singh, S., 2014b. Genetic variability and effect of postflowering drought on stalk sugar content in sorghum mini core collection. *Crop Sci.* 54(5), 2120–30.

Upadhyaya, H. D., Reddy, K. N., Vetriventhan, M., Murali, K. G., Irshad, A. M., Manyasa, E., Reddy M. T. and Singh, S. K., 2017a. Geographical distribution, diversity and gap analysis of East African sorghum collection conserved at the ICRISAT genebank. *Aust. J. Crop Sci.* 11(4): 424–37.

Upadhyaya, H. D., Dwivedi S. L., Vetriventhan M., Krishnamurthy L. and Singh S. K., 2017b. Post-flowering drought tolerance using managed stress trials, adjustment to flowering, and mini core collection in sorghum. *Crop Sci.* 57, 310–21.

Upadhyaya, H. D., Reddy, K. N., Vetriventhan, M., Irshad, A. M., Murali, K. N., Reddy T. M. and Singh, S. K., 2017c. Sorghum germplasm from West and Central Africa maintained in the ICRISAT genebank: Status, gaps, and diversity. *The Crop Journal.* doi:10.1016/j.cj.2017.07.002.

Wang, Y-H, Upadhyaya, H. D., Burrell, A. M., Sahraeian, S. M. E., Klein, R. R. and Klein, P. E., 2013. Genetic structure and linkage disequilibrium in a diverse, Representative collection of the C4 Model Plant, *Sorghum bicolor. G3 Genes Genomes Genet.* 3, 783–93.

Wang, Y-H., Bible, P., Loganantharaj, R. and Upadhyaya, H. D., 2012. Identification of SSR markers associated with height using pool-based genome-wide association mapping in sorghum. *Mol. Breed.* 30(1), 281–92.

Wang. Y-H., Upadhyaya, H. D. and Dweikat, I., 2015. Sorghum. In: M. Singh and H. D. Upadhyaya (Eds) *Genetic and Genomic Resources for Grain Cereals Improvement.* Academic Press, Oxford, pp. 227–51.

Westengen, O. T., Okongo, M. a., Onek, L., Berg, T., Upadhyaya, H. D, Birkeland, S., Kaur Khalsa, S. D., Ring, K. H., Stenseth, N. C. and Brysting, A. K., 2014. Ethnolinguistic structuring of sorghum genetic diversity in Africa and the role of local seed systems. *Proc. Natl. Acad. Sci. USA* 111, 14100–5.

Xu, F., Xinxiang, A., Zhang, F., Zhang, E., Tang, C., Dong, C., Yang, Y., Liu, X., Dai, L., 2014. On-farm conservation of 12 cereal crops among 15 ethnic groups in Yunnan (PR China). *Genet. Resour. Crop Evol.* 61, 423–34.

Zhang, D., Kong, W., Robertson, J., Goff, V. H., Epps, E., Kerr, A., Mills, G., Cromwell, J., Lugin, Y., Phillips, C., Paterson, A. H., 2015. Genetic analysis of inflorescence and plant height components in sorghum (Panicoidae) and comparative genetics with rice (Oryzoidae). *BMC Plant Biol.* 15, 107.

Zhang, H., Wang, J.-C., Wang, D.-J., Yao, F.-X., Xu, J.-F., Guo-An, S., Yan-An, G., Li, R.-Y., 2011. Assessment of genetic diversity in Chinese sorghum landraces using SSR markers as compared with foreign accessions. *Acta Agron. Sin.* 37, 224–34.

Part 2

Sorghum improvement techniques

Chapter 4

The principles and processes of sorghum breeding

Cleve D. Franks, DuPont Pioneer, USA

1 Introduction

New breeding technologies continue to become available at an ever-increasing pace, but the basic principles and processes involved in sorghum breeding have remained relatively unchanged. These fundamental practices apply to nearly all modern sorghum improvement programmes, and this chapter will seek to provide an overview of this process and relevant considerations. Since the widespread adoption of hybrid sorghum in the early 1960s, most public and private sorghum breeding programmes have been shaped by the need to conform to a cytoplasmic male sterility (CMS) seed production system. This system introduces a complexity to the breeding process that presents unique challenges and, in the end, determines the final form of sorghum germplasm breeders must deliver in order to be successful.

Sorghum's strength as a drought-tolerant crop frequently relegates it to more marginal and less intensively managed land, and this fact also plays a large role in governing breeding decisions such as population creation, choice of tester and locations of yield trials. Since nearly all sorghum is grown in water- and input-limited environments, it is imperative that sorghum breeding programmes evaluate their germplasm under similar production constraints.

Exotic germplasm has played an especially important role in sorghum improvement, and will undoubtedly serve as a source of favourable alleles as new challenges arise. The use of exotic germplasm has been approached via different strategies over the years, depending upon breeding objectives and programme location, but most sorghum researchers recognize the need to evaluate and incorporate some level of genetic diversity into their programmes. The proportion of exotic germplasm to be included in a sorghum

http://dx.doi.org/10.19103/AS.2017.0015.07

improvement project is one which should be weighed carefully, as it will have a profound impact upon the quality and form of lines eventually developed.

2 The structure of a sorghum breeding programme

In order to describe the structure of a traditional breeding programme, it is necessary to first describe the production system it seeks to develop lines for. In sorghum, this system has remained as described by Quinby et al. (1958), and most sorghum breeding programmes since that time have conformed to that model. Since sorghum is a predominantly self-pollinated species, to capitalize on heterosis, a means of forcing the hybrid parents to intercross is required. The CMS system was first described in onion by Jones and Clark (1943), but its utility in sorghum only became possible upon the discovery of sources of CMS in sorghum (Stephens and Holland, 1954). The original seed parents (A lines) in sorghum were created using Milo cytoplasm and restorer genes from Milo-type sorghums, while the nuclear male sterility genes came from Kafir sorghums.

Sorghum breeding programmes are normally divided quite distinctly into two separate pools of germplasm: seed parents (A/B) and restorers (R), and the two classes are normally not intercrossed except during the process of hybrid creation. In making breeding crosses between the two pools, it is difficult to predict the eventual fertility reaction of the resultant progeny, and very often lines derived from such crosses are partially fertile: neither suitable maintainers of sterility nor restorers of fertility. This tendency to only make intra-class breeding crosses, combined with the fact that most accessions in the world collection are A1 restorers, has resulted in a much more genetically diverse class of restorer lines, relative to the B-line germplasm pool in sorghum (Ahnert et al., 1996; Menz et al., 2003). Because two versions of each future A/B line pair must be carried forward at each stage after the initial sterilization cross, the cost and effort associated with the development of a female seed parent is much higher than that of a restorer line. Consequently, the relative proportion of unique lines within the two germplasm pools in a breeding programme is usually significantly shifted towards the restorer class.

To create an A line, a B line (maintainer) is first developed, and then crossed onto a suitable source of A1 cytoplasm. Prior to the initiation of sterilization, selection can be practised for traits of interest to the programme, such as disease resistance, plant height, maturity or general adaptation. Once the B line has attained a suitable level of homozygosity (usually at around the F_4 generation), crosses are made with an A line of known quality, usually a successful seed parent previously developed within the programme. The closer the cytoplasm donor is phenotypically to the B line being sterilized, the more uniform and similar the A and B pair will become in a shorter period of time. Individual panicles in the respective A and B rows are identified at crossing, so that 'paired' crosses can be identified at harvest and advanced in tandem as head-rows the following season. Throughout the process of sterilization, a few extra unpollinated bags are normally left on panicles within the A-line plot to serve as 'sterility checks', and any seed set observed under these bags prior to harvest results in the entire row being discarded. Phenotypic selection for traits important to the programme should be practised concurrently with the sterilization process, but it is important to remember that the traits of interest should be primarily selected within the segregating B line, especially in earlier generations of sterilization, since the A line will gradually conform to the phenotype of the maintainer with each generation of backcrossing. Lines with pedigrees containing only B-line parents normally produce reliably sterile A lines, but even in these cases, partially fertile plants can sometimes be found, and the attrition rate due to these discards can vary greatly across populations. The high cost of sterility breakage in production fields requires that breeders be ruthless in discarding any lines with even small frequencies of pollen shedders. Producibility is a trait that is of great value in hybrid production systems, and both seed parents and pollinators are evaluated for their ability to produce a high-yielding crop of hybrid seed. Inbreds are not normally yield tested over the course of the breeding process, but it is nevertheless important that sterile females produce enough seed when pollinated to justify their use in commercial production.

R-line development follows the standard pedigree breeding method used in other self-pollinated crops. It is important to note, however, that the time from the initial breeding cross to the first topcross hybrid is at least two generation cycles sooner for the restorer lines than for their B-line counterparts begun in the same year. Thus, the quickest route to incorporate a dominant trait into a finished hybrid is always via the male parent. Just as B lines are evaluated for producibility, R lines are scored for their capacity as pollinators, with important traits being anther extrusion, volume of pollen shed across a range of temperatures, duration of pollen shed and whether or not they are 'double-shedders' (shedding pollen over two time periods, thus extending the pollination interval for seed production).

Population improvement in sorghum has been proposed by a number of researchers, with the ultimate objective being to increase the frequency of favourable alleles in a random mating population and to break undesirable linkages. This is usually facilitated through the use of genetic male sterility, the identification and tagging of sterile segregates at flowering, and several cycles of random mating prior to selection. Doggett (1972, 1988) has provided a description of the methodologies and rationale associated with population improvement and recurrent selection in sorghum. Published results have usually shown favourable responses for the trait under selection, either in comparison with the initial population or a conventionally developed set of lines. Population improvement in sorghum has been used to breed for traits like iron chlorosis tolerance (Williams et al., 1987), downy mildew resistance (Craig et al., 1979), twin-seededness (Miller, 1977), fusarium head blight and stalk rot resistance (Duncan et al., 1987), anthracnose resistance (Duncan et al., 1991), acid soil tolerance (Duncan et al., 1989) and reduced dhurrin content (Gorz et al., 1990). Menkir et al. (1994) found more genetic variation and higher yields in *per se* F_3 lines derived from broad-based populations than from inbred lines when introgressing three exotic accessions via backcrossing. Despite these results, examples of commercial successes from population breeding efforts are scarce. Most often, populations are released as resources for other breeders, from which inbred lines would presumably be derived as in a conventional pedigree breeding scheme. In such cases, the recessive male sterility gene (usually *ms3*) must be eliminated during the selection process to ensure complete fertility in the subsequently developed B or R lines. Perumal et al. (2015), however, released nine elite A/B inbred pairs with stalk rot resistance and several grain quality characteristics which were derived from random mating populations.

Off-season winter nurseries are commonly used in sorghum improvement programmes to produce seed and advance breeding generations. Data collected and decisions made during the summer cycle are acted upon in the creation of the winter nursery to generate the next cycle of germplasm and hybrids for testing. New hybrid combinations are created in the off season, so that they can be yield tested the following summer. It is critical to have a trained team with technical expertise on the ground, especially during flowering and harvest, to avoid any contamination and carry out the detailed field operations that a breeding programme requires. Phytosanitary and import requirements should also be considered when choosing a location for a winter nursery, as the requirements for field or seed inspections and customs documentation are not uniformly easy to manage for all countries. Any difficulties in the movement of seed can result in lost or delayed shipments, both of which can have disastrous long-term consequences on the success of the breeding programme.

3 Field operations

3.1 Selfing and crossing

Since sorghum is a mostly self-pollinated crop, care must be taken to prevent outcrossing during the breeding process and in the maintenance and increase of inbred lines. This is normally accomplished through placing bags over the panicles prior to flowering and stapling them closed to ensure that they stay in place until harvest. To make a controlled cross between a sterile and a fertile row, the plants in the sterile row are typically bagged at or near the initiation of flowering, being careful to remove all florets that may have opened. After approximately seven days have elapsed, pollen is brought from the male parent row, the bags are removed from the sterile plants and they are then pollinated and immediately bagged again to prevent any outcrosses. Bagging, crossing and stapling are by far the most labour-intensive activities in a sorghum nursery, but the sheer volume of individual plants in early generations within a full-scale programme simply precludes bagging all panicles. Maunder and Sharp (1963) showed that the frequency of outcrossing is considerably higher in the upper quarter of the panicle than in the lower three quarters, so the removal of this portion of the panicle serves to greatly reduce the level of contamination when bagging is not feasible. To increase seed of fertile inbreds, the panicles are often bagged slightly prior to or shortly after their emergence from the boot. This ensures complete self-pollination and helps schedule nursery activities so that the workload can be better staged.

3.2 Emasculation

Hand emasculation remains a favoured method of emasculation for some breeders, and it is certainly the surest means of obtaining F_1 seed for breeding purposes. To be efficient on a large scale, it does require a level of expertise which can only be acquired through much practice. Probably the most widely used method for creating breeding populations in sorghum is the plastic bag emasculation, also known as 'pour' crosses (Schertz and Clark, 1967). Briefly, all florets that have already flowered are removed from the panicle and a plastic bag is placed over it, binding the bottom with a twist tie or pipe cleaner, so that the bag is sealed. Transpiration quickly increases within the bag, preventing the dehiscence of pollen from the anthers. On the third morning after the bag was placed on the panicle (depending upon the weather), pollen is collected from the male parent with a pollinating bag and brought to the female plant on which the cross was prepared. Without removing the plastic bag, any florets that have not extruded anthers are removed, and as many anthers as is possible are removed from the remaining florets. The plastic bag is removed, and the paper bag with the pollen is placed over the female panicle with vigorous shaking to agitate the pollen. While this method is less precise and yields lower crossing percentages compared with hand emasculations, the required level of expertise is lower and the overall number of breeding crosses one can make is usually much greater than that with any other method. There are varying numbers of self-pollinations within the F_1 progeny row, so it is customary to choose as a female a line with one or more clearly recognizable recessive phenotypic traits, such as white seed colour, tan plant colour or awns. The pollinator, or male parent, would carry the dominant allele, so that the F_1 plants can be easily distinguished from the self-pollinated female plants. The use of molecular markers can greatly facilitate F_1 validation and obviates the need for choosing breeding parents on phenotypic traits. Designing the nursery so that breeding crosses can be made easily, and planting the crossing block over several dates can greatly enhance the success rate of emasculated crosses. Hot water emasculation was used early in the era of sorghum hybrid development as a means of generating larger quantities of seed which could be used for varietal development and inheritance studies (Stephen and Quinby, 1933). The sorghum panicle is typically submersed in water at 48°C for 10 min, after which time it is bagged until complete flowering has occurred. The range of temperatures and length of exposure are quite exacting, and even small deviations often result in floret blasting or incomplete pollen kill. The proportion of hybrid seed generated can be as high as 100%, but success rates are greatly influenced by physical characteristics of the maternal plant, such as panicle openness and degree of anther extrusion. The use of hot water emasculation has been limited in applied breeding programmes by low throughput and the exacting requirements of the process. Embryo rescue has been proposed as a way to expedite the overall turnaround time on cycling breeding generations, and, by extension, to increase the number of generations per year that can be advanced. Rizal et al. (2014) were able to shorten the breeding cycle of sorghum from 17 to 11 weeks through the use of embryo rescue at 10–12 days after pollination. More often, however, the technique has been used to overcome natural barriers to interspecific hybridization (Kuhlman et al., 2010; Price et al., 2006; Wen et al., 1986).

3.3 Harvesting, threshing and processing

Harvesting can occur at any time after the seed reaches the black layer stage, although it does not usually take place until the seed reaches approximately 17% moisture. Harvesting at higher moisture content is sometimes necessary to expedite processing the seed, as to meet a winter nursery deadline, but this can result in shrunken seed and reduced germ, and requires a forced air seed dryer of the proper capacity to be efficient. A certain amount of forethought should go into the harvest, so that the processing of the seed for the next nursery can be done as efficiently as possible. There is often as little a few weeks between the harvest of one nursery and sowing of the next, so all selections, advancements, row tags and envelopes should be managed well before the actual harvest occurs, if at all possible.

4 Selection and topcrossing

4.1 Selecting breeding material

During the process of self-pollination, sorghum breeders are presented with several opportunities for selecting germplasm that meets the objectives of their programme. If the breeding material contains an appreciable measure of exotic germplasm, selection may be performed simply on the basis of highly heritable adaptation traits, such as maturity, height or the presence or absence of a testa. However, if more quantitative or difficult-to-phenotype traits are the breeding objectives, then more intensively managed screening nurseries are usually called for. Specialized screens have been described in sorghum for charcoal rot resistance (Rosenow, 1984), downy mildew (Naranyana et al., 1995), anthracnose (Pande et al., 1994), sugarcane aphid (Sharma et al., 2014), greenbug (Harvey et al., 1985), early season cold tolerance (Kapanigowda et al., 2013), aluminium tolerance (Hill et al., 1989) and many other traits. Except in the case of easily phenotyped, monogenic traits, selection is not normally practised in the F_2 generation due to the difficulty of imposing a uniform screen on individual plants. Selections made in tropical winter nurseries can also be misleading, especially for breeders in temperate zones, since disease pressure is often greater and earlier maturing phenotypes are very poorly adapted to the conditions there. One caveat of early generation selection is that when the material is to be used in developing quantitative trait loci (QTL) or as training sets for genomic prediction, phenotypic selection can have unintended downstream consequences, in terms of segregation distortion.

Whether public or private, the goals of a breeding programme are largely determined by the regional needs and preferences of growers. Diseases and insects are, with a few notable exceptions, mostly local menaces. Where common diseases are present across multiple areas, difference races and pathotypes can make breeding across zones difficult (Prom et al., 2012, 2015). Commercial entities usually prefer hybrids that perform well across broad geographical swaths, making hybrid seed production, marketing and sales simpler.

4.2 Topcrossing

At the point the breeder determines that sufficient homozygosity has been attained, the inbred is topcrossed to an elite tester to ascertain its value as a hybrid parent. Breeders must strike a balance between developing germplasm as quickly as possible, and waiting until their material is genetically stable enough to make meaningful inferences based on hybrid performance. Typically, the first hybrid topcross is made using inbreds between the F_4 and F_6 generation for R lines, and the BC_1 or BC_2 stage for A lines in sterilization.

Some workers have proposed the use of sterile versions of A1 restorers created by backcrossing into A3 cytoplasm in order to topcross early generation B lines prior to the completion of the sterilization process (Schertz et al., 1990; Lee et al., 1992). In this approach, males widely used in A1-based hybrids are backcrossed into a different cytoplasm for which they are a maintainer, usually A3. These can then be used as seed parents to create topcross hybrids with breeding material from a B-line pool sooner than could be done if the breeder had to sterilize all candidate females. While the procedure can work, it relies upon the availability of ample pollen during the time the hybrids are flowering, and can thus result in blasting and ergot in situations when pollen flow is less than optimal. Moran and Rooney (2003) found a significant reduction in the yield of A3 cytoplasm-based, versus A1 and A2 versions of the same hybrids, even under conditions favouring excellent seed set.

The purpose of the initial topcross is to identify those lines with greater general combining ability and eliminate those with little or no chance of giving rise to a commercial hybrid. The trait of interest is usually harvestable yield, but other important traits can be incorporated into the advancement decision as well. Published studies often use numerous testers to estimate variance components or heritabilities. In an applied breeding programme, however, these estimates are of secondary importance in comparison with the relative sorting of new lines for combining ability. It is obvious that the size of the testing programme must increase by the size of the entire inbred class for every tester added, so, given the inherent constraints in any research programme, most breeders elect to test more progeny with less precision than to test fewer inbreds. The sheer numbers involved at the earliest topcross generation preclude using more than one, or, at most, two testers. Abel and Pollack (1991) found differences among testers in their ranking of unadapted maize lines, and recommended the use of more than one tester when evaluating exotic accessions.

The decision of which tester to use is, therefore, of the utmost importance. The critical requirement of a tester is that it allow for the ranking of inbreds according to their combining ability in a target set of environments, and within the context of the breeding programme's germplasm. This usually means selecting a line or lines which have a proven record of producing high-performing hybrids in that zone, often parents of commercial hybrids. The tester is usually chosen with the intent of maximizing potential yield and exposing any weakness to regional breeding objectives. If traits like disease or insect resistance are a priority, then the tester should be deficient for that trait, so as to expose the weaknesses of the inbreds being evaluated. In order for the tester to be predictive of the performance of the inbred in future testing, it must also be representative of the germplasm in the rest of the programme. Packer and Rooney (2011) showed that an F_1 tester can be effectively utilized to discriminate early generation inbreds, thus buffering to some extent the specific combining ability effects of using a single tester. In practice, however, most programmes still use a single well-characterized inbred for their initial topcross.

If the programme spans a wide range of maturities, different testers are used for each maturity class so that seed production is feasible, and only material with the same tester is normally included in a given experiment. Hybrids including current elite inbreds crossed with the same tester should also be included in the trial, both to help evaluate the new material with respect to their absolute hybrid performance and put other agronomic information such as maturity and height into a relative context. These lines can be successful commercial parents or elite codes from other programmes, but in all cases, care must be taken to ensure that they are well representative of the breeders' objectives, since they are to be the yardstick by which the new testing class is measured.

At the second topcross, the breeding material is more homozygous and the number of inbreds has been reduced to the size where the inclusion of larger number of testers is feasible. The total number of hybrids created at the two topcross stages may be comparable, but at the second topcross, it is common to use four or more testers for each maturity class for a more in-depth understanding of the relative performance of the breeding material. While individual hybrids may be selected for further testing, the goal is to rank all the remaining lines on the basis of their breeding values to provide a representative set of hybrids for that region. It is only after the second topcross that the most elite female and male lines begin to be crossed with other elite lines for the purpose of designing hybrids that have the potential to become commercialized. The decision on which specific inbreds to cross has long relied on the breeder's intuition and knowledge of the strengths and weaknesses of the germplasm in his or her programme, but increasingly these decisions are being bolstered by the more widespread use of genomic predictions. The number of new inbreds under consideration can be in the thousands at the first topcross, to hundreds at the second topcross, and finally to perhaps a few dozen when the lines reach the most elite stage.

The criteria for choosing environments in which to test topcross hybrids are the same as those employed when choosing which testers to use, namely, their ability to allow the breeder to unambiguously discriminate among a large number of disparate hybrids. The consolidation and discontinuation of sorghum breeding programmes over the past few decades have, in some cases, resulted in those remaining programmes having to develop germplasm for much larger geographies. The breeder must then place trialling sites in strategic locations so that the entire range of target environments and markets is covered. Given that sorghum is grown in regions where water is often or usually limited, hybrid yield under drought conditions is an implicit requirement for any new sorghum inbred. Yield trials are normally conducted on growers' farms, where lack of available moisture leads to low yields, lodging and even crop failure at times. In arid regions, especially, the breeder must balance the need to collect 'real world' data with the likelihood of forfeiting yield data altogether when harvest is not possible. Often, breeders balance this by using a combination of dryland, irrigated and limited-irrigation yield locations.

4.3 The use of exotic germplasm

Sorghum has greatly benefited from the availability of a vast pool of diverse accessions as sources of favourable alleles for virtually all traits of importance to breeders, including yield. The USDA-ARS sorghum conversion programme as described by Stephens et al. (1967) has been widely recognized as a great success story for the impact it has had upon sorghum breeding programmes. Although there is considerable germplasm diversity within the world collection, most breeding programmes are reluctant to stray very far from within the narrow confines of their own breeding programmes and risk what genetic gains they have achieved. The lines released from the sorghum conversion programme have served as a bridge to allow breeders to incorporate a modicum of diversity into their programmes without the need to cope with the challenges of coping with tall, photoperiod-sensitive material. This important work has recently begun anew and updated as the reinstated sorghum conversion programme using marker-assisted backcrossing (Klein et al., 2016).

5 Conclusion and future trends

While the overall process described above will likely be standard for the foreseeable future, a number of other technologies are poised to have a profound impact on the ways that sorghum breeders make decisions within their programme. The use of molecular markers has become nearly universal across all crops, including sorghum, and whether the marker work itself is done in-house or contracted through a third party, the financial barrier to their availability is becoming less each year. As genotyping capacities increase, however, the primary constraint in the use of molecular markers will not be in their availability, but rather in their interpretation and application. Marker-assisted selection for traits governed by one or a few genes is feasible without extensive computational resources, given the availability of a reliable map for the gene or QTL. However, in order to bring more computationally demanding technologies such as marker-assisted backcrossing or genomic predictions to bear, a very different skill set than the one which most classically trained plant breeders possess is required. In crops such as sorghum, where resources are often limited in comparison with those available in major crops like corn, soybeans or cotton, it can be difficult to justify the added expense of a molecular breeding scientist. In conjunction with the proliferation of molecular marker technology has come the availability of high-throughput means of phenotyping plants. Unmanned aerial vehicles, for example, offer the ability to collect a tremendous volume of information, in the medium of digital images, in a fraction of the time such information would take to collect through conventional measurements, and can do so with added precision and in locations that would be difficult to evaluate under normal circumstances. These images have a permanence that allows their subsequent analysis to take place as resources become available, and can even be used to capture information that could not be collected through visual breeding scores, such as infrared reflectance or leaf area index values continuously over the course of a season. The conversion of these images to actionable data that are useful in a breeding programme, however, presents a significant challenge that is still being resolved. The coupling of the deluge of phenotypic and genotypic data will require a level of investment that will tend to provide a competitive advantage to larger institutions that are able to leverage the cost of developing these resources across crops and regions. Although classical breeding methodologies and knowledge will continue to be a requirement for future sorghum breeders, to be successful, they will also have to adapt to a range of constantly evolving technologies that will afford better data with which to make breeding decisions, and, in turn, better sorghum varieties in the fields of growers.

6 Where to look for further information

For an overview of the history of sorghum breeding, two excellent resources are *Sorghum Breeding and the Genetics of Growth*, by J. Roy Quinby and 'History of cultivar development in the United States: From Memoirs of A.B. Maunder – Sorghum breeder', In: *Sorghum: Origin, History, Technology, and Production*; see References for full details.

Despite the closure of several prominent public programmes in recent years, there nevertheless remain many excellent centres of excellence for sorghum genetics and improvement globally. Any of these programmes would serve as an outstanding starting point for learning more about the ways that new technologies are being incorporated into conventional sorghum breeding programmes.

Clemson University, South Carolina: http://www.clemson.edu/cafls/faculty_staff/profiles/skresov

ICRISAT: http://exploreit.icrisat.org/profile/Sorghum/193

Iowa State University: http://faculty.agron.iastate.edu/mgsalas/

Kansas State University, both Manhattan and Hays: https://www.hays.k-state.edu/programs/sorghum/index.html and https://www.agronomy.k-state.edu/people/faculty/tesso-tesfaye/index.html

Purdue University: https://ag.purdue.edu/agry/directory/Pages/gejeta.aspx and https://ag.purdue.edu/agry/directory/Pages/mtuinstr.aspx

Texas A&M University: http://soilcrop.tamu.edu/program/corn-sorghum/

USDA-Agricultural Research Service: https://www.ars.usda.gov/people-locations//people-list/?modeCode=30-96-05-10 and https://www.ars.usda.gov/people-locations/person/?person-id=48564

The University of Queensland Centre for Crop Science: https://qaafi.uq.edu.au/team/centre-crop-science

7 References

Abel, B. C. and Pollack, L. M. (1991), 'Rank comparisons of unadapted maize population by testers and per se evaluation', *Crop Sci.* 31:650–6.

Ahnert, D., Lee, M., Austin, D. F., Livini, C., Woodman, W. L., Openshaw, S. J., Smith, J. S. C., Porter, K. and Dalton, G. (1996), 'Genetic diversity among elite sorghum inbred lines assessed with DNA markers and pedigree information', *Crop Sci.* 36:1385–92.

Craig, J., Edmunds, L. K., Casady, A. J., Webster, O. J. and Sotomayor-Rios, A. (1979), 'Registration of RSP3BR sudangrass germplasm (Reg. No. GP34)', *Crop Sci.* 19:300.

Doggett, H. (1972), 'Recurrent selection in sorghum populations', *Heredity* 28, 9–29.

Doggett, H. (1988), *Sorghum*. Wiley, New York.

Duncan, R. R., Sotomayor-Rios, A. I., Hepperly, P. R., Rosenow, D. T., Miller, F. R., Narro, J., Forbes, G. A. and Frederiksen, R. A. (1987), 'Registration of GPTM3BR(H)C4 fusarium head blight/stalk rot resistant sorghum population', *Crop Sci.* 27:1321–2.

Duncan, R. R., Torres-Cardona, S., Goenaga, R., Sotomayor-Rios, A. and Shuman, L. M. (1989), 'Registration of GPP4BR(H)C5 acid-soil-tolerant sorghum germplasm population', *Crop Sci.* 29(6):1581–3

Duncan, R., Hepperly, P., Torres-Cardona, S. and Sotomayor-Rios, A. (1991), 'Registration of GPP5BR (M/H/F) C3 anthracnose-resistant sorghum germplasm population', *Crop Sci.* 31:241–2.

Gorz, H. J., Haskins, F. A., Johnson, B. E. and Sotomayor-Rios, A. (1990), 'Registration of NP36 and NP37, two random mating grain sorghum populations selected for reduced dhurrin content', *Crop Sci.* 30:761–2.

Harvey, T. L., Kofoid, K. D., Martin, T. J. and Sloderbeck, P. E. (1985), 'A new greenbug virulent to E-biotype resistant sorghum', *Crop Sci.* 31:1689–91.

Hill, P. R., Ahlrichs, J. L. and Ejeta, G. (1989), 'Rapid evaluation of sorghum for aluminum tolerance', *Plant Soil* 114:85–90.

Jones, H. A. and Clarke, A. E. (1943), 'Inheritance of male sterility in the onion and the production of hybrid seed', *Am. Soc. Hort. Sci. Proc.* 43:189–94.

Kapanigowda, M. H., Perumal, R., Aiken, R. M., Herald, T. J., Bean, S. R. and Little, C. R. (2013), 'Analyses of sorghum [*Sorghum bicolor* (L.) Moench] lines and hybrids in response to early-season planting and cool conditions', *Can. J. Plant Sci.* 93(5):773–84.

Klein, R. R., Miller, F. R., Bean, S. and Klein, P. E. (2016), 'Registration of 40 converted germplasm sources from the Reinstated Sorghum Conversion program', *J. Plant Regist.* 45:57–61.

Kuhlman, L. C., Burson, B. L., Stelly, D. M., Klein, P. E., Klein, R. R., Price, H. J. and Rooney, W. L. (2010), 'Early-generation germplasm introgression from *Sorghum macrospermum* into sorghum (*S. bicolor*)', *Genome* 53(6):419–29.

Lee, R. D., Johnson, B. E., Eskridge, K. M. and Pedersen, J. F. (1992), 'Selection of superior female parents in sorghum utilizing A3 cytoplasm', *Crop Sci.* 32:918–21.

Maunder, A. and Sharp, G. (1963). 'Localization of outcrosses within panicle of fertile sorghum', *Crop Sci.* 3:449.

Menkir, A., Bramel-Cox, P. J. and Witt, M. D. (1994), 'Comparisons of methods for introgressing exotic germplasm into adapted sorghum', *Theor. Appl. Genet.* 89(2):233–9.

Menz, M. A., Klein, R. R., Unruh, N. C., Rooney, W. L., Klein P. E. and Mullet, J. E. (2003), 'Genetic diversity of public inbreds of sorghum determined by mapped AFLP and SSR markers', *Crop Sci.* 44(4):1236–44.

Miller, F. R. (1977), 'Registration of TP11R sorghum germplasm population (Reg. No. GP 29)', *Crop Sci.* 17:676–7.

Moran, J. L. and Rooney, W. L. (2003), 'Effect of cytoplasm on the agronomic performance of grain sorghum hybrids', *Crop Sci.* 43:777–81.

Naranyana, Y. D., Mughogho, L. K. and Bandyopadhyay, R. (1995), 'Evaluation of greenhouse inoculation techniques to screen sorghum for resistance to downy mildew', *Euphytica* 86(1):49–53.

Packer, D. and Rooney, W. L. 2011. A comparison of inbred line and F1 testers for evaluating sorghum experimental lines in testcrosses. *Field Crops Res.* 123:47–50.

Pande, S., Thakur, R. P., Karunakar, R. I., Bandyopadhyay, R. and Reddy, B. V. S. (1994), 'Development of screening methods and identification of stable resistance to anthracnose in sorghum', *Field Crops Res.* 38(3):157–66.

Perumal, R., Tesso, T., Kofoid, K. D., Prasad, P. V. V., Aiken, R. M., Bean, S. R., Wilson, J. D., Herald, T. J. and Little, C. R. (2015), 'Registration of nine grain sorghum seed parent (A/B) lines', *J. Plant Regist.* 9:244–8.

Price, H. J., Hodnett, G. L., Burson, B. L., Dillon, S. L., Stelly, D. M. and Rooney, W. L. (2006), 'Genotype dependent interspecific hybridization of *Sorghum bicolor*', *Crop Sci.* 46(6):2617–22.

Prom, L. K., Perumal, R., Erattaimuthu, S. R., Little, C. R., No, E. G., Erpelding, J. E., Rooney, W. L., Odvody, G. N. and Magill, C. W. (2012), 'Genetic diversity and pathotype determination of *Colletotrichum sublineolum* isolates causing anthracnose in sorghum', *Eur. J. Plant Pathol.* 133:671–85.

Prom, L. K., Perumal, R., Montes-Garcia, N., Isakiet, T., Odvody, G. N., Rooney, W. L., Little, C. R. and Magill, C. (2015), 'Evaluation of Gambian and Malian sorghum germplasm against downy mildew pathogen, *Peronosclerospora sorghi*, in Mexico and the USA', *J. Gen. Plant Pathol.* 81:24–31.

Quinby, J. R., Kramer, N. W., Stephens, J. C., Lahr, K. A. and Karper, R. E. (1958), 'Grain sorghum production in Texas', Texas Agricultural Experiment Station Bulletin no. 912. Texas Agricultural Experiment Station, College Station, TX.

Rizal, G., Karki, S., Alcasid, M., Montecillo, F., Acebron, K., Larazo, N., Garcia, R., Slamet-Loedin, I. H. and Quick, W. P. (2014), 'Shortening the breeding cycle for sorghum, a model crop for research', *Crop Sci.* 54:520–9.

Rosenow, D. T. (1984), 'Breeding for resistance to root and stalk rots in Texas', in *Sorghum Root and Stalk Rots: A Critical Review: Proceedings of the Consultative Group Discussion on Research Needs and Strategies for Control of Sorghum Root and Stalk Rot Diseases, 27 November – 2 December 1983 Bellagio, Italy*. ICRISAT, Patancheru, AP 502 324, India.

Schertz, K. F. and Clark, L. E. (1967), 'Controlling dehiscence with plastic bags for hand crosses in sorghum', *Crop Sci.* 7:540–2.

Schertz, K. F., Clark, L. E. and Rosenow, D. T. (1990), 'Registration of A3Tx430 and A3Tx7000 sorghum genetic stocks', *Crop Sci.* 30:1163.

Sharma, H. C., Bhagwat, V. R., Daware, D. G., Pawar, D. B., Munghate, R. S., Sharma, S. P., Kumar, A. A., Reddy, B. V. S., Prabhakar, K. B., Ambekar, S. S. and Gadakh, S. R. (2014), 'Identification of sorghum genotypes with resistance to the sugarcane aphid *Melanaphis sacchari* under natural and artificial infestation', *Plant Breed.* 133(1):36–44.

Stephens, J. C. and Holland, R. F. (1954), 'Cytoplasmic male sterility for hybrid sorghum seed production', *Agron. J.* 46:20–3.

Stephens, J. C. and Quinby, J. R. (1933), 'Bulk emasculation of sorghum flowers', *J. Am. Soc. Agron.* 25(3):233–4.

Stephens, J. C., Miller, F. R. and Rosenow, D. T. (1967), 'Conversion of alien sorghums to early combine genotypes', *Crop Sci.* 7:396.

Wen, N. F., Lee, S. H. and Liang, G. H. (1986), 'Interploidy crosses in sorghums', *Sorghum Newslett.* 29:79–80.

Williams, E. P., Clark, R. B. and Ross, W. M. (1987), 'Variability and correlation of iron-deficiency symptoms in a sorghum population evaluated in the field and growth chamber', *Plant Soil* 99: 127–37.

Chapter 5

Advanced technologies to accelerate sorghum breeding

Joseph E. Knoll, USDA-ARS, USA

1 Introduction

Genetic variation is the foundation of all plant breeding programmes. As a species, *Sorghum bicolor* (L) Moench contains an incredible amount of diversity, and much good work has been done in collecting and characterizing this vast germplasm (Dahlberg and Madera-Torres, 1997; Morris et al., 2013; Cuevas and Prom, 2013; Cuevas et al., 2015). The task of the breeder is to effectively incorporate these diverse sources of genes into their crop improvement programmes, in order to develop cultivars with higher yields, greater stress tolerance or other novel traits. Accurate and rapid selection for certain traits that are recessive, environmentally variable or otherwise difficult to assess is challenging in a large breeding programme. The time, labour, field space and other expenses required to conduct selection are limiting factors in any plant breeding operation. Technologies such as molecular markers and HTP approaches can help breeders to address some of these limitations. Despite the great amount of diversity in cultivated sorghum, some genes

http://dx.doi.org/10.19103/AS.2017.0015.08
Published by Burleigh Dodds Science Publishing Limited, 2018.

or traits of interest are simply not found within this primary gene pool, and this presents another challenge for the breeder. Technologies for incorporating genetic diversity from outside the primary sorghum gene pool include wide crossing and transgenic technology. Wide crossing has been used for many years in sorghum, while transgenic technology has only recently been developed for this crop and is not yet widely used. The existing sorghum genome can also be altered to create *de novo* genetic diversity through mutagenesis in a somewhat haphazard manner, or more precisely through newer genome-editing technologies. The time required to develop new cultivars is also a challenge for any plant breeder. Commercial competition or an immediate need to release cultivars with resistance to the latest disease or insect outbreak necessitates that breeders develop new cultivars in a timely manner. With the use of an off-season nursery, a sorghum breeder can achieve two generations in a year, or perhaps three with fast-maturing lines in a greenhouse. Doubled haploid breeding is currently being used in maize (*Zea mays* L) and other crops to speed up the process of developing inbred lines. This technology is still in its infancy for sorghum, but there is good potential for it to develop into a mature technology to expedite sorghum improvement. This chapter first reviews some of the technologies that are currently in common use among the sorghum breeding community with specific examples of how these technologies have been used in cultivar development. This is followed by an exploration of some emerging technologies in various stages of development that have potential for direct application in sorghum improvement.

2 Molecular markers: overview and types

2.1 Overview

The ability to select desirable genotypes and eliminate undesirable ones is critical to any breeding programme, but phenotypic selection alone may not be sufficient. A desirable trait, disease resistance for example, may be conditioned by a simply inherited single gene, but it may be recessive (Singh et al., 2006; Fukuoka et al., 2009; Kawahigashi et al., 2011), so that heterozygous carriers would go unselected. Spatial effects in a field can confound the ability to select for certain traits, particularly in F_2 populations where each plant is unique. In an off-season nursery or greenhouse, it may be impossible or impractical to impose the proper biotic or abiotic stress for proper selection pressure. In these situations, molecular markers offer a means for selection. Molecular markers can also be used for selection of more complex or quantitative traits, for example disease resistance (Singh et al., 2006), insect resistance (Gorthy et al., 2017), cold temperature tolerance (Knoll and Ejeta, 2008), or drought tolerance (Kassahun et al., 2010), for which selection may be difficult, particularly in early generations or off-season nurseries. Molecular markers also offer a simple means of differentiating true crosses versus accidental selfs when breeding crosses are made following emasculation (Rooney, 2004).

2.2 Molecular marker types

Numerous molecular marker types are available to the sorghum breeder, each suited to specific applications. For routine genotyping at one or only a few marker loci, markers based on the polymerase chain reaction (PCR) are most commonly used. Arbitrary PCR markers

Published by Burleigh Dodds Science Publishing Limited, 2018.

that do not require prior sequence information include Random Amplified Polymorphic DNA (RAPD; Williams et al., 1990) and Amplified Fragment Length Polymorphism (AFLP; Vos et al., 1995). However, with the publication of the sorghum genome sequence (Paterson et al., 2009) such arbitrary markers have become less popular. Higher-quality PCR markers include sequence-tagged sites (STSs), cleaved amplified polymorphic sequences (CAPSs), and simple sequence repeats (SSRs). Any known single-copy DNA sequence that can be amplified can be considered an STS marker (Talbert et al., 1994), and they may be dominant or co-dominant, depending on the nature of the polymorphism (Wen et al., 2002). If a DNA polymorphism creates or eliminates a restriction site, a CAPS marker can be developed. Primers are designed to flank the target sequence containing the restriction site, which is then amplified by PCR. The product is then digested with the specific enzyme, and then separated on a gel by electrophoresis (Knoll, 2007). SSR markers are commonly used for routine MAS in sorghum. These markers rely on the variability in length of short repetitive DNA sequences. They are plentiful and scattered throughout the sorghum genome, are highly polymorphic and are also co-dominant, allowing the breeder to select heterozygotes. Several SSR markers can often be multiplexed within the same PCR, or individual products can be fluorescently labelled and loaded together on a capillary electrophoresis DNA sequencing instrument, allowing for semi-automated allele calling (Knoll, 2007). Several newer technologies are available to detect single nucleotide polymorphisms (SNPs) that do not require gel electrophoresis. One example is Kompetitive Allele-Specific PCR (KASP; LCG Genomics). Briefly, PCR is carried out with two allele-specific forward primers, each containing a specific tail sequence. The tail sequences are complementary to two dye-labelled cassettes. The dyes are quenched (unable to emit a fluorescent signal) until they are incorporated into a DNA molecule in subsequent rounds of PCR. The intensities of the fluorescent signals are quantified and used to infer genotypes. Because two different dyes are used, heterozygotes are detectable (Semagn et al., 2014; LGC Genomics, 2017).

Several hundred to a few thousand markers can be genotyped simultaneously using microarray-based methods such as Diversity Arrays Technology (DArT). Mace et al. (2008) developed the first DArT array for sorghum, which contains about 12 000 genomic clones. Briefly, the DNA to be genotyped undergoes a complexity reduction step using restriction enzymes, followed by fluorescent labelling. It is then hybridized to the array and scanned. Alleles are called based on the presence or absence of signal for each spot on the array. The initial development of the libraries for the DArT array did not require prior sequence information; however, the sequence of any clone on the array can be determined (Mace et al., 2009) and then converted into another type of marker if needed.

Next-generation sequencing (NGS) technologies have opened the door to even higher throughput generation of SNP marker data. Though several variations of the procedure have been developed (reviewed by Poland and Rife, 2012), typically genomic DNA samples are subjected to complexity reduction using restriction enzymes, and are then ligated to unique 'barcode' adapters. Multiple samples are then combined (multiplexed) and sequenced using a NGS platform. After sequencing, the 'barcodes' allow for the assignment of each sequence to its original sample, a process called 'demultiplexing'. These technologies are constantly evolving, but a few current examples of NGS platforms include the Ion Torrent systems from Applied Biosystems (Thermo Fisher Scientific, 2017), and several platforms from Illumina based on its sequencing-by-synthesis chemistry (Illumina, Inc., 2017). One NGS run will typically generate millions of sequence reads, from which tens to hundreds of thousands of SNPs can be identified, depending on the methods used. For example, in a large genome-wide association study of 971 sorghum accessions, Morris et al. (2013) used the Genotyping-by-Sequencing method (Elshire et al., 2011) on Illumina NGS platforms to generate over six million sequence tags, from which they identified over 260 000 SNPs. Such large volumes of data require bioinformatics software and sufficient computing power to analyse. Because the sorghum genome has a published reference sequence (Paterson et al., 2009), software such as SAMtools (Li et al., 2009) can be used to directly anchor SNPs to their physical locations in the genome. A disadvantage of NGS technologies is that breeders in developing countries or at small institutions may not be able to afford the instrumentation, or may not have ready access to it. In addition, many sequence-based genotyping technologies are currently protected by patents, so they must be outsourced to a commercial company or university core facility that is licensed to perform these services. A brief list of companies offering these services is given in Section 11 .

3 Marker-assisted selection in sorghum breeding

Molecular markers have been utilized in sorghum breeding programmes to simplify the selection for traits such as maintenance of male sterility. In sorghum, a cytoplasmic male sterility (CMS) system is used to produce hybrid seed. The development of new male-sterile seed parents typically begins by crossing a known B-line (sterility maintainer) as the female with another line that has desirable characteristics. Usually both parents in such a cross are known maintainer lines, so that the progeny will most likely be maintainers. However, increasing the diversity among seed parents is desirable, and certain traits may only be available in a known restorer line or in germplasm which has not been tested for sterility reaction. The identification of potential B-lines from large germplasm collections, or the recovery of maintainers from B-line by R-line crosses can be facilitated by the use of molecular markers. The A_1, or milo, CMS system is the most commonly used, but other male-sterile cytoplasms and their restorers have been characterized, and could be useful for increasing the diversity of a sorghum breeding programme. Several major fertility restorer (*Rf*) genes have in fact been mapped (Table 1), and linked molecular markers can be used to select for them. A collection of germplasm can be screened using one or more of these markers to identify potential B-lines, those carrying recessive *rf* alleles. These lines would still need to be crossed with A-lines to verify that they indeed maintain sterility, but this approach can dramatically reduce the number of candidate maintainer lines to be tested. These markers can also be used to genotype segregants in the F_2 and subsequent generations when a B-line and R-line are crossed and the goal is to recover a maintainer. Here, also, complete maintenance of sterility must be verified in crosses to A-lines, as partial restoration due to other loci often occurs (Jordan et al., 2010, 2011), but the number of required test crosses is reduced by first screening with markers.

Klein et al. (2001) mapped the first fertility restorer gene *Rf1* to linkage group H (Menz et al., 2002), which was later assigned to chromosome SBI-08 (Kim et al., 2005). In a later study Klein et al. (2005) refined the map of this region, added additional markers and identified the underlying candidate gene, which encodes a mitochondrial-targeted pentatricopeptide repeat (PPR) protein. Gao et al. (2013) demonstrated how the markers linked to *Rf1* can be used in a breeding programme to develop new B-lines. They tested five SSR markers in the region mapped by Klein et al. (2005) and selected *Xtxp18* (1.18 cM

Table 1 Major restorer of fertility (*Rf*) genes in sorghum with linked markers

Rf gene	Cytoplasm	Chromosome	Linked markers (<10 cM from *Rf* gene)[†]	References
Rf1	A_1	SBI-08	*Xtxp18, Xtxp407, Xtxp408, Xtxp406, Xtxp250*	Klein et al. (2001, 2005)
Rf2	A_1	SBI-02	*Xtxp211, Xtxp250, Xtxp304, Xtxp616, Xtxp654*	Jordan et al. (2010)
Rf3	A_3 (with *Rf4*)	Unknown	*orf107*[‡], *urf209*[‡]	Tang, Chang and Pring (1998); Tang and Pring (2003)
Rf4	A_3 (with *Rf3*)	SBI-07	LW7[§], LW8[¶], LW9[§]	Wen et al. (2002)
Rf5	A_1 and A_2	SBI-05	Xtxp65, Sb05QGM279, SB05-2429989, SB05-3693459, Sb05QGM273, Sb05QGM280	Jordan et al. (2011)
Rf6	A_1 and A_2	SBI-04	SB2385, SB2386, SB2387, SB2388	Praveen et al. (2015)

[†] Simple sequence repeat (SSR) markers, unless indicated otherwise.
[‡] Mitochondrial RNA transcript processing activity (TPA).
[§] Sequence tagged site (STS) marker.
[¶] Cleaved amplified polymorphic sequence (CAPS) marker.

from *Rf1*) because it showed a high degree of polymorphism among the materials surveyed (ten alleles among 43 sweet sorghum accessions), and because it amplified reliably. The same marker was then used to select potential B-lines from the F_2 generation of two B-line by R-line crosses.

A second locus *Rf2*, which also restores fertility in A_1 cytoplasm, was mapped to SBI-02 by Jordan et al. (2010). Later, Jordan et al. (2011) also mapped a major restorer gene, *Rf5*, to chromosome SBI-05, which restores fertility in both A_1 and A_2 cytoplasms. In addition, a quantitative trait locus (QTL) for partial fertility restoration was mapped to SBI-04 in the same study. It was shown that, in both A_1 and A_2 cytoplasms, the QTL alone partially restored fertility, resulting in about 5% seed set on bagged panicles. In A_1 cytoplasm the QTL increased seed set in combination with *Rf5*, but did not show any interaction with *Rf5* in A_2 cytoplasm. Most recently, Praveen et al. (2015) identified a new locus *Rf6*, which also restores fertility in both A_1 and A_2 cytoplasms. This gene was mapped to SBI-04, but not in the same region as the partial restoration of QTL identified by Jordan et al. (2011). *Rf1*, *Rf2*, *Rf5* and *Rf6* all map to regions of the genome that contain one or more PPR genes.

The A_3 cytoplasmic sterility system is not currently used for commercial hybrid production because restorers in A_3 are very rare (Dahlberg and Madera-Torres, 1997; Pedersen et al., 2003; Kuhlman et al., 2006), and because of possible detrimental effects on grain yield in A_3 hybrids compared with A_1 or A_2 hybrids (Moran and Rooney, 2003). Two genes, *Rf3* and *Rf4*, are responsible for fertility restoration in this system. To restore male fertility, both genes must be present in the dominant form, and unlike the A_1 and A_2 systems, restoration in A_3 is gametophytic, occurring within the pollen itself (Tang et al., 1998). This means that in a doubly heterozygous hybrid (*Rf3,rf3/Rf4,rf4*), only 25% of the pollen is viable, and in a single heterozygote (*Rf3,Rf3/Rf4,rf4* or *Rf3,rf3/Rf4Rf4*) only 50% viable pollen would be produced (Wen et al., 2002). Tang and Pring (2003) reported reduced levels of seed set in A_3 hybrids as a result of low pollen viability. However, this system could have applications for sweet and forage sorghum cultivars, for which sugar yield, biomass yield or forage digestibility are more important than grain yield. Using AFLP markers, Wen et al. (2002) mapped *Rf4* to LG E (Menz et al., 2002), which corresponds to SBI-07 (Kim et al., 2005). Three AFLP markers lined to *Rf4* were then converted to STS or CAPS markers. Tang and Pring (2003) used molecular markers to generate B-lines which are homozygous dominant at either *Rf3* or *Rf4* and which are also recessive at the other locus. The objective was to create a single-gene restoration system in A_3 cytoplasm, which is simpler to use and results in higher pollen viability compared to the two-gene system. Although a DNA marker for *Rf3* has not been reported, Tang and Pring (2003) used the transcript processing activity (TPA) of a linked gene, named *Mmt1* (*modification of mitochondrial transcript 1*), as a marker for *Rf3*. *Mmt1* TPA results in an 832 nucleotide mRNA transcripts of the mitochondrial gene *urf209*, while the recessive *mmt1* does not show this activity. The STS and CAPS markers (LW8 and LW9) developed by Wen et al. (2002) were then used to identify individuals carrying *Rf4*. Bunphan, Jaisil and Sanitchon (2013) used the STS marker LW7 developed by Wen et al. (2002) to screen a group of 25 sweet, forage and grain sorghum lines for potential B-lines, those carrying *rf4*. Two lines, which were already known to be maintainers, were correctly identified by the marker. One of these lines (03-B) was then crossed to a sweet sorghum line (SSV84) in order to transfer the sterility-maintaining trait to a sweet background. The LW7 marker was used to identify potential B-lines from segregating F_2 and backcross populations. The progeny carrying *rf4* would still need to be confirmed as maintainers by conventional crossing to an A-line. The role of the *Rf3* locus was not mentioned in this study.

Resistance to parasitic weeds of the genus *Striga* is one such trait for which traditional field-based phenotypic selection and breeding has been difficult. Using field-based observations, Haussmann et al. (2004) identified five QTLs associated with resistance to *Striga hermonthica* in a recombinant inbred population. The resistant QTL alleles all derived from the resistant parent, N13. Yohannes et al. (2015) later utilized N13 as a source of *Striga* resistance in their breeding programme in Eritrea. A farmer-preferred variety known as 'Hugartay', which is *Striga*-susceptible, was crossed to N13. Because genotyping facilities were not available in Eritrea, DNA samples were sent to collaborators in Kenya for analysis. Simple sequence repeat markers were used to introgress the QTL into 'Hugartay' through repeated backcrossing. In the BC_1 and BC_2 generations, plants carrying markers linked to the N13 alleles of the QTL were selected. This is called foreground selection. In the BC_3 generation, foreground selection was performed again to select for the desired QTL, but numerous markers scattered elsewhere in the genome were also used to select for recovery of the recurrent 'Hugartay' genome, in a process called background selection. Background selection helps to hasten recovery of the recurrent parent genome and thus eliminate several generations of backcrossing. Lines carrying from one to four of the QTLs were recovered and were screened in the field under controlled *Striga* infestation to confirm the effects of the introgressed QTL. Lines carrying three or four of the QTLs were found that supported less *Striga* growth than the recurrent parent, and also showed higher grain yield as a result.

The 'stay-green' trait, in which post-flowering senescence is delayed, has been associated with tolerance to terminal drought stress in sorghum (Rosenow et al., 1983; Borrell et al., 1999). It is a complex quantitative trait, but numerous studies have identified QTL associated with the stay-green phenotype (reviewed by Sanchez et al., 2002), and MAS has been used to develop near-isogenic lines to more closely analyse the effects of

Published by Burleigh Dodds Science Publishing Limited, 2018.

individual stay-green QTL (Tuinstra et al., 1998; Harris et al., 2007). Kassahun et al. (2010) demonstrated that stay-green QTL from a non-senescent line, B35, can be introgressed into a senescent line, R16, to improve post-flowering drought tolerance. B35 was crossed to R16, and then the F_1 was backcrossed to R16. Some of the BC_1F_1 plants were self-pollinated and others were backcrossed again to R16. In the BC_1F_3 and BC_2F_2 generations, plants were screened with markers for four stay-green QTLs: *StgB* (Crasta et al., 1999), *Stg1*, *Stg3* and *Stg4* (Subudhi et al., 2000; Xu et al., 2000), all from the B35 parent. Background selection was also performed to recover R16 alleles. Selections were advanced to the BC_1F_4 and BC_2F_3 for field evaluation. Nine partial introgression lines, three with only *StgB* and six with two or three QTLs, were evaluated in the field under different environments. In a well-watered environment, the onset of senescence, as evidenced by a decrease in per cent green leaf area (%GLA), began much sooner in R16 (15 days after anthesis) than in the introgression lines (average 25 days after anthesis), or the donor parent B35 (40 days after anthesis). Under a late-onset drought environment, senescence began about the same time in all lines, but the rate of decrease in %GLA was lowest in B35, intermediate in the introgression lines, and greatest in R16. Though none of the introgression lines expressed the stay-green trait to the degree of B35, they clearly demonstrated the utility of MAS to improve a complex trait like drought tolerance in sorghum.

4 Genomic selection

NGS approaches now make it possible to generate thousands of SNPs for each individual in a breeding population. An extension of MAS known as genomic selection utilizes all available markers in the genome, perhaps several thousand, to predict the breeding value of each individual, rather than identifying the most significant markers, as in traditional QTL analyses (Meuwissen et al., 2001; Jannink et al., 2010). In order to implement genomic selection, one must first assemble a training population, a collection of individuals which are both genotyped using high-density markers and phenotyped for the trait or traits of interest. The individuals in the training population can be highly related, such as progeny from a biparental cross (Lorenzana and Bernardo, 2009), or very diverse, such as a diversity panel (Riedelsheimer et al., 2012). Albrect et al. (2014) recommended using a training population that represents the diversity within an entire breeding programme. The data from the training population is used to build a statistical model which is then used to predict the breeding value of individuals in the breeding population, which are genotyped using the same marker set, but not phenotyped. Ordinary least squares (i.e. multiple regression or ANOVA) procedures are limited by degrees of freedom, so they are not suitable for genomic selection models (Meuwissen et al., 2001). Using simulated data, Meuwissen et al. (2001) first demonstrated the use of best linear unbiased prediction (BLUP) or Bayesian statistical approaches to estimate breeding values from dense marker data. Numerous methods have since been proposed for genomic selection models, some of which are reviewed by Jannink, Lorenz and Iwata (2010) and by Desta and Ortiz (2014). Heslot et al. (2012) compared 11 different methods using real datasets, and concluded that Bayesian LASSO (Park and Casella, 2008), weighted Bayesian shrinkage regression (Hayashi and Iwata, 2010), and ridge regression BLUP (RR-BLUP) had a good combination of accuracy and ease of implementation, and also limited the degree of model overfitting. They also concluded that Random Forests (Breiman, 2001), a machine-learning algorithm, may also be appropriate. Depending on the model used, accuracy of genomic selection generally increases with increasing trait heritability and a larger training population size (Daetwyler et al., 2010; Meuwissen, 2009). In general, more accurate predictions are also made when the training and breeding populations are closely related. As the two populations become more distant, linkage disequilibrium between markers and QTLs is expected to decrease so that more markers and a larger training population are required to maintain prediction accuracy (Meuwissen, 2009). Greater marker density should also improve the accuracy of genomic selection, but there is a plateau above which adding more markers has little effect (Lorenzana and Bernardo, 2009). Meuwissen (2009) notes that the training population must be quite large to take advantage of very high marker densities. As in traditional QTL mapping studies, the accuracy of genomic prediction is also strongly affected by the quality of the phenotypic data, and genotype by environment interactions (G × E) should not be ignored, especially for traits such as yield which are heavily influenced by environmental conditions. Ideally the training population should be phenotyped in several environments to account for G × E effects. Perhaps there is a need to develop genomic selection models that incorporate environmental co-variates.

Genomic selection can be used for population improvement, but an even more powerful application of this technique, particularly for sorghum breeders, is the prediction of hybrid performance from inbred genotypes (Zhao et al., 2012). This has been demonstrated in maize with mixed results (Schrag et al., 2009; Mendes and de Souza, 2016), though it has not yet been described in the literature for sorghum. As an example of how this might be applied in a sorghum breeding programme, if a sorghum breeder has developed ten A-lines and 84 R-lines, there are 840 possible hybrid combinations among them that could be tested. All 94 parental lines could be genotyped in one 96-well plate (typically one or two blank wells are included), and the genotypes of all 840 potential hybrids could then be constructed *in silico*. Co-dominant markers are not necessary in this scenario, and only a subset of hybrids would need to be generated and actually phenotyped in the field. This subset would serve as the training population. A genomic selection model such as RR-BLUP would then be used to determine which of the remaining hybrids should be generated for further field testing.

5 Wide crosses

Crossing with the secondary or tertiary gene pool, or wide crossing, is one means of introducing novel genes into a crop. Cultivated *Sorghum bicolor* ssp. *bicolor* is diploid ($2n = 2x = 20$) and will readily cross with many wild or weedy *Sorghum bicolor* subspecies, including sudangrass (ssp. *drummondii*), and shattercane (ssp. *arundinaceum*), which constitute the secondary gene pool. The tertiary gene pool of sorghum includes several rhizomatous *Sorghum* species, including *Sorghum propinquum* ($2n = 2x = 20$), Johnsongrass (*Sorghum halepense*, $2n = 4x = 40$), and Columbus grass (*Sorghum almum* Parodi, $2n = 4x = 40$). Wide crosses in sorghum are usually performed using the cultivated line as the female, because the flowers are larger and thus easier to emasculate, or because male sterility can be used. In addition, some sorghum lines have been identified that carry a recessive, non-functional allele of the *Inhibition of Alien Pollen* (*iap*) gene. This allele allows a higher rate of adhesion and pollen tube growth of foreign pollen, and thus makes the plant more conducive to interspecific (Price et al., 2006) and even intergeneric crosses (Bartek et al., 2012; Gill et al., 2014).

Johnsongrass is a perennial weedy relative of sorghum, but despite its reputation as an invasive pest (Follak and Essl, 2013; Rout et al., 2013; Werle et al., 2016), it may possess useful genes that can be exploited in cultivated sorghum. Johnsongrass is tetraploid, which somewhat prevents it from crossing with the diploid cultivated species. However, occasional crosses are successful, usually resulting in sterile triploids or fertile tetraploids, and very rarely, diploids. Dweikat (2005) reported that about 180 out of 18 000 male-sterile sorghum flowers pollinated with Johnsongrass pollen set small shrivelled seeds which did not grow, but one normal seed was recovered. This normal seed germinated and was found to be diploid by observation of its chromosomes under the microscope. The recovery of a normal diploid in this cross is unusual. Either the Johnsongrass parent produced a monohaploid (1*n*) pollen grain, or ten of the Johnsongrass chromosomes were somehow eliminated after fertilization. The phenotype of the diploid hybrid was intermediate between sorghum and Johnsongrass, and it was fully male- and female-fertile. This plant was self-pollinated, and its progeny showed normal Mendelian segregation for 34 of 36 polymorphic SSR markers tested.

Crosses between cultivated annual sorghum and its perennial weedy relatives are being used as the basic breeding material for the creation of perennial sorghum cultivars. The Land Institute in Salina, KS, for example, has been developing perennial grain cultivars, including sorghum, as well as perennial grain production systems as a means to make farming more environmentally sustainable (Glover et al., 2010; Van Tassel and DeHaan, 2013). Rhizomatous tetraploid hybrids were generated by crossing tetraploid grain sorghum with Johnsongrass (Piper and Kulakow, 1994). These were then backcrossed to grain sorghum to recover more of the domesticated traits such as larger grain size and higher grain yield. Progress was made after three subsequent cycles of selection for winter survival, grain yield and grain size. Currently grain quality and yields do not yet compare to those of annual grain sorghum, but further improvements should be possible through continued breeding and selection (Nabukalu and Cox, 2016). A few studies have examined the potential of Johnsongrass for cellulosic bioenergy production (Kering et al., 2012) or forage (McLaughlin, Fairbrother and Rowe, 2004; Adeli et al., 2005), but due to its invasive nature it is not likely to be widely adopted as a biomass crop. However, these rhizomatous hybrids may be useful for development of perennial bioenergy or forage sorghum cultivars with less invasive potential and better forage quality than wild Johnsongrass, but this has not yet been studied. Crosses between the diploids *Sorghum bicolor* and *Sorghum propinquum* have been used in QTL mapping of traits related to perenniality and weediness, such as rhizome number, rhizome length, and winter survival (Paterson et al., 1995; Washburn et al., 2013).

When wide crosses are attempted, normal fertilization may occur but often normal seeds fail to develop. The recovery of viable seedlings then requires embryo rescue, a procedure in which the developing embryo is excised from the seed and cultured on sterile media. Though *Sorghum bicolor* and *Sorghum propinquum* are reported to readily cross, Rizal et al. (2014) reported that many of the seeds desiccated before reaching maturity. To improve the rate of recovery, embryo rescue was attempted at varying times, with the greatest success at 10–12 days after pollination. After thorough surface sterilization of the fertilized spikelets, the embryos were excised from immature grains using forceps and a sharp blade. They were transferred to solid MS medium (Murashige and Skoog, 1962) with the scutellum (root) end of the embryo in contact with the surface. Within one day the embryos grew into normal seedlings. After seven days the seedlings may be transferred to soil, but a higher rate of survival was achieved by first transferring to Yoshida culture solution (Yoshida et al., 1972) for an additional week before transferring to soil. Price et al. (2006) reported that embryo rescue was necessary to recover seedlings from crosses between *Sorghum bicolor* and *Sorghum angustum* ($2n = 2x = 10$) and between *Sorghum bicolor* and *Sorghum nitidum* ($2n = 4x = 20$). Interestingly, normal viable seeds were recovered from crosses between *Sorghum bicolor* and *Sorghum macrospermum* ($2n = 4x = 40$); however, the resulting progeny were, as expected, sterile triploids. Embryo rescue is often necessary to recover plants from wide crosses, but Rizal et al. (2014) also demonstrated that it can be used to shorten the breeding cycle for regular breeding crosses by at least a month. With embryo rescue the breeder does not need to wait for the seed to mature, break dormancy, imbibe and germinate. The breeder would need to decide whether the time savings is worth the additional technical effort.

A few reports describe crosses between sorghum and the highly complex polyploids sugarcane (*Saccharum officinarum*) or other *Saccharum* species (Singh, 1934; Doggett, 1988; Nair, 1999; Hodnett et al., 2010). In earlier cases the sugarcane parent was used as the female because of its higher chromosome number, but sorghum has also been used as the female parent in such crosses, though the success rate was initially very low (five seeds out of 3670 pollinated florets; Nair, 1999). By taking advantage of the *iap* gene in the female sorghum parent (Tx3361), Hodnett et al. (2010) were able to achieve a much higher rate of seed set in crosses with *Saccharum* pollinators. Still, some seeds were found to be viviparous (germinating while still attached to the mother plant) or were unable to germinate. These problems were alleviated by harvesting the seed before full maturity, removing the seed coat and immediately germinating the seedlings on tissue culture media under sterile conditions. *Sorghum* × *Saccharum* hybrids vary greatly in their level of vigour, but generally resemble the sugarcane parent more so than the sorghum parent (Nair, 1999; Hodnett et al., 2010). The usefulness of such crosses as breeding material still needs to be fully evaluated, but they may be quite useful for development of biomass or sugar-based energy crops.

6 Mutagenesis breeding

Genetic variability is the foundation of any plant breeding effort, but sometimes the desired variation cannot be found naturally in the primary or secondary gene pool of a crop. In these cases it is possible to artificially create genetic variation through the use of mutagenesis. Mutations may be induced in DNA through various methods including radiation (Mehlo et al., 2013; Mizuno et al., 2013) and chemicals (Jenks et al., 1994). Radiation, such as X-rays, γ-rays, or fast neutrons, tends to induce greater damage to DNA than chemicals, sometimes resulting in dramatic changes such as chromosome breaks, deletions or inversions (Mizuno et al., 2013, Nawaz and Shu, 2014). In contrast, chemical mutagenesis typically results in numerous random point mutations or smaller insertion/deletion mutations (Ozias-Akins et al., 2010). Common chemical mutagens include ethyl methanesulfonate (EMS) and diethyl sulphate (DES). After recovering mutagenized plants, a breeder must screen through a large number of them to identify desirable mutants, but often the desired mutation does not have an easily discernible phenotype. Reverse genetic approaches like TILLING (Targeting Induced Local Lesions in Genomes) (McCallum et al., 2000; Till et al., 2003) allow the breeder to select for mutations within the gene or genes of interest. The original TILLING approach was developed by McCallum et al. (2000) to screen for induced mutations in *Arabidopsis*. Briefly, DNA from up to eight mutagenized individuals is pooled and subjected to PCR to amplify the target sequence. The PCR product is then denatured by heat and then cooled to allow it to renature. If a mutation is present, some heteroduplex DNA will be formed in this process. The original technique (McCallum et al., 2000) utilized denaturing HPLC to detect these heteroduplex DNA molecules. Colbert et al. (2001) and Till et al. (2003) refined the technique to use CEL1 nuclease to cleave the mismatched sites in the heteroduplex DNA, followed by polyacrylamide gel electrophoresis. The forward and reverse PCR primers are each labelled with different fluorescent IRDyes (Li-Cor, Lincoln, NE, USA), so that the fragments can be detected using a Li-Cor 4300 DNA Analyzer or similar instrument. The CEL1 nuclease can be extracted from fresh celery (*Apium graveolens* L.) stalks using the protocol of Till et al. (2006). The technique has since been applied to numerous crop species, including maize (Weil and Monde, 2007), rice (*Oryza sativa* L.) (Till et al., 2007), soybean [*Glycine max* (L.) Merr.] (Anai, 2011) and peanut (*Arachis hypogaea* L.) (Knoll et al., 2011). As the cost of high-throughput sequencing continues to decrease, this approach is being replaced by direct sequencing approaches (Guo et al., 2015; King et al., 2015; Nida et al., 2016). One major drawback to using mutagenesis for breeding is its very random nature. It is nearly impossible to predict where a mutation will occur, and there is no certainty that a mutation in the desired gene will be recovered at all. In addition, mutagenesis can produce unintended changes elsewhere in the genome that may go undetected. Nonetheless mutagenesis breeding is not subject to the regulatory hurdles of transgenic technology (Nida et al., 2016), which may be a distinct advantage to the sorghum breeder.

Xin et al. (2008; 2009) developed a large TILLING population in sorghum utilizing the cultivar BTx623, the same line that was used to develop the first sorghum genome sequence (Paterson et al., 2009). This population was developed by soaking seeds in aqueous solutions of EMS at various concentrations (ranging between 0.1 and 0.3% v/v) on a rotary shaker. After 16 h, the seeds were soaked in tap water for 5 h, with the water changed every half hour to remove the excess mutagen. These seeds were planted (M_1 generation) and it was observed that their viability decreased sharply above 0.25% EMS. Of about 10 000 seeds in the 0.25% EMS treatment, 1305 survived and flowered. These plants were self-pollinated to produce the M_2 generation. The M_2 seeds were planted in headrows, and leaf tissue from one M_2 plant from each headrow was used for DNA extraction and TILLING analysis. The same M_2 plants were self-pollinated to produce the M_3 generation. Ten M_3 plants from each row were selfed and bulked to create over 1600 mutant lines which are available to sorghum researchers through the USDA-ARS. The TILLING approach was used to screen a subset of this population for mutations in caffeic acid *O*-methyltransferase (*COMT*), a gene which is essential in the production of lignin. Sorghum plants with reduced lignin have a distinct brown midrib and are more digestible to livestock than normal plants (Porter et al., 1978; Yerka et al., 2015). Two missense mutations were identified in the *COMT* gene, and both of the lines carrying these mutations had the brown-midrib phenotype. Although the main focus of this TILLING population is for basic genetic studies, other interesting phenotypes were observed in this population which could be useful to breeders, including alterations in tillering, male sterility, plant height, early or late-maturity and erect leaves. In addition, a potentially useful multiseeded mutant was identified and named *msd1*. This mutant has fertile pedicellate spikelets, which results in more seeds per panicle than wild-type BTx623. Even though the individual seeds are smaller, the mutant also has longer and more numerous primary panicle branches, resulting in greater total yield per panicle (Burow et al., 2014). The *msd1* mutation is inherited as a single recessive gene, and could be very useful for increasing yields in grain sorghum.

Mutation breeding of sorghum has successfully resulted in the release of improved cultivars, at least since 1970. The Mutant Variety Database (International Atomic Energy Agency, 2017) lists 18 registered sorghum or sudangrass varieties that have been developed directly by mutation or by crossing with an induced mutant. For most of these, gamma radiation at dosages from 100 to 400 Gy is listed as the mutagen. Various traits were targeted in these cultivars, including grain quality, grain yield, maturity, lodging and plant height. Other examples of sorghum improvement via mutagenesis have been reported. Pahuja et al. (2012) used a combination of γ-rays (300 Gy) and EMS (0.2%) to induce mutations in the forage sorghum line SSG 59-3. Higher dosages of γ-rays with the 0.2% EMS treatment did not produce viable plants. Fifteen promising mutant lines were recovered and were then tested over multiple environments for two years. Although none of these mutant lines yielded greater than SSG 59-3, some had desirable traits such as higher digestibility, higher protein, higher leaf:stem ratio, reduced hydrocyanic acid or earlier maturity. Blomstedt et al. (2012) mutagenized a proprietary forage sorghum line using EMS, and then used a biochemical approach to identify mutants with reduced hydrocyanic acid potential (HCNp). Seeds were soaked in aqueous solutions ranging in concentration between 0.15 and 0.4% EMS for 16 h, followed by rinsing. From about 53000 treated seeds, about 16300 plants were recovered. Of these, about 4200 produced seeds, but the highest dose treatment (0.4% EMS) had severely reduced fertility. The biochemical screen identified 264 potential mutants, which were then subjected to TILLING to identify mutations in genes in the dhurrin biosynthesis pathway. One stable mutant, named *tcd1* (*totally cyanide deficient 1*), was recovered that completely lacks HCNp. TILLING and subsequent sequence analysis revealed a missense mutation in the gene *CYP79A1*, the first committed step in the dhurrin synthesis pathway. Three additional mutants with greatly reduced HCNp in the adult stage were identified, and named *acdc* (*adult cyanide deficient category*). The causal polymorphisms in the *acdc* mutants were not identified. Interestingly some other mutants with elevated HCNp were also identified in this study.

Mutagenesis frequently results in loss-of-function mutations. For example, Jenks et al. (1994) used chemical mutagenesis to generate a series of mutants with reduced epicuticular wax. Seeds were treated with 5.7–11.5 mM DES solution for 3 h or 4.7–9.4 mM EMS solution for 18 h. The original brown-midrib lines reported by Porter et al. (1978) were also generated using chemical mutagenesis. These mutations are all recessive, but occasionally a dominant, or gain of function, trait is created by mutagenesis. Oria et al. (2000) reported that a semi-dominant mutation for increased grain protein digestibility was induced in sorghum by chemical mutagenesis, and was later incorporated into the cultivar P851171. More recently, Petti et al. (2014) used EMS (0.75% for 12 h) to induce mutations in the sweet sorghum cultivar Della. Out of approximately 12 000 M_1 plants, they identified a single mutant with distinctly red leaves. The red colouration was found to be inherited as a single dominant gene, and was associated with increased production of 3-deoxyanthocyanidin pigments, which can be used as natural food colourants or antioxidants.

7 Emerging technologies: doubled haploids

The production of inbred lines typically takes five or more generations of self-pollination, but through the use of doubled haploid technology, the time for inbred development can be dramatically shortened to around one year. In maize, many breeding programmes are developing inbred lines through the use of *in vivo* maternal doubled haploids. Briefly, an

F_1 hybrid or other heterozygous plant is pollinated with pollen from a haploid inducer line. The haploid inducer carries one or more genes which cause the occasional elimination of the paternal genome in its progeny, resulting in maternal haploids. The exact mechanism by which this occurs is still unclear, but candidate genes have recently been identified in a region of maize chromosome 1 (Hu et al., 2016). Haploids are typically produced at rates ranging from 2.3 to 8.1% (Coe, 1959; Röber et al., 2005) depending on the inducer being used as well as the genotype of the maternal parent (Eder and Chalyk, 2002). Several improved derivatives of the original haploid inducer 'Stock6' (Coe, 1959; Hu et al., 2016) are available and many carry the dominant *R1-Navajo* (*R1-nj*) marker for purple pigmentation (Chang and Coe, 2014). With this marker, all crosses with the inducer will show a coloured crown of the endosperm, but normal diploid crosses will also have a purple embryo, while haploids will have a colourless embryo (Nanda and Chase, 1966; Yu and Birchler, 2016), allowing for visual selection. Inducers carrying the dominant trait of high seed oil content have been developed which allow for rapid haploid seed selection using nuclear magnetic resonance (NMR) technology (Melchinger et al., 2013). Haploid seeds are selected, germinated, treated with colchicine to double their chromosome number, and then grown to maturity. Colchicine treatment may be accomplished by soaking newly germinated seedlings in a colchicine solution or by injection of colchicine directly into growing seedlings at the 3–4 leaf stage just above the growing point (Eder and Chalyk, 2002). Seedlings may also be germinated in sand for ten days, after which the roots are washed clean and then the roots are soaked in colchicine solution (de Oliveira Couto et al., 2015).

A similar *in vivo* haploid induction system has not been fully developed in sorghum, but the potential for it does exist. The first major hurdle is the identification of a reliable haploid inducer. Chumley (2015) reported the discovery of a haploid-inducing pollinator (coded 'HIP1' to protect its identity) by crossing a large number of accessions with male-sterile A-lines. It was reported that haploid seeds were produced at a rate of 9–12% when using pollen from HIP1. A recent press release (United Sorghum Checkoff Program, 2017) also announced that researchers at DuPont Pioneer have discovered two haploid inducer lines of sorghum. Methods for rapidly identifying haploid sorghum seeds and for doubling their chromosome number are still in development. Chumley (2015) used male-sterile seed parents to identify a haploid inducer, but for this type of system to be practical in sorghum breeding, a means for emasculating the target germplasm is needed. Hand-emasculation would be far too labour-intensive, but the hot water method, first described in 1933 by Stephens and Quinby, could possibly be used. The treatment should kill the pollen while preserving the stigmas and ovaries, but the optimum time and temperature may be genotype-dependent, and may require some initial testing to determine. Stephens and Quinby (1933) reported that immersing the unopened panicle in water at 44°C, cooling to 42°C for 10 min resulted in 100% pollen kill and 90% seed set upon cross-pollination in the variety 'Schrock.' Hot water emasculation may be best accomplished in a greenhouse using a digital circulating water bath with exact temperature control, and potted plants that can be tipped on their sides, or even inverted over the water bath, in order to immerse the panicle for the specified amount of time. Downes and Marshall (1971) reported inducing male sterility in sorghum by exposing plants to cold temperatures (18°C d/13°C night) at about the time of flag leaf emergence, when meiosis occurs in pollen development. Preventing anther dehiscence by covering the panicle with a plastic bag to maintain high humidity (Schertz and Clark, 1968) may also be an option. With any form of bulk emasculation, accidental self-pollinations still occur, so these seeds would also need to be sorted out along with the crossed diploid seeds. In maize, because the *R1-nj* marker is unreliable in certain genetic backgrounds, Yu and Birchler (2016) used a transgenic approach to insert a dominant green fluorescent protein (GFP) marker into a haploid inducer line. If the GFP marker were used in a similar sorghum system, all crosses would display fluorescence in the endosperm, but accidental selfs would not. Upon germination of the crossed seeds, normal diploid crosses would show fluorescence in the radicle, but haploids would not. Though transgenic sorghum has not been approved for cultivation in the field, this technology could be used under containment in a greenhouse. The resulting haploid plants, being entirely maternal, should not retain any transgenic DNA in their genomes; this could be confirmed by Southern blot analysis.

An alternative method for producing doubled haploids through anther or microspore culture has been employed in barley (*Hordeum vulgare* L.) (Sriskandarajah et al., 2015), rice (*Oryza sativa* L.) (Bhattacharya et al., 2014), pepper (*Capsicum annuum* L.) (Bhattacharya et al., 2015; Keles et al., 2015), *Brassica* species (Shumilina et al., 2015; Zeng et al., 2015) and numerous other crops. This practice has not been widely adopted in sorghum, mostly because sorghum is notoriously difficult to grow in tissue culture. Callus induction and plant regeneration rates are generally low, and many genotypes have a tendency to produce phytotoxic phenolic compounds in tissue culture media (Guo and Liang, 1993; Gurel et al., 2009). Nonetheless an anther culture-based system for generating doubled haploid sorghum breeding lines has been reported (Kumaravadivel and Rangasamy, 1994). Anthers at the uninucleate stage of microspore development were cultured on N_6 medium (Chu et al., 1975) with 0.8% agar supplemented with 3% sucrose. Different concentrations of 2,4-dichlorophonoxyacetic acid (2,4-D) in the medium and different temperatures were tested to determine the optimum conditions for callus induction. The highest callus induction rates were achieved at 2.0 mg L^{-1} 2,4-D and an incubation temperature of 30°C. Callus formation and growth took up to 30 days before the calli were transferred to regeneration medium. Modified MS medium (Murashige and Skoog, 1962) supplemented with 6-benzylaminopurine (BAP) and indole acetic acid at 2.0 and 0.3 mg L^{-1}, respectively, yielded the greatest number of green plantlets, with a similar number of albinos also produced. Even at the optimum conditions, only 7.6 green plants were generated for every 100 anthers cultured. However, 95.3% of these plants were diploid, having spontaneously doubled at some point during the tissue culture process. Self-pollination of these plants and observation of their progeny in the field confirmed that they were indeed doubled haploids. It should be noted that only one genotype was used in this study, but it has been shown that callus induction and plant regeneration in sorghum are genotype- and media-dependent (Guo and Liang, 1993; Nakamura et al., 1997), so the culture conditions may need to be optimized for each specific genotype.

8 Emerging technologies: transgenic technology

Efforts to introduce foreign genes into sorghum via genetic transformation have lagged behind other major crops like maize, soybean and cotton (*Gossypium hirsutum* L.). The possibility of transgene escape from cultivated sorghum to its invasive weedy relatives, Johnsongrass (Arriola and Ellstrand, 1996) and shattercanes (Muraya et al., 2012), has perhaps stifled research on transgenic sorghum. In addition, sorghum is generally considered difficult to grow in tissue culture (Guo and Liang, 1993; Gurel et al., 2009), which also hinders research and development of transgenic protocols for this crop. Nonetheless, a few protocols have been developed and shown to be successful at introducing foreign genes into sorghum.

The first successful report of transformation and regeneration of viable sorghum plants was published in 1993 by Casas et al. Immature embryos were harvested and tissue-cultured to produce embryogenic callus. These were then subjected to bombardment with DNA-coated gold or tungsten particles. Though successful, only two stably transgenic plants were recovered from 600 embryos from one cultivar. Transformation of seven other cultivars was attempted but was not successful, suggesting that some genotypes may be more recalcitrant to transformation or regeneration than others. The same group later achieved a slight improvement in efficiency by using immature inflorescences as explants for bombardment (Casas et al., 1997). The first successful attempt at transforming sorghum using *Agrobacterium tumefaciens* was reported in 2000 by Zhao et al. Immature embryos from two different cultivars were used as explants, and various conditions were tested, including *Agrobacterium* concentration during the infection stage and tissue culture media composition. Averaged across all conditions tested, approximately 2.1 stably transformed plants were recovered for every 100 embryos cultured. Several other researchers have since reported improvements in transformation efficiency using *Agrobacterium*. For example Gurel et al. (2009) achieved transformation efficiency up to 8.3% by first applying a 3-min 43°C heat shock to immature embryos, and then cooling them to 25°C before co-cultivation with *Agrobacterium*. Wu et al. (2014) reported an optimized protocol using immature embryos of Tx430 in which cupric sulphate and BAP were added to the resting and selection media. Also, they noted significant differences in transformation efficiency between two *Agrobacterium* strains LBA4404 and AGL1 (average of 9.7% and 33.2%, respectively).

Recently, several researchers have demonstrated that sorghum transformation via *Agrobacterium* can be accomplished without the need for tissue culture, reminiscent of the 'floral dip' method described by Clough and Bent (1998) for *Arabidopsis thaliana*, which has now become a standard protocol for this species. Transformation can be accomplished by dipping sorghum panicles at about 30% anthesis in a culture solution of *Agrobacterium* carrying the gene of interest (Peng et al., 2012; Ordonio et al., 2016). The reported efficiency of this technique was less than 1%, but perhaps with some refinement the efficiency could be improved. Li et al. (2016) recently demonstrated direct transformation of intact sorghum seeds. After thorough surface sterilization and 12 h of soaking in water to imbibe the seed, the embryo was pierced with a sterile needle. The pierced seeds were then soaked in *Agrobacterium* (strain EHA105) cultures (LB medium with appropriate antibiotics), supplemented with 100 mM acetosyringone. The time of immersion was not reported, but the seeds were then transferred to wet papers in Petri dishes for four days and were kept moist in the dark. After four days they were soaked in carbenicillin solution for 1 h to kill the remaining *Agrobacterium*. The seeds were then sown in conventional potting media. The vectors used for transformation in this study carried genes for hygromycin resistance; thus, hygromycin solution was brushed onto the leaves at the three-leaf stage to screen for successful transformation. The rate of transformation was low, at most 4%, but given the simplicity and rapidity of this technique compared to tissue culture, it seems to be a practical means of transforming sorghum. In a similar study, Yellisetty, Reddy and Mandapaka (2015) used a hypodermic needle to inoculate *Agrobacterium* directly into the apical meristem of three-day old sorghum seedlings. The initial transformants (T_0) were chimeric, but many of these produced seed, and stable transgene integration was confirmed in the T_1 generation. High transformation efficiency (26–38%) was reported.

To date, most of the transgenes deployed commercially in major crops include those for herbicide resistance or insect resistance, traits which could create problems if accidentally introduced into sorghum's weedy relatives (Arriola and Ellstrand, 1997). However, some transgenes might not confer any specific advantage to these weeds, though such 'neutral' genes would likely persist in wild populations, so there may be possibilities for transgenic technology to be commercially viable in sorghum, depending on the traits involved. In addition, male sterility could be used to reduce, though not completely eliminate, the probability of accidental gene flow from transgenic sorghum to nearby weedy relatives, or to adjacent fields of non-transgenic sorghum. Seeds of a transgenic male-sterile hybrid could be mixed and planted with about 10% of a non-transgenic male-fertile version of the same or very similar hybrid. The male-fertile plants should produce enough pollen to fertilize the entire crop, and would also function as an in-field refuge for traits like insect resistance (Pedersen et al., 2003). Most commercial sorghum utilizes the A_1 CMS system, but Pedersen et al. (2003) proposed leveraging the A_3 system for controlling gene flow from transgenic sorghum. Because restorers in the A_3 system are extremely rare (Dahlberg and Madera-Torres, 1997; Pedersen, Marx and Funnell, 2003; Kuhlman et al., 2006), using this system, any seed that is accidentally left in the field to volunteer the next season should be male-sterile, further limiting accidental gene escape. To test this, open-pollinated seed was saved from F_1 hybrids in both A_1 and A_3 cytoplasms. These seeds were planted in two subsequent seasons, and panicles were bagged prior to anthesis. An average of 0.04% seed set was observed in the A_3 plants, compared to 74% in A_1 cytoplasm, indicating that volunteer plants with A_3 cytoplasm would have highly reduced pollen viability (Pedersen et al., 2003).

9 Emerging technologies: genome editing and high-throughput phenotyping

9.1 Genome editing

The ability to introduce foreign genes into sorghum opens up the possibility to precisely alter specific genes already present in the sorghum genome, in a process known as site-directed mutagenesis or genome editing. Genome-editing technology utilizes specifically targeted nucleases to introduce breaks in the target sequence, or sequences, of the host DNA. A gene expression cassette containing a nuclease domain, one or more DNA-binding domains, and, in some systems targeting sequences, is introduced into a plant, usually with *Agrobacterium*, though particle bombardment has also been used (Sun et al., 2016). The host cell will express the nuclease, which will cause targeted double-strand breaks in the host DNA. The host cell will naturally try to repair these breaks in a process known as non-homologous end-joining (NHEJ). Often the DNA is repaired to its original condition, but occasionally NHEJ results in small insertions or deletions, which can result in reading frame shifts or other changes that affect gene function. Alternatively, a process known as homology-directed repair can insert specific sequences into the target site if a

Published by Burleigh Dodds Science Publishing Limited, 2018.

'donor' template DNA sequence is available (Sun et al., 2016). After regeneration, single-copy transformants (T_0) are selected and self-pollinated to generate T_1 populations. In the T_1 generation, individuals are selected which have the correctly edited target sequences, and which have also lost the transgene through normal genetic segregation. In this way, genome-edited plants are produced which are not considered to be transgenic, and are thus not subject to the same regulation as transgenic, or so-called 'GMO', crops (Khatodia et al., 2016; Waltz, 2016).

Three different types of proteins have been developed for the purpose of genome editing: the zinc-finger nucleases (ZFNs), transcription activator-like effector nucleases (TALENs) and the clustered regularly interspaced short palindromic repeat (CRISPR)-associated protein (Cas) system. The ZFNs and TALENs are chimeric proteins consisting of several sequence-specific DNA-binding domains fused to a nonspecific endonuclease domain from *Fok1*. Several zinc-finger domains or transcription activator-like effector (TALE) domains can be linked together and thus engineered to recognize and bind to a specific sequence in the genome, based on the amino acid 'code' for that particular class of domain. An in-depth review of ZFNs and TALENs is provided by Gaj et al. (2013). Constructing an engineered protein is time-consuming and technically challenging, but the newly discovered CRISPR/*Cas9* system works through a different mechanism than the ZFNs and TALENs, and thus offers greater simplicity and flexibility in genome targeting. Rather than using sequence-specific DNA-binding domains, the Cas protein is guided by RNA that is complementary to the target DNA sequence. In the native CRISPR/*Cas9* system in the bacterium *Streptococcus pyogenes*, two RNA molecules are needed to guide the nuclease to its target: CRISPR RNA (crRNA), which is complementary to the 20-nucleotide target sequence, and trans-acting crRNA (tracrRNA). These two RNA molecules form a complex which is recognized by Cas9. For genome editing, these two RNA genes can be fused together in what is called single guide RNA (sgRNA), which further simplifies this system. The Cas9 nuclease, the most commonly used Cas protein for genome editing, requires a NGG nucleotide sequence, referred to as the protospacer-adjacent motif (PAM), adjacent to the 3′ end of the target site with which to bind. The PAM sequence can vary depending on the specific Cas protein being used. The CRISPR/*Cas9* system has been used to induce directed mutations in numerous plants, including important crop species like maize (Zhu et al., 2016). In rice it has been used to improve blast disease resistance (Wang et al., 2016), and to engineer herbicide resistance (Sun et al., 2016). The CRISPR/*Cas9* system has been shown to function in sorghum immature embryos (Jiang et al., 2013). A binary expression vector was constructed which contained the *Cas9* nuclease gene, the sgRNA gene and a GFP marker gene, all under monocot-specific promoters. The vector also contained the *dsRED2* gene which encodes a red fluorescent protein marker. The *dsRED2* gene was intentionally mutated to be complementary to the sgRNA target sequence. This mutation resulted in a frameshift and thus a non-functional gene. The vector was transformed into immature sorghum embryos using *Agrobacterium* via the protocol described by Howe et al. (2006). After two weeks, GFP expression could be observed, confirming successful transformation. Sectors that also showed red fluorescence from the restored *dsRED2* gene were observed, confirming that the CRISPR/*Cas9* system was indeed functioning, and suggesting that the edited *dsRED2* gene was stably integrated in the cells giving rise to these sectors. This is, to date, the only published report of the utilization of the CRISPR/*Cas9* system to edit genes in sorghum, but given the evidence so far, it should be a useful tool for sorghum crop improvement in the future.

9.2 High-throughput phenotyping

With NGS technologies it is now possible to generate massive amounts of genomic sequence data, and new software and high-capacity computing are making it possible to analyse all of that data. However, genomic sequence data are only useful to a breeder if they can be reliably linked to phenotypes in the field. HTP, or phenomics, refers to automated or semi-automated approaches to collect and process large amounts of phenotypic data in a relatively short amount of time compared to traditional approaches. At the cutting edge of HTP technology are several greenhouse-based facilities, including the Plant Accelerator at the University of Adelaide in Australia (Nielson et al., 2015); the Bellwether Foundation Phenotyping Facility at the Donald Danforth Plant Science Center in St. Louis, MO, USA; and the National Plant Phenomics Centre at Aberystwyth University in Wales. Scientists may hire these facilities for the purpose of running highly automated HTP experiments on plants, including sorghum. Potted plants are grown on a conveyor system which automatically moves them through an imaging station at various time points throughout the growing season, as frequently as several times per day if desired by the researcher. The imaging station uses one or more digital cameras, mounted overhead or on the side to capture images from which various types of data can be extracted by computer software. Depending on the phenotypes under study, the cameras may be sensitive to visible light, fluorescence or infrared wavelengths. Environmental parameters such as irrigation and fertilization are also fully automated, providing consistency between individual pots. Nielson et al. (2015) grew two sorghum varieties under well-watered and water-limiting conditions at the Plant Accelerator. The plants were photographed daily for several weeks, and they showed that differences in growth rate could be accurately measured by using non-destructive imaging. At 45 days, the plants were photographed in the late afternoon when they were most water-stressed, and again the following morning when water stress was less severe. It was shown that a decrease in leaf area between morning and evening due to leaf rolling can be documented using imaging. Leaf rolling is a common indicator of drought stress in sorghum, but accurately quantifying it using traditional methods is difficult. They also grew sorghum under varying levels of nitrogen fertilization. Colour imaging was able to differentiate well-fertilized and nitrogen-stressed plants based on the level of green colour in their leaves, and the percentage of senesced area on the leaves could also be quantified. An outdoor HTP facility has been constructed at ICRISAT in India, called the LeasyScan system (Vadez et al., 2015). In this system, rather than moving the plants, scanners are placed on a movable overhead irrigation boom, and the plants are grown in containers set atop scales. The scanners use near-infrared lasers to construct three-dimensional images of the plant canopy, while the scales monitor evapotranspiration in real time. The system has been designed such that additional sensors could be added later. Because this is an outdoor system, weather data are also collected throughout the day.

Such highly automated facilities are probably cost-prohibitive for most plant breeders, and they are still limited by the size of the pots and capacity of the facility. However, less expensive, more flexible field-based HTP methods are being developed which could be very useful to sorghum breeders. Various types of cameras or sensors can be mounted onto tractors, irrigation systems or other vehicles, which collect data as they drive through the field (Araus and Cairns, 2014). Sensors and cameras may also be mounted on towers overlooking the field (Araus and Cairns, 2014) or they can be deployed using unmanned aerial vehicles, popularly referred to as UAVs or 'drones' (Rasmussen et al., 2016). A less costly and highly portable option is to mount sensors onto small carts that are manually pushed or pulled through the field as they collect data. For a cost of approximately US$12 000, Crain et al. (2016) created a lightweight HTP pushcart using a wheel and portion of the frame from a bicycle. Four sensors were mounted to the cart: an infrared thermometer to record canopy temperature, a GreenSeeker spectral reflectance sensor to record the normalized difference vegetation index, a digital camera, and a global navigation system to georeference all of the data points so that they can be automatically assigned to a plot using computer software. These were all connected to an ordinary laptop computer which functions as the main controller and data recorder for the unit. Crain et al. (2016) demonstrated the utility of their 'Phenocart' in wheat plots, but such technology should be easily transferrable to other crops, and other types of sensors could be used, depending on the traits being studied. Compared to expensive tractor-mounted or aerial systems, technologies like the 'Phenocart' are more likely to enable sorghum breeders, particularly those in developing countries, to utilize HTP techniques.

10 Conclusions and future trends

These advanced technologies are not meant to replace conventional breeding techniques, but rather to enhance the efficiency of breeding programmes. The choice of which techniques to incorporate into a breeding programme will depend largely on the main objectives of the programme, as well as financial cost, ease of implementation and technical expertise of the breeder and staff. Though yield is always a major target for crop improvement, local biotic and abiotic constraints will dictate which other traits are most important. Traditional MAS is appropriate for major QTLs with relatively large effect, but selection for more complex traits may be better suited to genomic selection or HTP approaches. Traits which are not readily available in the cultivated sorghum germplasm can be generated in a variety of ways, including mutagenesis, genome editing or transgenic technology, though the latter is unlikely to be commonly used in sorghum in the near future. Once it is fully developed, doubled haploid breeding could be used to quickly combine multiple traits into inbred lines. Though much faster, one disadvantage of doubled haploid breeding is that there is no ability to eliminate inferior individuals in early generations, so a large number of doubled haploid lines must be screened. MAS can be used to select doubled haploids with a desirable combination of major traits, such as disease or insect resistance and non-restoration. After fixing these major traits, the breeder could then apply HTP and genomic selection methods to further select for traits like grain yield or to predict the best hybrid combinations. As an example, Watanabe et al. (2016) used a UAV equipped with cameras to measure plant height in a set of 151 sorghum accessions. These measurements were reasonably correlated ($r = 0.523$) with traditional ruler measurements. Each accession was genotyped for over 66 000 SNPs using RAD-Seq, and then genomic prediction models (RR-BLUP) were built for both UAV-generated plant height and for ruler-measured plant height. Interestingly, the model-predicted plant heights measured by the two methods were highly correlated ($r = 0.842$).

Climate change, land degradation and an ever-increasing human population will present the greatest overall challenge to sorghum breeders to develop cultivars that are ever more tolerant to stresses such as heat, drought or flood while also increasing yields.

Emerging pests and diseases will also continue to threaten the crop, requiring breeders to continually search for and incorporate new sources of genetic resistance. Additional challenges may come from the rapidly changing needs of growers or industry end-users of sorghum. In much the same way as dwarf grain sorghum cultivars were introduced to facilitate mechanized harvesting in the early twentieth century, the emerging biofuels and bioproducts industries may require the development of novel plant types (Anami et al., 2015) or the incorporation of new biochemical traits. Newer technologies such as sequence-based genotyping, genome editing and HTP will continue to improve and eventually become more widely used by those who have the means to access them. Thus, another challenge will be to get these technologies into the hands of scientists and breeders in developing countries where sorghum will continue to be a major source of sustenance, particularly for smallholder farmers.

11 Where to look for further information and resources

The National Plant Germplasm System holds over 52 500 accessions of *Sorghum bicolor*, and numerous accessions of related *Sorghum* species, most of which are publicly available at no cost to researchers. Germplasm requests may be made directly from the website (http://www.ars-grin.gov/npgs/index.html), but researchers outside the United States may need to provide an import permit to receive seed. The Annotated Individually-pedigreed Mutagenized Sorghum (AIMS) collection contains over 1600 sorghum TILLING mutants, developed by the USDA-ARS Plant Stress and Germplasm Development Unit. Information on specific mutant lines can be found at http://www.lbk.ars.usda.gov/psgd/index-sorghum.aspx.html.

Sequence-based genotyping services are available commercially though several companies including the Elshire Group, Ltd., in New Zealand (https://www.elshiregroup.co.nz/), which uses the GBS method described by Elshire et al. (2011). Keygene owns the patent for a similar procedure called Sequence-Based Genotyping (Truong et al., 2012). Customers may request services directly from Keygene (http://www.keygene.com/), and the company has also licensed this technology to other companies and university facilities (Keygene, 2017). Data2Bio, LLC, based in Iowa, offers genotyping services using the tGBS process (Ott et al., 2017) and can also assist in development of genomic selection models using tGBS data (http://www.data2bio.com/). In India, Xcelris Genomics offers genotyping by GBS/RAD-Seq (Davey et al., 2011). They also offer NGS-based TILLING services (http://www.xcelrisgenomics.com/Home.html).

Numerous genomics and computing resources are available on the internet. The *Sorghum bicolor* v. 3.1 genome sequence assembly can be accessed at the Phytozome website (https://phytozome.jgi.doe.gov/). Luo et al. (2016) recently described a database of SNPs, called SorGSD, created from resequencing data of 48 diverse sorghum accessions. It may be accessed at this website, which also contains links to numerous other sorghum genomics resources: http://sorgsd.big.ac.cn/snp/index.jsp. CyVerse, formerly the iPlant Collaborative, is a public computing resource for data storage, bioinformatics and image analysis; it can be accessed at http://www.iplantcollaborative.org. The Integrated Breeding Platform (IBP; https://www.integratedbreeding.net/) provides software tools and access to genotyping and other breeding-related services. Its main mission is to provide modern plant breeding tools to breeders in developing countries, though they are available to all scientists. Software subscription fees vary based on institution type and country, and are free to scientists in certain parts of the world. A CyVerse/iPlant account is required to access IBP services. Many genomic selection models and other bioinformatics analyses can be run using 'packages' within the R statistical software platform, available for free download from the R Project (https://www.r-project.org).

Information on access and resources available at HTP facilities can be found at the Donald Danforth Plant Science Center (https://www.danforthcenter.org/home), the Australian Plant Phenomics Facility (http://www.plantphenomics.org.au/) and the National Plant Phenomics Centre (http://www.plant-phenomics.ac.uk/en/).

Several conferences may be of interest to sorghum breeders. The Sorghum Improvement Conference of North America (SICNA) is held annually, usually in the fall, and usually in Kansas or Texas (http://sicna.net). Each December the American Seed Trade Association (ASTA) holds its Annual Corn and Sorghum Seed Research Conference in Chicago, IL (http://www.betterseed.org/). The Sweet Sorghum Association (formerly Sweet Sorghum Ethanol Association) annual meeting is usually held in January (http://www.sseassociation.org/). For the past several years it has been held in Orlando, FL, and is likely to remain at the same location for the foreseeable future. The Plant and Animal Genome (PAG) conference, billed as the largest agricultural genomics meeting in the world, is held each January in San Diego, CA (www.intlpag.org/).

Funding opportunities may be available through the Feed the Future Innovation Lab for Collaborative Research on Sorghum and Millet, headquartered at Kansas State University (https://www.k-state.edu/smil). Its purpose is to bring US and international researchers together to improve sorghum and millet production and markets in developing countries, primarily in Africa. Other funding opportunities for sorghum improvement in the United States may be available through the United Sorghum Checkoff Program (http://www.sorghumcheckoff.com/research).

12 Disclaimer

Mention of trade names is solely for providing specific information and does not imply recommendation or endorsement by the United States Department of Agriculture.

13 References

Adeli, A., Varco, J. J., Sistani, K. R. and Rowe, D. E. (2005), Effects of swine lagoon effluent relative to commercial fertilizer applications on warm-season forage nutritive value. *Agron. J.* 97(2):408–17.

Albrecht, T., Auinger, H.-J., Wimmer, V., Ogutu, J. O., Knaak, C., Ouzunova, M., Piepho, H. P. and Schön, C. C. (2014), Genome-based prediction of maize hybrid performance across genetic groups, testers, locations, and years. *Theor. Appl. Genet.* 127:1375–86.

Anai, T. (2011), Potential of a mutant-based reverse genetic approach for functional genomics and molecular breeding in soybean. *Breed. Sci.* 61(5):462–7.

Anami, S. E., Zhang, L.-M., Xia, Y., Zhang, Y.-M., Liu, Z.-Q. and Jing, H.-C. (2015). Sweet sorghum ideotypes: Genetic improvement of the biofuel syndrome. *Food Energy Secur.* 4(3):159–77.

Araus, J. L. and Cairns, J. E. (2014), Field high-throughput phenotyping: The new crop breeding frontier. *Trends Plant Sci.* 19(1):52–61.

Arriola, P. E. and Ellstrand, N. C. (1996), Crop-to-weed gene flow in the genus *Sorghum* (Poaceae): Spontaneous interspecific hybridization between johnsongrass, *Sorghum halepense*, and crop sorghum, *S. Bicolor. Am. J. Bot.* 83(9):1153–9.

Arriola, P. E. and Ellstrand, N. C. (1997), Fitness of interspecific hybrids in the genus *Sorghum*: Persistence of crop genes in wild populations. *Ecol. Appl.* 7(2):512–18.

Bartek, M. S., Hodnett, G. L., Burson, B. L., Stelly, D. M. and Rooney, W. L. (2012), Pollen tube growth after intergeneric pollinations of *iap*-homozygous sorghum. *Crop Sci.* 52(4):1553–60.

Bhattacharya, A., Mikkilineni, V., Verma, L., Palan, B., Mali, K. and Char, B. (2014), Evaluation of doubled haploid culture conditions and regeneration of an indica rice hybrid. *Ind. J. Genet. Plant Breed.* 74(3):384–6.

Bhattacharya, A., Palan, B., Mali, K. and Char, B. (2015), Assessment of double haploid culture conditions in bell pepper. (*Capsicum annuum* L.). *Ind. J. Genet. Plant Breed.* 75(4):529–31.

Blomstedt, C. K., Gleadow, R. M., O'Donnell, N., Naur, P., Jensen, K., Laursen, T., Olsen, C. E., Stuart, P., Hamill, J. D., Møller, B. L. and Neale, A. D. (2012), A combined biochemical screen and TILLING approach identifies mutations in *Sorghum bicolor* L. Moench resulting in acyanogenic forage production. *Plant Biotechnol. J.* 10:54–66.

Borrell, A. K., Bidinger, F. R. and Sunitha, K. (1999), Stay-green trait associated with yield in recombinant inbred sorghum lines varying in rate of leaf senescence. *Int. Sorghum Millets Newslett.* 40:31–4.

Breiman, L. (2001), Random forests. *Mach. Learn.* 45:5–32.

Bunphan, D., Jaisil, P. and Sanitchon, J. (2013), Application of LW7 marker for identification of progenies with male sterility gene in sweet sorghum population. *Crop Breed. Appl. Biotechnol.* 13:59–66.

Burow, G., Xin, Z., Hayes, C. and Burke, J. (2014), Characterization of a multiseeded (*msd1*) mutant of sorghum for increasing grain yield. *Crop Sci.* 54:2030–7.

Casas, A. M., Kononowicz, A. K., Zehr, U. B., Tomes, D. T., Axtell, J. D., Butler, L. G., Bressan, R. A. and Hasegawa, P. M. (1993), Transgenic sorghum plants via microprojectile bombardment. *Proc. Natl. Acad. Sci. USA* 90:11212–16.

Casas, A. M., Kononowicz, A. K., Haan, T. G., Zhang, L., Tomes, D. T., Bressan, R. A. and Hasegawa, P. M. (1997), Transgenic sorghum plants obtained after microprojectile bombardment of immature inflorescences. *In Vitro Cell. Dev. Biol.-Plant* 33:92–100.

Chang, M. T. and Coe, E. (2014), Procedures to improve Stock 6. *Maize Genet. Coop. Newslett.* 87:12–13.

Chu, C. C., Wang, C. C., Sun, C. S., Chen, H., Yin, K. C., Chu, C. Y. and Bi, F. Y. (1975), Establishment of an efficient medium for anther culture of rice through comparative experiments on nitrogen-sources. *Sci. Sinica* 18(5):659–68.

Chumley, F. G. (2015), Development of a sorghum doubled haploid breeding system. Small Business Grant project report no. KANW-2014-00225. USDA-NIFA. https://portal.nifa.usda.gov/web/crisprojectpages/1002865-development-of-a-sorghum-doubled-haploid-breeding-system.html (Accessed 23 March 2016).

Clough, S. J. and Bent, A. F. (1998), Floral dip: A simplified method for *Agrobacterium*-mediated transformation of *Arabidopsis thaliana*. *Plant J.* 16(6):735–43.

Coe Jr., E. H. (1959), A line of maize with high haploid frequency. *Am. Nat.* 93:381–2.

Colbert, T., Till, B. J., Tompa, R., Reynolds, S., Steine, M. N., Yeung, A. T., McCallum, C. M., Comai, L. and Henikoff, S. (2001), High-throughput screening for induced point mutations. *Plant Physiol.* 126(2):480–4.

Crain, J. L., Wei, Y., Barker III, J., Thompson, S. M., Alderman, P. D., Reynolds, M., Zhang, N. and Poland, J. (2016), Development and deployment of a portable field phenotyping platform. *Crop Sci.* 56:965–75.

Crasta, O. R., Xu, W. W., Rosenow, D. T., Mullet, J. and Nguyen, H. T. (1999), Mapping of post-flowering drought resistance traits in grain sorghum: Association between QTLs influencing premature senescence and maturity. *Mol. Gen. Genet.* 262:579–88.

Cuevas, H. E. and Prom, L. K. (2013), Assessment of molecular diversity and population structure of the Ethiopian sorghum [*Sorghum bicolor* (L.) Moench] germplasm collection maintained by the USDA-ARS National Plant Germplasm System using SSR markers. *Genet. Resour. Crop Evol.* 60(6):1817–30.

Cuevas, H. E., Prom, L. K., and Erpelding, J. E. (2015), Tapping the US sweet sorghum collection to identify biofuel germplasm. *Sugar Tech.* 17(4):428–38.

Daetwyler, H. D., Pong-Wong, R., Villanueva, B. and Woolliams, J. A. (2010), The impact of genetic architecture on genome-wide evaluation methods. *Genetics* 185:1021–31.

Dahlberg, J. A. and Madera-Torres, P. (1997), Restorer reaction in A_1 (ATx 623), A_2 (A_2Tx 632), and A_3 (A_3SC 103) cytoplasms to selected accessions from the Sudan sorghum collection. *Int. Sorghum Millets Newslett.* 38:43–58.

Davey, J. W., Hohenlohe, P. A., Etter, P. D., Boone, J. Q., Catchen, J. M. and Blaxter M. L. (2011), Genome-wide genetic marker discovery and genotyping using next-generation sequencing. *Nat. Rev. Genet.* 12:499–510.

de Oliveira Couto, E. G., de Resende Von Pinho, E. V., Von Pinho, R. G., Veiga, A. D., Bustamante, F. D. O. and das Graças Dias K. O. (2015), *In vivo* haploid induction and efficiency of two chromosome duplication protocols in tropical maize. *Cienc. Agrotec.* 39(5):435–42.

Desta, Z. A. and Ortiz, R. (2014), Genomic selection: Genome-wide prediction in plant improvement. *Trends Plant Sci.* 19(9):592–601.

Doggett, H. (1988), *Sorghum.* 2nd ed. John Wiley & Sons, New York.

Downes, R. W. and Marshall, D. R. (1971), Low temperature induced male sterility in *Sorghum bicolor. Aust. J. Exp. Agric. Animal Husbandry* 11(50):352–6.

Dweikat, I. (2005), A diploid, interspecific, fertile hybrid from cultivated sorghum, *Sorghum bicolor*, and the common Johnsongrass weed *Sorghum halepense. Mol. Breed.* 16:93–101.

Eder, J. and Chalyk, S. (2002), In vivo haploid induction in maize. *Theor. Appl. Genet.* 104:703–8.

Elshire, R. J., Glaubitz, J. C., Sun, Q., Poland, J. A., Kawamoto, K., Buckler, E. S. and Mitchell, S. E. (2011), A robust, simple Genotyping-by-Sequencing (GBS) approach for high diversity species. *PLoS ONE* 6(5):e19379.

Follak, S. and Essl, F. (2013), Spread dynamics and agricultural impact of *Sorghum halepense*, an emerging invasive species in Central Europe. *Weed Res.* 53(1):53–60.

Fukuoka, S., Saka, N., Koga, H., Ono, K., Shimizu, T., Ebana, K., Hayashi, N., Takahashi, A., Hirochika, H., Okuno, K. and Yano, M. (2009), Loss of function of a proline-containing protein confers durable disease resistance in rice. *Science* 325:998–1001.

Gaj, T., Gersbach, C. A. and Barbas III, C. F. (2013), ZFN, TALEN, and CRISPR/Cas-based methods for genome engineering. *Trends Biotechnol.* 31(7):397–405.

Gao, J., Xia, B., Luo, F., Sun, S., Pei, Z., Gui, Z., Yuan, Q. and Li, X. (2013), Marker-assisted breeding for *rf1*, a nuclear gene controlling A1 CMS in sorghum (*Sorghum bicolor* L. Moench). *Euphytica* 193:383–90.

Gill, J. R., Rooney, W. L. and Klein, P. E. (2014), Effect of humidity on intergeneric pollinations of *iap* (*Inhibition of Alien Pollen*) sorghum [*Sorghum bicolor* (L.) Moench]. *Euphytica* 198:381–7.

Glover, J. D., Reganold, J. P., Bell, L. W., Borevitz, J., Brummer, E. C., Buckler, E. S., Cox, C. M., Cox, T. S., Crews, T. E., Culman, S. W., DeHaan, L. R., Eriksson, D, Gill, B. S., Holland, J., Hu, F., Hulke, B. S., Ibrahim, A. M. H., Jackson, W., Jones, S. S., Murray, S. C., Paterson, A. H., Ploschuk, E., Sacks, E. J., Snapp, S., Tao, D., Van Tassel, D. L., Wade, L. J., Wyse, D. L., Xu, Y. (2010), Increased food and ecosystem security via perennial grains. *Science* 328:1638–9.

Gorthy, S., Narasu, L., Gaddameedi, A., Sharma, H. C., Kotla, A., Deshpande, S. P., and Are, A. K. (2017), Introgression of shoot fly (*Atherigona soccata* L. Moench) resistance QTLs into elite post-rainy season sorghum varieties using marker assisted backcrossing (MABC). *Front. Plant Sci.* 8:1494.

Guo, J. H. and Liang, G. H. (1993), Callus induction and plant regeneration of cultivated and wild sorghums. *Cytologia* 58:203–10.

Guo, Y., Abernathy, B., Zeng, Y. and Ozias-Akins, P. (2015), TILLING by sequencing to identify induced mutations in stress resistance genes of peanut (*Arachis hypogaea*). *BMC Genom.* 16:157.

Gurel, S., Gurel, E., Kaur, R., Wong, J., Meng, L., Tan, H.-Q. and Lemaux, P. G. (2009), Efficient, reproducible *Agrobacterium*-mediated transformation of sorghum using heat treatment of immature embryos. *Plant Cell Rep.* 28:429–44.

Harris, K., Subudhi, P. K., Borrell, A., Jordan, D., Rosenow, D., Nguyen, H., Klein, P., Klein, R. and Mullet, J. (2007), Sorghum stay-green QTL individually reduce post-flowering drought-induced leaf senescence. *J. Exp. Bot.* 58(2):327–38.

Haussmann, B. I. G., Hess, D. E., Omanya, G. O., Folkertsma, R. T. Reddy, B. V. S., Kayentao, M., Welz, H. G. and Geiger, H. H. (2004), Genomic regions influencing resistance to the parasitic weed *Striga hermonthica* in two recombinant inbred populations of sorghum. *Theor. Appl. Genet.* 109:1005–16.

Hayashi, T. and Iwata, H. (2010), EM algorithm for Bayesian estimation of genomic breeding values. *BMC Genet.* 11:1–9.

Heslot, N., Yang, H.-P., Sorrells, M. E. and Jannink, J.-L. (2012), Genomic selection in plant breeding: A comparison of models. *Crop Sci.* 52:146–60.

Hodnett, G. L., Hale, A. L., Packer, D. J., Stelly, D. M., da Silva, J and Rooney, W. L. (2010), Elimination of a reproductive barrier facilitates intergeneric hybridization of *Sorghum bicolor* and *Saccharum*. *Crop Sci.* 50:1188–95.

Howe, A., Sato, S., Dweikat, I., Fromm, M. and Clemente, T. (2006), Rapid and reproducible *Agrobacterium*-mediated transformation of sorghum. *Plant Cell Rep.* 25:784–91.

Hu, H., Schrag, T. A., Peis, R., Unterseer, S., Schipprack, W., Chen, S., Lai, J., Yan, J., Prasanna, B. M., Nair, S. K., Chaikam, V., Rotarenco, V., Shatskaya, O. A., Zavalishina, A., Scholten, S., Schön, C.-C. and Melchinger, A. E. (2016), The genetic basis of haploid induction in maize identified with a novel genome-wide association method. *Genetics* 202:1267–76.

Illumina, Inc. (2017), Illumina sequencing platforms. Online at: https://www.illumina.com/systems/sequencing-platforms.html.

International Atomic Energy Agency (2017) *Mutant Variety Database* [website]. Vienna, Austria: IAEA. Available at: https://mvd.iaea.org/(Accessed 3 October 2017).

Jannink, J.-L., Lorenz, A. J. and Iwata, H. (2010), Genomic selection in plant breeding: From theory to practice. *Brief. Func. Genom.* 9(2):166–77.

Jenks, M. A., Joly, R. J., Peters, P. J., Rich, P. J., Axtell, J. D. and Ashworth, E. N. (1994), Chemically induced cuticle mutation affecting epidermal conductance to water vapor and disease susceptibility in *Sorghum bicolor* (L.) Moench. *Plant Physiol.* 105:1239–45.

Jiang, W., Zhou, H., Bi, H., Fromm, M., Yang, B. and Weeks, D. P. (2013), Demonstration of CRISPR/Cas9/sgRNA-mediated targeted gene modification in Arabidopsis, tobacco, sorghum and rice. *Nucleic Acids Res.* 41(20):e188.

Jordan, D. R., Mace, E. S., Henzell, R. G., Klein, P. E. and Klein, R. R. (2010), Molecular mapping and candidate gene identification of the *Rf2* gene for pollen fertility restoration in sorghum [*Sorghum bicolor* (L.) Moench]. *Theor. Appl. Genet.* 120:1279–87.

Jordan, D. R., Klein, R. R., Sakrewski, K. G., Henzell, R. G., Klein, P. E. and Mace, E. S. (2011), Mapping and characterization of Rf_5: A new gene conditioning pollen fertility restoration in A_1 and A_2 cytoplasm in sorghum (*Sorghum bicolor* (L.) Moench). *Theor. Appl. Genet.* 123:383–96.

Kassahun, B., Bidinger, F. R., Hash, C. T. and Kuruvinashetti, M. S. (2010), Stay-green expression in early generation sorghum [*Sorghum bicolor* (L.) Moench] QTL introgression lines. *Euphytica* 172:351–62.

Kawahigashi, H., Kasuga, S., Ando, T., Kanamori, H., Wu, J. Z., Yonemaru, J., Sazuka, T. and Matsumoto, T. (2011), Positional cloning of *ds1*, the target leaf spot resistance gene against *Bipolaris sorghicola* in sorghum. *Theor. Appl. Genet.* 123(1):131–42.

Keles, D., Pinar, H., Ata, A., Taskin, H., Yildiz, S. and Buyukalaca, S. (2015), Effect of pepper types on obtaining spontaneous doubled haploid plants via anther culture. *Hortscience* 50(11):1671–6.

Kering, M. K., Butler, T. J., Biermacher, J. T. and Guretzky, J. A. (2012), Biomass yield and nutrient removal rates of perennial grasses under nitrogen fertilization. *Bioenerg. Res.* 5:61–70.

Published by Burleigh Dodds Science Publishing Limited, 2018.

Keygene (2017), Sequence-based genotyping. Online at: http://www.keygene.com/products-tech/seq-based-genotyping/.

Khatodia, S., Bhatotia, K., Passricha, N., Khurana, S. M. P. and Tuteja, N. (2016), The CRISPR/Cas genome-editing tool: Application in improvement of crops. *Front. Plant Sci.* 7:506.

Kim, J.-S., Klein, P. E., Klein, R. R., Price, H. J., Mullet, J. E. and Stelly, D. M. (2005), Chromosome identification and nomenclature of *Sorghum bicolor. Genetics* 169:1169–73.

King, R., Bird, N., Ramirez-Gonzalez, R., Coghill, J. A., Patil, A., Hassani-Pak, K., Uauy, C. and Phillips, A. L. (2015), Mutation scanning in wheat by exon capture and next-generation sequencing. *PLoS ONE* 10(9):e0137549.

Klein, R. R., Klein, P. E., Chhabra, A. K., Dong, J., Pammi, S., Childs, K. L., Mullet, J. E., Rooney, W. L. and Schertz, K. F. (2001), Molecular mapping of the *rf1* gene for pollen fertility restoration in sorghum (*Sorghum bicolor* L.). *Theor. Appl. Genet.* 102:1206–12.

Klein, R. R., Klein, P. E., Mullet, J. E., Minx, P., Rooney, W. L. and Schertz, K. F. (2005), Fertility restorer locus *Rf1* of sorghum (*Sorghum bicolor* L.) encodes a pentatricopeptide repeat protein not present in the colinear region of rice chromosome 12. *Theor. Appl. Genet.* 111:994–1012.

Knoll, J. E. (2007), Improvement of early-season cold tolerance in sorghum by introgression of favorable genes from Chinese kaoliang germplasm. Ph.D. Diss. Purdue Univ, West Lafayette, IN, USA.

Knoll, J. E. and Ejeta, G. (2008), Marker-assisted selection for early-season cold tolerance in sorghum: QTL validation across populations and environments. *Theor. Appl. Genet.* 116:541–53.

Knoll, J. E., Ramos, M. L., Zeng, Y., Holbrook, C. C., Chow, M., Chen, S., Maleki, S., Bhattacharya, A. and Ozias-Akins, P. (2011), TILLING for allergen reduction and improvement of quality traits in peanut (*Arachis hypogaea* L.). *BMC Plant Biol.* 11:81.

Kuhlman, L. C., Pring, D. R., Rooney, W. L. and Tang, H. V. (2006), Allelic frequency at the *Rf3* and *Rf4* loci and the genetics of A_3 cytoplasmic fertility restoration in converted sorghum lines. *Crop Sci.* 46:1576–80.

Kumaravadivel, N. and Rangasamy, S. R. S. (1994), Plant regeneration from sorghum anther cultures and field evaluation of progeny. *Plant Cell Rep.* 13:286–90.

LGC Genomics (2017), KASP genotyping technology. Online at: http://www.lgcgroup.com/LGCGroup/media/PDFs/Products/Genotyping/KASP-brochure.pdf.

Li, H., Handsaker, B., Wysoker, A., Fennell, T., Ruan, J., Homer, N., Marth, G., Abecasis, G. and Durbin, R. (2009), The Sequence Alignment/Map format and SAMtools. *Bioinformatics* 25(16):2078–9.

Li, J., Wang, L., Zhan, Q., Fan, F., Zhao, T. and Wan, H. (2016), Development of a simple and efficient method for *Agrobacterium*-mediated transformation in sorghum. *Int. J. Agric. Biol.* 18:134–8.

Lorenzana, R. E. and Bernardo, R. (2009), Accuracy of genotypic value predictions for marker-based selection in biparental plant populations. *Theor. Appl. Genet.* 120:151–61.

Luo, H., Zhao, W. M., Wang, Y. Q., Xia, Y., Wu, X. Y., Zhang, L. M., Tang, B. X., Zhu, J. W., Fang, L., Du, Z. L., Bekele, W. A., Tai, S. A., Jordan, D. R., Godwin, I. D., Snowdon, R. J., Mace, E. S., Jing, H. C. and Luo, J. C. (2016), SorGSD: A sorghum genome SNP database. *Biotechnol. Biofuels* 9:6.

Mace, E. S., Xia, L., Jordan, D. R., Halloran, K., Parh, D. K., Huttner, E., Wenzl, P. and Kilian, A. (2008), DArT markers: Diversity analyses and mapping in *Sorghum bicolor. BMC Genomics* 9:26.

Mace, E. S., Rami, J.-F., Bouchet, S., Klein, P. E., Klein, R. R., Kilian, A., Wenzl, P., Xia, L., Halloran, K. and Jordan, D. R. (2009), A consensus genetic map of sorghum that integrates multiple component maps and high-throughput Diversity Array Technology (DArT) markers. *BMC Plant Biol.* 9:13.

McCallum, C. M., Comai, L., Greene, E. A. and Henikoff, S. (2000), Targeting induced local lesions in genomes (TILLING) for plant functional genomics. *Plant Physiol.* 123(2):439–42.

McLaughlin, M. R., Fairbrother, T. E. and Rowe, D. E. (2004), Nutrient uptake by warm-season perennial grasses in a swine effluent spray field. *Agron. J.* 96(2):484–93.

Mehlo, L., Mbambo, Z., Bado, S., Lin, J., Moagi, S. M., Buthelezi, S., Stoychev, S. and Chikwamba, R. (2013), Induced protein polymorphisms and nutritional quality of gamma irradiation mutants of sorghum. *Mutation Res. -Fund. Mol. Mech. Mutagenesis* 748:66–72.

Melchinger, A. E., Schipprack, W., Würschum, T., Chen, S. and Technow, F. (2013), Rapid and accurate identification of *in vivo*-induced haploid seeds based on oil content in maize. *Sci. Rep.* 3:2129.

Mendes, M. P. and de Souza Jr., C. L. (2016), Genomewide prediction of tropical maize single-crosses. *Euphytica* 209:651–63.

Menz, M. A., Klein, R. R., Mullet, J. E., Obert, J. A., Unruh, N. C. and Klein P. E. (2002), A high-density genetic map of *Sorghum bicolor* (L.) Moench based on 2926 AFLP®, RFLP and SSR markers. *Plant Mol. Biol.* 48:483–99.

Meuwissen, T. H. E. (2009), Accuracy of breeding values of 'unrelated' individuals predicted by dense SNP genotyping. *Genet. Selection Evol.* 41:35.

Meuwissen, T. H. E., Hayes, B. J. and Goddard, M. E. (2001), Prediction of total genetic value using genome-wide dense marker maps. *Genetics* 157:1819–29.

Mizuno, H., Kawahigashi, H., Ogata, J., Minami, H., Kanamori, H., Nakagawa, H. and Matsumoto, T. (2013), Genomic inversion caused by gamma irradiation contributes to downregulation of a *WBC11* homolog in *bloomless* sorghum. *Theor. Appl. Genet.* 126:1513–20.

Moran, J. L. and Rooney, W. L. (2003), Effect of cytoplasm on the agronomic performance of grain sorghum hybrids. *Crop Sci.* 43(3):777–81.

Morris, G. P., Ramu, P., Deshpande, S. P., Hash, C. T., Shah, T., Upadhyaya, H. D., Riera-Lizarazu, O., Brown, P. J., Acharya, C. B., Mitchell, S. E., Harriman, J., Glaubitz, J. C., Buckler, E. S. and Kresovich S. (2013), Population genomic and genome-wide association studies of agroclimatic traits in sorghum. *Proc. Natl. Acad. Sci. USA* 110(2):453–8.

Murashige, T. and Skoog, F. (1962), A revised medium for rapid growth and bio assays with tobacco tissue cultures. *Physiol. Plant.* 15(3): 473–97.

Muraya, M. M., Geiger, H. H., Sagnard, F., Toure, L., Traore, P. C. S., Togola, S., de Villiers, S. and Parzies, H. K. (2012), Adaptive values of wild × cultivated sorghum (*Sorghum bicolor* (L.) Moench) hybrids in generations F_1, F_2, and F_3. *Genet. Resour. Crop Evol.* 59:83–93.

Nabukalu, P. and Cox, T. S. (2016), Response to selection in the initial stages of a perennial sorghum breeding program. *Euphytica* 209(1):103–11.

Nair, N. V. (1999), Production and cyto-morphological analysis of intergeneric hybrids of *Sorghum × Saccharum*. *Euphytica* 108:187–91.

Nakamura, S., Can, N. D. and Yoshida, T. (1997), Study on callus induction from anther and inflorescence culture of sorghum. *J. Fac. Agric. Kyushu Univ.* 42(1–2):1–9.

Nanda, D. K. and Chase, S. S. (1966), An embryo marker for detecting monoploids of maize (*Zea mays* L.). *Crop Sci.* 6:213–15.

Nawaz, Z. and Shu, Q. (2014), Molecular nature of chemically and physically induced mutants in plants: A review. *Plant Genet. Resour. Char. Utiliz.* 12(S1):S74–S78.

Nida, H., Blum, S., Zielinski, D., Srivastava, D. A., Elbaum, R., Xin, Z., Erlich, Y., Fridman, E. and Shental, N. (2016), Highly efficient *de novo* mutant identification in a *Sorghum bicolor* TILLING population using the ComSeq approach. *Plant J.* 86(4):349–59.

Neilson, E. H., Edwards, A. M., Blomstedt, C. K., Berger, B., Møller, B. L. and Gleadow R. M. (2015), Utilization of a high-throughput shoot imaging system to examine the dynamic phenotypic responses of a C_4 cereal crop plant to nitrogen and water deficiency over time. *J. Exp. Bot.* 66(7):1817–32.

Ordonio, R., Ito, Y., Morinaka, Y., Sazuka, T. and Matsuoka, M. (2016), Molecular breeding of *Sorghum bicolor*, a novel energy crop. *Int. Rev. Cell Mol. Biol.* 321:221–57.

Oria, M. P., Hamaker, B. R., Axtell, J. D. and Huang, C.-P. (2000), A highly digestible sorghum mutant cultivar exhibits a unique folded structure of endosperm protein bodies. *Proc. Natl. Acad. Sci. USA* 97(10):5065–70.

Ott, A., Liu, S., Schnable, J. C., Yeh, C.-T., Wang, C. and Schnable, P. S. (2017) Tunable Genotyping-By-Sequencing (tGBS®) enables reliable genotyping of heterozygous loci. *BioRχiv*. doi:http://dx.doi.org/10.1101/100461.

Ozias-Akins, P., Chu, Y., Knoll, J. and Bhattacharya, A. (2010), Mitigating allergenicity of crops. *Adv. Agron.* 107:93–121.

Pahuja, S. K., Aruna, C., Shrotria, P. K., Kaur, S., Ranwah, B. R. and Patil, J. V. (2012), Inducing variability in multi-cut forage sorghum through mutagenesis. *Plant Genet. Resour. Char. Utiliz.* 11(2):114–20.

Park, T. and Casella, G. (2008) The Bayesian Lasso. *J. Am. Stat. Assoc.* 103:681–6.

Paterson, A. H., Schertz, K. F., Lin, Y.-R., Liu, S.-C. and Chang, Y.-L. (1995), The weediness of wild plants: Molecular analysis of genes influencing dispersal and persistence of johnsongrass, *Sorghum halepense* (L.) Pers. *Proc. Natl. Acad. Sci. USA* 92:6127–31.

Paterson, A. H., Bowers, J. E., Bruggmann, R., Dubchak, I., Grimwood, J., Gundlach, H., Haberer, G., Hellsten, U., Mitros, T., Poliakov, A., Schmutz, J., Spannagl, M., Tang, H., Wang, X., Wicker, T., Bharti, A. K., Chapman, J., Feltus, F. A., Gowik, U., Grigoriev, I. V., Lyons, E., Maher, C. A., Martis, M., Narechania, A., Otillar, R. P., Penning, B. W., Salamov, A. A., Wang, Y., Zhang, L., Carpita, N. C., Freeling, M., Gingle, A. R., Hash, C. T., Keller, B., Klein, P., Kresovich, S., McCann, M. C., Ming, R., Peterson, D. G., Mehboob-ur-Rahman, Ware, D., Westhoff, P., Mayer, K. F. X., Messing, J. and Rokhsar, D. S. (2009), The *Sorghum bicolor* genome and the diversification of grasses. *Nature* 457, 551–6.

Pedersen, J. F., Marx, D. B. and Funnell, D. L. (2003) Use of A_3 cytoplasm to reduce risk of gene flow through sorghum pollen. *Crop Sci.* 43:1506–9.

Peng, L., Hou, S., Du, R., Li, S., Ji, G. and Liu, G. (2012), Floral dip transformation of sorghum. In: *2012 China Crop Science Society of Academic Annual Meeting Abstract Book.*

Petti, C., Kushwaha, R., Tateno, M., Harman-Ware, A. E., Crocker, M., Awika, J. and DeBolt, S. (2014), Mutagenesis breeding for increased 3-deoxyanthocyanidin accumulation in leaves of *Sorghum bicolor* (L.) Moench: A source of natural food pigment. *J. Agric. Food Chem.* 62:1227–1232.

Piper, J. K. and Kulakow, P. A. (1994), Seed yield and biomass allocation in *Sorghum bicolor* and F_1 and backcross generations of *Sorghum bicolor* × *Sorghum halepense* hybrids. *Can. J. Bot.* 72(4):468–74.

Poland, J. A. and Rife, T. W. (2012), Genotyping-by-Sequencing for plant breeding and genetics. *Plant Genome* 5:92–102.

Porter, K. S., Axtell, J. D., Lechtenberg, V. L. and Colenbrander, V. F. (1978), Phenotype, fiber composition, and *in vitro* dry matter disappearance of chemically induced brown midrib (*bmr*) mutants of sorghum. *Crop Sci.* 18:205–8.

Praveen, M., Anurag Uttam, G., Suneetha, N., Umakanth, A., Patil, J. V. and Madhusudhana, R.. (2015), Inheritance and molecular mapping of Rf_6 locus with pollen fertility restoration ability on A_1 and A_2 cytoplasms in sorghum. *Plant Sci.* 238:73–80.

Price, H. J., Hodnett, G. L., Burson, B. L., Dillon, S. L., Stelly, D. M., and Rooney, W. L. (2006), Genotype dependent interspecific hybridization of *Sorghum bicolor*. *Crop Sci.* 46(6):2617–22.

Rasmussen, J., Ntakos, G., Nielsen, J., Svensgaard, J., Poulsen, R. N. and Christensen, S. (2016), Are vegetation indices derived from consumer-grade cameras mounted on UAVs sufficiently reliable for assessing experimental plots? *Eur. J. Agron.* 74:75–92.

Riedelsheimer, C., Technow, F. and Melchinger, A. E. (2012), Comparison of whole-genome prediction models for traits with contrasting genetic architecture in a diversity panel of maize inbred lines. *BMC Genom.* 13:452.

Rizal, G., Karki, S., Alcasid, M., Montecillo, F., Acebron, K., Larazo, N., Garcia, R., Slamet-Loedin, I. H. and Quick, W. P. (2014), Shortening the breeding cycle of sorghum, a model crop for research. *Crop Sci.* 54:520–9.

Röber, F. K., Gordillo, G. A. and Geiger, H. H. (2005), *In vivo* haploid induction in maize – performance of new inducers and significance of doubled haploid lines in hybrid breeding. *Maydica* 50:275–83.

Rooney, W. L. (2004), Sorghum improvement – integrating traditional and new technology to produce improved genotypes. *Adv. Agron.* 83:109.

Rosenow, D., Quisenberry, J. E., Wendt, C. E. and Clark, L. E. (1983), Drought tolerant sorghum and cotton germplasm. *Agric. Water Manag.* 7:207–22.

Rout, M. E., Chrzanowski, T. H., Smith, W. K. and Gough, L. (2013), Ecological impacts of the invasive grass *Sorghum halepense* on native tallgrass prairie. *Biol. Invasions* 15(2):327–39.

Sanchez, A. C., Subudhi, P. K., Rosenow, D. T. and Nguyen, H. T. (2002), Mapping QTLs associated with drought resistance in sorghum (*Sorghum bicolor* L. Moench). *Plant Mol. Biol.* 48:713–26.

Schertz, K. F. and Clark, L. E. (1968), Controlling dehiscence with plastic bags for hand crosses in sorghum. *Crop Sci.* 7(5):540–2.

Schrag, T. A., Möhring, J., Maurer, H. P., Dhillon, B. S., Melchinger, A. E., Piepho, H.-P., Sørensen, A. P. and Frisch, M (2009), Molecular marker-based prediction of hybrid performance in maize using unbalanced data from multiple experiments with factorial crosses. *Theor. Appl. Genet.* 118:741–51.

Semagn, K., Babu, R., Hearne, S. and Olsen, M. (2014), Single nucleotide polymorphism genotyping using Kompetitive Allele Specific PCR (KASP): Overview of the technology and its application in crop improvement. *Mol. Breed.* 33:1–14.

Shumilina, D. V., Shmykova, N. A., Bondareva, L. L. and Suprunova, T. P. (2015), Effect of genotype and medium culture content on microspore-derived embryo formation in Chinese cabbage (*Brassica rapa* ssp *chinensis*) cv. Lastochka. *Biol. Bull.* 42(4):302–9.

Singh, M., Chaudhary, K., Singal, H. R., Magill, C. W. and Boora, K. S. (2006), Identification and characterization of RAPD and SCAR markers linked to anthracnose resistance gene in sorghum [*Sorghum bicolor* (L.) Moench]. *Euphytica* 149:179–87.

Singh, T. S. N.. (1934), Chromosome numbers in sugar cane × sorghum hybrids. *Ind. J. Agric. Sci.* 4:1050.

Sriskandarajah, S., Sameri, M., Lerceteau-Köhler, E. and Westerbergh, A. (2015), Increased recovery of green doubled haploid plants from barley anther culture. *Crop Sci.* 55:2806–12.

Stephens, J. C. and Quinby, J. R. (1933), Bulk emasculation of sorghum flowers. *J. Am. Soc. Agron.* 25(3):233–4.

Subudhi, P. K., Rosenow, D. T. and Nguyen H. T. (2000), Quantitative trait loci for the stay green trait in sorghum (*Sorghum bicolor* L. Moench): Consistency across genetic backgrounds and environments. *Theor. Appl. Genet.* 101:733–41.

Sun, Y. W., Zhang, X., Wu, C. Y., He, Y. B., Ma, Y. Z., Hou, H., Guo, X. P., Du, W. M., Zhao, Y. D. and Xia, L. Q. (2016), Engineering herbicide-resistant rice plants through CRISPR/*Cas9*-mediated homologous recombination of acetolactate synthase. *Mol. Plant* 9(4):628–31.

Talbert, L. E., Blake, N. K., Chee, P. W., Blake T. K. and Magyar, G. M. (1994), Evaluation of 'sequence-tagged-site' PCR products as molecular markers in wheat. *Theor. Appl. Genet.* 87:789–94.

Tang, H. V. and Pring, D. R. (2003), Conversion of fertility restoration of the sorghum IS1112C (A3) male-sterile cytoplasm from two genes to one gene. *Crop Sci.* 43:1747–53.

Tang, H. V., Chang, R. and Pring, D. R. (1998), Cosegregation of single genes associated with fertility restoration and transcript processing of sorghum mitochondrial *orf107* and *urf209*. *Genetics* 150:383–91.

Till, B. J., Cooper, J., Tai, T. H., Colowit, P., Greene, E. A., Henikoff, S. and Comai, L. (2007), Discovery of chemically induced mutations in rice by TILLING. *BMC Plant Biol.* 7:19.

Till, B. J., Reynolds, S. H., Greene, E. A., Codomo, C. A., Enns, L. C., Johnson, J. E., Burtner, C., Odden, A. R., Young, K., Taylor, N. E., Henikoff, J. G., Comai, L. and Henikoff, S. (2003), Large-scale discovery of induced point mutations with high-throughput TILLING. *Genome Res.* 13(3):524–30.

Till, B. J., Zerr, T., Comai, L. and Henikoff, S. (2006), A protocol for TILLING and Ecotilling in plants and animals. *Nature Protocols* 1(5):2465–77.

Thermo Fisher Scientific (2017), Next-generation sequencing (NGS). Online at: http://www.thermofisher.com/us/en/home/life-science/sequencing/next-generation-sequencing.html.

Truong, H. T., Ramos, A. M., Yalcin, F., de Ruiter, M., van der Poel, H. J. A., Huvenaars, K. H. J., Hogers, R. C. J., van Enckevort, L. J. G., Janssen, A., van Orsouw, N. J. and van Eijk, M. J. T. (2012),

Sequence-Based Genotyping for marker discovery and co-dominant scoring in germplasm and populations. *PLoS ONE*. 7(5):e37565.

Tuinstra, M. R., Ejeta, G. and Goldsbrough, P. (1998), Evaluation of near-isogenic sorghum lines contrasting for QTL markers associated with drought tolerance. *Crop Sci.* 38(3):835–42.

United Sorghum Checkoff Program (2017), Research funded by United Sorghum Checkoff Program advances sorghum breeding technology. Press release. Online at: http://www.sorghumcheckoff.com/news-and-media/newsroom/2017/03/06/research-funded-by-united-sorghum-checkoff-program-advances-sorghum-breeding-technology/.

Vadez, V., Kholová, J., Hummel, G., Zhokhavets, U., Gupta, S. K. and Hash, C. T. (2015), LeasyScan: A novel concept combining 3D imaging and lysimetry for high-throughput phenotyping of traits controlling plant water budget. *J. Exp. Bot.* 66(18):5581–93.

Van Tassel, D. and DeHaan, L. (2013), Wild plants to the rescue. *Am. Sci.* 101:218–25.

Vos, P., Hogers, R., Bleeker, M., Reijans, M., Vandelee, T., Hornes, M., Frijters, A., Pot, J., Peleman, J., Kuiper, M. and Zabeau, M. (1995), AFLP - a new technique for DNA-fingerprinting. *Nucl. Acids Res.* 23(21):4407–14.

Waltz, E. (2016), Gene-edited CRISPR mushroom escapes US regulation. *Nature* 532:293.

Wang, F. J., Wang, C. L., Liu, P. Q., Lei, C. L., Hao, W., Gao, Y., Liu, Y. G. and Zhao, K. J. (2016), Enhanced rice blast resistance by CRISPR/*Cas9*-targeted mutagenesis of the ERF transcription factor gene *OsERF922*. *PLoS ONE* 11(4):e0154027.

Washburn, J. D., Murray, S. C., Burson, B. L., Klein, R. R. and Jessup, R. W. (2013), Targeted mapping of quantitative trait locus regions for rhizomatousness in chromosome SBI-01 and analysis of overwintering in a *Sorghum bicolor* × *S. propinquum* population. *Mol. Breed.* 31:153–62.

Watanabe, K., Guo, W., Arai, K., Takanashi, H., Kajiya-Kanegae, H., Kobayashi, M., Yano, K., Tokunaga, T., Fujiwara, T., Tsutsumi, N. and Iwata, H. (2016), High-throughput phenotyping of sorghum plant height using an unmanned aerial vehicle and its application to genomic prediction modeling. *Front. Plant Sci.* 8:421.

Weil, C. F. and Monde, R. A. (2007), Getting the point - Mutations in maize. *Crop Sci.* 47(1):S60–S67.

Wen, L., Tang, H. V., Chen, W., Chang, R., Pring, D. R., Klein, P. E., Childs, K. L. and Klein, R. R. (2002), Development and mapping of AFLP markers linked to the sorghum fertility restorer gene *rf4*. *Theor. Appl. Genet.* 104:577–85.

Werle, R., Jhala, A. J., Yerka, M. K., Dille, J. A. and Lindquist, J. L. (2016), Distribution of herbicide-resistant shattercane and Johnsongrass populations in sorghum production areas of Nebraska and northern Kansas. *Agron. J.* 108(1):321–8.

Williams, J. G., Kubelik, A. R., Livak, K. J., Rafalski, J. A. and Tingey, S. V. (1990), DNA polymorphisms amplified by arbitrary primers are useful as genetic markers. *Nucl. Acids Res.* 18:6531–5

Wu, E., Lenderts, B., Glassman, K., Berezowska-Kaniewska, M., Christensen, H., Asmus, T., Zhen, S., Chu, U., Cho, M.-J. and Zhao, Z.-Y. (2014), Optimized *Agrobacterium*-mediated sorghum transformation protocol and molecular data of transgenic sorghum plants. *In Vitro Cell. Dev. Biol.-Plant* 50:9–18.

Xin, Z., Wang, M.-L., Barkley, N. A., Burow, G., Franks, C., Pederson, G. and Burke, J. (2008), Applying genotyping (TILLING) and phenotyping analyses to elucidate gene function in a chemically induced sorghum mutant population. *BMC Plant Biol.* 8:103.

Xin, Z., Wang, M.-L., Burow, G. and Burke, J. (2009), An induced sorghum mutant population suitable for bioenergy research. *Bioenerg. Res.* 2:10–16.

Xu, W., Subudhi, P. K., Crasta, O. R., Rosenow, D. T., Mullet, J. E. and Nguyen, H. T. (2000), Molecular mapping of QTLs conferring stay-green in grain sorghum (*Sorghum bicolor* L. Moench). *Genome* 43:461–9.

Yellisetty, V., Reddy, L. A. and Mandapaka, M. (2015), *In planta* transformation of sorghum (*Sorghum bicolor* (L.) Moench) using *TPS1* gene for enhancing tolerance to abiotic stresses. *J. Genet.* 94(3):425–34.

Yerka, M. K., Watson, A., Toy, J. J., Erickson, G., Pedersen, J. F., and Mitchell, R. (2015), Yield and forage value of a dual-purpose *bmr-12* sorghum hybrid. *Crop Sci.* 55(2):681–7.

Yohannes, T., Abraha, T., Kiambi, D., Folkertsma, R., Hash, C. T., Ngugi, K., Mutitu, E., Abraha, N., Weldetsion, M., Mugoya, C., Masiga, C. W. and de Villiers, S. (2015), Marker-assisted introgression improves *Striga* resistance in an Eritrean Farmer-Preferred Sorghum Variety. *Field Crops Res.* 173:22–9.

Yoshida, S., Forno, D. A., Cock, J. H. and Gomez, K. A. (1972), *Laboratory Manual for Physiological Studies of Rice*. International Rice Research Institute, Los Baños, Laguna, Philippines.

Yu, W. and Birchler, J. A. (2016), A green fluorescent protein-engineered haploid inducer line facilitates haploid mutant screens and doubled haploid breeding in maize. *Mol. Breed.* 36:5.

Zeng, A. S., Yan, Y. Y., Yan, J. Y., Song, L. X., Gao, B., Li, J. Q., Hou, X. L. and Li, Y. (2015) Microspore embryogenesis and plant regeneration in Brussels sprouts (*Brassica oleracea* L. var. *gemmifera*). *Sci. Hort.* 191:31–7.

Zhao, Z., Cai, T., Tagliani, L., Miller, M., Wang, N., Pang, H., Rudert, M., Schroeder, S., Hondred, D., Seltzer, J. and Pierce, D. (2000), *Agrobacterium*-mediated sorghum transformation. *Plant Mol. Biol.* 44:789–98.

Zhao, Y., Mette, M. and Reif, J. (2012), Genomic selection in hybrid breeding. *Plant Breed.* 134:1–10.

Zhu, J. J., Song, N., Sun, S. L., Yang, W. L., Zhao, H. M., Song, W. B. and Lai, J. S. (2016), Efficiency and inheritance of targeted mutagenesis in maize using CRISPR-*Cas9*. *J. Genet. Genom.* 43(1):25–36.

Published by Burleigh Dodds Science Publishing Limited, 2018.

Chapter 6

Harnessing genetic and genomic resources to transform the production and productivity of sorghum

Tesfaye T. Tesso, Dereje D. Gobena, Dechassa O. Duressa, Kraig Roozeboom and Krishna Jagadish, Kansas State University, USA; Ramasamy Perumal, Agricultural Research Center – Hays, Kansas State University, USA; and Desalegn D. Serba and Dilooshi Weerasooriya, Kansas State University, USA

1 Introduction

According to United Nations' projections, the world's population is expected to reach 9.7 billion by 2050, from the current estimate of about ~7.3 billion. The majority of the anticipated increase is expected to happen in developing countries, with African countries taking the lion's share, adding 1.3 billion more people. According to these estimates, the population in Africa by the mid-twenty-first century will be at least twice its current size of about 1.2 billion. The majority of the people that are chronically food insecure at present live in developing countries, with one in four people in Africa being food insecure (FAO, 2015). Even if the millennium development goal of reducing hunger by half by 2015 had been met, that still would have left about 600 000 people in chronic hunger. At the current rate of growth in agricultural production and productivity, African and other developing countries will see the numbers of their hungry population double by 2050. This is a formidable challenge, not only to the developing nations that will bear the brunt of the hunger, but also to the entire world.

http://dx.doi.org/10.19103/AS.2017.0015.09

The quantity of food produced at a global scale at present is sufficient to feed the current world population, which implies that one billion people are currently suffering from chronic food insecurity not due to the shortage of food but to the lack of equitable food distribution caused by disproportionate income, poor governance and a failing trade policy as well as the lack of an effective transport system. Every prediction indicates that these scenarios are expected to change for the worse. First, global food production is going to fall short of population growth, so no matter how conducive trade and administrative policies may be, there will be a segment of the world's population that will never have access to food in sufficient quantity and quality. Second, rising temperatures, drought and the degradation of natural ecosystems, aggravated by climate change, are expected to challenge agricultural productivity in most parts of the globe. The modest improvements in productivity that have been seen over recent decades will be far from enough to see a hunger-free world, and food security will continue to be the major global challenge in the twenty-first century.

Taking into account current production levels, projected population growth and changing food preferences associated with rising income, numerous research results have shown that food production has to double by 2050 in order to feed the nine billion plus people expected to be born in the next three decades. In order to achieve this, the productivity of the world's major crops has to increase by an average of 2.4% per year. According to estimates by USDA, yields of the four major world crops – maize, wheat, rice and soybean – are increasing at an average rate of 1.2%. At this rate, the production of these crops is expected to increase by an average of 51% in 2050, which is far less than the predicted demand (Ray et al., 2013). Much of the increase in global agricultural produce in past decades was the combined result of the increased use of chemical fertilizers, the development and adoption of high-yielding crop varieties and agronomic practices, the development of agricultural enterprises that facilitated the production and delivery of modern agricultural inputs, such as improved seeds and agro-chemicals, the expansion of irrigation systems, and acreage under cultivation. Not many of these elements can be repackaged to further increase productivity to meet the anticipated demand.

Moreover, due to the effects of impending climate change, the quality of the natural resource base on which agriculture hinges is expected to change, mostly in an unfavourable direction. Declining soil fertility, changes in rainfall distribution, dwindling water resources for irrigation and the rising temperatures are going to pose serious challenges to increasing agricultural productivity throughout the world. Poor governance and political instability in several African countries have turned some of the most productive lands in the continent into a desolate savannah, leaving their inhabitants helpless and destitute, and there is little sign for hope. The increasing use of food crops for biofuel production, or the use of crop land for growing biofuel feedstock, is another challenge to increasing food production. Thus, future efforts to increase agricultural production need to take these elements into account and identify new avenues and strategies by which crop productivity can be increased to meet the projected need.

The challenges that need to be confronted are complex and the solutions are not straightforward. It will require a comprehensive analysis of the existing national and global agricultural research and development policies and the design of new investment programmes and strategies that will take these challenges into consideration and spur agricultural productivity. This should include selection of strategic crop species to make sure that crops that have a better chance of success in a given environment are grown, and increasing research efforts to allow the development of tools, techniques and practices that can be deployed to maximize productivity.

In view of the expected impacts of climate change, crops such as sorghum and millet appear to have a greater potential, especially in geographies where the impact of impending climate change is expected to result in higher temperatures and drought or cause a degradation of soil fertility. Sorghum is considered to be one of the drought-hardy crops that can be grown successfully in conditions where other cereals have little or no chance of success.

2 Sorghum as a vital food grain of the twenty-first century

While sorghum in the developed world is largely used as animal feed, the crop remains the primary source of energy and protein for the majority of subsistence farmers living in the driest parts of the world. Although the overall global acreage has not increased over the last five decades, areas under sorghum cultivation in Africa have doubled during this period (FAO, 1995), attesting to its value as means of livelihood. As drought conditions become worse and soil fertility degradation continues, more and more acreage may come under sorghum cultivation since many other cereals do not produce good yields under extreme conditions. While the availability of some essential nutrients, such as protein, may continue to be an important issue of academic debate, sorghum is not inferior to any other cereals in terms of its nutritional composition. Some of the local food-processing methods used by traditional sorghum consumers have been reported to improve this situation though may not resolve the problem completely. Food diversification, which is usually practised by smallholder farmers to avert the risk of hunger, has been shown to contribute to reducing malnutrition in that it adds complementary diets to the menu. Moreover, the wealth of knowledge that farmers have developed over the years on selecting particular cultivars for certain purposes has a positive impact on nutrition. Local farmers have not only adopted unique landraces that are adapted to specific localities, but have also found specific uses for them and have developed the management skills for growing them under particular conditions. They have also developed distinct processing methods for making various traditional foods and have selectively adopted unique variants of the crop suited for these purposes. These include cultivars with high lysine content, selectively grown as a high-value food, cultivars specifically grown for traditional brewing, those adapted to cold temperatures (Stemler et al., 1977), transplant sorghums that are cultivated following retreating seasonal waters (Batello et al., 2004) and ratoon sorghums that are grown in pockets of areas in the tropics and subtropics (Escalada and Plucknett, 1975; Touchton and Martin, 1981).

Moreover, sorghum is increasingly gaining acceptance in many places where it has never been popular before. Partly because of its gluten-free property, the crop is gradually making its way into the food system of western society. In contrast to the 1970s, when only sorghum molasses were available, different food products and beverages, such as pop snacks, cookies, snack bars, beer and so forth, made from sorghum are showing up on grocery shelves throughout the United States. All these changes happened without any regard to its agronomic potential. As climate change continues to leave its mark on the environment (Schlenker and Roberts, 2009; Brevik, 2013), the popularity of sorghum as a food/feed cereal and industrial raw material is expected to increase further. The key elements signifying the value of sorghum as a major food-security cereal lies in its inherent ability to thrive in marginal environments, the health-promoting compounds packed in its grain and the potential for further improvement in its productivity.

2.1 Adaptation to marginal growing conditions

Domesticated under warm and dry conditions, sorghum is inherently resilient to harsh weather conditions such as drought and high temperature stress as well as low-nutrient soils. Such a unique adaptation of the crop makes it an ideal candidate as twenty-first century food and feed grain.

Because of its drought resilience, sorghum is a preferred crop in regions frequently suffering from water scarcity. In addition to originating in sub-Saharan Africa, sorghum is widely cultivated there because yields remain stable despite the low-moisture stress and the scorching sun that characterize much of the subcontinent. While maize is more productive and a preferred food crop in areas where drought is not the major constraint, sorghum remains the dominant cereal in the drier parts of Africa, with acreage under the crop continuously increasing (FAO, 2015). Despite disproportionate investment by most national programmes and donor agencies, and exclusion from development portfolios, sorghum's share of acreage in sub-Saharan Africa has continually increased over the last three decades.

In the United States, grain sorghum production is concentrated in the semi-arid environments extending from Nebraska to South Texas, mainly as rain-fed. Farmers in this belt prefer sorghum not because of cultural ties and food preference, unlike in Africa but because sorghum is more profitable than maize in these semi-arid conditions. In a recent study comparing sorghum and maize in conditions where water was supplied on a gradient, results indicated that while phenology, water uptake and biomass and grain yield were affected by water treatments in both crops, sorghum out-yielded maize under moderate to severe water deficits because of its superior ability to extract water from deeper soil layers (Farré and Faci, 2006).

In addition to its resilience to drought and high temperatures, sorghum also has a remarkable tolerance to low-nutrient supply. It can grow on marginal soils with little or no soil fertility and still produces a reasonable yield. In Africa, sorghum is grown not only in drought-stressed, high-temperature conditions, but also on highly degraded soils devoid of organic matter and essential nutrients and yet it produces enough grain to sustain life. Even under high-input mechanized agriculture, sorghum is grown on low-nutrient soils with only half the fertilizer rate of other cereals. Sorghum-growing areas in the United States are not only prone to drought and high temperature stress, but also have shallow soils and are low in nutrients compared to the deep, fertile soils often left aside for maize. But sorghum receives fertilizer at the rate of only 50–60% that of maize. Resilience to both drought and low-nutrient supply can be improved to further optimize production in marginal environments. Crop management approaches that improve soil nutrients and water availability have the potential to increase sorghum yields. Soil management practices that increase organic matter content and water harvesting systems such as the use of tied ridge to harvest and retain rainwater near crop roots have significantly improved both sorghum and maize yields in sub-Saharan Africa (Belay et al., 1998; Gebrekidan, 2003). Likewise, not only can drought tolerance be enhanced through genetic means, but adaptation to low nutrient levels, especially to limited phosphorus (P) supply, can also be improved using genetic approaches (Hetrick et al., 1993; Cobb et al., 2017).

2.2 Health benefits of sorghum

Over 60% of global grain sorghum is used as human food in the developing countries of Africa and Asia, while the remaining 40% is used primarily as animal feed or biofuel feedstock. In the United States, grain sorghum is grown as the second major feed source for the livestock and poultry industries. Sorghum is also a principal forage/silage crop, with the brown mid-rib sorghum silages being particularly important in the dairy industry (Rook et al., 1977; Oba and Allen, 1999).

Due to its gluten-free property and the host of health-promoting bioactive compounds and phytochemicals it contains (Awika and Rooney, 2004a; Stefoska-Needham, et al., 2015), the wrong perception regarding the food value of sorghum in the western world is gradually changing. Because of its low glycemic index, sorghum foods are recommended for fighting obesity and diabetes (Awika and Rooney, 2004). Its rich antioxidant content has added to its growing popularity as a health food that food scientists and health-conscious consumers are promoting the use of sorghum in various food applications (Hahn and Rooney, 1986; Singh et al., 2003; Awika et al., 2005). The chemical ingredients of sorghum, including phenols (phenolic acids, anthocyanins and tannins), phytosterol, policosanol, resistant starch, carotenoids and polyamines, have been reported to have significant health benefits, such as reducing bad cholesterol and countering the proliferation of cancer cells both in model animals and humans (Awika et al., 2009; Yang et al., 2009). Certain types of sorghum, especially the red and black types, are the only dietary source of 3-deoxyanthocyanidins, anthocyanin-type polyphenols with antimicrobial, antioxidant and anticancer properties (Awika and Rooney, 2004; Awika et al., 2004; Dykes and Rooney, 2007; Stefoska-Needham et al., 2015; Cardoso et al., 2016). Black sorghum bran is a superior source of anthocyanin compared to most fruits and vegetables, including blueberries (Awika et al., 2004). Due to its chemically stable anthocyanin 3-deoxyanthocyanidin, sorghum has been identified as a potential commercial source of natural food colourant with health-promoting properties (Awika et al., 2004; Dia et al., 2016).

The lipid profile of grain sorghum includes potent health-promoting phytochemicals that are present in the germ fraction (phytosterol) and in the seed surface wax (policosanol). Sorghum is one of the cereal grains that are rich in these two compounds (Table 1). Using male hamsters, Carr et al. (2005) demonstrated that sorghum phytosterol and policosanol are particularly important for lowering plasma and liver cholesterol levels, a major risk factor in coronary heart disease. The study showed diets containing lipid extracts from sorghum grain significantly reduced plasma non-HDL cholesterol concentrations in a dose-dependent manner, with reductions of 18%, 36% and 69% in hamsters fed on diets containing 0.5%, 1.0% and 5.0% grain sorghum lipids, respectively, compared to the control. Liver cholesterol ester concentration and cholesterol absorption efficiency were significantly reduced in the same way. A follow-up study conducted to determine the specific components of grain sorghum's crude lipid extract responsible for lowering cholesterol showed that the phytosterol containing oil fraction was the likely component, with the policosanol containing wax having a minimal impact (Lee et al., 2014). Phytosterols have a structural similarity to cholesterol and their robust cholesterol-lowering effect may be related to the competitive inhibition of cholesterol absorption (Carr et al., 2005; Lee et al., 2014). The suggested mechanism for the policosanol cholesterol-lowering effect is through the reduction of cholesterol synthesis in the liver (Carr et al., 2005), which may be vitally important for people who are genetically predisposed to high cholesterol. In

Table 1 Phytochemical content from whole grain samples of common cereals with elevated grain pigment

	Cereals			
Grain chemical concentration	Sorghum	Maize	Wheat	Rice
Carotenoid (mg/100 g)[1]	2.9–13.7	10.69–13.66	0.2–0.6	4
Phytosterol (mg/100 g)[2]	46–51	70–88	40–74	72
Policosanol (mg/100 g)[3]	74.5	0.2	Trace	1.9–3.7
Polyamines (mg/100 g)[4]	5.8–41.4	73.3	23.8	16.4
Total phenol (μg/g)[5]	1500–19000	2000–3000	1000–1500	500–10000
Phenolic acid (μg/g)[5]	385–746	601	1342	197–376
Anthocyanin (μg/g)[5]	944	93–965	13–153	2283
Resistance starch (%)[6]	45.6	32.3	26.8	14.8
Slowly digestible starch (%)[6]	17.6	20.0	14.4	51.4
Rapidly digestible starch(%)[6]	36.9	47.6	58.9	33.6
Total dietary fibre (g/100 g)[6]	4.6	19.6	17	1.5–5.7
Soluble fibre (g/100 g)[6]	0.4	3.6	2.3	0.3–1.4

Total phenol content is extrapolated (μg/g) from a figure in Dykes and Rooney (2007).
[1] Keana et al. (2011) and Mellado-Ortega et al. (2015); [2] Awika and Rooney (2004); [3] Hwang et al. (2004); [4] Ali et al. (2011) and Paiva et al. (2017); [5] Dykes and Rooney (2007); [6] Bednar et al. (2001).

addition to phytosterol and policosanol, the sorghum lipid profile is rich in the heart-healthy mono- and polyunsaturated fatty acids. All these findings are stimulating interest in developing sorghum as an alternative source of quality edible oil (Mahmood et al., 2008; Stefoska-Needham, et al., 2015).

Another class of sorghum phytochemicals with health benefits is resistant starch. Resistant starch is the portion of starch that is indigestible in the small intestine but is broken down by colon-resident bacteria. Like any other cereal crop, starch is the major component of sorghum grain, but sorghum has an unusually high fraction of resistant starch compared to other grains (Table 1). Resistant starch accounts for 45.6% of total starch in sorghum grain (Bednar et al., 2001). When resistant starch is broken down in the colon it provides many benefits, including improving gut health and maintaining intestinal pH levels, which helps to improve the bioavailability of minerals and prevent cancer (Champ, 2004; Shen et al., 2015).

2.3 Potential for improvement

As a C4 species, sorghum possesses efficient photosynthetic mechanisms that allow the accumulation of enormous biomass. Moreover, due to its adaptation to low inputs and drought stress, these vital processes tend to continue under conditions they normally halt in other crops. Sorghum also has great potential for further improvement, not only in increasing yields, but also in other added values, such as nutritional composition, agronomic adaptation and resilience to biotic and abiotic stresses, and attributes related to the utilization of the crop. Two important points attest to this potential: the strikingly large difference in yield figures in research fields compared to production fields, and the gigantic genetic resource of the crop, which is discussed elsewhere.

A brief glimpse at the current sorghum yield gap between research plots and actual production (~4 versus ~9 ton/ha) or between irrigated and rain-fed crop points to the extent to which sorghum yields can be improved. Irrigated sorghum yields from various areas, including from agriculturally challenging environments, indicate that the present global average yield of the crop can be doubled even with existing technologies, whereas sorghum yields in the largest sorghum-producing continent of Africa (0.9 ton/ha) are less than one-third of the global average. These differences in productivity show the massive potential for increasing global sorghum production. If the global community can come together to try to achieve what is most achievable, that is, closing the yield gap between the average yield of the crop in Africa and that of the global average, it is going to bring about a 2-ton/ha yield increase on 30 million hectares of land, which will increase global sorghum production by an average of 60 million tons. This may seem unrealistic, but past experiences and current data indicate that this is indeed possible (Ransom et al., 2007).

First, the yield gap between research-managed and farmer-managed plots in Africa shows a marked contrast. Yields from regional and national performance tests of experimental hybrids and varieties in Africa are always four to five times higher than yields harvested from farmers' fields. The primary reason for this is that over 80% of smallholder farmers grow ancient, low-yielding landraces without supplemental inputs. Although improved varieties and production recommendations have become available over the last few decades, their adoption remains very slow for various reasons. Some of the important causes include the high cost of inputs associated with the use of improved varieties, a lack of consistent supply of the technology and the technology not addressing the needs of the farmer. Since most varieties are developed under conditions of optimal nutrient supply, farmers are also required to use chemical fertilizers, which, in addition to seed cost, pushes up the cost of production. Most farmers cannot afford to pay for seeds and fertilizers at the same time, and hence resort to their traditional varieties. Even if farmers can afford these costs, they do not want to take the risk as the chances of crop failure are high in dryland sorghum production. Similarly, agronomic management practices such as the use of tied ridges have been shown to increase yields through harvesting and keeping rainwater near the root zone, but its adoption is again slow because the implement used to make tied ridges is often too heavy to manipulate. These and other problems are slowing the adoption of production recommendations. One needs to invest time and resources in identifying and investigating the key constraints along the value chain that can potentially harm productivity. Since circumstances in different countries or regions of a country can be different, so do the intervention mechanisims.

Second, as shown in Fig. 1, the sorghum yields in the pre-hybrid era in the United States were similar to the current sorghum yields in Africa. But science helped to change that course in the United States. Following the deployment of hybrid cultivars in the late 1950s, the sorghum yield, which was stagnant in preceding decades, began to show dramatic improvements to reach the current average of 4.5 ton/ha (Fig. 1). Although agriculture in Africa and the United States is based on a different set of operations, there is no reason why similar strides cannot be made if the right technological ingredients are deployed. Africa has an even better potential to achieve this for two major reasons: (1) the continent harbours the greatest variability of the crop for every trait, and this can be exploited to

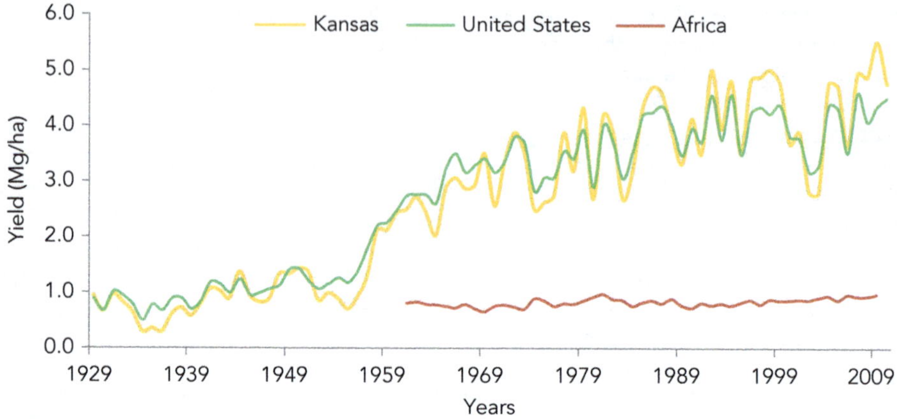

Figure 1 Pre- and post-hybrid era sorghum yields in the United States and the state of Kansas versus the current average yield in Africa.

develop unique varieties that are selectively adapted to a particular agro-ecology; and (2) unlike in temperate environments, sorghum production in Africa is less constrained by temperature and photoperiod so that all existing germplasm collections can be readily used in breeding programmes to develop new varieties or hybrids. Thus, investment to develop new varietal technologies and innovative crop management practices tailored to specific conditions and an increased emphasis on facilitating agricultural development, such as strengthening institutions for enhancing the production and delivery of improved seeds, fertilizers and various agrochemicals, can triple sorghum productivity in Africa and increase global sorghum production.

3 Major impediments to improved yield in sorghum

Despite its magnificent adaptation to different stress conditions and the wide genetic variability for almost every trait, the productivity of sorghum is considerably lower than its sister cereal, maize. In addition to the difference in the genetic architecture of the two crops and the fact that they are produced under a completely different set of environmental and management conditions, other factors continue to undermine the productivity of sorghum compared to that of maize.

3.1 Lack of technological breakthroughs

Sorghum is not particularly intractable to manipulation. The complete lack of or delay in deploying certain technologies in sorghum that have transformed the production of other crops is simply a reflection of the amount of financial and intellectual commitment to the improvement of the crop.

Although moving at a slow pace, research conducted over the last 50–60 years in the United States and elsewhere has generated fundamental knowledge about the biology of the crop. This knowledge was instrumental in the development of tools and techniques for enhancing the production and productivity of sorghum. The discovery of cytoplasmic male sterility, which facilitated hybrid seed production (Stephens et al., 1952), the change in plant architecture through the manipulation of dwarfing genes (Quinby and Karper, 1954) and the manoeuvring of the photoperiod response to alter the plant's adaptation to long-day conditions (Quinby, 1967) are among the early discoveries that expedited the adoption of sorghum as a commercial grain crop in the industrialized agriculture. The key changes made through these discoveries facilitated the adaptation of the crop to a narrow, long-day growing season in temperate environments and made it amenable to mechanized harvesting. These and numerous other discoveries, including the development of mutants with low levels of lignin and the identification of variants that are uniquely rich in antioxidant compounds, not only increased the value of sorghum, but also highlighted the adaptability of the crop to technological manipulation and its potential for improvement. However, the increasingly shrinking research funding from public institutions not only slowed down the pace of research and innovation that are needed to enhance the productivity of the crop, but also gradually led to the loss of a critical mass of sorghum scientists, as shown by the abandonment of many research programmes over the years. The impacts of these losses are being manifested through the pace at which the productivity of the crop is increasing.

Sorghum has not fully benefited from the discoveries that science has brought to other crops, primarily due to the limited investment in its improvement. In the mechanized agriculture of the developed world, weed control remains the major constraint limiting sorghum production (Al-Khatib et al., 2007; Norwood and Currie, 1997). Insect pests such as green bug (*Schizaphis graminum*) have been key constraints in the past (Thindwa and Teetes, 1994; Teetes and Pendelton, 2000). At present, a similar but more devastating pest, the sugar cane aphid (*Melanaphis sacchari* Zehntner), has appeared in the United States (Knusten et al., 2015). Other insects, such as stem borers (*Busseola fusca*, *Chilo partellus* and *Sesamia calamistis*) and sorghum midge (*Contarinia sorghicola*), are critical pests in Africa, the United States and Australia (Haile and Hofsvang, 2001; Tefera, 2004). Despite tremendous progress in understanding the biology of *Striga* and the genetics of host plant resistance (Ejeta, 2007), the pest remains the single most major biotic constraint to sorghum production in Africa. Similarly, abiotic stresses such as drought, high temperatures, nutrient deficiency and mineral toxicity continue to undermine sorghum production to a different degree, with drought being the most devastating problem at a global scale. Many of these constraints in other crops are either not very critical, or have been resolved through science. Both public and private investment in research and innovation are critical to exploiting the numerous potentials of the crop, from mitigating the effects of climate change addressing food security, to promoting health as a superfood crop.

3.2 Public perception

Contrary to common perceptions, numerous research findings show that sorghum is not inferior to other cereals in terms of its nutrient profile, and it is, in fact considered richer in certain minerals. Sorghum is also rich in health-promoting compounds that certain sorghum types have been shown to reduce the incidence of some of the most common types of cancer (Yang et al., 2009; Suganyadevi et al., 2013), reduce blood cholesterol (Hoi et al., 2009) and prevent blood sugar spikes in diabetic patients (Chung et al., 2011; Park et al., 2012). Nevertheless, the myth that sorghum is inferior to other cereals continues to negatively affect consumer preferences and its utilization in the food industry. As a result, the market price of sorghum grain in most countries is always less than that of corn. And yet, all the research indicates that sorghum has a similar biological value as other feed grains when it is used in cattle feedlot (Abdelhadi and Santini, 2006; Wood et al., 2011) or fed to swine and poultry (Douglas et al., 1990; Torres et al., 2013; Mohamed et al., 2015), provided that the right kinds of cultivars are used. Tannin-free sorghums have been shown to have a similar biological value to maize in both swine and poultry rations. Silage from selected low-lignin sorghum hybrids have super dry matter digestibility and are highly desired by the dairy industry (Rook et al., 1977; Oba and Allen, 1999). The substitution of ground sorghum for ground maize on an equal weight basis did not affect the weight gain of lambs, but reduced *Haemonchus contortus* infection and improved the meat colour by increasing the accumulation of antioxidants in the meat (Zhong et al., 2017). A similar study showed identical growth performance and ruminal fermentation in steers that were fed sorghum and maize-dried distiller grains (Abdelhadi and Santini, 2006).

Studies on the industrial application of sorghum show that it is well suited for various purposes. When used as a biofuel feedstock, sorghum grain produces as many gallons of ethanol as maize and fermentation efficiency can be improved with the use of the right genetics (Wu et al., 2007; Yan et al., 2011). Because of its sugar-rich stalks and ability to accumulate massive biomass, sorghum has great potential as a source of renewable energy feedstock (Shoemaker and Bransby, 2010; Ratnavathi et al., 2011). Likewise, research results have demonstrated the potential of sorghum as a replacement for barley in lager beer production with the light and yellow pericarp cultivars having the greatest potential (Palmer et al., 1989; Owuama, 1999). Sorghum beer is common elsewhere; the Eagle brand from the Nile brewery in east Africa, which is made from a white sorghum variety (Epuripur), has been offering consumers a unique clear beer for nearly two decades. Similarly, a sorghum-based beer and liquor revolution is unfolding in China, Nigeria and elsewhere (Ilori, 2002). Because sorghum is gluten-free grain, numerous local breweries have emerged in the United States offering gluten-free, sorghum beer, not only to consumers with celiac disease, but also to the larger community who prefer the distinct sorghum flavour. Thus, in addition to its numerous agronomic advantages, especially its magnificent tolerance to environmental stresses, sorghum offers numerous benefits to consumers both as a health-food ingredient and through its use in different kinds of food and non-food applications. Changes are gradually coming and consumers are becoming more open to including sorghum products in their menus. This is evident from the successes of local food companies, which are changing the market landscape.

In addition to increasing support for sorghum research and innovation of new cultivars, and for a fair and equitable agricultural development policy for all crops, significant effort is needed to educate communities to change their erroneous perceptions of sorghum and to communicate the many benefits of the crop (especially as climate change begins to leave its mark on agricultural production) and the health benefits of including sorghum in their diet.

3.3 The poverty factor

Sorghum research in Africa, which started as agronomy testing during the colonial period, has gradually expanded to cover broader research disciplines, including the development of different varieties. Efforts over several decades of research, including on-farm tests, were highly promising. Yet, sorghum and other crop yields in Africa remain unacceptably low. Multitudes of interrelated problems may be responsible for this and it is difficult to understand what is at the core of this complex. A lack of ready cash and the absence of flexible credit services to acquire modern inputs and the lack of a functional market system for farm outputs have limited the capacity of farmers to deploy yield enhancement inputs. The chronic shortage of public investment on research and innovation has also undermined the development of technological options, forcing farmers to stick to their traditional farming practices. Similarly, the absence of private/public enterprises devoted to serving agriculture, ineffective public intervention strategies, civil conflicts, can be cited among the causes. All these can be summed up as what is called the 'poverty factor'.

Poverty has a far-reaching impact on agricultural production in general and on sorghum production in particular, not only because smallholder sorghum farmers are the most cash-strapped communities, but also because they farm in a difficult terrain, on highly degraded soils, parched by high temperatures and drought. Existing technologies are input-intensive and sorghum varieties are rarely tailored to local circumstances. As a result, the majority of farmers grow local landraces using century-old local implements and family labour. These problems are not just limited to sorghum, but sorghum suffers the most due to the poor economic situation of the communities growing the crop.

While poverty may be to blame for all these problems, the greatest culprit is the reluctance of the national and international communities. The majority of international aid advanced to developing countries is not meant to address development issues, and those that are targeted at development are often not carefully monitored and accounted for. These practices not only have failed to curb poverty in the recipient countries, but have also propped up corruption, which has trickled down to grassroot levels, jeopardizing development efforts and installing a culture of perversion. Research institutions in the continent have been weakened due to either an insufficient allocation of resources for research and innovation or a massive exodus of trained manpower in search of better opportunities overseas. Private seed enterprises are reluctant to make large-scale investments due to the low purchasing potential of the smallholder farmers to acquire seeds and associated inputs, and due to unpredictable government policies, which can affect both the input and output market.

4 Deploying science to enhance sorghum productivity

Advances in science have changed the state of agricultural production in the world. Developments in agricultural mechanization, soil fertility management and crop protection technologies as well as the development of high-yielding and stress-resistant cultivars are among the major breakthroughs that transformed agricultural production in the twentieth century. The emerging discoveries in genetics and genomics, and the development of various tools, techniques and platforms for studying genotypes at the genomic level, holds enormous promise for further improvements in yield. Given the broad, untapped genetic diversity of sorghum, the use of these tools and techniques to mine novel high-yield and stress-resistant alleles and to modify plant architecture to maximize resource capture and conversion may lead to increased productivity in sorghum.

4.1 Mining novel alleles

The greatest opportunity for improving sorghum productivity and enhancing its utilization lies on the broad genetic resource base of the crop. Sorghum is one of the most diverse species, with magnificent phenotypic differences in plant height, maturity, panicle and leaf architecture, grain colour, size and shape (Grenier et al., 2000; Abu Assar et al., 2005; Perumal et al., 2007). It also displays remarkable variability for numerous agronomic characteristics, stress tolerance responses, utilization attributes and nutritional quality traits (Duncan, 1988; Sun et al., 2014; Sakhi et al., 2014). According to the USDA Germplasm Resource Information Network (GRIN) database, over 40000 sorghum germplasm accessions are currently maintained in the United States and a similar size is held at the International Crop Research Institute for Semi-Arid Tropics (ICRISAT) (Upadhyaya et al., 2016). Likewise, most countries collect and preserve germplasm materials either by their respective national commodity research programmes or through their national gene banks. The widespread exchange of germplasm between countries and among researchers in the 1970s and 1980s may have resulted in some duplicate samples stored across these repositories. Regardless, these are invaluable resources, without which improvement of productivity cannot be contemplated.

Advances in other fields of science over the past decades have had a positive spillover effect on other scientific disciplines, including agriculture. The numerous tools and technologies that were developed to understand and improve human health or build modern defence power have been successfully adopted to study and understand the biology of animals and crop plants. The emerging high-throughput phenotyping platforms, new genomic tools and techniques as well as statistical tools and computing devices are among the key discoveries adopted by the agricultural science community. These tools and platforms have dramatically increased the capacity of scientists not only to conduct phenotypic and genotypic characterization more efficiently, but also to discover the key genes and genomic regions affecting agriculturally important traits. Designing the next generation of sorghum cultivars, whether it is for the large-scale commercial agriculture of the west or for smallholder farmers in Africa, will require the use of these tools and techniques to exhaustively characterize the germplasm resources for the key plant traits associated with improved productivity, stress tolerance and utilization attributes. In other words, the next generation of plant breeding will need to mine novel alleles found in the natural population or generate new beneficial mutations for use in breeding programmes. The new cultivars should combine all novel alleles for key plant traits into the most suitable and productive background to make sure that the new cultivars are high yielders and tolerant to biotic and abiotic stresses and produce the right quality grain for specific end uses. Moreover, knowledge of the genetics of these traits is equally essential to predict the expression and inheritance of the traits and to determine the right breeding approach to improve them.

The systematic use of tropical germplasm sources in breeding programmes has been extremely useful not only for introducing major genes for stress tolerance, but also for extracting fresh alleles for increasing yield potential. The continued appearance of more productive commercial hybrids on the market and the improved productivity of experimental hybrid yields (see Table 2) indicate that sorghum is not nearing its yield

Table 2 Tests of cross-hybrid performance from one of the experimental sets from newly developed parental lines indicate opportunities for further enhancing yield potentials through deploying novel alleles

Hybrid entry	Days to flowering	Grain yield (ton/ha)*	Plant height
KSU2015-527 × 618	61	9.39	55.25
KSU2015-539 × 589*	65	8.61	56
KSU2015-531 × 611	60	8,30	59.5
KSU2015-505 ×	59	8.27	55.75
KSU2015-531 ×	62	8.25	54.5
KSU2015-531 × Tx430	60	8.18	55
KSU2015-557 × 582	61	8.08	57.25
KSU2015-553 × 653	60	8.0	56
KSU2015-555 × Tx430	62	8.0	52.75
KSU2015-501 × 592	61	7.92	57
Commercial check 1	64	7.79	50.5
KSU2015-507 × 589	–	7.68	58
KSU2015-503 × 608	62	7.61	55
KSU2015-507 × 588	58	7.60	56.5
KSU2015-527 × 651	63	7.59	58
KSU2015-501 × 685	64	7.57	64.5
KSU2015-505 × 687	63	7.56	54.5
KSU2015-549 × Tx430	65	7.54	55.5
KSU2015-505 × 587	62	7.52	52
Commercial check 2	58	7.37	55
KSU2015-555 × 617	61	7.25	56
KSU2015-503 × R-45	63	7.18	48.5
KSU2015-531 × 633	62	7.15	53

* Data based on 2016 replicated hybrid yield trail conducted at Manhattan KS on a 5-m long two-row plot.

plateau. Thus, the systematic exploration of new alleles, and their deployment in ways that can be most useful, will enhance global sorghum production and contribute to tackling food security concerns.

In addition to discovering and deploying novel alleles, the development of the next generation cultivars will also require a thorough understanding of the culture and socio-economic circumstances of the target communities and the physical resource base (the production environment) of the target regions. Societies are diverse in terms of language, culture, economic status, food habits and geography, as well as in their preference for technologies and their ability to adopt them. Thus, just as much as the biology of the target traits, knowledge of the socio-economic conditions of the target communities and the natural resource base is a prerequisite for developing cultivars that are tailored to specific conditions of production. The one-size-fits-all approach will never work, especially in geographies that are characterized by diverse topography, soil types and weather phenomena. Past efforts by international and national research organizations have not made much impact on the lives of poor communities, largely because the technologies were high cost. Most high-yielding cultivars released in Africa were not adopted as they require an optimal supply of inputs or supplemental irrigation which farmers do not have access to.

Thus, future efforts to increase sorghum productivity need to explore the power of modern science to identify and manipulate novel alleles in breeding programmes while at the same time paying attention to the circumstances of target beneficiaries and their environment, so that the research outputs are useable by the community. Only those cultivars that combine the right traits for target environment have the potential to bring about meaningful change in the lives of the people.

4.2 Tackling constraints undermining productivity

Like many other crops, sorghum suffers from a myriad of constraints that in addition to developing and deploying a genetically high-yielding hybrids/varieties, enhancing productivity requires that the major constraints undermining productivity are thoroughly understood. The most outstanding problems include drought stress, low nutrient supply, various types of disease and insect pests as well as infestation by the parasitic weed *Striga*. Past research has significantly reduced the negative impact of some of these constraints, except in sub-Saharan Africa, where almost all of them remain the major challenges. More efforts are needed to better understand the dynamics of these constraints in order to develop technologies that can help reduce their impact. The research information generated in the past may serve as the basis for venturing into a robust and comprehensive effort to reduce the impact of these stresses on sorghum production.

Although sorghum is relatively tolerant to drought, it suffers most from this stress because its production is largely concentrated in regions where the problem is rampant. This may be evident from the marked differences in yields obtained from genotypes subjected to the stress as well as through physiological indicators such as transpiration efficiency (Xin et al., 2009; Gholipoor et al., 2010), rate of fixation of carbon dioxide and non-senescence or stay-green (Rosenow et al., 1983). Some of these traits, notably the non-senescence, have been widely exploited both in public and private breeding programmes and the stay-green trait has been deployed in commercial hybrids, contributing to significant improvements in yield under drought stress (Borrell et al., 2000; Jordan et al., 2012). Genomic regions associated with the expression of the stay-green trait have been mapped (Sanchez et al., 2002; Haussmann et al., 2003) and the major QTLs validated through field evaluations (Harris et al., 2007). A decade ago, efforts were initiated by ICRISAT to incorporate the stay-green traits into local varieties from a handful of countries in Africa. While this is significant, this alone cannot solve the complex problem of drought. Efforts should continue to identify alternative drought tolerance mechanisms to incorporate into breeding programmes.

A recent effort to this end has generated some interest. Preliminary results from physiological studies have indicated the presence of a genetic difference in transpiration regulation as a likely response to drought stress (Xin et al., 2009). Certain tolerant genotypes appear to save water by slowing water loss through transpiration, regardless of the vapour pressure deficit, as opposed to susceptible genotypes that continue to transpire at a high rate until the water runs out (Gholipoor et al., 2010). Other tolerant types transpire at a high rate when moisture is available and then slow down transpiration as stress sets in. Although physiological screening was not carried out, tolerant sources that appear to express a similar mechanism have been observed in the field. These genotypes express little or no sign of wilting or chlorophyll loss under severe moisture stress, but growth is visibly arrested for the duration of the stress. Once the stress has subsided, normal physiological functions resume and growth occurs rapidly, allowing the plants to produce normal grains. This mechanism resembles what is known as 'slow wilting', a drought response mechanism that is widely exploited in soybeans (Sadok et al., 2012). In addition to such visible growth plasticity, the latter group of genotypes also has stable stay-green expression and characteristic narrow leaf angles, and erect and slender leaves with a low tillering tendency. This seems to offer a good opportunity for exploring sorghum germplasm resources to identify additional tolerant genotypes that express similar or different patterns of drought response.

Just like drought, sorghum also has a remarkable tolerance to low nutrient supply, and hence has superior nutrient use efficiency compared to its related cereal, maize (Lemaire et al., 1996; Klocke et al., 2014). In the United States, sorghum is produced with just 50–60% of the fertilizer nutrient normally supplied to maize. Numerous studies indicate that sorghum can be improved further for tolerance to low-nutrient soils. A preliminary study in Oklahoma indicated that sorghum cultivars of African origin tend to thrive better on low-nutrient soils compared to US hybrids bred under high-input conditions. African cultivars readily form a symbiotic association with arbuscular mycorrhizal (AM) fungi when grown in low-nutrient prairie soils, and produce more grain compared to US-adapted commercial hybrids which seem to be unfriendly to AM fungi. The AM fungal hyphae have been reported to improve nutrient and water uptake of most terrestrial plant species grown in low-nutrient soils (Smith and Read, 2008). The formation of a symbiotic association with mycorrhiza appears to be an important structural and biochemical adaptation of the roots, which is under genetic control. Cultivars have been shown to vary considerably in the extent to which they form symbiotic associations, as well as in the degree to which they benefit from such associations (Hetrick et al., 1995). Analysis of sorghum plant and grain samples harvested from this study indicated that an improved association of tropical cultivars with AM fungi resulted in a greater uptake of inorganic phosphorus (P) and formation of grain protein compared to commercial hybrids grown under similar conditions. Similar results have been reported in other agriculturally important species (Hetrick et al., 1993), leading to speculation that selection under an optimum input condition (high P input) may have led to the elimination of alleles or allelic functions responsible for AM fungal symbiosis.

4.3 Optimizing plant architecture

There is little information on the genetics of variability in both the above- and below-ground morphological traits of sorghum and their translation to enhancing crop performance. In corn, leaf angle and stalk strength have been effectively exploited to create hybrids tolerant to lodging and high plant populations (Pepper et al., 1977; Li et al., 2011). As a result, maize seeding rates over the last several decades have been increasing significantly and yields have been increasing proportionally. Although the modern hybrids may not necessarily yield more than the older hybrids when planted at low population levels, their enhanced root and stalk strength and the erect leaf structure have allowed closer plant spacing and standability without sacrificing the yield per plant, thus ensuring higher yields per unit area (Duvick, 2005). In addition to the flexibility it may afford in managing plant populations, erect leaf architecture has been reported to increase radiation use efficiency in several crops, including maize, rice and sorghum (Pepper et al., 1977). Tomato cultivars with a more spacious canopy architecture, due to long internodes and long and narrow leaves, led to an increase in crop photosynthesis by up to 10% through facilitating deeper light penetration (Sarlikioti et al., 2011). In rice, hybrids with an erect leaf had a lower leaf degradation rate and higher photosynthesis, dry matter production and yield under high light intensity than cultivars with flat leaves (Lang et al., 2004). As a result, the carbon assimilation rate on a whole-plant basis and biomass accumulation tends to be higher in cereals with narrow leaf angles compared to regular-type cultivars.

Differences in root morphology could have a similar effect on crop productivity. In addition to its structural integrity, the length, density and angle of the roots and the distribution of fine roots in the soil profile are critical for successful crop production (Comas et al., 2013; Hochholdinger et al., 2004; Ristova and Busch, 2014), while their role varies according to soil types and the availability of growth resources. Model-based simulation studies and validation exercises indicated that yield improvements in maize over successive decades were mainly associated with changes in root architecture, which facilitated water uptake and biomass accumulation, than with changes made to the leaf angles (Hammer et al., 2009). However, an erect leaf angle may have made an indirect contribution through increasing leaf area retention and the partitioning of carbohydrates to the ear. A recent study in China comparing root morphology and the physiology of maize hybrids with variable yield potentials indicated that the high-yielding hybrid, which produced 20% and 19% more biomass and grain yield, respectively, had a faster root growth rate. Root growth in the high-yielding hybrid reached 160–200 cm at the VT stage, while the other hybrid reached that depth at the R3 stage (Qui et al., 2012). In a study on the founder parents of maize nested association mapping population highlighted a robust variability for root traits (Manavalan et al., 2011). This was confirmed by other similar studies that reported at different times on diverse groups of germplasm (Li et al., 2015). A similar variation was reported in several cereal species, including wheat (Narayanan and Prasad, 2014; Manschadi et al., 2008) and barley (Robinson et al., 2016), as well as legumes such as lentil (Gahoonia et al., 2006), lupins (Chen et al., 2006), chickpea (Lalitha et al., 2015) and common bean (de Melo et al., 2016).

Sorghum displays immense above-ground morphological variation including leaf architecture, tillering property, plant height, and panicle and grain characteristics. Many of these traits have been widely exploited for improving the crop both as a grain crop and a forage crop. Root characteristics of sorghum have not been widely studied, but the existing limited information reveals that the crop harbours significant variation in root traits (Singh et al., 2011). The future high-yielding cultivars need to exploit this diversity and contain the most desirable combination of these traits to enhance productivity and utilization. Changes made to the above-ground plant characteristics in the past, such as the manipulation of plant height and maturity that facilitated the management and adaptation of the crop to western mechanized agriculture, may have negatively affected yield potentials. A study comparing 1-dwarf and 3-dwarf sorghum isogenic lines under different water management conditions revealed marked differences in yields, with the 1-dwarf cultivar having a markedly higher yield than the 3-dwarf line (Blum et al., 1997).

This was hypothesized to be due to a higher stem sugar reserve maintained in the taller cultivars, which was later remobilized to fill grains under moisture stress. Another study comparing a series of 2-dwarf and 3-dwarf lines, varying for the Dw3 locus, showed that the 2-dwarf lines produced more biomass and grain yields than their corresponding 3-dwarf lines (George-Jaeggli et al., 2011) and this was reported to be partly attributed to better light interception and radiation use efficiency in the taller genotypes (George-Jaeggli et al., 2013). This concurs with reports on other crops, where shorter internodes caused by dwarfing genes resulted in a close overlap of leaves, preventing light penetration to the lower leaves. The effect may be even larger if terminal moisture stress is imposed as better stem sugar reserves in taller cultivars can be remobilized to fill grains. In addition to its role in maintaining higher stalk sugar for remobilization, the taller cultivars may have other physiological benefits. Comparisons of the morphological features of the highest yielding (5%) and lowest yielding (5%) accessions among a total of 200 accessions evaluated at multiple locations in Ethiopia showed that the top-yielding accessions took a significantly longer time to reach flowering, were taller and had remarkably longer and wider leaves, a greater number of leaves and a larger leaf area than the lowest yielding accessions (Tesso et al., 2011).

Thus, it is imperative that future efforts to improve sorghum productivity take a serious look into both above-ground and below-ground plant architecture. Root depth has been cited as an important trait for improving crop performance under drought conditions. Given that sorghum is primarily a dryland crop, an increasing emphasis on root system research has the potential to provide insight for designing the next-generation sorghum cultivars. Moreover, the above-ground plant architecture also needs to be revisited. The development of 3-dwarf lines combined with tall hybrids has undoubtedly facilitated the mechanization of the crop, but it has also hurt yield potential. With the current state of mechanization technology, farmers can harvest taller hybrids, so it is no longer necessary to grow very short 3-dwarf hybrids. Based on the existing data, combining stalk strength, tallness and erect plant architecture with a high yield potential can significantly transform productivity without hindering mechanization and affecting standability.

5 Bridging the gap to improve yield potential

5.1 Breeding and genetics – unlocking heterosis

Both the science and the art of plant breeding saw dramatic advance during the twentieth century with numerous discoveries and innovations contributing to the current state of agricultural production. One outstanding advancement was the advent of hybrid technology. The superiority of F_1 hybrids in crops has been recognized since Mendel's time, but its commercial exploitation did not start until the mid-1920s when the first hybrid maize made it to the market (Crow, 1998). The discovery of cytoplasmic male sterility facilitated the commercial exploitation of heterosis in naturally self-pollinated crops through simplifying the commercial production of hybrid seeds. Like other major crops, the hybrid technology in sorghum was the major driver for revolutionizing sorghum yields throughout the world. As presented in Fig. 1, pre-hybrid era sorghum yields in the United States were only about 1 ton/ha. As soon as the hybrid seed market started, the average yield of the crop dramatically increased. Today, sorghum yield in all parts of the world where hybrid technology has been fully adopted is two to three times higher than in areas where the technology is not widely used. Although developments in crop and soil fertility management had a significant effect in increasing sorghum yields, this level of yield was never achieved until hybrids were deployed. In addition to enhancing yield potential, breeding also allowed the manipulation of several plant characteristics, including plant architecture and physiology, to enhance the adaptation of the crop to temperate climates and mechanized harvesting. Likewise, developments in breeding for tolerance to biotic and abiotic stress were critical for facilitating the management of pests and diseases. Nevertheless, despite significant strides over the last 60 years, sorghum productivity is still far from reaching its achievable potential. Breeding and genetics still have a tremendous role to play in continuing to push sorghum productivity up. Two fundamental research areas need to be addressed in order to further exploit the power of heterosis for increasing productivity in sorghum: (1) understanding the genetic causes of heterosis and developing genetic tools to predict it and (2) establishing sorghum heterotic pools.

Hybrid breeding is a tedious process. It involves the development and selection of parental lines and the test cross evaluation of experimental hybrids to determine their combining ability and the further testing of test crosses from the most elite parents. It is expensive and time-consuming. Any tool that can help to predict heterosis can simplify this process. The genetic distance between parental inbreds has been traditionally considered as a prerequisite for heterosis in several species. Several studies have been conducted to determine the association between molecular marker-based diversity with hybrid performance and heterosis in different crop species, including rice (Tao et al., 2010; Singh et al., 2011), sorghum (Jordan et al., 2003) and maize (Smith et al., 2000; Xu et al., 2004; Drinic et al., 2002). The results have been mixed. Significant and positive correlations between molecular marker-based genetic distance and heterosis have been reported for grain yield and yield components in rice (Cai and Lan, 2005; Zha et al., 2008), although an earlier study on the same crop reported otherwise (Zhang et al., 1994; Xiao et al., 1995). However, studies on other species did not find a correlation between heterosis and the genetic distance between parents (Bernardo, 1992; Munhoz et al., 2009; Liao et al., 1998; Zhang et al., 2007). Some of the possible reasons for the lack of a strong correlation may be due to the absence of a strong linkage between a large proportion of the markers used to study diversity and the QTLs controlling the trait of interest. In other words, a majority of the markers used to determine the genetic distance between the parents have no association with the trait under consideration, such that any two parents that are genetically distant from each other can carry the same or similar alleles at a particular locus or loci affecting the trait. Thus, any effort to develop a predictive model for heterosis should first determine the key loci associated with the trait of interest and the mode of action of the alleles at these loci.

In addition to mapping the major loci and haplotypes contributing to heterosis, it is critically important that the germplasm systematically sorting the germplasm according to the heterotic pools is critical for facilitating hybrid breeding. Unlike maize, sorghum has no defined heterotic pools from which to breed complementary seed and pollinator parents. The current designation of pollinator and seed parent pools is simply based on whether or not a genotype carries a fertility restorer gene (*Rf*) for a given type of male sterility system. Based on these, sorghum germplasm is sorted into a B (seed parent) and an R (pollinator parent) group. These pools, however, cannot be designated as heterotic, that is, a B × B or an R × R cross can be as vigorous as a B × R cross. In other words, genotypes can have a more or less similar overall genetic structure but vary in the fertility restoration loci to be wrongly perceived as heterotic, or they can be genetically distant from each other but carry similar alleles at the fertility restoration locus and again be wrongly regarded as belonging to the same pool. Thus, in order to exploit heterosis to its maximum, it is important to have a robust genetic screen of germplasm and their test cross-hybrid performance to generate a heterosis map to serve as a tool to sort the germplasm according to the complementary haplotypes they harbour. The biggest challenge here is that this cannot be done for all traits. In other words, two parents that are heterotic for yield may not necessarily be so for other traits. It is important that priority is given to the most critical traits, such as those related to productivity or yield formation.

5.2 Production and management

Sorghum production systems are often derived from those of maize in the United States and other countries where maize was historically more common. As a result, field operations for sorghum production have been optimized based on technologies developed primarily, or at least initially, for maize. Fertility recommendations are perhaps the most glaring example of using recommendations developed for maize and adjusting them for sorghum, rather than developing sorghum-specific recommendations (Leikam et al., 2003). This is particularly important for soil-mobile nutrients like nitrogen and sulphur, given the differences in the timing of planting and nutrient uptake between maize and sorghum (Abendroth et al., 2011; Roozeboom and Prasad, 2016). From the fact that only about half or less of applied nitrogen fertilizer is used by the crop in current production systems (Raun and Johnson, 1999), and given the potential for excess nitrogen to cause environmental problems (Feinerman et al., 1990), it is clear that there is a need not only to optimize nitrogen management for improved yields, but also to reduce the environmental problems caused by excess nitrogen in the soil. Other production practices adopted from maize cultivation include tillage operations, plant and row spacing and herbicide use. For both crops, the goal is to utilize the available environmental resources (precipitation, season length) to maximum effect by eliminating the yield limitations that are amenable to management. Developing production systems specifically intended to optimize sorghum productivity may facilitate a greater expression of the genetic potential of this crop.

The context of the cropping system is critical for the successful management of any crop, and is no less so for sorghum. Cropping systems can include a wide range of factors (Connor et al., 2011), but two fundamental components are crop sequencing (rotations) and tillage systems. These two components are so intertwined that their interaction is likely to be of greater importance than either one independently (Peterson and Roozeboom, 2007). High-residue, reduced tillage or no-till systems are becoming popular in many countries and can double precipitation storage efficiency (Nielsen et al., 2005), depending on the amount and orientation of residue, soil depth, texture and structure. This additional stored soil moisture is particularly critical considering that sorghum is usually targeted for water-limited environments and has the potential to produce an additional 184 kg/ha of grain yield for every additional centimetre of stored soil water (Stone and Schlegel, 2006). Longer intervals between the same crop or crop type (e.g. summer annual grass) have been shown to improve yields of that crop (Kelly, 2005), and this becomes even more important in high-residue cropping. A side effect of high-residue systems, particularly in temperate environments, is that the soil remains cooler, potentially slowing the emergence of early seedling growth (Unger, 1978). Mechanical solutions, such as strip tillage or in-row residue management, have been used extensively to address this issue. Planting sorghum after legume or broadleaf crops or cover crops (Kelley, 2005; Williams et al., 2012) and having access to hybrids that emerge and grow vigorously with relatively cool soil temperatures are necessary to fully realize the benefits of high-residue production systems.

A foundational element of optimizing the sorghum production system is hybrid selection. Yield differentials between commercial hybrids in a given production environment can be as great as two- or three-fold (Lingenfelser et al., 2015). Several traits must be considered when selecting a hybrid appropriate for a given environment and production system, but the most essential is hybrid maturity. A short-season hybrid may transition to reproductive growth too soon if season length and/or water supply is not limiting, reducing the source capacity and eventually the yield (Roozeboom and Prasad, 2016). A full-season hybrid may use the available water and/or growing season to grow a larger canopy with tremendous source capacity, but may have a reduced sink capacity due to a reduced seed set or grain fill duration. The challenge is predicting water availability and season length in the variable environments where sorghum is typically grown. As site-specific sorghum management becomes more broadly adopted, developing algorithms to optimize management factors for specific zones will become increasingly important (Robert, 2002). A major knowledge gap for optimizing sorghum production is understanding the complex interactions of environment, genetics and management (Hatfield and Walthall, 2015).

Planting management, such as the planting date, density and planting geometry (row spacing and arrangement within rows), can be optimized to maximize resource use efficiency, particularly water, light and temperature. The planting date should balance the minimum temperature requirements with hybrid maturity, season length and water supply. Temperatures during vegetative growth can influence the number of fertile tillers (Lafarge et al., 2002), requiring adjustments to the seeding rate, depending on the planting date and the genetic potential for tillering. The typical row spacing of 76 cm in the United States may limit yields in environments with greater production potential (Fig. 2a), possibly due to light limitation during early vegetative growth (Fig. 2b). Research exploring alternative row spacing (skip rows) and plant arrangement within the row (clump or section) shows promise for managing plant carbohydrate partitioning to increase the harvest index and conserve stored soil moisture for later stages of grain filling in environments where water is the most limiting factor (Akwasi et al., 2011; Prasad et al., 2006). A heavy residue layer to minimize evaporation must be paired with aggressive weed control to minimize soil profile water reductions for these alternative planting geometry systems to be successful. Most of this research has been done with a limited number of genotypes, revealing an opportunity to further explore the complex interactions of genetics with planting management.

Sorghum is particularly sensitive to early season weed competition because of the relative lack of size and vigour of sorghum seedlings relative to other crop species (Knezevic et al., 1997). Herbicide options for sorghum are another example of 'trickle down' from maize or other crop research (Thompson et al., 2017). Few, if any, post-emergence herbicide options are available for grass weed control, highlighting the importance of adequate weed control during the seedling stages coupled with developing and maintaining a healthy leaf canopy to minimize in-season weed emergence and growth (Burnside, 1977). To date, no transgenic herbicide tolerance traits have been released in sorghum, creating a potential opportunity for future research, but also maintaining a unique, non-transgenic market niche for sorghum grain products.

Although stay-green hybrids have shown greater tolerance to post-anthesis drought stress, the moisture-limited environments where sorghum is often grown can lead to stress-induced stalk rots that make the plants susceptible to yield-reducing lodging (Jardine,

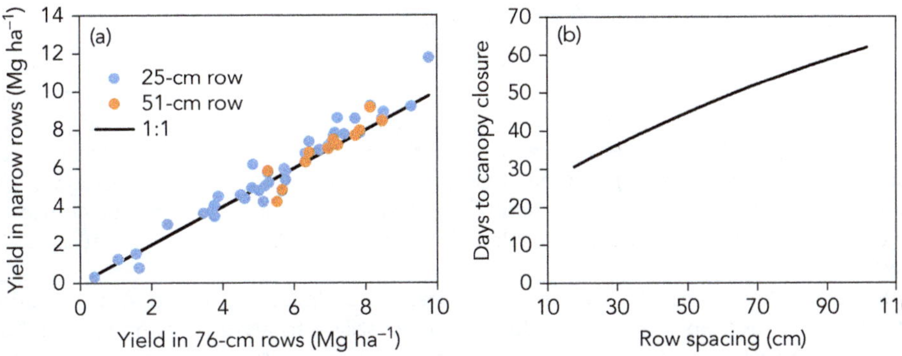

Figure 2 Yield response to row spacing in 53 experiments comparing 76-cm row spacing to narrower row spacing (a), and the effect of row spacing on time to closure of the crop canopy in sorghum (b).

2006). The management of cropping systems that limits post-anthesis moisture stress, coupled with improved genetics, can be further improved to minimize these potentially devastating yield losses.

Given the number of cropping systems and in-season production factors that affect sorghum yields in any one field or in any one management zone within a field, the potential for the further optimization of the interactions of genetics, the environment and management is huge. Future sorghum genetic improvements and crop management efforts must account for these complex interactions to fully realize the productive potential of this crop.

Crop management in smallholder production situations in the semi-arid regions is primarily guided by the prevailing weather conditions, especially drought, and by traditional practices handed down through the generations. This is evident in countries like Ethiopia where the crop is grown by culturally and ethnically diverse communities in various agro-ecologies. In southern Ethiopia, where drought is widespread, local farmers have adopted a mitigation mechanism where the crop is clump planted at the centre of a ridge constructed to capture and retain rainwater. Mulching using crop residues and grass straws is common to prevent water loss through evaporation. Likewise, in eastern parts of the country, terracing and intercropping sorghum with other crops is common, with the primary aim not only of maximizing productivity, but also of minimizing water and nutrient loss through runoff and for the efficient use of the land. These practices have found a scientific basis and research is ongoing to refine these practices and simplify their use.

One important drought mitigation practice that has been widely studied and has also shown to have a significant effect is planting sorghum in furrows tied at intervals (tied ridge) to harvest rainwater. The practice has almost doubled grain yields compared to the normal planting methods. In addition to increasing yields through improving moisture availability, the practice has also been shown to reduce *Striga* pressure in infested fields (Tesso and Ejeta, 2011), either by enhancing crop growth or affecting the germination and establishment of *Striga*. However, the technology is not being picked up as a desirable practice due to various socio-economic constraints. In addition to the cost, the weight of the implement has limited its use by farmers. It is important that further research aimed at developing innovative crop, soil and moisture management practices takes into account the farmers' situations to make sure the technology is not only effective but can also be readily used by the target communities.

5.3 Physiology and plant adaptation

Similar to other field crops, sorghum has been assumed to be highly tolerant to heat and drought stress during the vegetative stage. However, all crops have physiological limits, and with sorghum being pushed to harsher environments as a result of changing climate, an opportunity could be missed if we overlook the need to enhance the resilience of the source (photosynthetic tissue) for maintaining the efficient production and storage of pre-flowering assimilates. These stored assimilates have been shown to be crucial in other cereals, such as wheat and rice, particularly in stressed conditions (McIntyre et al., 2012; Slewinski, 2012).

Sorghum is known for its adaptability to heat and drought stress during flowering, but the actual mechanisms which induce its adaptability to adverse climatic conditions has not been systematically investigated. A recent review of the response of field crops to heat stress indicated the role of different response mechanisms, such as escape, avoidance and true tolerance, particularly during the reproductive stages (Prasad et al., 2016). These different adaptation mechanisms need to be systematically investigated to develop a strategic breeding programme to enhance resilience to heat stress in sorghum. Temperatures of about 30°C during reproductive development have been reported as optimal for sorghum, while 38°C has been found to be damaging (Nguyen et al., 2013). Although the male reproductive organ has been identified to be highly sensitive to heat stress (Djanaguiraman et al., 2014), the female reproductive organs can also be affected by heat stress, leading to a dysfunctional stigma, desiccated style and damaged ovary (Prasad et al., 2016). This indicates that heat stress and/or drought stress can simultaneously impact the viability of both male and female reproductive organs and this adds complexity to stress-tolerance breeding efforts.

Sorghum heads generally appear to fill normally under mild stress conditions, although the actual gradient in grain weight along the head may be affected, so there may also be potential to increase sink size. Also, there may be sterile florets that in some genotypes result in smaller seed numbers regardless of panicle size. So there are two options to increase the panicle sink: (1) increasing the number of reproductive units on a head by exploring the existing diversity for this trait, and (2) exploring options to introduce panicle architectural modifications to accommodate additional grains, such as the multi-seed approach (ref). While past efforts have not succeeded due to trade-offs between the number of grains and grain weight, some emerging tools may be useful to minimize this trade-off. Moreover, as indicated above, hybrids that are slightly taller may provide additional stem storage capacity for more assimilates, which can later be used to fill the additional reproductive units, and this minimizes the trade-off. This approach can therefore increase the yield potential, especially under harsher heat and drought stress conditions where current assimilates may be low due to the effect of these stresses.

The importance of the availability of higher water-soluble carbohydrates in the stem is well documented in studies on wheat (Beed et al., 2007), with the increase in height allowing the additional penetration of light (Miralles and Slafer, 1997) to the lower leaves to enhance radiation use efficiency. The stay-green trait is a well-established phenomenon in sorghum and is another key aspect that allows sorghum grains (highest in cereals – about 3–4 K) on a single head to be filled. Thus, a programmed senescence to balance the stay-greenness at the top leaves with a senescence in the lower leaves could provide additional assimilate transport efficiency (Jagadish et al., 2016) that can support grain development from additional florets, which genetic modification of the panicle architecture may bring.

In addition to stay-green, efforts to enhance drought tolerance related to 'stress recovery efficiency' are well underway. Although sorghum as a species is believed to have an improved recovery rate, it may vary between genotypes according to their adaptation locale. Since the sorghum plant has a robust, sturdy stem, it can be hypothesized to support additional grains, but this warrants investigation, particularly under stress-prone conditions which may weaken the stem. Thus, exploiting these traits and combining them into one background trait may provide positive synergy and enhance stress adaptation.

6 Conclusion and future trends

6.1 Genetic resources

In the preceding sections we highlighted the plant characteristics that can be exploited to maximize productivity, including root and shoot architecture, water and nutrient uptake, photosynthesis, assimilation and translocation, light interception and radiation use efficiency. Combining these inherent plant characteristics with traits related to improved productivity through mining and deploying high-yield alleles, unlocking key physiological mechanisms related to yield formation and optimizing source–sink balance may be used to tailor the next generation cultivars to improve productivity in a given production scenario. One important resource that the sorghum community has at hand to achieve this is the large germplasm base that nature has endowed in the crop. All of the key traits named above and several others are harboured in this vast germplasm resource.

In addition to the roughly 40 000 accessions maintained at the USDA gene bank and the similar number held at ICRISAT, almost all of the national programmes or gene banks of major sorghum-producing countries maintain substantial numbers of accessions from local collections in their respective countries. These resources can offer even more novel traits in addition to those associated with production and productivity, including unique physical and chemical grain traits related to nutrition and utilization attributes, processing qualities and so forth. It is important that the global sorghum research community works together to designate pathways for efficiently exploiting these resources for the public good. The key task ahead should be to focus on deploying all relevant fields of science, genetics, genomics, physiology and soil science to thoroughly understand the value of the different plant characteristics for yield improvement and to determine the interaction between them to help design plant ideotypes that maximize resource utilization and translate them to productivity. This is a monumental task that takes a great deal of time and resources to implement. However, it is essential that the public sorghum research community finds a common purpose and agree on the priority researchable areas that are most relevant to enhancing productivity and adapting to the climatological and ecological phenomena witnessed in the major sorghum-growing areas of the world. Enhancing the record-keeping of this information, such as through publication or creating user-friendly public databases, to inform present and future scientists, is critical.

Although sorghum has relatively rich genetic resources, the accessions so far captured by the global gene banks represent only a small fraction of the *in situ* sorghum genetic variability, and efforts are still needed to collect more germplasm variants, especially from areas that were inaccessible to early explorers. Therefore, it is very important that new collection initiatives are undertaken. This requires local administrations, farmers and germplasm curators to educate one another on the value of maintaining these important resources so that they can capture them before they disappear permanently. Africa, where the majority of the sorghum germplasm originates, has been experiencing tremendous natural and man-made disasters and instabilities, many of which have been violent and have caused the massive displacement of rural communities. The result has been genetic erosion. It is important that whatever is remaining *in situ* be collected and preserved before the next catastrophes. Even in areas where earlier collections have been made, or where natural and man-made calamities are not so widespread, conducting additional rounds of collection may be important in order to capture fresh germplasm that may have been introduced to the area either through natural evolution or the adoption of new cultivars.

Finally, the research community and the agriculture industry in general, while utilizing the germplasm resources, should take a moment not only to recognize the contribution of smallholder farming households and communities who have maintained these resources and accord them appropriate rewards based on the value of the materials they may offer.

6.2 Emerging tools and technologies for accelerating innovation

Modern plant breeding and genetics research have been at work on different species for over a century. Fundamental breakthroughs have been made in research on almost all crops, and this is evident from the productivity improvements that have been made over the last 100 years. In sorghum, the discovery of the cytoplasmic male sterility system to facilitate the commercial production of hybrid seeds, of maturity genes for manipulating photoperiod sensitivity, and of the alteration of plant height to develop combined height hybrids are among the key genetic discoveries that have facilitated the production of the crop. The discovery of the molecular marker systems and sequencing technology over the last few decades and the continued refinement of these tools bring enormous promise to further enhance crop productivity. The completion of the whole genome sequence of a reference sorghum genotype in the last few years (Paterson et al., 2009) was a new genetic breakthrough and a significant milestone in sorghum research. It presents new opportunities to further improve the crop through allowing the efficient characterization of genotypes and may facilitate the improvement of complex and economically important traits, such as drought tolerance, yield potential and other quantitative traits. With the sequencing technology increasingly improving and the costs continuously coming down, multiple reference genotypes can be re-sequenced to provide a more complete and accurate genomic catalogue for an array of genotypes representing the diverse phenotypes of interest. Such developments have significant implications in our understanding of the biology and evolution of the species as well as for marker-trait association studies which facilitate genetic improvements. However, despite the advances in sequencing technology, exploiting and filtering the massive sequence data and applying them in plant breeding programmes remain a challenge. Effective and user-friendly bio-informatics resources and tools are needed to exploit the genomic information being generated so it can be efficiently utilized in breeding programmes. Moreover, for such developments to be successfully applied in solving practical agricultural problems, future cloning and sequencing efforts should focus on the genetic events associated with the traits of economic value, such as those responsible for improved agronomic adaptation, yield potential and enhanced utilization attributes.

Today, there are more technologies, innovations, knowledge and research tools than there were 50 years ago. Advancements made over the last two decades have enhanced the scientific knowledge on sorghum and other crops and new tools have become available that can be used to enhance the capability of researchers to better target critical problems. Molecular tools that can aid breeding for complex traits, such as yield, drought tolerance and insect and disease resistance, have become available (Xu et al. 2000; Sanchez et al., 2002; Satish et al., 2009). Biotechnological capabilities for breeding *Striga* resistance have significantly increased in past decades (Ejeta, 2007). The complete genome sequence of sorghum has become available and can expedite the development of unique traits and also help take advantage of developments in other species (Paterson et al., 2009). Protocols for the genetic transformation of sorghum have improved over the last decade (Wang et al., 2007; Gurel et al., 2009). Knowledge and technologies on the improved utilization of sorghum as food, feed and industrial raw material have been accumulating (Dicko et al., 2006; Rooney et al., 2007). These new tools and innovations have the power to change the existing production scenario in developing countries and elsewhere if they are intelligently integrated and packaged into mature technologies for deployment. Such information and tools can also be exploited to design new research approaches to develop new technologies to tackle the specific needs of communities that are dependent on the crop. This, however, requires consistent and stable support to build the research and educational institutions of the developing countries from the bottom up, and thereby to build local capacity to address domestic problems. Poor nations obviously do not have the resources to build these foundations, but a modest support from the global community can bring this to reality.

6.3 Future research needs

Future studies to improve sorghum production and productivity for improving food or feed grain quality need to exploit all community resources that are available to look in to all aspects of production. The growing literature resources on sorghum germplasm, new genetic and genomic tools and statistical methods may be utilized to address key constraints undermining the productivity of the crop. Deploying such tools and information for unraveling the physiology of yield, genetics of heterosis, below and above-ground plant architecture and its impact on resource use efficiency are critical for enhancing yield potential. New production practices and soil and water management techniques are key elements that should go in parallel with genetic improvements.

7 References

Abdelhadi, L. O. and F. J. Santini. 2006. Corn silage versus grain sorghum silage as a supplement to growing steers grazing high quality pastures: Effects on performance and ruminal fermentation. *Animal Feed Sci. Tech.* 127: 33–43.

Abendroth, L. J., R. W. Elmore, M. J. Boyer and S. K. Marlay. 2011. Corn growth and development. IA St. Univ. Ext. PMR1009. Iowa State University, Ames, IA.

Abu Assar, A. H., R. Uptmoor, A. A. Abdelmula, M. Salih, F. Ordon and W. Friedt. 2005. Genetic variation in sorghum germplasm from Sudan, ICRISAT, and USA Assessed by Simple Sequence Repeats (SSRs). *Crop Sci.* 45(4): 1636–44.

Akwasi, A. A., R. B. Ferguson, C. S. Wortmann, D. J. Lyon, S. C. Mason, S. Irmak and R. N. Klein. 2011. Grain sorghum water use with skip-row configuration in the Central Great Plains of the USA. *Afr. J. Agr. Res.* 6: 5328–38.

Ali, M. A., E. Poortvliet, R. Stromberg and A. Yngve. 2011. Polyamines in foods: Development of a food database. *Food Nutr. Res.* 55, 5572–87.

Al-Khatib, K., K. S. Kershner and M. Tuinstra. 2007. Resistance to acetolactate synthase- inhibiting herbicides in grain sorghum. *North Central Weed Sci. Soc. Proc.* 62:139.

Awika, J. M. 2011. Sorghum flavonoids: Unusual compounds with promising implications for health. In: J. Awika (Ed.), *Advances in Cereal Science: Implications to Food Processing and Health Promotion* (pp. 171–200). American Chemical Society, Washington, DC.

Awika, J. M. and L. W. Rooney. 2004. Sorghum phytochemicals and their potential impact on human health. *Phytochemistry* 65: 1199–221

Awika, J. M., L. W. Rooney and R. D. Waniska 2004a. Properties of 3-deoxyanthocyanins from sorghum. *J. Agric. Food Chem.* 52(14):4388–94.

Awika, J. M., L. W. Rooney and R. D. Waniska. 2004b. Anthocyanins from black sorghum and their antioxidant properties. *Food Chem.* 90: 293–301.

Awika, J. M., C. M. McDonough and L. W. Rooney. 2005. Decorticating sorghum to concentrate healthy phytochemicals. *J. Agric. Food Chem.* 53: 6230–4.

Awika, J. M., L. Yang, J. D. Browning and A. Faraj. 2009. Comparative antioxidant, antiproliferative and phase II enzyme inducing potential of sorghum (Sorghum bicolor) varieties. *Food Sci. Tech.* 42: 1041–6.

Batello, C., M. Marzot and A. H. Toure. 2004. The Future is an ancient lake- traditional knowledge, biodiversity and genetic resources for food and agriculture in lake Chad basin ecosystems. FAO Interdepartmental Working Group on Biological Diversity, FAO, Rome.

Bednar, G. E., A. R. Patil, S. M. Murray, C. M. Grieshop, N. R. Merchen and G. C. Fahey. 2001. Starch and fiber fractions in selected food and feed ingredients affect their small intestinal digestibility and fermentability and their large bowel fermentability in vitro in a canine model. *J. Nutr.* 131(2): 276–86.

Beed, F. D., N. D. Paveley and R. Sylvester-Bradley. 2007. Predictability of wheat growth and yield in light-limited conditions. *J. Agric. Sc., Cambridge* 145: 63–79.

Belay, H., H. Gebrekidan and Y. Uloro. 1998. Effect of tied ridges on yield response of Maize (*Zea mays* L.) to application of crop residue and residual N and P on two soil types at Alemaya, Ethiopia. *S. Afr. J. Plant Soil* 15(4): 123–9.

Bernardo, R. 1992. Relationship between single-cross performance and molecular marker heterozygosity. *Theor. Appl. Genet.* 83: 628–34

Blum, A., G. Golan, J. Mayer and B. Sinmena. 1997. The effect of dwarfing genes on sorghum grain filling from remobilized stem reserves, under stress author links open the overlay panel. *Field Crops Res.* 52:43–54

Borrell, A. K., G. L. Hammer and R. G. Henzell. 2000. Does maintaining green leaf area in sorghum improve yield under drought? II. Dry matter production and yield. *Crop Sci.* 40: 1037–48.

Brevik, E. C. 2013. The potential impact of climate change on soil properties and processes and corresponding influence on food security. *Agriculture* 3: 398–417.

Burnside, O. C. 1977. Control of weeds in non-cultivated, narrow-row sorghum. *Agron J.* 69: 851–4.

Cai, J. and W. Lan. 2005. Using of AFLP marker to predict the hybrid yield and yield heterosis in rice. *Chin. Agric. Sci. Bull.* 21(4): 39–43.

Cardoso, L. de M., S. S. Pinheiro, H. S. D. Martino and H. M. Pinheiro-Sant'Ana. 2016. Sorghum (*Sorghum bicolor* L.): Nutrients, Bioactive Compounds, and Potential Impact on Human Health. *Critical Reviews in Food Science and Nutrition*, 57:372–90.

Carr, T. P., C. L. Weller, V. L. Schlegel, S. L. Cuppett, D. M. Guderian and K. R. Johnson. 2005. Grain sorghum lipid extract reduces cholesterol absorption and plasma non-HDL cholesterol concentration in hamsters. *J. Nutr.* 135, 2236–40.

Champ, M. M. 2004. Physiological aspects of resistant starch and in vivo measurements. *J. AOAC Int.* 87(3): 749–55.

Chen, Y., F. Shan, M. N. Nelson, K. H. M. Siddique and Z. Rengel. 2006. Root trait diversity, molecular marker diversity, and trait-marker associations in a core collection of *Lupinus angustifolius*. *J. Exp. Bot.* 67: 3683–97.

Chung, I. M., E. H. Kim, M. A. Yeo, S. J. Kim, M. C. Seo and H. I. Moon. 2011. Antidiabetic effects of three Korean sorghum phenolic extracts in normal and streptozotocin-induced diabetic rats. *Food Res. Int.* 44: 127–32.

Cobb, A. B., G. W. T. Wilson, C. L. Goad, S. R. Bean, T. T. Tesso and J. D. Wilson. 2017. Assessing the influence of farm fertility amendments, field management, and sorghum genotypes on soil microbial communities and grain quality. *Appl. Soil Ecol.* 119: 367–74.

Comas, L. H., S. R. Becker, V. M. V. Cruz, P. F. Byrne and D. A. Dierig. 2013. Root traits contributing to plant productivity under drought. *Front Plant Sci.* 4: 442.

Connor, D. J., R. S. Loomis and K. G. Cassman. 2011. Agricultural systems. In: D. J. Connor, R. S. Loomis and K. G. Cassman (Eds), *Crop Ecology: Productivity and Management in Agricultural Systems*. Cambridge University Press, New York.

Crow, F. 1998. 90 years ago: The beginning of hybrid maize. *Genetics* 148(3): 923–8.

de Melo, R. C., R. S. Paulo Henrique Cerutti, A. F. Guidolin and J. L. M. Coimbra. 2016. Genetic variation in the trait root distribution over segregating generations of common bean *Euphytica* 207(3): 665–74.

Dia, V. P., P. Pangloli, L. Jones, A. McClure and A. Patel. 2016. Phytochemical concentrations and biological activities of Sorghum bicolor alcoholic extracts *Food Funct.* 7: 3410.

Dicko, M. H., H. Gruppen, A. S. Traoré, A. G. J. Voragen and W. J. H. van Berkel. 2006. Sorghum grain as human food in Africa: Relevance of content of starch and amylase activities African *J. Biotechnol.* 5(5): 384–95.

Djanaguiraman, M., P. V. V. Prasad, M. Murugan and U. K. Reddy. 2014. Physiological differences Among sorghum (*Sorghum bicolor* L. Moench) genotypes under high temperature stress. *Environ. Exp. Bot.* 100: 43–54.

Douglas, J. H., T. W. Sullivan, P. L. Bond and F. J. Struwe. 1990. Nutrient composition and metabolizable energy values of selected grain sorghum varieties and yellow corn. *Poultry Sci.* 69(7): 1147–55.

Drinic, S. M., S. Trifunovic and G. Drinic. 2002. Genetic divergence and its correlation to heterosis in maize as revealed by SSR-based makers. *Maydica* 47: 1–8.

Duncan, R. R. 1988. Sequential development of acid soil tolerant sorghum genotypes under field stress conditions. *Commun. Soil Sci. Plant Anal.* 19: 1295–305.

Duvick, D. N. 2005. The contribution of breeding to yield advances in maize (*Zea mays* L.). *Adv. Agron.* 86: 83–145.

Dykes, L. and L. W. Rooney. 2007. Phenolic compounds in cereal grains and their health benefits. *Cereal Foods World.* 52(3): 105–11.

Ejeta, G. 2007. Breeding for *striga* resistance in sorghum: Exploitation of an intricate host–parasite biology. *Crop Sci.* 47: S-216–27.

Elkonin, L. A., J. V. Italianskaya, I. V. Domanina, N. Yu. Selivanov, A. L. Rakitin and N. V. Ravin. 2016. Transgenic sorghum with improved digestibility of storage proteins obtained by Agrobacterium-mediated transformation *Russ. J. Plant Physiol.* 63(5): 678–89.

Escalada, R. G. and D. L. Plucknett. 1975. Ratoon cropping of sorghum: I. Origin, time of appearance, and fate of tillers. *Agron J.* 67: 243–478.

FAO, 1995. *Sorghum and Millets in Human Nutrition*. Food and nutrition series, No. 27. FAO, Rome. ISBN: 92-103381-1. http://www.fao.org/docrep/T0818E/T0818E00.htm.

FAO, 2015. *Regional overview of food insecurity: African food security prospects brighter than ever.* Accra.

Farré, I. and J. M. Faci. 2006. Comparative response of maize (*Zea mays* L.) and sorghum (*Sorghum bicolor* L. Moench) to deficit irrigation in a Mediterranean environment. *Agric. Water Manag.* 83: 135–43.

Feinerman, E., E. K. Choi and S. R. Johnson. 1990. Uncertainty and split nitrogen application in crop production. *Am. J. Agr. Econ.* 72: 975–84.

Frey, S. D., E. T. Elliott and K. Paustian. 1999. Bacterial and fungal abundance and biomass in conventional and no-tillage agroecosystems along two climatic gradients. *Soil Biol. Biochem.* 31: 573–85.

Gahoonia, T. S., O. Ali, A. Sarker, N. E. Nielsen and M. M. Rahman. 2006. Genetic variation in root traits and nutrient acquisition of lentil genotypes *J. Plant Nutr.* 29: 643–55.

Gebrekidan, H. 2003. Grain yield response of sorghum (*Sorghum bicolor*) to tied ridges and planting methods on entisols and vertisols of Alemaya Area, Eastern Ethiopian highlands. *J. Agric. Rural Dev. Trop. Subtrop.* 104(20): 113–28.

George-Jaeggli, B., D. R. Jordan, E. J. van Oosterom and G. L. Hammer. 2011. Decrease in sorghum grain yield due to the dw3 dwarfing gene is caused by reduction in shoot biomass. *Field Crops Res.* 124: 231–9.

George-Jaeggli, B., D. R. Jordan, E. J. Van Oosterom, I. J. Broad and G. L. Hammer. 2013. Sorghum dwarfing genes can affect radiation capture and radiation use efficiency. *Field Crops Res.* 149, 283–90.

Gholipoor, M. P., V. V. Prasad, R. N. Mutava and T. R. Sinclair. 2010. Genetic variability of transpiration response to vapor pressure deficit among sorghum genotypes. *Field Crops Res.* 19: 85–90

Grenier, C., M. Deu, S. Kresovich, P. J. Bramel-Cox and P. Hamon 2000. Assessment of genetic diversity in three subsets constituted from the ICRISAT sorghum collection using random vs non-random sampling procedures B. Using molecular markers. *Theor. Appl. Genet.* 101: 197–202.

Gurel, S., E. Gurel, R. Kaur, J. Wong, L. Meng, H. Q. Tan and P. G. Lemaux. 2009. Efficient, reproducible Agrobacterium-mediated transformation of sorghum using heat treatment of immature embryos. *Plant Cell Rep.* 28(3): 429–44. doi: 10.1007/s00299-008-0655-1.

Hahn, D. H. and L. W. Rooney. 1986. Effects of genotype on tannins and phenols of sorghum. *Cereal Chem.* 63: 4–8.

Haile, A. and T. Hofsvang. 2001. Survey of lepidopterous stem borer pests of sorghum, maize and pearl millet in Eritrea. *Crop Prot.* 20(2): 151–7

Hammer, G. L., Z. Dong, G. McLean, A. Doherty, C. Messina, J. Schussler, C. Zinselmeier, S. Paszkiewicz and M. Cooper. 2009. Can changes in canopy and/or root system architecture explain historical maize yield trends in the U.S. corn belt? *Crop Sci.* 49(1): 299–312.

Harris, K., P. K. Subudhi, A. Borrell, D. Jordan, D. Rosenow, H. Nguyen, P. Klein, R. Klein and J. Mullet. 2007. Sorghum stay-green QTL individually reduce post-flowering drought-induced leaf senescence. *J. Exp. Bot.* 58(2): 327–38.

Hatfield, J. L. and C. L. Walthall. 2015. Meeting global food needs: Realizing the potential via genetics × environment × management interactions. *Agron. J.* 107: 1215–26.

Haussmann, B. I. G., V. Mahalakshmi, B. V. S. Reddy, N. Seetharama, C. T. Hash and H. H. Geiger. 2003. QTL mapping of stay-green in two sorghum recombinant inbred populations. *Theor. Appl. Genet.* 106: 133–42.

Hetrick, B. A. D., G. W. T. Wilson and T. S. Cox.1993. Mycorrhizal dependence of modern what cultivars and ancestors: A synthesis. *Can. J. Bot.* 71: 512–18.

Hetrick, B. A. D., G. W. T. Wilson, B. S. Gill and T. S. Cox. 1995. Chromosome location of mycorrhizal responsive genes in wheat. *Can. J. Bot.* 73: 891–7.

Hochholdinger, F., W. J. Park, M. Sauer and K. Woll. 2004. From weeds to crops: Genetic analysis of root development in cereals. *Trends Plant Sci.* 9: 42–8.

Hoi, J. T., C. L. Weller, V. S. Schlegel S. L. Cuppett, J.-Y. Lee and T. P. Carr. 2009. Sorghum distillers dried grain lipid extract increases cholesterol excretion and decreases plasma and liver cholesterol concentration in hamsters *J. Funct. Foods* 1: 381–6.

Hwang, K. T., C. L. Weller, S. L. Cuppett and M. A. Hanna. 2004. Policosanol contents and composition of grain sorghum kernels and dried distillers grains. *Cereal Chem.* 81(3): 345–9.

Ilori, M. O. 2002. The use of sorghum for beer brewing in Nigeria. *Technovation* 11: 27–37.

Jagadish, S. V. K., R. N. Bahuguna, M. Djanaguiraman, R. Gamuyao, P. V. V. Prasad and P. Q. Craufurd. 2016. Implications of high temperature and elevated CO_2 on flowering time in plants. *Front Plant Sci*. 7: 913.

Jardine, D. 2006. Stalk rots of corn and sorghum. Ag. Exp. Sta. L-741. Kansas State Univ., Manhattan, KS.

Jordan, D., Y. Tao, I. Godwin, R. Henzell, M. Cooper and C. McIntyre. 2003. Prediction of hybrid performance in grain sorghum using RFLP markers. *Theor. Appl. Genet*. 106: 559–67.

Jordan, D. R., H. Hunt, A. W. Cruickshank, A. K. Borrell and R. G. Henzell. 2012. The relationship between the stay-green trait and grain yield in elite sorghum hybrids grown in a range of environments. *Crop sci*. 52: 1153–61.

Keana, E. G., N. Bordenavea, G. Ejeta, B. R. Hamaker and M. G. Ferruzzi. 2011. Carotenoid bioaccessibility from whole grain and decorticated yellow endosperm sorghum porridge. *J. Cereal Sci*. 54: 450–45

Kelley, K. 2005. Grain sorghum and soybean cropping sequence affect yield and fertilizer n requirement. *Crop Manag*. 4(1): 105–12.

Klocke, N. L., R. S. Currie, I. Kisekka and L. R. Stone. 2014. Corn and grain sorghum response to limited irrigation, drought, and hail. *Appl. Eng. Agric*. 30: 915–24.

Knezevic, S. Z., M. J. Horak and R. L. Vanderlip. 1997. Relative time of redroot pigweed (*Amaranthus retroflexus* L.) emergence is critical in pigweed–sorghum [*Sorghum bicolor* (L.) Moench] competition. *Weed Sci*. 45: 502–8.

Knutson, A., R. Bowling, P. Porter, E. Bynum, R. Villanueva, C. Allen and S. Biles. 2015. The sugarcane aphid: A new pest of grain and forage sorghum. Texas AgriLife Research and Extension.

Lafarge, T. A., I. J. Broad and G. L. Hammer. 2002. Tillering in grain sorghum over a wide range of population densities: Identification of a common hierarchy for tiller emergence, leaf area development and fertility. *Ann. Bot. (Lond.)* 90: 87–98.

Lalitha, N., H. D. Upadhyaya, L. Krishnamurthy, J. Kashiwagi, P. B. Kavikishor and S. Singh. 2015. Assessing genetic variability for root traits and identification of trait-specific germplasm in chickpea reference set. *Crop Sci*. 55: 5: 2034–45

Lang, Y., Z. Zhang, X. Gu, J. Yang and Q. Zhu. 2004. Physiological and ecological effects of crimpy leaf character in rice (*Oryza sativa* L.) II. Photosynthetic character, dry mass production and yield forming. *Zuo Wu Xue Bao* 30, 883–7.

Lee, B. H., T. P. Carr, C. L. Weller, S. Cuppett, I. M. Dweikat and V. Schlegel. 2014. Grain sorghum whole kernel oil lowers plasma and liver cholesterol in male hamsters with minimal wax involvement. *J. Funct. Foods* 7: 709–18.

Leikam, D. F., R. E. Lamond and D. B. Mengel. 2003. Soil test interpretations and fertilizer recommendations. KS St. Univ. Ag. Exp. Sta. and Coop. Ext. Service. MF-2586. Kansas State Univ., Manhattan, KS.

Lemaire, G., X. Charrier and Y. Hébert. 1996. Nitrogen uptake capacities of maize and sorghum crops in different nitrogen and water supply conditions. *Agronomie*. 16:231–46.

Lemlioglu-Austin, D. 2014. Sorghum: Obliging alternative and ancient grain. *Cereal Foods World* 59(1): 12–20.

Li, Y, X. L. Ma, T. Y. Wang, Y. X. Li, C. Liu and Z. Liu. 2011. Increasing maize productivity in China by planting hybrids with germplasm that responds favorably to higher planting densities. *Crop Sci*. 51: 2391–400.

Li, R., Y. Zeng, J. Xu, Q. Wang, F. Wu, M. Cao, H. Lan, Y. Liu and Y. Lu. 2015. Genetic variation for maize root architecture in response to drought stress at the seedling stage. *Breed Sci*. 65(4): 298–307.

Liao, F. M., K. L. Zhou, H. H. Yang and Q. S. Xu. 1998. Genetic difference of parents and its relation to heterosis in hybrid rice. *Chin. J. Rice Sci*. 12(4): 193–9.

Lingenfelser, J., D. Jardine, M. Knapp, E. Adee, G. Cramer, P. Evans and A. Schlegel. 2015. Ag. Exp. Sta. SRP1122. Kansas State Univ., Manhattan, KS.

Manavalan, L. P., T. Musket and H. T. Nguyen. 2011. Natural genetic variation for root traits among diversity lines of maize (Zea Mays L.). *Maydica* 56: 1707.

Manschadi, A. M., G. L. Hammer, J. T. Christopher and P. deVoil. 2008. Genotypic variation in seedling root architectural traits and implications for drought adaptation in wheat (*Triticum aestivum* L.). *Plant Soil* 303: 115–29.

McIntyre, C. L., D. Seung, R. E. Casu, G. J. Rebetzke, R. Shorter and G. P. Xue. 2012. Genotypic variation in the accumulation of water soluble carbohydrates in wheat. *Funct. Plant Biol.* 39(7): 560–8.

Mehmood, S., I. Orhan, Z. Ahsan, S. Aslan and M. Gulfraz 2008. Fatty acid composition of seed oil of different Sorghum bicolor varieties. *Food Chem.* 109(4): 855–9 ·

Mellado-Ortega, E. and D. Hornero-Me´ndez. 2015. Carotenoids in cereals: An ancient resource with present and future applications. *Phytochem. Rev.* 14: 473–890.

Miralles, D. J. and G. A. Slafer. 1997. Radiation interception and radiation use efficiency of near isogenic wheat lines with different height. *Euphytica* 97: 201–8.

Mohamed, A., M. Urge and K. Gebeyew. 2015. Effects of replacing maize with sorghum on growth and feed efficiency of commercial broiler chicken. *J. Vet. Sci. Technol.* 6: 224.

Munhoz, R. E. F., A. J. Prioli, A. T. Amaral Junior, C. A. Scapim and G. A. Simon. 2009. Genetic distance between popcorn populations based on molecular markers and correlations with heterosis estimates made by diallel analysis of hybrids. *Genet. Mol. Res.* 8(3): 951–62.

Narayanan, S. and P. V. V. Prasad. 2014. Characterization of a spring wheat association mapping panel for root traits. *Crop Sci.* 106: 1593–604.

Nguyen, T. C., V. Singh, E. J. van Oosterom, S. C. Chapman, D. R. Jordan and G. L. Hammer. 2013. Genetic variability in high temperature Effects on seed-set in sorghum. *Funct. Plant Biol.* 40: 439–48.

Nielsen, D. C., P. W. Unger and P. R. Miller. 2005. Efficient water use in dryland cropping systems in the great plains. *Agron. J.* 97: 364–72.

Norwood, C. A. and R. Currie. 1997. Dryland corn vs. grain sorghum in Western Kansas. *J. Prod. Agric.* 10(1): 152–7.

Oba, M. and M. S. Allen. 1999. Effects of brown midrib-3 mutation in maize silage on dry matter intake and productivity of high yielding dairy cows. *J. Dairy Sci.* 82, 135–42.

Owuama, C. I. 1999. Brewing beer with sorghum. *J. Inst. Brew.* 105: 23–34.

Paiva, C. L., V. A. V. Queiroz, M. L. F. Simeone, R. E. Schaffert, A. C. de Oliveira and C. S. da Silva. 2017. Mineral content of sorghum genotypes and the influence of water stress. *Food Chem.* 14: 400–5.

Palmer, G. H., O. U. Etokakpan and M. A. Igyor. 1989. Sorghum as brewing materials. *World J. Microbiol. Biotechnol.* 5: 265.

Park, J. H., S. H. Lee, I-M. Chung and Y. Park. 2012. Sorghum extract exerts an anti-diabetic effect by improving insulin sensitivity via PPAR-γ in mice fed a high-fat diet. *Nutr. Res. Pract.* 6(4): 322–7.

Paterson, A. H., J. E. Bowers, R. Bruggmann, I. Dubchak, J. Grimwood, H. Gundlach, G. Haberer, U. Hellsten, T. Mitros, A. Poliakov, J. Schmutz, M. Spannagl, H. Tang, X. Wang, T. Wicker, A. K. Bharti, J. Chapman, F. A. Feltus, U. Gowik, I. V. Grigoriev, E. Lyons, C. A. Maher, M. Martis, A. Narechania, R. P. Otillar, B. W. Penning, A. A. Salamov, Y. Wang, L. Zhang, N. C. Carpita, M. Freeling, A. R. Gingle, C. T. Hash, B. Keller, P. Klein, S. Kresovich, M. C. McCann, R. Ming, D. G. Peterson, Mehboob-ur-Rahman, D. Ware, P. Westhoff, K. F. Mayer, J. Messing and D. S. Rokhsar. 2009. The *Sorghum bicolor* genome and the diversification of grasses. *Nature* 457, 551–6.

Pepper, G. E., R. B. Pearce and J. J. Mock. 1977. Leaf orientation and yield of maize. *Crop Sci.* 17: 883–6.

Perumal, P., P. Krishnaramanujam and M. A. Menz 2007. Genetic diversity among sorghum races and working groups based on AFLPs and SSRs. *Crop Sci.* 47: 1375–83.

Peterson, D. and K. Roozeboom. 2007. Long-term yield responses to crop rotation and tillage in northeast Kansas. Agronomy Abstracts. ASA, Madison, WI.

Prasad, V. B., B. A. Stewart, R. L. Baumhardt, S. Ambati, C. A. Robinson and A. Schlegel. 2006. Growing dryland grain sorghum in clumps to reduce vegetative growth and increase yield. *Agron. J.* 98:1109–20.

Prasad, P. V. V., R. Bhemanahalli and S. V. K. Jagadish. 2016. Field crops and the fear of heat stress - opportunities, challenges and future directions. *Field Crops Res.* 200, 114–21.

Qui, W.-Z., H.-H. Liu, P. Liu and B. Zhao. 2012. Morphological and physiological characteristics of corn (Zea mays L.) roots from cultivars with different yield potentials. *Eur. J. Agron.* 38(1): 54–63

Quinby, J. R. 1967. The maturity genes of sorghum. In: A. G. Norman (Ed.), *Advances in Agronomy*, Volume 19. Academic Press, New York, pp. 267–305.

Quinby, J. R. and R. E. Karper. 1954. Inheritance of height in sorghum. *Agron. J.* 46: 211–16.

Ransom, J. K., A. G. Babiker and G. D. Odhiambo. 2007. Integrating Crop management practices for Striga Control, pp. 213–28. In: G. Ejeta and J. Gressel. *Integrating New Technologies for Striga Control: Towards Ending the Witch-hunt.* World Scientific Publishing, Mainland Press Pte Ltd, Singapore.

Ratnavathi, C. V., S. K. Chakravarthy, V. V. Komala, U. D. Chavan and J. V. Patil. 2011. Sweet sorghum as feedstock for biofuel production: A review. *Sugar Tech.* 13(4): 399–407.

Raun, W. R. and G. V. Johnson. 1999. Improving nitrogen use efficiency for cereal production. *Agron. J.* 91: 357–63.

Ray, D. K., N. D. Mueller, P. C. West and J. A. Foley. 2013. Yield trends are insufficient to double global crop production by 2050. *Plos One* 8(6): e66428.

Ristova, D. and W. Busch. 2014. Natural variation of root traits: From development to nutrient uptake. *Plant Physiol.* 166(2): 518–27.

Robert, P. C. 2002. Precision agriculture: A challenge for crop nutrition management. *Plant Soil* 247: 143–9.

Robinson, H., L. Hickey, C. Richard, E. Mace, A. Kelly, A. Borrell, J. Franckowiak and G. Fox. 2016. Genomic regions influencing seminal root traits in barley. *Plant Genome* 9(1). doi: 10.3835/plantgenome2015.03.0012.

Rook, J. A., L. D. Muller and D. B. Shank. 1977. Intake and digestibility of brown-midrib corn silage by lactating dairy cows. *J. Dairy Sci.* 60: 1894–904.

Rooney, W. L., J. Blumenthal, B. Bean and J. E. Mullet. 2007. Designing sorghum as a dedicated bioenergy feedstock. *Biof. Biop. Bioref.* 1(2): 147–57.

Roozeboom, K. and P. V. V. Prasad. 2016. Sorghum growth and development. In: I. Ciampitti, V. Prasad (Eds), *Sorghum: State of the Art and Future Perspectives.* ASA monograph 58. ASA and CSSA, Madison, WI. doi:10.2134/agronmonogr58.2014.0062.

Rosenow, D. T., J. E. Quisenberry, C. W. Wendt and L. E. Clark. 1983. Drought tolerant sorghum and cotton germplasm. *Agric. Water Manag.* 7: 207–22.

Sadok, W. and T. R. Sinclair. 2009. Genetic variability of transpiration response to vapor pressure de cit among soybean [*Glycine max* (L.) Merr.] cultivars. *Crop Sci.* 49: 955–60.

Sadok, W., G. E. Matthew, A. W. S. Raza and T. R. Sinclair. 2012. Basis of slow wilting phenotype in soybean PI 471938. *Crop Sci.* 52: 1261–9.

Sakhi, S., S. Rehman, K. Okuno, A. Shahzad and M. Jamil. 2014. Evaluation of sorghum (*Sorghum bicolor*) core collection for drought tolerance: Pollen fertility and mean performance of yield traits and its components at reproductive stage. *Int. J. Agric. Biol.* 16(2): 251. ISSN Print: 1560–8530; ISSN Online: 1814–9596 13S–038/2014/16–2–251–260 (http://www.fspublishers.org).

Sanchez, A. C., P. K. Subudhi, D. T. Rosenow, and H. T. Nguyen. (2002). Mapping QTLs associated with drought resistance in sorghum (*Sorghum bicolor* L. Moench). *Plant Mol. Biol.* 48: 713–26.

Sarlikioti, V., P. H. B. de Visser, G. H. Buck-Sorlin and L. F. M. Marcelis. 2011. How plant architecture affects light absorption and photosynthesis in tomato: Towards an ideotype for plant architecture using a functional–structural plant model. *Ann. Bot.* 108: 1065–73.

Satish, K., G. Srinivas, R. Madhusudhana, P. G. Padmaja, R. N. Reddy, S. M. Mohan, N. and Seetharama. 2009. Identification of quantitative trait loci for resistance to shoot fly in sorghum [*Sorghum bicolor* (L.) Moench]. *Theor. Applied Genet.* 119(8): 1425–39.

Schlenker, W. and M. J. Roberts. 2009. Nonlinear temperature effects indicate severe damages to U.S. crop yields under climate change. *PNAS* 106(37): 15594–8.

Shen, R. L., W. L. Zhang, J. L. Dong, G. X. Ren and M. Chen. 2015. Sorghum resistant starch reduces adiposity in high-fat diet-induced overweight and obese rats via mechanisms involving adipokines and intestinal flora. *Food Agric. Immunol.* 26, 120–30.

Shoemaker, C. E. and D. Bransby. 2010. The role of sorghum as a bioenergy feedstock. In: R. Braun, D. Karlen and D. Johnson (Eds), *Sustainable Alternative Fuel Feedstock Opportunities, Challenges and Roadmaps for Six U.S. Regions, Proceedings of the Sustainable Feedstocks for Advance Biofuels Workshop*. Soil and water conservation society, Ankeny, IA, pp. 149–59.

Singh, V., E. J. van Oosterom, D. R. Jordan, C. H. Hunt and G. L. Hammer. 2011. Genetic variability and control of nodal root angle in sorghum. *Crop Sci.* 51: 5: 2011–20

Singh, V., R. A. Moreau and K. B. Hicks. 2003. Yield and phytosterol composition of oil extracted from grain sorghum and its wet-milled fractions. *Cereal Chem.* 80: 126–9.

Slewinski, T. L. 2012. Non-structural carbohydrate partitioning in grass stems: A target to increase yield stability, stress tolerance, and biofuel production. *J. Exp. Bot.* 63(13):4647–70. doi: 10.1093/jxb/ers124

Smith, O. S., H. Sullivan, B. Hobart and S. J. Wall. 2000. Evaluation of a divergent set of SSR markers to predict F1 grain yield performance and grain yield heterosis in maize. *Maydica* 45: 235–41.

Smith, S. E. and D. J. Reed. 2008. Mycorrhizal symbiosis, 3rd ed. Academic Press, London.

Stefoska-Needham, A., E. J. Beck, S. K. Johnson and L. C. Tapsel. 2015. Sorghum: An underutilized cereal whole grain with the potential to assist in the prevention of chronic disease. *Food Rev. Int.* 31: 401–37.

Stemler, A. B. L., J. R. Harlan and J. M. J. de Wet. 1977. The sorghums of Ethiopia. *Econ. Bot.* 31: 446–60.

Stephens, J. C., G. H. Kuykendall and D. W. George. 1952. Experimental production of hybrid sorghum seed with a three-way cross. *Agron J.* 44:369.

Stone, L. R. and A. J. Schlegel. 2006. Yield–water supply relationships of grain sorghum and winter wheat. *Agron. J.* 98: 1359–66

Suganyadevi, P. K., M. Saravanakumar and S. Mohandas. 2013. The antiproliferative activity of 3-deoxyanthocyanins extracted from red sorghum (Sorghum bicolor) bran through P53-dependent and Bcl-2 gene expression in breast cancer cell line. *Life Sci.* 92: 379–82.

Sun, Y., N. G. Niu, P. Osuna-avila, L. Zhao, G. Ganjegunte, G. Peterson, J. R. Peralta-Videa and J. L. Gardea-Torresdey. 2014. Variability in Salt Tolerance of *Sorghum bicolor* (L.). *Agric. Sci.* 2(2): 9–21.

Tao, Z., X. L., Ni, K. F. Jiang, H. F. Deng, Q. He, Q. H. Yang, L. Yang, X. Q. Wang, Y. J. Cao and J. K. Zheng. 2010. Relationship between heterosis and parental genetic distance based on molecular markers for functional genes related to yield traits in rice. *Rice Sci.* 17(4): 288–95.

Teetes, G. L. and Pendleton, B. P. 2000. Insect pests of sorghum, pp. 443–96. In: Smith, C. W. and Fredriksen, R. A. (Eds), *Sorghum: Origin, History, Technology and Production*. Wiley & Sons, New York, NY, p. 825.

Tefera, T. 2004. Lepidopterous stem borers of sorghum and their natural enemies in eastern Ethiopia. *Trop. Sci.* 44(3): 128–30.

Tesso, T. and G. Ejeta. 2011. Integrating multiple control options enhances striga management and sorghum yield on heavily infested soils. *Agron J.* 103: 1464–71.

Tesso, T., A. Tirfessa and H. Mohammed. 2011. Association between morphological traits and yield components in the durra sorghums of Ethiopia. *Hereditas* 148: 98–109.

Thindwa, H. P. and T. L. Teetes 1994. Effect of temperature and photoperiod on sorghum resistance to biotype C and E greenbug (Homoptera: Aphididae). *J. Econ. Entomol.* 87: 1366–72.

Thompson, C. R., D. E. Peterson, W. H. Fick, P. W. Stahlman and J. W. Slocombe. 2017. Chemical weed control for field crops, pastures, rangeland, and noncropland. Ag. Exp. Sta. SRP1132. Kansas State Univ., Manhattan, KS.

Torres, K. A. A., J. M. Pizauro Jr., C. P. Soares, T. G. A. Silva, W. C. L. Nogueira, D. M. B. Campos, R. L. Furlan and M. Macari. 2013. Effects of corn replacement by sorghum in broiler diets on performance and intestinal mucosa integrity. *Poultry Sci.* 92(6): 1564–71.

Touchton, J. T. and P. B. Martin. 1981. Response of Ratooning Grain Sorghum to Nitrogen Fertilizer and Insecticides. *Agron J.* 73: 298–300.

Unger, P. 1978. Straw mulch effects on soil temperatures and sorghum germination and growth. *Agron. J.* 70: 858–64.

Upadhyaya, H. D., S. L. Dwivedi, Yi-Hong Wang and M. Vetriventhan. 2016. Sorghum genetic resources. In: Ciampitti, I. and P. V. V. Prasad (Eds), *Sorghum: State of the Art and Future Perspectives*. Agronomy monographs 58. ASA and CSSA, Madison, WI.

Wang, W., J. Wang, C. Yang, Y. Li, L. Liu and J. Xu. 2007. Pollen-mediated transformation of *Sorghum bicolor* plants. *Biotechnol. Appl. Biochem.* 48(2): 79–83

Williams, J. R., M. J. Pachta, K. L. Roozeboom, R. V. Llewelyn, M. M. Claassen and J. S. Bergtold. Risk analysis of tillage and crop rotation alternatives with winter wheat. *J. Agri. Appl. Econ.* 44(4): 561–76.

Wood, K. M., H. Salim, P. L. McEwen, I. B. Mandell, S. P. Miller and K. C. Swanson. 2011. The effect of corn or sorghum dried distillers grains plus solubles on growth performance and carcass characteristics of cross-bred beef steers. *Anim. Feed Sci. Technol.* 165: 23–30.

Wu, X., R. Zhao, S. R. Bean, P. A. Seib, J. S. McLaren, R. L. Madl, M. Tuinstra, M. C. Lenz and D. Wang. 2007. Factors impacting ethanol production from grain sorghum in the dry-grind process. *Cereal Chem.* 84: 130–6.

Xiao, J., J. Li, L. Yuan and S. D. Tanksley. 1995. Dominance is the major genetic basis of heterosis in rice as revealed by QTL analysis using molecular markers. *Genetics* 140: 745–54.

Xin, Z., R. Aiken, J. Burke. 2009. Genetic diversity of transpiration efficiency in sorghum. *Field Crops Res.* 111: 74–80

Xu, S.-X., J. Liu and G.-S. Liu. 2004. The use of SSRs for predicting the hybrid yield and yield heterosis in 15 key inbred lines of Chinese maize. *Hereditas* 141: 207–15.

Xu, W, P. K. Subudhi, O. R. Crasta, D. T. Rosenow, J. E. Mullet and H. T. Nguyen. 2000. Molecular mapping of QTLs conferring stay-green in grain sorghum (Sorghum bicolor L. Moench). *Genome* 43(3): 461–9

Yan, S., X. Wu, S. R. Bean, J. F. Pedersen, T. Tesso, Y. Chen and D. Wang. 2011. Evaluation of waxy grain sorghum for ethanol production. *Cereal Chem.* 88(6): 589–95.

Yang, L., J. D. Browning and J. M. Awika. 2009. Sorghum 3-deoxyanthocyanins possess strong phase II enzyme inducer activity and cancer cell growth inhibition properties. *J. Agric. Food Chem.* 57(5): 1797–804.

Zha, R. M., Y. H. Ling, Z. L. Yang, F. M. Zhao, B. Q. Zhong, R. Xie, X. C. Sang and G. H. He. 2008. Prediction of hybrid grain yield performances in Indica rice (*Oryza sativa* L.) with effect-increasing loci. *Mol. Breed.* 22: 467–76.

Zhang Q. F., Y. J. Gao, S. H. Yang, R. A. Ragab, M. A. Saghai-Maroof and Z. B. Li. 1994. A diallel analysis of heterosis in elite hybrid rice based on RFLPs and microsatellites. *Theor. Appl. Genet.* 89: 185–92.

Zhang, Q. F., Z. Q. Zhou, G. P. Yang, C. G. Xu, K. D. Liu and M. A. Saghai-Maroof. 1996. Molecular marker heterozygosity and hybrid performance in indica and japonica rice. *Theor. Appl. Genet.* 93(8): 1218–24.

Zhang, X. Q., X. D. Wang, P. D. Jiang, S. J. Hua, H. P. Zhang and Y. Dutt. 2007. Relationship between molecular marker heterozygosity and hybrid performance in intra- and interspecific hybrids of cotton. *Plant Breed.* 126: 385–91.

Zhong, R. Z., Y. Fang, Y. Q. Wang, H. X. Sun and D. W. Zhou. 2017. Effects of substituting finely ground sorghum for finely ground corn on feed digestion and meat quality in lambs infected with *Haemonchus contortus*. *Animal Feed Sci. Technol.* 211: 31–40.

Chapter 7

Sorghum breeding for biotic stress tolerance

Ramasamy Perumal, Agricultural Research Center – Hays, Kansas State University, USA; Clint W. Magill, Texas A&M University, USA; Gary C. Peterson, Texas A&M AgriLife Research and Extension Center, USA; Louis K. Prom, USDA-ARS, USA; and Elfadil M. Bashir, Tesfaye T. Tesso, Desalegn D. Serba and Christopher R. Little, Kansas State University, USA

1 Introduction

Biotic and abiotic constraints can play a major role in limiting productivity and grain quality in sorghum. Host plant resistance should be the most effective way to reduce losses incurred by biotic stresses. However, classical breeding approaches for the development of resistant cultivars have been complicated by the fact that the disease and pest interaction often involves: a) complex polygenic inheritance, b) continuous development of new races or biotypes and c) enormous environmental interactions that affect pest/disease incidence and severity. These complications when combined with tedious and time-consuming inoculation procedures for sorghum diseases or insects that mimic natural conditions have presented challenges to screening genotypes under field conditions and, hence, the development of new germplasm tolerant/resistant to biotic stresses is critical for a thriving sorghum industry. Molecular diagnostics offers a more reliable method to screen large numbers of germplasm to identify potential resistance sources in a short time, but useful markers must be identified and verified before the technology can be applied.

Biotic stresses are one of the major constraints in sorghum production, and the problems with disease and insect pests are likely to increase due to climate change and associated

http://dx.doi.org/10.19103/AS.2017.0015.10

changes in rainfall and abiotic stresses such as drought and cold. The first step in any of the resistance breeding programme is to rapidly screen available genetic resources, including exotic germplasm accessions using controlled environment screening and field tests. The available well-established inoculation techniques for various pathogens and pests of sorghum (Table 1) are currently being used to identify resistant cultivars and to develop many bi-parental, recombinant inbred line (RIL) populations for detailed molecular studies. Resistance sources identified for different sorghum diseases and pests are presented in Tables 2 and 3, respectively. Molecular markers developed for diseases and pests are detailed in Tables 4 and 5, respectively.

2 Fungal diseases of sorghum

2.1 Anthracnose

Anthracnose caused by *Colletotrichum sublineolum* seems to be a universal problem in sorghum. In part, this reflects observations that the pathogen is quite variable, with races constantly being discovered (Prom et al. 2012b). Consequently, there are also many reports of germplasm screening to identify sources of resistance. Examples include germplasm originating from Burkina Faso and South Africa (Cuevas et al. 2016) and Ethiopia (Chala and Tronsmo 2012). Also, several resistance genes have been tagged using a variety of methods.

Table 1 Established screening methods to identify sources of resistance against different sorghum diseases and pests

Biotic stress	Screening method	References
Disease		
Anthracnose	Colonized sorghum grain and spray inoculation	Prom et al. (2009)
Head smut	Seedling syringe inoculation	Perumal et al. (2007)
Stalk rots	Tooth pick and basal stalk syringe inoculation	Das et al. (2007); Adeyanju et al. (2015)
Ergot	Panicle (boot leaf stage) spray inoculation	Thakur and Rao (2007)
Grain mould	Panicle spray inoculation	Thakur et al. (2007a); Katilé et al. (2010)
Rust	Urediniospore spray inoculation	Thakur et al. (2007c)
SDM	Sandwich conidia inoculation	Thakur et al. (2007b); Radwan et al. (2011)
Pest		
Midge	No-choice headcage technique	Sharma et al. (2003)
Shoot fly	Cage screening	Sharma et al. (2003)
Stem borer	Mass rearing and artificial infestation	Sharma et al. (2003)
Greenbug	Cage screening	Deol et al. (1997)
Sugar-cane aphids	Plastic cone screening	Bayoumy et al. (2015)

Table 2 Sources of resistance to sorghum diseases and parasitic weed (Striga)

Disease	Resistant source and origin	Reference
Anthracnose	PI525629, PI585711, PI585852, PI608874, PI610036, PI562783 (Mali)	Erpelding and Prom (2004)
	PI481828, PI481830, PI481834, PI481838, PI481843, PI481850 (Mozambique)	Erpelding and Prom (2006)
	PI563924, PI563905, PI430471, PI563960 (China)	Prom et al. (2007)
	PI297210, PI533833, PI576337, PI534117, PI534144, PI297199 (Uganda)	Prom et al. (2011)
	PI608974, PI608986, PI608990, PI608992, PI609009, PI609015, PI609947, PI609991 (Mali); PI154748, PI154788, PI154966, PI297093, PI297215, PI584033, PI584283, PI330983 (Uganda); PI568373, PI568533, PI568637, PI570743, PI570873 (Sudan)	Prom et al. (2012b)
	PI533991 (Sudan); PI534131 (Ethiopia); PI595735 (United States); PI482718, PI482820 (Zimbabwe)	Cuevas et al. (2014a)
	PI154758, PI584214, PI584284, PI330983 (Uganda); PI563301, PI217674 (Sudan); PI609008 (Mali)	Prom et al. (2014)
	PI586191, PI586186, PI586178, PI586183, PI586187, PI647711, PI576369 (Burkina Faso); PI576422, PI533976, (South Africa); PI656070 (Sudan)	Cuevas et al. (2016)
Head smut	Female 46038 and males Tx437 and LRB-204 and their hybrids RB-5x204, RB-118x437, RB-118x204, RB-119x430 in Mexico	Williams-Alanis et al. (2009)
	BTx635, Tx2962 through Tx2978 (Texas, USA)	Little et al. (2012)
Fusarium and charcoal rot	SC33, SC35 and SC599, SC134, SC1030 (sorghum converted lines)	Tesso et al. (2012)
	IS30562 (ICRISAT, India) and 1790E R (USA adapted restorer line)	Kapanigowda et al. (2013)
Ergot	IS8525D (Ethiopia)	Prom and Erpelding (2006)
	PI63923, PI511832, PI610749, PI610688 (China)	Prom et al. (2008)
Grain mould	PI267548, PI570011, PI570027, PI569992, PI569882, PI571312, PI570759 (Sudan)	Prom and Erpelding (2009)
	PI534117, PI297156, PI297134 (Uganda)	Prom et al. (2011)
	ICSB400, ICSB52, PVK801, ICSR89058, ICSR91011, IS41675 (ICRISAT, India)	Ashok Kumar et al. (2011)
	PI525954 (Mali); PI276840, PI276841 (Ethiopia)	Prom and Erpelding (2013)
	Sureno (Honduras); RTx2911 (United States); IS18760 (Sudan); SC719-11E (Uganda)	Prom et al. (2014)
	PI586182, PI586186, PI647705, PI647706, PI647707, PI647708, PI647710, PI647712, PI586183, PI647704 (Burkina Faso); PI61666, PI644521, PI644283, PI644304, PI19770 (South Africa)	Cuevas et al. (2016)

(*Continued*)

Table 2 (*Continued*)

Disease	Resistant source and origin	Reference
Downy mildew Pathotype - P3	PI511832, PI563519, PI563521, PI563850, PI610677, PI610724 (China)	Prom et al. (2007)
	PI609151, PI609442, PI585852 (Mali)	Prom et al. (2015)
Pathotype - P6	PI282860, PI282864, PI563505, PI282843, PI282877, PI549196, PI563438, PI282842, PI549189, PI282870, PI282829 (Chad); PI297210, PI576386, PI576395, PI576435, PI533832, (Uganda)	Prom et al. (2010)
	IS4060, IS4360, IS4372, IS4631, IS5094 (India); IS20625, IS20632 (United States); IS1212 (China); IS7305 (Nigeria); IS12804 (Turkey); IS13549 (Mexico); IS21083 (Kenya); IS23992 (Yemen); IS24463 (South Africa)	Radwan et al. (2011)
	PI609151, PI609442, PI612815 (Mali); PI522111, PI522160 (Gambia)	Prom et al. (2015b)
Pathotype P3 & P6	Tx3301 through Tx3360 (Texas, USA)	Rosenow et al. (2014b)
Rust	PI482787, PI482709, PI482718, PI482802, PI482813, PI482820, PI482727, PI482784, PI482788, PI482809, PI482801, PI482804 (Zimbabwe)	Cuevas et al. (2012)
Potyvirus spp.	IS7679 (Nigeria) and IS20740 (ICRISAT, India) - Resistance to *Maize dwarf mosaic virus* (MDMV), *Sugarcane mosaic virus* strain MDB (SCMV-MDB), *Sorghum mosaic virus* (SrMV), *Johnsongrass mosaic virus*, Nigeria (JGMV-N) and *Zea mosaic virus* (ZeMV)	Seifers et al. (2012)
Striga	PI189950 'N13' (France)	Haussmann et al. (2004)
	WSA-1, WSE-1 and WSA-2 (Sudan)	Mbuvi et al. (2017)

The approach used by Biruma et al. (2012) was to convert mRNAs present at 1, 2 and 3 days after inoculation with *Colletotrichum sublineolum* of a resistant (BS04/05 from Uganda) and a susceptible cultivar (MU07/193D) into cDNA and then to search for differences in expressed genes via amplified fragment length polymorphism (AFLP) analysis among the cDNAs produced. The sequences of two genes more highly expressed in the response of the resistant cultivar revealed nucleotide binding (NB) and leucine-rich repeat (LRR) motifs which are typical of R genes. Verification that the two genes, both on chromosome 9, have a role in resistance was confirmed when virus-induced gene silencing or VIGS led to susceptibility.

By taking advantage of the sorghum minicore diversity panel, Upadhyaya et al. (2013a) used association mapping with data from over 14 000 single nucleotide polymorphisms (SNPs) to locate eight loci common to anthracnose resistance. At least seven of these were associated with sequences previously associated with disease resistance, such as two genes on chromosome 10 with homology to the *Pib* gene for blast resistance in rice. Other defence-related genes were identified on chromosomes 1, 4 and 8. By using a panel of simple sequence repeat (SSR) primers with segregating F_2 and RIL populations, Cuevas et al. (2014b) tagged resistance in cultivar SC112-14 with closely associated markers. The

Table 3 Sources of resistance to sorghum pests

Pest	Resistant source and origin	Reference
Midge	DJ6514 (ICRISAT, India), TAM2566 (Texas, USA)	Agrawal et al. (1988)
	IS7005 (Sudan); AF28, IS10712, IS8918, ICSV197, ICSV745, ICSV843, ICSV88013, ICSV88032 (ICRISAT, India)	Sharma et al. (1992)
	Testers IS15107, TAM2566 and DJ6514 resistant in India and susceptible in Kenya	Sharma et al. (2004)
Shoot fly	ICSV700, ICSV701, ICSV705, ICSV714, ICSV717 (ICRISAT, India)	Sharma et al. (1992)
	RHRB12, ICSV713, ICSV93046, ICSV25027, IS33844-5, ICSB463, RHRB19, Parbhani Moti, ICSV705, IS5480, IS5622, IS17726, IS18368, IS34722, RVRT1, ICSR93031, Dagidi Solapur, ICSB461, ICSB463, Phule Yasodha, ICSV711, ICSV25010, ICSV93089, IS18662, Phule Vasudha, IS18551, IS33844-5, Barsizoot (ICRISAT, India)	Riyazaddin et al. (2015)
	Hybrids involving A4(M) cytoplasm (ICRISAT, India)	Sanjana Reddy et al. (2015)
	CSV22, ICSB422, ICSB425, ICSB428, ICSB432, ICSB458, ICSB463, IS2312, IS5480, IS18662, Phule Chitra, RSV1093, IS18551, RSV1235 (ICRISAT, India)	Sharma et al. (2015)
	Maulee, Phule Anuradha, M 35-1, CSV18R, Giddi Maldandi, RVRT3, ICSB433, ICSV700, ICSV25019, ICSV25022, ICSV25026, ICSV25039, PS35805, Akola Kranti, IS18551 (ICRISAT, India); IS2312 (Sudan)	Mohammed et al. (2016)
	ICSV700, Phule Anuradha, ICSV25019, PS35805, IS2123, IS2146, IS18551 (ICRISAT, India)	Riyazaddin et al. (2016)
Stem borer	IS5470, IS5604, IS8320 (India); IS18573, ICSV443, ICSV700, PB12779-1, IS2205 (ICRISAT, India)	Sharma et al. (1992)
	IS1044, IS2123, IS1054, ICSV714 (ICRISAT, India)	Kishore Kumar et al. (2006)
	PSFR94011A, SPSFR94006A, SP55299A, SP55301A, ICSV705, ICSV700, ICSV708, PS30710, IS18551, SFCR151, SFCR125 (ICRISAT, India)	Sharma et al. (2006)
	ICSA467, ICSA473, MACIA, ICSB464 (ICRISAT, India)	Muturi et al. (2012b)
	ICSA472, ICSV700, ICSA464 (ICRISAT, India)	Muturi et al. (2012a)
Greenbug	PI266965 (Soviet Union, Former) – Biotype E	Harvey et al. (1991)
	PI297181, PI297212 (Uganda); PI302137, PI302178, PI302236 (Portugal); PI302231 (Australia); PI302233 (Soviet Union, Former); PI308442, PI229828 (South Africa); PI308976 (Kenya); PI220248 (Italy); PI494893, PI524770, PI500963 (Zambia); PI514604 (Senegal); IS809, Capbam (Sarvasi) and KS 30 (United States) – Biotype C and E	Peterson (1996)

(Continued)

Table 3 (*Continued*)

Pest	Resistant source and origin	Reference
	PI266965 (Soviet Union, Former); PI550607 (China); PI221719, PI560387 (South Africa); PI550610 (Syria); PI607900, PI596542, PI535779, PI591008 (United States); PI452752, PI455203, PI455512, PI455812, PI456490, PI456504, PI457212, PI457314 (Ethiopia); PI482903 (Zimbabwe); PI515888 (Togo); PI536594 (Honduras); PI545501 (Sudan); PI562891 (India); PI585393 (Nigeria); IS25246, IS27002, IS27834, IS27903 (ICRISAT, India) – Biotype I	Peterson (1996)
	KS99A/B to KS107A/B (Kansas State University, USA) – Biotype E	Kofoid and Harvey (2000a)
	KS97 (Kansas State University, USA) – Biotype I	Tuinstra et al. (2001)
	KS108R to KS114R (Kansas State University, USA) – Biotype I	Kofoid and Harvey (2000b)
	KS116A/B to KS120A/B (Kansas State University, USA) – Biotype I	Kofoid and Harvey (2005)
	KS121R to KS132R (Kansas State University, USA) – Biotypes E, I and K	Kofoid et al. (2012)
Sugarcane aphid	*Germplasm accessions*: IS44, IS718, IS2312 and IS3796 (USA); IS84(Mexico); IS1063, IS1117, IS1133C (SC202), IS1134C (SC203), IS1139C (SC205), IS1144C (SC451), IS1366C (SC210), IS1598C (SC214), IS1840, IS4657, IS5188C (SC245), IS5490, IS5887C (SC 248), IS6389C (SC489), IS6416C, IS6426C (SC497), IS6962C, IS8100C and IS33843 (India); IS8100C (Japan); IS12599C (SC90) (Congo); IS14048 (Malawi); IS12158C (SC984), IS12551C (SC31), IS12608C (PI257 595,SC108-14-E), IS12609C (SC109), IS12610C (SC110-14), IS12611C (SC111 or SC170), IS12612C, IS12637C (SC146), IS12645C, IS12661C, IS12664C and IS12667C (Ethiopia) *Parental lines (A/B)*: 2B, 9B, 27A, 33B, 42B, 53B, 104A/B, 116A/B, 117A/B, 205B, 296B, 2077A/B, 2219A/B, 36227A/B, AKMS14A, ICSB # 53, 70, 84, 90, 101, 2730, 2731, 88001 and 88014 (India), 129-3A (Taiwan) *Parental lines (R)*: 2R, 5R, R128, R131 and R132, 354R, 1413R, RS29, RS67, RS291, RS530, RS649, C43, C81, ICSR # 160, 161, 162, 165, 172, 174, 194, 38111, 89008, and 89009, M 148-138 and SB1085 (India); RTAM428 (USA) *Agronomic elite lines/experimental varieties*: Agr4S2560 Shallu (Taiwan); 7511, HB37 (China); NK349, PE954177 (Japan); PNR8537 (Botswana); CE151, Ent.62SADC, FGYQ336, FGYQ353, SA967 and SDSL89426 (South Africa); BPT28, E108, E119, E303, Hami, PVR10, ICSV # 9, 148, 197, 743, 745, 89013, 93028, 93038, 93093, 95081, 95085, 95093, 95099, 95110 and 96081, SPV # 97, 101, 102, 224, 232, 236, 243, 245, 257, 291, 303, 378, 386, 492, 504, 969, 991, 998, 1002, 1004, 1051, 1053, 1054, 1055, 1178, 1359, 1380, 1413, 1450, 1452, 1453, 1462 and 1465 (India)	*Reviewed by* Singh et al. (2004)
	PSFR94011A, SPSFR94006A, SP55299A, SP55301A, ICSV705, ICSV700, ICSV708, PS30710, IS18 551, SFCR151, SFCR125 (ICRISAT)	Sharma et al. (2006)

Pest	Resistant source and origin	Reference
	ICSB215, ICSB323, ICSB724, ICSR165, ICSV12001, ICSV12004, IS40615 (ICRISAT)	Sharma et al. (2014)
	RTx2783 (Texas, USA)	Armstrong et al. (2015)
	PI550610 (Syria)	Bayoumy et al. (2015)

resistance on chromosome 5 was effective against *Colletotrichum sublineolum* pathotypes from Puerto Rico and Texas, whereas a nearby but non-identical gene was effective for Arkansas isolates. However, neither of these genes applied to Georgia isolates and they are not the same as *Cg*1, a gene also on chromosome 5 that had been tagged previously to linkage group 5 (Perumal et al. 2009). More extensive genotyping by sequencing (GBS) methodology has been used to show that *Cg*1 from resistant variety SC748-5 falls in a cluster of defence-related genes with several nucleotide substitutions that differ from BTx623, the susceptible parent used for sequencing and to establish the mapping RIL progeny (Burrell et al. 2015). Consequently, the authors suggest that there may be several closely linked genes involved. The GBS concept was also used to identify prospective resistance genes in BK7 in $F_{3:4}$ and $F_{4:5}$ populations from a cross to susceptible Early Hegari Sart (Felderhoff et al. 2016). In these crosses over 5000 SNPs were used to associate quantitative trait loci (QTLs) for resistance from BK7 on chromosomes 7 and 9, again in regions with well-defined defence response genes. In another SSR-based tagging effort, Beshir et al. (2016) identified two QTLs segregating in 126 $F_{8:9}$ individuals that contributed to both anthracnose and leaf blight resistance: one on chromosome 2 and the other on chromosome 5.

2.2 Head smut

Head smut disease is caused by the soil-borne pathogen *Sporisorium reilianum* (Kühn) Langdon & Fullerton, a facultative biotrophic basidiomycete fungus. This disease is reported to be found in all sorghum-growing areas of the world. Once infected, the pathogen survives in the soil for several years and even proper crop rotation does not help to remove the pathogen from the soil. *Sporisorium reilianum* infects sorghum at the seedling stage, with the symptoms appearing in the flowering stage and replacing the grain completely with black sori. Disease incidence has recently increased dramatically in sorghum fields in the Coastal Bend area of Texas (G.N. Odvody, pers. comm.).

Host resistance is the most effective way to control *Sporisorium reilianum* infection. Identification of resistance sources is a slow process due to observing disease expression in late-maturing photoperiod-sensitive sorghums (Craig and Frederiksen 1992) and due to several plant and environmental factors that can interfere with natural infection events (Edmunds 1963). Zou et al. (2010) conducted an inheritance and molecular study on sorghum head smut and inferred that the resistance to sorghum head smut physiological race 3 is controlled by two mutually independent non-allelic genes, and there is an interaction between them. The identified molecular markers are easier to find in B-lines (maintainer or seed parent) than R-lines (pollinator) during the screening of microsatellite markers and showed different resistance mechanisms between restorer and maintainer lines.

Table 4 Developed molecular markers for sorghum diseases

Disease	Markers/genes	Source	Reference
Anthracnose	RAPD markers linked to a recessive gene conditioning anthracnose resistance	115 F_2 segregated individuals SC326-6 × BTx623	Boora et al. (1998)
	Identified two RAPD-based DNA markers (*OPI*16 and *OPD*12) linked to anthracnose disease resistance	Bulked segregant analysis using 110 F_2 populations from the cross G73 × HC136	Panday et al. (2002)
	Resistance gene maps to the long arm of Chr-8 on the basis of linked RAPD (OPJ01) and SCAR (SCJ01-1 and SCJ01-2) markers	110 F_2 population from HC136 × G73	Singh et al. (2006)
	*OPA*12, *OPJ*01, *OPF*07 and *OPL*04, *OPI*12 and *OPD*12 (RAPD markers), *Xtxp*61 and *Xtxp*212 (SSRs) and *SCA*12 and *SCJ*01 (SCARs)	Bulked segregant analysis using two $F_{2:3}$ populations: HC136 (susceptible) x G73 (resistant) and SC326-6 (resistant) x BTx623 (susceptible)	Singh and Boora (2008)
	AFLP marker *Xtxa*6227 and SSR marker *Xtxp*549 on LG-05 for anthracnose resistance locus *Cg1*	71 F_{2-3} families from SC748-5 x BTx623	Perumal et al. (2009)
	Eight linked loci: 4 on Chr-1, 2 on Chr-6, 1 on Chr-8, and 1 on Chr-10	Mini core (242 accessions) partially sequenced and aligned to the sorghum reference genome	Upadhyaya et al. (2009)
	One QTL on chromosome 4, and three on Chr-6	168 RILs from 296B x IS18551	Mohan et al. (2010)
	Two R-genes encoding proteins with NBS-LRR domains, Cs1A and Cs2A, as the primary sources of resistance.	Two Ugandan sorghum genotypes BS04/05 and MU07/193D	Biruma et al. (2012)
	One marker on chromosome 5 linked to one or several closely linked resistance loci	50 RILs from SC112-14 × PI 609251	Cuevas et al. (2014b)
	One QTL on Chr-5	117 RILs from SC748-5 × BTx623	Burrell et al. (2015)
	Resistance loci on Chr-7 and Chr-9	Bi-parental (Bk7 × Early Hegari-Sart) mapping population with 135 lines (28 $F_{3:4}$ and 107 $F_{4:5}$)	Felderhoff et al. (2016)
Head smut	RFLP loci detected by probes *pSbTXS*560 and *pSbTXS*1294 and one RAPD locus from primer *OPG*5 were linked to head smut resistance gene (*Shs*)	52 F_2 plants from SC325 x RTx7078	Oh et al. (1994)

Disease	Markers/genes	Source	Reference
	*Xtxp*13 located in LG- B in the R (restorer) population, and *Xtxp*145 located in LG- I in B (seed parent) population was linked to resistance	Bulk segregant analysis using two F_2 populations: R (2381R × Aisi) and B (BTx622B × 7050B)	Zou et al. (2010)
	One pair of primers IS10264, showed consistent polymorphism bands between parents and bulks located in LG- I, with a recombination percentage between marker and resistant gene of 9.8% and the genetic distances to resistant gene are 9.9 cM	R population (2381R × Aisi) with F_2 (114 resistant and 60 susceptible) plants	Li et al. (2011)
Stalk rot (Charcoal rot)	Five QTLs at Dharwad and 4 QTL at Bijapur locations for the component traits of charcoal rot disease resistance. QTL for number of internodes crossed, length of infection and per cent lodging accounted phenotypic variation with 31.8, 10.8 and 18.9% at Dharwad and 14.9, 10.5 and 26.4 at Bijapur, respectively	RILs from IS22380 (susceptible) × E-36-1 (resistant)	Reddy et al. (2008)
	Nine stable QTLs for morphological trait along with three QTLs for biochemical trait influencing charcoal rot disease incidence	93 RILs from E 36-1 (resistant) × SPV 86 (susceptible)	Patil et al. (2012)
	Fourteen loci and a set of candidate genes appear to be involved in connected functions controlling plant defence response	Sorghum diversity panel with 300 genotypes	Adeyanju et al. (2015)
Ergot	Identified 9, 5 and 4 QTLs linked to molecular markers for percentage ergot infection, pollen quantity and pollen viability, respectively	146 RILs from R931945-2-2 × IS 8525 (resistant)	Parh et al. (2008)
Grain mould	Five QTLs in linkage groups D, E, F, G and I, each accounting for between 10 and 23% of the phenotypic variation	125 RILs from Sureno × RTx430	Klein et al. (2001)

(*Continued*)

Table 4 (*Continued*)

Disease	Markers/genes	Source	Reference
	Two linked loci; one contained a homologue of the maize non-host resistance gene *Rxo1*	Mini core (242 accessions)	Upadhyaya et al. (2013b)
Rust	Four regions, each in a separate linkage group, linkage group 10; it accounted for 40% of the total phenotypic variation	160 RILs from QL39 x QL41	Tao et al. (1998)
	Five linked loci. Two of rust-linked loci each contained the rust resistance gene homologous to the maize rust resistance gene *Rp1-D* which is the B locus and to the wheat rust resistance gene *Lr1*	Mini core (242 accessions)	Upadhyaya et al. (2013b)
	Sixty-four significant QTLs and 21 suggestive rust resistance QTLs were identified representing 55 unique genomic regions (54 rust resistance QTLs, and 31 rust QTL as multi-trait effect QTLs)	3 bi-parental RILs; 119 from ICSV745 × R890562-1-2, 246 from IS8525 × R931945-2-2 and 214 from AusTRCF317961 x R931945-2-2	Wang et al. (2014)
Sorghum downy mildew (SDM)	The clone *txs*552 revealed an RFLP linked to resistance	BTx623 (resistant) × IS3620C (susceptible)	Gowda et al. (1995)
	Identified two RFLPs that segregated with a resistance gene, the markers were 5–7 map units from the gene	F_2 plants from SC325 (resistant to P1, P2 and P3 pathotypes) × RTx7078 (susceptible to all pathotypes)	Oh et al. (1996)
	Three QTLs (1 each on Chr-2, Chr-3 and Chr-6) and two QTLs (1 each on Chr-3 and Chr-6), all of which were contributed by NAI116. The significance of the major QTL on Chr-6 (bin 6.05) that confers resistance to diverse DMs in tropical Asia, including SDM and RDM (Rajasthan Downy Mildew) in India	151 BC_1F_1 progeny from the cross CM139 (susceptible) x NAI116 (resistant to both SDM and RDM)	Nair et al. (2005)

Disease	Markers/genes	Source	Reference
Striga	Eleven QTLs and nine QTLs in RIP-1 and RIP-2 populations, respectively	RIP-1: 116 $F_{3:5}$ lines from the cross IS9830 × E36-1 RIP-2: 110 $F_{3:5}$ lines from the cross N13 × E36-1	Haussmann et al. (2004)
	Zero to five QTLs	BC_2 lines of farmer preferred sorghum cultivars from Eritrea, Kenya, Mali and Sudan	Muth et al. (2012)
	Two QTLs from group 5 exhibit close association with Striga resistance levels using SSRs markers *Xtxp*201, *Xtxp*15, *Xtxp*225, *Xtxp*65, *Xtxp*302, *Xtxp*303, *Xtxp*145, *Xtxp*57, *Xtxp*208, *Xtxp*50 and *Xtxp*304	Backcross-derived progenies BC1S1-67-H, BC1S1-110-H, and BC1S1-16-H, from 'Tabat' x N13	Gamar and Mohamed (2013)
	Two or more major QTLs	BC_3S_4 populations from N13 with Tabat, Wad Ahmed and AG-8	Mohamed et al. (2014)

2.3 Stalk rots

Several pathogens may be associated with stalk rot infection, but *Macrophomina phaseolina* and *Fusarium thapsinum* are recognized as the major causal organisms. Severe drought and high temperature at the grain-filling stage favour Fusarium stalk rot (*Fusarium* spp.) infection. Of the various *Fusarium* species, *F. thapsinum* was the most virulent against sorghum under both greenhouse and field conditions (Tesso et al. 2010). Pathogens causing Fusarium stalk rot can be found in host tissues at various growth stages, but disease symptoms are visible only after flowering and during the grain-filling stage. Access to plants is gained via injury to tissues or damage by insects. The infection causes reddish pith, and the upper internodes have discoloured vascular bundles. Under severe conditions, the disease may lead to premature plant death, poor grain development, complete stalk collapse and crop lodging. Charcoal rot [*Macrophomina phaseolina* (Tassi) Goidanich] incidence is reported the most widespread in the tropics, subtropics and temperate regions. The higher the height the more the plant is exposed to drought stress during grain development and the more the disease incidence.

Recent studies have indicated increased incidence of stalk rot worldwide (Khangura and Aberra 2009). Yield is drastically reduced due to stalk rot infection followed by lodging and it is reported in all sorghum-growing regions including Africa (Hulluka and Esele 1992), India (Seetharama et al. 1987), Australia (Trimboli and Burgess 1982) and the United States (Jardine and Leslie 1992). The extent of the reduction, however, may vary from region to region depending on disease severity. Many of the known sources of resistance to *Fusarium* are not resistant to charcoal rot (Tesso et al. 2010). Adeyanju (2015) conducted a genome-wide association study using a 300-entry sorghum diversity panel in three environments at Kansas State University. Following the standard field inoculation procedures for both pathogens, phenotyping data were integrated with 79132 SNP

Table 5 Developed molecular markers for sorghum pests

Pest	Markers/genes	Source	Reference
Midge	Two genetic regions located on separate linkage groups associated with antixenosis and explained respectively 12 and 15%, of the total variation in egg numbers/spikelet laid in a cage experiment. One region was significantly associated with antibiosis and explained 34.5% of the variation of the difference of egg and pupal counts in the RILs	RILs from the cross ICSV745 × 90562	Tao et al. (2003)
Shoot fly	Twenty-nine QTLs: 4 each for leaf glossiness and seedling vigour, 7 for oviposition, 6 for dead hearts, 2 for adaxial trichome density and 6 for abaxial trichome density	168 RILs from the cross 296B (susceptible) × IS18551 (resistant)	Satish et al. (2009)
	Twenty-five QTLs: 5 each for leaf glossiness and seedling vigour, 10 for deadhearts, 2 for adaxial trichome density and 3 for abaxial trichome density. Stable QTLs were identified for the traits across different environments	210 RILs from the cross 27B (susceptible) × IS2122 (resistant)	Aruna et al. (2011)
	QTL which is common for glossiness, oviposition and dead hearts was detected on linkage group A. Phenotypic variation by this QTL ranged from 3.8 to 6.3%. Two more QTLs were detected following multi-trait composite interval mapping (MCIM), one each on LG-A and C	135 RILs from IS18551 (resistant) × 296B (susceptible)	Apotikar et al. (2011)
	QTLs were mapped to a 37-72 cM or 5-15 Mb interval on the long arm of sorghum SBI-10L with flanking markers *Xgap*001 and *Xtxp*141. One QTL each for glossiness (*QGls*10) and trichome density (*QTd*10) was mapped in marker interval *Xgap*001-*Xnhsbm*1044 and *Xisep*0630-*Xtxp*141, confirming their loose linkage, for which phenotypic variation ranged from 2.29 to 11.37%	Three RILs: i) 1894 F_2s, ii) 369 F_2 and iii) 369 $F_{2:3}$ progenies, from RSG04008-6 (susceptible) × J2614-11 (resistant)	Kiranmayee et al. (2016)
Stem borer	Twenty-nine QTLs associated with different traits such as stem tunnelling, deadheart, leaf feeding, recovery resistance and overall resistance. A stable putative QTL on SBI-07 was associated with stem tunnelling	266 RILs from susceptible parent ICSV 745 and resistant parent PB 15520	Vinayan (2010)
	One hundred and four SSR markers exhibited polymorphism among individuals of the mapping population	94 RILs from ICSV745 (susceptible) × PB15881-3 (resistant)	YueYing et al. (2010)
	QTL on sorghum Chr-1,3,4,5,6,8,9 and Chr-10 controlling *Busseola fusca* and *Chilo partellus* resistance against leaf damage, deadheart, exit holes and stem tunnelling	243 RILs from ICSV 745 (susceptible) x PB 15520-1 (resistant)	Muturi et al. (2013)
Greenbug	Nine QTLs were detected affecting both resistance and tolerance to greenbug biotypes I and K. with phenotypic variation by each QTL ranging from 5.6 to 38.4%	93 RILs from GBIK x Redlan	Agrama et al. (2002)

Pest	Markers/genes	Source	Reference
	Three QTLs associated with biotype I and 5 QTLs associated with biotype K were identified. The phenotypic variation by these QTLs ranged from 9 to 19.6%	RILs from the cross 96-4121 (tolerant) x Redlan (susceptible)	Nagaraj et al. (2005)
	Sixteen SSRs spread over 5 chromosomes were significant for greenbug resistance. A major QTL resided in the interval of 6.8 cM between SSR markers *Xtxp*358 and *Xtxp*289 on SBI-09	F_2s with 277 plants and $F_{2:3}$ with 233 families derived from Westland A line x PI550610	Wu et al. (2007)
	Two QTLs on sorghum SBI-09 consistently conditioned the resistance to greenbug and were designated as *QSsgr*-09-01 (major QTL) and *QSsgr*-09-02 (minor QTL), accounting respectively for 55-80% and 1–6% of the phenotypic variation for the resistance to greenbug feeding. These QTLs appeared to have additive and partially dominant gene effects. The markers *Xtxp*358, *Xtxp*289, *Xtxp*67 and *Xtxp*230 closely flanked the respective QTLs	277 F_2 progeny and their $F_{2:3}$ families from Westland A line (susceptible parent) and PI550610 (resistant parent)	Wu and Huang (2008)
	Twenty-nine markers significantly associated with the plant response to greenbug feeding damage. Four major QTLs for greenbug resistance on SBI-09 collectively accounted for 34–82% of the phenotypic variation in greenbug resistance	371F_2 plants from the cross PI607900 (resistant) x BTx623 (susceptible)	Punnuri et al. (2013)
Sugarcane aphids	A dominant gene, *RMES1*, had previously been identified to confer sugarcane aphid resistance in a Chinese line (Chr-6). Three pairs of AFLP markers (*EcoR* I-*CTG/Mse* I-*CCT*, *EcoR* I-*CTG/ Mse* I-*CAT* and *EcoR* I-*AGT/Mse* I-*CCC*) showed linkage with aphis resistance. A specific band of 300 bp was amplified by a pair of SCAR primers designed based on the sequence	Bulked segregant analysis using F_2 population from the cross Qiansan (aphid sensitive) x Henong (aphid resistant)	Chang Jin-hua et al. (2012)
	One of the mapping populations is progressing well towards the placement of the QTLs on sorghum chromosomes	Core collection (~280 sorghum lines) was evaluated for their responses to sugarcane aphids (SCA), identify > 10 sources with various levels of resistance. Several mapping populations were developed with the resistance lines	Huang (2015)
Nematodes	A major QTL on SBI-03 for total eggs and eggs per gram of root. SSR markers were developed in the total eggs and eggs per gram of root QTL region and the markers flanking the resistance gene are 4.7 and 2.4 cM away	130 RILs from the cross Honey Drip (resistant) x Collier (susceptible)	Harris-Shultz et al. (2015b)

markers developed through GBS. Fourteen loci and a set of candidate genes involved in controlling plant defence response were identified. SNP markers' effects were small, accounting only for 19–30% of the total phenotypic variation. All SNPs were genetically independent as evinced from the linkage disequilibrium analysis.

2.4 Rust

Sorghum rust, caused by *Puccinia purpurea* is especially a problem in Australia (White et al. 2012). Though less studied than anthracnose, there is convincing evidence that different resistance genes are available. An early study using restriction fragment length polymorphism (RFLP) in an RIL population from a cross between QL41 and QL39 located QTLs on four linkage groups, with high and moderate resistance, respectively (Tao et al. 1998). Evidence of different genes needed for resistance was also confirmed by demonstrating that ten different host cultivars showed differential responses to isolates of *Puccinia purpurea* (White et al. 2015). In that test, four different rust pathotypes were identified among 28 rust isolates. By taking advantage of newer GBS technology, specifically, the ability to identify SNPs scattered throughout the genome, tags associated with resistance genes were identified. For example, Upadhyaya et al. (2013b) were able to identify five SNPs associated with rust resistance among 242 minicore accessions; two of these showed sequence homology to previously defined rust resistance genes from maize and wheat and the others to typical host defence response genes. By combining DarTseq™, a process designed to improve the detection of polymorphisms in active genes, along with SSR and RFLP markers on three RIL populations, each with different parents with various levels of rust resistance, QTLs were detected in a total of 55 genome areas (Wang et al. 2014). As with the previous case, some of these were associated with clusters of 'R' genes that contain NB and LRR motifs that function to initiate signalling pathways to induce host defences. Additional data collected in the analysis showed a correlation between the level of rust resistance with earliness and short stature, likely accounting for some of the many QTLs detected.

2.5 Ergot

Sorghum ergot is caused by *Claviceps africana* (most common species in the Americas, Australia, Asia and Africa), *Claviceps sorghi* (limited to Asia) and *Claviceps sorghicola* (reported only in Japan). Ergot was reported in the United States, Brazil and Australia in 1997 (Isakeit et al. 1998; Casela et al. 1999; Ryley and Henzell 1999) and is now endemic in most sorghum-producing states. Male sterile sorghum lines used in hybrid seed production are highly susceptible, especially when pollination is delayed due to environmental or genetic factors (Futrell and Webster 1965). Yield losses due to ergot in hybrid seed production fields ranges from 10% to 80% (Sangitrao et al. 1999). Also, the sugary exudates produced by the pathogen reduce seed quality and make harvesting and seed processing more difficult (Bandyopadhyay et al. 1998).

The type of resistance in male fertile plants would be functional in varieties and hybrids, but would be ineffective for hybrid seed production based on the use of male sterile lines. Two new resistant sources (IS14131 and IS14257) were identified by Reed et al. (2002) and this emphasized the need to study the physiological basis for ergot resistance. Also, when the identified genetic resistance sources are crossed to diversified cytoplasmic (A3) male sterile lines, hybrid progenies were sterile, which favours ergot infection. Results indicated that pollen-based fertile hybrids are more tolerant to ergot infection than male sterile hybrids irrespective of any resistance source background (Prom and Erpelding 2006). Overall, stable genetic resistance in sorghum is lacking and special studies are needed to develop male sterile lines and hybrids with ergot resistance (McLaren 1992).

Tooley et al. (2010) characterized the species of sorghum ergot using real-time polymerase chain reaction (PCR) analysis with the primers and probes designed from the intron 3 region of the β-tubulin gene (for *Claviceps africana* and *Claviceps sorghi*) and the intron 4 region of EF-1α (for *Claviceps sorghicola*). The developed molecular probes provide useful tools for detecting ergot pathogens in seed and grain shipments to determine the species, thereby aiding in the regulatory decision-making process. Parth et al. (2008) identified molecular markers linked to ergot resistance and two pollen traits, pollen quantity (PQ) and pollen viability (PV) and assessed the relationship between the two pollen traits and ergot resistance in sorghum. The study revealed that two of the QTLs for PQ and PV are located on SBI-01 and SBI-06. Identification and analysis of candidate genes to deduce information about the nature and function of the detected loci for these traits will be helpful in determining the genetic architecture of resistance to sorghum ergot.

2.6 Grain mould

One of the major obstacles to sorghum productivity and seed quality is grain mould, a complex fungal disease. Over 40 fungal genera are capable of infecting and colonizing sorghum grain at all levels of maturity and in storage. Of these, *F. thapsinum* and *Cochliobolus lunata* are considered the most common grain-moulding fungi (Prom et al. 2015). The disease is most prevalent when moist weather conditions prevail after flowering and continue until harvest. Damage caused by complex fungal pathogens includes reduction of yield and grain density, and poor germination. Diseased grain may not be suitable for food or feed since some of the infecting fungi have the capacity to produce mycotoxins (Little et al. 2012) capable of causing disease and even death to human and animals.

The inheritance of grain mould resistance was studied by Rodriguez-Herrera et al. (2000) using F_1, F_2 and backcross generations derived from the cross Sureño (dual-purpose food grain and forage variety, resistant to grain mould) and RTx430 (adapted inbred line, susceptible to grain mould). Additive, dominance and epistatic gene effects were detected. Four to ten genes were estimated to contribute to grain mould resistance. The study indicated that selection in specific environments is useful for enhancing resistance to mould as the inheritance of these complex fungi is not stable for a wide range of environments.

Grain mould fungi colonize the internal tissues of the floret and developing grain (i.e. living tissue). Grain weathering fungi primarily colonize non-living tissue, that is, mostly external surfaces of the grain (Bandyopadhyay et al. 2000). Forty-four grain weathering-resistant sorghum lines (Tx3364 through Tx3407) were developed and released by the Texas A&M AgriLife Research and Extension Center, Lubbock, Texas as sources of resistance to grain mould and grain weathering in improved agronomic backgrounds (Rosenow et al. 2014a). Further, based on the *in vitro* screening of these 48 sorghum lines, seven lines, Tx3371, Tx3373, Tx3374, Tx3376, Tx3407, Tx3400 and Tx3402, were identified as potential sources of grain mould resistance as each showed minimal fungal infection and higher grain quality traits (Tomar et al. *manuscript in preparation*).

2.7 Sorghum downy mildew (SDM)

SDM is often grouped with fungal diseases but is caused by a fungal-like oomycete, *Peronosclerospora sorghi*. Since systemically infected plants produce little if any seed, losses can be estimated by the relative fraction of infected plants, which can be high in susceptible varieties under optimal conditions for the pathogen. Although less variability has been demonstrated for SDM, at least six pathotypes have been identified based on differential host responses in the United States (Isakeit and Jaster 2005). Although the latest emerging pathotype, described in 2005, was also resistant to metalaxyl-based seed treatment, alternative sources of genetic resistance have been identified (Rosenow et al. 2014b; Prom et al. 2015) and it is likely that commercial pathotype 6 resistant hybrids have been released based on observations by G. Odvody (*pers. comm.*) that isolates of that pathotype are no longer being detected in the affected area. Genetic diversity of *Peronosclerospora sorghi* isolates was studied using AFLP (Perumal et al. 2006) and microsatellite (Perumal et al. 2008) markers to differentiate isolates of P3 and P6 pathotypes and different *Peronosclerospora* species. SDM also infects maize, and the re exist reports on resistance genes (Wongkaew et al. 2014) and QTL (Jampatong et al. 2013; Lohithaswa et al. 2015) identification only with regard to maize.

3 Sorghum pests

3.1 Midge

Sorghum midge [*Stenodiplosis sorghicola* (Coquillett)] is a serious pest and its infestation is reported worldwide where sorghum is grown (Teetes and Pendleton 2000). The insect was first reported as a sorghum pest in Queensland, Australia, in 1894 (Tyron 1895). The insect only damages sorghum at anthesis with female midges ovipositing in flowering florets. Before the discovery of genetic resistance, the pest could be partially controlled by the uniform planting of sorghum over a wide geographic area, resulting in the crop flowering within a relatively short period, or by application of insecticides to suppress the adult female midge population. Insecticides are not effective against the damaging larval stage.

Breeding for genetic resistance in sorghum did not begin until the discovery of resistance (Johnson et al. 1973) in partially converted lines from the Texas A&M University/USDA-ARS Sorghum Conversion Program (Stephens et al. 1967). Evaluation of the converted exotic sorghum collection resulted in the identification of 31 resistant lines (Peterson et al. 1994). These lines formed the basis for the sorghum midge-resistant breeding programmes in Texas and Australia. In Brazil, the line 'AF28' was identified as resistant. ICRISAT evaluated at least 15 000 accessions from the world sorghum collection (Sharma 1985; Sharma et al. 1993). The evaluations resulted in the identification of 19 additional sources of resistance. Biotypes of sorghum midge have not been reported.

Studies on insect resistance have reported disease resistance due to the extensive use of insecticides and the complex nature of host plant resistance and insect interactions (Muturi et al. 2012a). The genetics of resistance to sorghum midge are complex. Several studies concluded resistance was controlled by an unknown number of recessive to partially dominant genes (Widstrom et al. 1972; Bergquist et al. 1974; Teetes and Johnson

1978; Rossetto and Igue 1983; Boozaya-Angoon et al. 1984; Widstrom et al. 1984; Agrawal et al. 1988). Resistance is controlled by mostly additive gene action (Sharma et al. 1996) with significant general and specific combining ability effects (Page 1979; Patil and Thombre 1985; Agrawal et al. 1988). Tao et al. (2003) used one RIL population and identified QTLs for antixenosis and antibiosis. Two QTL regions, each identified on separate linkage groups for antixenosis, explained respectively 12% and 15% of the total variation. For antibiosis, one QTL explained 34.5% of the variation of the difference of egg and pupal counts. Diarisso et al. (1998a,b) studied the floral structure and time of flowering of sorghum midge-resistant and susceptible lines. They concluded that while floral structures (stigmas, filaments, anthers and ovaries) differed between resistant and susceptible lines the differences were not consistent. Resistant genotypes completed early flowering were no longer vulnerable to oviposition when female midges were present. Thus, tightness of glume closure and time of flowering could be a major cause of sorghum resistance to sorghum midge.

3.2 Shoot fly

The worldwide yield loss due to shoot fly has been estimated to be over US $ 274 million (Sharma 2006). The pest causes serious problems especially in the late-sown crops and also infests early sowings when the preceding dry season is interrupted by frequent rain showers (Nimbalkar and Bapat 1987). Numerous papers suggest that shoot fly (*Atherigona soccata*) resistance is a critical need for sorghum production in India. Post-rainy season sorghums grown under receding moisture conditions are exposed to peak shoot fly populations between September and October (Mohammed et al. 2016). Shoot fly infestation decreases plant stand and causes severe losses in grain and fodder yield. The increase in shoot fly dead hearts by 1% results in a loss of 143 kg grain yield/ha, and an overall loss of 90–100% was reported under delayed sowings (Dhaliwal et al. 2004).

High heritability and genetic advance were reported for shoot fly resistance and morphological traits suggested that these traits were under the control of additive genes, and these traits can be used in selecting genotypes for use in sorghum improvement programmes (Mohammed et al. 2016). While much of the research has dealt with screening to identify sources of resistance, QTL mapping has also been conducted in progeny of some susceptible by resistant crosses. Several physiological factors that contribute to resistance have been identified such as leaf glossiness and density of trichomes (Kiranmayee et al. 2016; Riyazaddin et al. 2016).

3.3 Stem borer

Spotted stem borer, *Chilo partellus* (Lepidoptera: Pyralidae) is predominant in Asia, and eastern and southern Africa, causing an estimated loss of US $1.0 billion in the semi-arid tropical region annually (ICRISAT 1992). Pest infection was reported in all growing stages starting from two-week-old seedlings to harvest. Larvae start feeding on the seedling leaf whorl and start boring the stem with extensive tunnelling, thus causing 'deadheart' symptom which interrupts nutrient supply, finally leading to crop lodging. Spraying insecticide controls the infestation but it is not affordable to poor farmers (Kishore Kumar et al. 2006).

Another predominant species is African stem borer *Busseola fusca* Fuller (Lepidopteran: Noctuidae), which reduces grain yield through leaf feeding, dead heart formation and stem damage (Karaya et al. 2009). The larvae of *Busseola fusca* infest sorghum at the seedling stage and thrive till maturity, resulting in a substantial loss in grain yield (Sally et al. 2007). The larvae remain protected inside the stems, and, thus, are less vulnerable to insecticides and natural enemies (Muhammad et al. 2009). *Busseola fusca* reduces grain yield by more than 15% depending on the pest population density at the time of the attack, crop age and variety (Karaya et al. 2009).

In the case of the spotted stem borer (*Chilo partellus*), transgenic sorghum based on the use of *Bt* toxin has already been produced (Girijashankar et al. 2005; Visarada et al. 2014). Although neither transgenic line gave complete protection, both studies showed significantly decreased leaf damage in transgenic lines versus untransformed controls, and, in the latter report, methods for use in previously difficult-to-transform accessions are described.

3.4 Greenbug

Greenbug [*Schizaphis graminum* (Rondani)] was first reported to infest Johnsongrass [*Sorghum halepense* (L.) Pers.] as early as 1909 (Webster and Phillips 1912) and serious damage to sorghum in Kansas was first reported in 1916 (Hayes 1922) and the pest became virulent on sorghum in 1968 (Harvey and Hackerott 1969). So far, over twenty greenbug biotypes have been identified (Burd and Porter 2006) of which only four (C, E, I and K) attack sorghum. The first greenbug biotype causing widespread damage to sorghum in the US Great Plains was designated biotype C (Harvey and Hackerott 1969). The second greenbug biotype to damage sorghum, biotype E, was identified in 1979 (Porter et al. 1982). Biotype I was identified in 1990 (Harvey et al. 1991) and biotype K in 1992 (Harvey et al. 1997). Biotypes C and E were the most widespread and caused the most damage to sorghum production. Damage to farmers' fields from biotypes I and K was less except for an occasional, isolate outbreak. The biotype I and K-resistant hybrids were not in much demand as compared to biotypes C and E.

Numerous germplasm lines were screened for resistance to each biotype (Table 2). Identified and widely used sources of resistance to biotype C greenbug were SA7536-1, KS30, IS809 and PI264453 (Schuster and Starks 1973; Teetes et al. 1974a,b). Resistance to biotype E was originally discovered in the sorghum line 'Capbam'. Other lines resistant to biotype E were also resistant to biotype C. Initial sources of resistance to biotype I were identified in introductions from Russia (Andrews et al. 1993). Later systematic evaluation of the US sorghum germplasm collection resulted in the identification of 21 lines resistant to biotype I (Huang 2011). Resistance to biotype K was identified in 'Cargill 607E' and PI550607 (Kofoid and Harvey 1992). Resistance to biotype C was determined to be dominant or incompletely dominant (Hackerott et al. 1969; Johnson 1971; Johnson and Teetes 1972; Weibel et al. 1972). Resistance to biotype E was reported to be inherited as a dominant character (Johnson et al. 1981), or complexly inherited and partially dependent on the resistant and susceptible parents (Dixon et al. 1991). Resistance in one source was conditioned by two genes (Kofoid and Harvey 1992). One source of biotype I resistance was found to be incompletely dominant with two dominant genes requiring complementary gene action controlling resistance.

In the case of greenbug, resistance is challenged by constantly evolving biotypes. Currently, biotype I is the most impactful on sorghum. Studies of several QTL mapping experiments using different sources of resistance have been published. The research identified multiple regions conditioning resistance to biotypes C, E, I and K. Katsar et al. (2002) identified resistance to biotypes E, C, I and K from four resistance sources. At least nine loci on eight linkage groups were determined to condition resistance to greenbug. None of the loci were effective against all biotypes. Agrama et al. (2002) evaluated resistance to biotypes I and K. Nine QTLs were detected affecting resistance to biotypes I and K. Four SSRs and one RAPD marker were associated with all resistance or tolerance traits. Nagaraj et al. (2005) identified three QTLs associated with biotype I and five QTLs associated with biotype K. Wu and Huang (2008) identified two QTLs for resistance to biotype I. Punnuri et al. (2013) made crosses between a line with relatively good resistance (PI607900) and BTx623, which is very susceptible. Testing for co-segregation of resistance among 371 F_2 progeny and 102 polymorphic SSRs scattered throughout the genome revealed that resistance is quantitative. In all, 29 markers were associated with some level of resistance; 4 of these accounting for much of the resistance were mapped to chromosome 9 near a receptor-like kinase with homology to a rice gene *Xa*21 (Punnuri et al. 2013). The rice gene codes for a protein that includes a leucine-rich recognition sequence and is involved in conferring resistance to many races of *Xanthomonas oryzae* (bacterial blight) (Jiang et al. 2013b). While it has not been verified that the sorghum gene is involved in protection against greenbug, the CRISPR technology offers an excellent opportunity to test this hypothesis.

In an earlier study by Wu et al. (2006), a total of 30 000 sorghum accessions were tested, and 26 were identified as having weak (8), moderate (13) and strong (5) resistance to greenbug biotype I as compared to a susceptible control. When these accessions, which originated in 10 different countries, were analysed using AFLP, resistant lines fell into two different clusters, suggesting they represent different sources of resistance. In wheat, which is also susceptible to greenbug, a gene (Gb3) for resistance to biotype E has been mapped to a narrow region of chromosome 7D (Azhaguvel et al. 2012). That location has been verified using SNP array analysis (Liu et al. 2016a), so the identification of the *Gb*3 gene is likely to come soon and equivalent gene, knowledge of its function, availability and potential use in sorghum are also possible through transgene technology in the years to come.

3.5 Sugarcane aphid (SCA)

The sugarcane aphid [*Melanaphis sacchari* (Zehntner)] has been reported worldwide in all areas of sorghum production (Eastop 1955, 1965; Mead 1978; CIE 1981) and is considered an economically important pest in China (Wang 1961), India (Young 1970), Japan (Setokuchi 1973), South Africa (van Rensburg 1973), Taiwan (Chang 1981a,b; Pi and Hsieh 1982) and Botswana (Flattery 1982). The sugarcane aphid was reported on sugarcane in the United States in Florida as early as 1922 (Wilbrink 1922) and confirmed in 1978 (Mead 1978; Denmark 1988). The aphid was later confirmed in Louisiana (White 2001). In 2013, SCA infested grain sorghum in Southwest Louisiana, South and East Texas, southern Oklahoma, eastern Mississippi and northeastern Mexico (Villanueva et al. 2014). Ironically, though SCA was reported in mid-1980 in South Africa, it became a serious problem on sugarcane in 2013 (Way et al. 2014). Even with the use of insecticidal compounds, over the last two years, the infestation area enlarged significantly. It was calculated that growers in the lower Rio Grande Valley (South Texas) lost over US $31 million in 2014 and 2015 due to reduced yields and increased expenses for insecticide and cost of applications. By the end of 2016, the aphid had been confirmed in 18 states, basically all areas in the United States where sorghum is grown.

Evaluation of 483 converted exotic sorghum genotypes and elite lines in Botswana and Zimbabwe identified 41 lines with low plant damage and aphid abundance ratings (Teetes et al. 1995). Further evaluation of sorghum accessions, parental lines, elite lines/ experimental varieties and hybrids and local cultivars for resistance has been conducted at many locations (Singh et al. 2004). In the United States none of the commercial cultivars being grown in affected areas when the SCA became virulent on sorghum showed resistance (Armstrong et al. 2015). Nibouche et al. (2014) assessed genetic diversity among 1333 accessions from 15 countries on sugarcane and sorghum. The identified mitochondrial COI gene (658 bp) grouped 96 individuals into five haplotypes with a narrow genetic divergence of 19%. Continuous evaluations identified a diverse array of resistant germplasm. Subsequent evaluation of additional lines at ICRISAT identified seven lines with moderate levels of resistance (Sharma et al. 2014). Seven other lines exhibited susceptibility to damage but suffered from relatively low grain yield loss, indicating these lines may have tolerance to the aphid. Following the sugarcane aphid becoming virulent on sorghum in the United States an evaluation of several lines identified one line that was selected for resistance to two greenbug biotypes (Tx2783) also showed resistance to SCA (Armstrong et al. 2015). Tx2783 was identified as resistant in Botswana and Zimbabwe (Teetes et al. 1995). Currently, biotypes that would be virulent on some resistant lines and not others have not been reported.

Several studies have reported the inheritance of genetic resistance to SCA. Crosses between PI 257595 (highly resistant), 129-3A (moderately resistant) and RTx430 (susceptible) in greenhouse and field studies led to the conclusion that resistance is monogenic and controlled by a single dominant gene (Hsieh and Pi 1982; Pi and Hsieh 1982; Tan et al. 1985). Studies also showed that PI 257595 and 129-3A have the same gene for resistance, and the 129-3A resistance gene has modifiers (Pi and Hsieh 1982). Additive gene action accounts for the resistance expression although dominant and additive gene action is involved (Hsieh and Pi 1988). A cross between RTx430 and 129-3A showed the presence of complimentary gene action. Manthe (1992) studied the expression of resistance in crosses of TAM428 and SC170 (both resistant) crossed to susceptible lines. Segregation ratios indicated the presence of a single dominant gene for resistance. In all crosses, more resistant progeny were identified than expected to indicate the presence of modifiers tightly linked to major resistance genes. Pi and Hsieh (1982) also concluded that TAM428 contains a single dominant gene for resistance. They also concluded that while 129-3A and TAM428 contain the same resistance gene, the 129-3A resistance gene is conditioned by modifiers.

SCA had clearly been a problem in other sorghum-producing areas before its sudden explosion in the United States. Harris-Shultz et al. (2015a) crossed a sweet sorghum-resistant line with a susceptible line AN109, evaluated F_1 and F_2 populations and reported that the resistance is likely conferred by a major recessive gene. Chang et al. (2012) showed that resistance in Chinese grain sorghum cultivar HN16 segregated as a single dominant gene (RMES1) when crossed to a susceptible cultivar and further, they tagged the gene using bulked segregant analysis to identify three linked AFLP markers which were converted to SCARs. Wang et al. (2013) took it one step further by repeating the cross using BTx623, the original line used for sequencing the sorghum genome as a susceptible parent and then examining the published sorghum genome near the SCAR markers to identify microsatellites (SSRs) which were then successfully used to develop PCR primers to amplify markers more closely linked to RMSE1.

4 Other threats: bacterial diseases, *Potyvirus* spp., *Striga* and nematodes

4.1 Bacterial diseases

During the 1900s, literature dealing with bacterial pathogens of sorghum centred on three diseases: bacterial leaf stripe caused by *Pseudomonas* (now *Burkholderia*) *andropogonis*, bacterial leaf streak caused by *X. campestris* and bacterial leaf spot caused by *Pseudomonas syringae*. However, even when present the bacteria caused so little yield loss and were not considered a serious threat. Perhaps for that reason, or perhaps because sorghum breeders have been extremely effective in incorporating resistance, no recent mentions of losses or resistance screening for bacterial pathogens have been reported. However, based on personal observations in the coastal bend area of Texas (Odvody and Magill, unpublished) some stands of Johnsongrass (*Sorghum halepense*) growing near grain sorghum production fields showed high levels of bacterial streak disease, so a reservoir of the pathogen is present should a susceptible cultivar be grown.

4.2 *Potyvirus* spp.

Although many viruses have been found in sorghum, they seem to have minimal effect on yield, so relatively little recent information on resistance is available. A specific gene, 'Krish', originally from *Sorghum halepense* has been widely effective for many years against various types of mosaic viruses that are members of the Potyvirus family (Gordon and Thottappilly 2013). Another recent study identified IS 7679 and IS 20740 as sources of resistance to MDMV (*Maize dwarf mosaic virus*) and several others, but not all, mosaic viruses (Seifers et al. 2012). The most prevalent viral diseases in India recently have been reported to be stripe disease of maize caused by *Maize stripe virus* (MSpV-Sorg) and sorghum mosaic caused by *Sugarcane streak mosaic virus* (SStMV-Sorg) and a rhabdovirus (Indian sorghum mosaic virus-ISMV) (Srinivas et al. 2013). Ironically, 'sorghum' mosaic virus seems to be much more damaging to sugarcane than sorghum.

4.3 *Striga*

Where *Striga* is endemic, it creates one of the worst biotic stresses on sorghum and other crops. Seeds of this parasitic weed can survive long periods in the soil and are stimulated to germinate by exudate from host roots that include a strigolactone. Inbred lines that differ in resistance have been discovered, and at least two of these (IS9830 and N13) were used to create segregating RILs for use in QTL mapping (Haussmann et al. 2004). Overall, 11 and 9 QTLs were identified as contributing alleles from each resistant parent when the progeny were tested across 5 locations in multiple years. Five of the QTLs were shared by both resistant parents. More recently, a project has successfully used marker-assisted selection (MAS) in a backcrossing scheme to incorporate these QTLs into a variety favoured by sorghum growers in Eritrea (Yohannes et al. 2015). While not as resistant as the N13 parent, progeny lines with three or four of the QTLs showed significantly improved production in test plots. A single gene (*lgs* for low germination stimulant) that reduces the amount of strigolactone secreted has also been tagged with SSR markers to a 2 cM region (Satish et al. 2012), which contains several candidate genes. If the gene (genes) from cultivar SRN39 was (were) not included in the previous groups, its addition to the improved lines could provide added protection against *Striga*. If *lgs* can be identified and cloned, it also becomes a candidate for use in both transgenic resistance and MAS.

4.4 Nematodes

Although sorghums, in general, do not seem to suffer losses to nematodes, some of the sweet sorghum varieties that are being assessed for use in producing biofuels are susceptible. In South Africa, sweet *Sorghum bicolor* cv. 'Ndendane-X1' was identified as resistant to both *Meloidogyne incognita* and *Meloidogyne javanica* (http://transylvanianreview.org/index.php/Trr/article/view/551). Likewise, Harris-Shultz et al. (2015b) working in Georgia (USA) have been able to identify a major QTL on chromosome 3 in the resistant cultivar 'Honey Drip', and tagged it between two SSR markers around 6 cM apart. The ability to transfer a cotton gene for resistance to *Meloidogyne incognita* to tobacco (GHNTR1) and the fact that it is a typical NB-LRR-based 'R' gene (Zhang et al. 2015) suggests that transformation may be a rapid alternative for incorporation in sorghum once the identified gene has been cloned.

5 Resistance genes

5.1 Structure and function

Over 16000 plant genes that individually confer resistance to specific pathogens have now been cloned or identified via sequence homology (Sanseverino et al. 2012). The great majority of these 'R' genes have been found to encode proteins that share two conserved motifs, namely an NB site motif, and an LRR (McHale et al. 2006). Many of these also have conserved N-terminal motifs. In plants, these are either of two types, CC (coiled coil) or TIR (Toll, interleukin-1 receptor), although the TIR class is rare or absent in monocots such as sorghum (Ve et al. 2015). Consequently, as full genome sequences are released for a plant species, it is possible to use refined searches to estimate the number of resistance gene analogue (RGA) family members that are present. Searches of the sorghum genome have revealed 228-245 RGAs in this class (Cheng et al. 2010; Mace et al. 2014), far fewer than the 600 (Linden et al. 2004) to 900 (Ballini et al. 2008) were detected in rice.

Under current models, members of the NB-LRR class of R genes (also often referred to as NLR genes for NB-LRR Receptor genes) are expressed in host cells and 'recognize' invasion by a pathogen. The interaction initiates a signal transduction path that leads to expression of genes that function in host defence. Most recent models assume that the R gene-encoded proteins are maintained in the cell at a steady state involving auto-inactivated complexes of self or paired gene products (Eitas 2010; Takken and Goverse 2012). Steady-state levels of each complex require not only mRNA transcription and translation but also interactions with chaperones and other proteins and ubiquitin-induced proteasome degradation (Cheng and Li 2012; Takken and Goverse 2012). A further complication involves the discovery that the amount of mRNA present from NB-LRR genes can be altered by microRNAs (Shivaprasad et al. 2012). However, there is also evidence from some hosts, including sorghum, that expression of specific NB-LRR genes is induced in pathogen-resistant cultivars by exposure to the pathogen (Biruma et al. 2012; Martin et al. 2011).

Whether constitutive or induced, each NB-LRR complex acts as a sensor for the detection of a potentially pathogenic microbe through molecular interactions. Consequently, if detected, a factor (now often referred to as an effector) that may otherwise increase pathogen virulence becomes an avirulence factor. CC domains, when present, allow homo- or hetero-aggregates to assemble, which may be critical for function. An example of a hetero-protein complex being involved in pathogen resistance comes from rice: interaction of RGA4 and RGA5 lead to blast resistance (Cesari et al. 2013). The LRR motif allows for interactions of the R gene product with other proteins, possibly of pathogen origin, and possibly in combination with host components (the guard hypothesis) to activate NB activity. NB site changes, involving addition or removal of a phosphate, serve to alter the R protein conformation (Takken and Goverse 2012). The new conformation provides the ability to initiate signal transduction pathways, typically involving plant 'WRKY' or 'SPL' transcription factors that then interact with enhancer elements in the promoters of defence response genes to activate transcription (Liu et al. 2016b). Since transcription factors with the WRKY amino acid sequence are often involved in activating stress response genes, it should not be surprising that pathogens have evolved effectors that target the same sequence to prevent active defence, or that plants, in turn, have evolved 'decoys' that can trap such effectors (Nishimura et al. 2015).

Common induced defence responses include expression of genes whose products have antimicrobial effects. Examples include creation of reactive oxygen species, chitinases and glucanases that destroy fungal cell walls, genes in the PAL or terpenoid pathways that lead to compounds with antimicrobial activity (phytoalexins), or genes that trigger apoptosis (Ingle et al. 2006; Katilé 2008; Spoel and Dong 2012). It should be noted that NB-LRR-mediated resistance is not limited to fungal or bacterial pathogens but can also be involved in virus and insect resistance (Dogimont et al. 2014).

Host defence mechanisms are typically activated in both susceptible and resistant interactions, meaning that the speed and level of response may be crucial in determining the outcome. The extremely rapid response can trigger a hypersensitive response (HR), in which host cells die to prevent the spread of a biotrophic pathogen. This HR defence mechanism can even extend to what are referred to as non-host reactions, meaning that microbial species pathogenic to other plant hosts are recognized and prevented from ever gaining a foothold in the 'non-host' (Huitema et al. 2003; Mysore and Ryu 2004). In this case, the recognition often occurs via pathogen recognition receptors (PRRs) that span the host cell membrane with the recognition component external rather than via the internal NB-LRR gene products (Dangl et al. 2013). While non-host resistance may occur by other mechanisms including structural features such as thick cuticles or through the presence of transmembrane PRRs (Lipka et al. 2008), some non-host interactions directly involve classic NB-LRR genes (Sarris Panagiotis et al. 2011). Both PRR- and NB-LRR-based resistance mechanisms involve induction of the same active host defence pathways.

5.2 Gene deployment

The most direct method of using defined R gene alleles to enhance durable resistance in crop plants is the use of MAS. Since even a SNP or larger change that differentiates an R from an S allele can be detected with appropriate PCR primers or from sequencing data, the progeny of a cross carrying the desired NB-LRR allele(s) can be identified. In theory, with appropriate crosses, it would be possible to combine a series of R genes that interact with different effectors to confer multiple layers of resistance. Since multiple changes would then be required for the pathogen to overcome resistance, stacking genes in this manner is expected to provide durable resistance. Next-generation sequencing technology has developed to the point that whole genome sequencing is a realistic option for use in MAS, especially if reduced representation is sufficient to identify polymorphisms (Elshire et al. 2011). Also, data handling progress and bioinformatics capacity to create and align base sequences in all chromosomes in a manner that allows customers to explore polymorphisms of all types down to the level of genes will permit tracking multiple NB-LRR genes simultaneously. The software company known as NRGene process was credited with the rapid sequencing of the genomes of hexaploid wheat based on ILLUMINA sequencing data in 2016; (see Wheat Sequencing Consortium Releases Key Resource to the Scientific Community press release http://www.wheatgenome.org/News/Press-releases/Wheat-Sequencing-Consortium-Releases-Key-Resource-to-the-Scientific-Community).

However, there are potential problems associated with MAS, even when progeny with the desired combination of R genes can be identified. In addition to the several generations of breeding involved, there is a good chance that other undesirable alleles may be introduced, that is, linkage drag. In part, this is because many of the R genes occur in closely linked clusters (Yang and Wang 2016) with no guarantee that all nearby alleles will be beneficial and that close linkage may make desired combinations rare. Wheat stem rust including *Puccinia graminis* f. sp. *tritici* strain UG99 that spread rapidly from Africa into Asia. Wheat resistance gene *Sr22* is a good example as it was found to confer broad spectrum of resistance (Steuernagel et al. 2016).

Thus, indirect methods of gene transfer may prove to be more efficient than MAS, again by taking advantage of recent advances in technology. A key element is the ability to identify and clone specific R genes. Sequence capture technology based on hybridization to conserved motifs has proven to be effective in cloning genomic fragments that include R genes (Jupe et al. 2013; Steuernagel et al. 2016; Witek et al. 2016). An excellent example of stacking using cloned R genes comes from potato where three unique R genes on a single binary vector were added to a late blight-susceptible cultivar using Agroinfection (Haverkort et al. 2016; Zhu et al. 2012). While the added genes are functional and provide resistance, this method of transformation generally leads to random insertion of the transgenes in the host genome. In many cases, targeted gene replacement may be a better option. The ability of CRISPR/Cas9 systems to create targeted double-strand breaks in DNA is already proving valuable, not simply a means to disrupt genes to verify function, but also for targeted gene replacement. While the system has been demonstrated to function in sorghum and rice (Jiang et al. 2013a), the best current example of the possibilities for CRISPR-Cas9 in a cereal crop comes from maize. Svitashev et al. (2015) bombarded immature embryos with RNA guides and a *Cas9* gene to create double strand breaks at 12 specific sites at five locations in maize chromosomes. By also adding regions of homology they created specified missense mutations in an *ALS* gene and by including a sequence with outside homology to a target gene that was removed as directed by two nearby guide RNAs, they accomplished targeted gene insertion. In these studies, selectable markers were used so that transformed callus could be selected and grown into plants. Thus, for any system where callus regeneration is feasible, the ability to add specific R genes should be practical. This includes creating promoter changes or exchanges for enhanced expression as well. If the transferred genes can be directly used for selection as is often the case for host plant resistance, the need for added selectable markers would be negated. In any case, the CRISPR-based system offers an excellent mechanism for verifying gene function through the creation of knockouts, and in many cases may be more practical and informative than the use of VIGS.

6 Summary

Genes that offer resistance to the major biotic stresses, diseases and insects that affect sorghum production have been identified but for the most part only by studies based on mapping or co-segregation with DNA-based polymorphisms of one sort or another. Through primarily classical breeding approaches, plant breeders working with plant pathologists and entomologists have achieved success in developing improved parental lines and thus hybrids resistant to one or more biotic stress(es). This success was achieved through knowledge of germplasm and proper methodology combined with field and laboratory activity to select resistant genotypes along with the myriad of other traits required to produce superior hybrids or cultivars. A partial list of successes will include hybrids resistant to anthracnose, head smut, downy mildew, foliar diseases, grain mould and grain weathering, *Striga*, greenbug and sorghum midge. Deployment of resistant hybrids or cultivars has placed selection pressure on the stress-causing organism resulting in new disease pathotypes (races) or insect biotypes. The process to identify new resistance genes and incorporating the genes into elite germplasm must then be repeated and constantly be continued as an integral part of the classical breeding programmes. Also, the identified exotic photosensitive resistant sources in sorghum for both diseases and pests are needed to be converted into adapted breeding line(s) background continuously through partial conversion programmes integrating both classical breeding and molecular marker technology as followed by Klein et al. (2016).

The new molecular biological technologies offer the opportunity to accelerate progress and improve efficiency in understanding resistance genes and incorporating the genes into germplasm for subsequent deployment. These technologies provide the opportunity to determine exact chromosomal location of the resistance genes and determine the relationship to other resistance genes. If the resistance genes are different, MAS will enable more than one resistance gene to be incorporated into the genome of elite germplasm for subsequent deployment. Although new races or biotypes will still become virulent, the time interval between the old and new types may be increased. Also, as knowledge of the structure of R genes increases and the techniques for tracking or transferring them become simpler and less costly, prospects for creating more durable forms of resistance seem near at hand. In this chapter, we have attempted to describe and reference many of the R genes that have already been identified and tagged. While these genes will not eliminate the need for the discovery of additional sources of resistance in the future, we hope the information will be useful in future efforts to develop sorghums with combined resistance to the same as well as multiple threats.

Taken together, a coordinated programme incorporating classical and molecular approaches combined with access to field evaluation environments will result in great progress to understanding biotic stress(es) on sorghum and the genetic technology required to alleviate the stress. Increased reduction in sustainability can potentially reduce environmental disruption and will be achieved at little if any cost to farmers.

7 Where to look for further information

A wealth of genetic resources of more than 30 000 exotic photo-period-sensitive sorghum germplasms still remain unexploited. Enormous research efforts are needed to develop many new sorghum conversion lines integrating both classical and molecular breeding approaches. Several identified exotic resistance sources for most of the sorghum diseases and pests respectively for many pathotypes and biotypes are to be given top priority in sorghum conversion programmes. This would help to develop several new breeding populations by introgressing biotic stress tolerance/resistance traits into adapted breeding lines and to develop potential high-yielding hybrids for these biotic stresses.

8 Acknowledgements

This publication is contribution no. 17-300-B-B from the Kansas Agricultural Experiment Station, Manhattan, KS.

9 References

Adeyanju, A., Little, C., Yu, J. and Tesso, T. 2015. Genome-wide association study on resistance to stalk rot diseases in grain sorghum. *G3: Genes Genom Genet.* 5:1165–75.

Agrama, H. A., Wilde, G. E., Reese, J. C., Campbell, L. R. and Tuinstra, M. R. 2002. Genetic mapping of QTLs associated with greenbug resistance and tolerance in *Sorghum bicolor*. *Theor. Appl. Genet.* 104:1373–8.

Agrawal, B. L., Abraham, C. V. and House, L. R. 1988. Inheritance of resistance to midge, *Contarinia sorghicola* Coq., in sorghum, *Sorghum bicolor* (L.) Moench. *Insect Sci. Appl.* 9:43–5.

Andrews, D. J., Bramel-Cox P. J. and Wilde, G. E.. 1993. New sources of resistance to greenbug, biotype I, in sorghum. *Crop Sci.* 33:198–9.

Apotikar, D. B., Venkateswarlu, D., Ghorade, R. B., Wadaskar, R. M., Patil, J. V. and Kulwal, P. L. 2011. Mapping of shoot fly tolerance loci in sorghum using SSR markers. *J. Genet.* 90:59–66.

Armstrong, J. S., Rooney, W. L., Peterson, G. C., Villenueva, R. T., Brewer, M. J. and Sekula-Ortiz, B. 2015. Sugarcane aphid (Hemiptera: Aphididae): host range and sorghum resistance including cross-resistance from greenbug sources. *J. Econ. Entomol.* 108:576–82.

Aruna, C., Bhagwat, V. R., Madhusudhana, R., Sharma, V., Hussain, T., Ghorade, R. B., Khandalkar, H. G., Audilakshmi, S. and Seetharama, N. 2011. Identification and validation of genomic regions that affect shoot fly resistance in sorghum [*Sorghum bicolor* (L.) Moench]. *Theor. Appl. Genet.* 122(8):1617–30.

Ashok Kumar, A., Reddy, B. V. S., Thakur, R. P., Sharma, R., Ramaiah, B. and Rao, V. P. 2011. Genetic enhancement for grain mold resistance and grain yield in white pericarp sorghum. *Indian J. Dryland Agric. Res. Devel.* 26:21–7.

Azhaguve, I. P., Rudd, J. C., Ma, Y., Luo, M. C. and Weng, Y. 2012. Fine genetic mapping of greenbug aphid-resistance gene Gb3 in *Aegilops tauschii*. *Theor. Appl. Genet.* 124:555–64.

Ballini, E., Morel, J. B., Droc, G., Price, A., Courtois, B., Notteghem J. L. and Tharreau, D. 2008. A genome-wide meta-analysis of rice blast resistance genes and quantitative trait loci provides new insights into partial and complete resistance. *Mol. Plant Microbe Interact.* 21:859–68.

Bandyopadhyay, R., Frederickson, D. E., McLaren., N, Odvody, G. N. and Ryley, M. J. 1998. Ergot: A new disease threat to sorghum in the Americas and Australia. *Plant Dis.* 82:356–67.

Bandyopadhyay, R., Butler, D. R., Chandrashekar, A., Kanaka Reddy, R. and Navi, S. S. 2000. Biology, epidemiology, and management of sorghum grain mold, pp. 34–71. In: Chandrashekar, A., Bandyopadhyay, R. and Hall, A. J. (Eds), *Technical and Institutional Options for Sorghum Grain Mold Management: Proceeding of an International Consultation*, May 2000, ICRISAT, Patancheru, India.

Bayoumy, M. H., Perumal, R. and Michaud, J. P. 2015. Comparative life histories of greenbugs and sugarcane aphids (Hemiptera: Aphididae) co-infesting susceptible and resistant sorghums. *J. Econ. Entomol.* 109:385–91.

Bergquist, R. R., Rotar, P. and Mitchell, W. C. 1974. Midge and anthracnose head blight resistance in sorghum. *Trop. Agric.* 51:431–5.

Beshir, M., Ali, A., Rubaihayo, P., Ahmed, N. and Okori, P. 2016. Simple sequence repeat markers associated with anthracnose and Turcicum leaf blight resistance in sorghum. *Afr. Crop Sci. J.* 24:97–107.

Biruma, M., Martin, T., Fridborg, I., Okori, P. and Dixelius, C. 2012. Two loci in sorghum with NB-LRR encoding genes confer resistance to *Colletotrichum sublineolum*. *Theor. Appl. Genet.* 124:1005–15.

Boora, K. S., Frederiksen, R. and Magill, C. 1998. DNA-based markers for a recessive gene conferring anthracnose resistance in sorghum. *Crop Sci.* 38:1708–9.

Boozaya-Angoon, D., Starks, K. J., Weibel, D. E. and Teetes, G. L. 1984. Inheritance of resistance in sorghum, *Sorghum bicolor*, to the sorghum midge, *Contarinia sorghicola* (Diptera: Cecidomyiidae). *Environ. Entomol.* 13:1541–3.

Burd, J. and Porter, D. 2006. Biotypic diversity in greenbug (Hemiptera: Aphididae): Characterizing new virulence and host associations. *J. Econ. Entomol.* 99:959–65.

Burrell, A. M., Sharma, A., Patil, N. Y., Collins, S. D., Anderson, W. F., Rooney, W. L. and Klein, P. E. 2015. Sequencing of an anthracnose-resistant sorghum genotype and mapping of a major QTL reveal strong candidate genes for anthracnose resistance. *Crop Sci.* 55:790–9.

Casela, R., Ferreira, A. S. and Pinto, N. F. J. A. 1999. Sugary disease of sorghum in Brazil: An overview, pp. 23–5. In: Casela, C. R. and J. A. Dahlberg (Eds), *Proc. Global Conf. Ergot Sorghum (1–8 June 1997)*. EMBRAPA, Sete Lagaos, Brazil.

Cesari, S., Thilliez, G., Ribot, C., Chalvon, V., Michel, C., Jauneau, A., Rivas, S., Alaux, L., Kanzaki, H., Okuyama, Y., Morel, J.-B., Fournier, E., Tharreau, D., Terauchi, R. and Kroj, T. 2013. The rice resistance protein pair RGA4/RGA5 recognizes the *Magnaporthe oryzae* effectors AVR-Pia and AVR1-CO39 by direct binding. *Plant Cell* 25:1463–81.

Chala, A. and Tronsmo, A. M. 2012. Evaluation of Ethiopian sorghum accessions for resistance against *Colletotrichum sublineolum*. *Eur. J. Plant Pathol.* 132:179–89.

Chang, N. T. 1981a. Resistance of some grain sorghum cultivars to sorghum aphid injury. *Plant Prot. Bull. (Taiwan)* 23:35–41.

Chang, S. C. 1981b. Sources of resistance in sorghum to sugarcane aphid, *Melanaphis sacchari* (Zehntner). *Rep. Corn Res. Cent.*, Taiwan DAIS 15:10–14.

Chang, J.-H., Xue, W. and Zhang, Q. 2012. Identification of molecular markers for a aphid resistance gene in sorghum and selective efficiency using these markers. *J. Integrative Agric.* 11:1086–92.

Cheng, Y. T. and Li, X. 2012. Ubiquitination in NB-LRR-mediated immunity. *Curr. Opin. Plant Biol.* 15:392–9.

Cheng, X., Jiang, H., Zhao, Y., Qian, Y., Zhu, S. and Cheng, B. 2010. A genomic analysis of disease-resistance genes encoding nucleotide binding sites in *Sorghum bicolor*. *Genet. Mol. Biol.* 33:292–7.

CIE (Commonwealth Institute of Entomology). 1981. Sugarcane aphid (*Melanaphis sacchari*: Zehntner). Distribution Maps of Pests. Series A (Agriculture). Map No. 420, Farnham Royal, UK.

Craig, J. and Frederiksen, R. A. 1992. Comparison of sorghum seedling reactions to *Sporisorium reilianum* in relation to sorghum head smut resistant classes. *Plant Dis.* 76:314–18.

Cuevas, H., Prom, L. K. and Magill, C. 2012. Reaction to rust by a subset of sorghum accessions from Zimbabwe. *Asian J. Plant Pathol.* 6:33–40.

Cuevas, H. E., Prom, L. K., Erpelding, J. E. and Brotons, V. 2014a. Assessment of genetic diversity and anthracnose disease response among Zimbabwe sorghum germplasm. *Plant Breed.* 133(2):234–42.

Cuevas, H. E., Prom, L. K. and Erpelding, J. E. 2014b. Inheritance and molecular mapping of anthracnose resistance genes present in sorghum line SC112-14. *Mol. Breed.* 34:1943–53.

Cuevas, H. E., Prom, L. K., Isakeit, T. and Radwan, G. 2016. Assessment of sorghum germplasm from Burkina Faso and South Africa to identify new sources of resistance to grain mold and anthracnose. *Crop Prot.* 79:43–50.

Dangl, J. L., Horvath, D. M. and Staskawicz, B. J. 2013. Pivoting the plant immune system from dissection to deployment. *Science* 341(6147). doi: 10.1126/science.1236011.

Das, I. K., Rao, S. V. and Sanjana Reddy, P. 2007. Charcoal/Stalk rot. In: *Screening Techniques for Sorghum Diseases*. Information Bulletin # 76. International Crops Research Institute for the Semi-Arid Tropics (ICRISAT), Pantacheru - 502 324, India, pp. 40–6.

Denmark, H. A. 1988. Sugarcane aphids in Florida (Homoptera: Aphididae). Entomology Circular No. 302. Division of Plant Industry, Florida Department of Agriculture and Consumer Services, p. 2.

Deol, G. S., Reese, J. C. and Gill, B. S. 1997. A rapid nondestructive technique for assessing chlorophyll loss from greenbug (Homoptera:Aphididae) feeding damage on sorghum leaves. *J. Kans. Entomol. Soc.* 70:305–12.

Dhaliwal, G. S., Arora, R. and Dhavan, A. K. 2004. Crop losses due to insect pests in Indian agriculture. An update. *Indian J. Ecol.* 31:1–7.

Diarisso, N. Y., Pendleton, B. B., Teetes, G. L., Peterson, G. C. and Anderson, R. M. 1998a. Floret morphology of sorghum midge-resistant sorghum. *Southwest. Entomol.* 23:67–75.

Diarisso, N. Y., Pendleton, B. B., Teetes, G. L., Peterson, G. C. and Anderson, R. M. 1998b. Spikelet flowering time: Cause of sorghum resistance to sorghum midge (Diptera: Cecidomyiidae). *J. Econ. Entomol.* 91(6):1464–70.

Dixon, A. G., Bramel-Cox, P. J. and Harvey, T. L. 1991. Complementarity of genes for resistance to greenbug [*Schizaphis graminum* (Rondani)], biotype E in sorghum [*Sorghum bicolor* (L.) Moench]. *Theor. Appl. Genet.* 81:105–10.

Dogimont, C., Chovelon, V., Pauquet, J., Boualem, A. and Bendahmane, A. 2014. The Vat locus encodes for a CC-NBS-LRR protein that confers resistance to *Aphis gossypii* infestation and *A. gossypii*-mediated virus resistance. *Plant J.* 80, 993–1004.

Eastop, V. F. 1955. Notes on East African aphids. VI. Cereal and grass root feeding species. *East Afr. Agric. J.* 20:209–12.

Eastop, V. F. 1965. A taxonomic study of *Australina Aphidoidea* (Homoptera). *Aust. J. Zool.* 14:399–592.

Edmunds, L. K. 1963. Use of sporidial hypodermic injection to test sorghum for head smut resistance. *Plant Dis. Rep.* 47:909–13.

Eitas, T. and Dangl, J. L. 2010. NB-LRR proteins: Pairs, pieces, perception, partners, and pathways. *Curr. Opin. Plant Biol.* 13:1–6.

Elshire, R. J., Glaubitz, J. C., Sun, Q., Poland, J. A., Kawamoto, K., Buckler, E. S. and Mitchell, S. E. 2011. A robust, simple genotyping-by-sequencing (GBS) approach for high diversity species. *PLoS ONE* 6:e19379.

Erpelding, J. E. and Prom, L. K. 2004. Evaluation of Malian sorghum germplasm for resistance against anthracnose. *Plant Pathol. J.* 3:65–71.

Erpelding, J. E. and Prom, L. K. 2006. Variation for anthracnose resistance within sorghum germplasm collection from Mozambique, Africa. *Plant Pathol. J.* 5:28–34.

Felderhoff, T., McIntyre, L., Saballos, A. and Vermerris, W. 2016. Using genotyping by sequencing to map two novel anthracnose resistance loci in *Sorghum bicolor*. *G3: Genes Genom. Genet.* 6:1935–46.

Flattery, K. E. 1982. An assessment of pest damage of grain sorghum in Botswana. *Exp. Agric.* 18:319–28.

Futrell, M. C. and Webster, O. J. 1965. Ergot infection and sterility in grain sorghum. *Plant Dis. Rep.* 49:680–3.

Gamar, Y. A. and Mohamed, A. H. 2013. Introgression of Striga resistance genes into a Sudanese sorghum cultivar, Tabat, using marker assisted selection (MAS). *Greener J. Agric. Sci.* 3:550–6.
Girijashankar, V., Sharma, H. C., Sharma, K. K., Swathisree, V., Prasad, L. S., Bhat, B. V., Royer, M., Secundo, B. S., Narasu, M. L., Altosaar, I. and Seetharama, N. 2005. Development of transgenic sorghum for insect resistance against the spotted stem borer (*Chilo partellus*). *Plant Cell Rep.* 24:513–22.
Gordon, D. and Thottappilly, G. 2013. Maize and sorghum. In: Loebenstein, G. and Thottappilly, G. (Eds), *Virus and Virus-like Diseases of Major Crops in Developing Countries*. Springer-Verlag Berlin Heidelberg, USA, XLVII, pp. 295–336. ISBN: 978-94-010-3729-7.
Gowda, B., Xu, G.-X., Frederiksen, R. and Magill, C. W. 1995. DNA markers for downy mildew resistance in sorghum. *Genome* 38:823–6.
Hackerott, H. L., Harvey, T. L. and Ross, W. M. 1969. Greenbug resistance in sorghum. *Crop Sci.* 9:656–8.
Harris-Shultz, K., Ni, X., Knoll, J. and Anderson, W. 2015a. Genetic studies on sugarcane aphid resistance in sorghum. In: Jugulam, M., Burow, G. B. and Everhart-Valentin, Kira (Eds), *Proceedings of the 2015 Sorghum Improvement Conference of North America (SICNA)*, 1–3 September 2015. Manhattan, Kansas USA, p. 15.
Harris-Shultz, K. R., Davis, R. F., Knoll, J. E., Anderson, W. and Wang, H. 2015b. Inheritance and identification of a major quantitative trait locus (QTL) that confers resistance to *Meloidogyne incognita* and a novel QTL for plant height in sweet sorghum. *Phytopathol.* 105:1522–8.
Harvey, T. L. and Hackerott, H. L. 1969. Recognition of a greenbug biotype injurious to sorghum. *J. Econ. Entomol.* 62:776–9.
Harvey, T. L., Kofoid, K. D., Martin, T. J. and Sloderbeck, P. E. 1991. A new greenbug virulent to E-biotype resistant sorghum. *Crop Sci.* 31:1689–991.
Harvey, T. L., Wilde, G. E. and Kofoid, K. D. 1997. Designation of a new greenbug biotype, biotype K, injurious to resistant sorghum. *Crop Sci.* 37:989–91.
Haussmann, B. I., Hess, D. E., Omanya, G. O., Folkertsma, R. T., Reddy, B. V., Kayentao, M., Welz, H. G. and Geiger, H. H. 2004. Genomic regions influencing resistance to the parasitic weed *Striga hermonthica* in two recombinant inbred populations of sorghum. *Theor. Appl. Genet.* 109:1005–16.
Haverkort, A. J., Boonekamp, P. M., Hutten, R., Jacobsen, E., Lotz, L. A. P., Kessel, G. J. T., Vossen, J. H. and Visser, R. G. F. 2016. Durable late blight resistance in potato through dynamic varieties obtained by cisgenesis: Scientific and societal advances in the DuRPh project. *Potato Res.* 59:35–66.
Hayes, W. P. 1922. Observations on insects attacking sorghum. *J. Econ. Entomol.* 51:341–56.
Hsieh, J. and Pi, C. P. 1982. Genetic study on aphid resistance in sorghum. *J. Agric. Assoc. China New Ser.* 117:6–14.
Hsieh, J. and Pi, C. P. 1988. Diallel analysis of resistance to aphid in sorghum. *J. Agric. Assoc. China* 152:67–84.
Huang, Y. 2011. Improvement of crop protection against greenbug using the worldwide sorghum germplasm collection and genomics-based approaches. *Plant Genet. Resour.* 9:317–20.
Huang, Y. 2015. Identification, characterization, and development of insect resistant wheat, barley, and sorghum germplasm. USDA Agricultural Research Service Plains Area, Sorghum Research Accomplishments, USDA-ARS-PA-NRRC, Office of the Director, 2150 Centre Avenue, Building D, Suite 300, Fort Collins, CO, 80526 USA, p. 4.
Huitema, E., Vleeshouwers, V. G., Francis, D. M. and Kamoun, S. 2003. Active defence responses associated with non-host resistance of *Arabidopsis thaliana* to the oomycete pathogen *Phytophthora infestans*. *Mol. Plant Pathol.* 4:487–500.
Hulluka, M. and Esele, J. P. E. 1992. Sorghum diseases in East Africa, pp. 21–4. In: de Millano, W. A. J., Frederiksen, R. A. and Bengston, G. D (Eds), *Sorghum and Millet Diseases: A Second World Review*. ICRISAT, Patancheru, AP, India.
ICRISAT 1992. *The Medium Term plan. Annual Progress Report*, Volume II. International Crop Research Institute for the Semi-Arid Tropics, Patancheru - 502 324, Andhra Pradesh, India, p. 312.

Ingle, R. A., Carstens, M. and Denby, K. B. 2006. PAMP recognition and the plant-pathogen arms race. *BioEssays* 28:880–9.

Isakeit, T. and Jaster, J. 2005. Texas has a new pathotype of *Peronosclerospora sorghi*, the cause of sorghum downy mildew. *Plant Dis.* 89:529.

Isakeit, T., Odvody, G. N. and Shelby, R. A. 1998. First report of sorghum ergot caused by *Claviceps africana* in the United States. *Plant Dis.* 82:592.

Jampatong, C., Jampatong, S., Jompuk, C., Sreewongchai, T., Grudloyma, P., Balla, C. and Prodmatee, N. 2013. Mapping of QTL affecting resistance against sorghum downy mildew (*Peronosclerospora sorghi*) in maize (*Zea mays* L). *Maydica.* 58:119–26

Jardine, D. J. and Leslie, J. F. 1992. Aggressiveness of *Gibberella fujikuroi* (*Fusarium moniliforme*) isolates to grain-sorghum under greenhouse conditions. *Plant Dis.* 76:897–900.

Jiang, W., Zhou, H., Bi, H., Fromm, M., Yang, B. and Weeks, D. P. 2013a. Demonstration of CRISPR/Cas9/sgRNA-mediated targeted gene modification in Arabidopsis, tobacco, sorghum and rice. *Nucleic Acids Res.* 41:e188.

Jiang, Y., Chen, X., Ding, X., Wang, Y., Chen, Q. and Song, W. Y. 2013b. The XA21 binding protein XB25 is required for maintaining XA21-mediated disease resistance. *Plant J.* 73:814–23.

Johnson, J. W. 1971. Evaluation of sorghum for greenbug resistance. *Sorghum Newsl.* 14:114–16.

Johnson, J. W. and Teetes, G. L. 1972. Evaluation of sorghum hybrids and lines for adult plant greenbug resistance. *Sorghum Newsl.* 15:137.

Johnson, J. W., Rosenow, D. T. and Teetes, G. L. 1973. Resistance to the sorghum midge in converted exotic sorghum cultivars. *Crop Sci.* 13:174–5.

Johnson, J. W., Teetes, G. L., Phillips, J. M. and Riggs, V. 1981. Resistance to biotype E greenbug in sorghum. *Sorghum Newsl.* 24:84.

Jupe, F., Witek, K., Verweij, W., Sliwka, J., Pritchard, L., Etherington, G. J., Maclean, D., Cock, P. J., Leggett, R. M., Bryan, G. J., Cardle, L., Hein, I. and Jones, J. D. 2013. Resistance gene enrichment sequencing (RenSeq) enables reannotation of the NB-LRR gene family from sequenced plant genomes and rapid mapping of resistance loci in segregating populations. *Plant J.* 76:530–44.

Kapanigowda, M., Perumal, R., Maduraimuthu, D., Aiken, M. R., Prasad P. V. V., Tesso. T. and Little, C. R. 2013. Genotypic variation in sorghum [*Sorghum bicolor* (L.) Moench] exotic germplasm collection for drought and disease tolerance. *SpringerPlus.* 2:650. DOI:10.1186/2193-1801-2-650. URL: http://www.springerplus.com/content/2/1/650

Karaya, H., Njoroge, K., Mugo, S. and Nderitu, H. 2009. Combining ability among twenty insect resistant maize inbred lines resistant to *Chilo partellus* and *Busseola fusca* Stem borers. *J. Plant Prod.* 3(1):115–26.

Katilé, S. O. 2008. Expression of defence genes in sorghum grain mold and tagging and mapping a sorghum anthracnose resistance gene. PhD Thesis Dissertation, Department of Plant Pathology & Microbiology, Texas A&M University, College Station, Texas, USA, p. 168.

Katilé, S., Perumal, R., Rooney, W. L., Prom, L. K. and Magill, C. W. 2010. Expression of pathogenesis-related protein PR-10 in sorghum floral tissues following inoculation with *Fusarium thapsinum* and *Curvularia lunata*. *Mol. Plant Pathol.* 11:93–03.

Katsar, C. S., Peterson, A. H., Teetes, G. L. and Peterson, G. C. 2002. Molecular analysis of sorghum resistance to the greenbug (Homoptera; Aphididae). *J. Econ. Entomol.* 95:448–57.

Khangura, R. and Aberra, M. 2009. First report of charcoal rot on canola caused by *Macrophomina phaseolina* in Western Australia. *Plant Dis.* 93:666–7.

Kiranmayee, K. N. S. U., Kishor, P. B. K., Hash, C. T. and Deshpande, S. P. 2016. Evaluation of QTLs for shoot fly (*Atherigona soccata*) resistance component traits of seedling leaf blade glossiness and trichome density on sorghum (*Sorghum bicolor*) chromosome SBI-10L. *Trop. Plant Biol.* 9:12–28.

Kishore Kumar, V., Sharma, H. C. and Dharma Reddy, K. 2006. Antibiosis mechanism of resistance to spotted stem borer, *Chilo partellus* in sorghum, *Sorghum bicolor*. *Crop Prot.* 25:66–72.

Klein, R. R., Rodriguez-Herrera, R., Schlueter, J. A., Klein P. E., Yu, Z. H. and Rooney, W. L. 2001. Identification of genomic regions that affect grain-mould incidence and other traits of agronomic importance in sorghum. *Theor. Appl. Genet.* 102:307–19.

Klein, R. R., Miller, F. R., Bean, S. and Klein, P. E. 2016. Registration of 40 converted germplasm sources from the reinstated Sorghum Conversion Program. *J. Plant Reg.* 10:57–61.

Kofoid, K. D. and Harvey, T. L. 1992. Inheritance of resistance to greenbug biotypes E and I in the sorghum hybrid 'Cargill 607E'. Agronomy Abstracts. ACSESS Books Series. In: *ASA/CSSA/SSSA Annual Meetings*, 1–6 November 1992. Minneapolis, MN, USA.

Kofoid, K. D. and Harvey, T. L. 2000a. Registration of biotype E greenbug resistant grain sorghum germplasm. *Crop Sci.* 40:1509–10.

Kofoid, K. D. and Harvey, T. L. 2000b. Registration of seven biotype I greenbug resistant restorer germplasm lines of sorghum. *Crop Sci.* 40:1510–11.

Kofoid, K. D. and Harvey, T. L. 2005. Registration of greenbug resistant germplasm lines KS 116 A/B through KS 120 A/B. *Crop Sci.* 45:802.

Kofoid, K. D., Perumal, R., Reese, J. C. and Campbell, L. R. 2012. Registration of twelve sorghum germplasm lines tolerant to greenbug feeding damage. *J. Plant Regis.* 6:1–3.

Li, Y. Y., Ma, L. L., Lu, S. Y., Zhang, Y. H., Lin, F., Hussain, K and Li, X. M. 2011. SSR analysis of resistance gene to head smut physiological strain no.3 on sorghum. *J. Sci. Res.* 7:392–6

Linden, C., Wouters, D., Mihalka, V., Kochieva, E., Smulders, M. and Vosman, B. 2004. Efficient targeting of plant disease resistance loci using NBS profiling. *Theor. Appl. Genet.* 109:384–93.

Lipka, U., Fuchs, R. and Lipka, V. 2008. Arabidopsis non-host resistance to powdery mildews. *Curr. Opin. Plant Biol.* 11:404–11.

Little, C. R., Perumal, R., Tesso T., Prom, L. K. and Magill, C. W. 2012. Sorghum pathology and biotechnology - A fungal disease perspective: Part I. Grain mold, head smut, and ergot. *Eur. J. Plant Sci. Biotech.*, Global Science Books 6:10–30.

Liu, S., Assanga, S. O., Dhakal, S., Gu, X., Tan, C. T., Yang, Y., Rudd, J., Hays, D., Ibrahim, A., Xue, Q., Chao, S., Devkota, R., Shachter, C., Huggins, T., Mohammed, S. and Fuentealba, M. P. 2016a. Validation of chromosomal locations of 90K array single nucleotide polymorphisms in US wheat. *Crop Sci.* 56:364–73.

Liu, X., Inoue, H., Hayashi, N., Jiang, C. J. and Takatsuji, H. 2016b. CC-NBS-LRR-Type R proteins for rice blast commonly interact with specific WRKY transcription factors. *Plant Mol. Biol. Rep.* 34:533–7.

Lohithaswa, H. C., Jyothi, K., Sunil Kumar, K. R. and Puttaramanaik, H. S. 2015. Identification and introgression of QTLs implicated in resistance to sorghum downy mildew (*Peronosclerospora sorghi* (Weston and Uppal, C. G. Shaw) in maize through marker-assisted selection. *J. Genet.* 94:741–8.

Mace, E., Tai, S., Innes, D., Godwin, I., Hu, W., Campbell, B., Gilding, E., Cruickshank, A., Prentis, P., Wang, J. and Jordan, D. 2014. The plasticity of NBS resistance genes in sorghum is driven by multiple evolutionary processes. *BMC Plant Biol.* 14:1–14.

Manthe, C. S. 1992. Sorghum resistance to the sugarcane aphid (Homoptera: Aphididae). Ph.D. dissertation, Texas A&M University, College Station, Texas.

Martin, T., Biruma, M., Fridborg, I., Okori, P. and Dixelius, C. 2011. A highly conserved NB-LRR encoding gene cluster effective against *Setosphaeria turcica* in sorghum. *BMC Plant Biol.* 1:151.

Mbuvi, D. A., Masiga, C. W., Kuria, E., Masanga, J., Wamalwa, M., Mohamed, A., Odeny, D. A. Hamza, N., Timko, M. P. and Runo, S. 2017. Novel sources of witchweed (Striga) resistance from wild sorghum accessions. *Front. Plant Sci.* 8:1–15.

McHale, L., Tan, X., Koehl, P. and Michelmore, R. W. 2006. Plant NBS-LRR proteins: Adaptable guards. *Genome Biol.* 7:212.

McLaren, N. W. 1992. Quantifying resistance of sorghum genotypes to the sugary disease pathogen (*Claviceps africana*). *Plant Dis.* 76:986–8.

Mead, F. W. 1978. Sugarcane aphid, *Melanaphis sacchari* (Zehntner) – Florida – New Continental United States record. *Coop. Plant Pest Rep.* 3: 475.

Mohamed, A., Ali, R., Elhassan, O., Suliman, E., Mugoya, C., Masiga, C. W., Elhusien, A. and Hash C. T. 2014. First products of DNA marker-assisted selection in sorghum released for cultivation by farmers in sub-Saharan Africa. *J. Plant Sci. Mol. Breed.* 3:1–10.

Mohammed, R., Munghate, R. S., Are, A. K., Polavarapu, K. K. B., Reddy, B. V. S. and Sharma, H. C. 2016. Components of resistance to sorghum shoot fly, *Atherigona soccata*. *Euphytica* 207:419–38.

Mohan, S. M., Madhusudhana, R., Mathur, K., Chakravarthi, D., Rathore, S., Reddy, R. N., Sathish, K., Srinivas, G., Mani, N. S. and Seetharama, N. 2010. Identification of quantitative trait loci associated with resistance to foliar diseases in sorghum [*Sorghum bicolor* (L.) Moench]. *Euphytica* 176:199–211.

Muhammad, A., Zahid, N., Muhammad, H. B. and Bilal, S. K. 2009. Analysis of host plant resistance in some genotypes of maize against *Chilo partellus* (swinhoe) (pyralidae: lepidoptera). *Pak. J. Bot.* 41:421–8.

Muth, P., Elhassan, O., Mohammed, A., Rattunde, H. F. W., Toure, A. and Haussmann, B. I. G. 2012. Effect of marker-assisted backcrossing to introgress resistance to *Striga hermonthica* into African sorghum varieties. In: Eric T. (Ed.), *Resilience of Agricultural Systems Against Crises. Tropentag 2012: International Research on Food Security, Natural Resource Management and Rural Development*, Georg-August-Universität, Göttingen, 19–21 September. Cuvilier Verlag, Gottingen, Germany.

Muturi, P. W. 2013. Resistance to the African and spotted stem borers in sorghum in Kenya. PhD thesis, University of Makerere, Kampala, Uganda.

Muturi, P. W., Rubaihayo, P., Mgonja, M., Kyamanywa, S., Sharma, H. C. and Hash, C. T. 2012a. Novel source of sorghum tolerance to the African stem borer, *Busseola fusca*. *Afr. J. Plant Sci.* 6:295–302.

Muturi, P. W., Rubaihayo, P., Mgonja, M. A., Kyamanywa, S., Kibuka, J. and Sharma, H. C. 2012b. New sources of resistance to spotted stem borer, *Chilo partellus* in sorghum. *Int. J. Agron. Agric. Res.* 2:18–28.

Mysore, K. S. and Ryu, C. M. 2004. Non-host resistance: How much do we know? *Trends Plant Sci.* 9:97–104.

Nagaraj, N., Reese, J. C., Tuinstra, M. R., Smith, C. M., Amand, P. S., Kirkham, M. B., Kofoid, K. D., Campbell, L. R. and Wilde, G. E. 2005. Molecular mapping of sorghum genes express tolerance to damage by greenbug (Homoptera: Aphididae). *J. Econ. Entomol.* 98:595–602.

Nair, S. K., Prasanna, B. M, Garg, A., Rathore, R. S., Setty, T. A. S. and Singh, N. N. 2005. Identification and validation of QTLs conferring resistance to sorghum downy mildew (*Peronosclerospora sorghi*) and Rajasthan downy mildew (*P. heteropogoni*) in maize. *Theor. Appl. Genet.* 110:1384–92.

Nibouche, S., Fartek, B., Mississipi, S., Delatte, H., Reynaud, B. and Costet, L. 2014. Low Genetic diversity in *Melanaphis sacchari* aphid populations at the worldwide scale. *PLoS ONE* 9:e106067. doi:10.1371/journal.pone.0106067.

Nimbalkar, V. S and Bapat, D. R. 1987. Genetic analyses of shoot fly resistance under high level of shoot fly infestation in sorghum. *J. Maharashtra Agric. Univ.* 12:331–4.

Nishimura Marc, T., Monteiro, F. and Dangl Jeffery, L. 2015. Treasure your exceptions: unusual domains in immune receptors reveal host virulence targets. *Cell* 161:957–60.

Oh, B. J., Frederiksen, R. A. and Magill, C. W. 1994. Identification of molecular markers linked to head smut resistance gene (*Shs*) in sorghum by RFLP and RAPD analyses. *Phytopathol.* 84:830–3.

Oh, B. J., Frederiksen, R. A. and Magill, C. W. 1996. Identification of RFLP markers linked to a gene for downy mildew resistance (*Sdm*) in sorghum. *Can. J. Bot.* 74:315–17.

Page, F. D. 1979. Resistance to sorghum midge (*Contarinia sorghicola* Coquillet) in grain sorghum. *Aust. J. Exp. Agric. Animal Hus.* 19:97–101.

Panday, S., Sindhu, A. and Boora, K. S. 2002. RAPD based DNA markers linked to anthracnose disease resistance in *Sorghum bicolor* (L.) Moench. *Indian J. Exp. Biol.* 40:20–211.

Parh, D. K., Jordan, D. R., Aitken, E. A. B., Mac, E. S., Jun-ai, P., McIntyre, C. L. and Godwin, I. D. 2008. QTL analysis of ergot resistance in sorghum. *Theor. Appl. Genet.* 117:369–82.

Patil, R. C. and Thombre, M. V. 1985. Inheritance of earhead midge resistance in sorghum. *Sorghum Newsl.* 27:90.

Patil, A., Fakrudin, B., Salimath, P. M. and Rajkumar. 2012. Genome-wide molecular mapping and QTL analysis, validated across locations and years for charcoal rot disease incidence traits in *Sorghum bicolor* (L.) Moench. *Indian J. Genet. Plant Breed.* 72:296–302.

Perumal, R., Isakei, T., Menz, M. A., Katile, S., No, E. G. and Magill, C. W. 2006. Characterization and genetic distance analysis of isolates of *Peronosclerospora sorghi* using AFLP fingerprinting. *Mycol. Res.* 110:471–8.

Perumal, R., Frederiksen, R. A., Prom, L. K. and Magill, C. W. 2007. Head Smut. In: Screening Techniques for Sorghum Diseases, Information Bulletin #76. International Crops Research Institute for the Semi-Arid Tropics (ICRISAT), Pantacheru - 502 324, India, pp. 58–63.

Perumal, R., Nimmakayala, P., Erattaimuthu, S., No, E. G., Reddy, U. K., Prom, L. K., Odvody, G. N., Luster, D. G. and Magill, C. W. 2008. Simple sequence repeat markers useful for sorghum downy mildew (*P. sorghi*) and related species. *BMC Genet.* 9:77.

Perumal, R., Menz, M. A., Mehta, P. J., Katilé, S., Gutierrez-Rojas, L. A., Klein, R. R., Klein, P. E., Prom, L. K., Schlueter, J. A., Rooney, W. L. and Magill, C. W. 2009. Molecular mapping of *Cg*1, a gene for resistance to anthracnose (*Colletotrichum sublineolum*) in sorghum. *Euphytica* 165:597–606.

Peterson, G. C. 1996. Greenbug resistance in sorghum. In: *Proceedings of the International Conference on Genetic Improvement of Sorghum and Pearl Millet*, 22–27 September 1996. Holiday Inn Plaza, Lubbock, Texas, USA.

Peterson, G. C., Teetes, G. L. and Pendleton, B. B. 1994. Resistance to the sorghum midge in the United States. *Int. Sorghum Millets Newsl.* 35:48–63.

Pi, C. and Hsieh, J. S. 1982. Preliminary studies on aphid resistance in sorghum. *Natl. Sci. Counc. Month. Repub. China* 10:153–60.

Porter, K. B., Peterson, G. L. and Vise, O. 1982. A new greenbug biotype. *Crop Sci.* 22:847–50.

Prom, L. K. and Erpelding, J. E. 2006. Ergot response for the sorghum genotype IS8525. *Plant Pathol. J.* 5:216–20.

Prom, L. K. and Erpelding, J. E. 2009. New sources of grain mold resistance among sorghum accessions from Sudan. *J. Trop. Subtrop. Agroecosys.* 10:457–63.

Prom, L. K. and Erpelding, J. E. 2013. Evaluation of sorghum accessions from Ethiopia and Mali against *Fusarium thapsinum*. *J. Trop. Agric.* 51:92–7.

Prom, L. K., Erpelding, J. E. and Montes-Garcia, N. 2007. Chinese sorghum germplasm evaluated for resistance to downy mildew and anthracnose. *Commun. Biometry Crop Sci.* 2:26–31.

Prom, L. K., Erpelding, J. E. and Montes-Garcia, N. 2008. Evaluation of sorghum germplasm from China against *Claviceps africana*, causal agent of sorghum ergot. *Plant Health Prog.* (online). https://www.plantmanagementnetwork.org/pub/php/research/2008/sorghum/.

Prom, L. K., Perumal, R., Erpelding, J. E., Isakeit, T., and Montes, N. and Magill, C. W. 2009. A pictorial technique for mass screening of sorghum germplasm for anthracnose resistance. *Open Agric. J.* 3:20–5.

Prom, L. K., Montes-Garcia, N., Erpelding, J. E., Perumal, R. and Medina-Ocegueda, S. 2010. Response of sorghum accessions from Chad and Uganda to natural infection by the downy mildew pathogen, *Peronosclerospora sorghi* in Mexico and the USA. *J. Plant Dis. Prot.* 117:2–8.

Prom, L. K., Isakeit, T., Perumal, R., Erpelding, J. E., Rooney, W. and Magill, C. W. 2011. Evaluation of the Ugandan sorghum accessions for grain mold and anthracnose resistance. *Crop Prot.* 30:566–71.

Prom, L. K., Erpelding, J., Perumal, R., Isakeit, T. and Cuevas, H. 2012a. Response of sorghum accessions from four African countries against *Colletotrichum sublineolum*, causal agent of sorghum anthracnose. *Am. J. Plant Sci.* 3:125–9.

Prom, L. K., Perumal, R., Erattaimuthu, S. R., Little, C. R., No, E. G., Erpelding, J. E., Rooney, W. L., Odvody, G. N. and Magill, C. W. 2012b. Genetic diversity and pathotype determination of *Colletotrichum sublineolum* isolates causing anthracnose in sorghum. *Eur. J. Plant Pathol.* 133:671–85.

Prom, L. K., Perumal, R., Cisse, N., and Little, C. R. 2014. Evaluation of selected sorghum lines and hybrids for resistance against grain mold and long smut fungi in Senegal, West Africa. *Plant Health Prog.* 15:74–7.

Prom, L. K., Perumal, R., Montes-Garcia, N., Isakeit, T., Odvody, G. N., Rooney, W. L., Little, C. R. and Magill, C. 2015. Evaluation of Gambian and Malian sorghum germplasm against downy mildew pathogen, *Peronosclerospora sorghi*, in Mexico and the USA. *J. Gen. Plant Pathol.* 81:24–31.

Punnuri, S., Huang, Y., Steets, J. and Wu, Y. 2013. Developing new markers and QTL mapping for greenbug resistance in sorghum [*Sorghum bicolor* (L.) Moench]. *Euphytica* 191:191–203.

Radwan, G. L., Perumal, R., Isakeit, T., Magill, C. W., Prom, L. K., Little, C. R. 2011. Screening exotic sorghum germplasm, hybrids and elite lines for resistance to a new virulent pathotype (P6) of *Peronosclerospora sorghi* causing downy mildew. *Plant Health Prog.* (Online). https://www.plantmanagementnetwork.org/pub/php/research/2011/p6/p6.pdf.

Reddy, P. S., Fakrudin-Rajkumar, B., Punnuri, S. M., Arun, S. S., Kuruvinashetti, M. S., Das, I. K. and Seetharama, N. 2008. Molecular mapping of genomic regions harboring QTLs for stalk rot resistance in sorghum. *Euphytica* 159:191–8.

Reed, J. D., Ramundo, B. A., Claflin, L. E. and Tuinstra, M. R. 2002. Analysis of resistance to ergot in sorghum and potential alternate hosts. *Crop Sci.* 42:1135–8.

Riyazaddin, M., Kavi Kishor, P. B. Ashok Kumar, A., Reddy, B. V. S. Munghate, R. and Sharma, H. C. 2015. Mechanisms and diversity of resistance to sorghum shoot fly, *Atherigona soccata*. Plant Breed. 134:423–36.

Riyazaddin, M., Kumar, A. A., Munghate, R. S., Ramaiah, B., Kavi Kishor, P. B. and Sharma, H. C. 2016. Inheritance of resistance to sorghum shoot fly, *Atherigona soccata* in sorghum, *Sorghum bicolor* (L.) Moench. *Front. Plant Sci.* 7:01–18.

Rodriguez-Herrera., R., Rooney W. L., Rosenow, D. T. and Frederiksen, R. A. 2000. Inheritance of grain mold resistance in grain sorghum without a pigmented testa. *Crop Sci.* 40:1573–8.

Rosenow, D. T., Collins, S. D., Odvody, G. N., Schaefer, K., Rooney, W. L. and Peterson, G. C. 2014a. Registration of Tx3364 through Tx3407 grain weathering-resistant sorghum germplasm. *J. Plant Reg.* 8:324–8.

Rosenow, D. T., Odvody, G. N., Frederiksen, R. A., Schaefer, K., Collins, S. D., Remmers, J., Peterson, G. C., Woodfin, C. A. and Rooney, W. L. 2014b. Registration of Tx3301 through Tx3360 sorghum germplasms with resistance to downy mildew. *J. Plant Reg.* 8:90–5.

Rossetto, C. J. and Igue, T. 1983. Inheritance of resistance to *Contarinia sorghicola* (Coquillett) of sorghum variety AF 28. *Brag.* 42:211–19.

Ryley, M. J. and Henzell, B. 1999. Sorghum ergot in Australia, pp. 182–90. In: Casela, C. R. and Dahlberg, J. A. (Eds), *Proc. Global Conf. Ergot Sorghum (1–8 June 1997)*. EMBRAPA, Sete Lagaos, Brazil.

Sally, L. D., Frances, M. S., Robert, J. H., Giovanni, C. and Liz, I. 2007. Domestication to crop improvement: Genetic resources for *Sorghum* and *Saccharum* (Andropogoneae). *Ann. Bot.* 100:975–89.

Sangitrao, C. S., Indira, S. and Bandyopadhyay, R. 1999. Sorghum ergot in India, pp. 41–54. In: Casela, C. R. and Dahlberg, J. A. (Eds), *Proc. Global Conf. Ergot Sorghum (1–8 June 1997)*. EMBRAPA, Sete Lagaos, Brazil.

Sanjana Reddy, P., Reddy, B. V. S. Ashok Kumar, A. and Sharma, H. C. 2015. Performance of A1, A2, A3, A4(M), A4(G) and A4(VZM) cytoplasms based iso-nuclear sorghum hybrids for shoot fly resistance across-rainy and post-rainy seasons. *Indian J. Genet. Plant Breed.* 75:324–9.

Sanseverino, W., Roma, G., De Simone, M., Faino, L., Melito, S., Stupka, E., Frusciante, L. and Ercolano, M. R. 2012. PRGdb: A bioinformatics platform for plant resistance gene analysis. *Nucleic Acids Res.* 38:D814–21.

Sarris Panagiotis, F., Duxbury, Z., Huh Sung, U., Ma, Y., Segonzac, C., Sklenar, J., Derbyshire, P., Cevik, V., Rallapalli, G., Saucet Simon, B., Wirthmueller, L., Menke Frank, L. H., Sohn Kee, H., Schulze-Lefert, P. and Panstruga, R. 2011. A molecular evolutionary concept connecting nonhost resistance, pathogen host range, and pathogen speciation. *Trends Plant Sci.* 16:117–25.

Satish, K., Srinivas, G., Madhusudhana, R., Padmaja, P. G., Nagaraja, Reddy, R., Murali Mohan, S. and Seetharama, N. 2009. Identification of quantitative trait loci for resistance to shoot fly in sorghum [*Sorghum bicolor* (L.) Moench]. *Theor. Appl. Genet.* 119:1425–39.

Satish, K., Gutema, Z., Grenier, C., Rich, P. J. and Ejeta, G. 2012. Molecular tagging and validation of microsatellite markers linked to the low germination stimulant gene (*lgs*) for Striga resistance in sorghum [*Sorghum bicolor* (L.) Moench]. *Theor. Appl. Genet.* 124:989–1003.

Schuster, D. J. and Starks, K. J. 1973. Greenbugs: Components of host-plant resistance in sorghum. *J. Econ. Entomol.* 66:1131–4.

Seetharama, N., Bidinger, F. R., Rao, K. N., Gill, K. S. and Mulgund, M. 1987. Effect of pattern and severity of moisture deficit stress on stalk rot incidence in sorghum: I. Use of line source irrigation technique, and the effect of time of inoculation. *Field Crops Res.* 15:289–308.

Seifers, D. L., Perumal, R. and Little, C. R. 2012. New sources of resistance in sorghum (*Sorghum bicolor*) germplasm are effective against a diverse array of *Potyvirus* spp. *Plant Dis.* 96:1775–9.

Setokuchi, O. 1973. Ecology of *Longiunguis sacchari* infesting sorghum. I. Nymphal period and fecundity of apterous viviparous females. *Proc. Assoc. Plant Prot. Kyushu* 19:95–7.

Sharma, H. C. 1985. Screening for sorghum midge resistance and resistance mechanisms, pp. 275–92. In: *Proc. Int. Sorghum Ent. Workshop*, 15–21 July 1984, College Station, Texas, USA. International Crops Research Institute for the Semi-Arid Tropics (ICRISAT), Patancheru 502 - 324, Andhra Pradesh, India.

Sharma, H. C. 2006. *Integrated Pest Management Research at ICRISAT: Present Status and Future Priorities*. International Crops Research Institute for the Semi-Arid Tropics (ICRISAT), Patancheru, Andhra Pradesh, India, 48p.

Sharma, H. C., Taneja, S. L., Leuschner, K. and Nwanze, K. F. 1992. Techniques to screen sorghums for resistance to insects. Information Bulletin #32. International Crops Research Institute for the Semi-Arid Tropics (ICRISAT), Pantacheru - 502 324, India.

Sharma, H. C., Agrawal, B. L., Vidyasagar, P., Abraham, C. V. and Nwanze, K. F. 1993. Identification and utilization of resistance to sorghum midge, *Contarinia sorghicola* (Coquillett) in India. *Crop Prot.* 12:343–51.

Sharma, H. C., Mukuru, S. Z. and Nwanze, K. F. 1996. Stability of resistance to the sorghum midge *Contarinia sorghicola*, pp. 429–35. In: Foale, M. A., Henzell, R. G. and Kneipp, J. (Eds), *Proc. Third Aust. Sorghum Conf.*, 20–22 Fegryary 1996, Tamworth, Australia. Australian Institute of Agricultural Science, Melbourne, Australia. Occasional Publication No. 93.

Sharma, H. C., Taneja, S. L., Kameswara Rao, N. and Prasada Rao, K. E. 2003. Evaluation of sorghum germplasm for resistance to insect pests. Information Bulletin #63. International Crops Research Institute for the Semi-Arid Tropics (ICRISAT), Pantacheru - 502 324, India.

Sharma, H. C., Mukuru, S. Z. and Stenhouse J. W. 2004. Variation in inheritance to sorghum midge, *Stenodiplosis sorghicola* across locations in India and Kenya. *Euphytica* 138:219–25

Sharma, H. C., Dhillon, M. K. and Pampapathy, G. 2006. Multiple-resistance to sorghum shoot fly, spotted stemborer and sugarcane aphid in sorghum. *Int. J. Trop. Insect Sci.* 26:239–45.

Sharma, H. C., Bhagwat, V. R., Daware, D. G., Pawar, D. B., Munghate, R. S., Sharma, S. P., Kumar, A. A., Reddy, B. V. S., Prabhakar, K. B., Ambekar, S. S. and Gadakh, S. R. 2014. Identification of sorghum genotypes with resistance to the sugarcane aphid *Melanaphis sacchari* under natural and artificial infestation. *Plant Breed.* 133:36–44.

Sharma, H. C., Bhagwat, V. R., Munghate, R. S., Sharma, S. P., Daware, D. G., Pawar, D. B., Ashok Kumar, A., Reddy, B. V. S., Prabhakar, K. B., Mehtre, S. P., Kalpande, H. V. and Gadakh, S. R. 2015. Stability of resistance to sorghum shoot fly, *Atherigona soccata*. *Field Crops Res.* 178:34–41.

Shivaprasad, P. V., Chen, H. M., Patel, K., Bond, D. M., Santos, B. A. and Baulcombe, D. C. 2012. A microRNA superfamily regulates nucleotide binding site-leucine-rich repeats and other mRNAs. *Plant Cell* 24:859–74.

Singh, M. and Boora, K. S. 2008. Molecular characterization of anthracnose resistance gene in sorghum. In: Rao, G. P., Wagner, C., Singh, R. K. and Sharma, M. L. (Eds), *Plant Genomics and Bioinformatics*. Stadium Press LLC, Houston, Texas, USA, pp. 373–88.

Singh, B. U., Padmaja, P. G. and Seetharama, M. 2004. Biology and management of the sugarcane aphid, *Melanpahis sacchari* (Zehntner) (Homoptera: Aphididae), in sorghum: A review. *Crop Prot.* 23:739–55.

Singh, M., Chaudhary, K., Singal, H. R., Magill, C. W. and Boora, K. S. 2006. Identification and characterization of RAPD and SCAR markers linked to anthracnose resistance gene in sorghum [*Sorghum bicolor* (L.) Moench]. *Euphytica* 152: 139 .

Spoel, S. H. and Dong, X. 2012. How do plants achieve immunity? Defence without specialized immune cells. *Nat. Rev. Immunol.* 12:89–100.

Srinivas, K. P., Sreekanth Reddy, M., Hema, M. and Sreenivasulu, P. 2013. Multiplex RT-PCR for parallel detection of three RNA viruses associated with stripe and mosaic diseases of sorghum in India. *Eur. J. Plant Pathol.* 137:211–15.

Steuernagel, B., Periyannan, S. K., Hernandez-Pinzon, I., Witek, K., Rouse, M. N., Yu, G., Hatta, A., Ayliffe, M., Bariana, H., Jones, J. D. G., Lagudah, E. S. and Wulff, B. B. H. 2016. Rapid cloning of disease-resistance genes in plants using mutagenesis and sequence capture. *Nat. Biotech.* 34:652–5.

Stephens, J. C., Miller, F. R. and Rosenow, D. T. 1967. Conversion of alien sorghums to early combine genotypes. *Crop Sci.* 7:396.

Svitashev, S., Young, J. K., Schwartz, C., Gao, H., Falco, S. C. and Cigan, A. M. 2015. Targeted mutagenesis, precise gene editing and site-specific gene insertion in maize using *Cas9* and guide RNA. *Plant Physiol.* 169:931–45.

Takken, F. L. W. and Goverse, A. 2012. How to build a pathogen detector: structural basis of NB-LRR function. *Curr. Opin. Plant Biol.* 15:375–84.

Tan, W. Q., Li, S. M., Gua, H. P. and Gao, R. P. 1985. A study of the inheritance of aphid resistance in sorghum. *Shanxi Agric. Sci.* 8:12–14.

Tao, Y. Z., Jordan, D. R., Henzell, R. G. and McIntyre, C. L. 1998. Identification of genomic regions for rust resistance in sorghum. *Euphytica* 103:287–92.

Tao, Y. Z., Hardy, A., Drenth, J., Henzell, R. G., Franzmann, B. A., Jordan, D. R., Butler, D. G. and McIntyre, C. L. 2003. Identifications of two different mechanisms for sorghum midge resistance through QTL mapping. *Theor. Appl. Genet.* 107:116–22.

Teetes, G. L. and Johnson, J. W. 1978. Insect resistance in sorghum, pp. 167–89. In: *Proc. of the 33rd Annual Corn and Sorghum Res. Conf.*, Chicago, Illinois, USA.

Teetes, G. L. and Pendleton, B. B. 2000. Insect pests of sorghum, pp. 443–95. In: Wayne Smith, C. and Frederiksen, R. A. (Eds), *Sorghum: Origin, History, Technology, and Production.* Wiley Series in Crop Science, John Wiley & Sons, Inc., New York, USA.

Teetes, G. L., Schaefer, C. A. and Johnson, J. W. 1974a. Resistance in sorghums to the greenbug; laboratory determination of mechanisms of resistance. *J. Econ. Entomol.* 67:383–96.

Teetes, G. L., Schaefer, C. A., Johnson, J. W. and Rosenow, D. T. 1974b. Resistance in sorghum to the greenbug: Field evaluation. *Crop Sci.* 14:706–8.

Teetes, G. L., Manthe, C. S., Peterson, G. C., Leuschner, K., and Pendleton, B. B. 1995. Sorghum resistant to the sugarcane aphid *Melanaphis sacchari* (Zehntner) (Homoptera: Aphididae). *Insect. Sci. Appl.* 16: 63–71.

Tesso, T. T, Perumal R., Little C. R., Adeyanju A., Radwan G. L., Prom L. K. and Magill. C. W. 2012. Sorghum pathology and biotechnology - A fungal disease perspective: Part II. Anthracnose, stalk rot, and downy mildew. *Eur. J. Plant Sci. Biotech.*, Global Science Books 6(1):31–41.

Tesso, T. T., Ochanda, N., Little, C. R., Claflin, L. and Tuinstra, M. R. 2010. Analysis of host plant resistance to multiple *Fusarium* species associated with stalk rot disease in sorghum [*Sorghum bicolor* (L.) Moench]. *Field Crops Res.* 118:177–82.

Thakur, R. P. and Rao, P. 2007. Ergot. In: *Screening Techniques for Sorghum Diseases*, Information Bulletin #76. International Crops Research Institute for the Semi-Arid Tropics (ICRISAT), Pantacheru - 502 324, India, pp. 47–52.

Thakur, R. P., Rao, V. P., Reddy, B. V. S. and Sanjana Reddy, P. 2007a. Grain mold. In: *Screening Techniques for Sorghum Diseases*, Information Bulletin # 76. International Crops Research Institute for the Semi-Arid Tropics (ICRISAT), Pantacheru - 502 324, India, pp. 5–14.

Thakur, R. P., Rao, V. P. and Sanjana Reddy. P. 2007b. Downy mildew. In: *Screening Techniques for Sorghum Diseases*, Information Bulletin # 76. International Crops Research Institute for the Semi-Arid Tropics (ICRISAT), Pantacheru - 502 324, India, pp. 31–9.

Thakur, R. P., Rao, V. P. and Sanjana Reddy. P. 2007c. Rust. In: *Screening Techniques for Sorghum Diseases*, Information Bulletin #76. International Crops Research Institute for the Semi-Arid Tropics (ICRISAT), Pantacheru - 502 324, India, pp. 53–7.

Tooley, P. W., Carras, M. M., Sechler, A. and Rajasab, A. H. 2010. Real-time PCR detection of sorghum ergot pathogens *Claviceps africana, Claviceps sorghi* and *Claviceps sorghicola*. *J. Phytopathol.* 158:698–704.

Trimboli, D. S. and Burgess, L. W. 1982. The fungi associated with stalk and root rot of grain sorghum in New South Wales. *Sorghum Newsl.* 25:105.

Tuinstra, M. R., Wilde, G. E. and Kriegshauser, T. 2001. Genetic analysis of biotype I resistance in sorghum. *Euphytica* 121:87–91.

Tyron, H. 1895. The insect enemies of cereals belonging to the genus *Cecidomyia*. *Trans. Nat. Hist. Soc. Queensland* 1:21–9.

Upadhyaya, H., Pundir, R. P. S., Dwivedi, S. L, Gowda, C. L. L., Reddy, V. G. and Singh, S. 2009. Developing a mini core collection of sorghum for diversified utilization of germplasm. *Crop Sci.* 49:1769–80.

Upadhyaya, H. D., Wang, Y. H., Sharmam R. and Sharma, S. 2013a. Identification of genetic markers linked to anthracnose resistance in sorghum using association analysis. *Theor. Appl. Genet.* 126:1649–57.

Upadhyaya, H. D., Wang, Y.-H., Sharma, R. and Sharma, S. 2013b. SNP markers linked to leaf rust and grain mold resistance in sorghum. *Mol. Breed.* 32:451–62.

Van Rensburg, N. J. 1973. Notes on the occurrence and biology of the sorghum aphid in South Africa. *J. Entomol. Soc. S. Afr.* 36:293–8.

Ve, T., Williams, S. J. and Kobe, B. 2015. Structure and function of Toll/interleukin-1 receptor/ resistance protein (TIR) domains. *Apoptosis* 20:250–61.

Villanueva, R. T., Brewer, M., Way, M. O., Biles, S., Sekula, D., Bynum, E., Swart, J., Crumley C., Knutson, A., Porter, P., Parker, R., Odvody, G., Allen, C., Ragsdale, D., Rooney, W., Peterson G., Kerns, D., Royer, T. and Armstrong, S. 2014. Sugarcane aphid: A New Pest of Sorghum. Texas A&M AgriLife Extension. ENTO-35.

Vinayan, M. T. 2010. Genetic architecture of spotted stem borer resistance in sorghum as inferred from QTL mapping and synteny with the maize genome. PhD thesis, Tamil Nadu Agricultural University, India.

Visarada, K. B. R. S., Padmaja, P. G., Saikishore, N., Pashupatinath, E., Royer, M., Seetharama, N. and Patil, J. V. 2014. Production and evaluation of transgenic sorghum for resistance to stem borer. *In Vitro Cellular and Developmental Biology - Plant* 50:176–89.

Wang, Y. S. 1961. Studies on the sorghum aphid, *Aphis sacchari* Zehntner. *Acta Entomol. Sin.* 10:363–80.

Wang, F., Zhao, S., Han, Y., Shao, Y., Dong, Z., Gao, Y., Zhang, K., Liu, X., Li, D., Chang, J. and Wang, D. 2013. Efficient and fine mapping of RMES1 conferring resistance to sorghum aphid *Melanaphis sacchari*. *Mol. Breed.* 31:777–84.

Wang, X., Mace, E., Hunt, C., Cruickshank, A., Henzell, R., Parkes, H. and Jordan, D. 2014. Two distinct classes of QTL determine rust resistance in sorghum. *BMC Plant Biol.* 14:366.

Way, M., Conlong, D., Martin, L., Mcfarlane, S. R. S., Keeping, M. and Rutherford, R. 2014. First record of yellow sugarcane aphid, *Sipha flava* (Homoptera: Aphididae), in the South African sugarcane industry. In: *Proc. S. Afr. Sug. Technol. Ass.*, pp. 53–7.

Webster, F. M and Phillips, W. J. 1912. The spring grain-aphis or 'green bug'. *USDA Bur. Entomol. Bull.* 110:1–153.

Weibel, D. E., Starks, K. J., Wood, E. A. Jr. and Morrison, R. D. 1972. Sorghum cultivars and progenies rated for resistance to greenbugs. *Crop Sci.* 12:334–6.

White, W. H., Reagan, T. E. and Hall, D. G. 2001. *Melanaphis sacchari* (Homoptera: Aphididae), a sugarcane pest new to Louisiana. *Florida Entomol.* 84:435–6.

White, J. A., Ryley, M. J., George, D. L., Kong, G. A. and White, S. C. 2012. Yield losses in grain sorghum due to rust infection. *Australasian Plant Pathol.* 41:85–91.

White, J., Ryley, M., George, D. and Kong, G. 2015. Identification of pathotypes of the sorghum rust pathogen, *Puccinia purpurea*, in Australia. *Australasian Plant Pathol.* 44:1–4.

Widstrom, N. W., Wiseman, B. R. and McMillian, W. W. 1972. Some gene effects conditioning resistance to midge and webworm injury in sorghum. *Sorghum Newsl.* 15:22–3.

Widstrom, N. W., Wiseman, B. R. and McMillian, W. W. 1984. Patterns of resistance in sorghum to the sorghum midge. *Crop Sci.* 24:791–3.

Wilbrink, G. 1922. An investigation on spread of the mosaic disease of sugarcane by aphids. *Medid Procfst. Java Suikerind.* 10:413–56.

Williams-Alanis, H., Pecina-Quintero, V., Montes-Garcia, N., Arcos-Cavazos, G., Zavala-Garcia, F. and Gamez-Vazquez, J. A. 2009. Incidence of head smut *Sporisorium reilianum* (Kuhn) Langdon and Fullerton in grain sorghum [*Sorghum bicolor* (L.) Moench] hybrids. *Revista Mexicana de Fitopathologia* 27:36–44.

Witek, K., Jupe, F., Witek, A. I., Baker, D., Clark, M. D. and Jones, J. D. G. 2016. Accelerated cloning of a potato late blight-resistance gene using RenSeq and SMRT sequencing. *Nat. Biotech.* 34:656–60.

Wongkaew, A., Phumichai, C., Chunwongse, J., Jampatong, S., Grudloyma, P., Pulam, T. and Doungchan, W. 2014. Detection of candidate R genes and single nucleotide polymorphisms for downy mildew resistance in maize inbred lines by association analysis. *Euphytica* 197:109–18.

Wu, Y. and Huang, Y. 2008. Molecular mapping of QTLs for resistance to the greenbug *Schizaphis graminum* (Rondani) in *Sorghum bicolor* (Moench). *Theor. Appl. Genet.* 117:117–24.

Wu, Y. Q., Huang, Y., Tauer, C. G. and Porter, D. R. 2006. Genetic diversity of sorghum accessions resistant to greenbugs as assessed with AFLP markers. *Genome* 49:143–9.

Wu, Y. Q., Huang, Y., Porter, D. R., Tauer, C. G. and Hollaway, L. 2007. Identification of a major quantitative trait locus conditioning resistance to greenbug biotype E in sorghum PI 550610 using simple sequence repeat markers. *J. Econ. Entomol.* 100:1672–8.

Yang, X. and Wang, J. 2016. Genome-wide analysis of NBS-LRR genes in sorghum genome revealed several events contributing to NBS-LRR gene evolution in grass species. *Evol. Bioinform..* 12:9–21.

Yohannes, T., Abraha, T., Kiambi, D., Folkertsma, R., Tom Hash, C., Ngugi, K., Mutitu, E., Abraha, N., Weldetsion, M., Mugoya, C., Masiga, C. W. and de Villiers, S. 2015. Marker-assisted introgression improves Striga resistance in an Eritrean Farmer-Preferred Sorghum Variety. *Field Crops Res.* 173:22–9.

Young, W. R. 1970. Sorghum insects. In: Wall, J. S. and Ross, W. M. (Eds), *Sorghum Production and Utilization*. AVI Publishing Co., Westport, CT, USA, pp. 235–87.

YueYing, L., LiJun, C., Feng, L., Hussain, K. and JianQiu, Z. 2010. Identification of SSR markers linked to resistance against the spotted stem borer in sorghum. *Middle East J. Sci. Res.* 6:505–11.

Zhang, B., Yang, Y., Wang, J., Ling, X., Hu, Z., Liu, T., Chen, T. and Zhang, W. 2015. A CC-NBS-LRR type gene GHNTR1 confers resistance to southern root-knot nematode in *Nicotiana benthamiana* and *Nicotiana tabacum*. *Eur. J. Plant Pathol.* 142:715–29.

Zhu, S., Li, Y., Vossen, J. H., Visser, R. G. F. and Jacobsen, E. 2012. Functional stacking of three resistance genes against *Phytophthora infestans* in potato. *Transgenic Res.* 21(1):89–99.

Zou, J. Q., Li, Y. Y., Zhu, K. and Wang, Y. Q. 2010. Study on inheritance and molecular markers of sorghum resistance to head smut physiological race 3. *Scientia Agricultura Sinica* 43:713–20.

Chapter 8

Sorghum breeding for abiotic stress tolerance

John J. Burke, USDA-ARS, USA

1 Introduction

As a food, feed and biofuel crop, sorghum is an essential component of global crop production. Worldwide, sorghum is the fifth most important grain crop, and annual world production and the area planted to sorghum have both decreased marginally over the past 30 years from 62.8 to 59.3 million metric tons and from 44.5 to 41.9 million hectares (http://www.cgiar.org/our-strategy/crop-factsheets/sorghum/). Recently, sorghum has become the number two crop for grain-based ethanol in the United States after maize. Sorghum (*Sorghum bicolor* (L.) Moench) is believed to have been first domesticated in northeastern Africa (Southern Sudan and Ethiopia) (De Wet and Huckabay, 1967; Doggett, 1988) and has spread to tropical and subtropical regions of all continents. This crop has enormous diversity within the National Genetic Resources Program (NGRP), maintaining 43,077, accessions of *S. bicolor* subsp. *bicolor* that can be found in the Germplasm Resource Information System (GRIN). Because sorghum is rich in genetic diversity it is employed for various purposes. In addition to serving as an important dietary staple for more than 500 million people, sorghum is used for animal forage, shelter and recently as a feedstock for biofuel production. One of the key characteristics of sorghum is its ability to yield across a range of environments. Primarily grown in arid and semi-arid regions in the United States, Asia and sub-Saharan Africa, sorghum has been adapted for these harsh environments, representing perhaps the best cereal model for abiotic stress tolerance studies (Borrell et al., 1999; Doggett, 1988; Rosenow and Clark, 1982; Sanchez et al.,

http://dx.doi.org/10.19103/AS.2017.0015.11
Published by Burleigh Dodds Science Publishing Limited, 2018.

2002; Walulu et al., 1994). Although sorghum exhibits better abiotic stress tolerances than many crops, significant sorghum yield losses are experienced annually in response to unfavourable environments. Globally, domesticated sorghum is often grown on marginal lands and in environments exposing the plant to temperature extremes and water deficits. There are numerous reviews on sorghum abiotic stress tolerance (Mickelbart et al., 2015; Mittler, 2006; Prasad et al., 2015; Sanchez et al., 2002; Wang, 2015), and so much of that information will not be repeated here. This chapter will focus on (1) germplasm diversity in response to low temperatures and water-deficit stress, and (2) methods for identifying diversity among germplasm collections.

2 Exploiting genetic diversity: cold tolerance

Early and late season low temperatures are commonly experienced by sorghum throughout the United States and in mountainous regions of the world. Cool temperatures during the early growing season are a major limitation to growing sorghum in temperate areas. Singh (1985) referred to the term 'cold tolerance' as the ability of sorghum genotypes to germinate, grow and produce satisfactory grain yields under conditions of relatively cold (but above freezing) air and soil temperatures. Singh (1985) investigated 380 excessively tall, photoperiod-sensitive and late-maturing accessions obtained from China, Ethiopia and Uganda. Of these, 39 were tolerant, 157 were partially tolerant and the remainder were susceptible to cold when rated at physiological maturity. The study reported that the highest levels of cold tolerance were found among accessions from Ethiopia (e.g. Alemaya 70, Je-jewegere 935 and Muyra) and Uganda (Mabere, Magune and Nyundo). Cold tolerance appeared to be a dominant trait. However, cold-susceptible x cold-tolerant crosses (Singh, 1985) showed that there was a wide range of variation among cold-tolerant genotypes, and the frequency of highly tolerant segregates was less than 10% in the F2, suggesting that several genes controlled the inheritance of the character and that their effects were largely cumulative.

Tiryaki and Andrews (2001a,b) evaluated the rate of germination, final germination percentage and the rate of early shoot growth as estimators of cold tolerance using 12 diverse genotypes from China, Russia, West Africa and the University of Nebraska and their test crosses. The largest difference among genotypes under both normal [25–30°C (day) and 20–25°C (night)] and cold stress conditions (constant 15°C) was shown by the rate of germination in petri dishes. The rate of germination under controlled cold conditions was strongly correlated ($P < 0.01$) with the rate of emergence in a field experiment seeded early in cold soil.

Franks et al. (2006) evaluated early season cold tolerance in ten Chinese Kaoliang, ten U.S. inbred parental lines and ten U.S. commercial hybrids for a range of cold tolerance traits under laboratory, growth chamber and field settings. They reported that Chinese lines were superior to both the inbred and hybrid classes in laboratory germination rates and field-based rates of emergence. At the cooler temperature, Kaoliangs were significantly greater than U.S. inbreds for fresh shoot weight; when tested at the warmer temperature, Kaoliangs had higher dry root weight, fresh and dry shoot weights, and fresh and dry whole plant weights, relative to the U.S. inbred class. They concluded that Chinese accessions from this working group would serve as a source of favourable genes primarily for tolerance to low temperatures during the germination and emergence phase of growth in the breeding of cold-tolerant sorghum lines.

Knoll et al. (2008) used a population of 153 RI lines, developed from a cross between Chinese landrace 'Shan Qui Red' (SQR, cold-tolerant) and SRN39 (cold-sensitive), to identify quantitative trait loci (QTLs) associated with early season performance under both cold and optimal conditions by single marker analysis, simple interval mapping (SIM) and composite interval mapping (CIM). Two QTLs for germination were identified: one on linkage group SBI-03a, derived from SRN39, was significant under cold and optimal temperatures. The other, on group SBI-07b, showed greater significance under cold temperatures and was contributed by SQR. They reported that a region of group SBI-01a, derived from SQR, showed strong associations with the seedling emergence and seedling vigour scores under early and late field plantings. A QTL for both early and late emergence was identified by CIM on SBI-02 that favoured the SRN39 allele. SIM identified a QTL for early vigour on SBI-04 favouring the SQR genotype. These findings represented the first step in the development of a marker-assisted breeding effort to improve early season performance in sorghum. Several landraces from China have been found to exhibit higher emergence and greater seedling vigour under cool conditions than most breeding lines currently available but tended to lack desirable agronomic characteristics. The introgression of desirable genes from Chinese landraces into elite lines could be expedited by marker-assisted selection.

Burow et al. (2011) used a mapping population consisting of 171 F7–F8 recombinant inbred lines (RILs) derived from the cross between RTx430 (cold-sensitive) and PI610727 (cold-tolerant) and a genetic map was constructed using 141 microsatellites or simple sequence repeat (SSR) markers. They evaluated RILs for cold and optimal temperature germinability in the laboratory, field emergence and seedling vigour in two locations during early season planting. They detected two or more QTLs for all traits, except for seedling vigour, with only one QTL being detected in the population. A QTL for cold germinability (Germ 12-2.1) showed the highest LOD value and was also associated with optimal germinability. One of the QTL for field emergence, Fearlygerm-9.3, a contribution from PI610727, was found significant in both locations used for the study. This study showed the alignment of QTLs in SBi1 (Fearlygerm-1.2 and FGerm30-1.2) with previously reported QTLs associated with late field emergence identified from a different mapping population. They concluded that PI617027 shares some common loci with other known early season cold-tolerant sorghum germplasm but also harbours novel QTLs that could be useful in introgression of enhanced laboratory germination and early season field emergence.

Knoll and Ejeta (2008) used three SSR markers (Xtxp43, Xtxp51 and Xtxp211), each representing a QTL, tested in two new populations: (Tx2794 x SQR F(3)) and (Wheatland x SQR BC(1)F(3)) to genotype individual families and early season field performance. They reported that the SQR allele of Xtxp43 had favourable effects on seedling vigour in both populations, and on emergence in the Tx2794 population. They also found a large positive effect of the SQR allele of Xtxp51 in the Tx2794 population for vigour and emergence. Knoll and Ejeta (2008) validated QTL markers in various genetic backgrounds and demonstrated the utility of marker-assisted selection for a quantitative trait, early season cold tolerance, evaluated in the field.

Bekele et al. (2014) collected multi-trait phenotype data under optimal and suboptimal temperatures in a genetically diverse RIL mapping population showing contrasting segregation patterns for chilling tolerance. Emergence and root establishment were found to be the key determinants of development and survival under chilling stress. They reported that highly interactive epistatic QTL hot spots, including a previously unknown QTL on Sb06 with a significant effect on prolonged chilling survival, were found to regulate different physiological mechanisms contributing to the maintenance of growth and development despite the chilling temperatures. The major QTL regions harbour promising candidate genes with known roles in abiotic stress tolerance. They concluded that identification of loci in the QTL hot spot regions conferring maintenance of cell division and growth under early chilling stress represents a promising step towards breeding for successful establishment of sorghum in temperate climates.

Chopra et al. (2015) identified differentially expressed genes (DEGs) between contrasting sorghum genotypes, Hong Ke Zi (cold tolerant) and BTx623 (cold sensitive), under cool and control temperatures. They used an RNAseq approach to elucidate the molecular basis of sorghum response to cold stress. An analysis of the transcriptome profile showed a total of 1910 DEGs under cold and control temperatures. Chopra identified a subset of genes under cold stress for downstream analysis, including transcription factors that exhibit differential abundance between the sensitive and tolerant genotypes. Transcription factors including dehydration-responsive element-binding factors, C-repeat binding factors, and ethylene-responsive transcription factors were significantly upregulated during cold stress in the cold-tolerant HKZ. Additionally, specific genes such as plant cytochromes, glutathione *S*-transferases and heat shock proteins were found to be differentially regulated under cold stress between cold-tolerant and -susceptible genotypes of sorghum. A total of 41,603 single-nucleotide polymorphisms (SNPs) were identified between the cold-sensitive and -tolerant genotypes. This report represented a new strategy that involved an integrated analysis of differential gene expression and identification of bi-allelic SNP to determine and analyse DEGs and variation involved in the cold stress response of sorghum. Chopra concluded that the results gathered provide an insight into the complex mechanisms associated with cold response in sorghum, which involves an array of transcription factors and genes that were previously related to abiotic stress responses.

Upadhyaya et al. (2016) evaluated the 242 accessions of the ICRISAT sorghum mini-core collection for seed germination and seedling vigour at 12°C as a measure of cold tolerance. Genome-wide association analysis was performed with approximately 162,177 SNP markers. They reported that only one marker locus (Locus 7-2) was significantly associated with low-temperature germination and none with vigour.

Maulana and Tesso (2013) investigated the impact of cold temperature during plant development and identified genetic sources of tolerance for use in breeding programmes. Plants were either subjected to two temperature regimes, cold temperature [15°C (day) and 13°C (night)] and control temperature [25°C (day) and 23°C (night)], at seedling and flowering stages using growth chambers, and the stresses lasted 10 d, or evaluated in the field under cold and control temperatures imposed by varying planting dates. Their results showed that early season cold stress reduced seedling vigour (3.4 vs. 1.2) and leaf chlorophyll content (28.83 vs. 36.59) and delayed flowering and maturity by 8 and 4 d, respectively, but had no effect on yield components and plant height. They also showed that stress at the flowering stage delayed maturity by 9 d and significantly reduced all yield components. They concluded that the effect of cold stress at seedling stage may not be critical to yield provided that emergence is not affected, but stress at flowering, however, may be detrimental and the effects seem irreversible.

Published by Burleigh Dodds Science Publishing Limited, 2018.

2.1 Case study 1

Souza and Marcos Filho (1975) evaluated sorghum seed vigour using standard germination, rate of germination, rapid ageing, ammonium chloride stress test and field emergence. They concluded that the results were highly correlated with seed quality; and, among the laboratory tests, the best predictors of seed vigour were germination velocity and ammonium chloride stress test. Sorghum seeds from the Association Panel (Casa et al., 2008) were evaluated for early season germination and vigour in April 2008 in Lubbock, Texas. Seed obtained from the NGRP were planted on April 1 and evaluated for germination weekly and for vigour 30 days after planting. A poor relationship between the per cent germination and seedling vigour was observed (Fig. 1). An R2 of only 0.10691 was obtained for this data set. The poor germination percentages observed in this study were most likely associated with seed quality as the tested seed came directly from the collection and were not grown out to evaluate seed from a single location and age prior to performing the field cold test. Our more recent studies of the relationship between sorghum seed germination and seedling vigour exhibited a very significant relationship between the two parameters (Fig. 2). This study evaluated advanced sorghum breeding lines selected for either cold tolerance or post-flowering drought tolerance in April 2105 and 2016. Low average daily air temperatures were observed in both years compared with the optimum sorghum temperature of 27°C (Fig. 2, horizontal line in the temperature graphs). Seed germination (measured as emergence) and seedling vigour were highly correlated with a R2 of 0.94031 in 2015 and 0.80813 in 2016. The seed planted in 2015 were obtained from well-watered plants harvested six months earlier from the field, while the seeds planted in 2016 were the same seed lot as those planted in 2015 (18 months after harvest) that had been stored at 25°C. The increased scatter in the 2016 cold test most likely resulted from seed quality issues arising from the storage method. These findings emphasize the need for evaluating only high-quality seeds in these stress bioassays. Figure 3 is a photograph of a line from the Ethiopian collection thirty days after

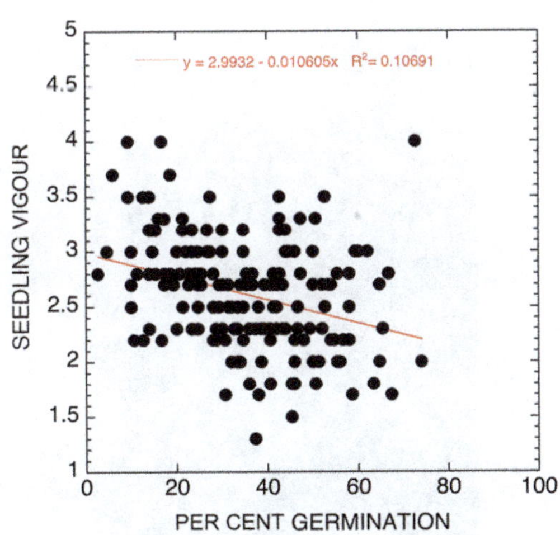

Figure 1 Graph of a 2008 field cold test comparing per cent germination and seedling vigour for the Sorghum Association Panel.

Published by Burleigh Dodds Science Publishing Limited, 2018.

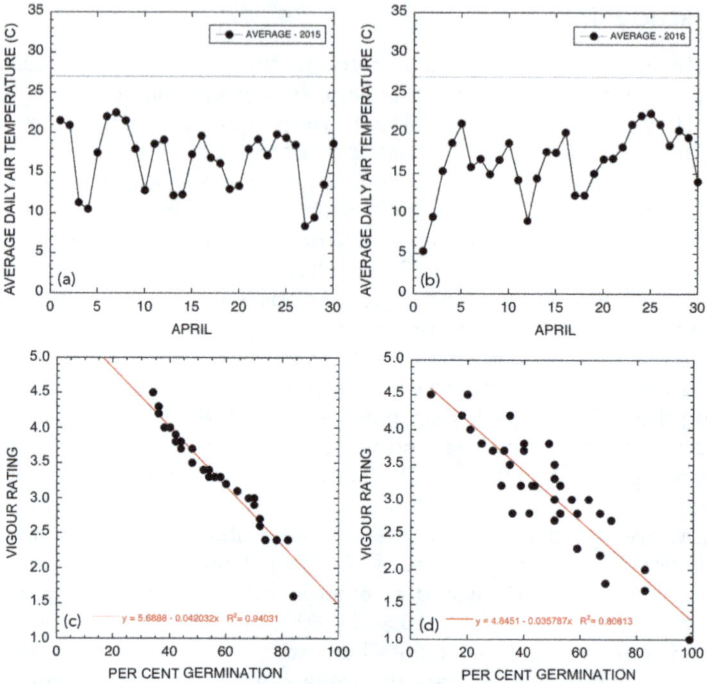

Figure 2 Graphs of average air temperatures during April cold test in Lubbock, Texas during the 2015 and 2016 growing seasons (a and b), and graphs of the per cent germination versus seedling vigour during the field cold tests.

Figure 3 Photograph of a cold-tolerant sorghum line from the Ethiopian collection at the end of the 30-day cold test.

planting in the 2016 field cold test study. Excellent early season germination and seedling vigour can be obtained from plant introductions maintained from the NGRP.

In summary, significant variability exists for early seedling germination and vigour. Although the majority of sorghum accessions are sensitive to low temperatures, selections from central Africa, Asia and Russia exhibited excellent early season germination and vigour. The advancement of molecular marker development for QTLs associated with cold tolerance is enhancing the development for cold-tolerant hybrids. Although technically difficult, more studies are needed to identify germplasm with late season cold tolerance as low-temperature stress injury at flowering may be irreversible (Maulana and Tesso, 2013).

3 Exploiting genetic diversity: drought tolerance

Drought stress at any stage of development can directly affect grain yield, but the effects of stress are manifested differently depending on developmental stage. Rosenow and colleagues described three growth stages in sorghum that are critical in understanding drought tolerance: GS1, seedling establishment (early vegetative stage); GS2, pre-flowering (panicle differentiation to flowering); and GS3, post-flowering (grain fill to physiological maturity of grain) (Rosenow and Clark, 1982; Rosenow et al., 1983). Most of the sorghum drought research in the past 30 years has focused on the GS2 and GS3 growth stages (Borrell et al., 1999, 2000b,c, 2001; Borrell and Hammer, 2000a; Burgess, et al., 2002; Burke et al., 2010, 2013, 2015; Chapman, et al., 2000; Crasta et al., 1999; Dev Sharma et al., 2006; Fracasso et al., 2016; Garrity et al., 1984; Girma and Krieg, 1992a,b; Harris et al., 2007; Haussmann et al., 1998, 1999, 2000, 2002; Henzel et al., 1976; Jedmowski et al., 2014; Johnson et al., 2014; Kannangara et al., 1982; Kapanigowda et al., 2013; Kassahun et al., 2010; Kebede et al., 2001; Mahalakshmi and Bidinger, 2002; Manjarrez-Sandoval et al., 1989; Menezes et al., 2015; Muchow and Sinclair, 1989; Peng and Krieg, 1992; Prasad et al., 2015; Rosenow et al., 1996; Rosenow and Clark, 1982; Sanchez et al., 2002; Stout et al., 1978; Tenkouano et al., 1993; Tuinstra et al., 1997; Wang, 2015; Wright and Smith, 1983; Wright et al., 1983; Xu et al., 2000). Plants under stress from panicle differentiation to flowering will exhibit a pre-flowering response. This type of stress directly affects panicle size, grain number and grain yield. Symptoms include leaf rolling, uncharacteristic leaf erectness, leaf bleaching, leaf tip and margin burn, delayed flowering, saddle effect, poor panicle exertion, panicle blasting and floret abortion, and reduced panicle size (Rosenow et al., 1996). Symptoms of post-flowering drought stress, which occurs from flowering through maturity (GS3), include premature plant death, increased susceptibility to charcoal rot of sorghum, stalk collapse and lodging, and a significant reduction in seed size, all of which result in decreased yield potential (Rosenow et al., 1996). Identification of the genetic factors underlying drought tolerance would provide a solid foundation for improving drought tolerance.

Retention of green leaf area at maturity (GLAM), known as stay-green, has been used as an indicator of post-anthesis drought resistance in sorghum programmes in the United States and Australia (Borrell et al., 2000b). Borrell et al. reported that under terminal water deficit, grain yield was correlated positively with GLAM ($r = 0.75$, $P < 0.01$) and negatively with the rate of leaf senescence ($r = -0.74$, $P < 0.01$) (Borrell et al., 2000c). Their results indicated that sorghum hybrids possessing the stay-green (SG) trait have a significant yield advantage under post-anthesis drought compared with hybrids not possessing this trait. Sorghum containing the SG trait retain more photosynthetically active leaves under drought than do hybrids that do not contain this trait (Borrell et al., 2000b). Nine hybrids varying in the B35 and KS19 sources of SG were grown under a fully irrigated control, post-flowering water deficit and terminal water deficit. Genotypic differences in delayed onset and reduced rate of leaf senescence were explained by differences in specific leaf N and N uptake during grain filling. Borrell et al. reported that leaf N concentration at anthesis was correlated with onset ($r = 0.751$, $P < 0.01$, $n = 27$) and rate ($r = -0.783$, $P < 0.01$, $n = 27$) of leaf senescence under terminal water deficit (Borrell and Hammer, 2000a).

Burke et al. (2013) reported a correlation between leaf dhurrin [(*S*)-*p*-hydroxyman delonitrile-β-D-glucopyranoside] contents and the level of SG based on leaf and plant death (LPD) ratings during post-flowering drought stresses. Post-flowering drought-tolerant SG sorghums had elevated dhurrin contents while pre-flowering drought-tolerant sorghum lines exhibited lower dhurrin contents. Lines with intermediate LPD ratings exhibited intermediate dhurrin levels. They concluded that dhurrin levels determined before flowering from sorghum grown under irrigated or dry land conditions were related to the degree of pre- or post-flowering drought tolerance. Burke et al. (2013) concluded that leaf dhurrin levels provide a measurement of the level of post-flowering drought stress tolerance and are highly correlated with the SG phenotype. Additionally, relationship between leaf dhurrin content and SG drought tolerance is stable across irrigation regimes and may be used as a screening tool during pre-flowering sorghum growth, eliminating the need for field-based screening and difficult visual ratings when environmental conditions are variable.

4 Identifying QTL for post-flowering drought tolerance

Numerous studies have identified QTLs associated with post-flowering drought tolerance. Tuinstra et al. (1997) used QTL analysis to identify 13 regions of the genome associated with one or more measures of post-flowering drought tolerance. They reported that two QTLs were identified with major effects on yield and 'stay-green' under post-flowering drought. These loci were also associated with yield under fully irrigated conditions suggesting that these tolerance loci have pleiotropic effects on yield under non-drought conditions. They reported that regions of linkage groups F (Chr 9) and I (Chr 6) were associated with grain yield in the post-flowering drought trial. They determined that these loci account for about 12% of the variability in grain yield under post-flowering drought.

Crasta et al. (1999) identified three major QTLs (SGA, SGD and SGG) influencing the SG trait, each contributing to more than 10% of the phenotypic variability with LOD scores of more than 2.0. They concluded that together these three major QTLs contributed to 41.2% of the phenotypic variability with a cumulative LOD score of 9.0. Two major (SGA and SGD) and three minor QTLs (SGI.1, SGI.2 and SGJ) influenced the SG response.

Xu et al. (2000) used a restriction fragment length polymorphism (RFLP) map, developed from a RIL population, to identify four SG QTLs, located on three linkage groups. The QTLs (Stg1 and Stg2) were reported to be on linkage group A, with the other two, Stg3 and Stg4, on linkage groups D and J, respectively. They stated that two SG QTLs, Stg1 and Stg2, explained 13–20% and 20–30% of the phenotypic variability, respectively. Three QTLs for chlorophyll content (Chl1, Chl2 and Chl3), explaining 25–30% of the phenotypic variability, were also identified under post-flowering drought stress. All of the three QTLs for chlorophyll content (Chl1, Chl2 and Chl3) coincided with the three SG QTL regions (Stg1, Stg2 and Stg3) accounting for 46% of the phenotypic variation. They concluded that such spatial arrangement showed that linkage group A was important for drought- and heat-stress tolerance and yield production in sorghum.

Kebede et al. (2001) compared the SG QTL locations with the earlier reports and indicated that the SG QTLs Stg A, Stg G and Stg J of the current map were consistent with the QTLs identified earlier in different populations (Crasta et al., 1999; Tuinstra et al., 1997; Xu et al., 2000). Kebede et al. reported that Stg A and Stg J were located in identical genomic locations in the B35 x Tx430 and B35 x Tx7000 populations, and Stg G might be the same QTL as reported by Crasta et al. (1999) and Tuinstra et al. (1997) on linkage groups G and F, respectively.

In a study by Haussmann et al. (2002), the three measures of SG mapped were the green leaf area percentages at 15, 30 and 45 days after flowering (% GL15, % GL30 and % GL45, respectively). Across the three measures of the SG trait, three QTLs on linkage groups A, E and G were common to both of the sorghum recombinant inbred populations developed from crosses between (1) IS9830 x E36-1 and (2) N13 x E36-1, with the SG alleles originating from E36-1. These QTLs were therefore consistent across the tested genetic backgrounds and years.

Harris et al. (2007) mapped alleles that contribute to the SG trait to four major QTLs, Stg1–Stg4, using a population derived from BTx642 and RTx7000. They analysed the size and location of BTx642 DNA regions in each RTx7000 NIL using 62 DNA markers spanning the four SG QTLs. RTx7000 NILs were identified that contained BTx642 DNA completely or partially spanning Stg1, Stg2, Stg3 or Stg4. Physiological analysis of four RTx7000 NILs containing only Stg1, Stg2, Stg3 or Stg4 showed that Stg2 retained more GLAM under terminal drought conditions than RTx7000 or the other RTx7000 NILs. They also reported that under post-anthesis water deficit the Stg2, Stg3 and Stg4 NILs exhibited a trend for delayed onset of leaf senescence compared with RTx7000.

Borrell et al. (2014) reported that the positive effects of Stg QTLs on grain yield under drought can be explained as emergent consequences of their effects on temporal and spatial water-use patterns that result from changes in leaf-area dynamics. They showed that the four Stg QTLs regulate canopy size by (1) reducing tillering via increased size of lower leaves, (2) constraining the size of the upper leaves and (3) in some cases, decreasing the number of leaves per culm. In addition, they variously affect leaf anatomy and root growth. They concluded that the multiple pathways by which Stg QTLs modulate canopy development can result in considerable developmental plasticity. The reduction in canopy size associated with Stg QTLs reduced pre-flowering water demand, thereby increasing water availability during grain filling and, ultimately, grain yield. George-Jaeggli et al. (2017) stated that by restricting their leaf area growth, and even shedding some leaf area early during the vegetative growth phase via senescence, sorghum lines with SG introgressions use soil moisture more slowly, effectively shifting water use to the later crop stages. They concluded that this ensures that they remain productive for longer periods during the critical grain-filling period.

A recent report by Hayes et al. (2015) using a RIL mapping population derived from BTx642 and RTx7000 identified a previously unidentified SG QTL (Stg5) on SBI01 that is associated with a dhurrin QTL (Dhu1). Dhu1 is aligned with genes involved in dhurrin biosynthesis (CYP79A1, CYP71E1 and UGT85B1), consistent with prior studies showing an association between high leaf dhurrin levels and this region of SBI01 (Hayes et al., 2015). High levels of leaf dhurrin have been found in grain sorghum genotypes that also exhibit SG during post-anthesis water deficit (Burke et al., 2013). Leaf dhurrin was highly heritable and Dhu1 explained a large percentage of the variation of leaf dhurrin in the population. Hayes et al. suggested that protein sequence variants found in CYP71E1 and UGT85B1 could be the cause of the observed differences in leaf dhurrin levels in BTx642, Tx7000 and other sorghum lines that vary in leaf dhurrin content (Hayes et al., 2015).

Adeyanju et al. (2016) investigated the link between the post-flowering drought response and stalk rot resistance, especially charcoal rot. Fourteen genotypic groups derived from the Tx642 × Tx7000 RIL population carrying a combination stg QTL were evaluated under three environments in four replications. They reported that stg QTL have variable effects on severity of stalk rot diseases, showing that genotypes carrying either stg1 or stg3, or their combination stg1+3 expressed high levels of resistance to *Macrophomina phaseolina*, while resistance to *Fusarium thapsinum* required a combination of stg1 and stg3. They found that other stg QTL blocks, such as stg2 and stg4, did not have an effect on stalk rot resistance caused by either pathogen. Furthermore, stalk rot resistance had no significant association with levels of dhurrin and soluble sugars in the leaves.

Johnson et al. (2015) used a transcriptomic approach to compare gene expression between SG (B35) and senescent (R16) sorghum varieties. They reported that ontological analysis of the differentially expressed transcripts identified an enrichment of genes involved with the 'response to osmotic stress' Gene Ontology (GO) category. In particular, delta-1-pyrroline-5-carboxylate synthase 2 (P5CS2) was highly expressed in the SG line compared with the senescent line, and this high expression was correlated with higher proline levels under a PEG-induced osmotic stress. Comparisons of the DEGs with those that lie in known SG QTLs revealed that P5CS2 lies within the Stg1 QTL. They identified polymorphisms in known *cis*-elements identified in the putative promoter region of P5CS2 and these could be responsible for the differences in the expression of this gene. Burke et al. (2015) measured stress-induced changes in relative water content, abscisic acid (ABA), proline, dhurrin, sucrose and carbon assimilation during the onset of water-deficit stress in the pre-flowering drought-tolerant line SC1211–11E and the post-flowering drought-tolerant line BTx642 to determine if there were differential responses to the onset of soil drying. In both greenhouse and field studies, it was found that SC1211–11E had lower relative water contents and accumulated higher levels of ABA and proline than the BTx642. Clearly additional research into the metabolic differences between senescent and SG sorghums is needed to advance our understanding of the sorghum stress responses.

Sukumaran et al. (2016) used genetic mapping in a sorghum RIL population developed from Tx436 (a non-SG (NSG) high food quality inbred) and 00MN7645 (a SG high yield inbred) and evaluated in eight environments (location and year combination) in a hybrid background of Tx3042 (a NSG A-line). They evaluated agronomic traits (grain yield and flowering time), physiological traits of SG (chlorophyll content [SPAD] and chlorophyll fluorescence [Fv/Fm] measured on the leaves), and green leaf area visual score (GLAVS). Using genotyping-by-sequencing (GBS) technology, they identified 42 QTLs. A consistent QTL for grain yield under normal and stressed conditions was identified in chromosome 1 that explained 8–16% of the phenotypic variation. QTLs for flowering time were identified in chromosomes 2, 6 and 9 that explained 6–11% of the phenotypic variation. They concluded that these identified QTLs with flanking SNPs of known genomic positions could be used to improve grain yield, flowering time and SG in sorghum molecular breeding programmes.

5 Case studies in improving drought tolerance

5.1 Case study 2

Rosenow et al. (1983) emphasized the importance of knowing the major physiological and morphological traits associated with drought tolerance in order to develop effective screening techniques and to know which traits may be utilized in a breeding programme. Retention of GLAM is a visual evaluation for the SG trait under field conditions to obtain LPD ratings (Wanous et al., 1991; Xu et al., 2000). LPD ratings have been used as an indicator of post-flowering drought resistance in sorghum programmes in the United States, Ethiopia, India and Australia (Borrell et al., 1999, 2000b,c, 2001; Borrell and Hammer, 2000a; Burke et al., 2013; Harris et al., 2007; Henzell et al., 1997; Kassahun et al., 2010; Mahalakshmi and Bidinger, 2002; McIntyre et al., 1997; Rosenow et al., 1983; Subudhi et al., 2000; Tao et al., 2000; Tuinstra et al., 1996, 1997). Unfortunately, year-to-year environmental variability results in a range of LPD ratings for a given line making reliable and quantitative evaluation of this trait difficult. Henzell et al. (1997) conducted multi-environment tests for the selection of SG line in developing superior drought-resistant genotypes in Australian sorghum breeding programmes. However, Tao et al. (2000) stated that as SG is expressed only in those tests in which terminal stress occurred, neither the efficiency nor the reliability of selection is high when only conventional breeding approaches are used for selection of this trait.

A method was described for identifying sorghum lines having the SG trait under well-watered field conditions (Burke et al., 2010). By means of a 30-min high-temperature challenge to leaf tissue during flowering of well-watered sorghum and a 30-min room temperature recovery, Burke et al. (2010) showed that SG lines can be readily identified. This procedure was a modification of a previously published technique for evaluation of water-deficit stress responses in cotton (Burke, 2007). The chlorophyll fluorescence bioassay paper (Burke et al., 2010) warned that because of the assay's sensitivity, plants cannot be under any water stress or SG false positives would be obtained. Figure 4 shows

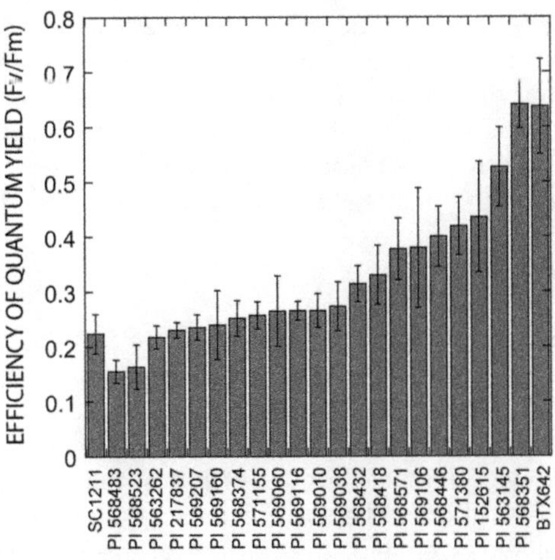

Figure 4 Graph of the efficiency of quantum yield (Fv/Fm) for SC1211 (senescent), BTx642 (stay-green), and sorghum lines from the Sudan sorghum collection following the stress bioassay described by Burke (2010).

Table 1 Peak amplitudes of dhurrin in BTx406 (senescent), BTX642 (stay-green) and sorghum plant introductions from Sudan and Ethiopia. Average peak amplitudes are from a VP Series HPLC system fitted with an evaporative light scattering detector-LT (Shimadzu, Columbia, MD)

Line	Country of origin	Relative dhurrin content average peak amplitude [mVolts]
BTx406		82
PI 568351* BTx406	Sudan	364
PI 571380* BTx406	Sudan	478
PI 563145* BTx406	Sudan	737
PI 452707* BTx406	Ethiopia	257
PI 452610* BTx406	Ethiopia	295
PI 452808* BTx406	Ethiopia	436
PI 452605* BTx406	Ethiopia	442
BTx642	Ethiopia	485
PI 452706* BTx406	Ethiopia	575
PI 452830* BTx406	Ethiopia	693

the efficiency of quantum yield (Fv/Fm) for a senescent sorghum line (SC1211), a SG sorghum line (BTx642) and a series of sorghum accessions from the Sudan collection. Most of the plant introductions exhibited low efficiency of quantum yield values similar to that of the SC1211. A few plant introductions, however, did exhibit elevated Fv/Fm levels similar to those of the SG line BTx642. Three of these introductions (PI 568351, PI 563145 and PI 571380) were advanced into our breeding programme as potential sources of the SG trait. The PIs were crossed with BTx406 and F1 offspring were evaluated for dhurrin levels to determine if the three Sudan lines identified by the efficiency of quantum yield method exhibited elevated dhurrin levels similar to the known SG lines described by Burke et al. (2013). Table 1 shows that the F1 plants from the crosses between BTx406 and the three selected lines exhibited dhurrin levels three to seven times greater than BTx406. These results support the usefulness of the chlorophyll fluorescence technique as a selection tool for SG breeding lines when the plants are grown in the absence of water stress.

5.2 Case study 3

A second method for identifying potential sources of the SG trait was evaluated in Burke's lab. This method is based upon the observation of morphological differences among photoperiod-sensitive lines grown on 40-inch rows under long day lengths in Lubbock, Texas. Under well-watered conditions, these plants develop canopies 10–20 feet in height and experience low-light levels at the bottom of the canopy. Initial studies of 500 lines from the Ethiopian sorghum collection revealed that over 95% of those evaluated exhibited significant leaf senescence at the bottom of the canopy (right side of Fig. 5c). This senescence of the lower leaves is morphologically similar to the lower leaf senescence of the post-flowering drought-sensitive line TX7000 under late season water-deficit stress shown in Fig. 5b. About 5% of the Ethiopian lines exhibited maintenance of lower leaves

Figure 5 Photographs of (a) BTx642 and (b) TX7000 following a post-flowering drought exposure, and (c) a photograph of photoperiod sensitive Ethiopian lines grown under well-watered conditions.

even under low-light conditions (left side of Fig. 5c). This maintenance of lower leaves under the low-light stress is morphologically similar to the maintenance of lower leaves in the post-flowering drought-tolerant line BTx642 TX7000 under late season water-deficit stress (Fig. 5a). Six Ethiopian lines that maintained lower leaves under low-light conditions were advanced in our breeding programme and crossed to BTx406 to convert the lines into photoperiod-insensitive lines. Similar to the lines identified in Case Study 1, these lines were evaluated for dhurrin levels (Table 1). The F1 lines exhibited dhurrin levels three to seven times that of the BTx406 breeding line used to convert the lines to photoperiod-insensitive lines. The results of the dhurrin analyses suggested that these lines might provide additional sources of the SG trait. Subsequent selections and analysis of field LPD ratings of these lines under post-flowering drought stress confirmed the presence of the SG trait.

5.3 Case study 4

Burke et al. (2015) recently reported the findings of a study investigating the metabolic responses of two contrasting sorghums to water-deficit stress. In that study the canopy temperatures were measured from three to twelve days following irrigation in the SG (BTx642) and the senescent lines (SC1211-11E). The field-grown BTx642 and SC1211-11E

Published by Burleigh Dodds Science Publishing Limited, 2018.

exhibited a gradual rise in maximum day temperature from day three to day nine as the soil water was being depleted. Burke et al. (2015) reported that the daily swing between minimum and maximum temperatures increased from approximately 16°C for days three to eight to approximately 22°C between days nine and twelve. From day eight to twelve, BTx642 exhibited higher temperatures than the SC1211-11E in the morning, and in the afternoons the SC1211-11E had higher temperatures than the BTx642. The magnitude of this difference increased as stress progressed. They observed that the differences between the lines appeared to become greater at temperatures below 27°C and above 30°C. Because only one SG and one senescent line were evaluated during the first twelve days after initiating water-deficit stress by Burke et al. (2015), additional SG and senescent

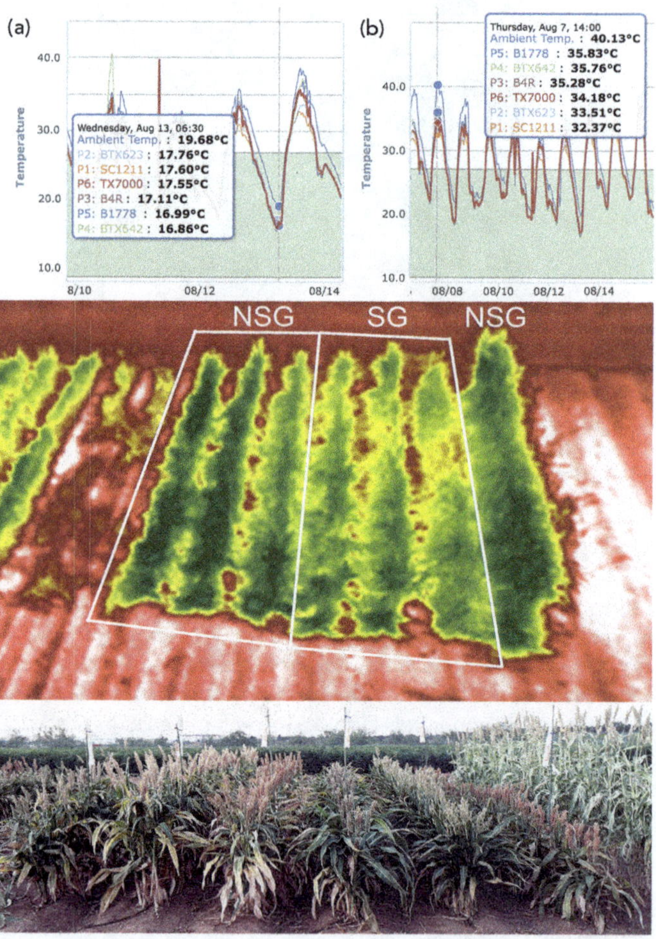

Figure 6 Graphs of (a) morning and (b) afternoon canopy temperatures of three senescent sorghums (BTx623, SC1211, and TX7000) and three stay-green sorghums (BTX642, B1778, and B4R). (c) A thermal image of the test plots at mid-day (NSG, non-stay-green; SG, stay-green) and (d) a photograph of the senescent and stay-green lines showing the SmartCrop sensors above the canopies.

lines needed to be evaluated further into water-deficit stress to determine if this is a characteristic of these two lines or whether similar responses are observed in other SG and senescent lines. Figure 6 shows canopy temperatures of three senescent (TX7000, BTx623 and SC1211) and three SG (B1778, BTx642 and B4R) sorghum lines under prolonged post-flowering water-deficit stress. Unlike the thermal signatures described by Burke et al. (2015) for SC1211 and BTx642 early in their stress responses, the results of multiple SG and senescent lines later in the stress exposure showed opposite results. Figure 6a shows morning canopy temperatures and Fig. 6b shows afternoon canopy temperatures obtained from SmartCrop thermometers shown in Fig. 6d. This data showed lower morning canopy temperatures and warmer afternoon canopy temperatures for the SG lines compared with the three senescent lines. The warmer afternoon canopy temperatures in the SG lines were further supported by thermal imaging of the plots (Fig. 6c). The senescent (NSG) lines exhibited cooler mid-day canopy temperatures than the SG lines as illustrated by the green versus yellow colour. Additional studies are needed to better understand the relationship between the timing of water-deficit stress and thermal canopy signatures in senescent and SG sorghums. These preliminary studies suggest that thermal stress signatures of sorghum lines under post-flowering water-deficit stresses may be useful in future germplasm screening for the SG trait.

In summary, significant progress has been made in understanding the genetics associated with the post-flowering drought tolerance trait (SG). Numerous lines possessing this trait have been identified and many of these have been used in the development of recombinant inbred populations needed for molecular mapping of the genes associated with this trait. The SG trait is morphologically associated with reduced leaf area prior to flowering that results in the conservation of soil water for use during the grain-filling period. Metabolically, much remains to be known about the biochemical basis for this trait, and more research is needed to elucidate the range of metabolic changes occurring in SG accession.

6 Summary

Sorghum genetic diversity existing within the NGRP provides a foundation for the development of hybrids with enhanced performance across a range of temperatures and water availabilities. New molecular technologies and germplasm screening techniques are expediting discovery of germplasm with improved abiotic stress tolerances. These traits are being introgressed into elite germplasm through the use of marker-assisted selection. The newly developed lines with the improved abiotic stress tolerances will help stabilize yields across the range of climates in which sorghum is grown.

Declining water availability from aquifers and more extreme weather events will position sorghum as the crop of choice in the future. Its ability to grow and yield in marginal agricultural regions, combined with the many uses of the crop, will increase sorghum's importance in global agriculture. The advancement of molecular techniques has outpaced the development of screening techniques needed to mine the germplasm collections. Future emphasis on the development of reliable high-throughput screening technologies should result in the development of hybrids with enhanced abiotic stress tolerance.

7 Where to look for further information

In the next few years, new information on the molecular and evolutionary mechanisms of cuticular wax in plants in response to drought stress will become available. Additionally, the importance of genetic diversity in the size of sorghum root systems and root angles in response drought will be better understood. The significant influx of sorghum research funding sponsored by the Department of Energy will help us better understand the role of epigenetics in allowing plants to survive drought conditions, and the development of high-throughput screening technologies to identify sorghum germplasm with improved abiotic stress tolerances. Germplasm releases with improved pre-and post-flowering drought tolerances will become more frequent because of the use of marker-assisted breeding. Finally, germplasm with improved cold tolerances both early and late in the sorghum plants' development will become available.

8 References

Adeyanju, A., J. Yu, C. Little, W. Rooney, P. Klein, J. J. Burke and et al. 2016. Sorghum RILs segregating for stay-green QTL and leaf dhurrin content show differential reaction to stalk rot diseases. *Crop Science* 56: 2895–903.

Bekele, W. A., K. Fiedler, A. Shiringani, D. Schnaubelt, S. Windpassinger, R. Uptmoor and et al. 2014. Unravelling the genetic complexity of sorghum seedling development under low temperature conditions. *Plant, Cell & Environment* 37. (no 3): 707–23https://doi.org/10.1111/pce.12189.

Borrell, A., G. Hammer and E.v. Oosterom. 2001. Stay-green: A consequence of the balance between supply and demand for nitrogen during grain fillings? *Annals of Applied Biology* 138(1): 91–5.

Borrell, A. K., F. R. Bidinger and K. Sunitha. 1999. Stay-green trait associated with yield in recombinant inbred sorghum lines varying in rate of leaf senescence. *International Sorghum and Millets Newsletter* 40: 31–4.

Borrell, A. K. and G. L. Hammer. 2000a. Nitrogen dynamics and the physiological basis of stay-green in sorghum. *Crop Science* 40: 1295–307.

Borrell, A. K., G. L. Hammer and A. C. L. Douglas. 2000b. Does maintaining green leaf area in sorghum improve yield under drought? I. Leaf growth and senescence. *Crop Science* 40: 1026–37.

Borrell, A. K., G. L. Hammer and R. G. Henzell. 2000c. Does maintaining green leaf area in sorghum improve yield under drought? II. Dry matter production and yield. *Crop Science* 40: 1037–48.

Borrell, A. K., J. E. Mullet, B. George-Jaeggli, E. J. van Oosterom, G. L. Hammer, P. E. Klein and et al. 2014. Drought adaptation of stay-green sorghum is associated with canopy development, leaf anatomy, root growth, and water uptake. *Journal of Experimental Botony* 65: 6251–63. doi:10.1093/jxb/eru232.

Burgess, M. G., C. M. Rush, G. Piccinni and G. Schuster. 2002. Relationship between charcoal rot, the stay-green trait, and irrigation in grain sorghum. *Phytopathology* 92: S10.

Burke, J. J. 2007. Evaluation of source leaf responses to water-deficit stresses in cotton using a novel stress bioassay. *Plant Physiology* 143: 108–21.

Burke, J. J., J. Chen, G. Burow, Y. Mecherf, D. Rosenow, P. Payton and et al. 2013. Leaf dhurrin content is a quantitative measure of the level of pre- and postflowering drought tolerance in sorghum. *Crop Science* 53: 1056–65.

Burke, J. J., C. D. Franks, G. Burow and Z. Xin. 2010. Selection system for the stay-green drought tolerance trait in sorghum germplasm. *Agronomy Journal* 102. (no 4): 1118–22.

Burke, J. J., P. Payton, J. Chen, Z. Xin, G. Burow and C. Hayes. 2015. Metabolic responses of two contrasting sorghums to water-deficit stress. *Crop Science* 55: 344–53.

Burow, G., J. J. Burke and Z. Xin. 2011. Genetic dissection of early-season cold tolerance in sorghum (Sorghum bicolor (L.) Moench). *Molecular Breeding* 28: 391–402.

Casa, A. M., G. Pressoir, P. J. Brown, S. E. Mitchell, W. L. Rooney, M. R. Tuinstra and et al. 2008. Community resources and strategies for association mapping in sorghum. *Crop Science* 48. (no 1): 30–40.

Chapman, S. C., M. Cooper, G. L. Hammer and D. G. Butler. 2000. Genotype by environment interactions affecting grain sorghum. II. Frequencies of different seasonal patterns of drought stress are related to location effects on hybrid yields. *Australian Journal of Agricultural Research* 51(2): 209–21.

Chopra, R., G. Burow, C. Hayes, Y. Emendack, Z. Xin and J. Burke. 2015. Transcriptome profiling and validation of gene based single nucleotide polymorphisms (SNPs) in sorghum genotypes with contrasting responses to cold stress. BMC Genomics 16: 1040. doi:10.1186/s12864-015-2268-8.

Crasta, O. R., W. W. Xu, D. T. Rosenow, J. Mullet and H. T. Nguyen. 1999. Mapping of post-flowering drought resistance traits in grain sorghum: Association between QTLs influencing premature senescence and maturity. *MGG, Molecular Genetics and Genomics* 262: 579–88.

De Wet, J. M. J. and J. P. Huckabay. 1967. The origin of sorghum bicolor. II. Distribution and domestication. *Evolution* 21: 787–802.

Dev Sharma, A., S. Kumar and P. Singh. 2006. Expression analysis of a stress-modulated transcript in drought tolerant and susceptible cultivars of sorghum (Sorghum bicolor). *Journal of Plant Physiology* 163: 570–6. doi:10.1016/j.jplph.2005.06.011.

Doggett, H. 1988. *Sorghum*. John Wiley, New York.

Fracasso, A., L. M. Trindade and S. Amaducci. 2016. Drought stress tolerance strategies revealed by RNA-Seq in two sorghum genotypes with contrasting WUE. *BMC Plant Biology* 16: 115. doi:10.1186/s12870-016-0800-x.

Franks, C. D., G. B. Burow and J. J. Burke. 2006. A Comparison of U.S. and Chinese sorghum germplasm for early season cold tolerance. *Crop Science* 46. (no 3): 1371–6.

Garrity, D. P., C. Y. Sullivan and D. G. Watts. 1984. Changes in grain sorghum stomatal and photosynthetic response to moisture stress across growth stages. *Crop Science* 24: 441–6.

George-Jaeggli, B., M. Y. Mortlock and A. K. Borrell. 2017. Bigger is not always better: Reducing leaf area helps stay-green sorghum use soil water more slowly. *Environmental and Experimental Botony* 138: 119–29.

Girma, F. S. and D. R. Krieg. 1992a. Osmotic adjustment in sorghum. I. Mechanisms of diurnal osmotic potential changes. *Plant Physiology* 99: 577–82.

Girma, F. S. and D. R. Krieg. 1992b. Osmotic adjustment in sorghum. II. Relationship to gas exchange rates. *Plant Physiology* 99: 583–8.

Harris, K., P. K. Subudhi, A. Borrell, D. Jordan, D. Rosenow, H. Nguyen and et al. 2007. Sorghum stay-green QTL individually reduce post-flowering drought-induced leaf senescence. *Journal of Experimental Botany* 58. (no 2): 327–38.

Haussmann, B. I. G., V. Mahalakshmi, B. V. S. Reddy, N. Seetharama, C. T. Hash and H. H. Geiger. 2002. QTL mapping of stay-green in two sorghum recombinant inbred populations. *Theoretical and Applied Genetics* 106: 133–42.

Haussmann, B. I. G., A. B. Obilana, P. O. Ayiecho, A. Blum, W. Schipprack and H. H. Geiger. 1999. Quantitative-genetic parameters of sorghum Sorghum bicolor (L.) Moench grown in semi-arid areas of Kenya. *Euphytica: Netherlands Journal of Plant Breeding* 105(2): 109–18.

Haussmann, B. I. G., A. B. Obilana, P. O. Ayiecho, A. Blum, W. Schipprack and H. H. Geiger. 2000. Yield and Yield stability of four population types of grain sorghum in a semi-arid area of Kenya. *Crop Science* 40: 319–29.

Haussmann, B. I. G., A. B. Obilana, A. Blum, P. O. Ayiecho, W. Schipprack and H. H. Geiger. 1998. Hybrid performance of sorghum and its relationship to morphological and physiological traits under variable drought stress in Kenya. *Plant Breed Z Pflanzenzucht* 117: 223–9.

Hayes, C. M., G. Burow, P. Brown, C. Thurber, Z. Xin and J. J. Burke. 2015. Natural variation in synthesis and catabolism influence dhurrin content in sorghum (Sorghum bicolor L. Moench). *Plant Genome* 8: 1–9. doi: 10.3835/plantgenome2014.3809.0048.

Hayes, C. M., B. W. Weers, M. Thakran, G. B. Burow, Z. Xin, Y. Emendack, et al. 2015. Discovery of a dhurrin QTL in sorghum: Co-localization of dhurrin biosynthesis and a novel stay-green QTL. *Crop Science* 56: 104–12.
Henzell, R. G., K. J. McCree, C. H. M. Van Bavel and K. F. Schertz. 1976. Sorghum genotype variation in stomatal sensitivity to leaf water deficit. *Crop Science* 16: 660–2.
Henzell, R. G., G. L. Hammer, A. K. Borrell, C. L. McIntyre and S. C. Chapman. 1997. Research on drought resistance in grain sorghum in Australia. *International Sorghum and Millets Newsletter*: 1–9.
Jedmowski, C., A. Ashoub, T. Beckhaus, T. Berberich, M. Karas and W. Bruggemann. 2014. Comparative analysis of sorghum bicolor proteome in response to drought stress and following recovery. *International Journal of Proteomics* 2014: 395905. doi:10.1155/2014/395905.
Johnson, S. M., I. Cummins, F. L. Lim, A. R. Slabas and M. R. Knight. 2015. Transcriptomic analysis comparing stay-green and senescent Sorghum bicolor lines identifies a role for proline biosynthesis in the stay-green trait. *Journal of Experimental Botony* 66: 7061–73. doi:10.1093/jxb/erv405.
Johnson, S. M., F. L. Lim, A. Finkler, H. Fromm, A. R. Slabas and M. R. Knight. 2014. Transcriptomic analysis of Sorghum bicolor responding to combined heat and drought stress. *BMC Genomics* 15: 456. doi:10.1186/1471-2164-15-456.
Kannangara, T., R. C. Durley, G. M. Simpson and D. G. Stout. 1982. Drought resistance of Sorghum bicolor. 4. Hormal changes in relation to drought stress in field-grown plants. *Canadian Journal of Plant Science* 62: 317–30.
Kapanigowda, M. H., R. Perumal, M. Djanaguiraman, R. M. Aiken, T. Tesso, P. V. Prasad and et al. 2013. Genotypic variation in sorghum [Sorghum bicolor (L.) Moench] exotic germplasm collections for drought and disease tolerance. *Springerplus* 2: 650. doi:10.1186/2193-1801-2-650.
Kassahun, B., F. R. Bidinger, C. T. Hash and M. S. Kuruvinashetti. 2010. Stay-green expression in early generation sorghum Sorghum bicolor (L.) Moench QTL introgression lines. *Euphytica* 172. (no 3): 351–62.
Kebede, H., P. K. Subudhi, D. T. Rosenow and H. T. Nguyen. 2001. Quantitative trait loci influencing drought tolerance in grain sorghum (Sorghum bicolor L. Moench). *Theoretical and Applied Genetics* 103: 266–76.
Knoll, J. and G. Ejeta. 2008. Marker-assisted selection for early-season cold tolerance in sorghum: QTL validation across populations and environments. *Theoretical and Applied Genetics* 116: 541–53. doi:10.1007/s00122-007-0689-8.
Knoll, J., N. Gunaratna and G. Ejeta. 2008. QTL analysis of early-season cold tolerance in sorghum. *Theoretical and Applied Genetics* 116: 577–87. doi:10.1007/s00122-007-0692-0.
Mahalakshmi, V. and F. R. Bidinger. 2002. Evaluation of stay-green sorghum germplasm lines at ICRISAT. *Crop Science* 42: 965–74.
Manjarrez-Sandoval, P., V. A. Gonzalez-Hernandez, L. E. Mendoza-Onofre and E. M. Engleman. 1989. Drought stress effects on the grain yield and panicle development of sorghum. *Canadian Journal of Plant Science Revue Canadienne de Phytotechnie* 69: 631–41.
Maulana, F. and T. T. Tesso. 2013. Cold temperature episode at seedling and flowering stages reduces growth and yield components in sorghum. *Crop Science* 53: 564–74.
McIntyre, C. L., Y. Tao, D. R. Jordan and R. G. Henzell. 1997. Application of molecular markers to sorghum breeding in Australia. *International Sorghum and Millets Newsletter*: 11–15.
Menezes, C. B., D. C. Saldanha, C. V. Santos, L. C. Andrade, M. P. Mingote Julio, A. F. Portugal and et al. 2015. Evaluation of grain yield in sorghum hybrids under water stress. *Genetics and Molecular Research* 14: 12675–83. doi:10.4238/2015.October.19.11.
Mickelbart, M. V., P. M. Hasegawa and J. Bailey-Serres. 2015. Genetic mechanisms of abiotic stress tolerance that translate to crop yield stability. *Nature Review Genetics* 16: 237–51.
Mittler, R. 2006. Abiotic Stress, the field environment and stress combination. *Trends in Plant Science* 11: 15–19.
Muchow, R. C. and T. R. Sinclair. 1989. Epidermal conductance, stomatal density and stomatal size among genotypes of Sorghum bicolor (L.) Moench. *Plant, Cell and Environment* 12: 425–31.

Peng, S. and D. R. Krieg. 1992. Gas exchange traits and their relationship to water use efficiency of grain sorghum. *Crop Science* 32: 386–91.

Prasad, P. V., M. Djanaguiraman, R. Perumal and I. A. Ciampitti. 2015. Impact of high temperature stress on floret fertility and individual grain weight of grain sorghum: Sensitive stages and thresholds for temperature and duration. *Frontiers in Plant Science* 6: 820. doi:10.3389/fpls.2015.00820.

Rosenow, D., G. Ejeta, L. E. Clark, M. L. Gilbert, R. G. Henzell, A. K. Borrell and et al. 1996. Breeding for pre- and post-flowering drought stress resistance in sorghum. In D. Rosenow (Ed.) *The International Conference on Genetic Improvement of Sorghum and Millet.* INTSORMIL, Lubbock, pp. 400–24.

Rosenow, D. T. and L. E. Clark. 1982. Drought tolerance in sorghum. Proceeding of Annu Corn Sorghum Ind Res Conf Am Seed Trade Assoc Corn Sorghum Div Corn, pp. 18–30.

Rosenow, D. T., G. Ejeta, L. E. Clark, M. L. Gilbert, R. G. Henzell, A. K. Borrell and et al. 1996. Breeding for pre- and post-flowering drought stress resistance in sorghum. In D. Rosenow (Ed.) *The International Conference on Genetic Improvement of Sorghum and Millet.* INTSORMIL, Lubbock, TX, pp. 400–24.

Rosenow, D. T., J. E. Quisenberry, C. W. Wendt and L. E. Clark. 1983. Drought tolerant sorghum and cotton germplasm. *Agricultural Water Management* 7: 207–22.

Sanchez, A. C., P. K. Subudhi, D. T. Rosenow and H. T. Nguyen. 2002. Mapping QTLs associated with drought resistance in sorghum (Sorghum bicolor L. Moench). *Plant Molecular Biology* 48: 713–26.

Singh, S. P. 1985. Sources of cold tolerance in grain sorghum. *Canadian Joutrnal of Plant Science Revue Canadienne de Phytotechnie* 65: 251–7.

Souza, F. H. and J. Marcos Filho. 1975. Estudi comparativo de metodos para a avaliacao do vigor de sementes de sorgo (Sorghum vulgare Pers). *Anais da Escola Superior Luiz de Queiroz* 33: 369–83.

Stout, D. G., T. Kannangara and G. M. Simpson. 1978. Drought resistance of Sorghum bicolor. 2. Water stress effects on growth. *Canadian Journal of Plant Science* 58: 225–33.

Subudhi, P. K., D. T. Rosenow and H. T. Nguyen. 2000. Quantitative trait loci for the stay green trait in sorghum (Sorghum bicolor L. Moench): Consistency across genetic backgrounds and environments. *Theoretical and Applied Genetics* 101: 733–41.

Sukumaran, S., X. Li, X. Li, C. Zhu, G. Bai, R. Perumal and et al. 2016. QTL mapping for grain yield, flowering time, and stay-green traits in sorghum with genotyping-by-sequencing markers. *Crop Science* 56(no. 4): 1429–42.

Tao, Y. Z., R. G. Henzell, D. R. Jordan, D. G. Butler, A. M. Kelly and C. L. McIntyre. 2000. Identification of genomic regions associated with stay green in sorghum by testing RILs in multiple environments. *Theoretical and Applied Genetics* 100: 1225–32.

Tenkouano, A., F. R. Miller, R. A. Frederiksen and D. T. Rosenow. 1993. Genetics of nonsenescence and charcoal rot resistance in sorghum. *Theoretical and Applied Genetics* 85: 644–8.

Tiryaki, I. and D. J. Andrews. 2001a. Germination and seedling cold tolerance in sorghum. I. Evaluation of rapid screening methods. *Agronomy Journal* 93: 1386–91.

Tiryaki, I. and D. J. Andrews. 2001b. Germination and seedling cold tolerance in sorghum. II. Parental lines and hybrids. *Agronomy Journal* 93: 1391–7.

Tuinstra, M. R., E. M. Grote, P. B. Goldsbrough and G. Ejeta. 1996. Identification of quantitative trait loci associated with pre-flowering drought tolerance in sorghum. *Crop Science* 36: 1337–44.

Tuinstra, M. R., E. M. Grote, P. B. Goldsborough and G. Ejeta. 1997. Genetic analysis of post-flowering drought tolerance and components of grain development in Sorghum bicolor (L.) Moench. *Molecular Breeding* 3 (6): 439–48.

Upadhyaya, H. D., Y. H. Wang, D. V. Sastry, S. L. Dwivedi, P. V. Prasad, A. M. Burrell, and et al. 2016. Association mapping of germinability and seedling vigor in sorghum under controlled low-temperature conditions. *Genome* 59: 137–45. doi:10.1139/gen-2015-0122.

Walulu, R. S., D. T. Rosenow, D. B. Wester and H. T. Nguyen. 1994. Inheritance of the stay green trait in sorghum. *Crop Science* 34: 970–2.

Wang, Y.-H., Burow, G. B., Burke, J. J. 2015. *Genetic Mapping of Abiotic Stress Responses in Sorghum.* CRC Press, Boca Raton, Florida.

Wanous, M. K., F. R. Miller and D. T. Rosenow. 1991. Evaluation of visual rating scales for green leaf retention in sorghum. *Crop Science* 31: 1691–4.

Wright, G. C. and R. C. G. Smith. 1983. Differences between two grain sorghum genotypes in adaptation to drought stress. II. Root water uptake and water use. *Australian Journal of Agricultural Research* 34(6): 627–36.

Wright, G. C., R. C. G. Smith and J. R. McWilliam. 1983. Differences between two grain sorghum genotypes in adaptation to drought stress. I. Crop growth and yield responses. *Australian Journal of Agricultural Research* 34(6): 615–26.

Wright, G. C., R. C. G. Smith and J. M. Morgan. 1983. Differences between two grain sorghum genotypes in adaptation to drought stress. III. Physiological responses. *Australian Journal of Agricultural Research* 34(6): 637–51.

Xu, W., D. T. Rosenow and H. T. Nguyen. 2000. Stay green trait in grain sorghum: Relationship between visual rating and leaf chlorophyll concentration. *Plant Breed Z Pflanzenzucht* 119: 365–7.

Xu, W. W., P. K. Subudhi, O. R. Crasta, D. T. Rosenow, J. E. Mullet and H. T. Nguyen. 2000. Molecular mapping of QTLs conferring stay-green in grain sorghum (Sorghum bicolor L. Moench). *Genome* 43: 461–9.

Published by Burleigh Dodds Science Publishing Limited, 2018.

Chapter 9

Sorghum tolerance to low-phosphorus soil conditions

Willmar L. Leiser, University of Hohenheim, Germany; H. Frederick Weltzien-Rattunde and Eva Weltzien-Rattunde, University of Wisconsin-Madison, USA; and Bettina I.G. Haussmann, University of Hohenheim, Germany

1 Introduction

Phosphorus (P) is an essential plant nutrient. Phosphate rock, the main source of P fertilizer, is finite and just five countries control an estimated 88% of the global phosphate rock reserves. Many sorghum-growing areas in Africa are affected by P scarcity, and in these areas small-scale farmers depend on sorghum for food security and income but have difficulty accessing or paying for P fertilizer. With the expected future price increases for P fertilizer, the major but hidden P-deficiency problem will only worsen. Therefore, enhancing sorghum tolerance to low-P soil conditions is of great importance both now and for the future. This chapter first gives an overview of the worldwide P challenge, and then focuses on sorghum breeding for low soil P conditions, taking breeding for smallholder farmers in West Africa as a case study. The issues and challenges presented are:

- Breeding sorghum for smallholder farmers in West Africa;
- Genotype selection under low-input conditions – challenges and solutions;
- Quantitative genetics – a basis for setting up a specific breeding programme;
- Adaptation of plants to low-P soil conditions;
- Genetics of low-P adaptation.

http://dx.doi.org/10.19103/AS.2017.0015.30

For each point, a general review of related literature is followed by concrete results from West Africa. This is then consolidated in a summary of implications for applied sorghum breeding in West Africa. We continue with future trends and challenges (both breeding and agronomic/socio-economic) in bringing benefits of sorghum low-P tolerance to West African farmers, and close with some hints regarding where to look for further information.

2 Phosphorus: a worldwide future challenge

Phosphorus is a key component of DNA, cell membranes and cellular energy, hence vital to every form of life. There is no substitute for P in food production and it may be the most limiting mineral nutrient for plants across all arable land (Kochian, 2012). Currently 90% of all mined rock phosphate is used in food production and its worldwide use is constantly increasing due to a higher demand for food, feed and fuel production (Cordell et al., 2009). Estimates of when worldwide P reserves will be exhausted range from 40 to 400 years, depending on the source of information (Vaccari, 2009; Cooper et al., 2011; Obersteiner et al., 2013; Cordell and White, 2013) and the estimated worldwide demand for this scarce mineral (MacDonald et al., 2011; Sattari et al., 2012). Geopolitical conflicts over control of mineable P reserves are likely to continue, since these are heavily concentrated with Morocco and Western Sahara holding about 74% of the global share, followed by China 6% and Algeria 3% (Jasinski, 2013; Fig. 1).

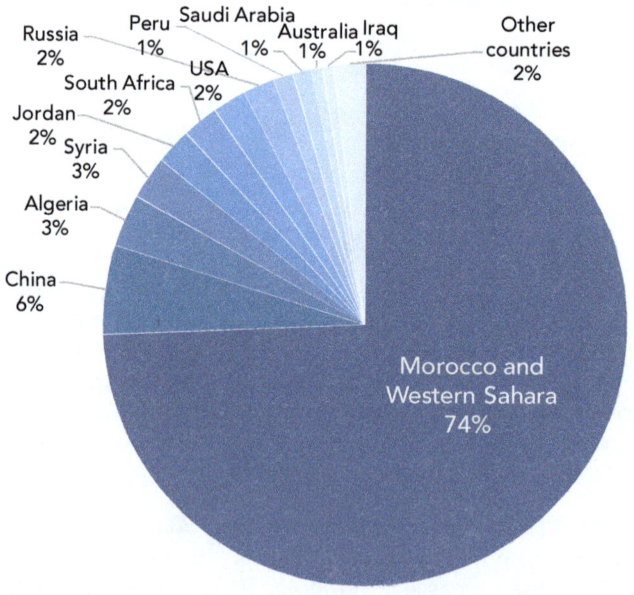

Figure 1 Phosphate rock reserves by country, indicating that the top 5 countries (Morocco and Western Sahara, China, Algeria, Syria and Jordan) control 88% of remaining reserves (Data based on Jasinski, 2013).

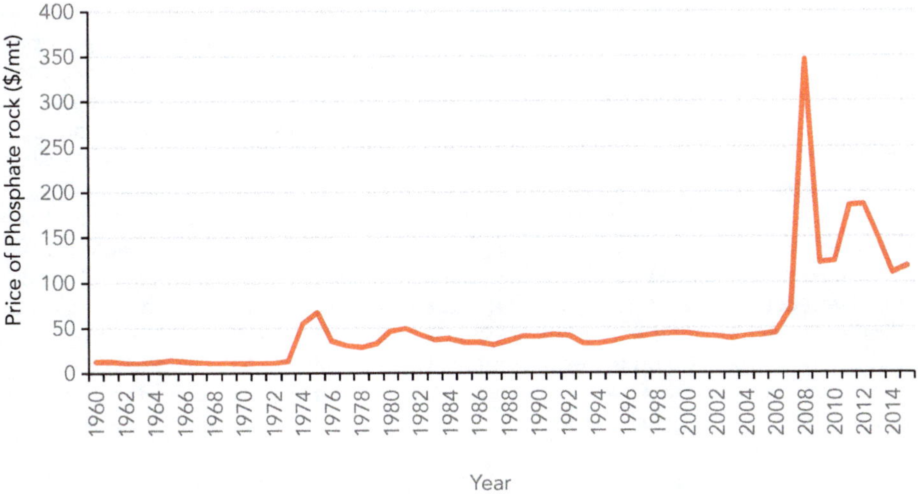

Figure 2 Phosphate rock price ($/mt) development in the years 1960–2015 based on commodity price data from World Bank (http://www.worldbank.org/en/research/commodity-markets).

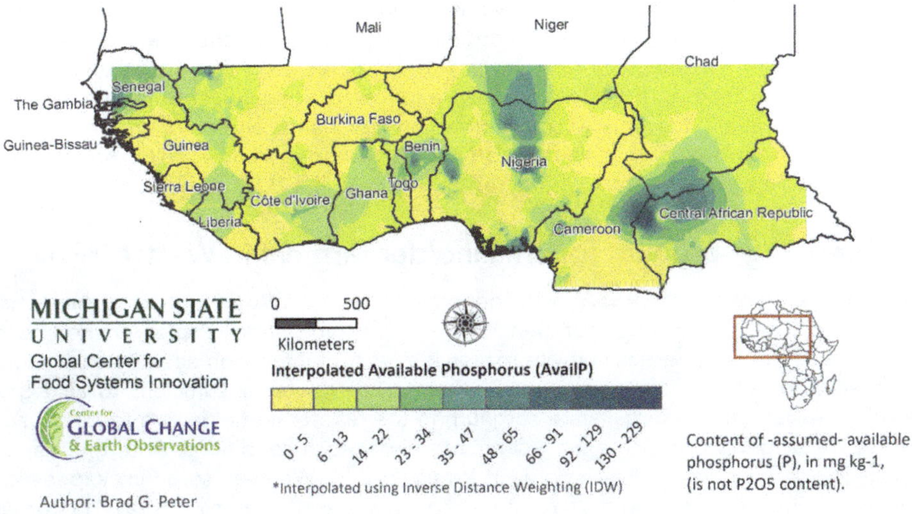

Figure 3 Interpolated plant-available phosphorus levels in West and Central Africa.

The P fertilizer prices skyrocketed in 2008 and since then have stayed above 2.6 times higher than in 2006 (Fig. 2). While farmers worldwide are economically affected by increasing fertilizer prices, smallholder farmers in sub-Saharan Africa (SSA) are hardest struck by such strong increases. Fertilizer prices in SSA are often higher, up to three times the world market price, due to limited infrastructure, inefficient supply chains and dealer margins (Cordell et al., 2015).

Most of the SSA soils are highly weathered low pH soils with a high-P retention level (Kochian, 2012) that fix most (70–90%) of the applied P as plant-unavailable phosphate (Holford, 1997) (Fig. 3). Current fertilizer application rates are very low in SSA, mostly below 5 kg P ha^{-1} (Obersteiner et al., 2013), resulting in P deficits for the agricultural production system of certain regions such as West Africa (WA) (MacDonald et al., 2011). Therefore, there is an urgent need for crop varieties that can cope with low plant-available P soils in these regions.

In contrast, industrialized countries, for example, the United States, Europe and many Asian countries, tend to have soils with high-P levels and eutrophication of sea bays (e.g. Gulf of Mexico, Baltic Sea) and lakes (e.g. Lake Erie) due to excessive use of P resulting in negative ecosystem and economic consequences. In these countries, a reduction of P fertilization is needed in order to meet specific regulations and reduce the negative impact of agricultural production on the environment. Therefore, here, we will need plants, which can efficiently take up the excess P present in the soil and use it very efficiently internally to produce the highest possible biomass.

3 Breeding sorghum for low-P soil conditions

Although adaptation to low-P soil conditions has been studied for many crops such as rice and maize, the only extensive research on sorghum adaptation to low-P soils that we know of is from WA. Field testing of diverse sorghum materials under low-P conditions was started in 2006 by the Institut d'Economie Rurale (IER) and the International Crops Research Institute for the Semi-Arid Tropics (ICRISAT) in Mali. Since that time several hundred to a few thousand progenies have been tested annually under long-term low-P conditions on-station as well as in smallholder farmers' fields. Therefore, in the following sections we present general challenges on breeding crops for low-P soils and then give specific examples from the work conducted on sorghum in WA.

3.1 Breeding sorghum for smallholder farmers in West Africa

Sorghum is a staple crop in SSA, with the largest share in WA (42%) across the African continent (Fig. 4). Its good adaptation to the challenging and complex environmental conditions of SSA and particularly WA makes it a crop vital for food security and income for smallholder farmers of this region. Further, where rainfall is sufficient to also grow maize, these farmers often cultivate sorghum in the less fertile fields due to its capacity to more dependably produce grain than maize under such conditions. And, the majority of soils in WA have low plant-available P levels (Fig. 3). Whereas sorghum experiences P-deficiency below a threshold of 7–10 mg kg^{-1} plant-available soil P (Bray-1P) content (Doumbia et al., 1993), most farmers' fields have P levels below this threshold (mean = 7.4, median = 5.5; unpublished data). Women's fields tend to have even lower P levels (Fig. 5), as they are typically allocated fields at the end of a crop rotation cycle when the fertility is deemed too low to be of interest for use by men. Limited soil P availability is therefore a frequent and serious constraint to sorghum growth and productivity across the range of environments in WA (Buerkert et al., 2001). Although sorghum has a grain yield potential of several tons per hectare in WA, current average grain yields are only about 1t ha^{-1} since 1960 (FAO, 2012), due in part to low soil fertility and low-input production systems (Vom Brocke et al., 2010).

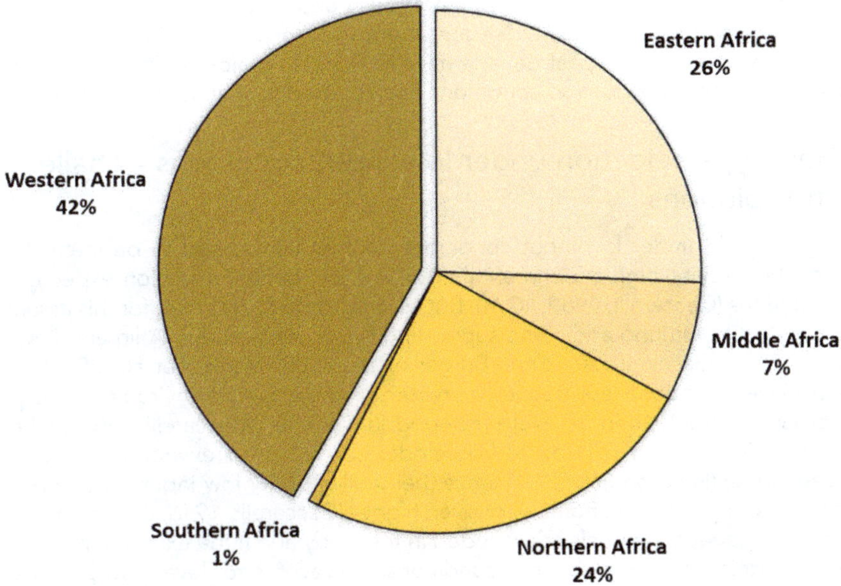

Figure 4 Proportional sorghum production in Africa in 2014 based on FAOSTAT 2014 data.

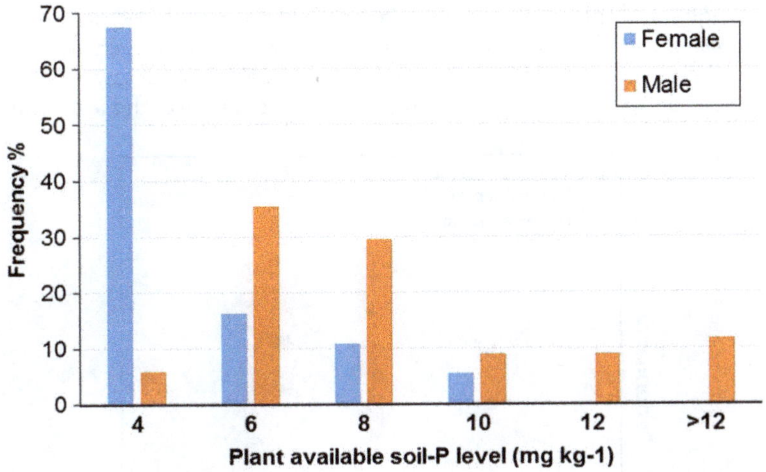

Figure 5 The frequency distribution of a sample of 91 Malian farmers' sorghum production fields for level of plant available P (Bray-1 in ppm).

Increasing sorghum productivity by applying mineral P fertilizer is currently only a limited option for most smallholder farmers in WA. The lack of financial resources, high prices, risk aversion and inadequate rural infrastructure hinder their annual fertilizer application rate to below 5 kg P ha^{-1} (MacDonald et al., 2011; Obersteiner et al., 2013). Breeding sorghum for adaptation to low-P conditions is, therefore, needed to address one of the most critical challenges for WA smallholder sorghum producers. Researchers have taken this need seriously and over the past decade have launched a major effort to breed sorghum adapted to low-input production conditions as described further in case studies below.

3.2 Genotype selection under low-input conditions – challenges and solutions

Selecting for yield under low-input conditions is often considered to be ineffective due to the low heritability, high experimental error and low genetic variation expected under these conditions (Ceccarelli, 1989, 1994). But contradictory results exist for this assumption, with some studies refuting and some supporting these assumptions (Atlin and Frey, 1989; Ceccarelli, 1994; Gallais et al., 2008; Burger et al., 2008; Mandal et al., 2010). Genetic variation for yield has been reported to be greater in highly stressed environments, especially when locally adapted landraces were included in the trials (Ceccarelli, 1996), whereas a smaller genetic variation is expected in moderately stressed environments (cross-over point; see Fig. 6) (Simmonds, 1991). Hence, heritability under low-input conditions can be comparable to high-input conditions or even higher (Ceccarelli, 1996; Burger et al., 2008), if appropriate genetic materials are included in the study and if the experimental error is of the same magnitude as in high-input conditions. However, error levels typically are much higher under low-input management since spatial variation of environmental factors, for example, soil fertility, is not compensated by external inputs, and small fertility differences can have very large effects on plant growth (Marschner, 1995: 184–6; Voortman and Brouwer, 2003). Genotypic variation and selection therefore can be biased by environmental factors leading to a higher unexplained residual variance (Grondona et al., 1996).

Therefore, control of the environmental heterogeneity needs to be pursued by use of appropriate design and analysis options to successfully select under low-input conditions. The various field designs and corresponding analyses created over the last century mostly

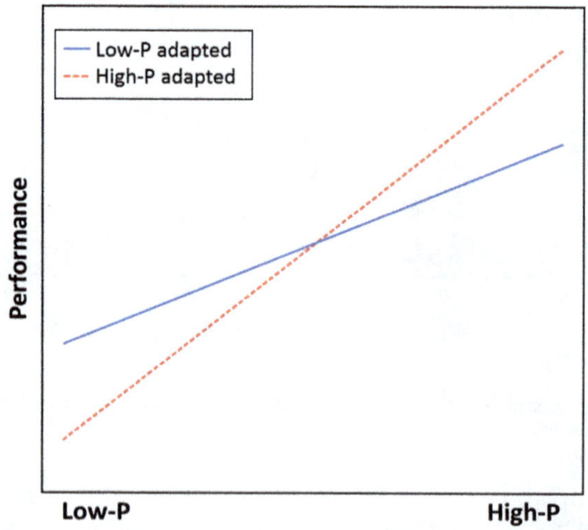

Figure 6 Genotype-by-environment cross-over interaction for low- versus high-P soil conditions.

control environmental heterogeneity by blocking structures and replications (Edmondson, 2005). These techniques have their limitations if spatial variation cannot be captured by the applied design. Various spatial adjustment techniques have been developed (e.g. autoregressive models) and have been shown to significantly reduce residual error, hence increase heritability and therefore make selection more efficient especially in abiotic stress environments (Gilmour et al., 1997; Singh et al., 2003).

Case study results from West Africa

A study to investigate the potential of spatial models to increase the efficiency of low-P field trials for grain yield selection was conducted in Mali with a series of 17 experiments testing 70 sorghum genotypes (Leiser et al., 2012a). All trials were analysed for grain yield using different mixed models, including models with autoregressive spatial correlation terms. Although addition of soil or plant growth parameters as covariates in a standard analysis can increase heritability (Mühleisen et al., 2013), these parameters, and particularly soil parameters, accounting mostly only for a small proportion of genotypic performance (Voortman and Brouwer, 2003; Leiser et al., 2012a), are laborious to collect and expensive to analyse. Hence, several spatial models (AR1, AR2) were applied that generally improved broad-sense heritability estimates for grain yield, averaging gains by 10 and 6 percentage points relative to RCB and lattice models, respectively (Leiser et al., 2012a). The heritability estimate gains were even higher under low-P conditions and in less replicated (even two replicate) situations. Further, the use of spatial models to analyse low-input trials resulted in important changes in genotype ranking for grain yield, indicating that the better discrimination among genotypes provided by these models can lead to different, more effective, selection outcomes. Although no single model was best for all environments, the AR1xAR1 model captured most of the gains for heritability and relative efficiency.

In this study, mostly spatial trends occurred along rows and columns, with spatial models accounting for both trends being most effective (Leiser et al., 2012a). Thus, many sorghum trials conducted on-station and on-farm by the national programmes and ICRISAT are now laid out as incomplete blocks in two dimensions comprising both rows and columns in consecutive field trials. This practice generally gave similar high h^2 estimates using either incomplete block or spatial model analyses (data not shown). Thus, proper blocking based on prior knowledge rather than the spatial adjustment methods proposed by Leiser et al. (2012a) can be relied upon for routine testing. The two-dimensional blocking incorporated in the R package DiGGer can be easily applied in any breeding programme to model spatial covariance structures in a mixed model framework without investment in expensive software.

No matter which approach is finally chosen, it is essential to correct properly for environmental errors by either design or analysis for effective low-P field testing. The use of spatial models, especially under low-P conditions, will lead to substantially higher h^2 estimates, more precise genotype estimates and different genotypic rankings, hence influencing genotype selection (Leiser et al., 2012a,b). Further, the best possible phenotypic data are essential to identify marker–trait associations to enable molecular breeding using either QTL-based or genomic selection methods. The power of detecting QTL (Riedelsheimer et al., 2012a) and the accuracy of genomic prediction (Lado et al., 2013) increase with increasing h^2. Hence, spatial models can be of great importance also for molecular breeding efforts targeting stress-prone environments.

3.3 Quantitative genetics: a basis for setting up a specific breeding programme

Knowledge about the prevailing stresses and the amount and type of genotype-by-environment interaction (GEI) within the target region is essential to guide the breeder in designing any breeding programme, including one targeting yield under low-P conditions in WA. If the GEI is of the non-cross-over type, a breeder may choose to select for a wide adaptation across many environmental conditions within the target region. Whereas, if there is a significant and large GEI of cross-over type where genotype ranking changes in different environmental conditions (e.g. in low-P vs high-P conditions; Fig. 6 or within specific low-P environments), selection targeting specific conditions should be pursued. Knowledge of type and extent of GEI within the target production system is necessary to determine appropriate focus for selection and allocation of testing resources. Appropriate to consider would be GEI of geographic (e.g. genotype x rainfall zones) and of sowing date origins. Furthermore, when targeting a specific environmental condition (e.g. adaptation to low-P soils), the question arises whether direct selection under these specific conditions (e.g. low-P) or indirect selection under normal conditions (e.g. high-P) is superior. The response to direct versus indirect selection depends on the heritabilities of the target trait (e.g. grain yield) under each condition and on the genetic correlation of the target trait between both conditions (Atlin and Frey, 1989). Therefore, if genotypic performance in both conditions is highly correlated and the heritability estimates in the direct selection environment (e.g. low-P) are lower, then indirect selection would be more efficient and should be the method of choice.

Case study results from West Africa

A diversity panel of 70 diverse West African sorghum lines and varieties was evaluated for yield performance in low-P and high-P soil conditions from 2006 to 2010 by Leiser et al. (2012b). Sorghum growth was found to be severely reduced under the P-deficient conditions, resulting in large delays of flowering and reductions of grain yield and plant height. Similar responses to P-deficiency were observed in other plant species (Rossiter, 1978; Fageria et al., 1988; Atlin and Frey, 1989; Wissuwa and Ae, 2001; Manske et al., 2001; Turk et al., 2003; Chen et al., 2008; Cichy et al., 2008; Parentoni et al., 2010).

The WA sorghum diversity panel showed a relatively good relationship between performance under high-P and low-P conditions as compared to other crops (Atlin and Frey, 1989; Beebe et al., 1997; Hammond et al., 2009; Parentoni et al., 2010; Ding et al., 2012), suggesting that WA sorghum germplasm as a group possess relatively good adaptation to low-P soil conditions (Leiser et al., 2012b). Yet this WA germplasm exhibited substantial genetic variation for yield response to level of soil P, with some genotypes, particularly landrace varieties, showing specific adaptation to low-P conditions with significant genotype-by-P cross-over type interaction.

Whereas genetic variation for yield was detectable under both low-P and high-P conditions, there was actually less GEI observed over low-P environments as compared to the high-P environments. This indicates that genotypic selection for yield could be made across the full range of low-P environments without regard to specific mega-environments (such as degree of environmental P-deficiency, geographic or rainfall levels) among the Sudanian zone test environments. Also, the resulting high broad-sense heritabilities under both low- and high-P conditions showed that direct selection under low-P conditions is possible and would be more efficient than indirect selection under high-P conditions for enhancing grain yield in low-P environments (<7 ppm Bray-1 P) (Leiser et al., 2012b). Direct selection under low-P was predicted to be 18% more effective for achieving yield gains under low-P conditions based on the WA sorghum diversity panel (Leiser et al. 2012b). This prediction was validated by large-scale evaluations of early generation progenies from a random-mating population established specifically for adaptation to low-P conditions. Based on evaluations of these population progenies, direct selection for yield under low-P was estimated to be 62% more efficient than indirect selection under high-P conditions, with only a weak genotypic relationship between low-P and high-P grain yield performance (R^2 = 0.13; Fig. 7) and heritability under low-P (h^2 = 0.66) that was nearly as high as under high-P (h^2 = 0.76).

Further validation comes from a selection methods study. Here, the on-farm performance of progenies derived from a dwarf Guinea population revealed that on-station testing under low-P conditions better predicted the multi-environment on-farm performance than did on-station testing under high-P conditions (Rattunde et al., 2016). Although direct yield testing on-farm was found to be more effective than low-P on-station testing for predicting sorghum yield performance under smallholder farmers' conditions, empirical results showed that combining on-farm and on-station yield performance could provide the greatest gains.

Various studies in WA (Leiser et al., 2012b, 2014a,b, 2015) also showed that the choice of sorghum germplasm is important for optimizing gains for low-P tolerance. The sorghum racial groups showed differing response to P level, with Guinea and Durra races showing more specific adaptation to low-P conditions, whereas the Caudatum race showed more adaptation to high-P conditions (Leiser et al., 2014b, 2015). The Guinea race sorghums, the predominant germplasm in the wetter Sudanian zone, and the Durra race sorghums, cultivated in the drier Sahelian zone (Zongo et al., 1993; Deu et al., 2008; Sagnard et al., 2011), thus represent promising germplasm pools for breeding sorghum targeting P-limited conditions in their respective agro-climatic zones in WA. Furthermore, landrace varieties showed more specific adaptation to low-P conditions than lines from formal breeding programmes. This may reflect the consequence of landrace varieties being the product of farmers' selection under low-input conditions, whereas formal

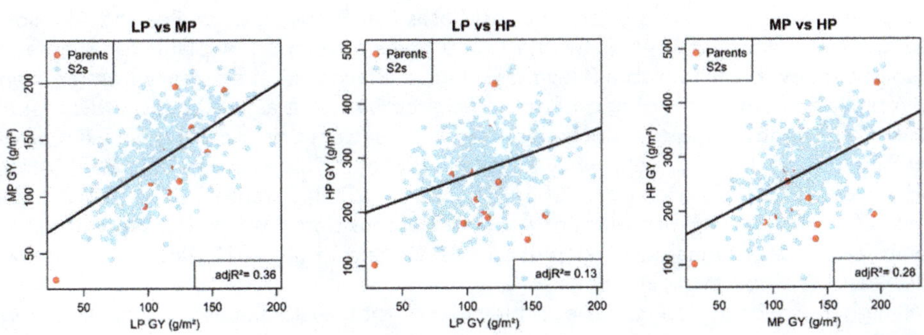

Figure 7 Relationship of grain yield performance of 435 S2 bulks (mixtures of S2 plants derived from the same S1 plant) and 14 parental lines in three different P treatments (low-P (LP), medium-P (MP) and high-P (HP) conditions).

sorghum breeding programmes were traditionally mostly conducted under well-fertilized conditions. However, both landrace and formally bred materials showed significant within-group variation for grain yield under low-P conditions and thus both pools can be used to enhance grain yield production under farmer's low-input field conditions.

Also, photoperiod-sensitive genotypes proved to be more adapted to low-P conditions, exhibiting less delay to flowering under low-P conditions and having a higher P uptake rate than photoperiod-insensitive varieties (Leiser et al., 2014b, 2015). The early sown photoperiod-sensitive sorghums were found to exhibit bilinear rates of above-ground growth, with growth slowed after initiation of the 20th leaf (Clerget et al., 2008) but root growth, assessed by rooting depth, continued unaltered (B. Clerget, pers. comm. as per other Leiser paper). Thus, photoperiod-sensitive sorghums may, in effect, have the capacity to go into 'low gear' with relatively greater root versus above-ground growth that can maintain adequate P uptake at later growth stages. Similar findings on phenological delay, root growth duration and adaptation to low-P conditions were also observed in *Arabidopsis thaliana* (Nord and Lynch, 2008), pointing to the need of further research on this adaptation mechanism.

Recommendations for allocating testing resources to optimize yield gains by sorghum breeding programmes targeting performance under farmers' low-P conditions, therefore, include (i) emphasis of directly testing under lower P conditions on-station; (ii) opportunity of directly testing for yield on-farm; and (iii) possibility of maximizing gains by combining on-station low-P and on-farm yield testing. The germplasm base for the breeding programme would benefit from inclusion of (i) landrace materials appropriate for the targeted agro-ecology and (ii) photoperiod-sensitive materials.

4 Plant adaptation to low-P conditions

4.1 Adaptation of plants to low-P soil conditions

Phosphorus in soils occurs mostly as orthophosphate and can be divided into mineral and organic P. More than 90% of the total soil P is inaccessible to plants since it is fixed in organic matter or as Al-, Fe- or Ca-phosphates (i.e. P adsorbed on Fe- and Al-oxides) (Mengel and Kirkby, 2001: 453). Two basic strategies developed by plants for adaptation to soils with low plant-available P levels: (1) higher P acquisition efficiency from soils and (2) improved internal physiological P use efficiency (Vance et al., 2003; Richardson et al., 2011). Traits contributing to P uptake are generally better at predicting grain yield across low-P conditions than P use efficiency traits (Jones et al., 1989; Manske et al., 2001; Araújo and Teixeira, 2003; Ozturk et al., 2005; Cichy et al., 2008; Parentoni et al., 2010; Leiser et al., 2014b). Higher P use efficiency should be considered where increased P uptake would lead to substantial soil P depletion (MacDonald et al., 2011; Rose and Wissuwa, 2012; van der Velde et al., 2014).

A higher P acquisition can be achieved by root exudates (e.g. organic anions, phosphatases), greater root biomass, altered root architecture (e.g. root angle, root hair, aerenchyma and finer roots) and by symbioses with mycorrhiza (Lynch and Brown, 2008; Lynch, 2011; Richardson et al., 2011). A higher internal P use efficiency (e.g. relatively more plant biomass per unit of P uptake) is characterized by reduced growth rate, better internal translocation, alternative respiration pathways and modified carbon cycles (Vance et al., 2003).

Although both adaptation strategies show large genotypic variation, most of the low-P breeding efforts have been devoted to improve P acquisition efficiency since it showed higher correlations to final grain yield production (Rose and Wissuwa, 2012). Whether P acquisition or internal P use efficiency should be considered in a breeding programme depends on the soil P status and the agricultural production system targeted (Richardson et al., 2011; Simpson et al., 2011). In low-P soils with a high-P retention potential, as predominant in SSA (Kochian, 2012), P acquisition efficiency is considered to be the more promising approach, whereas in high-input cropping systems, internal P use efficiency is regarded as more important (Wang et al., 2010). Nevertheless, both adaptation strategies should be considered in SSA to enhance adaptation and minimize soil P mining (Stoorvogel et al., 1993). Most of SSA soils are characterized by low pH values, low organic matter content (C_{org}) and higher aluminium values (Al^{3+}) (Eswaran et al., 1993; Buerkert et al., 2001; Kochian et al., 2004; FAO/IIASA/ISRIC/ISS-CAS/JRC, 2012). Especially under these harsh conditions root exudates may play a major role for crop growth. Release of citrate and other organic anions into the rhizosphere prevents root damage by chelating Al^{3+} and can lead to more plant-available P by mobilizing previously bound P mainly by ligand exchange, dissolution and occupation of P sorption sites (Zhang et al., 1997; Neumann and Römheld, 1999; Ma et al., 2001). Thus, exudation of organic anions can lead to a higher P acquisition rate by releasing more bound P and keeping roots intact for further soil exploration (Richardson et al., 2011). Plant growth under such stressful conditions as prevalent in WA is not merely restricted to a sole abiotic or biotic stress (Haussmann et al., 2012). Attention to genetic variation for resistance to constraints other than low-P *per se*, such as foliar and root diseases, other mineral deficiencies and water stagnation or drought, can contribute to increasing yield under low-P conditions without directly targeting higher P acquisition or internal P use efficiency (Simpson et al., 2011). Therefore, crop adaptation to low-P soil conditions is complex and is best examined in a holistic manner, attentive to the targeted production system, soil type and risk of soil mining, in order to close the yield gaps in SSA (Mueller et al., 2012) and enhance production system nutrient efficiency in a sustainable manner.

Case study results from West Africa

Large genetic variation for P uptake was found among WA sorghum genotypes under both low-P and high-P field conditions (Leiser et al., 2014b). Landrace genotypes and photoperiod-sensitive materials generally showed higher P uptake and grain P concentration. Mycorrhiza colonized root length and crown root angle, two factors generally influencing P uptake in plants (Richardson et al., 2011), were examined in a diverse set of 187 sorghum genotypes from West and Central Africa in pots with low-P soil (<6ppm Bray-1 P). Neither of these traits showed correlation to grain yield production across 15 independent low-P field trials (Leiser et al., 2016). Mycorrhiza colonization in fact had a negative influence on biomass production and P uptake at early developmental stages; probably reflecting 'carbohydrate costs' of initial mycorrhiza colonization. Thus, no single mechanism primarily responsible for variation in P uptake could be identified. Several mechanisms are likely involved in P acquisition of sorghum and further research to understand how different mechanisms interact and complement each other in specific genetic materials and WA environments would be useful.

Large genetic variation for P use efficiency was also found among WA sorghum genotypes (Leiser et al., 2014b). The variation for P use efficiency among WA sorghums was positively correlated to grain yield productivity, and thus selection for higher P use efficiency may be pursued without adversely impacting grain yield in P-limited conditions (Leiser et al., 2015). Higher P use efficiency was exhibited by the formally bred sorghum materials that tended to have lower P content in the grain. Leiser et al. (2014) showed that P concentration in the grain is a reliable, simple and rather inexpensive measure of P use efficiency that fulfils the requirement of not being confounded with harvest index (Rose et al., 2011; Rose and Wissuwa, 2012). Selection for low grain P concentration could, therefore, be used to enhance P use efficiency and help minimize soil P mining (Leiser et al., 2015). A concurrent reduction of phytic acid content in the grain could increase Zn and Fe bio-availability (Hurrell et al., 2003), with nutritional benefit in WA where sorghum is a staple food and micronutrient deficiencies are alarmingly high (Birner et al., 2007). However, reduction of phytic acid might reduce early plant vigour, with negative impact on biomass production under low-P conditions (Veneklaas et al., 2012).

One remarkable WA sorghum genotype, NafalenP6, combined both superior P acquisition and P use efficiency (Leiser et al., 2014b). This genotype was derived from a random-mating population based on approximately 75% WA Guinea landrace germplasm and 25% breeding materials. The selection process involved farmer participatory on-farm selection and testing as well as on-station testing. NafalenP6 shows the possibility of combining both high-P uptake and P use efficiency and suggests promising pathways for doing so. First, by intermating Guinea landrace and elite breeding materials which have shown higher potential for P uptake and P use efficiency (Leiser et al., 2014b), respectively. Second, using population improvement methods to achieve effective recombination and concentration of favourable genes through multiple cycles of selection. And third, through a testing and selection process that involves farmer and research assessments in multiple environments that represent the targeted smallholder production conditions.

4.2 Genetics of low-P adaptation

Extensive study of the genetics of low-P adaptation in *A. thaliana* (for review, see Nilsson et al., 2010; Rouached et al., 2010; Hammond and White, 2011) reveal a complex adaptation and response network. Nevertheless, most of the genes identified could not contribute to higher yields in crop plants under field conditions (Wissuwa et al., 2016). Rice is the crop that has been most extensively studied for low-P adaptation (Lafitte et al., 2007; Shimizu et al., 2008; Chin et al., 2009, 2011; Panigrahy et al., 2009; Torabi et al., 2009; Li et al., 2009, 2010; Secco et al., 2010; Park et al., 2010; Famoso et al., 2011; Gamuyao et al., 2012; Topp et al., 2013). *PUP-1* is the most important quantitative trait locus in rice for increasing grain yield under low-P field conditions and was first mapped in 1998 (Wissuwa et al., 1998). Underlying *PUP-1* is a single kinase gene, *PSTOL1*, which increases early root growth and P acquisition efficiency under low-P conditions in several different genetic backgrounds (Gamuyao et al., 2012). *PSTOL1* is one of the few genes proven to contribute to productivity under low-P field conditions. Since the rice genome exhibits substantial synteny to other grasses including sorghum (Soderlund et al., 2006; Ramu et al., 2009), it might be possible to exploit the genetic knowledge from rice and find genes similar to *PSTOL1* in sorghum which contribute to root growth and P uptake under low-P conditions.

Case study results from West Africa

Sorghum, with its African origin (Wet and Harlan, 1971), its extensive cultivation in WA under P-limited conditions and its genome being fully sequenced (Paterson et al., 2009), makes the most promising crop for studying the genetic architecture of traits contributing to low-P adaptation. To date, only studies in Brazil (Hufnagel et al., 2014) and WA (Leiser et al., 2014a) have begun to explore the genetic variation and underlying genetics of low-P adaptation in sorghum.

Two approaches were pursued in WA to examine the genetics underlying low-P adaptation in sorghum. First, SNPs within candidate gene homologues from *A. thaliana* and rice were identified and associated to low-P yield performance. Second, a whole-genome scan using 220,934 SNPs derived from genotyping-by-sequencing (Elshire et al., 2011) was conducted.

Although many SNPs were identified and studied in several different genes (homologues of *OsPHR2, SIZ1, PHO2, OsSPX1, PHT1 and PSTOL1*) involved in P starvation response, with functions ranging from P transporter, transcription factor, mycorrhiza inducer, root growth and P signalling (Fang et al., 2009; Nilsson et al., 2010; Rouached et al., 2010b; Gamuyao et al., 2012), no strong associations to traits related to low-P adaptation were identified, except for *PSTOL1* homologues. SNPs within two *PSTOL1* homologues located on chromosomes 3 and 7 showed significant associations to early shoot P uptake and shoot biomass production, and to grain yield under low-P conditions in the WA sorghum diversity panel (Hufnagel et al., 2014). The same SNPs were also associated to grain yield and P uptake under low-P field conditions and to several root architectural traits in nutrient solution in a genetically independent mapping population evaluated in Brazil (Hufnagel et al., 2014). These results indicate that *PSTOL1* homologues in sorghum have the ability to stably enhance P uptake and crop performance under low-P soil conditions by a mechanism related to early root growth enhancement, similar to *PSTOL1* in rice (Gamuyao et al., 2012). Although significant associations of the *PSTOL1* homologues to grain yield production under low-P field conditions were found, they generally explained only a small proportion ($r^2 < 7\%$) of the genotypic variance.

A genome-wide association study using 220 934 SNPs identified one specific region among the WA material that was highly associated to grain yield production under P-limited conditions. The Al-tolerance locus *AltSB* and its underlying gene *SbMATE* in sorghum were collocated in this region. The *AltSB*-specific SNPs showed very high associations to grain yield production, especially under low-P conditions, and explained up to 16% of the genotypic variance (Leiser et al., 2014a). Thus *AltSB* appears to have a pleiotropic role in providing tolerance to two of the most serious abiotic stresses for sorghum in WA, Al-toxicity and P-deficiency. It was suggested that Al-tolerance and P uptake under low-P conditions in sorghum can be regulated by similar processes, since *AltSB* was associated to a higher citrate release of sorghum roots (Magalhaes et al., 2007). Although the frequency of the positive low-P-specific and Al-tolerant SNPs was low (9%) in the evaluated WA sorghum panel, the frequency is much higher in the Guinea race germplasm, with 95% of the genotypes carrying the positive *SbMATE* alleles. A similar allele frequency of these SNPs and a higher frequency in WA Guinea sorghums was also found in a worldwide sorghum collection (Caniato et al., 2011), hence pointing to this germplasm group as an important source for Al-tolerance in sorghum (Caniato et al., 2014). Since both P-limitation and Al-toxicity show a great spatial variation (Voortman et al., 2004) and have been reported to have a major influence on crop growth in WA (Manu et al., 1996), marker-assisted selection targeting both traits simultaneously can and should be carried out to achieve genotypes with a stable performance across multiple soil conditions in WA.

Although a highly associated genomic region with a plausible molecular link to low-P adaptation was identified, only a small proportion ($r^2 < 20\%$) of the total genotypic variance could be explained by these regions (Leiser et al., 2014a; Hufnagel et al., 2014) and no single adaptation strategy (e.g. P uptake, P use efficiency) could be identified (Leiser et al., 2014b). Hence, adaptation of sorghum to P-limited soils is highly polygenic and complex, encompassing different adaptation mechanisms involving P acquisition and or internal P use efficiency. A method to overcome these hurdles of complexity may be genomic selection. In contrast to QTL-based selection strategies, genomic selection does not involve significance tests and uses all available markers to predict genotypic performance (Meuwissen et al., 2001), hence minor QTL will not be neglected in genotypic selection. It gained recently much more attention in plant breeding, especially for highly polygenic traits, due to its higher response to selection compared to QTL-based or phenotypic selection (Heffner et al., 2010; Jannink et al., 2010; Massman et al., 2012). The predictive ability of genomic selection was estimated within the WA diversity panel of 187 genotypes for grain yield under low-P, high-P, across both low- and high-P conditions and for grain yield ratios, as a measure for low-P specificity (Fig. 8). Pearson's correlation coefficient between predicted genotypic values and observed phenotypic values in the validation set show that a high predictive ability for grain yield under low-P, high-P and across both fertility conditions was attainable, and a moderate predictive ability for grain yield ratios (Fig. 8). This points to the possibility of using genomic prediction as a tool for breeding

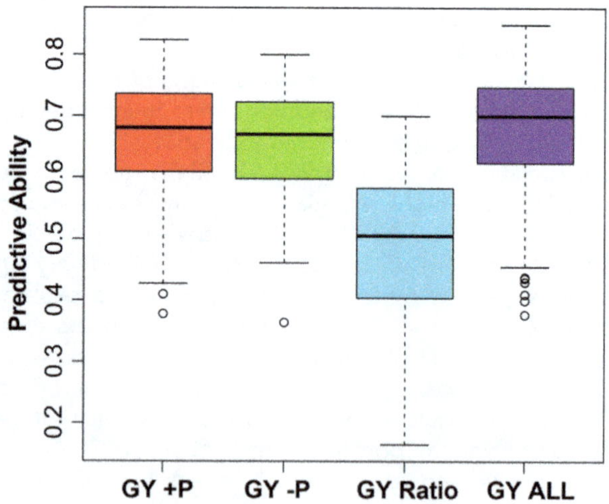

Figure 8 Genomic selection predictive ability of grain yield BLUPs across +P, –P sites separately and combined (GY ALL) and grain yield ratios as –P/+P performance. Values derived from 5-fold cross validation with 100 replications using RR-BLUP.

for low-P adaptation. Whether genome-wide approaches can be chosen in the regional breeding programmes depends on the costs of DNA extraction and marker analyses. To date, these methods are still too costly for the regional programmes and phenotypic selection on-station in lower P fields and on-farm are more promising approaches and should be favoured.

5 Summary: implications for applied sorghum breeding in West Africa

In summary, the following conclusions can be drawn, which are of direct relevance for sorghum breeding targeting P-limited environments in WA:

- WA sorghum is generally well adapted to low-P soil conditions and shows a large exploitable genetic variation for P-efficiency.
- Adequate field designs and/or spatial analyses are necessary tools for efficient genotype selection under low-input conditions.
- Direct selection under low-P conditions is feasible, necessary and more efficient than indirect selection under high-P conditions.
- On-farm yield testing is feasible and can help increase genetic gains for yield for smallholder sorghum producers.
- Landrace genotypes from WA show a higher P acquisition capacity and a specific adaptation to low-P conditions, with Guinea and Durra race sorghums offering promising source material for low-P adaptation for the Sudanian and Sahelian zones, respectively.
- Photoperiod-sensitive genotypes show less delay in heading under low-P conditions, a higher P acquisition rate and are more specifically adapted to low-P conditions, hence providing a valuable germplasm source for climate and low-P resilience breeding.
- WA sorghum breeding materials with some introduced Caudatum germplasm tend to have lower grain P concentration and higher P use efficiency.
- Selection for low P concentration of grain can be used to enhance internal P use efficiency, without any negative impact on grain yield production.
- Combining superior P uptake and P use efficiency in a single sorghum genotype is feasible.
- Population improvement is a promising approach for enhancing the highly polygenic low-P adaptation complex.
- SNPs in *PSTOL1* homologues of rice stably enhance P uptake and crop performance of sorghum under low-P soil conditions and can be used in marker-assisted selection to increase root growth and final grain yield production.
- P-efficiency and Al-tolerance are pleiotropically regulated by the same genomic region. SNPs of this region can be used for marker-assisted selection.
- WA Guinea race sorghum is a major source for Al-tolerant and low-P-specific SNPs and can serve as source germplasm for allele mining and marker-assisted selection.
- Genomic selection appears to be a very promising approach to increase genetic gains for low-P adaptation.

6 Future trends and challenges in bringing benefits of sorghum low-P tolerance to West African farmers

The presented findings show that breeding sorghum for P-limited soils is necessary and possible using different approaches. Although the conclusions presented here are based on extensive data, many questions and new challenges are to be addressed to increase sorghum production on a large scale across millions of smallholder farmer's fields in WA. Breeding challenges directly arising from previous work with potential to be addressed in the near future will be presented first. Subsequently, issues of major agronomic and socio-economic importance that must be considered and integrated to tackle the issue of food in the context of P scarcity in WA will be highlighted.

6.1 Breeding challenges

WA germplasm with specific adaptation to low-P have been identified and are available as parental materials for breeding. A random-mating low-P adaptation population, based on several of these landrace and breeding lines exhibiting superior low-P adaptation, has been created and well random-mated with very low linkage disequilibrium. This source material is available for population improvement. Deriving moderate to highly inbred single seed decent progenies from this population would facilitate selection methodology studies.

The identified SNP markers for grain yield production under low-P conditions will need to be further validated to draw final conclusions and give final advice on how to proceed. Furthermore, expression or knock-out studies of the identified genes will be necessary to prove the postulated mechanisms. Once validated, it should be possible to incorporate the desired loci for both traits into elite breeding materials with minimum linkage drag using marker-assisted selection. Furthermore, the already validated gene-specific markers for Al-tolerance will make allele mining possible in large germplasm collections, hence increase allelic frequency and Al-tolerance of the currently used germplasm in WA. These markers have been converted to the cost-efficient KASP system of LGC Genomics and are available on the Integrated Breeding Platform of the Generation Challenge Programme (https://www.integratedbreeding.net/), hence can be directly used by all breeders in WA. Once validation steps are completed all other markers will be converted to the KASP system and made accessible on the same open breeding platform.

Genomic selection provides a promising tool to select for grain yield under low-P. This potential is based on data derived *in silico* using cross validations. Therefore, further validation steps with an independent set of genotypes will be required to prove the effectiveness of genome-wide versus QTL-based selection among different sets of genotypes.

Adaptation to low-P appears to involve multiple mechanisms. Using a less diverse set or even a bi-parental mapping population derived from contrasting genotypes might be an option to detect single contributing mechanisms.

Sorghum hybrids are rapidly gaining momentum in WA, with hybrid seed production and sales doubling every year. Some of the first released hybrid cultivars show farmer-preferred characteristics and higher than 30% yield superiorities relative to an excellent Guinea landrace variety across the full range of farmers' field conditions (Rattunde et al.,

2013; Kante et al., 2017). Examination of what contributes to these hybrids' superior growth and yield under low-P conditions would be revealing. Also, designing the hybrid sorghum breeding programme to maximize yield gains for low-P production conditions is needed and currently being addressed in collaborative work by the Malian Sorghum Program of the IER, ICRISAT-Mali and the University of Hohenheim.

6.2 Agronomic and socio-economic challenges

Agronomic practices and socio-economic factors and interventions are vital in achieving sustainable and P efficient production systems and for adoption of newly developed technologies (Cordell and White, 2013).

Agronomic practices such as intercropping (Henry et al., 2008), crop rotations (Horst et al., 2001), livestock integration, manure and compost applications (Simpson et al., 2011) offer great promise for smallholder farmers to address the constraint of limited P, as these practices are less dependent on external inputs. However, labour is one of the most limiting factors in these production systems, and appropriate smaller-scale labour-saving equipment would be most useful.

Farmers' use of inorganic fertilization is constrained by availability and feasibility of purchase. However, the methods of applying fertilizer to sorghum offers great opportunities for increasing effectiveness. The timing of fertilizer application is a major factor for P, since sorghum farmers typically apply fertilizer only after the first weeding, at three to five weeks after sowing. However, for P applications to benefit sorghum yield, P fertilizer needs to be applied latest by two weeks after sowing. Also, the method of applying fertilizer, broadcast or incorporated, is a major determinant of effectiveness of P fertilization. Only soil-incorporated P-containing fertilizer will benefit the sorghum crop since P is not mobile. Small-scale mechanized options for fertilizer application are desperately needed and are under development but not yet satisfactory for wide-scale adoption for sorghum cultivation.

Phosphorus deficiency in sorghum production is not only smallholder farmers' issue but also a women's issue. Women are typically allocated fields at the end of a rotation when the soil fertility is lowest, resulting in extremely low plant-available P levels (Fig. 5). The P levels are so low that simply application of ash from the cooking fire at the time of sowing increased sorghum yield by 30% in women's fields. Although this result was only from a single year it highlights the need to identify options to address the low-P issue that are feasible for women.

Although the mentioned agronomic measures can help enhance crop production under P-limited conditions, their usefulness to improve Al-tolerance or decrease Al-toxicity is rather limited. It was shown that Al-toxicity and P-deficiency can be reduced by incorporating crop residues, hence increasing soil organic matter and reducing P mining from the soils (Haynes and Mokolobate, 2001); nevertheless, applying lime ($CaCO_3$) is generally the method of choice to ameliorate soils, which show high levels of Al-toxicity (Pavan et al., 1984; Moody et al., 1998; Ila'ava et al., 2000). To date, liming is no viable option to smallholder farmers in WA due to the lack of and access to lime. Therefore, the only sustainable solution for WA is to increase soil organic matter through agronomic measures and improve Al-tolerance within the current breeding germplasm.

In recent decades, several methods have been developed and proven to be effective in enhancing crop production under low-input conditions and many high-yielding sorghum varieties have been released in WA. Nevertheless, the adoption rates of these new tools have been limited on farm level. Many factors were identified for being responsible for the observed lack of adoption among farmers, ranging from political and socio-economic to agronomic reasons (CGIAR and Stoop, 2002). By proper identification of the farmers' needs using participatory research methodologies (e.g. participatory plant breeding), on-farm validation and farmer-led testing of new technologies under minimal risk (e.g. seed mini-packs), strengthening of local seed production systems that work for smallholder farmers, and broad-based information distribution (e.g. learning DVDs, http://www.agroinsight.com/resources.php), it is possible to increase adoption rates and reach a much larger scale of smallholder farmers.

7 Where to look for further information

From the references listed below, we recommend especially the following articles:

- Cordell, D., A. Turner and J. Chong. 2015. The hidden cost of phosphate fertilizers: mapping multi-stakeholder supply chain risks and impacts from mine to fork. *Glob. Change Peace Secur.* 27: 323–43. doi:10.1080/14781158.2015.1083540.
- Gamuyao, R., J. H. Chin, J. Pariasca-Tanaka, P. Pesaresi, S. Catausan, C. Dalid, I. Slamet- Loedin, E. M. Tecson-Mendoza, M. Wissuwa and S. Heuer. 2012. The protein kinase Pstol1 from traditional rice confers tolerance of phosphorus deficiency. *Nature.* 488: 535–9. doi:10.1038/nature11346.
- Gemenet, D. C., W. L. Leiser, F. Beggi, L. H. Herrmann, V. Vadez, H. F. W. Rattunde, E. Weltzien, C. T. Hash, A. Buerkert and B. I. G. Haussmann. 2016. Overcoming phosphorus deficiency in West African pearl millet and sorghum production systems: Promising options for crop improvement. *Front. Plant Sci.* 7 doi:10.3389/fpls.2016.01389.
- Leiser, W. L., H. F. W. Rattunde, H.-P. Piepho, E. Weltzien, A. Diallo, A. E. Melchinger, H. K. Parzies and B. I. G. Haussmann. 2012b. Selection strategy for sorghum targeting phosphorus-limited environments in West Africa: analysis of multi-environment experiments. *Crop Sci.* 52: 2517–27. doi:10.2135/cropsci2012.02.0139.
- Leiser, W. L., H. F. Rattunde, E. Weltzien, N. Cisse, M. Abdou, A. Diallo, A. O. Tourè, J. V. Magalhaes and B. I. Haussmann. 2014a. Two in one sweep: aluminum tolerance and grain yield in P-limited soils are associated to the same genomic region in West African Sorghum. *BMC Plant Biol.* 14: 206. doi:10.1186/s12870-014-0206-6.
- Leiser, W. L., H. F. W. Rattunde, E. Weltzien and B. I. G. Haussmann. 2014b. Phosphorus uptake and use efficiency of diverse West and Central African sorghum genotypes under field conditions in Mali. *Plant Soil.* 377: 383–94. doi:10.1007/s11104-013-1978-4.
- Rose, T. J., M. T. Rose, J. Pariasca-Tanaka, S. Heuer and M. Wissuwa. 2011. The frustration with utilization: Why have improvements in internal phosphorus utilization efficiency in crops remained so elusive? *Front. Plant Sci.* 2 doi:10.3389/fpls.2011.00073.
- Rose, T. J., and M. Wissuwa. 2012. Rethinking internal phosphorus utilization efficiency. In *Advances in Agronomy*. Amsterdam: Elsevier, pp. 185–217.

- Wissuwa, M., T. Kretzschmar and T. J. Rose. 2016. From promise to application: root traits for enhanced nutrient capture in rice breeding. *J. Exp. Bot.*: erw061. doi:10.1093/jxb/erw061.
- Vance, C. P., C. Uhde-Stone and D. L. Allan. 2003. Phosphorus acquisition and use: critical adaptations by plants for securing a nonrenewable resource. *New Phytol.* 157: 423–47.

Furthermore, the following webpages or conferences might be monitored for up-to-date information:

- Sustainable Phosphorus Summit (SPS): A global multidisciplinary event held every two years to discuss the phosphorus production and utilization, management and sustainability. Each event is announced on several websites, and more information could be found on these websites: http://phosphorusfutures.net/and http://www.phosphorusplatform.eu/.
- European Sustainable Phosphorus Platform (EPSP): EPSP brings together companies and stakeholders to address the Phosphorus Challenge and its opportunities (http://www.phosphorusplatform.eu/).
- Global Phosphorus Research Initiative (GPRI): The GPRI is a collaboration between six independent research institutes in Europe, Australia and North America. Their work is the first major research globally to investigate how phosphorus security can ensure global food security. The GPRI undertakes independent, quality, interdisciplinary research on global phosphorus security for future food production. In addition to research, the GPRI also facilitates networking, dialogue and awareness-raising among policy makers, industry, scientists and the community on the implications of global phosphorus scarcity and possible sustainable solutions (http://phosphorusfutures.net/).
- International Plant Nutrition Institute is a not-for-profit, science-based organization dedicated to the responsible management of plant nutrition for the benefit of the human race (http://www.ipni.net/).
- Roots Lab: It is located at Penn State College of Agricultural Science, Pennsylvania. They have a strong focus on root research and conduct several projects related to nutrient use efficiency (http://plantscience.psu.edu/research/labs/roots).
- Matthias Wissuwa Research Group: This work group is working on rice adaptation to low-P conditions for several decades and keeps on working on this topic. Major research findings in breeding crops for low-P soils came out of this group (https://sites.google.com/site/wissuwaresearchgroup/home).

8 References

Altmann, M. Stitt, L. Willmitzer and A. E. Melchinger. 2012a. Genome-wide association mapping of leaf metabolic profiles for dissecting complex traits in maize. *Proc. Natl. Acad. Sci.* 109: 8872–7. doi:10.1073/pnas.1120813109.

Araújo, A. P., and M. G. Teixeira. 2003. Nitrogen and phosphorus harvest indices of common bean cultivars: Implications for yield quantity and quality. *Plant Soil.* 257: 425–33. doi:10.1023/A:1027353822088.

Atlin, G. N., and K. J. Frey. 1989. Predicting the relative effectiveness of direct versus indirect selection for oat yield in three types of stress environments. *Euphytica*. 44: 137–42. doi:10.1007/BF00022608.

Beebe, S., J. Lynch, N. Galwey, J. Tohme and I. Ochoa. 1997. A geographical approach to identify phosphorus-efficient genotypes among landraces and wild ancestors of common bean. *Euphytica*. 95: 325–38.

Birner, R., S. A. Kone, N. Linacre and D. Resnick. 2007. Biofortified Foods and Crops in West Africa: Mali and Burkina Faso. AgBioForum Vol 10 No 3 2007.

Buerkert, A., A. Bationo and H.-P. Piepho. 2001. Efficient phosphorus application strategies for increased crop production in sub-Saharan West Africa. *Field Crops Res*. 72: 1–15. doi:10.1016/S0378-4290(01)00166-6.

Burger, H., M. Schloen, W. Schmidt and H. H. Geiger. 2008. Quantitative genetic studies on breeding maize for adaptation to organic farming. *Euphytica*. 163: 501–10. doi:10.1007/s10681-008-9723-4.

Caniato, F. F., C. T. Guimarães, M. Hamblin, C. Billot, J. F. Rami, B. Hufnagel, L. V. Kochian, J. Liu, A. A. F. Garcia and C. T. Hash. 2011. The relationship between population structure and aluminum tolerance in cultivated sorghum. *PLoS ONE*. 6: e20830.

Caniato, F. F., M. T. Hamblin, C. T. Guimaraes, Z. Zhang, R. E. Schaffert, L. V. Kochian and J. V. Magalhaes. 2014. Association mapping provides insights into the origin and the fine structure of the Sorghum Aluminum Tolerance Locus, AltSB. *PLoS ONE*. 9: e87438. doi:10.1371/journal.pone.0087438.

Ceccarelli, S. 1989. Wide adaptation: how wide? *Euphytica*. 40: 197–205.

Ceccarelli, S. 1994. Specific adaptation and breeding for marginal conditions. *Euphytica*. 77: 205–19. doi:10.1007/BF02262633.

Ceccarelli, S. 1996. Adaptation to low/high input cultivation. *Euphytica*. 92: 203–14.

CGIAR, and W. A. Stoop. 2002. A Study and Comprehensive Analysis of the Causes for Low Adoption Rates of Agricultural Research Results in West and Central Africa : Possible Solutions Leading to Greater Future Impacts – the Mali and Guinea Case Studies. Available at: http://library.cgiar.org/handle/10947/84 (verified 24 February 2014).

Chen, J., L. Xu, Y. Cai and J. Xu. 2008. QTL mapping of phosphorus efficiency and relative biologic characteristics in maize (Zea mays L.) at two sites. *Plant Soil*. 313: 251–66. doi:10.1007/s11104-008-9698-x.

Chin, J. H., R. Gamuyao, C. Dalid, M. Bustamam, J. Prasetiyono, S. Moeljopawiro, M. Wissuwa and S. Heuer. 2011. Developing rice with high yield under phosphorus deficiency: Pup1 sequence to application. *Plant Physiol*. 156: 1202–16. doi:10.1104/pp.111.175471.

Chin, J. H., X. Lu, S. M. Haefele, R. Gamuyao, A. Ismail, M. Wissuwa and S. Heuer. 2009. Development and application of gene-based markers for the major rice QTL Phosphorus uptake 1. *Theor. Appl. Genet*. 120: 1073–86. doi:10.1007/s00122-009-1235-7.

Cichy, K. A., S. S. Snapp and M. W. Blair. 2008. Plant growth habit, root architecture traits and tolerance to low soil phosphorus in an Andean bean population. *Euphytica*. 165: 257–68. doi:10.1007/s10681-008-9778-2.

Clerget, B., M. Dingkuhn, E. Goze, H. F. W. Rattunde and B. Ney. 2008. Variability of phyllochron, plastochron and rate of increase in height in photoperiod-sensitive sorghum varieties. *Ann. Bot*. 101: 579–94. doi:10.1093/aob/mcm327.

Cooper, J., R. Lombardi, D. Boardman and C. Carliell-Marquet. 2011. The future distribution and production of global phosphate rock reserves. *Resour. Conserv. Recycl*. 57: 78–86.

Cordell, D., J.-O. Drangert and S. White. 2009. The story of phosphorus: Global food security and food for thought. *Glob. Environ. Change* 19: 292–305. doi:10.1016/j.gloenvcha.2008.10.009.

Cordell, D., A. Turner and J. Chong. 2015. The hidden cost of phosphate fertilizers: mapping multi-stakeholder supply chain risks and impacts from mine to fork. *Glob. Change Peace Secur*. 27: 323–43. doi:10.1080/14781158.2015.1083540.

Cordell, D., and S. White. 2013. Sustainable phosphorus measures: Strategies and technologies for achieving phosphorus security. *Agronomy*. 3: 86–116. doi:10.3390/agronomy3010086.

Delhaize, E., P. Taylor, P. J. Hocking, R. J. Simpson, P. R. Ryan and A. E. Richardson. 2009. Transgenic barley (Hordeum vulgare L.) expressing the wheat aluminium resistance gene (TaALMT1) shows enhanced phosphorus nutrition and grain production when grown on an acid soil. *Plant Biotechnol. J.* 7: 391–400. doi:10.1111/j.1467-7652.2009.00403.x.

Deu, M., F. Sagnard, J. Chantereau, C. Calatayud, D. Hérault, C. Mariac, J.-L. Pham, Y. Vigouroux, I. Kapran, P. S. Traore, A. Mamadou, B. Gerard, J. Ndjeunga and G. Bezançon. 2008. Niger-wide assessment of in situ sorghum genetic diversity with microsatellite markers. *Theor. Appl. Genet.* 116: 903–13. doi:10.1007/s00122-008-0721-7.

Ding, G., Z. Zhao, Y. Liao, Y. Hu, L. Shi, Y. Long and F. Xu. 2012. Quantitative trait loci for seed yield and yield-related traits, and their responses to reduced phosphorus supply in Brassica napus. *Ann. Bot.* 109: 747–59. doi:10.1093/aob/mcr323.

Doumbia, M. D., L. R. Hossner and A. B. Onken. 1993. Variable sorghum growth in acid soils of subhumid West Africa. *Arid Soil Res. Rehabil.* 7: 335–46. doi:10.1080/15324989309381366.

Edmondson, R. N. 2005. Past developments and future opportunities in the design and analysis of crop experiments. *J. Agric. Sci.* 143: 27–33. doi:10.1017/S0021859604004472.

Elshire, R. J., J. C. Glaubitz, Q. Sun, J. A. Poland, K. Kawamoto, E. S. Buckler and S. E. Mitchell. 2011. A robust, simple genotyping-by-sequencing (GBS) approach for high diversity species. *PLoS ONE.* 6: e19379. doi:10.1371/journal.pone.0019379.

Endelman, J. B. 2011. Ridge regression and other kernels for genomic selection with R package rrBLUP. *Plant Genome J.* 4: 250. doi:10.3835/plantgenome2011.08.0024.

Eswaran, H., E. van den Berg and P. Reich. 1993. Organic carbon in soils of the world. *Soil Sci. Soc. Am. J.* 57: 192–4.

Fageria, N. K., R. J. Wright and V. C. Baligar. 1988. Rice cultivar evaluation for phosphorus use efficiency. *Plant Soil* 111: 105–9. doi:10.1007/BF02182043.

Famoso, A. N., K. Zhao, R. T. Clark, C.-W. Tung, M. H. Wright, C. Bustamante, L. V. Kochian and S. R. McCouch. 2011. Genetic architecture of aluminum tolerance in rice (*Oryza sativa*) determined through genome-wide association analysis and QTL mapping. *PLoS Genet.* 7: e1002221. doi:10.1371/journal.pgen.1002221.

Fang, Z., C. Shao, Y. Meng, P. Wu and M. Chen. 2009. Phosphate signaling in Arabidopsis and Oryza sativa. *Plant Sci.* 176: 170–80. doi:10.1016/j.plantsci.2008.09.007.

FAO. 2012. FAO-Statistics. Available at: http://faostat.fao.org/site/567/DesktopDefault.aspx?PageID=567#ancor

FAO/IIASA/ISRIC/ISS-CAS/JRC. 2012. Harmonized World Soil Database (version 1.2). FAO, Rome, Italy and IIASA, Laxenburg, Austria. FAO, Rome, Italy and IIASA, Laxenburg, Austria.

Gallais, A., M. Coque and P. Bertin. 2008. Response to selection of a maize population for adaptation to high or low nitrogen fertilization. *Maydica.* 53: 21–8.

Gamuyao, R., J. H. Chin, J. Pariasca-Tanaka, P. Pesaresi, S. Catausan, C. Dalid, I. Slamet-Loedin, E. M. Tecson-Mendoza, M. Wissuwa and S. Heuer. 2012. The protein kinase Pstol1 from traditional rice confers tolerance of phosphorus deficiency. *Nature.* 488: 535–9. doi:10.1038/nature11346.

Gilmour, A. R., B. R. Cullis and A. P. Verbyla. 1997. Accounting for natural and extraneous variation in the analysis of field experiments. *J. Agric. Biol. Environ. Stat.* 2: 269–93.

Grondona, M. O., J. Crossa, P. N. Fox and W. H. Pfeiffer. 1996. Analysis of variety yield trials using two-dimensional separable ARIMA processes. *Biometrics* 52: 763–70.

Hammond, J. P., M. R. Broadley, P. J. White, G. J. King, H. C. Bowen, R. Hayden, M. C. Meacham, A. Mead, T. Overs, W. P. Spracklen and D. J. Greenwood. 2009. Shoot yield drives phosphorus use efficiency in Brassica oleracea and correlates with root architecture traits. *J. Exp. Bot.* 60: 1953–68. doi:10.1093/jxb/erp083.

Hammond, J. P., and P. J. White. 2011. Sugar signaling in root responses to low phosphorus availability. *Plant Physiol.* 156: 1033–40. doi:10.1104/pp.111.175380.

Haussmann, B. I. G., H. Fred Rattunde, E. Weltzien-Rattunde, P. S. C. Traoré, K. vom Brocke and H. K. Parzies. 2012. Breeding strategies for adaptation of pearl millet and sorghum to climate variability and change in West Africa. *J. Agron. Crop Sci.* 198: 327–39. doi:10.1111/j.1439-037X.2012.00526.x.

Haynes, R. J., and M. S. Mokolobate. 2001. Amelioration of Al toxicity and P deficiency in acid soils by additions of organic residues: A critical review of the phenomenon and the mechanisms involved. *Nutr. Cycl. Agroecosyst* 59: 47–63. doi:10.1023/A:1009823600950.

Heffner, E. L., A. J. Lorenz, J.-L. Jannink and M. E. Sorrells. 2010. Plant breeding with genomic selection: Gain per unit time and cost. *Crop Sci.* 50: 1681. doi:10.2135/cropsci2009.11.0662.

Henry, A., P. J. A. Kleinman and J. P. Lynch. 2008. Phosphorus runoff from a phosphorus deficient soil under common bean (Phaseolus vulgaris L.) and soybean (Glycine max L.) genotypes with contrasting root architecture. *Plant Soil.* 317: 1–16. doi:10.1007/s11104-008-9784-0.

Holford, I. C. R. 1997. Soil phosphorus: its measurement, and its uptake by plants. *Aust. J. Soil Res.* 35: 227–40.

Horst, W. J., M. Kamh, J. M. Jibrin and V. O. Chude. 2001. Agronomic measures for increasing P availability to crops. *Plant Soil* 237: 211–23. doi:10.1023/A:1013353610570.

Hufnagel, B., S. M. de Sousa, L. Assis, C. T. Guimaraes, W. Leiser, G. C. Azevedo, B. Negri, B. G. Larson, J. E. Shaff, M. M. Pastina, B. A. Barros, E. Weltzien, H. F. W. Rattunde, J. H. Viana, R. T. Clark, A. Falcão, R. Gazaffi, A. A. F. Garcia, R. E. Schaffert, L. V. Kochian and J. V. Magalhaes. 2014. Duplicate and conquer: Multiple homologs of PHOSPHORUS- STARVATION TOLERANCE1 enhance phosphorus acquisition and sorghum performance on low-phosphorus soils. *Plant Physiol.* 166: 659–77. doi:10.1104/pp.114.243949.

Hurrell, R. F., M. B. Reddy, M.-A. Juillerat and J. D. Cook. 2003. Degradation of phytic acid in cereal porridges improves iron absorption by human subjects. *Am. J. Clin. Nutr.* 77: 1213–19.

Ila'ava, V. P., P. Blamey and C. J. Asher. 2000. Effects of lime and gypsum on growth of sweet potato in two strongly acid soils. *Aust. J. Agric. Res.* 51: 1031–8.

Jannink, J.-L., A. J. Lorenz and H. Iwata. 2010. Genomic selection in plant breeding: from theory to practice. *Brief. Funct. Genomics.* 9: 166–77. doi:10.1093/bfgp/elq001.

Jasinski, S. M. 2013. Phosphat Rock. Available at: http://minerals.usgs.gov/minerals/pubs/commodity/phosphate_rock/mcs-2012- phosp.pdf (verified 3 January 2014)

Jones, G. P. D., G. J. Blair and R. S. Jessop. 1989. Phosphorus efficiency in wheat – A useful selection criterion? *Field Crops Res.* 21: 257–64. doi:10.1016/0378-4290(89)90007-5.

Kante, M., H.F.W. Rattunde, W.L. Leiser, B. Nebié, B. Diallo, A. Diallo, A.O. Touré, E. Weltzien, and B.I.G. Haussmann. 2017. Can Tall Guinea-Race Sorghum Hybrids Deliver Yield Advantage to Smallholder Farmers in West and Central Africa? *Crop Sci.* 57: 833–842. doi:10.2135/cropsci2016.09.0765.

Kochian, L. V. 2012. Plant nutrition: Rooting for more phosphorus. *Nature.* 488: 466–7. doi:10.1038/488466a.

Kochian, L. V., O. A. Hoekenga and M. A. Pineros. 2004. How do crop plants tolerate acid soils? Mechanism of aluminum tolerance and phosphorus efficiency. *Annu. Rev. Plant Biol.* 55: 459–93. doi:10.1146/annurev.arplant.55.031903.141655.

Lado, B., I. Matus, A. Rodríguez, L. Inostroza, J. Poland, F. Belzile, A. del Pozo, M. Quincke, M. Castro and J. von Zitzewitz. 2013. Increased genomic prediction accuracy in wheat breeding through spatial adjustment of field trial data. G3 *Genes Genomes Genetics* 3: 2105–14. doi:10.1534/g3.113.007807.

Lafitte, H., G. Yongsheng, S. Yan and Z.-K. Li. 2007. Whole plant responses, key processes, and adaptation to drought stress: the case of rice. *J. Exp. Bot.* 58: 169–75. doi:10.1093/jxb/erl101.

Leiser, W. L., M. O. Olatoye, H. F. W. Rattunde, G. Neumann, E. Weltzien and B. I. G. Haussmann. 2016. No need to breed for enhanced colonization by arbuscular mycorrhizal fungi to improve low-P adaptation of West African sorghums. *Plant Soil* 401: 51–64. doi:10.1007/s11104-015-2437-1.

Leiser, W. L., H. F. Rattunde, H.-P. Piepho and H. K. Parzies. 2012a. Getting the most out of sorghum low-input field trials in West Africa using spatial adjustment. *J. Agron. Crop Sci.* 198: 349–59. doi:10.1111/j.1439-037X.2012.00529.x.

Leiser, W. L., H. F. W. Rattunde, H.-P. Piepho, E. Weltzien, A. Diallo, A. E. Melchinger, H. K. Parzies and B. I. G. Haussmann. 2012b. Selection strategy for sorghum targeting phosphorus-limited environments in West Africa: Analysis of multi-environment experiments. *Crop Sci.* 52: 2517–27. doi:10.2135/cropsci2012.02.0139.

Leiser, W. L., H. F. W. Rattunde, H.-P. Piepho, E. Weltzien, A. Diallo, A. Toure and B. I. G. Haussmann. 2015. Phosphorus efficiency and tolerance traits for selection of sorghum for performance in phosphorous-limited environments. *Crop Sci.* 55: 1–11. doi:10.2135/cropsci2014.05.0392.

Leiser, W. L., H. F. Rattunde, E. Weltzien, N. Cisse, M. Abdou, A. Diallo, A. O. Tourè, J. V. Magalhaes and B. I. Haussmann. 2014a. Two in one sweep: Aluminum tolerance and grain yield in P-limited soils are associated to the same genomic region in West African Sorghum. *BMC Plant Biol.* 14: 206. doi:10.1186/s12870-014-0206-6.

Leiser, W. L., H. F. W. Rattunde, E. Weltzien and B. I. G. Haussmann. 2014b. Phosphorus uptake and use efficiency of diverse West and Central African sorghum genotypes under field conditions in Mali. *Plant Soil* 377: 383–94. doi:10.1007/s11104-013-1978-4.

Li, L., C. Liu, and X. Lian. 2010. Gene expression profiles in rice roots under low phosphorus stress. *Plant Mol. Biol.* 72: 423–32. doi:10.1007/s11103-009-9580-0.

Li, J., Y. Xie, A. Dai, L. Liu and Z. Li. 2009. Root and shoot traits responses to phosphorus deficiency and QTL analysis at seedling stage using introgression lines of rice. *J. Genet. Genomics Yi Chuan Xue Bao.* 36: 173–83. doi:10.1016/S1673-8527(08)60104-6.

Liang, C., M. A. Piñeros, J. Tian, Z. Yao, L. Sun, J. Liu, J. Shaff, A. Coluccio, L. V. Kochian and H. Liao. 2013. Low pH, Aluminum, and phosphorus coordinately regulate malate exudation through GmALMT1 to improve soybean adaptation to acid soils. *Plant Physiol.* 161: 1347–61. doi:10.1104/pp.112.208934.

Lynch, J. P. 2011. Root Phenes for enhanced soil exploration and phosphorus acquisition: Tools for future crops. *Plant Physiol.* 156: 1041–9. doi:10.1104/pp.111.175414.

Lynch, J. P., and K. M. Brown. 2008. Root strategies for phosphorus acquisition. In *The Ecophysiology of Plant-Phosphorus Interactions*. Springer Netherlands, pp. 83–116.

Ma, J. F., P. R. Ryan and E. Delhaize. 2001. Aluminium tolerance in plants and the complexing role of organic acids. *Trends Plant Sci.* 6: 273–8. doi:10.1016/S1360-1385(01)01961-6.

MacDonald, G. K., E. M. Bennett, P. A. Potter and N. Ramankutty. 2011. Agronomic phosphorus imbalances across the world's croplands. *Proc. Natl. Acad. Sci.* 108: 3086–91. doi:10.1073/pnas.1010808108.

Magalhaes, J. V., J. Liu, C. T. Guimarães, U. G. P. Lana, V. M. C. Alves, Y.-H. Wang, R. E. Schaffert, O. A. Hoekenga, M. A. Piñeros, J. E. Shaff, P. E. Klein, N. P. Carneiro, C. M. Coelho, H. N. Trick and L. V. Kochian. 2007. A gene in the multidrug and toxic compound extrusion (MATE) family confers aluminum tolerance in sorghum. *Nat. Genet.* 39: 1156–61. doi:10.1038/ng2074.

Mandal, N. P., P. K. Sinha, M. Variar, V. D. Shukla, P. Perraju, A. Mehta, A. R. Pathak, J. L. Dwivedi, S. P. S. Rathi, S. Bhandarkar, B. N. Singh, D. N. Singh, S. Panda, N. C. Mishra, Y. V. Singh, R. Pandya, M. K. Singh, R. B. S. Sanger, J. C. Bhatt, R. K. Sharma, A. Raman, A. Kumar and G. Atlin. 2010. Implications of genotype × input interactions in breeding superior genotypes for favorable and unfavorable rainfed upland environments. *Field Crops Res.* 118: 135–44. doi:10.1016/j.fcr.2010.05.004.

Manske, G. G. B., J. I. Ortiz-Monasterio, M. van Ginkel, R. M. González, R. A. Fischer, S. Rajaram and P. L. G. Vlek. 2001. Importance of P uptake efficiency versus P utilization for wheat yield in acid and calcareous soils in Mexico. *Eur. J. Agron.* 14: 261–74. doi:10.1016/S1161-0301(00)00099-X.

Manu, A., L. P. Wilding, L. R. Hossner, A. A. Pfordresher and S. C. Geiger. 1996. Soil parameters related to crop growth variability in Western Niger, West Africa. *Soil Sci. Soc. Am. J.* 60: 283–8. doi:10.2136/sssaj1996.03615995006000010043x.

Marschner, H. 1995. *Mineral Nutrition of Higher Plants*. London: Academic Press.

Massman, J. M., H.-J. G. Jung and R. Bernardo. 2012. Genomewide selection versus marker- assisted recurrent selection to improve grain yield and stover-quality traits for cellulosic ethanol in maize. *Crop Sci.* 0: 0. doi:10.2135/cropsci2012.02.0112.

Mengel, K., and E. A. Kirkby. 2001. *Principles of Plant Nutrition*. 5th edn. Berlin: Springer.

Meuwissen, T. H. E., B. J. Hayes and M. E. Goddard. 2001. Prediction of total genetic value using genome-wide dense marker maps. *Genetics*. 157: 1819–29.

Moody, P. W., R. L. Aitken and T. Dickson. 1998. Field amelioration of acidic soils in south-east Queensland. III. Relationships of maize yield response to lime and unamended soil properties. *Aust. J. Agric. Res.* 49: 649–56.

Mueller, N. D., J. S. Gerber, M. Johnston, D. K. Ray, N. Ramankutty and J. A. Foley. 2012. Closing yield gaps through nutrient and water management. *Nature*. 490: 254–7. doi:10.1038/nature11420.

Mühleisen, J., J. C. Reif, H. P. Maurer, J. Möhring and H.-P. Piepho. 2013. Visual scorings of drought stress intensity as covariates for improved variety trial analysis. *J. Agron. Crop Sci.* 199: 321–30. doi:10.1111/jac.12025.

Neumann, G., and V. Römheld. 1999. Root excretion of carboxylic acids and protons in phosphorus-deficient plants. *Plant Soil.* 211: 121–30. doi:10.1023/A:1004380832118.

Nilsson, L., R. Müller and T. H. Nielsen. 2010. Dissecting the plant transcriptome and the regulatory responses to phosphate deprivation. *Physiol. Plant.* 139: 129–43. doi:10.1111/j.1399-3054.2010.01356.x.

Nord, E. A., and J. P. Lynch. 2008. Delayed reproduction in Arabidopsis thaliana improves fitness in soil with suboptimal phosphorus availability. *Plant Cell Environ.* 31: 1432–41. doi:10.1111/j.1365-3040.2008.01857.x.

Obersteiner, M., J. Peñuelas, P. Ciais, M. van der Velde and I. A. Janssens. 2013. The phosphorus trilemma. *Nat. Geosci.* 6: 897–8.

Ozturk, L., S. Eker, B. Torun and I. Cakmak. 2005. Variation in phosphorus efficiency among 73 bread and durum wheat genotypes grown in a phosphorus-deficient calcareous soil. *Plant Soil* 269: 69–80. doi:10.1007/s11104-004-0469-z.

Panigrahy, M., D. N. Rao and N. Sarla. 2009. Molecular mechanisms in response to phosphate starvation in rice. *Biotechnol. Adv.* 27: 389–97. doi:10.1016/j.biotechadv.2009.02.006.

Parentoni, S. N., and C. L. Souza Jr. 2008. Phosphorus acquisition and internal utilization efficiency in tropical maize genotypes. *Pesqui. Agropecuária Bras.* 43: 893–901.

Parentoni, S., C. de Souza Jr, V. de Carvalho Alves, E. Gama, A. Coelho, A. de Oliveira, P. Guimarães, C. Guimarães, M. Vasconcelos, C. A. P. Pacheco, W. F. Meirelles, J. V. de Magalhães, L. J. Moreira Guimarães, A. R. da Silva, F. Ferreira Mendes and R. E. Schaffert. 2010. Inheritance and breeding strategies for phosphorus efficiency in tropical maize (Zea Mays L.). *Maydica.* 55: 1–15.

Park, G.-G., J.-J. Park, J. Yoon, S.-N. Yu and G. An. 2010 A RING finger E3 ligase gene, Oryza sativa Delayed Seed Germination 1 (OsDSG1), controls seed germination and stress responses in rice. *Plant Mol. Biol.* 74: 467–78. doi:10.1007/s11103-010-9687-3.

Paterson, A. H., J. E. Bowers, R. Bruggmann, I. Dubchak, J. Grimwood, H. Gundlach, G. Haberer, U. Hellsten, T. Mitros, A. Poliakov, J. Schmutz, M. Spannagl, H. Tang, X. Wang, T. Wicker, A. K. Bharti, J. Chapman, F. A. Feltus, U. Gowik, I. V. Grigoriev, E. Lyons, C. A. Maher, M. Martis, A. Narechania, R. P. Otillar, B. W. Penning, A. A. Salamov, Y. Wang, L. Zhang, N. C. Carpita, M. Freeling, A. R. Gingle, C. T. Hash, B. Keller, P. Klein, S. Kresovich, M. C. McCann, R. Ming, D. G. Peterson, Mehboob-ur-Rahman, D. Ware, P. Westhoff, K. F. X. Mayer, J. Messing and D. S. Rokhsar. 2009. The Sorghum bicolor genome and the diversification of grasses. *Nature* 457: 551–6. doi:10.1038/nature07723.

Pavan, M. A., F. T. Bingham and P. F. Pratt. 1984. Redistribution of exchangeable calcium, magnesium, and aluminum following lime or gypsum applications to a Brazilian Oxisol. *Soil Sci. Soc. Am. J.* 48: 33–8.

Ramu, P., B. Kassahun, S. Senthilvel, C. Ashok Kumar, B. Jayashree, R. T. Folkertsma, L. A. Reddy, M. S. Kuruvinashetti, B. I. G. Haussmann and C. T. Hash. 2009. Exploiting rice–sorghum synteny for targeted development of EST-SSRs to enrich the sorghum genetic linkage map. *Theor. Appl. Genet.* 119: 1193–1204. doi:10.1007/s00122-009-1120-4.

Rattunde, H. F. W., E. Weltzien, B. Diallo, A. G. Diallo, M. Sidibe, A. O. Touré, A. Rathore, R. R. Das, W. L. Leiser and A. Touré. 2013. Yield of photoperiod-sensitive sorghum hybrids based on Guinea-race germplasm under farmers' field conditions in Mali. *Crop Sci.* 53: 2454. doi:10.2135/cropsci2013.03.0182.

Rattunde, H. F. W, S. Michel, W. L. Leiser, H.-P. Piepho, C. Diallo, K. vom Brocke, B. Diallo, B. I. G. Haussmann, E. Weltzien. 2016. Farmer participatory early-generation yield testing of sorghum in West Africa: possibilities to optimize genetic gains for yield in farmers' fields. *Crop Science* 56: 1–13. (open access)

Richardson, A. E., J. P. Lynch, P. R. Ryan, E. Delhaize, F. A. Smith, S. E. Smith, P. R. Harvey, M. H. Ryan, E. J. Veneklaas, H. Lambers, A. Oberson, R. A. Culvenor and R. J. Simpson. 2011. Plant and microbial strategies to improve the phosphorus efficiency of agriculture. *Plant Soil* 349: 121–56. doi:10.1007/s11104-011-0950-4.

Riedelsheimer, C., J. Lisec, A. Czedik-Eysenberg, R. Sulpice, A. Flis, C. Grieder, T. Altmann, M. Stitt, L. Willmitzer, and A.E. Melchinger. 2012a. Genome-wide association mapping of leaf metabolic profiles for dissecting complex traits in maize. *Proc. Natl. Acad. Sci.* 109: 8872– 8877. doi:10.1073/pnas.1120813109.

Riedelsheimer, C., F. Technow and A. E. Melchinger. 2012b. Comparison of whole-genome prediction models for traits with contrasting genetic architecture in a diversity panel of maize inbred lines. *BMC Genomics*. 13: 452. doi:10.1186/1471-2164-13-452.

Rose, T. J., M. T. Rose, J. Pariasca-Tanaka, S. Heuer and M. Wissuwa. 2011. The frustration with utilization: Why have improvements in internal phosphorus utilization efficiency in crops remained so elusive? *Front. Plant Sci.* 2 doi:10.3389/fpls.2011.00073.

Rose, T. J., and M. Wissuwa. 2012. Rethinking internal phosphorus utilization efficiency. In *Advances in Agronomy*. Amsterdam: Elsevier, pp. 185–217.

Rossiter, R. C. 1978. Phosphorus deficiency and flowering in subterranean clover (T. subterraneum L.). *Ann. Bot.* 42: 325–9.

Rouached, H., A. B. Arpat and Y. Poirier. 2010a. Regulation of phosphate starvation responses in plants: Signaling players and cross-talks. *Mol. Plant.* 3: 288–99. doi:10.1093/mp/ssp120.

Rouached, H., A. B. Arpat and Y. Poirier. 2010b. Regulation of phosphate starvation responses in plants: Signaling players and cross-talks. *Mol. Plant.* 3: 288–99. doi:10.1093/mp/ssp120.

Sagnard, F., M. Deu, D. Dembélé, R. Leblois, L. Touré, M. Diakité, C. Calatayud, M. Vaksmann, S. Bouchet, Y. Mallé, S. Togola and P. C. S. Traoré. 2011. Genetic diversity, structure, gene flow and evolutionary relationships within the Sorghum bicolor wild–weedy–crop complex in a western African region. *Theor. Appl. Genet.* 123: 1231–46. doi:10.1007/s00122-011-1662-0.

Sattari, S. Z., A. F. Bouwman, K. E. Giller and M. K. van Ittersum. 2012. Residual soil phosphorus as the missing piece in the global phosphorus crisis puzzle. *Proc. Natl. Acad. Sci.* 109: 6348–53. doi:10.1073/pnas.1113675109.

Secco, D., A. Baumann and Y. Poirier. 2010. Characterization of the Rice PHO1 Gene family reveals a key role for OsPHO1;2 in phosphate homeostasis and the evolution of a distinct clade in dicotyledons. *Plant Physiol.* 152: 1693–704. doi:10.1104/pp.109.149872.

Shimizu, A., K. Kato, A. Komatsu, K. Motomura and H. Ikehashi. 2008. Genetic analysis of root elongation induced by phosphorus deficiency in rice (Oryza sativa L.): fine QTL mapping and multivariate analysis of related traits. *Theor. Appl. Genet.* 117: 987–96. doi:10.1007/s00122-008-0838-8.

Simmonds, N. 1991. Selection for local adaptation in a plant breeding programme. *Theor. Appl. Genet.* 82: 363–7. doi:10.1007/BF02190624.

Simpson, R. J., A. Oberson, R. A. Culvenor, M. H. Ryan, E. J. Veneklaas, H. Lambers, J. P. Lynch, P. R. Ryan, E. Delhaize, F. A. Smith, S. E. Smith, P. R. Harvey and A. E. Richardson. 2011. Strategies and agronomic interventions to improve the phosphorus-use efficiency of farming systems. *Plant Soil* 349: 89–120. doi:10.1007/s11104-011-0880-1.

Singh, M., R. S. Malhotra, S. Ceccarelli, A. Sarker, S. Grando and W. Erskine. 2003. Spatial variability models to improve dryland field trials. *Exp. Agric.* 39: 151–60. doi:10.1017/S0014479702001175.

Soderlund, C., W. Nelson, A. Shoemaker and A. Paterson. 2006. SyMAP: A system for discovering and viewing syntenic regions of FPC maps. *Genome Res.* 16: 1159–68. doi:10.1101/gr.5396706.

Stoorvogel, J. J., E. M. A. Smaling and B. H. Janssen. 1993. Calculating soil nutrient balances in Africa at different scales. *Fertil. Res.* 35: 227–35. doi:10.1007/BF00750641.

Topp, C. N., A. S. Iyer-Pascuzzi, J. T. Anderson, C.-R. Lee, P. R. Zurek, O. Symonova, Y. Zheng, A. Bucksch, Y. Mileyko, T. Galkovskyi, B. T. Moore, J. Harer, H. Edelsbrunner, T. Mitchell- Olds,

J. S. Weitz and P. N. Benfey. 2013. 3D phenotyping and quantitative trait locus mapping identify core regions of the rice genome controlling root architecture. *Proc. Natl. Acad. Sci.* 110: E1695–704. doi:10.1073/pnas.1304354110.

Torabi, S., M. Wissuwa, M. Heidari, M.-R. Naghavi, K. Gilany, M.-R. Hajirezaei, M. Omidi, B. Yazdi-Samadi, A. M. Ismail and G. H. Salekdeh. 2009. A comparative proteome approach to decipher the mechanism of rice adaptation to phosphorus deficiency. *PROTEOMICS.* 9: 159–70. doi:10.1002/pmic.200800350.

Turk, M. A., A. M. Tawaha and M. K. J. El Shatnawi. 2003. Response of lentil (Lens culinaris Medik) to plant density, sowing date, phosphorus fertilization and ethephon application in the absence of moisture stress. *J. Agron. Crop Sci.* 189: 1–6. doi:10.1046/j.1439- 037X.2003.00002.x.

Vaccari, D. A. 2009. Phosphorus: A looming crisis. *Sci. Am.* 300: 54–9.

Vance, C. P., C. Uhde-Stone and D. L. Allan. 2003. Phosphorus acquisition and use: critical adaptations by plants for securing a nonrenewable resource. *New Phytol.* 157: 423–47.

van der Velde, M., C. Folberth, J. Balkoviç, P. Ciais, S. Fritz, I. A. Janssens, M. Obersteiner, L. See, R. Skalský, W. Xiong and J. Peñuelas. 2014. African crop yield reductions due to increasingly unbalanced Nitrogen and Phosphorus consumption. *Glob. Change Biol.* : n/a– n/a. doi:10.1111/gcb.12481.

Veneklaas, E. J., H. Lambers, J. Bragg, P. M. Finnegan, C. E. Lovelock, W. C. Plaxton, C. A. Price, W.-R. Scheible, M. W. Shane, P. J. White and J. A. Raven. 2012. Opportunities for improving phosphorus-use efficiency in crop plants. *New Phytol.* 195: 306–20. doi:10.1111/j.1469- 8137.2012.04190.x.

Vom Brocke, K., G. Trouche, E. Weltzien, C. P. Barro-Kondombo, E. Gozé, and J. Chantereau. 2010. Participatory variety development for sorghum in Burkina Faso: Farmers' selection and farmers' criteria. *Field Crops Res.* 119: 183–94. doi:10.1016/j.fcr.2010.07.005.

Voortman, R. L., and J. Brouwer. 2003. An empirical analysis of the simultaneous effects of nitrogen, phosphorus and potassium in millet production on spatially variable fields in SW Niger. *Nutr. Cycl. Agroecosyst.* 66: 143–64.

Voortman, R. L., J. Brouwer and P. J. Albersen. 2004. Characterization of spatial soil variability and its effect on millet yield on Sudano-Sahelian coversands in SW Niger. *Geoderma* 121: 65–82.

Wang, X., J. Shen, and H. Liao. 2010. Acquisition or utilization, which is more critical for enhancing phosphorus efficiency in modern crops? *Plant Sci.* 179: 302–6. doi:10.1016/j.plantsci.2010.06.007.

Wang, Y., H. Xu, J. Kou, L. Shi, C. Zhang and F. Xu. 2013. Dual effects of transgenic Brassica napus overexpressing CS gene on tolerances to aluminum toxicity and phosphorus deficiency. *Plant Soil* 362: 231–46. doi:10.1007/s11104-012-1289-1.

Wet, J. M. J. D., and J. R. Harlan. 1971. The origin and domestication of Sorghum bicolor. *Econ. Bot.* 25: 128–35. doi:10.1007/BF02860074.

Wimmer, V., C. Lehermeier, T. Albrecht, H.-J. Auinger, Y. Wang and C.-C. Schön. 2013. Genome-wide prediction of traits with different genetic architecture through efficient variable selection. *Genetics.* 195: 573–87. doi:10.1534/genetics.113.150078.

Wissuwa, M., and N. Ae. 2001. Genotypic variation for tolerance to phosphorus deficiency in rice and the potential for its exploitation in rice improvement. *Plant Breed.* 120: 43–8. doi:10.1111/j.1439-0523.2001.tb01698.x.

Wissuwa, M., T. Kretzschmar and T. J. Rose. 2016. From promise to application: root traits for enhanced nutrient capture in rice breeding. *J. Exp. Bot.* erw061. doi:10.1093/jxb/erw061.

Wissuwa, M., M. Yano and N. Ae. 1998. Mapping of QTLs for phosphorus-deficiency tolerance in rice (Oryza sativa L.). *Theor. Appl. Genet.* 97: 777–83. doi:10.1007/s001220050955.

Zhang, F. S., J. Ma and Y. P. Cao. 1997. Phosphorus deficiency enhances root exudation of low-molecular weight organic acids and utilization of sparingly soluble inorganic phosphates by radish (Raghanus satiuvs L.) and rape (Brassica napus L.) plants. *Plant Soil* 196: 261–4. doi:10.1 023/A:1004214410785.

Zongo, J.-D., P. H. Gouyon and M. Sandmeier. 1993. Genetic variability among sorghum accessions from the Sahelian agroecological region of Burkina Faso. *Biodivers. Conserv.* 2: 627–36. doi:10.1007/BF00051963.

Chapter 10

Advanced testing, multiplication and release of new sorghum varieties

Roger L. Monk, formerly DuPont Pioneer, USA

1 Introduction

The release of a new sorghum variety or hybrid is the culmination of many years of breeding and evaluation. Whether the release is by a public institution or a commercial enterprise, there are a number of steps to ensure a successful launch. The hybrid advancement process depends on accurate testing and analysis across a range of environments to ensure the product meets expectations. Increasing the parent lines for hybrid production and the hybrid for marketing must follow established guidelines to assure high-quality seed. Finally, there are decisions concerning registration, certification, plant variety protection (PVP) or plant patenting that must be determined based on the target markets and expected value. While many of the steps are similar for both varieties and hybrids, the primary focus of this chapter will be on hybrid evaluation and release.

2 Hybrid advancement considerations

In another chapter of this book, Franks examined the pre-breeding, development and testing of inbred through the level of topcross testing on F4/F5 inbreds. As he discussed, the development of A/B-lines and R-lines differ due to the need to develop a cytoplasmic male sterile through backcrossing. Topcross tests are often based on a single uniform tester, with the goal of selecting among siblings within a population and in identifying superior populations. The focus is generally on identifying yield potential, but the hybrids may also be evaluated for disease, insect or abiotic tolerance characteristics if the target

http://dx.doi.org/10.19103/AS.2017.0015.12

environment is limited by these issues. These topics are also discussed in a review by Rooney (2004). While the development of inbreds is central to the creation of unique hybrids, once the inbreds are homozygous enough for testing, continued advancement is based on hybrid performance if the goal is a commercial hybrid. Various schemes have been used to advance inbreds, some programmes name or code inbred lines directly at the topcross level, others may not name the inbred until after a second year of testing on multiple testers. The naming of an inbred merely simplifies the pedigree of the hybrid as it moves through the testing process. Depending on the resources available, an advancement cycle may follow a procedure similar to that outlined in Table 1. For commercial production, it helps to identify inbreds with good general combining ability as opposed to lines with high specific combining ability as a good general combining line may find use across a number of products. As more materials are genetically fingerprinted and phenotypically evaluated, it is also possible to use genomic prediction tools to plan new combinations for wider testing (Heffner et al., 2009).

The goal is to identify a stable, widely adapted hybrid. In sorghum, this can be a significant challenge, due to the diversity of environments where sorghum is grown. The interaction between the genotype and the environment (GXE) can be a significant constraint to making the appropriate decision. Sampling over seasons is vital to success as year-by-year variation is often larger than within-year variation across locations. One of the early goals of statistical procedures was to help make decisions in agricultural field trials and the tools and techniques have followed a progression in power with modern computer technology. Through a progression from randomized complete blocks and lattice designs to augmented designs, then to two-stage multiplicative models that account for spatial variability, these methods all try to ensure that the correct advancement decisions are made (Smith et al., 2001). At the same time, advances in field equipment and agronomic practices allow better population control, harvest accuracy and lower variation in the field data. The use of seed treatments similar to those used in commercial production can also help achieve more uniform results. Moving to larger equipment and data handling systems has allowed many programmes to expand the size of their hybrid testing programmes in recent years.

An entire science has developed along the lines of ensuring that testing is conducted in proper environments (Bramel-Cox et al., 1996; Muchow et al., 1996). Crop simulation models that use plant phenotypic information, geographic information and historical weather data have been used to classify environments for maize (Loffler et al., 2005). In

Table 1 Advancement and release scheme

Year	Testing stage	Multiplication stage
Year 1	Topcross – single tester	
Year 2	Multi-tester topcross	Breeders seed increase parental lines/Growouts
Year 3	Wide Area Testing	Foundation parental increase/Growouts
Year 4	Wide Area Testing	Hybrid increase/Growout
Year 5	On-Farm Strip Trials	
Year 6	On-Farm Trials	Hybrid increase – commercial quantities
Tear 7	Commercial	

order to be successful, a breeder must have a target environment in mind, and then ensure that the testing environments represent that target. One example of the methods that can be used was offered by Muchow et al. (1996). In Queensland, Australia, the growers expect a terminal drought environment, and that requires a different genetic response than environments where drought can occur earlier in the season. Using historic environmental records they were able to classify drought periods into five distinct scenarios at one location in Queensland, whereas another region showed four general schemes. It was possible to estimate the frequency of each climate and it suggests that testing programme needed to model these environments in order to be successful. This approach has been used in the United States to classify environments so that the new generation of maize hybrids can be evaluated at specific yield levels to ascertain the value of drought tolerance traits (Gaffney et al., 2016; Messina et al., 2015).

Since sorghum is often grown as a low cost, low profit margin crop, producers historically rely on native host plant resistance bred into the improved crop to protect against the myriad diseases, insects and abiotic challenges to successful production. Disease tolerance characterization is critical in areas where disease pressure is yield limiting. Each pathogen is unique in the speed which it takes to overcome host plant resistance mechanisms. *Colletotrichum graminicola* (anthracnose) is a diverse collection of strains and resistance genes are only effective against certain strains so that field tolerance is fleeting at best (Moore et al., 2008). On the other hand, diseases such as head smut are not as diverse and good levels of tolerance can be longer lasting. Field screens in the target environment are the most dependable method of ensuring that the breeder is addressing the current races or pathotypes. Identifying 'hot spots' and planting spreader rows to increase the inoculum load can help ensure a good screening environment. The problem with natural screening in disease-prone fields is the potential for escapes, which can be addressed in some situations with inoculation by the pathogen. This may require access to a pathology laboratory or other means to grow the organisms and prepare the inoculation media. Methods of growing the organisms and inoculation procedures for many common diseases are outlined in Fredericksen and Rosenow (1979), and the procedures and methods for screening a number of other diseases are described by Rooney (2004). The need to screen accurately for marker prediction has required the development of more precise screening techniques (Prom et al., 2011). Nevertheless, using these techniques, disease tolerance QTLs have been identified for a number of key pathogens and marker-assisted selection is being used to expedite the development of disease tolerance. Insect pests can be even more difficult to address in-field environments but effective greenhouse and field procedures have been developed to screen for key pests (Sharma et al., 1997). A recent example of an insect pest dramatically altering both production and breeding programmes is the rise of the sugarcane aphid (*Melanaphis sacchari*) across key sorghum growing areas of Mexico and the United States since its emergence in 2013 (Bowling et al., 2016). The aphid can rapidly increase in numbers under favourable conditions causing slow growth, reduced yields and large amounts of honeydew which is then covered in sooty mould. The honeydew can be so pronounced as to interfere with harvest in a way similar to that expressed with ergot infection. The aphid is not a new pest, but its affinity for sorghum is a recent development. Management of the pest is confounded by the fact that few insecticides are effective for control. Fortunately, USDA and the State entomologists and extension agents rapidly developed effective screening methods to identify tolerance and it was found that some of the older sources of greenbug (*Schizaphis graminum*) tolerance had cross tolerance to the sugarcane aphid (Armstrong e al., 2015). Tolerant inbred lines with dominant resistance have also been recently released (Mbulwe et al., 2016). An updated list on the National Sorghum Producers website lists 21 hybrids from nine commercial companies as having some level of tolerance (Anonymous, 2016). Combatting this pest will undoubtedly continue to be a major focus of both public and private breeding programmes as the pest continued to expand its range into Kansas in 2016.

One of the factors in determining the successful introduction is the realization that the hybrid or variety needs to match the target environment and market. Programmes whose focus is on identifying unique traits and ensure germplasm diversity is maintained are important to the long-term success of a breeding programme, but are somewhat separate from programmes aimed primarily to develop commercial hybrids. Commercial companies are at a distinct advantage in this regard as their sales and marketing staffs are closely attuned to the needs of their customers, however, they need to have a regular procedure for communicating with research personnel to ensure that everyone is working towards similar goals. However, research personnel should follow long-term, well-defined strategies since the targets of a breeding programme cannot change annually depending on the particular weather or specific problems that may occur.

Testing generally remains a function of the research department until it is transferred to sales and marketing teams. Research teams are limited in how many environments they can realistically sample during the development process. Once the agronomy or sales teams are involved, however, the product can be evaluated in 50–100 on-farm environments and compared to standard benchmark checks. If these trials are carefully placed and well conducted, then the entire team will gain confidence and information to aid in product placement for a successful launch. There are a number of difficulties that arise when the testing is transferred to the sales and marketing teams, however. These trials take time to place and conduct and often the farmer/cooperator is busy during that time of year with his own operations which means there can be mistakes in field selection, planting dates and entry placement. They tend to be placed with familiar farmers so that data will be collected, but the trial may not be placed in an area of sufficient stress to challenge the entries. All of these issues can make data analysis and interpretation difficult. It is imperative that the breeder or trained agronomist verifies the accuracy of the data before the results are compiled and used for advancement decisions. Since such strip trials are seldom replicated, the variation across the field needs to be considered and sometimes repeated benchmark checks are used to adjust for spatial problems. Hopefully this data will be compared or combined with ongoing multiyear research data for final considerations. Again, various companies differ as to whether they advance to commercial status based on one or two years of strip testing. The decision to advance a hybrid can also depend on the urgency of the new product's needs and the variation expected in the target environment. However, it is important that a potential release is fully characterized before it is introduced into the marketplace. Each company differs in the amount of information provided, but maturity, height, seed colour and other agronomic characteristics such as head type, exertion and lodging tolerance are usually described along with disease, insect and drought tolerance. The company will generally make a recommendation as to where the hybrid is best adapted for maximum production and reliability since the successful introduction often depends on success in the first year of introduction. Most companies limit the number of products that they produce and market in a given geography and they are very conscious of the product's life cycle. In the United States and Mexico, successful sorghum hybrids can have a long life cycle in the market. It is not uncommon for a hybrid to be sold for 15–20 years.

3 Inbred and hybrid increase

The use of cytoplasmic male sterility systems for the production of F1 hybrids creates a difference in the speed at which inbred parents can be developed. The selection of R-lines from an initial cross to an F3 or F4 selection that is uniform enough for evaluation is straightforward, but the development of the female (A-line) involves backcrossing to create the A/B pair, which slows down progress. As the materials progress through the process, it is critical that 100% sterility is maintained in the A-line and that the B-line produces adequate pollen for high yields on the A-line, in addition to being uniform for phenotypic traits. The R -line should also be phenotypically uniform and provide adequate pollen shed for high production yields (Rooney, 2004).

Once an inbred is identified that produces commercially acceptable hybrids, the inbred must be increased to a quantity that allows large-scale production. This first increase is sometimes the responsibility of the research team and sometimes is transferred to a parent seed group. If the inbreds or hybrids are to be certified, an application must be filed with the state agency responsible and inspections must be conducted throughout the growing season.

The initial parent increase is critical to determine genetic purity, quality and produceability. Often a research group may produce 20–50 pounds of an inbred through hand pollination which goes to the field scale of open pollination production with the goal of producing 30–50 (50 lb) units in anticipation of commercial production. The second increases are generally grown in an isolated area separate from the principal hybrid production area and are essentially small-scale production fields (1–3 acres). Careful roguing and isolation maintenance are critical. A pre-harvest roguing is also conducted as a final check to remove any questionable types. This breeders seed and parent increase should undergo a 'grow out' to ensure genetic purity and a low frequency of off-types. These off-types can be from a myriad of sources: they can be fertile plants in an A-line, they can be off-type for obvious phenotypic characteristics (head type, colour, height, awns, etc.) or they can be outcrosses with a foreign pollen source (forage, other grain, Johnsongrass, etc.). Such grow outs are often conducted during the off-cycle in a winter production environment so that the purity can be evaluated before the next growing season. Seed that is certified must be evaluated in grow outs under the auspices of the certifying agency. Trained personnel need to evaluate the materials throughout the growing season and record issues that they find. If the inbred is judged to be of preferred quality, then the source can be transferred to the production group for growing at a production scale. The seed certification authority sets standards for the number of different types of off-types allowed; an example is listed in Texas Department of Agriculture's website (https:texreg.sos.state.tx.us). The quality that is assessed will determine whether or not the lot must be discarded, can be used for hybrid production or is of sufficient quality for additional parent seed increases. Historically, these procedures have been followed to create the following four classes of seed (Poehlman and Sleper, 1995; Simmons, 1979):

1 Breeders seed – the seed produced by hand by the breeder.
2 Foundation seed – the direct increase of breeder's seed.
3 Registered seed – the generation of increase of breeder's seed or foundation seed.
4 Certified seed – first generation increase of any of the other classes. Certification requires repeated inspection by public officials and sets requirements for growing conditions, in-season inspections and laboratory evaluations.

Once the inbreds are deemed homozygous, they can be used to produce the experimental hybrids in anticipation of strip testing or can be used for commercial production. The hybrid seed is produced under similar procedures of isolation and roguing, with another grow out before strip testing or sale. The importance of fully characterizing the inbred during the increase period becomes apparent when planning for hybrid production. A-lines and R-lines may not be of the exact maturity and it is critical that they be planted in a manner that will ensure they are blooming at the same time for maximum seed set. Often the pollinator is planted on multiple dates or with differences in population to try to ensure adequate pollen coverage for the entire period that the A-line is receptive. Inadequate pollination can be a source of contamination if foreign pollen is available; splits longer that three weeks are risky and can lead to poor production. The cost of production obviously impacts the profit margin of each hybrid so that hybrids that are predictable, high yielding and consistent are preferred by production teams. It is good for breeders to be conscious of these issues throughout the advancement process to ensure a product is not dropped for reasons other than performance. It is apparent that the commercial launch of a hybrid must be anticipated at least two years ahead of launch to give time for the inbred increases, parent seed grow outs, hybrid production and hybrid grow outs. Off-cycle nurseries can help move the product forward, but most of this process is being conducted on lines as they work their way through the advancement cycle, so that the inbreds are in position if the hybrid advances. Obviously, expenses increase dramatically as an inbred advances to the level of parent seed increase, so discipline is required on all parties to only increase lines that have a good chance for success.

4 Registration and plant variety protection

While the principal discussion has been aimed at producing commercial products, the release of materials by public institutions has also become more complicated as PVP and intellectual property rights expanded over the last 40 years. Public and commercial products can be registered which is process used to describe the variety and secure a seed sample with the National Germplasm Collection (Ellis et al., 2010). Materials from public sources can often be acquired through the use of materials transfer agreements which specify the exact uses that are permitted by the granting agency. These documents are prepared in cooperation by the licensing and legal teams of the institutions and the requesting party. If the materials being transferred are finished inbreds, the agreement may allow the testing of the materials for a specified period of time, but may not allow the use of the materials in breeding crosses or genotyping. They will include statutes that indicate an additional agreement will need to be negotiated if the line is to be commercialized, which may include a royalty schedule. Some institutions (DAF, Australia) allow the use of the materials either directly or as parents in new inbreds, with a sliding royalty scale based on the percentage of the inbred in the final product (12.5%, 25%, 50%, 100%). Other institutions may have similar systems or different agreements in place. Although these procedures may apply to finished inbreds, public institutions may also release germplasm either freely or for a fee, without restrictions on use.

Although many legal channels are open to exploration for materials developed by a breeder, the certification process for the inbreds and hybrids is required for seed to be grown or sold in some countries. Each state where seed is produced can have its own requirements for both growing the seed and for the number of contaminants allowed.

For example, Arizona requires twice the isolation distance as required by Oklahoma (www.arizonacrop.org/crop standard, www. okcrop/sorghum.htm). If the seed meets the criteria, it can be issued official tags which serve as a record that the lot has met the standards. In the United States, this whole system is under the direction of the Association of Official Seed Certifying Agencies; this group was established by the Federal Seed Act. A second group, the Organization for Economic Cooperation and Development (OECD) was established in 1961 to assure that seed produced to standards in one country can be shipped to another country. An OECD tag is used to designate acceptable lots, if phytosanitary requirements are also met.

Inbreds can be protected from use by another entity through PVP, established in 1970 and amended in 1994. These regulations provide 20 years of protection. The PVP system requires submission of an application and fee prior to the commercial sale. The application includes a detailed phenotypic characterization, history of development, statement of distinctness, statement of ownership, photographs and a seed deposit to be submitted and is available on the internet at www.ams.usda.gov. A PVP examiner reviews the documentation and determines if the new variety warrants protection. If the PVP application is approved, a second sample is required to be deposited at the National Center for Genetic Resources Preservation facility in Fort Collins, CO. The United States system is based on the system established for international registrations (UPOV) in 1961. A standard form is used which again describes the phenotypic characteristic of the variety; such a characterization is needed before sales in the European Union, Mexico and many other countries. The UPOV sorghum testing guidelines and definitions form is available at www.upov.int.

An inbred or hybrid can also be protected through a patent or through trade secret statutes. The application process to file and receive a patent is lengthy, involved and expensive, requiring trained legal personnel to channel the application through the process. The patent process uses some of the same information as the UPOV and PVP applications, but also includes a great deal of other information.

In the United States, there is not a formal registration requirement before a hybrid can be offered for commercial sales. However, in many countries, that is not the situation. Mexico, Argentina, India and the European Union among others all require one or two years of government evaluation before a hybrid can be sold. The process can begin while a hybrid still has an experimental designation, but companies incur costs and must release significant information about the products as part of the process. These trials are designed to verify both the descriptors and uniqueness of the products and performance. Seed protection laws may not always be enforced effectively, so the risks associated with potentially losing control of germplasm must be considered.

Registration is only one aspect of shipping seed across country borders. There are phytosanitary regulations in place in each country and import permit requirements vary widely. For multinational companies, there are often logistical experts who track the requirements for shipping to each country and mage the process in a timely basis. Delays in shipping may cause a testing season to be missed, which slows the process. As part of the phytosanitary requirements, a given lot of seed must undergo field inspections during the growing season to identify possible diseases or insects that could prevent shipping to some countries. If a regulated disease or insect is noted on the inspection report, then the lot must be sampled and sent to a laboratory for additional testing, which may affect the shipping schedule. There is also a list of approved seed treatments that is unique to each country, so that each sample must be verified for treatments and rates before shipping. Certain countries require a quarantine grow out for seed entering the country.

5 Conclusion

While developing new varieties and hybrids is a significant challenge, carrying those products through to the market also requires diligence and careful attention to detail. There are challenges in ensuring that the product meets the requirements of the market through thorough testing and analysis. As the product is advancing through the final years of testing, the parental lines and the hybrid must be increased in quantity to allow large-scale testing and release. The path through certification and legal protection must be planned and documentation completed in a timely manner based on the intended markets. Procuring a team of people with expertise in all of these areas can help ensure that a successful introduction can be achieved.

6 References

Agricultural Marketing Service, United States Dept. of Agriculture. 2016. Plant Variety Protection. https://www.ams.usda.gov/services/plant-variety-protection (Doa 4/26/2016).

Anonymous. 2016. Sugarcane aphid. http:www.sorghumcheckoff.com/newsroom/2016/03/28/sugarcane-aphid. (Doa 2/23/2017).

Arizona Crop Improvement Association, Inc. 2016. Sorghum hybrids. http://www.arizonacrop.org/crop standard.html (Doa 5/30/2016).

Armstrong, J. S., W. L Rooney, G. C. Peterson, R. T. Villenueva, M. J. Brewer and D. Sekula-Ortiz. 2015. Sugarcane aphid (Hemiptera: Aphididae): Host range and sorghum resistance including cross- resistance from greenbug sources. *J. Econ. Entomol.* 108:576–82.

Bowling, R. D., M. J. Brewer, D. L. Kerns, J. Gordy, N. Seifer, N. Elliot, G. D Buntin, M. O. Way, T. A. Royer, S. Bites and E. Maxson. 2016. Sugarcane aphid (Hemiptera.Aphididae):a new pest on sorghum in North America. *J. Integr. Pest Manag.* 7:1–13. Doi10.1093/jipm/pmw011.

Bramel-Cox, P. J., G. A. Manziel, F. Zavala-Garcia, M. Chisi, E. A. Weltzein and E. O. Mongo. 1996. Breeding for diverse target environments. In: Proceedings of the International Conference on the Genetic Improvement of Sorghum and Pearl Millet. 22–27 September 1996. Lubbock, TX.

Ellis, D. D., K. A. Garland-Campbell, J. A. Grotenhuis, M. M. Jenderek and J. F. Pedersen. 2010. Crop registration: the pathway to public access of plant genetic materials to build crops for the future. *Crop Sci.* 50:1151–60.

Fredericksen, R. A., and D. T. Rosenow. 1979. Breding for disease resistance in sorghum. In *Biology and Breeding for Resistance to Arthropods and Pathogens in Agricultural Plants.* MP-1451. Texas Agricultural Experiment Station, College Station, TX.

Gaffney, J., J. Schussler, C. Loffler, W. Cai, S. Puszkiewicz, C. Messina, J. Groeteke, J. Keaschall and M. Cooper. 2015. Industry-scale evaluation of maize hybrids selected for increasing yield in drought stress conditions of the US Corn Belt. *Crop Sci.* 55:1608–18.

Heffner, E. L., M. E. Sorrells and J. Jannink. 2009. Genomic selection for crop improvement. *Crop Sci.* 49:1–12.

Heffner, E. L., A. J Lorenz, J. Jannink and M. E. Sorrells. 2010. Plant breeding with genomic selection: gain per unit time and cost. *Crop Sci.* 50:1681–90.

International Union for the Protection of New Plant Varieities. 2011. Test guidelines. http://www.upov.int/portal/index.html/.en (DOA 4/26/2016)

Loffler, C. M., J. Wei, T. Fast, J. Gogerty, S. Langton, M. Bergman, B. Merrill and M. Cooper. 2005. Classification of maize environments using crop simulation and geographic information systems. *Crop Sci.* 45:1708–16.

Mbulwe, L., G. C. Peterson, J. S. Armstrong and W. L Rooney. 2016. Registration of sorghum germplasm TX3408 and TX3409 with tolerance to sugarcane aphid [Melanaphis saccari (Zehnter)] *J. Plant Registrations* 10:51–56. Doi:10.3198/jpr2015.04.0025crg.

Messina, C. D., T. R. Sinclair, G. L. Hammer, D. Curan, J. Thompson, Z. Oler, C. Gho and M. Cooper. 2015. Limited transpiration trait may increase maize drought tolerance in the US Corn Belt. *Agron. J.* 107:1978–86.

Moore, J. W. M. Ditmore and D. O. TeBeest. 2008. Pathotypes of *Colletotrichum sublineolum* in Arkansas. *Plant Dis.* 92:1415–20.

Muchow, R. C., M. Cooper, F. R. Bidinger, G. L. Hammer, A. K. Borrell and S. C. Chapman. 1996. Targeting sorghum improvement in drought-prone environments: approaches and progress. In: Proceedings of the International Conference on the Genetic Improvement of Sorghum and Pearl Millet. 22–27 September 1996. Lubbock, TX.

Oklahoma Seed Association. 2016. Sorghum – Open Pollinated – Seed Certification Standard. www. okcrop/sorghum.htm (DOA 5/31/2016).

Podlich, D. W., M. Cooper and K. E. Basford. 1999. Computer simulation of a selection strategy to accommodate genotype-environment interactions in a wheat recurrent selection programme. *Plant Breed.* 18:17–28.

Poehlman, J. M., and D. A. Sleper. 1995. *Breeding Field Crops*. Iowa State University Press, Ames, Iowa.

Prom, L. K., R. Perumal, S. R. Erattaimuthu, J. E. Erpelding, N. Montes, G. N. Odvody, C. Greenwald, Z. Jin, R. Fredericksen, and C. W. Magill. 2011. Virulence and molecular genotyping studies of *Sporisorium reilianum* isolates in sorghum. *Plant Dis.* 95:523–9.

Rooney, W. L. 2004. Sorghum improvement – integrating traditional and new technology to produce improved genotypes. *Adv. Agron.* 83:37–107.

Sharma, H. C., F. Singh and K. F. Nwanze. 1997. Plant resistance to insects in sorghum. International Crops Research Institute for the Semi-Arid Tropics. Patancheru 502 324, Andhra Pradesh, India.

Simmons, N. W. 1979. *Principals of Crop Improvement*. Longman House, Burnt Mill, Harlow, Essex CM20 2JE, England.

Smith, A., B. Cullis and R. Thompson. 2001. Analyzing variety by environment data using multiplicative mixed models and adjustments for spatial field trend. *Biometrics* 57:1138–47.

Texas Department of Agriculture. 2016. Texas Administrative Code. https://texreg.sos.state.tx.us (DOA 5/31/2016).

Part 3

Sorghum production techniques

Chapter 11

Crop management practices for grain sorghum: an overview

M. Djanaguiraman, P. V. V. Prasad and I. A. Ciampitti, Kansas State University, USA

1 Introduction

Sorghum is grown in the tropical, subtropical or temperate regions of Africa, Asia and the Americas. Across the globe, sorghum production during 2016 was 63 million tons. Sorghum is produced in more than 100 countries; however, only 14 countries have an annual production of more than a million tons. The top ten sorghum producers in the world are: United States (8.4 million ton), Nigeria (6.5 million ton), Mexico (6.0 million ton), India (4.5 million ton), Sudan (4.5 million ton), China (3.8 million ton), Ethiopia (3.7 million ton), Argentina (3.5 million ton), Australia (1.9 million ton) and Burkina Faso (1.8 million ton). Africa produces 40% of the global sorghum crop, with West Africa's production being just over half of Africa's. The rest of the crop is produced in about 37% in the Americas and 20% in Asia and Oceania. For the United States, average sorghum yields are above 4 Mg ha^{-1}, which is higher than those of developing countries. India is a major sorghum producer, but its production is declining due to competition for resources (particularly irrigation to high-value crops) and other social/economic factors (Rao et al., 2006). Sorghum yields in China have improved compared to previous decades; however, it is still below 4 Mg ha^{-1}. Sorghum production in Mexico has increased over the past years but appears to have stabilized, with average yields of 3.7 Mg ha^{-1}. Sorghum yields in Argentina and Brazil were 4.5 and 2.4 Mg ha^{-1}, respectively (FAO, 2014), while in Australia sorghum production was significantly higher than maize production (FAO, 2014).

http://dx.doi.org/10.19103/AS.2017.0015.13

Sorghum is a versatile crop because it adapts well to different seasons, weather and cropping patterns. It is cultivated as a monocrop, intercrop or relay crop in different regions of the world. In general, sorghum is grown in rotation with legumes, cotton, oilseeds or vegetables. In Argentina, sorghum–groundnut, sorghum–wheat or maize, and sorghum–sunflower are most frequently adopted rotations. Mostly, in sub-Saharan Africa, sorghum is grown as a subsistence crop, as a monocrop or often mixed or intercropped with legumes or oilseeds. In semi-arid regions of India, sorghum-based crop rotation depends on soil fertility, local agro-climate and genotype. Two-year rotation is most common in India, especially on Vertisols (cotton–sorghum–groundnut) and Alfisols (cotton–sorghum–cowpea, sorghum–groundnut or sorghum–legume sequences). Sorghum monocropping is also practised on the Vertisols of Peninsular India.

Developing countries account for roughly 90% of the global sorghum area and 70% of the total production, primarily dominated by traditional varieties. However, in developed countries, most producers grow high-yielding hybrids. The major challenge of sorghum research systems is to provide technologies that can transform sorghum from a subsistence crop to a commercial crop. Adoption of improved production technologies can help enhance productivity of grain sorghum. In this chapter, we highlight some standard crop production technologies that should be considered. Our discussion is limited to crop management practices and does not describe the management of other yield limiting factors (e.g. insects and diseases). Production technologies of sorghum vary with regions; hence, regional educational resources or extension experts should be consulted for crop management options like time of sowing, seeding rate and nutrient and weed management to enhance the grain yield. The variation in the production technologies is due to changes in variety, soil, water and climatic conditions.

2 Planting

2.1 Planting time

Sowing of sorghum in India should be undertaken with the onset of the monsoon though sorghum is grown in two different seasons, namely rainy (Kharif) and post-rainy (Rabi). Kharif sorghum is sown with the onset of rains (from last week of May to mid-July) and the Rabi sorghum is sown from mid-September to mid-October. Dry sowing about one week in advance of monsoon set (firm forecast) is also practised in India. Crop failures are likely in northern India due to failure of the monsoon. Delay in sowing due to delay of the monsoon or other factors (e.g. non-availability of seeds) can result in substantial yield reduction and an increase in shoot fly infestation.

In West Africa, sorghum planting date is determined by the first rain, approximately June through July. In East Africa, sorghum is commonly planted from mid-October to mid-December. The October planting time has recorded better yields due to crop escape from midge damage and ergot disease. When sorghum was planted from mid-October to mid-January, the flowering will occur between mid-December and mid-March so that ergot development in florets is lower.

In Australia, planting time usually varies from September to January depending on the rainfall pattern and soil temperatures. Under moisture unlimited conditions, sorghum crop establishment and stand are dependent on air and soil temperature. Sorghum is a warm-season crop that requires a soil temperature of 15°C to start the germination process, below which the germination process will stop or slow down (ICRISAT, 1984). Early planting in August to September may result in crop failure due to low soil temperature.

The optimum planting date for higher yield in the United States is around early June for the states of Kansas and Oklahoma ('Northern Great Plains'). For the 'Southern Great Plains' (including the states of Texas, Louisiana, New Mexico and Mississippi), planting date depends on the region of the state, for example, planting in the Panhandle of Texas occurs from mid-April through June. On the other side of the state, recommended planting dates for Texas are late January to late February in the Lower Rio Grande Valley, late February to mid-March in the Coastal Bend and Upper Gulf Coast, and mid-March to mid-April in Central and North Central Texas (Ciampitti et al., 2017a). Research from Kansas, USA, has indicated that planting date also influences the final seeding rate. The number of tillers per plant was deceased in late plantings (late June or July) compared to normal date of planting in Kansas, USA, resulting in decreased grain yield (Ciampitti et al., 2017a). It is important to choose the right planting date, so that the reproductive stages (critical growth stages) have reduced exposure to abiotic stresses such as drought or high temperatures. A typical crop calendar for sorghum production in different regions of the world is provided in Fig. 1.

2.2 Soil preparation

Sorghum is generally planted at 3–4 cm depth depending on the soil texture and soil moisture. Grain sorghum is adapted to be grown on a wide range of soils. Like many crops, grain sorghum will likely produce greatest yields on deep, fertile, well-drained loamy soils. Relative to other cereals, sorghum can also tolerate mildly acidic to mild alkaline conditions with a pH range between 5.5 and 8.0 units and can also tolerate waterlogging for shorter periods. Shallower or deeper planting depths can affect the duration between

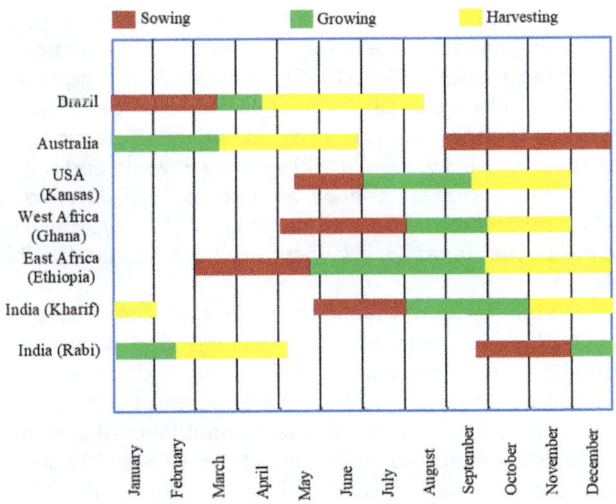

Figure 1 Crop calendar for sorghum production in different regions of the world. Data compiled from USDA and FAO and country profiles.

planting and emergence, affecting early season plant uniformity. In Kansas, research with three different sowing depths (1.2, 3.8 and 7.6 cm) has indicated that 3.8 and 7.6-cm deep placement resulted in similar shoot growth. However, the shallow seed placement (1.2 cm) has resulted in delayed development, fewer leaves and less shoot dry weight than optimum or deep placement (Ciampitti et al., 2014).

Grain sorghum needs a warm, moist soil for rapid germination. Tillage and planting systems can help to improve soil moisture and reduce soil and wind erosion. Conservation tillage is a common term used to describe reduced-till, mulch-till, strip-till, ridge-till, zero-till and no-till. It includes minimal mechanical soil disturbance (i.e. no tillage and direct seeding); maintenance of a mulch of carbon-rich organic matter (>30% residue cover after planting) and crop residues including cover crops; and rotations or sequences and associations of nitrogen-fixing legume crops, including trees. No-till is also known as direct seeding. In this system, the soil is left undisturbed from harvest to planting except for strips up to one-third of the row width. Planting or drilling is accomplished using machine with specially equipped no-till planters. In ridge-till the soil is left undisturbed from harvest to planting except for strips up to one-third of the row width. Planting is done using disk openers and usually involves the removal of the top of the ridge. Residue is left on the surface between ridges. Ridges are rebuilt during row cultivation. In mulch-till the full-width tillage involves one or more tillage trips which disturb the entire soil surface. Tillage tools such as chisels, field cultivators, disks or blades may be used. This is done prior to and/or during planting. Most of the sorghum grown in the United States was grown under conservative tillage.

2.3 Plant population

The optimum plant density for high yields depends on different factors such as available nutrients, soil moisture content and rainfall. Plant population and row spacing are variables that can have a significant impact on the yield of grain sorghum. Though optimal plant densities for grain sorghum differ from region to region, previous research has indicated that grain yields generally increase as plant populations increase (Jones and Johnson, 1991; Staggenborg et al., 1999). At lower than suggested plant densities, grain sorghum head number per plant or seed number per head increased when compared to the recommended plant density (Lafarge and Hammer, 2002). Under dry conditions, the seeding rate is 49 000–74 000 seed ha^{-1}, while under irrigation or higher moisture conditions, seeding rates can be as high as 272 000 seed ha^{-1}, depending on the planting date recommended (Fromme et al., 2012). Typically, seeding rates for producers along the upper Texas Gulf Coast vary from 124 000 to 247 000 seed ha^{-1} (Fromme et al., 2012).

Under irrigated or high levels of management or high productive soils, a final plant density of 200 000–275 000 plants ha^{-1} is recommended (Ciampitti et al., 2017b). Under rainfed conditions, recommended plant density is dependent on annual rainfall level. If the annual rainfall level is <500 mm, the recommended plant density level is 57 500–67 500 plants ha^{-1}. Similarly, for annual rainfall level of 500–660, 660–810 and above 810 mm, the recommended plant density is 62 500–112 500, 87 500–137 500 and 125 000–225 000 plants ha^{-1}, respectively (Ciampitti et al., 2017b). Lower plant densities minimize the risk of crop failure in dry environments by developing additional tillers and increasing panicle size (Myers and Foale, 1981; Thomas et al., 1981). In addition, final yields will be reduced in a dry year if plant density is too high. To achieve a plant density of 172 000 plants ha^{-1}, approximately 8 kg of seed is required (Borikar et al., 2007). However, the weight of the seed planted is not a good measure of plant density because seed size varies among the varieties or hybrid. If the seed germination percentage is >70, then lower seed rate is recommended. Most of the commercial hybrids and sorghum cultivars in the United States have a germination percentage of more than 70. The seeding rate can be calculated by using the following formula: desired planting rate divided by per cent actual germination equals number of seeds to plant for the desired population. For example: 30 000/95% gives a planting rate of 31 579 seeds acre^{-1}, that is, 30 000 germinating seeds per acre. An additional 10% can be added to the seeding rate to account for field mortality unrelated to germination percentages.

Studies have demonstrated that lower plant densities (<160 000 plants ha^{-1}) were not as sensitive to below or above normal amounts of precipitation compared to high plant densities (>209 000 plants ha^{-1}). In other words, lower plant densities maintained yield under less favourable moisture conditions, whereas yields declined slightly for excessive number of plants (Ciampitti et al., 2017b). Seeding rates between 65 000 and 80 000 seeds acre^{-1} (with about 90% stand) will provide good yield potential while reducing risks associated with higher seeding rates. In drier years, the lower end would likely be appropriate. On-farm research at Kansas State University, Manhattan, Kansas, has indicated an optimal response to plant density at around 74–110 000 seeds ha^{-1} in Kansas, USA (Ciampitti, 2017; Fig. 2).

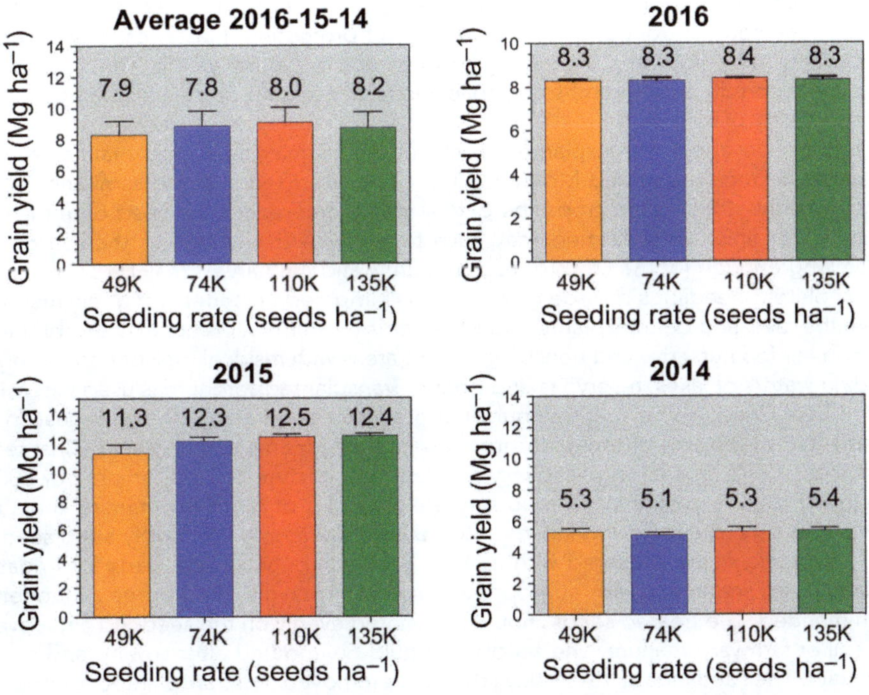

Figure 2 On-farm sorghum seeding rate experiment at Kansas, USA. Source: Ciampitti (2017).

Sorghum under irrigated condition can be raised both as a direct-sown and transplanted crop. In some areas of Africa and Asia, transplanting sorghum is followed, either to fill gaps after crop emergence and thinning or to compensate for a growth period that is too short for a complete crop cycle (Rehm, 1989). The quantity of seed required for transplanted sorghum is 3.0 kg $acre^{-1}$. The transplanted crop has advantages like (i) reduction in main field duration by ten days; (ii) effective and economical control of shoot fly, which attacks direct-sown crops during the first three weeks and which is difficult to control, in the nursery itself; (iii) elimination of seedlings which show chlorotic and downy mildew symptoms by hand removal, thereby minimizing incidence of downy mildew in the main field and (iv) maintenance of optimum plant density due to transplantation of only healthy seedlings.

Row spacing studies have indicated that grain sorghum yields will generally be higher with narrow (60 or 76 cm) rows compared to wider (101 cm) rows if other growing conditions are favourable, in most of the sorghum-growing regions of the United States, India and Africa. In the United States, most grain sorghum is planted in 76-cm rows because that is the row width of most equipment. Historically, 76-cm rows performed better than wider rows, and narrower rows have not consistently yielded better than 76-cm rows (Shroyer et al., 1998). When crop management requires weed management, the row width is limited by the cultivation equipment. When mechanical weed control practices are involved a minimum row spacing of 60 or 76 cm is allowed. Plants in narrower rows shade the soil quicker, improving weed control and reducing soil erosion; however, due to competition between the plants the yield will be decreased. For high-yielding sorghum, narrower row spacing (38-cm vs. 76-cm) can act not only as a best management practice for weed control but also as a practice for promoting a faster canopy coverage, intercepting more light and increasing biomass and potential yields. The row spacing range (between 38 and 76 cm) to be followed is dependent on rainfall, soil conditions and availability of irrigation.

Sorghum can also be transplanted from nursery to main field and maintained as a transplanted crop (Young and Mottram, 2003). It is practised in some semi-arid regions to minimize the risk of failed crops and patchy stands and reduce the costs of re-planting. If rains fail or finish early, farmers may have to re-sow – this is risky as the season may not be long enough for the crop to reach maturity, and the harvest may be small or even fail completely. Seedlings may be grown in small irrigated nurseries using minimal water before the rains and be transplanted into the fields when the rains are fully established or grown in rainfed nurseries and transplanted into areas with residual moisture following the receding waters of lakes, reservoirs and ponds. Transplanted sorghum is used extensively around Lake Chad, where sorghum nurseries are established during the rainy season and transplanted at the end of the rains into the sandy clay loams and clay soils (Chantereau and Nicou, 1994). For raising seedlings to plant one hectare of land, an area of 300 m^2 is required. Before ploughing the nursery area, 750 kg of farmyard manure should be added, and spread evenly and incorporated during ploughing. Uniformly sized seedbed (2 × 1.5 m) should be prepared with 30-cm spacing in between the beds for irrigation. Similarly, seed treatment with appropriate fungicide to avoid seed-borne pathogens is recommended. The treated seeds should be planted evenly on the seedbed and covered with soil or farmyard manure. The seedbed should be irrigated after sowing and once in three days. The seedlings at 15–18 days should be removed and transplanted for optimum growth (Anon., 2012).

3 Nutrient management

Sorghum grows best on soils with good tilth, balanced nutrients and a neutral pH. Sorghum can tolerate considerable variation in soil fertility; however, plant growth and grain yield will be reduced by both low levels of fertility and nutrient imbalance. In developing countries of Asia and Africa, traditionally little or no fertilizer is added to sorghum. However, with the development of high-yielding sorghum genotypes and improved crop production practices, today farmers are applying fertilizer to sorghum. Due to increasing fertilizer prices and decreasing purchasing power of sorghum farmers, attempts to identify sorghum genotypes with high nutrient use efficiency will be important. Improving plant efficiency for fertilizer or nutrient use is important to reduce costs of crop production (Mishra and Patil, 2015; Bernal et al., 2002). In India, under rainfed conditions, a fertilizer/nutrient level of 5 tonnes of farmyard manure, 40 kg N and 20 kg P_2O_5 ha^{-1} of fertilizers is generally followed. However, for irrigated sorghum, 80 kg of N and 40 kg P_2O_5 ha^{-1} is followed for higher productivity. One-half, that is, 40 kg N and full P_2O_5 is applied at sowing, while the remaining 40 kg N is applied at 40–50 days after sowing or panicle initiation stage. In light soils, split application of fertilizer is recommended, while in heavy soils, single application may be beneficial. The integrated nutrient management in sorghum improves the fertilizer use efficiency and maintains soil productivity.

Sorghum is relatively tolerant to low soil fertility in comparison to other cereals because it has finer and extensive fibrous root system. In sub-Saharan Africa, land degradation is acute because of long-term overuse of soil without addition of fertilizer to soil. The sorghum growers are economically weak and cannot afford to invest in fertilizer and also the growing season is risky with environmental extremes and unpredictable rainfall. The failure to replenish the soil nutrients leads to decreased sorghum yield. To address this problem and to enhance the fertilizer use efficiency, particularly for resource-poor smallholder subsistence farmers, micro-dosing is recommended. Micro-dosing involves the application of small, affordable quantities of fertilizer with the seed at planting time or as top dressing at 3–4 weeks after emergence. In micro-dosing 6-g doses of fertilizer are added to a single hole in which the seed is sown (Aune and Ousman, 2011). With the micro-dosing technique, between 4 and 9 kg of N, 4 kg of P and 4–7 kg of K to give a total of 20 kg of nutrients is added to sorghum field. The farmers dig a small hole before the start of rain, then fill it with manure, if available. When the rain begins, farmers put fertilizer and seed in the hole to avoid runoff of soil and to improve the root growth and nutrient use efficiency. In some other areas, micro-fertilization applies 0.3 g of NPK 16-16-16 $pocket^{-1}$ (seed sowing hole) of sorghum. This quantity corresponds to mixing seeds and fertilizer in a 1:1 ratio (Aune et al., 2007) and is equivalent to adding 3–10 kg mineral fertilizer ha^{-1}, depending on planting density. Seeds and fertilizer are mixed shortly before sowing in order to avoid prolonged contact between seeds and fertilizer; usually diammonium phosphate (DAP) or NPK 16-16-16 is used (Aune and Bationo, 2008). Nutrient mining occurs when cropped soils experience negative nutrient balances, with extraction losses occurring faster than the replacement of nutrients. Effects may be especially severe when food crops are integrated into intensive repeated sequences and rotations with inadequate nutrient management.

Meta-analysis indicates that nutrient application in African soil will increase sorghum grain yield by 66% relative to no nutrient application (Tonitto and Ricker-Gilbert, 2016). Sorghum grain yield gain by the addition of nutrients in African soil is illustrated in Fig. 3 (Tonitto and Ricker-Gilbert, 2016). Among the various fertilizers, N and P fertilizer application resulted in

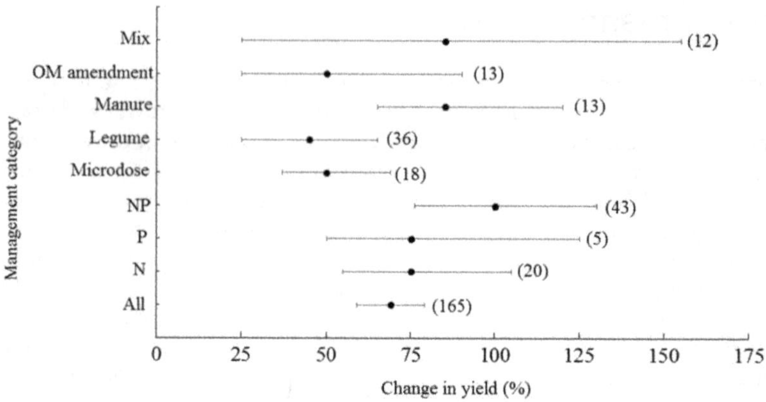

Figure 3 Percent change in yield by various nutrient management techniques in Africa. Effect of nutrient amendment on sorghum yield in units of percent change from the control (no nutrient inputs). Mean values and 95% confidence intervals of the back-transformed response ratios are shown (number of comparisons in parentheses). Source: Tonitto and Ricker-Gilbert (2016).

the highest yield improvement (98%) compared to control. Application of N fertilizer alone, P only and manure improved grain yield by 72%, 72% and 87%, respectively (Tonitto and Ricker-Gilbert, 2016). Micro-dose application improved yield by 47% relative to no nutrient inputs. Mixed treatment category (organic, inorganic and all nutrients) exhibited a large range in outcomes (Tonitto and Ricker-Gilbert, 2016).

In the United States, grain sorghum is mostly grown under reduced tillage for erosion control and more efficient use of available soil water. One of the most important issues under no-till condition is nitrogen (N) management, because of reduced N availability. This occurs due to slower N mineralization, greater N immobilization, denitrification and ammonia volatilization since N must be mostly surface applied. Low soil temperatures under no-till environments compared to high tillage leads to lower nutrient availability to the crop during the early crop season. The most common method of N fertilizer application under no-till condition is broadcasting either solid ammonium nitrate or urea, or spraying urea ammonium nitrate solutions on the soil surface immediately before or after planting. Surface application of N fertilizer can result in significant N losses through ammonia volatilization. The amount of N fertilizer required for the sorghum plants depends on the crop yield potential and the amount of residual soil N prior to planting. Nitrogen management is dependent on several factors such as water management, soil texture, history of manure application, type of N fertilizer, soil organic content, residue management, previous crop and tillage system adjustment. Factoring in these criteria, the typical sorghum N requirement in grain sorghum is about 1.2–1.5 kg ha^{-1} (Griffin et al., 2011). In most of the sorghum-producing areas in the United States, sorghum is grown as a rainfed crop where denitrification is the main cause of nitrogen loss. The first step in fertility management is soil testing. For sorghum following another cereal crop such as sorghum, corn or wheat in rotation, testing for residual nitrate-nitrogen in the soil to a depth of 0–60 or 0–90 cm is recommended; however, soil testing is not required if sorghum following soybean or alfalfa is followed. Separate equations are available to determine the economically optimal N rate for sorghum following a cereal crop and for sorghum following soybean in rotation. Nitrogen fertilizer following alfalfa will not be economical unless the alfalfa crop stand is poor (Wortmann et al., 2013). To determine phosphorus (P), potassium (K) and micronutrient requirements, soil samples from 0 to 20 cm depth at every four to five years have to be analysed. Phosphorus is the second most likely nutrient to be deficient in the soil. Phosphorus acquisition plays an important role for crop adaptation to low P soils (Wang et al., 2010) and a higher internal P use efficiency could help to limit soil nutrient mining (Rose and Wissuwa, 2012), especially in low-input farming conditions.

The uptake pattern of N, P and K in sorghum is presented in Fig. 4 (Vanderlip, 1993). In sorghum the uptake of N is very high during the reproductive growth preceding the panicle emergence and also during grain development. However, P accumulation is high during early vegetative stage and grain formation. Potassium uptake is the highest at vegetative growth prior to panicle emergence in soils with high K supply. To improve nutrient use efficiency, split application of N fertilizers is beneficial. The fertilizer should be applied at sowing and be placed 5 cm below the seed. Nitrogen fertilizer can be applied in split applications, but full doses of P_2O_5 and K_2O are recommended to be applied basally at the time of sowing.

Sorghum crop producing 5.5 Mg ha^{-1} of grain yield will remove 335 kg of nutrients ha^{-1} (149 kg N, 61 kg P and 125 kg K) from soil (Mishra and Patil, 2015). Sorghum grown in India removes on an average 22 kg N, 13.3 kg P_2O_5 and 34 kg K_2O to produce one tonne of grains (Kaore, 2006). A sorghum hybrid yielding approximately 8 Mg ha^{-1} removes about 250 kg N, 60 kg P, 160 kg K, 45 kg Mg and 40 kg S ha^{-1} (Maiti, 1996).

The amount of N fertilizer required for sorghum crop will vary depending on the yield potential of the cultivar and the amount of residual soil N available prior to planting. Traditionally, sorghum is known for its nutrient use efficiency, especially N. However, sorghum yield can be increased by addition of N fertilizer (Mahama et al., 2014). For modern sorghum hybrids, N application improves yields via modification of above-ground biomass, seed number and grain harvest index (Mahama et al., 2014). Sorghum grown in Africa, India and the United States showed a good response to N fertilizer (Fig. 5).

Sorghum yield can be increased by addition of N fertilizer up to 180 kg ha^{-1} (in contrast to total N uptake). However, 90 kg ha^{-1} gave profitable economic net returns. Similar results were observed by Mahama et al. (2014). The response of sorghum to N fertilizer depends on factors such as amount of available and residual soil N, physiochemical property of soil, soil moisture, rainfall pattern and cultivar (genotype or hybrid). Sorghum N use efficiency (NUE) significantly varied within its growing environment (Fig. 6).

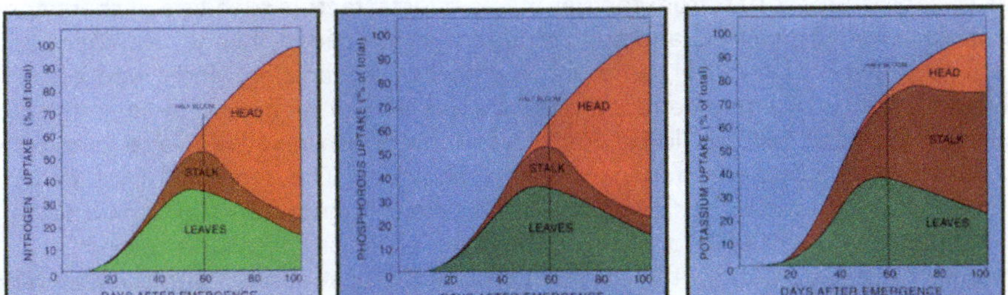

Figure 4 Uptake pattern of nitrogen (N), phosphorus (P) and potassium (K) in grain sorghum. Source: Vanderlip (1993).

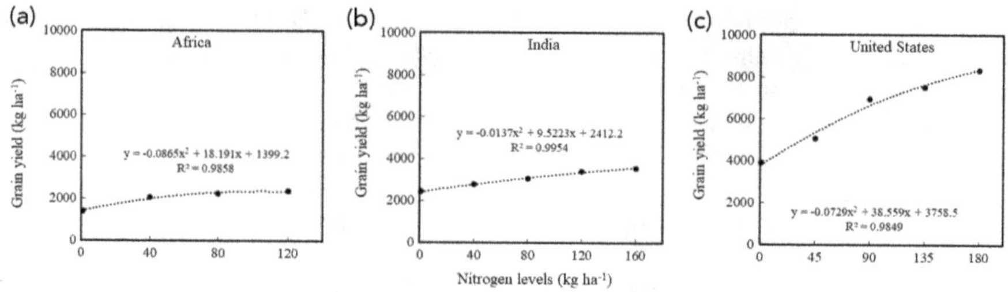

Figure 5 Response curves showing the grain yield as a function of nitrogenous fertilizer in (a) Africa, (b) India and (c) the United States. The response equations and coefficients of determination (R^2) are shown in each panel. The graph was drawn from the data of (a) Buah et al. (2012), (b) Pal et al. (1983) and (c) Mahama et al. (2016).

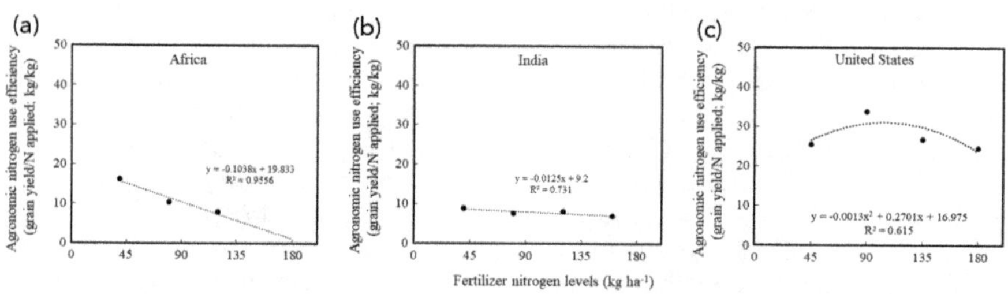

Figure 6 Agronomic nitrogen use efficiency (kg kg^{-1}) of sorghum genotypes grown in (a) Africa, (b) India and (c) the United States. The response equations and coefficients of determination (R^2) are shown in each panel. Agronomic nitrogen use efficiency was calculated using the formula NUE = $(Y_T - Y_O)/F_N$; where Y_T is grain yield (kg h^{-1}) in treatment with total N (soil +fertilizer), Y_O is crop yield (kg ha^{-1}) measured in a control treatment with no fertilizer N application (soil alone) and F_N is the amount of nitrogenous fertilizer (kg ha^{-1}) applied to the soil. The graph was drawn from the data of (a) Buah et al. (2012), (b) Pal et al. (1983) and (c) Mahama et al. (2016).

Among the different environments, sorghum grown in the United States had a higher NUE compared with Africa and India (Fig. 5). It could be due to the efficiency of genotypes. Efficient plants absorb and utilize the nutrients and increase the efficiency of applied fertilizers, resulting in reduced cost of inputs and also preventing the nutrient losses to ecosystems. Apart from the genotypic effects, factors such as soil characteristics (type of soil, soil pH, soil EC and soil management practices), tillage, weather, fertilizer management (type of fertilizer, time and method of application) and weed population (Mishra and Patil, 2015) may be responsible for low NUE in Africa and India. In the United States, the improved NUE could be due to adoption of conservative tillage, crop rotation and proper nutrient and weed management techniques. The overall NUE of a cropping system can be increased by achieving greater uptake efficiency from applied N inputs, and by reducing the amount of N lost from soil inorganic and organic pools or both.

The major nutrients deficient in sorghum crop are N, P, K and S. Nitrogen deficiency is first expressed in older leaves; the deficiency symptom includes yellowing of older leaves, leaf dropping and stunted growth. The deficiency can be corrected by application of 8–10 kg ha^{-1} of N as top dressing or 2–3% of urea as foliar spray. Phosphorus deficiency includes stunted growth, dark green leaves with red pigmentation and small root system. Potassium deficiency will be seen in older leaves, necrotic patterns with red pigmentation and streaked patterns on interveinal tissue at tips and margin of leaves. Sulphur deficiency appears in younger leaves as pale yellow.

Among micronutrients, sorghum is sensitive to zinc and iron deficiencies. Zinc deficiency affects stem elongation, resulting in stunted growth. Zinc deficiency usually occurs during early stages of crop growth, and when soil pH is greater than 7.5. Iron deficiency occurs in high pH and calcareous soils, causing reduced availability of soluble iron in the soil. Iron chlorosis causes stunted growth in the plant with interveinal chlorosis. The iron chlorosis will be aggravated by high rate of application of phosphatic fertilizer, because it forms iron phosphates which are less soluble in soil solution (Obour and Perumal, 2015).

4 Water requirement and management

Sorghum requires at least 500 mm of water per growing season for optimum production, that is, 200–250 mm from emergence to flowering, and another 250 mm from flowering to physiological maturity. Daily water use of grain sorghum cultivar varies with growth stage and can range from 3 to 7 mm ha^{-1}. Daily water use gradually increases during vegetative stages, further increases during growing point differentiation and peaks during booting and flowering; thereafter, it gradually decreases during grain filling and physiological maturity (Fig. 7).

Water use efficiency (WUE) can be defined as the above-ground biomass produced per unit area per unit of water used in evapotranspiration. Research at the University of Nebraska indicated that sorghum had higher WUE (372 kg cm^{-1}) than corn (272 kg cm^{-1}) and soybean (135 kg cm^{-1}), an important benefit of growing sorghum in water-limited environments. Research at Kansas State University showed that sorghum takes less water to initiate grain yield, which was about 152 mm for sorghum and about 280 mm for corn. The yield increases linearly and is proportional to water use up to about 1 kg mm^{-1} provided other inputs and agronomic practices are optimal. Tenhunen et al. (2002) have reported that soil moisture conditions have a significant influence on WUE of crops grown in arid and semi-arid regions.

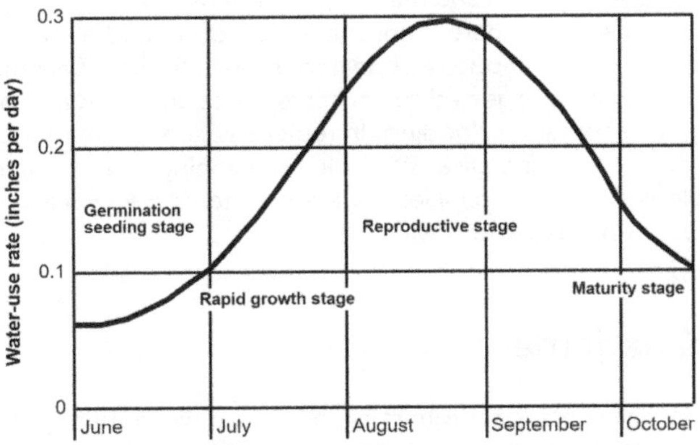

Figure 7 Characteristic water use pattern of grain sorghum in Kansas.

Khosla (1995) has reported that the grain yield per plant decreased at a decreasing rate with increase in plant density. However, the WUE increased at a decreasing rate with increase in plant density. This implies that crop water use per plant decreased at a faster rate than grain yield per plant over the range of plant density evaluated.

Sorghum responds positively to supplemental irrigation during dry periods. Usually, cultivars grown in rainy season respond favourably to irrigation or rainfall, but produce low yields exposed to post-rainy terminal drought conditions. Similarly, response of post-rainy season cultivars to irrigation during post-rainy season is low because of their photoperiod sensitivity and specific localized adaptation. In general, sorghum plants take >75% of water and nutrients from the top 90 cm of soil profile. To reduce soil evaporation during seedling stage, adopting water-conserving practices such as stubble mulching, narrow crop spacing (45 cm between rows) and appropriate planting date for rapid canopy establishment is ideal. Over-planting reduces panicle size, increases vulnerability to charcoal rot and lodging and increases competition for water and nutrients with little increase in yield. The most critical period for water availability begins about one week before panicle emergence (or boot) and continues through two weeks post-flowering (Stichler and Fipps, 2003). Sorghum plants experience high water demand at boot stage and the plant response to irrigation is greatest at this stage. Average daily water use of sorghum is <2.5 mm up to 7-leaf stage and thereafter, water use increases rapidly and an average of 7.5 mm is used per day through boot to milk stage (Stichler and Fipps, 2003). Irrigation at panicle initiation, boot and 50% flowering stages produces higher yields.

Water stress during pre-boot stage reduces the seeds per panicle. Crop water use is influenced by prevailing weather, available soil water, crop species and growth stage. At full cover, the crop canopy intercepts most of the incoming solar radiation, thereby reducing the amount of energy reaching the soil surface. Sorghum is said to be at full cover when the leaf area is three times the soil surface area under the canopy (i.e. LAI = 3), which usually occurs around boot to flowering stage. Medium-to-late maturing sorghum cultivar requires approximately 450–650 mm of water during a growing season. However, the daily water requirement depends on the growth stage. Early in the growing season (up to approximately 7-leaf stage), average daily water use is low. The water requirement then increases to around 7–10 mm day^{-1} until the boot stage. Maximum daily water use occurs from the boot stage until after anthesis. The daily water requirement then decreases gradually during grain fill as the crop begins to senesce leaves and mature (McWilliams, 2003).

During the growing season where monsoonal rains occur and early onset of monsoon followed by a long dry spell happens, repeated inter-cultivation operations like hoeing, weeding and mulching are suggested to conserve moisture. In India, a common method of irrigation for grain sorghum is flooding. In regions where sorghum is planted in broad beds furrow irrigation can increase WUE. In developing countries, more advanced irrigation technologies such as sprinkler irrigation systems are widely used. Sprinkler or drip irrigation systems have greater WUEs.

5 Weed management

One of the major constraints for sorghum production is weed management. Sorghum is commonly infested with both grassy (e.g. crabgrass, shattercane, panicum and witchgrass) and broadleaf (e.g. kochia, pigweed and cocklebur) weeds. In the United States, weeds

[Johnson grass (*S. halepense*); Shattercane (*S. bicolor*); Common cocklebur (*Xanthium strumarium*); Fall panicum (*Panicum dichotomiflorum*); Broadleaf signalgrass (*Brachiaria platyphylla*)] decreased yield of sorghum by about 26% (Bunce and Ziska, 2000; Gianessi and Sankula, 2003). One pigweed plant per metre of row left uncontrolled can reduce sorghum yields by 10%. In grain sorghum, weed control is best achieved by integrated approaches such as crop rotations, production practices (narrow rows), cover crop, herbicides and tillage to avoid herbicide resistance in weeds. Crop rotation also reduces weed pressure by varying the timing and types of tillage and herbicides. Fallow periods and rotation with soybeans during summer, or wheat during winter, will greatly reduce weed numbers. Weed control needs to be practised in sorghum before sowing, likely through tillage or chemical burn-down, so that sorghum plants can grow without any weed competition. There are four methods of weed control, namely cultural, manual or mechanical, biological and chemical methods. Effective weed control can be achieved by combining these methods. For example, weed control can be achieved by harrowing and ploughing after weeds are germinated and/or inter-cultivation using a cultivator at sorghum seedling stage, mulching and intercropping. Manual weeding is a very common practice to control weeds in sorghum crop in developing countries. Cultural and manual weed control measures are adopted in developing countries such as India and Africa. However, in developed countries such as the United States, chemical weed control method is more common. Herbicides account for nearly 47% and 25% of the total pesticide use in the United States and worldwide, respectively (Grube et al., 2011). Chemical control of weeds by herbicide application is mostly adopted because of its ease of use and greater efficiency (McErlich and Boydston, 2013).

Grain sorghum lacks tolerance to many of the commonly used grass and broadleaf herbicides. For one to apply herbicides, some knowledge about these chemicals is required, because improper use of herbicides may lead to the complete loss of the crop. Depending on the time of application, herbicides are grouped into three types (i) pre-planting, the herbicides are applied to the soil before sowing sorghum crop; (ii) pre-emergence, herbicides are applied after sowing of sorghum, however, before the emergence of sorghum seedling; and (iii) post-emergence, herbicides are applied after the emergence of weeds and sorghum. Herbicides should be selected based on the weed species present in the field. Control of grass type weeds using post-emergence herbicide will be very difficult than when using pre-emergence herbicides. If the field is weedy with annual grasses and broad-leaved weeds before sowing, burn-down herbicide (glyphosate, paraquat) can be used. There are a few options for herbicides applied at planting for grass and pigweed including Dual (S-metolachlor), Frontier, Partner and Ramrod. Appropriate rates of herbicides and calibrated sprayers and nozzles should be used, including use of safener for efficient weed control. A common post-emergence herbicide used is atrazine for control of annual grasses, while herbicides such as bromoxynil in combination with atrazine and other combinations are also available. Producers should follow the rates and timing of herbicides carefully. Non-atrazine options are available which include sulfonylurea types, which can be tank-mixed with 2,4-D or dicamba. New chemistries of herbicides are now available for efficient management of weeds to avoid building of herbicide tolerance. One of the newest post-emergence herbicides available for use in sorghum for control of broadleaf weeds such as kochia, pigweeds, is Huskie (Pyrasulfotole).

Of late, herbicide resistance in weeds has become widespread. In subtropical cropping areas of Argentina, populations of the C_4 grass *Sorghum halepense* have evolved glyphosate resistance. Similarly, glyphosate resistance evolution has been documented in

C_3 *Lolium rigidum* populations from temperate areas of Australia. In general, 372 unique, herbicide-resistant biotypes (in all crops) have been confirmed worldwide. The United States has 139 of these biotypes, Australia 60, Canada 52, France and Spain 33 each, Brazil 25, Germany 26, Israel 27, the United Kingdom 24, and there are from 1 to 19 in most other countries with intensive agriculture. About 21 glyphosate resistance biotypes were identified, out of which 13 were recorded in the United States (Vencill et al., 2012).

One of the strategies to control weeds in sorghum crop grown following wheat is through glyphosate and 2,4-D or dicamba followed by atrazine application in late summer. Sorghum planted into soybean or row-crop stubble normally requires no seedbed preparation other than weed control. Application of atrazine with crop-oil concentrate is recommended where mainly broadleaf weed pressure is low, and 2,4-D is recommended to control winter weeds and most germinated weeds through planting. Glyphosate and 2,4-D ester can be applied about a week before planting to control emerged annual weeds and established perennials. The rate of atrazine applied to the soil surface at planting should not exceed 1.1 kg ha^{-1} because surface-applied atrazine is vulnerable to loss in surface-water runoff.

Striga spp. (witch weed) is a root parasitic flowering plant, which affects the sorghum crop, particularly in Asia and Africa. *Striga* haustorium attaches to the root of sorghum plants to obtain water, nutrients and carbohydrates from the host plant (Van Ast, 2006), leading to wilting and rolling of leaves. *Striga* must be controlled before flowering because the tiny seeds of *Striga* are viable for 15–20 years in soil. It is particularly more prevalent in sorghum crop that is poorly managed (low-input fertilizer) and grown under rainfed or dry soil conditions (prolonged period of drought). Agronomic practices such as use of crop rotations, managing soil fertility, efficient water management, use of *Striga*-tolerant cultivars and seed treatment options are now available for management of *Striga* over several years, minimizing the seed bank in the soil. Application of high rates of atrazine as pre-emergence or post-emergence application controls *Striga*. However, low rates of atrazine were ineffective. Oxyfluorfen as an early post-emergence directed spray can/will control *Striga*. Herbicide seed treatment is a relatively new approach to herbicidal control of parasitic weeds. In this approach, low rates of herbicides are applied and it does not rely on specialized application equipment. Seed treatment can be given as seed coating or imbibing or priming seeds in a herbicide solution. The herbicide absorbed into seed will be subsequently transferred to an attached parasite or kill nearby *Striga* plants through diffusion into the surrounding soil. Gworgwor et al. (2002) have found that 1.5M brine (NaCl) solution can effectively control *Striga* emergence in sorghum. Tuinstra et al. (2009) demonstrated that the herbicide imazapyr and metsulfuron-methyl seed treatments cause few *Striga* attachments to the sorghum plant and also delay in attachment in acetolactate synthase herbicide-tolerant sorghum genotypes.

6 Conclusion

Sorghum yields have increased over time in most growing regions. However, the increase was greater in developed countries like the United States than in developing countries like India and those in Africa. In India, the area under sorghum cultivation can be improved, if sorghum is considered as a major cereal crop and industrial demand for grain-based starch or value-added products is developed. Research indicates that decreased productivity is due to non-adoption of proper production techniques because sorghum is mostly grown by subsistence farmers. Improving sorghum production potential is very critical because it is connected to food security in Africa and other regions. It is very clear that sorghum production can be improved by maintaining optimum plant population, split application of recommended N fertilizer, irrigation (if available) at critical growth stages and adoption of proper weed management strategies. However, the low sorghum yield in Africa can be improved by introducing high-yielding genotypes or hybrids, making inputs (quality seeds, fertilizer and plant protection chemicals) available to farmers at affordable price and correct time. It is important to match the genotype well with the environment and best management practices along with creation of enabling conditions for input supply chain and market opportunities. According to International Food Policy Research Institute (IFPRI), model demand for the cereals in semi-arid regions of Africa is forecast to increase by 40% by 2020. Hence, improving the production potential of sorghum by proper adoption of crop management practices is critical.

7 Acknowledgements

We thank Kansas Grain Sorghum Commission, United Sorghum Checkoff Program, K-State Center for Sorghum Improvement and Feed the Future Innovation Labs for Collaborative Research on Sorghum and Millet; and on Sustainable Intensification for supporting sorghum research. Contribution no. 18-306-B from Kansas Agricultural Experiment Station.

8 Where to look for further information

Understanding the impacts of different crop managements on sorghum grain yield will help in identification of best management practices. The newly released varieties or hybrids should be tested under different crop management practices and environments, so that the full genetic potential can be exploited. In addition, the social preferences and nutritional qualities and needs should be considered. In future, sorghum will be a potential alternative crop in environments facing water limitation or high temperatures, and hence, crop management practices under various abiotic and biotic stresses should be developed. Several universities (State Agricultural and National universities), national organizations (National Agricultural Research and Extension System; e.g. Indian Institute of Millet Research in India; Ethiopian Institute of Agricultural Research in Ethiopia) and international organizations (e.g. International Crops Research Institute for the Semi-Arid Tropics) are active in sorghum research and contribute to innovations. The results from these organizations and private industry will help improve productivity of sorghum. Continued funding from donor agencies, such as United States Agency for International Development, Australian Centre for International Agricultural Research, French Agricultural Research Centre for International Development, Department for International Development and Bill and Melinda Gates Foundation, will provide critical support and will enhance sorghum research to address the needs of farming communities and industry.

9 References

Anon. 2012. Sorghum. In: *Crop Production Guide 2012*. Department of Agriculture, Government of Tamil Nadu, Chennai, India and Tamil Nadu Agricultural University, Coimbatore, India, pp. 55–71.

Aune, J. B. and Bationo, A. 2008. Agricultural intensification in the Sahel – The ladder approach. *Agric. Syst.* 98:119–25.

Aune, J. B. and Ousman, A. 2011. Effect of seed priming and micro-dosing of fertilizer on sorghum and pearl millet in Western Sudan. *Expl. Agric.* 47:419–30.

Aune, J. B., Doumbia, M. and Berthe, A. 2007. Microfertilisation in Mali. *Outlook Agric.* 36:199–203.

Bernal, J. H., Navas, G. E. and Clark, R. B. 2002. Sorghum nitrogen use efficiency in Colombia. *Dev. Plant Soil Sci.* 92:66–7.

Borikar, S. T., Reddy, B. V. S., Alur, A. S., Reddy R. C., Birajadar, S. N., Kalapande, H. V. and Tripathi, C. M. S. 2007. *Rainy Season Sorghum Production Technologies for Dryland Areas of Maharashtra. Global Theme on Crop Improvement.* International Crops Research Institute for the Semi-Arid Tropics, Patancheru 502 324, Andhra Pradesh, India, pp. 20.

Buah, S. S. J., Kombiok, J. M. and Abatania, L. N. 2012. Grain sorghum response to NPK fertilizer in the guinea savanna of Ghana. *J. Crop Improv.* 26:101–15.

Bunce, J. A. and Ziska, L. H. 2000. Crop ecosystem responses to climatic change: Crop/weed interactions. In: Reddy, K. R. and Hodges, F. (Eds), *Climate Change and Global Crop Productivity*. CABI, New York, NY, pp. 333–48.

Chantereau, J. and Nicou, R. 1994. *Sorghum. The Tropical Agriculture Series*. MacMillan, London, pp. 4–11.

Ciampitti, I. A. 2017. Sorghum management consideration. Kansas State Research & Extension Bulletin, Issue 635, pp. 3–8.

Ciampitti, I. A., Diaz, D. R., Jardine, D., Thompson, C. R., Whitworth, R. J. and Rogers, D. H. 2014. Kansas Sorghum Management 2014. Kansas State Research & Extension Bulletin, MF3046. https://www.bookstore.ksre.ksu.edu/pubs/MF3046.pdf.

Ciampitti, I. A., Prasad, P. V. V., Schlegel, A. J., Haag, L., Schnell, R. W., Arnall, B. and Lofton, J. 2017a. Genotype × Environment × Management interactions: US sorghum cropping systems. In: Ciampitti, I. A. and Prasad, P. V. V. (Eds), *Sorghum: State of the Art and Future Perspectives*. Agronomy Monographs 58. American Society of Agronomy and Crop Science Society of America, Inc., Madison, WI.

Ciampitti, I. A., Diaz, D. R., Jardine, D., Thompson, C. R., Whitworth, R. J. and Rogers, D. H. 2017b. Kansas Sorghum Management 2017. Kansas State University Agricultural Experiment Station and Cooperative Extension Service, Publication number MF3046. Kansas State University, January 2017.

FAO. 2014. FAOSTAT. www.fao.org/faostat/en/#data. Visited on 8 February 2017.

Fromme, D. D., Fernandez, C. J., Grichar, W. J. and Jahn, R. L. 2012. Grain sorghum response to hybrid, row spacing, and plant populations along the upper Texas gulf coast. *Int. J. Agr.* 2012:5pp, Article ID 930630.

Gianessi, L. P. and Sankula, S. 2003. *The Value of Herbicides in U.S. Crop Production*. National Center for Food & Agricultural Policy, Washington DC. Retrieved from http://www.ncfap.org/benefits.htm.

Griffin, G., Jokela, W., Ross, D., Pettrinelli, D., Morris, T. and Wolf, A. 2011. Recommended soil nitrate-N tests. In: Sims, J. T. and Wolf, A. (Eds), *Recommended Soil Testing Procedures for the Northeastern United States*. Northeast Regional Bulletin #493. 3rd edition. Agricultural Experiment Station, University of Delaware, Newark, DE, pp. 27–38.

Grube, A., Donaldson, D., Kiely, T. and Wu, L. 2011. *Pesticide Industry Sales and Usage: 2006 and 2007 Market Estimates*. EPA's Biological and Economic Analysis Division, Office of Pesticide Programs, Office of Chemical Safety and Pollution Prevention, US EPA, Washington DC.

Gworgwor, N. A., Hudu, A. I. and Joshua, S. D. 2002. Seed treatment of sorghum varieties with brine (NaCl) solution for control of *Striga hermonthica* in sorghum. *Crop Prot.* 21:1015–21.

International Crops Research Institute for the Semi-Arid Tropics (ICRISAT). 1984. *Agrometeorology of Sorghum and Millet in the Semi-Arid Tropics: Proceedings of the International Symposium*, 15–20 November 1982. ICRISAT Center, Patancheru, Andhra Pradesh, India.

Jones, O. R. and Johnson, G. L. 1991. Row width and plant density effects on Texas High Plains sorghum. *J. Prod. Agric.* 4:613–21.

Kaore, S. V. 2006. An approach for crop wise plant nutrition prescription. *Indian J. Fertl.* 1:57–62.

Khosla, R. 1995. Yield and water use efficiency of different plant populations of dryland corn and sorghum. M.Sc. Thesis submitted to Department of Crop and Soil Environmental Science. Virginia Polytechnic Institute and State University, Blacksburg, VA.

Lafarge, T. A. and Hammer, G. L. 2002. Tillering in grain sorghum over a wide range of population densities: Modelling dynamics of tiller fertility. *Ann. Bot.* 90:99–110.

Mahama, G. Y., Prasad, P. V. V., Mengel, D. B. and Tesso, T. T. 2014. Influence of nitrogen fertilizer on growth and yield of grain sorghum hybrids and inbred lines. *Agron. J.* 106:1623–30.

Mahama, G. Y., Prasad, P. V. V., Roozeboom, K. L., Nippert, J. B. and Rice, C. W. 2016. Cover crops, fertilizer nitrogen rates, and economic return of grain sorghum. *Agron. J.* 108:1–16.

Maiti, R. K. 1996. Panicle development and productivity. In: *Sorghum Science*. Oxford and IBH Publishing Co., New Delhi, India, pp. 140–81.

McErlich, A. F. and Boydston, R. A. 2013. Current state of weed management in organic and conventional cropping systems. Publications from USDA-ARS/UNL Faculty Paper 1387.

McWilliams, D. 2003. Drought strategies for corn and grain sorghum. Coop Ext. Circ. 580, New Mexico State University, Las Cruces, NM.

Mishra, J. S. and Patil, J. V. 2015. Nutrient-use efficiency in sorghum. In: Rakshit, A., Singh, H. B. and Sen, A. (Eds), *Nutrient Use Efficiency: From Basics to Advances*. Springer Publication, India, pp. 297–315.

Myers, R. J. K. M. and Foale, M. A. 1981. Row spacing and population density in grain sorghum. A simple analysis. *Field Crops Res.* 4:147–54.

Obour, A. and Perumal, R. 2015. Evaluating the effectiveness of iron chelates in managing iron deficiency chlorosis in grain sorghum. *Kansas Agric. Exp. Sta. Res. Rep.* 2(6):3.

Pal, U. R., Singh, V. P., Singh, R. and Verma, S. S. 1983. Growth rate, yield and nitrogen uptake response of grain sorghum (*Sorghum bicolor* (L) Moench) to nitrogen rates in humid subtropics. *Fert. Res.* 4:3–12.

Rao, P. P., Birthal, P. S., Reddy, B. V. S., Rai, K. N. and Ramesh, S. 2006. Diagnostics of sorghum and pearl millet grains-based nutrition in India. *ISMN* 47:93–6.

Rehm, S. 1989. Hirsen. In: Rehm, S. (Ed.), *Spezieller Pflanzenbau in den Tropen und Subtropen*. Eugen Ulmer Verlag, Stuttgart, Germany, pp. 79–86.

Rose, T. J. and Wissuwa, M. 2012. Rethinking internal phosphorus utilization efficiency: A new approach is needed to improve PUE in grain crops. *Adv. Agr.*, 116:185–217.

Shroyer, J., Kok, H. and Fjell, D. 1998. Seedbed preparation and planting practices. In: *Grain Sorghum Production Handbook*. Kansas State University Agricultural Experiment Station and Cooperative Extension Service, Manhattan, KS, C-687, pp. 5–10.

Staggenborg, S. A., Fjell, D. L., Devlin, D. L., Gordon, W. B. and Marsh, B. H. 1999. Grain sorghum response to row spacings and seeding rates in Kansas. *J. Prod. Agric.* 12:390–5.

Stichler, C. and Fipps, G. 2003. Irrigating sorghum in South and South Central Texas. L-5434. Texas A&M University, College Station, TX.

Tenhunen, J., Roupsard, O. and Rambal, S. 2002. Severe drought effects on ecosystem CO_2 and H_2O fluxes at three Mediterranean evergreen sites: Revision of current hypotheses. *Global Change Biol.* 8:999–1017.

Thomas, G. A., Myers, R. J. K. M., French, M. U., Hall, B., Laewing, J. H., Dove, A. A., Taylor, G. K., Lehroy, E., Wylie, P. and Sterling, G. D. 1981. Evaluation of row spacing and population density effects of grain sorghum over a range of Australia environments. *Aust. J. Exp. Agric. Anim. Husb.* 21:210–17.

Tonitto, C. and Ricker-Gilbert, J. E. 2016. Nutrient management in African sorghum cropping systems: Applying meta-analysis to assess yield and probability. *Agron. Sustain. Dev.* 36:10.

Tuinstra, M. R., Soumana, S., Al-Khatib, K., Kapran, I., Toure, A., Ast, A., Bastiaans, L., Ochanda, N. W., Salami, I., Kayentao, M. and Dembele, S. 2009. Efficacy of herbicide seed treatments for controlling Striga infestation of sorghum. *Crop Sci.* 49:923–9.

Van Ast, A. 2006. The influence of time and severity of Striga on the *Sorghum bicolor- Striga hermonthica* association. Tropical Resource Management Papers, Wageningen University and Research Centre, Department of Plant Science, Wageningen, the Netherlands.

Vanderlip, R. L. 1993. *How A Sorghum Plant Develops.* Kansas State University, Manhattan, KS, January 1993. (http://www.nasecoseeds.com/products/sorghum/47.html).

Vencill, W. K., Nichols, R. L., Webster, T. M., Soteres, J. K., Mallory-Smith, C., Burgos, N. R., Johnson, W. G. and McClelland, M. R.2012. Herbicide resistance: Toward an understanding of resistance development and the impact of herbicide-resistant crops. *Weed Sci.* 60:2–30.

Wang, X., Shen, J. and Liao, H. 2010. Acquisition or utilization, which is more critical for enhancing phosphorus efficiency in modern crops? *Plant Sci.* 179:302–6.

Wortmann, C. S., Ferguson, R. B., Hergert, G. W., Shapiro, C. A. and Shaver, T. M. 2013. Nutrient management suggestions for grain sorghum. www.ianrpubs.unl.edu/pages/publicationD.jsp?publicationId=671.

Young, E. M. and Mottram, A. 2003. Transplanting sorghum and millet – a key to risk management. CAZS Natural Resources, University of Wales, UK, pp. 7.

Chapter 12

Improving soil and crop nutrition management in sorghum cultivation

David Mengel, Kansas State University, USA

1 Introduction

Sorghum (*Sorghum bicolor* L. Moench) ranks fifth in world production among cereals, and is an extremely important crop in Africa, where the crop originated, Australia, parts of the Americas, India, China and Southeast Asia, especially in hot, dry areas within these countries or regions where the production of other important cereals such as corn (*Zea mays* L.), wheat (*Triticum sativa*) or rice (*Oryza sativa*) is limited by the harsh climates or soil conditions (FAO, 1995). An ancient grain utilized by humans for centuries, and grown under quite variable soil and climatic conditions, great differences have been observed between different sorghum cultivars in their ability to survive adverse soil conditions or respond differently to varying levels of nutrient supply or deficiency (Baligar et al., 1995; Gemenet, et al., 2016; Mahama et al., 2014; Maritim et al., 2013).

While sorghum can utilize relatively large amounts of nutrients under high yielding conditions, similar to maize (*Zea mays* L.), and receives them in the form of manures or fertilizers in the developed countries, the crop does not always receive the nutrients needed, especially in areas of subsistence farming in Africa, Asia or Central America, where the required fertilizers or the money to buy them may not be available to local producers. The social and political issues that underlie these situations and require an

http://dx.doi.org/10.19103/AS.2017.0015.14

integrated systems approach to resolve are beyond the scope of this chapter. Thus, this chapter will focus on the science and technology used in commercial sorghum production in developed countries, with emphasis on the production systems and technologies used by sorghum farmers, the recommendations made to those farmers by both industries and governmental agencies and the underlying science and research on which these recommendations are based.

In countries such as the United States, where both corn and sorghum are grown, but substantially more corn is produced, it is often thought that since the crops appear to grow similarly, the nutrient requirements, both in the amounts needed and how they should be supplied, will be similar, and corn fertilizer recommendations are often used for sorghum. However, differences in area of crop growth, soil conditions and rainfall and temperature regimes and differences in cropping systems such as crop rotations, tillage systems and planting dates can significantly impact how the two crops respond to soil fertility practices and their nutrient or fertilizer needs. Consequently, efforts will be made to point out some of these differences when possible.

2 Quantities of the essential elements utilized by sorghum

There are 17 elements considered essential to all plants (Barker and Pilbeam, 2007). Three of these, carbon (C), oxygen (O) and hydrogen (H), are normally supplied from the atmosphere as gases or through water and metabolic reactions, while the remaining 14 essential mineral elements are classified into various groups. For an element to be considered essential for plant growth it must meet three basic criteria of essentiality proposed by Arnon and Stout (1939): all plants must have access to this nutrient to complete their life cycle; the nutrient must play a definite role in the metabolism of the plant and no other element may fully substitute for it; and all plants must require the element to meet their nutritional requirements.

One common method by which the essential elements are classified is based on the relative amounts of nutrients taken up by a crop. Nitrogen (N), phosphorus (P), potassium (K), calcium (Ca), magnesium (Mg) and sulphur (S) are commonly referred to as the macronutrients, since they are normally utilized in large amounts, >0.1% by weight. Iron (Fe), copper (Cu), manganese (Mn), zinc (Zn), molybdenum (Mo), boron (B), chlorine (Cl) and nickel (Ni) are referred to as micronutrients as they are generally used in small amounts, <0.01%, with the exception of Cl, which in grasses, such as sorghum, can be found in excess of 0.3% (Barker and Pilbeam, 2007; Mengel et al., 2009).

However, depending on the amounts and forms of these nutrients present in the soil prior to fertilization, the addition of any one of these elements as fertilizer may or may not result in increases in sorghum yield. Some of the factors which determine the amounts of these nutrients present and available in a certain field include: how long the area has been cropped, how intensive the cropping system used is, crop rotation utilized, geology of the region, past fertilizer usage; and many other factors. For example, in the most recent addition to the list of elements essential to all crops, nickel was established as an essential element in 1987. However, to date, no deficiencies of nickel have been diagnosed in the field, due to the extremely small amounts of this mineral needed for the crop to complete its life cycle, and adequate quantities present in seeds and soils.

Deficiencies of several of the essential mineral nutrients have been observed in areas of sorghum production. Among them nitrogen and phosphorus deficiencies are the most common, with N deficiency and fertilization being most common at higher yields in commercial production and P deficiency very common in lower yielding subsistence production areas such as West Africa (Gemenet et al., 2016). Other nutrients commonly observed to be deficient in sorghum production include: K, Mg, S, Fe, Zn, Mn and Cl (Grundon et al., 1987; Martens and Westermann, 1991; Mengel et al., 2009; Ruiz Diaz et al., 2011; Warrick et al., 2002; Whitney, 1998). Table 1 provides the amounts of several of the essential elements commonly found in high-yielding sorghum varieties in Kansas.

Another valuable way to classify the essential elements is by the degree to which they are retained or stored in soils for future use. A number of essential elements, including macronutrients such as P and K, are *immobile* in soil and react with the soil to enhance their retention or storage in the root zone. Other nutrients such as nitrate-N and chloride-Cl are highly water soluble, *mobile* in soil and move through the soil with drainage water and are not retained or stored for future use in humid regions. Unfortunately, the classification of elements as mobile or immobile is not clear-cut. Some nutrients such as sulphur and boron are considered slightly mobile or partially immobile in many soils. Other nutrients such as nitrogen are often present in a number of forms in soils and undergo chemical and biological transformations, which will alter both their availability to crops and their relative degree of retention. Thus, it is important to understand if a nutrient is retained in the root zone of soils, and also the retention mechanisms' need to be understood along with the degree to which an individual soil can retain a nutrient. This difference in retention of specific nutrients is extremely important in designing nutrient management programmes.

3 Tools commonly used in developing nutrient management programmes

3.1 Soil testing

The most important, and widely used, tool to help a sorghum grower plan a nutrient management programme is soil testing. Soil testing is the first step most sorghum producers should use to determine the relative availability of nutrients in soil and to determine the soil acidity level. Soil tests (STs) have been developed for use in most areas of the world to determine the fertilizer needs for many nutrients. However, all soils are not created equal, and for a specific test to be helpful it must meet two specific criteria with respect to the specific nutrient, crop and region for which it is being used. First, an ST must provide a good *correlation* between the nutrient level present in the soil as measured by the test, and crop growth and yield; and second, it must be *calibrated* to provide guidance on the specific rate or amount of fertilizer or soil amendment which will be needed at a given test level for a specific crop.

An excellent example of this situation is the use of different P STs in the Great Plains region of the United States based on soil pH. The Bray P1 P test was developed in the 1940s in Illinois to determine the P fertility status and fertilizer needs (Bray and Kurtz, 1945). The soils in Illinois at that time were primarily slightly acidic, and the test was based

Table 1 Total nutrient uptake in the above-ground portions of the sorghum plant at selected environments in Kansas, 2013[1]

	Yield	Nitrogen	Phosphorus	Potassium	Calcium	Magnesium	Sulphur	Zinc	Iron	Manganese	Copper
	Mg ha^{-1}					kg ha^{-1}					
High-yielding rainfed site, silty loam soil											
Grain[2]	9.5	107	23.4	31	1.1	9.6	6.5	0.13	0.26	0.13	0.015
Stover[3]	12.2	67	21.8	324	57.1	20.5	10.7	0.13	1.43	0.95	0.053
Total uptake	–	174	44.2	355	58.2	30.1	17.2	0.26	1.69	1.08	0.068
High-yielding irrigated site, sandy loam soil											
Grain[2]	10.0	130	29.2	34	1.4	12.5	7.5	0.17	0.31	0.13	0.020
Stover	14.4	96	24.3	383	78.9	28.1	12.8	0.21	1.43	0.71	0.099
Total uptake	–	226	53.5	417	80.3	40.6	20.3	0.38	1.74	0.84	0.119
Moderate-yielding supplemental irrigation site, loamy sand-textured soil											
Grain[2]	9.3	103	20.1	23	1.5	9.4	6.1	0.18	0.27	0.11	0.020
Stover	15.1	77	10.4	339	74.6	31.2	11.7	0.49	1.60	0.66	0.076
Total uptake	–	180	30.5	362	76.1	40.6	18.8	0.67	1.87	0.77	0.096

Mengel, Unpublished data.
[1] All nutrient uptake values reported on a dry weight basis.
[2] Grain yields corrected to 150 g/kg moisture content.
[3] Stover yields reported on dry weight basis.

on the use of a dilute acid to bring a portion of the sorbed, plant-available P in the soil into solution for measurement. However, when applied to higher pH soils, the dilute acid is quickly neutralized and little or no P is extracted. So while the Bray test works extremely well and is widely used and often considered the 'Gold Standard' for P soil testing on acid soils in much of North America and some other parts of the world, it fails to provide any useful information on P fertilizer needs for growers whose soil pH is high, above pH 7, and especially on calcareous soils, even in Illinois. In the higher pH region of the Great Plains, an alternative test was developed specifically for soils with pH >7 and containing free calcium carbonate, the Olsen Bicarbonate test (Olsen et al., 1954). The Olsen test is the 'Gold Standard' for sorghum growers in Western Kansas and Eastern Colorado where soil pH is commonly >7.5 and soils may contain as much as 10–20% free calcium carbonate (free lime). Many soil testing labs working in the Great Plains would determine the soil pH of a sample first, and then decide what is the appropriate test to use in that specific situation. In recent years a third soil P test has been developed, the Mehlich 3, which can work across a broad range of soil pH conditions, and is effective on moderately calcareous soils, with free calcium carbonate <10% (Mehlich, 1984). Many soil testing labs adopt this procedure once the appropriate correlation and calibration work is completed in the region. This P testing problem is an excellent example of why growers should use a test which has been developed specifically for their region, or one which has been shown to be well calibrated and correlated for their specific soil and climate situation. For a more detailed coverage of the ST correlation and calibration process, see Dahnke and Olson (1990).

It is equally important to recognize that soil sampling depth is also an important issue when sampling for immobile versus mobile nutrients. When sampling for immobile nutrients such as P and K, and for measuring soil acidity/pH levels, a sampling depth of 15–20 cm is normally recommended. The depth recommended generally reflects the sampling depth used in the supporting correlation and calibration research. The type and amount of tillage used will also impact where nutrients and acidity will accumulate in soils and can have an effect on sampling depth recommended. However, when sampling to estimate the availability of mobile nutrients such as nitrate-N, sulphate-S and chloride-Cl, a deeper sample of 60 cm to 1 m is often recommended, again reflecting the correlation and calibration data used to develop the fertilizer recommendations being made from those samples.

Unfortunately, well-correlated and calibrated STs have not been developed which can provide highly accurate information for all nutrients in all situations. In some cases, this is due either to the relatively low frequency of a deficiency of that nutrient, nickel for example, or to the complex relationships between other soil characteristics which influence the availability of a nutrient in the soil, such as soil pH, organic matter content and redox which can influence Fe and/or Mn availability, or environmental factors which can influence the plant's ability to take up a nutrient which can lead to deficiency such as blossom end rot in tomato. Many government organizations and universities provide guidance to growers on the relative value of individual STs as a sole source of information for making fertilizer recommendations for most crops.

3.2 Plant analysis

An additional tool available to growers to evaluate the nutrient status of their sorghum crop is plant analysis. While soil testing is commonly done before planting and is the primary tool used for making fertilizer recommendations, plant analysis is done in-season on the growing crop and may or may not allow effective corrections on the current growing crop, depending on the nutrient and when the plant samples are taken. Plant analysis is especially valuable as a tool to help manage nutrients which may not have dependable or reliable STs, for identifying low-level 'Hidden Hunger' deficiencies where deficiency symptoms may not be visible, for diagnosing in-season problems or as a quality control tool to ensure that your nutrient management programme is addressing all the factors impacting nutrient availability.

When utilizing plant analysis, one needs to understand the basic relationships between nutrient concentration in specific plant parts at specific growth stages and the correlation between those values and crop growth and yield to make effective management decisions. In most crops, including sorghum, nutrients are accumulated in the leaves and stems during vegetative growth, and transported to the developing seed or grain after flowering. Thus the nutrient levels in the leaves change over time. Nutrient content is normally highest in sorghum leaves as seedlings and tends to decline over time as the plant develops. As the canopy develops, the lower leaves are shaded and lose value to the plant as sites for photosynthesis. These older leaves gradually send nutrients and other valuable compounds out to the growing points for use in producing new leaves or stems, or to fill developing grain.

Due to these changes in nutrient content in many plant parts over time, specific instructions are usually provided as to the optimum growth stage for sampling to assess nutrient content, and the specific plant parts to collect. In grain sorghum production the second or third leaf below the head at heading or flowering is generally used for routine monitoring of the nutritional health of the plant.

Plant analysis is also a valuable tool for diagnosing suspected nutrient problems in the field. For this purpose, samples should be collected as soon as a problem is seen in the field, regardless of growth stage. Comparison plant and soil samples should be taken from both the 'problem' and an adjacent normal appearing area in the field to help with problem identification. As a rule of thumb, whole plant samples should be collected when the plants are <30 cm tall, and the top, fully developed leaves collected when >30 cm tall.

When interpreting plant nutrient content, one generally finds a rapid increase in yield as nutrient concentration increases from very low levels where deficiency symptoms may be present, up to some adequate or optimum level, where yield is maximized. The nutrient content where yield is optimized is generally referred to as the Critical Nutrient Level, or Critical level. At nutrient concentrations above the critical level, there is generally a broad yield plateau or 'Sufficiency Range' at which further increase in nutrient content does not increase yield. This is sometimes referred to as a range of 'Luxury Consumption'. As nutrient content continues to rise, yields may begin to decline due to toxic effects of excess nutrient levels in the leaf. This toxic effect is especially observed with nutrients such as manganese (Mn) and iron (Fe). One important point that growers need to be aware of is that it is very common to observe modest nutrient levels in most crops, at or slightly above the critical level, at high yield levels. Adding additional fertilizers to raise the nutrient concentrations further into the sufficiency range will not likely result in increased yield or a positive return on investment.

Table 2 provides reference sufficiency values for the essential elements most likely to be found deficient in sorghum.

Table 2 Nutrient sufficiency ranges for the essential elements commonly found deficient in grain sorghum[1,2]

Nitrogen ($g\ kg^{-1}$)	Phosphorus ($g\ kg^{-1}$)	Potassium ($g\ kg^{-1}$)	Calcium ($g\ kg^{-1}$)	Magnesium ($g\ kg^{-1}$)	Sulphur ($g\ kg^{-1}$)	Chloride ($g\ kg^{-1}$)
33–40	2.0–3.5	14–30	3.0–6.0	2.0–5.0	2.4–5.0	2.0–5.0
	Boron ($mg\ kg^{-1}$)	Copper ($mg\ kg^{-1}$)	Iron ($mg\ kg^{-1}$)	Manganese ($mg\ kg^{-1}$)	Zinc ($mg\ kg^{-1}$)	
	1.0–10	2.0–7.0	65–150	8.0–190	15–30	

[1] Values shown are for the nutrient content in the second or third leaf below the head, at heading to bloom.
[2] Values are compiled from several published sources, including: Lockman (1972); Mills and Jones (1996); Schwab et al. (2007); Zhang and Raun (2006) and Mengel, unpublished data.

4 Managing soil acidity in sorghum

4.1 Liming

A number of issues in acidic soils can have negative effects on the growth and yield of sorghum. These include the toxicity of soluble aluminium and manganese and deficiency of Ca and Mg (Foy, 1992). Sorghum is grown in many regions in the world, and as a result local varieties have been selected, directly or indirectly, over the centuries for conditions common to that region. Thus while soil acidity is a common issue limiting sorghum growth and yield in many areas where sorghum is grown, specific pH and liming recommendations can vary across regions due to soil conditions and crop rotations and locally used hybrid or cultivar adaptation. In general, pH and liming recommendations used for grain sorghum tend to be made for a pH range of 5.5–7.5, though the specific pH recommended will vary depending on the pH and Ca and Mg availability in the subsoil, Al contents of the soil, soil organic matter, acidity tolerance of the hybrid or cultivars used and the required pH of rotation crops. In Kansas, in the Central Great Plains of the United States, specific pH recommendations vary from 5.5 to 6.5 as a primary function of subsoil pH, cost of liming materials and the potential return on that investment, and the production of legumes in rotation. Soils with high-pH subsoils allow the maintenance of slightly lower surface soil pH since essential nutrients such as calcium and magnesium can be provided as the roots exploit the subsoil. When sorghum is grown in rotation with soybeans, clovers or alfalfa, these legumes require maintenance of higher pH than sorghum, but since pH changes slowly in many soils, maintaining soil pH at higher levels for all crops in the rotation is normally recommended (Whitney, 1998). Espinoza (2004) in Arkansas recommends a pH of 5.5 for sorghum production when grown in rotation with rice, and a minimum pH of 6.0, when grown with other crops such as soybeans or cotton. A recent study in Oklahoma across three locations and two years found that increasing soil pH above a critical value of 5.4 removed toxic effects of soluble Al and provided economic yield and optimum sorghum growth (Butchee et al., 2012). However, studies in Alabama have shown grain sorghum to be very responsive to lime at low soil pH, more so than corn and millet, and recommendations are made to lime soils to pH 6.5 for sorghum production (Adams, 1968; Mask et al., 1988). Thus while specific recommendations may vary, in general maintaining a slightly acidic soil pH is adequate for sorghum production.

4.2 Alternative approaches to dealing with low soil pH

Agricultural limestone may not always be available, or available at economical cost, to allow a producer to make the investment in lime to raise soil pH. A number of alternative acid-neutralizing sources have been used.

Wood ashes have been used as an alternative and are a viable option in many subsistence farming regions where sorghum is grown, or in developed regions where wood products or by-products are burned to produce electricity. The comparative value of wood ash to limestone will depend in part on the type of wood being burned and the degree of combustion. The calcium carbonate equivalent (CCE), a measure used to compare liming materials to pure, finely ground, calcitic limestone, of wood ash will vary, but is commonly measured at 15–40%. Agricultural limestone in the United States is commonly sold at an effective calcium carbonate equivalence of 40–60%. Wood ash also contains significant quantities of essential plant nutrients such as Ca, Mg, K and P, which can also increase their value to subsistence farmers (Griffin, 2006; Kopecky et al., 1995; Risse and Gaskin, 2013).

Another alternative which has been used successfully in the Southern and Central Great Plains of the United States, where ag lime is not readily available and the subsoil can provide adequate quantities of Ca and Mg, is to band apply 10–20 kg P ha^{-1} fertilizer with, or in close proximity to the seed of both sorghum and wheat, to reduce the toxic effects of Al on seedling root growth and allow the plants to root through the surface low-pH zone created with continued shallow or surface applications of N. The P fertilizer reacts with the soluble Al, forming insoluble Al-P complexes, reducing the toxic effects of the Al on root growth (Lamond et al., 2007).

Genetic differences in Al tolerance have also been identified and are being exploited by plant breeders, as a means of minimizing soil acidity effects on sorghum production (Baligar et al., 1995; Bernal and Clark, 1998). While the identification of acid-tolerant cultivars by seed suppliers is being done in other crops such as wheat, this is not currently a common practice in sorghum seed marketing in the United States at this point, in part due to limited demand. However, this technology is of high potential value to farmers in less developed countries where ag limestone and fertilizers may be unavailable or expensive.

5 Nitrogen fertilization and management

In many environments, particularly high-yielding environments, next to water, N deficiency is the most common and significant yield-limiting factor in sorghum production. That is not to say that productive agricultural soils will not supply significant amounts of N to a sorghum crop, because most productive soils do. However, to take advantage of the yield potential of modern sorghum hybrids and production practices, significant amounts of N fertilizer will likely be needed in many production settings.

Long-term sorghum N fertilizer studies can provide some important insights into the role of soils in storing N and providing N for crops over time. One Kansas study was initiated in 1982 and continued through 2008. The first 11 years are described in Matowo et al. (1997), with regular annual updates published, in the Kansas Fertilizer Research Reports, by the Kansas Agricultural Experiment Station and Cooperative Extension Service and available online at the KSU Agronomy Department Website (www.agronomy.k-state.edu/research/research-reports/). This study consisted of two tillage systems, four N rates and three N application systems. The study was continued until 2004, comparing different N fertilizer practices across tilled and no-till management systems, with N rates, 0, 34, 67 and 134 kg N ha^{-1}. The same N rates were applied to each plot every year, though N practices used were changed in 1995 and again in 2000. In 2005, the main study was terminated, and the area was used for a new study focused on the impact of the previous 23 years of sorghum production at three N application rates on the subsequent soil productivity, the effect of the previous N rates on the availability of N supplied by the soil and the responsiveness of sorghum grown on these plots to applied N for the next three years, from 2005 through 2007.

The revised experimental design used consisted of three residual N levels (34, 67 or 134 kg N ha^{-1} applied for 23 years) × three applied N rates (0, 67 and 134 kg N ha^{-1}) in a complete factorial design, with three replications. The residual N treatments consisted of plots which had received 34, 67 or 134 kg N ha^{-1} annually for the previous 23 years. Applied N rates were 0, 67 and 134 kg N ha^{-1} and these were applied as granular urea post emergence (2nd leaf). Nitrate-N levels in the top 60 cm of soil were measured prior to planting. Total above-ground N uptake, in both the vegetation and grain, was measured at harvest. The results from the previously tilled plots, using no-till planting, are summarized in Table 3.

Nitrate-N levels present in the soil were measured across all plots receiving the same residual N treatments in 2005, and for each combination of residual and applied N treatments in 2006 and 2007. The soil nitrate values for 2005 reflect a constant rate of N applied from 1982 through 2004 and reflect both the carry-over of unused fertilizer N applied at that rate from the previous crop and differences in N mineralization from soil organic matter and organic residues present where that rate of N had been applied from 1982 through 2004. The amount of ST N present in the soil observed in the spring of 2005, prior to planting and fertilization, increased as N rate applied during the previous years increased, particularly when the N rate applied exceeded the amount needed by previous crops. This is typical in the Central and Southern Plains, which are characterized by dry winters with low levels of N loss from harvest to subsequent planting. Throughout this region, winter or spring nitrate profile STs are recommended to measure 'carry-over' nitrogen and make appropriate adjustments in N fertilizer recommendations and applications.

Spring nitrate ST levels in 2006 and 2007 reflect both the long-term residual impacts of past N fertilization prior to 2005 on nitrogen cycling through soil organic matter and crop residues, and the carry-over of any unused fertilizer N from the rate of applied N in 2005 and 2006, when the study was redesigned. As one would expect, spring soil nitrate levels are lowest where suboptimal N rates were applied in the past, and in the previous crop year, and highest where the highest N rates were applied previously and in the past two years. Precipitation also has a significant impact on carry-over nitrate levels, both in terms of the total rainfall and when it occurs. In the Kansas study, annual precipitation received during the growing season was 729 mm in 2005, 654 mm in 2006 and 784 mm in 2007. Differences between years in carry-over N can be seen in the annual ST data, with carry-over N significantly lower in years of higher rainfall.

The total above-ground N uptake by the sorghum was influenced by both the levels of N applied in the first 23 years of the study and the direct N applications made to that crop. Highest N uptake by sorghum occurred where the highest N rate was applied, with a positive interaction seen between residual N effects and applied rate on N uptake. Optimum N rate for grain yield was found to be the highest applied rate, 134 kg ha^{-1},

Table 3 Impact of 23 years of nitrogen application rates on subsequent response of grain sorghum to applied N fertilizers, Manhattan, KS, USA

N rate applied 1982–2004 Kg N ha^{-1}	Grain yield 2002–04 Kg N ha^{-1}	Spring NO_3-N Soil test, 2005 Kg N ha^{-1}	N rate applied 2005–07 Kg N ha^{-1}	Spring NO_3-N Soil test, 2006–07 Kg N ha^{-1}	Total N uptake Kg N ha^{-1}	Grain yield 2005–07 MG ha^{-1}	NUE by recovery percent
34	2.72	40	0	22	34	2.19	–
			67	35	82	4.70	71.6%
			134	59	118	5.10	62.7%
67	3.78	45	0	37	42	2.73	–
			67	35	90	4.43	71.6%
			134	43	116	4.56	55.2%
134	4.43	69	0	48	60	3.22	–
			67	47	101	4.20	76.1%
			134	66	124	4.18	47.8%

Mengel, unpublished data.

where 34 kg N ha^{-1} was applied from 1982 to 2004, and at the intermediate applied N rate of 67 kg N ha^{-1} where the higher optimum N rates were applied previously. Though N uptake was increased by the highest N rate and residual N combination, this did not provide any additional yield increase. Throughout the full 26-year life of the study, optimum N rates varied between 67 and 134 kg N ha^{-1} with higher rates required during high-yielding, high-moisture years, and lower responses seen in the drier years (Mawoto et al., 1997). One important point to be considered is that a significant portion of the N required for optimum yield could be supplied by the soil (as measured by the unfertilized check plot N uptake and yield) where there has been a history of adequate N application over time. Moll et al. (1982) found a similar relationship in corn – that is, of the total N taken up by corn, 48–55% was supplied by mineralization of N from the soil.

5.1 The concept of nitrogen use efficiency, NUE

One measure of the effectiveness of N fertilization practices used is Nitrogen Use Efficiency (NUE). NUE can be measured in a number of ways. The components of NUE, as suggested by Moll et al. (1982), include the efficiency of N uptake, that is, how efficiently the plant takes up N from the soil against competing loss mechanisms, and the efficiency of N utilization, that is, how effectively the plant utilizes the N taken up to produce grain. Though both are important, the efficiency of uptake receives more attention by agronomists and the environmental community.

Varvel and Peterson (1990) used the same concepts to calculate the NUE of fertilizer recovery (NUE_{rec}) by difference. They used the following formula to calculate NUE using field experiment data:

$$\% \text{ fertilizer N recovery } (NUE_{rec}) = \frac{(NF)-(NC)}{R} \times 100$$

where NF = total N uptake in plants from fertilized plots,
NC = total N uptake in plants from unfertilized plots,
R = rate of fertilizer rate applied.

Applying the Varvel and Peterson formula for calculating NUE_{rec} to the data from the long-term Kansas study, 71.6–76.1% of the N fertilizer applied at the intermediate 67-kg N rate was recovered in the vegetation and grain of the sorghum, indicating relatively high levels of recovery, but only 47.8–62.7% recovery was found at the 134-kg N rate.

Cassman et al. (2002) take the concept further, to the cropping system levels, and discuss the effect of both soil N supplies and N rates on the overall NUE and NUE_{recv} of a cropping system. They define the NUE_{rec} of a cropping system as the portion of the total N supply available to the crop that is removed by the harvested crop biomass, the N contained in the recycled crop residue, and the N incorporated into soil organic matter and inorganic N pools. They consider the N lost to include any reductions in the soil organic matter and other organic and inorganic pools which supply N to crops over time. Thus they consider using suboptimal N rates which reduce the N balance in these soil organic and inorganic pools, as non-sustainable, and N extractive. These pools supply significant amounts of N to crops, and if reduced, which often occurs as significant shifts or changes in management systems occur, N and other nutrient deficiencies will likely occur over time, unless additional N is supplied in some form. These authors point out that N applications in management systems which reduce the size of the overall soil N pool, such as below optimum N rates, will result, over time, in lower yields, and NUE, as soil biota will compete with the crop for the fertilizer N. Thus applying suboptimal N rates to enhance NUE recovery as a means of reducing the amount of N being released into surface and ground water from crop production by runoff or leaching, or to minimize N_2O losses from incomplete denitrification, is likely not a sustainable practice. Rather efforts to optimize both crop yield and NUE will likely be more effective in the long run.

Raun and Johnson (1999) estimated that NUE_{rec} worldwide is only 33% for cereal grain production. They further estimate that in the developed world cereal NUE is approximately 42% and in the developing world it is 29%. The question we as an agricultural community must address is: What can we do to enhance NUE_{rec}, reduce the losses of reactive N from cropping systems, avoid the contamination of world water supplies and accelerate global warming, while sustaining, and/or the increasing production of important food crops such as sorghum?

In recent years, the concept of a 4-R's system of fertilizer management has been the focus of the fertilizer industry in North America as a means of enhancing fertilizer use efficiency (The Fertilizer Institute, 2017, www.nutrientstewardship.com). The system suggests that a farmer should focus on four basic ideas when developing a fertilizer programme: the Right rate of fertilizer to apply, the Right fertilizer source or combination of fertilizer source and additive to use, the Right place or method of application and the Right time of application. By optimizing these four ideas, costs can be minimized, yields can be enhanced and environmental impacts be minimized.

5.2 Right N rate

A number of different approaches are often used to develop and express the optimum or right N rate to use for cereal crops. Simplistically, the more variation in the soils and climate across the region, the more complex the nitrogen fertilizer recommendations appear to become, as more components of the soils, climate and management system tend to be included in the recommendation equation or formula. Essentially all N recommendation systems consider three common components, though they may not be visible to the end user. These include the amount of N which will be utilized by the sorghum crop, the amount of N which will be supplied by the soil and cropping system, and some estimate of the amount of N which will likely be lost from the system over the course of the growing season, NUE.

A number of sources across the sorghum-growing regions of the world will cite as general guidelines the need to apply 2 kg N per 100 kg grain yield ha^{-1}, or 90–110 kg fertilizer N ha^{-1} under good growing conditions for grain sorghum (Espinoza, 2004; Gangaiah, 2008; Mask et al., 1988; Stickler et al., 1997). These general guidelines are a starting point for developing a nitrogen rate recommendation. Most of the publications which cite these guidelines will go on to modify these recommendations based on soil characteristics, crop rotations and how long the particular field may have been in production.

The online Sorghum Production Guide for Queensland (www.daf.qld.gov.au/plants/field-crops-and-pastures/broadacres-field-crops/sorghum/) provides a formula to calculate the N removed in sorghum grain at harvest (N removed, kg N ha^{-1} =Yield (t ha^{-1}) × protein % × 1.6) and recommends doubling that value to estimate the total N supply needed by the crop. It further states that the N requirement needed can be met by some combination of soil nitrate (estimated by soil testing), soil mineralization (effected by how long a field has been cultivated, soil organic matter content, previous crop and rainfall) and fertilizers.

Although the goal of this formula is of course to provide a more precise recommendation to the farmer, that considers soil components such as residual nitrate and soil organic matter, it has increased the complexity of the system substantially.

A third level of complexity is the mechanistic formulas or equations used in some sorghum-growing states in the United States. Both Nebraska and Kansas use approaches which include direct measurements of residual nitrate-N, soil organic matter (SOM) and a credit or debit for the previous crop grown, and include a framework to consider other potential factors such as animal manure, N content of irrigation water, and in the case of Nebraska, the relative price of sorghum compared to the price of N fertilizers. Descriptions of the formulas and resulting recommendations can be found in Leikam et al. (2003) and Wortman et al. (2013).

The formula Kansas uses to make N recommendations for grain sorghum is as follows:

$$\text{N fertilizer rate, kg N ha}^{-1} = (\text{Available N supply needed}) - (\text{Measureable Soil N supply: SOM + ST nitrate}) - (\text{Previous Crop credit}) - (\text{Other credits})$$

where:
Available N supply needed = crop yield potential (t ha^{-1}) × 28.6;
SOM contribution (kg N ha^{-1}) = SOM content in g kg^{-1} soil × 2.2.
If no SOM test is available default to 44 kg N ha^{-1},
ST NO_3^- contribution = kg nitrate-N ha^{-1} in soil profile.
If no ST is available default to 34 kg N ha^{-1},
Previous crop credits = adjustment to N rate based on the relative contribution of a mineralized N from the previous crop grownprevious crop to the N available from soil using wide C:N ration residue crops as a base of 0.
Corn, sorghum, wheat = 0 credit;
Soybeans = 45 kg N ha^{-1} credit;
Alfalfa = 45–135 kg N ha^{-1} credit depending on alfalfa stand at time of termination;
Fallow = 22 if no profile N test is used, 0 if profile N test is used.
Other credits = reductions in N fertilizer needed to reflect N added as animal manure, in irrigation water, or from other sources such as N content of P fertilizers.

This formula works reasonably well, but is not perfect, primarily because it makes some major assumptions regarding NUE. In Kansas, it is assumed that the NUE_{rec} for all crops grown in all locations is 50%. Unfortunately, this is not always true. Research in Kansas has shown that the NUE_{rec} normally found in sorghum production can vary dramatically from year to year and location to location due to differences in soils, climate, cropping system, N source, inhibitor use and how N is applied. Our research has shown NUE_{rec} to vary, primarily with the intensity, timing and distribution of rainfall, and the methods by which N is managed. NUE_{rec} was found to vary in sorghum from 18 to 80% from 2006 to 2014 (unpublished).

Another critical assumption is that farmers can substitute default values for actual ST results for SOM and profile N tests. Again there is so much variability across fields, cropping systems and climate in most sorghum-growing regions that this is a major problem. The first thing a person can do to improve their N rate recommendations is to collect good ST data for SOM and residual nitrate-N present in most sorghum-growing areas. Even when the best soil testing and yield potential information are provided, N recommendations can only be a basic starting point. The N recommendation is generally made before planting. But sorghum is often grown under dry conditions in hot, dry regions where water, not N, is often the most limiting yield-determining factor. The final N rate one decides on should reflect the total situation: climate, soil, cropping system, N source used, method of application, time of application and other management tools applied. So using the current technology, N recommendations are normally a starting point for enhancing yield. Retaining the option of applying a portion of the N recommended later in the growing season is an excellent way to enhance yield in good years or reduce cost in bad years.

5.3 Right N source

There are a number of N sources available to farmers in many sorghum-growing regions. Granular urea is the most widely used source today, with anhydrous ammonia, UAN (urea-ammonium nitrate) solutions, ammonium sulphate and ammonium nitrate also commonly used. All are valuable products and each has particular advantages and disadvantages that can influence the efficiency of their use. Unfortunately, many of these products are not available to growers in all markets. When choosing a specific product from those available in your location, one should consider product cost, including transportation and storage, safety concerns, application costs and specific factors which could influence NUE.

Urea is normally marketed for sorghum as a dry granular product with an N content of 45–46%. One potential issue with urea-based N fertilizers is the potential for ammonia (NH_3) volatilization loss. When applied to soil, and in the presence of the urease enzyme and water, the urea molecule hydrolyzes and breaks down into ammonia and bicarbonate (Kissel et al., 1988). This process consumes H^+, temporarily raises the soil pH and promotes the volatilization of NH_3 (Fenn and Hosner,1985). If this process occurs on the soil surface, particularly at high pH and when water is evaporating from the soil surface, a portion of the ammonia will move off or volatilize to the atmosphere. When urea is incorporated into the soil by water or tillage, the free ammonia reacts with soil water to form ammonium, and is retained in the soil (Mengel and Wilson, 1988). However when on the surface, which will likely occur in surface applications of urea, losses of 15–30%, or higher, of the applied N through volatilization can occur (Engel et al., 2011; Keller and Mengel, 1986; Lightner et al., 1990).

Urea is very soluble in water, so as little as 10 mm of rainfall or irrigation can move the urea into the soil and stop further volatilization threat. One valuable management tool is to apply irrigation water immediately after urea application to incorporate the N (Holcomb et al., 2011). Another option in dryland production is to use a urease inhibitor with the urea to slow hydrolysis and increase the potential for a rain to incorporate the urea to occur before volatilization losses become critical (Engel et al., 2011; Hendrickson, 1992). A review of the impact of NBPT on corn yield showed that the addition of NBPT to urea surface applied for corn increased corn yield by an average of 267 kg ha^{-1} over 295 comparisons (Hendrickson, 1992).

Anhydrous ammonia is the initial product formed from the Haber–Bosch process of fixing and producing N. Essentially all other synthetic N products are produced from ammonia, so additional cost is required in producing urea, ammonium nitrate or UAN from ammonia.

Ammonia is a pungent and toxic gas under normal temperatures and atmospheric pressure. However, when cooled and kept in a pressurized tank, the gas becomes a liquid, which is how ammonia is normally handled and applied. As applied, anhydrous ammonia has an N content of 82%, making it the highest analysis N fertilizer product available. The main advantage of using ammonia is its low cost. The high cost of infrastructure for the transport, distribution and application of this product limits its use to developed areas of the sorghum-producing regions. Ammonia has an extremely high affinity for water. When placed in the soil, ammonia quickly reacts with water to form ammonium hydroxide. If the soils are dry, and adequate water is not present at application, the ammonia can volatilize or escape from the soil as a gas, resulting in N loss and damage to growing crops. Wet soils which prevent sealing of the application slot can also result in direct volatile loss of ammonia and crop injury. Losses as high as 25% have been documented under both wet and dry soil conditions. However, under good field conditions direct leakage loss at application is generally in the 1–2% range (Blue and Eno, 1954; Stamper, 2009).

Ammonia is an especially useful N source in poorly drained soils prone to denitrification. The nature of the material requires knifing or banding into soils, which has certain benefits. The toxic properties of the ammonia reduce microbial populations in the application zones, inhibiting the conversion of ammonia/ammonium to nitrate. This is unique to this product and can lower the potential loss of the ammonia from denitrification for 3–4 weeks. When the nitrification inhibitor nitrapyrin is added to the ammonia, this inhibition of nitrification can be extended to as much as 40–60 days for spring ammonia applications and 150 days for fall applications (Touchton et al., 1978).

A major concern with regard to ammonia use is safety. The high affinity of ammonia for water will cause it to react with living tissue, leading to severe chemical burns. Burns to the skin, eyes and lungs can occur. A number of people are affected in the United States each year from accidental release of ammonia during handling, transport and application. Improvements in application equipment have reduced this potential danger, but safety is still a major reason why farmers and fertilizer dealers choose other N sources today.

UAN solutions are produced using a mixture of urea and ammonium nitrate. These two products react and in combination form a very high N-content solution. Depending on temperature, products of 28, 30 of 32% N are commonly produced. All UAN products will 'salt out' at low temperatures, but the higher concentrations salt out at higher temperatures. UAN is a convenient product to handle and use. It is commonly used as a carrier for herbicides to allow combining fertilization and weed control trips. However, UAN is the most expensive product to manufacture, so normally the most expensive one to use.

One key issue which can reduce NUE with UAN solutions is the process of immobilization. Immobilization is the process of soil organisms utilizing N fertilizers to facilitate the decomposition of high carbon: low nitrogen crop residues. This problem can occur with all N fertilizers including ammonia, if the right circumstances are present. However, it is particularly problematic with UAN in no-till systems where UAN is sprayed onto surface residues such as wheat straw, corn stalks or wide C:N ratio cover crops. This uniform application, especially desirable when applied in combination with herbicides, creates conditions for immobilization. Alternative placement techniques such as coulter banding or surface banding can help overcome this loss.

An excellent review of nitrogen additives and specialty N fertilizers such as coated ureas can be found in Franzen (2017).

5.4 Right N placement

How any N source is applied in relation to the soil surface and crop residues is also critical to determining NUE. Placement and source interact so that each common source has specific reactions which can occur with the atmosphere and crop residue and impact the utilization efficiency of the fertilizer with different placement techniques.

Placing N sources such as urea or UAN solutions into the soil, rather than on the soil surface, will limit potential N losses from mechanisms such as immobilization or ammonia volatilization (Mengel et al., 1982). Developing methods to move N fertilizers into the soil can also reduce runoff losses of N. Subsurface banding of N fertilizers is an extremely effective tool for reducing N losses and enhancing NUE. How deep the fertilizer is placed in the soil can be important when using N products such as anhydrous ammonia to ensure sealing, but is not critical with urea or UAN to reduce volatilization or immobilization. Placing UAN 25–50 mm below the soil surface and crop residue can greatly enhance efficiency and N uptake.

Incorporation with tillage or water can also be effective in incorporating urea or UAN into the soil and reducing loss potential. A shallow tillage pass or an application of 10–20 mm of irrigation water can move the N off the surface and minimize immobilization or volatilization. One problem with incorporation using tillage is that there are chances that surface residue may also be incorporated, thus increasing the potential for immobilization. Surface banding of UAN, the process of concentrating the N solution in a narrow band 35–70 cm apart, is an alternative to subsurface or coulter banding. But, like many alternatives, this probably will not be as efficient, and equally as important, the performance will be more variable. Surface banding is most effective in reducing immobilization, and limiting contact of fertilizer with residue. The higher concentration in the band increases the N rate in the band to much more than the organisms responsible for decomposition can utilize, allowing a greater percentage of the applied N to move into the soil and be available to the crop. Some volatilization will likely still occur, but the overall efficiency will likely be increased over a surface broadcast application in high-residue systems.

5.5 Right time of application

Application timing is an especially valuable tool to fight N loss from denitrification and/or N leaching. The closer the N can be applied to the time it will be utilized by the sorghum crop (or any crop), the less opportunity for N loss from mechanisms such as leaching or denitrification to occur. Figure 1 adapted from Vanderlip (1993) shows that sorghum takes up only about 20% of the total N accumulated in the first 30 days after emergence, but nearly half of the total N uptake during the second month after emergence. The maximum uptake rates, indicated by the slope of the accumulation line, are found prior to heading. By bloom, as much as 70% of the total N uptake will be in the plant. Vanderlip also points out the importance of the period from growth stage 2 (five leaves) to growth stage 3 (7–10 leaves) for plant growth. Nutrient stress during this period can have significant negative effects on yield by limiting the number of seed-bearing florets formed. Consequently, it is important that adequate N be available to support this key growth stage. Some combination of residual nitrate-N in the soil, starter fertilizer N applied with the planter or a pre-plant or pre-emerge broadcast application of N fertilizer totaling 40–50 kg N ha^{-1} will assure adequate N to support the development of the head. In sandy soils prone to leaching, or heavy clay or silty clay loam soils prone to denitrification, losses of preplant N can limit the N supply later in the growing season. Retaining a portion of the N recommendation for a sidedress application at growth stage 3 will support the high rates of N uptake during the grand growth period from stage 3 through heading and bloom. If the soil

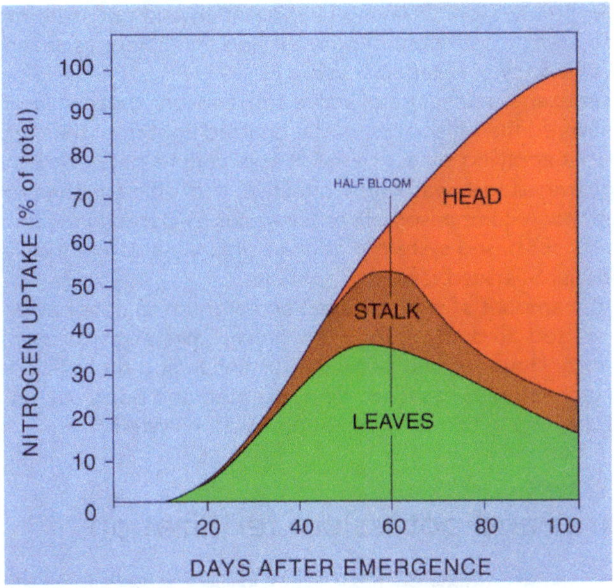

Figure 1 Cumulative N uptake by sorghum over the growing season. From Vanderlip (1993).

and climate, or irrigation practices, lead to a high probability of N loss, retaining a portion of the total N for a third application at bloom could also prove advantageous.

If the probability for N loss due to leaching or denitrification is low, such as on a deep, well-drained silt loam soil with low rainfall, then the return to delayed or multiple applications may also be low. However, also consider that a delayed or split application N plan can be a good way to adjust N rate for changes in yield potential. If rainfall is plentiful and yield potential is high, a delayed application plan will allow increasing the N rate to capture more of the increased yield potential. However, if rainfall is below normal and yield potential has decreased, deferring a sidedress N application, or reducing the planned rate of application may be appropriate.

The key places to get the greatest return from application timing are poorly drained soils in higher rainfall climates, prone to denitrification loss, or sandy, excessively well-drained soils in the same climates or where irrigation is used, which are prone to leaching loss. Limiting the exposure of soil and fertilizer nitrate N to loss conditions by adjusting when N is applied can pay big dividends by increasing yield and/or reducing N costs.

5.6 Sensor-based N applications

The use of crop sensors and remote sensing is currently being introduced in many areas with crops such as corn and wheat, to make later N applications based on the current status of the crop. There is considerable interest in these technologies as a way to enhance NUE, and to minimize pollution caused to both ground and surface waters by unused residual nitrogen fertilizers. With sorghum being grown in drier regions, yields can be variable from season to season. Delaying N application and rate determination until later growth stages and using the amount of biomass and leaf colour as an indicator of N need can enhance the efficiency of N fertilizer use.

Studies in Kansas examined the use of active crop sensors, those utilizing an internal light source, which measured the reflectance of the emitted light from the crop canopy (Tucker, 2009) and found that applying sensor-based N applications between 28 and 42 days after planting (GS-3 growth stage) was more effective and efficient than standard preplant ST-based applications. A comparison of sensor versus ST-based N recommendations at 10 sites showed the sensor-based system produced similar yields while calling for 30 kg ha^{-1} less N than traditional ST-based recommendations.

Unfortunately, the amount of work focused on sorghum to date has been minimal, due to the relatively limited amount of sorghum grown in relation to corn or wheat in the developed countries. However, the potential for enhancing the efficiency of N use with different sensing technologies has been demonstrated and hopefully will be examined as the technologies develop in corn and wheat production systems.

6 Phosphorus and potassium fertilization

Unlike N which is, when applied, or likely will become, a mobile nutrient in soils, both phosphorus (P) and potassium (K) are relatively immobile. Once placed in the soil they are unlikely to be lost in significant quantities. Thus the management of P and K is often quite different from N.

STs are the tool of choice for measuring the available P or K status in soils to make fertilizer recommendations. As discussed earlier, STs are designed and developed for specific soil situations and must be correlated and calibrated with a specific crop to be useful. Thus, different STs, or extractant solutions, are often used in different sorghum-growing regions.

In the primary sorghum-growing regions of the world, several different P STs are used. The underlying philosophies of making fertilizer recommendations and the resulting interpretations of the tests and fertilizer recommendations can also be quite different. In the United States, the Mehlich 3 ST is used for determining available P content in both Arkansas and Kansas (Espinoza et al., 2007; Leikam et al., 2003). In Arkansas, the P content of the Mehlich 3 extract from the soil is determined by ICP, integrated coupled plasma spectroscopy. In Kansas, the P content of the extract is determined colorimetrically. As a consequence of the method used to determine the P content of the extract, the values reported by the Arkansas Soil Testing Lab will be higher.

In Arkansas the fertilization philosophy used is one of Build and Maintain. The goal of this philosophy of fertilization is to build the ST value on low testing soils to some critical value, above which there is little or no chance of seeing a yield response to adding more fertilizer. This will allow the farmer to invest in his soil to gain future flexibility on when he has to fertilize. This flexibility can pay significant dividends in years when weather reduces the ability to make timely field operations and fertilizer can't be applied without seriously delaying planting and reducing yield, or when fertilizer price spikes or sorghum price drops. At ST levels below the critical level, <36 ppm in AR, the low to medium ranges, following the Arkansas fertilizer recommendations for an eight-year period will build the ST levels to the medium range, 26–35 ppm, the upper end of which is considered the critical level where no response to fertilizer would be expected. This Build recommendation will be for a large amount of fertilizer, and regular soil testing will be suggested to minimize excess fertilizer application. Once above the critical level, >36 ppm but <50 ppm, the optimum range, fertilizer recommendations would be equivalent to the amount of P that a sorghum crop of an expected yield potential would remove. This is the desired Maintain range of the Build and Maintain fertilization system. Above 50 ppm is considered above optimum and no fertilizer is recommended.

The advantage of the Build and Maintain system to a farmer is that over time he will reach a ST level where the soil will be able to supply the P needed, and fertilizer application will be optional for any given year. So, in years of anticipated low yields due to weather, or low commodity prices, he can defer P or K fertilizer application until his financial situation improves.

In Kansas, and in most of the Central and Northern Great Plains where the climate is hot and dry and weather more erratic, traditionally a different fertilization philosophy is followed, called the Nutrient Sufficiency system. This system only recommends the amount of fertilizer which will be needed at a given ST level, to meet the needs of the crop for that year. In this system annual fertilizer rates are considerably lower, but fertilizer will likely be required every year in perpetuity to avoid P deficiency, and STs will generally equilibrate at some level below the critical level. The advantage of this system is that annual fertilizer costs will be considerably lower on fields with low ST levels. Farmers who rent land on a cash rent basis, particularly short-term leases, common throughout the United States, will not be investing in improving land they do not own. But they will have to fertilize every year, regardless of weather, time demands, commodity prices and current cash flow.

In 2003, Kansas State University changed from only offering their traditional Nutrient Sufficiency fertilizer recommendation to providing farmers a choice of which system they wished to use. The logic for the change was that many farmers owned a portion of (or all) the land they farmed, it was very productive and could have irrigation, and they were in a financial condition where they wanted to invest in their farm to obtain the flexibility of using the Build and Maintain recommendation system. In some cases, this gave the farm business more available time to add additional acres, or it allowed them to better manage cash flow or better utilize animal waste products from livestock enterprises as fertilizer sources. Also, private ST labs were already providing Build and Maintain recommendations to these farmers. The university also felt that the appropriate person to decide which system should be used was the farmer, not an agronomist at a university. To illustrate the difference in the recommendation between the two systems, Table 4, adapted from Leikam et al. (2003), is used as an example.

Do these systems of making fertilizer recommendations work everywhere? Unfortunately, no. In the case of P, P fixation, or high levels of P adsorption in soils, can make the practice of Build and Maintain recommendations impractical in many places. In most areas of the United States, a broadcast application of 6–10 kg P ha^{-1} above the amount of P removed by the crop, will raise the P ST 1ppm. So, increasing the ST P level 10 or 15 ppm will require an investment of approximately 7.5–19 kg P $year^{-1}$ above crop removal for eight years to accomplish that task, which is expensive, but manageable. However, to do that on soils where it may take 5–10 times as much P to make that change, such as oxisols, ultisols or andepts in South America and Africa, especially for subsistence farmers, the cost may make it impossible (Sanchez and Uehara, 1980). However, systematic addition of locally available, finely ground phosphate rock and lime, together with using local cultivars more tolerant to acid soils and lower P levels, may be a more affordable option.

Table 4 Grain sorghum P and K recommendations for Kansas, showing both nutrient sufficiency and Build and Maintain recommendations

Mehlich 3	Potential sorghum yield			Ammonium acetate	Potential sorghum yield		
P test	5 t	7.5 t	10 t	Exchangeable K test	5 t	7.5 t	10 t
ppm		kg P ha^{-1}		ppm		kg K ha^{-1}	
Phosphorus sufficiency recommendations				***Potassium sufficiency recommendations***			
0–5 V low	27	29	32	0–40 V low	74	79	83
5–10 V low	20	22	22	40–80 V low	46	51	55
10–15 low	12	14	15	80–120 low	19	23	23
15–20 medium	7	7	7	120–130 medium	14	14	14
20+ high	0	0	0	130+ high	0	0	0
Crop removal	16	23	31	Crop removal	19	29	39
Phosphorus 8-year Build and Maintain recommendations				***Potassium 8-year Build and Maintain recommendations***			
0–5 V low	29	43	57	0–40 V low	94	142	188
5–10 V low	25	37	49	40–80 V low	68	102	136
10–15 low	21	32	43	80–120 low	37	55	73
15–20 medium	17	26	35	120–130 medium	18	27	36
20–30 optimum	16	23	31	130–160 optimum	19	29	39
>30 V high	0	0	0	160+ very high	0	0	0

Similar situations can also occur with K. Potassium fixation occurs in several areas of the world, including in the Corn Belt and Plains regions of the United States. As with P, large amounts of K are required to change low STs on these soils, making a Build and Maintain recommendation system expensive and inefficient for farmers. In these practices, nutrient sufficiency systems which utilize P and K placement techniques are primarily used.

6.1 Starter fertilizer

One commonly used tool at low ST levels to enhance P and K availability is placing a small quantity of fertilizer near the seed to enhance availability of the nutrient to the limited root system of seedling sorghum plants. By concentrating a relatively small quantity of fertilizer with limited volumes of soil, the concentration of P in the soil solution around this fertilizer will be higher, enhancing the nutrient uptake process. Generally most organizations making fertilizer recommendations will recommend that at very low or low ST levels for both P and K, 25–50% of the applied P and/or K should be banded close to the seed (Liekam et al., 2003). Care must be taken not to injure the seedlings through salt or ammonia injection when doing this. Having a minimum of 5 cm of soil between the fertilizer and seed will reduce the potential for seedling injury. When placing the fertilizer with or close to the seed one should minimize the quantity of salt, normally expressed as N + K in the fertilizer. With sorghum and corn in 76-cm rows, a maximum rate of 6–10 kg N + K ha^{-1} close to the seed is generally considered the maximum allowable rate.

6.2 Time of application

Unlike N, which is mobile and is often applied later in the growing season, the time of application for P and K is normally not critical. Unless working with a soil with high P or K fixation, preplant or planting time applications are normally considered optimum. In soils where the ST P levels are at or above the critical level, multiyear applications of P to maintain ST levels can be made.

7 Sulphur

Sulphur deficiency is commonly observed in many older, highly weathered, acid soils where sorghum is grown. Recently, S deficiencies have been noted in many areas in the developed countries as air pollution control measures and limits on the S content of P fertilizers has reduced the deposition of S to soils and the incidental addition of S through P fertilizers (Jeske, Diedrick and Cloyer, 2017).

Sulphur is taken up by plants as sulphate (SO_4^{2-}). Sorghum will commonly take up 15–20 kg S as sulphate per ha (Table 1). Sulphate is relatively mobile in soils and will leach down in the soil with drainage water and accumulate in zones with reduced water conductivity. Consequently, traditional surface (15–20 cm) soil samples will often underestimate the sulphur supply in many soils and profile or deep (60–90 cm) soil samples are commonly recommended to access S availability. Sulphur is also found in crop residues and SOM, and becomes available to crops through mineralization of organic S. Most organizations providing S recommendations to farmers will consider a S profile or deep ST and surface SOM content when determining sulphur needs (Lamond, 1997).

Since S is taken up by plants, including sorghum as sulphate (SO_4^{2-}), sources such as ammonium sulphate ($(NH_4)_2SO_4$), calcium sulphate or gypsum ($CaSO_4$) or liquid thio sulphate sources are commonly recommended. Elemental S can also be applied prior to planting if adequate time is allowed for the S to be oxidized to SO_4^{2-} for utilization.

8 Micronutrients

The micronutrients most commonly applied to sorghum are iron (Fe), zinc (Zn), manganese (Mn) and chloride. As discussed earlier, deficiencies of several others have been reported on sorghum, but the incidence of deficiency is limited.

8.1 Iron deficiency

Iron availability to plants is controlled by a number of soil factors including pH, free CaCO3 content, SOM content and redox potential. Reviews by Moraghan and Mascagni (1991) and Lindsay (1991) provide detailed discussions of these factors. Iron deficiency of sorghum occurs in many areas of the world, most commonly, on high-pH, calcareous soils (Martens and Westerman, 1991; Welch et al., 1991). Usually, Fe deficiency occurs not due to a shortage of Fe in the soil, but rather because of factors limiting its availability and uptake by plants, or limiting its use in metabolic processes in the plant (National Research Council, 1979 as cited by Welch et al. (1991)).

The addition of iron fertilizers such as ferrous sulphate at high rates to soils, or by adding acidifying materials by broadcasting elemental sulphur or sulphuric acid, has been shown to be expensive and relatively ineffective. Foliar applications of $FeSO_4$ or Fe chelates such as FeEDDHA are commonly recommended for correcting Fe deficiencies. Multiple applications may be required. The application of animal manure, or applying Fe in a band with acid forming N or N-P fertilizers, together with chlorosis-resistant varieties or hybrids can also be effective in overcoming Fe deficiency.

8.2 Zinc deficiency

Sorghum is one of several crops sensitive to zinc deficiency (Martens and Westermann, 1991). Zinc deficiencies are common on sandy or low-organic matter soils formed in low-zinc parent material (Krauskopf, 1972), highly weathered tropical soils, high-pH calcareous soils and organic soils. Over liming, high rates of P and removal of surface soil by erosion, land levelling or terracing are all management practices that can enhance or create zinc deficiency.

Soil testing is a good tool to predict areas of potential zinc deficiency in the field. Several zinc STs such as the DTPA or dilute acid tests are used and provide good predictions of zinc needs (Westfall and Bauder, 2011).

Zinc deficiencies can be corrected in several ways. Soil applications of Zn either as broadcast or band placed are commonly used for zinc fertilization to correct deficiencies. Broadcast applications of $ZnSO_4$ or some other soluble inorganic source at relatively high rates can be made to increase ST levels and provide a correction of the problem for several years. Higher rates are often needed in situations where soil properties such as high pH, low organic matter or calcareous conditions reduce Zn availability. The application of Zn in animal manures is also effective in correcting Zn deficiencies (Sutradhar et al., 2016). Band applications of organic Zn sources such as Zn EDTA, Zn lignosulphanate, Zn acetate, Zn citrate or Zn-ammonia complexes in combination with an acid forming N or N-P starter fertilizer are also effective. Much lower rates of Zn are normally applied in starter fertilizers, and will have little effect on subsequent Zn ST levels from single applications. Consequently, these applications will need to be made annually on zinc-sensitive crops.

Zinc can also be applied foliar for in-season correction of deficiencies of zinc in sorghum and other crops. $ZnSO_4$ or Zn chelates such as Zn EDTA can be applied to plants directly as a rescue application. Multiple applications may be needed.

8.3 Manganese deficiency

Sorghum is considered highly sensitive to manganese deficiency (Martens and Westermann, 1991). Manganese deficiency is commonly seen in high-organic matter soils, high-pH/calcareous soils and slightly acid, poorly drained sandy soils. Soil pH is the soil property which most significantly impacts Mn availability; as soil pH increases, Mn availability decreases. As a result, care must be taken when liming soils to avoid creating Mn deficiency (Gilbert et al., 1926; Blair and Prince, 1936; Sanchez and Kamprath, 1959). Manganese availability is also strongly impacted by changes in oxidation state. Mn^{2+} is the primary form taken up by plants and the soluble form commonly found in soil solutions. Other more oxidized forms of Mn are much less soluble and precipitate in various oxide and hydroxide forms. Manganese availability can also be enhanced by natural chelates formed in the soil through microbial processes.

Common methods used to correct Mn deficiency include banding Mn as $MnSO_4$ or a Mn chelate or complex with or without an acid forming N or N-P fertilizer near the seed, banding an acid-forming starter fertilizer alone without added Mn or foliar application of $MnSO_4$ or Mn chelate. Broadcast applications of Mn to soil are relatively ineffective and must be made at extremely high, uneconomical rates. Banding rates of 0.5–5 kg $MnSO_4$ ha^{-1} (Mascagni and Cox, 1985; Randall et al., 1975) have been effective in correcting Mn deficiencies. The use of similar rates of Mn in combination with an acid-forming N or N-P fertilizer, or banding the acid-forming fertilizer alone near the seed, has also been shown to be effective (Eck, 1995; Petrie and Jackson, 1984; Steckel, 1946).

Foliar application of Mn is also commonly used to correct Mn deficiency. The number of foliar applications of Mn may vary, depending on the severity of the deficiency.

8.4 Chloride deficiency

Chloride was established relatively recently as an essential element for plant growth (Boyer et al., 1954). In the 1980s the responsiveness of some crops to this nutrient became more widely established (Fixen, 1993). The wide use of KCl as a potassium fertilizer source in many crops satisfied the need for Cl in many areas. Chlorine is taken up by plants as the Cl^{1-} ion (Hockman, 2007).

Standard STs have been developed for chlorine and are commonly used in areas such as the Great Plains of the United States (Gelderman et al., 1998). As a mobile element in soil, a deep or profile sample is used to assess the Cl supply. Sampling depth is generally recommended to be from 60 cm to 1 m. Interpretation of the test generally falls into three classes: low or deficient, <4 mg kg^{-1}; medium or marginal, 4–6 mg kg^{-1} and high or adequate, >6 mg kg^{-1} of soil (Lamond, 2002).

A series of chloride studies were conducted in Kansas from 1996 through 2006 with grain sorghum. Twenty of the sites with common application rates of 0, 22 and 44 kg Cl as KCl or NH_4Cl broadcast applied at or before planting were examined in a combined analysis. Nineteen of the sites showed a significant response to chloride fertilization, with 22 kg Cl increasing yields 608 kg ha^{-1}, and no additional increase in response to the higher 44 kg ha^{-1} rate (Mengel et al., 2009). No difference in response was seen in these studies between $CaCl_2$, $MgCl_2$, KCl and NH_4Cl fertilizer sources.

9 Conclusion

The tools and procedures for developing a nutrient management plan for commercial grain sorghum production have been well defined in the developed countries for some time. The following are some of the tools and practices a grower, regardless of where they may be producing sorghum, should consider today:

First, realistically assess the soil and climate to determine the sorghum yield potential of their field. Key factors to consider are soil texture, internal drainage and water-holding capacity of the soil, and quantity and distribution of rainfall during the growing season. Yield potential may be limited for example, on a sandy soil with limited water-holding capacity in regions where rainfall comes in scattered heavy storms. Poorly drained, heavy-textured soils may also create yield-limiting conditions due to saturation and low oxygen in the root zone. Nitrogen management may also require special considerations to limit leaching losses in these types of environments. However on deep, well-drained silty loam-textured soils which can retain as much as 5 cm of available water per 30 cm of soil, yield potential in the same environment would be much higher and N loss potential would be much lower.

Second, ST the field prior to planting to determine the available nutrient levels, soil acidity and basic fertilizer needs. Basic tests would include determination of soil pH and lime requirement, and available P and K. Other tests which could be helpful depending on the region would include available aluminium, exchangeable magnesium, sulphur, zinc, iron, manganese or chloride plus SOM. Local extension or agricultural agency staff or industry agronomists can provide assistance in deciding which tests may be helpful in a specific area.

Third, develop specific fertilizer recommendations for your field. Most fertilizer recommendations are yield potential driven. As yield goal or potential goes up the recommended fertilizer dose also goes up. So using a realistic yield goal based on soil, climate and the grower's experience is important. It is also important to decide which type of fertilization system one wishes to use – a nutrient sufficiency system, which focuses on short-term goals and is best suited for many of the harsh environments where sorghum is commonly grown, or a Build and Maintain system which builds STs to levels where fertilizers are not needed for immobile nutrients like P and K in any given year. This type of system is best suited to high-value crops or large farms where the flexibility of only having to fertilize a portion of the farm each year is valuable.

Fourth, develop an N application plan. Key will be the N rate needed, primarily a function of yield potential, and time of application, a function of yield stability or variability across years and N loss potential from mechanisms such as leaching or denitrification. For many sorghum growers a N application system which uses multiple N applications, some at or before planting and some later, allows adjusting N rates and expenses, to the impacts of weather on yield and N loss.

Fifth, use plant analysis as a quality control tool to monitor the effectiveness of your fertilization/nutrient management programme. Are there nutrients limiting yield? Are nutrient levels very high, indicating over fertilization. Plant analysis can provide insights to help enhance yield and/or efficiency.

Following these basic five steps utilizing currently available methods and research and time-proven technology can lead sorghum to more efficient sorghum production.

10 References

Adams, F. 1968. *Relative Response of Grain and Annual Forage Crops to Lime Phosphorus and Potassium on Norfolk Sandy Loam*. Circular 161. Auburn University Agricultural Experiment Station, Auburn, Alabama. https://aurora.auburn.edu/bitstream/handle/1120/1959/1159CIRC.pdf;sequence-1.

Arnon, D. J. and P. R. Stout. 1939. The essentiality of certain elements in minute quantities for plants with special reference to copper. *Plant Physiol.* 14: 371–5.

Baligar, V. C., I. Anghimoni, G. V. E. Pitta, H. L. dos Santos, E. Cunha Filho and R. E. Schaffert. 1995. Aluminum effects on plant and aluminum uptake parameters of soil and solution grown sorghum genotypes. *J. Plant Nutr.* 18(11): 2325–38.

Barker, A. V. and D. J. Pilbeam 2007. *Handbook of Plant Nutrition*. Taylor & Francis, Boca Raton, FL.

Bernal, J. H. and R. B. Clark. 1998. Growth traits among sorghum genotypes in response to aluminum. *J. Plant Nutr.* 21(3): 297–303.

Blair, S. and A. L. Prince. 1936. Manganese in New Jersey soils. *Soil Sci.* 42: 327–33.

Blue, W. G. and C. F. Eno. 1954. Distribution and retention of anhydrous ammonia in sandy soils. *Soil Sci. Soc. Am. Proc.* 18: 420–34.

Boyer, T. C., A. B. Carlton, C. M. Johnson and P. R. Stout. 1954. Chlorine- a micronutrient element for higher plants. *Plant Physiol.* 29: 526–32.

Bray, R. H. and L. T. Kurtz. 1945. Determination of total, organic and available forms of phosphorus in soil. *Soil Soc.* 59: 39–45.

Butchee, K., D. B.Arnall, A. Sutradhar, C. Godsey, H. Zhang and C. Penn. 2012. Determining the critical soil pH for grain sorghum production. *Int. J. Agron.* 2012, Article ID 130254, 6. doi:10.1155/2012/130254.

Cassman, K. G., A. Dobermann and D. T. Walters. 2002. Agroecosystems, nitrogen use efficiency, and nitrogen management. *Ambio* 31(2): 132–40.

Dahnke, W. C. and R. A. Olson. 1990. Soil test correlation, calibration and interpretation, pp. 45–72. In: R. L. Westerman (Ed.), *Soil Testing and Plant Analysis*, 3rd Edition. Soil Science Society of America, Madison, WS.

Eck, K. J. 1995. *Diagnosing and Correcting Manganese Deficiencies in Soybeans in Indiana*. M.S. Thesis, Purdue University, West Lafayette, IN.

Engel, R., C. Jones and R. Wallander. 2011. Ammonia volatilization from urea and mitigation by NBPT following surface application to cold soils. *Soil Sci. Soc. Am. J.* 75(6): 2348–57.

Espinoza, L. 2004. Soil fertility and liming, pp. 21–4. In: *Grain Sorghum Production Handbook*. Bulletin MP 297. University of Arkansas, Division of Agriculture and Cooperative Extension Service. www.uaex.edu/publications/pdf/mp297.pdf.

Espinoza, L., N. Slaton and M. Morzaffari. 2007. *Understanding the Numbers on your Soil Test Report*. Bulletin FSA2118. University of Arkansas Division of Agriculture and Cooperative Extension Service. www.uaex.edu/publications/PDF/FSA-2118.pdf.

FAO. 1995. *Sorghum and Millets in Human Nutrition. FAO Food and Nutrition Series No. 27*. Food and Agricultural Organization of the United Nations, Rome. www.fao.org/docrep/TO0818e/T0818E00.htm

Fenn, L. B and L. R. Hossner. 1985. Ammonia volatilization from ammonium or ammonium forming nitrogen fertilizers. *Adv. Soil. Sci.* 1: 123–69.

Fertilizer Institute. 2017. http://www.nutrientstewardship.com.

Fixen, P. E. 1993. Crop responses to chloride. *Adv. Agron.* 50: 107–50.

Foy, C. D. 1992. Soil chemical factors limiting plant root growth. *Adv. Soil Sci.* 19: 97–149.

Franzen, D. W. 2017. *Nitrogen Extenders and Additives for Field Crops*. Bulletin SF1581 (revised). NDSU Extension Service, North Dakota State University. www.ag.ndsu.edu/publications/crops/nitrogen-extenders-and-additives-for-field-crops.

Gelderman, R. H., J. L. Denning and R. J. Goos. 1998. Chlorides. In: M. V. Nathan and R. Gelderman (Eds), *Recommended Soil Test Procedures in the North Central Region*, online version. Bulletin 221, Revised. msue.anr.msu.edu/uploads/234/68557/Rec_Chem_Soil_proce55c.pdf.

Gemenet, D. C., W. L. Leiser, F. Beggi, L. H. Herrmann, V. Vadez, H. F. W. Rattunde, E. Weltzien, C. T. Hash, A. Buerkert and B. I. Haussmann. 2016. Overcoming phosphorus deficiency in West African pearl millet and sorghum production systems: Promising options for crop improvement. *Front. Plant Sci.* 7: 1389. doi:10.3389/fpls.2016.01389.

Gilbert, B. E., E. T. McLean and L. J. Hardin. 1926. The relation of manganese and iron to lime-induced chlorosis. *Soil Sci.* 22: 437–46.

Griffin, T. S. 2006. *Using Wood Ash on your Farm.* Bulletin #2279. University of Maine Cooperative Extension. https://extension.umaine.edu/publications/2279e/.

Grundon, N. J., D. G. Edwards, P. N. Takkar, C. J. Asher and R. B. Clark. 1987. *Nutritional Disorders of Grain Sorghum.* Australian Centre for International Agricultural Research, Canberra. https://espace.library.uq.edu.au/data/UQ_520329/Nutritional_Disorders_of_Grain_Sorghum.pdf.

Heckman, J. R.2007. Chlorine, pp. 279–91. In: A. V. Barker and D. J. Pilbeam (Eds), *Handbook of Plant Nutrition.* Taylor & Francis, Boca Raton, FL.

Hendrickson, L. L. 1992. Corn yield response to the urease inhibitor NBPT: Five-year summary. *J. Prod. Agric.* 5: 131–7.

Holcomb, J. C., D. A. Sullivan, D. A. Horneck and G. H. Clough. 2011. Effect of irrigation rate on ammonia volatilization. *Soil Sci. Soc. Am. J.* 75(6): 2341–7.

Jeschke, M. and K. Diedrick. *Sulfur Fertility for Crop Production. DuPont-Pioneer On-line Crop Insights.* https://www.pioneer.com/home/site/us/agronomy/sulfur-fertility-crop-production/.

Keller, G. D. and D. B. Mengel. 1986. Ammonia volatilization from nitrogen fertilizers surface applied to no-till corn. *Soil Sci. Soc. Am. J.* 50: 1060–3.

Kissell, D. E., M. L. Cabrera, and R. B. Ferguson. 1988. Reactions of ammonia and urea hydrolysis products with soil. *Soil Sci. Soc. Am. J.* 52: 1793–6.

Kopecky, M. J., N. L. Meyers and W. Wasko. 1995. *Using Industrial Wood Ash as a Soil Amendment.* Bulletin A3635. University of Wisconsin-Extension, Cooperative Extension. https://learningstore.uwex.edu/assets/pdfs/A3635.pdf.

Lamond, R. E. 1997. *Sulphur in Kansas: Plant, Soil and Fertilizer Considerations.* Bulletin MF2264. Kansas State University Agricultural Experiment Station and Cooperative Extension Service. https://www.bookstore.ksre.ksu.edu/pubs/mf2264.pdf.

Lamond, R. E. 2002. *Chloride in Kansas: Plant, Soil and Fertilizer Considerations.* Bulletin MF2570. Kansas State University Agricultural Experiment Station and Cooperative Extension Service. https://www.bookstore.ksre.ksu.edu/pubs/mf2570.pdf.

Lamond, R. E., A. J. Sunderman, D. A. Whitney, S. R. Duncan and T. L. Wesley. 2007. Phosphorus benefits grain sorghum on acid soils. *Better Crops* 81(1): 10–11.

Leikam, D. F., R. E. Lamond and D. B. Mengel. 2003. *Soil Test Interpretations and Recommendations.* Bulletin MF-2586. Kansas State University Agricultural Experiment Station and Cooperative Extension Service. https://www.bookstore.ksre.ksu.edu/pubs/mf2586.pdf.

Lightner, J. W., D. B. Mengel and C. L. Rhykerd. 1990. Ammonia volatilization form nitrogen fertilizer applied to orchard grass sod. *Soil Sci. Soc. Am. J.* 54: 1478–82.

Lindsey, W. L. 1991. Inorganic equilibria affecting micronutrients in soils, pp. 89–112. In: J. J. Mortvedt, F. R. Cox, L. M. Shuman and R. M. Welch (Eds), *Micronutrients in Agriculture*, 2nd Edition. SSSA Book Series, no. 4.

Lockman, R. B. 1972. Mineral composition of grain sorghum samples, Part III. Suggested nutrient sufficiency limits at various stages of growth. *Comm. Soil Sci. Plant Anal.* 3: 295–304.

Mahama, G. Y., P. V. Prasad, D. B. Mengel and T. T. Tesso. 2014. Influence of nitrogen fertilizer on growth and yield of grain sorghum hybrids and inbred lines. *Agron. J.* 106: 1623–30.

Maritim, K. K., S. Gudu, B. A. Were, A. O. Onkware and E. J. Too. 2013. Phosphorus efficiency among selected sorghum (Sorghum bicolor L. Moench) lines and segregating families. *Afr. J. Educ. Tech.* 1(2): 158–62.

Martens, D. C. and D. T. Westermann. 1991. Fertilizer applications for correcting micronutrient deficiencies, pp. 549–92. In: J. J. Mortvedt, F. R. Cox, L. M. Shuman and R. M. Welch (Eds), *Micronutrients in Agriculture*, 2nd Edition. SSSA Book Series, no. 4.

Mascagni Jr., H. J. and F. R. Cox. 1985. Effective rates of fertilization for correcting manganese deficiency in soybeans. *Agron. J.* 77: 363–6.

Mask, P. L, A. Hagan and C. L. Mitchell, Jr. 1988. *Production Guide for Grain Sorghum.* A Circular ANR-502. Alabama Cooperative Extension Service, Auburn University, Auburn, AL. https://www.aces.edu/edu/pubs/docs/A/ANR-0502.pdf.

Matowo, P. R, G. M. Pierzynski, D. A. Whitney and R. E. Lamond. 1997. Long term effects of till age and nitrogen source, rate and placement on grain sorghum production. *J. Prod. Agric.* 10(1): 141–6.

Mehlich, A. 1984. Mehlich 3 soil test extractant: A modification of the Mehlich 2 extractant. *Comm. Soil Sci. Plant Anal.* 15: 1409–16.

Mengel, D. B., D. W. Nelson and D. M. Huber. 1982. Placement of nitrogen fertilizers for no-till and conventional till corn. *Agron. J.* 74: 515–18.

Mengel, D. B. and F. E. Wilson. 1988. Timing of nitrogen fertilizer for rice in relation to paddy flooding. *J. Prod. Agric.* 1: 90–2.

Mengel, D., R. Lamond, V. Martin, S. Duncan, D. Whitney and B. Gordan. 2009. Chloride fertilization and soil testing – Update for major crops in Kansas. *Better Crops* 93(4): 20–2.

Mills, H. A. and J. Benton. 1996. *Plant Analysis Handbook II.* Micro Macro Publishing, Inc., Athens, GA.

Moll, R. H., E. J. Kamprath and W. A. Jackson. 1982. Analysis and interpretation of factors which contribute to efficiency in nitrogen utilization. *Agron. J.* 74: 562–4.

Moraghan, J. T. and H. J. Mascagni, Jr. 1991. Environmental and soil factors affecting micronutrient deficiencies and toxicities, pp. 371–426. In: J. J. Mortvedt, F. R. Cox, L. M. Shuman and R. M. Welch (Eds), *Micronutrients in Agriculture*, 2nd Edition. SSSA Book Series, no. 4.

Olsen, S. R., C. V. Cole, E. S. Watanabe and L. A. Dean. 1954. *Estimation of Available Phosphorus in Soils by Extraction with Sodium Bicarbonate*. USDA Circular 939. US Government Printing Office, Washington DC. https://archive.org/details/estimationofavai939olse.

Petrie, S. E. and T. L. Jackson. 1984. Effect of fertilization on soil solution pH and manganese concentration. *Soil Sci. Amer. J.* 48: 315–18.

Randall, G. W., E. E. Schulte and R. B.Corey. 1975. Effect of soil and fertilizer applied manganese on the micronutrient content and yield of soybeans. *Agron. J.* 67: 502–7.

Raun, W. R. and G. V. Johnson. 1999. Improving nitrogen use efficiency for cereal production. *Agron. J.* 91: 357–61.

Risse, M. and J. Gaskin. 2013. *Best Management Practices for Wood Ash as Agricultural Soil Amendment.* Bulletin 1142. University of Georgia Cooperative Extension Service. https://www.uga.edu/publications/details.html?number B1142.

Ruiz Diaz, D. A., K. L. Martin and D. B. Mengel. 2011. *Diagnosing Nutrient Deficiencies in the Field.* Bulletin MF3028. Kansas state University Agricultural Experiment Station and Cooperative Extension Service. https://www.bookstore.ksre.ksu.edu/pubs/mf3028.pdf.

Sanchez, C. and E. J. Kamprath. 1959. Effect of liming and organic matter content on the availability of native and applied manganese. *Soil Sci. Soc. Amer. Proc.* 23: 302–4.

Sanchez, P. A. and G. Uehara.1980. Management considerations for acid soils with high P fixation capacity, pp. 471–514. In: F. E. Khasawneh, E. C. Sample and E. J. Kamprath (Eds), *The Role of Phosphorus in Agriculture*. ASA, CSSA, SSSA, Madison, WI.

Schwab, G. J., C. D. Lee, R. Pierce and W. O. Thom. 2007. *Sampling Plant Tissue for Nutrient Analysis.* Bulletin AGR-92. University of Kentucky Cooperative Extension Service. https://simpson.ca.uky.edu/files/sampling_plant_tissue_for_nutrient_analysis.pdf.

Stamper, J. D. 2009. *Evaluation of Method of Placement, Timing and Rate of Application for Anhydrous Ammonia in No-till Corn Production*. M.S. Thesis, Kansas State University.

Steckel, J. E. 1946. Manganese fertilization of soybeans in Indiana. *Soil Sci. Soc. Am. Proc.* 11: 346–8.

Sutradhar, A. K., D. E. Kaiser, C. J. Rosen and J. A. Lamb. 2016. *Zinc for Crop Production*. Bulletin FO-0720-C-(Revised 2016). University of Minnesota Extension. https://www.extension.umn.edu/agriculture/nutient-management/micronutrients/zinc-for-crop-production/.

Touchton, J. T., R. G. Hoeft, and L. F. Welch. 1978. Effect of nitrapyrin on nitrification of fall and spring applied anhydrous ammonia. *Agron. J.* 70: 811–16.

Tucker, A. N. 2009. *Managing Nitrogen in Grain Sorghum to Maximize N Use Efficiency and Yield While Minimizing Producer Risk*. M.S. Thesis, Kansas State University.

Vanderlip, R. L.1993. *How a Sorghum Plant Develops*. Bulletin S-3. Kansas State University Experiment Station and Cooperative Extension Service. https://www.bookstore.ksre.ksu.edu/pubs/s3.pdf.

Varvel, G. E. and T. A. Peterson. 1990. Nitrogen fertilizer recovery by corn in monoculture and rotation systems. *Agron. J.* 82: 935–8.

Warrick, B. E., Sansone, C. and Johnson, J. 2002. *Grain Sorghum Production in West Central Texas*. Texas Cooperative Extension, The Texas A&M University System. https://sanangelo.tamu.edu/extension/agronomy/agronomy-publications/grain-sorghum-production-in-west-central-texas/.

Whitney, D. 1998. Fertilizer requirements, pp. 12–15. In: *Grain Sorghum Production Handbook*, C-687 (revised). Kansas State University Agricultural Experiment Station and Cooperative Extension Service, Manhattan, KS. https://www.bookstore.ksre.ksu.edu/pubs/c687.pdf.

Welch, R. M., W. H. Allaway, W. A. House and J. Kubota. 1991. Geographic distribution of trace elements problems, pp. 31–57. In: J. J. Mortvedt, F. R. Cox, L. M. Shuman and R. M. Welch (Eds), *Micronutrients in Agriculture*, 2nd Edition. SSSA Book Series, no. 4.

Westfall, D. G. and T. A. Bauder. 2011. *Zinc and Iron Deficiencies*. Colorado State University Extension. https://extension.colostate.edu/docs/pubs/crops/00545.pdf.

Wortman, C. S., R. A. Ferguson, G. W. Hergert, C. A. Shapiro and T. M. Shaver. 2013. *Nutrient Management Suggestions for Grain Sorghum*. Bulletin G1669. University of Nebraska-Lincoln Extension, Institute of Agriculture and Natural Resources. extensionpublications.unl.edu/assets/pdf/g1669.pdf.

Zhang, H., and W. R. Raun. 2006. *Oklahoma Soil Fertility Handbook*, 6th Edition. Department of Plant and Soil Sciences, Oklahoma State University, Stillwater, Oklahoma. https://www.soiltesting.okstate.edu/extension-fact-sheets/SFHB2006C.pdf.

Chapter 13

Improving water management in sorghum cultivation

Jourdan Bell, Texas A&M AgriLife Research and Extension Center, USA; Robert C. Schwartz, USDA-ARS Conservation and Production Research Laboratory, USA; Kevin McInnes, Texas A&M University, USA; Qingwu Xue and Dana Porter, Texas A&M AgriLife Research and Extension Center, USA

1 Introduction

Grain sorghum (*Sorghum bicolor* [L.] Moench) is a drought-tolerant crop that has been grown across the American Great Plains since the early twentieth century, primarily as a feed crop. Physiologically, grain sorghum is well adapted to the semi-arid conditions of the American Great Plains because it can withstand periods of water stress. Under ideal conditions and management practices, modern hybrids can yield over 11 000 kg ha^{-1}, but in semi-arid regions, plant available water is a major factor limiting crop production (Abunyewa et al., 2011). While early cultivation of grain sorghum was under dryland production, it was recognized that the efficient use of water, including precipitation and stored soil moisture, was critical for improved performance. Evaluation of sorghum water use efficiencies (WUEs; biomass production including grain yield per unit of water used) was reported as early as 1914 by Briggs and Shantz (1914), who noted improved WUEs of sorghum in comparison to other grain crops. Numerous authors have since defined the improvements in WUEs as a result of sorghum being a C4 plant with a high transpiration efficiency (Xin, 2009; Rooney, 2004; Mortlock and Hammer, 1999; Boyer, 1996). As a C4 plant, sorghum produces more biomass per unit of water transpired than cool-season or C3 plants because C4 plants have a more efficient photosynthetic rate. This enables C4 plants such as sorghum to lose less CO_2 during photorespiration, resulting in greater conversion of CO_2 to biomass. In addition to

http://dx.doi.org/10.19103/AS.2017.0015.15

being a C4 plant, sorghum is more drought tolerant than other C4 plants such as corn due to a deep root system and epicuticular wax to minimize transpirational water losses. Sorghum's deep, fibrous root system provides greater root volume and therefore potential access to a greater soil water volume, enabling sorghum to withstand greater periods of water stress if there is ample stored soil water. Additionally, sorghum is able to maintain turgor pressure in cells through the regulated accumulation of solutes (osmotic adjustment) (Ferres et al., 1978; Assefa, 2010). Under water deficits, the low osmotic potential in the roots reduces the water movement through the plant, resulting in less water loss through the stomatal openings (Fekade and Daniel, 1992). Furthermore, Sanchez-Diaz and Kramer (1971) concluded that after periods of water stress, the plant water potential and saturation potential recovered more quickly than corn. While sorghum is physiologically adapted to withstand prolonged periods of water stress, such tolerances can come at the expense of reduced yield (Assefa, 2010; Peacock, 1982; Fereres et al., 1978).

With the expansion of the fed beef cattle industry on the Southern High Plains in the 1970s as well as corn's greater yield potential and WUE under full irrigation, corn replaced sorghum on much of the US irrigated acreage. However, grain sorghum is still the predominant summer grain crop under dryland and limited irrigation. In the last 25 years, US sorghum production has ranged from 1733 to 7160 MT, Mg (214–884 million bushels) with 75% of the US sorghum production occurring in Kansas and Texas (USDA-NASS, 2017). Consequently, the two lowest production years (2011 and 2012) coincided with record droughts in Kansas and Texas (Fig. 1); during this period, annual precipitation ranged from 102 to 260 mm. Regionally, mean annual precipitation across the US sorghum-producing regions (Fig. 2) ranges from 380 to 800 mm from the Texas Southern High Plains to central Kansas. To minimize yield losses accrued during periods of water stress for optimal production, adoption of improved strategies under both dryland and irrigated systems is essential.

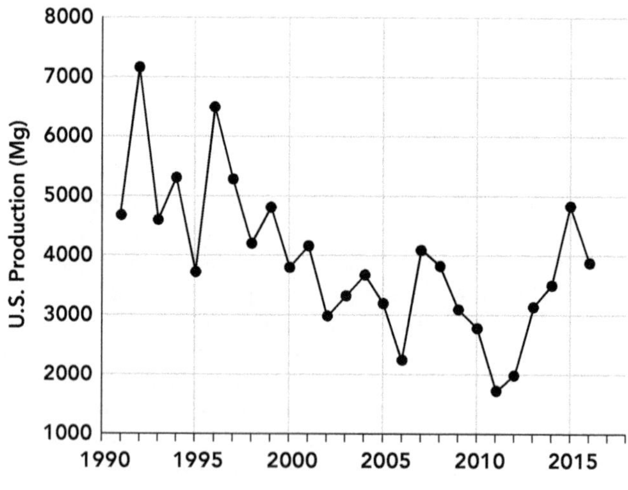

Figure 1 25 years of US grain sorghum production (USDA-NASS, 2017).

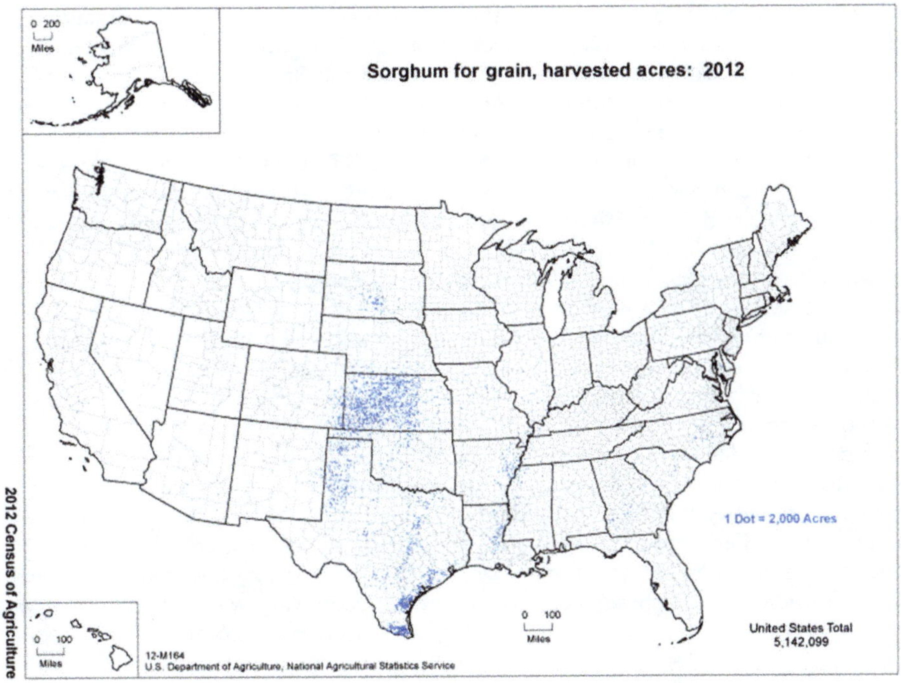

Figure 2 US 2012 grain sorghum harvested acres (USDA Census of Agriculture, 2012).

2 Dryland production

Historical grain sorghum yield data at the USDA-ARS Conservation and Production Research Laboratory in Bushland revealed that sorghum yields increased 139% between 1939 and 1997 (Unger and Baumhardt, 1999). While the authors recognized the importance of hybrid improvement in the yield increase, 67% of the increase was attributed to improved soil water content at planting because of reduced tillage and residue management.

In dryland production systems, conservation tillage and residue management have been essential to yield stabilization by increasing stored soil water. On the Great Plains, the wheat–fallow–sorghum rotation is widely adopted under both dryland and limited irrigation. This rotation sequence balances production and the fallow period efficiency to increase stored soil water at planting. By definition, conservation tillage leaves 30% of the crop residue on the soil surface after tillage and after the subsequent crop is planted (Unger and Baumhardt, 1999). Improvements in residue management were initially in response to widespread soil erosion resulting from the drought of the 1930s and the Great American Dust Bowl, but scientists and farmers quickly recognized that maintaining soil residue also minimized evaporative (E) losses from the soil surface and increased soil water contents at planting. Conservation tillage practices commonly used in dryland sorghum production are no-till (NT) and stubble-mulch tillage (SMT). Because weeds utilize valuable soil water, weed control is critical to the success of the conservation tillage practices (Wise and Army, 1960), especially in NT dryland systems. NT fully relies on chemical herbicides for weed control because tillage is eliminated. Unger and Baumhardt (1999) reported that NT is often considered the 'ultimate' conservation tillage method, but they recognize that other conservation tillage methods such as SMT leave sufficient residue on the soil surface to minimize soil erosion and evaporation. Previous NT research has demonstrated that yields are often comparable to yields with other conservation practices on fine-textured soils in the semi-arid Great Plains of Texas (Baumhardt et al., 2017; Jones et al., 1994), but in dry years with low biomass production, yields are reduced under NT. This is often pronounced in semi-arid dryland systems with fine-textured soils due to soil crust formation and surface sealing (Baumhardt et al., 2017; Jones et al., 1994). Schwartz et al. (2003) reported that crust formation decreased infiltration and soil water storage in these soils. With SMT, V-shaped blades cut weed roots approximately 10 cm below the soil surface without inversion and therefore retain residue on the soil surface. In a long-term research study conducted by Baumhardt et al. (2017) from 1983 to 2013, the annualized cumulative run-off during the fallow period averaged 27.4 mm for NT and 17.3 mm for SMT; however, evaporative losses were lower under NT, resulting in increased precipitation storage.

Planting geometry has been evaluated as a method to improve crop water use and mitigate stress. Early research indicated that grain sorghum planted in narrow rows produced greater biomass and grain yields in humid and subhumid climates as well as under full irrigation, but in semi-arid regions or in years with water stress, there was a greater risk for yield reduction as a result of increased plant competition for spatially variable stored soil water (Fernandez et al., 2012; Jones and Johnson, 1991; Staggenborg et al., 1999; Steiner, 1986; Grimes and Musick, 1960).

3 Irrigation

The expansion of internal combustion engines and turbine pumps led to widespread adoption of irrigation that stabilized and increased production in the semi-arid Great Plains (Colaizzi et al., 2009; Howell, 2006; Rhoades, 1997; Musick et al., 1988). Grain sorghum acreage peaked during the 1950s as a result of extensive irrigation, but yields continued to increase with advances in agronomic and irrigation management in addition to improved grain sorghum hybrids. Improvements in irrigation management including the transition from gravity-fed irrigation with graded furrows to sprinkler irrigation, and improved application efficiencies, further increased sorghum yield per acre as well as yield per unit of water applied (irrigation water use efficiency, IWUE). While improvements in irrigation technology have greatly increased IWUE, advancements in irrigation management strategies will likely provide the most extensive water savings (Garces-Restrepo et al., 2007). The lack of precipitation can potentially magnify yield responses to irrigation treatments (Allen and Musick, 1990; Howell, 2007). Under water-limited conditions, improvements in IWUE coupled with practices to improve precipitation and soil water management are necessary to optimize the crop WUEs (crop production per unit of water consumed). Crop WUE accounts for crop utilization of stored soil water, precipitation and irrigation.

In much of the US Great Plains, irrigation water is supplied by the Ogallala Aquifer (Fig. 3). The Ogallala underlies portions of eight states, with the saturated thickness currently in

Figure 3 Distribution of the Ogallala Aquifer in the Great Plains (https://en.wikipedia.org/wiki/File:Ogallala_Aquifer_map.png).

decline from Kansas south through Texas because of significant withdrawals and negligible recharge rates. Consequently, irrigation water from the Ogallala Aquifer is becoming less reliable due to the subsequent decline in well capacities in addition to increased competition for water supplies between urban and rural sectors. Traditionally, irrigation has been applied to fully meet the seasonal crop water demand; however, future water use will be limited either by restricted pumping or by reduced well capacities. To sustain domestic and industrial requirements, agricultural water use is projected to decrease by up to 34% by 2060 in the Texas High Plains due to ensuing policies that will restrict pumping (Wagner, 2012). Utilization of dryland management strategies with limited irrigation rather than irrigating to meet the full crop water demand could potentially stabilize crop production and support an agriculturally driven economy for an extended period (Bordovsky et al., 2011).

Combining conservation tillage practices and improved irrigation management strategies often allows producers to maximize the soil water storage from variable precipitation to improve WUE under conditions of limited irrigation water.

With limited water for irrigation, deficit irrigation (DI) is the most common irrigation strategy used for sorghum production. Howell et al. (2007) demonstrated that DI grain sorghum established on a nearly full soil water profile permitted enhanced soil water extraction and only marginal declines in yield compared to fully irrigated sorghum. However, frequent and shallow applications of irrigation water (e.g. $\leq$20 mm) during the growing season can result in shallow sorghum root development (Myers et al., 1984), which may reduce water uptake deeper in the profile. Moreover, both the timing and amount of irrigation during the growing season are critical to the success of DI strategies (Bell, 2014).

Irrigation scheduling often utilizes external estimates of crop evapotranspiration (ETc) and/or measurements of soil water depletion to estimate crop water requirements, while ETc equates the crop water use as a combination of evaporation (E) from the soil surface and transpiration (T) for plant growth. ETc is calculated using an appropriate crop-specific coefficient (Kc) and the reference evapotranspiration (ET_o). ET_o is the calculated evapotranspiration using weather station data including air temperature, relative humidity, precipitation, wind speed and solar irradiance. Oftentimes, local weather data are unavailable and therefore ET-based scheduling is not practical, so ET can be calculated directly through measurements in the changes in water content within the root zone. In practice, however, soil water sensors are not normally installed throughout the rooting zone and often are of insufficient accuracy to estimate ETc and hence irrigation is scheduled based on the attainment of established soil water depletion levels or water potentials at one or more depths in the most active regions of the root zone. Advancements in electronic sensing of soil moisture provides irrigators information they can use to assess the level of soil water depletion or increase and to determine when to initiate irrigation as well as how much water to apply to avoid water stress and minimize drainage below the root zone. Irrigation can be scheduled based on ETc and soil water monitoring or a combination of the two methods for successful grain sorghum yield (Bell, 2014).

Utilization of ETc is facilitated through on-farm weather stations or regional ET networks that distribute data on both the daily weather and ETc for irrigation scheduling. Daily ETc data permit producers to make daily adjustments based on the changes in ETc rather than irrigating at a constant rate. Wireless technologies and Internet integration enhance the application of ETc for irrigation scheduling by providing producers access to real-time data to improve irrigation efficiency and integrate irrigation with precision technologies (Pierce and Elliott, 2008). Real-time data permit the producer to balance the daily environmental demands with available soil moisture and irrigation capacity to minimize overwatering the crop, thus improving economic returns by reducing irrigation costs. Online ET-based irrigation scheduling tools, downloadable ET calculators and smartphone apps for irrigation scheduling utilize ET_o data from regional ET networks along with producer inputs to determine ETc (Rogers, 2012; Migliaccio et al., 2015). Just a few of the many user inputs are crop type, irrigated area, irrigation system type and efficiency, row spacing, rooting depth, soil type and water conservation mode. Migliaccio et al. (2015) noted that user-friendly apps can provide producers notifications on factors such as the daily water balance and weather forecasts that prompt a response to a critical event.

4 Deficit irrigation

Comprehensive irrigation strategies are necessary to optimize limited water supplies. Irrigation management should supplement seasonal precipitation and stored soil water to a degree sufficient to meet crop requirements while maximizing net returns and minimizing run-off and drainage below the root zone. Irrigation scheduling strives to manage irrigation timing and rates to efficiently satisfy crop water demand such that periods of excessive water stress associated with yield reductions are minimized. Regulated or managed DI enables the irrigator to apply water at a reduced rate throughout selected periods of the growing season in an attempt to optimize the rate of return. DI may increase the WUE of some crops (English, 1990; Fereres and Soriano, 2007; Musick et al., 1994); however, judicious scheduling of irrigation combined with appropriate and well-maintained technology is necessary to supply water at deficit levels during non-critical growth stages while ensuring adequate irrigation during critical growth stages to mitigate water stress (Perry et al., 2009). Vadez et al. (2013) concluded that research results from many crops grown under DI strategies suggest that incorporation of managed deficit irrigation strategies would shift crop water use from the vegetative to the reproductive stage to increase yield under water-limited conditions.

Howell and Hiller (1975) concluded that the yield response of grain sorghum to irrigation was most sensitive to having water available at critical times. Seasonal crop water requirements coincide with the growth patterns (Fig. 4) tempered by the atmospheric water demands during the growing season. Traditionally, DI entailed irrigation throughout the entire growing season targeting a fraction of crop water requirements compared with full irrigation (FI) (Howell, 2007). However, Bell (2014) evaluated a managed deficit irrigation (MDI) strategy whereby a fraction of the full irrigation requirement was applied at critical growth stages. Using MDI, early season irrigation was either reduced or eliminated; but a greater percentage of the daily crop water demand (ETc) was applied during the reproductive stages from growing point differentiation to half-bloom. Although the greatest grain yields and WUEs were achieved with FI, irrigation under MDI reduced WUE in only one of the three years. In contrast, yields and WUEs of DI were reduced in all years compared to FI.

With grain sorghum, it has been well documented (Van Oosterom and Hammer, 2008; Tolk et al., 2013; Prasad et al., 2008; Blum, 1996; Crauford and Peacock, 1993; Peacock, 1982; Eck and Musick, 1979) that water stress extending from growing point differentiation through half-bloom suppresses grain yield due to reduced seeds per panicle. Quantifying the effects of the magnitude of water stress at critical growth stages on the resultant ETc and grain yield is essential to evaluate irrigation strategies and improve WUEs. It is also imperative that under conditions of limited water, precipitation is considered in designing the irrigation management plan.

Growing point differentiation is the initial stage of reproductive development in grain sorghum (Vanderlip, 1979). At growing point differentiation, the crop begins to develop rapidly – 7–10 leaves have expanded and rapid nutrient uptake has begun. Water stress during growing point differentiation can reduce the number of seeds developed per panicle. Water stress during the boot stage minimizes head extension from the flag leaf sheath, which reduces pollination (Gerick et al., 2003). Half-bloom is the final stage of reproductive development where approximately half of the sorghum plants in a field have begun flowering. During half-bloom, water stress may induce floral abortion and decrease

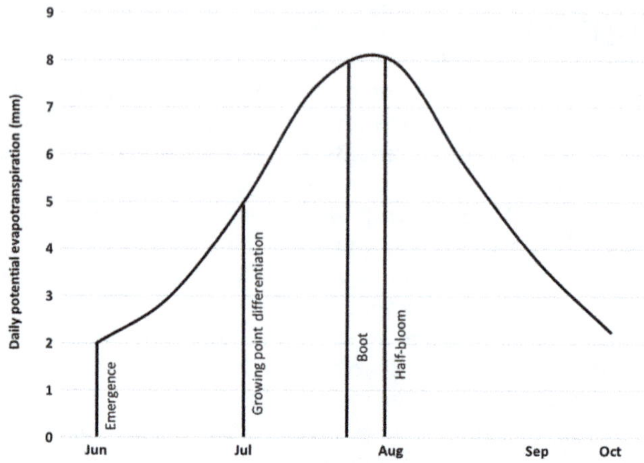

Figure 4 Generalized pattern of grain sorghum that coincides with periods of crop water demand.

grain yield. In contrast, water stress from anthesis through the dough stage may reduce grain/seed mass (Ockerby et al., 2001; Maman et al., 2004).

Irrigation scheduling should also utilize knowledge of the crop rooting depth, soil physical properties, irrigation system capabilities/constraints and irrigation water quality to optimize yield. Knowledge of the crop rooting depth can aid the irrigator in determining the potential volume of plant available water and the allowable depletion when scheduling irrigation. Under dryland, research has shown that the sorghum root

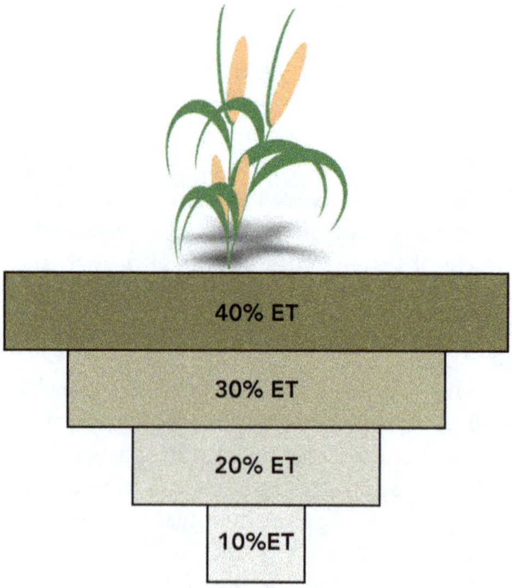

Figure 5 Generalized depiction of soil water extraction by rooting depth.

system often reaches depths over six feet (Moroke et al., 2005). However, it is important for producers to recognize that while deep soil moisture is available, research has shown that approximately 40% of water used originates in the upper 25% of the root zone (Fig. 5).

5 Soils and irrigation management

Soil properties such as texture and structure directly affect irrigation. The soil water-holding capacity as well as water flow into and through the soil and the rate and frequency at which irrigation is applied are functions of the soil type. Consequently, knowledge of soil physical properties is essential for successful irrigation scheduling. Soil textural classes are based on the fractions of sand, silt and clay particles. Soil structure is the arrangement of the sand, silt and clay particles into aggregates defined as granular, platy, blocky, columnar and prismatic. Aggregates are formed and their properties are altered through both natural and human activities such as wetting, freezing, thawing, organic matter decomposition and tillage. The aggregate arrangement affects the soil porosity (pore spaces in the soil) and thus water storage and flow in the soil. Coarse-textured soils with a high percentage of sand have large pores, while fine-textured soils with a high percentage of silts and clays have many small pores. Water drains freely through large pores. In comparison, capillary forces retain water in small pores; consequently, the clay fraction is the most important component affecting a soil's water-holding capacity due to its influence on structure and porosity. As the clay fraction increases, the quantity of water held in the soil pores after drainage increases. Soil texture and porosity can be directly measured by the soil bulk density, the weight of soil including solids and pores per unit volume. Soil texture and structure affect the infiltration rate and the permeability of water into the soil. Infiltration is the rate at which water moves into the soil, while permeability is the rate at which water moves through the soil. Fine-textured soils have a low infiltration rate. In comparison, coarse soils have a high infiltration rate. The infiltration rate directly affects the rate of run-off. If the irrigation is applied at a rate greater than the infiltration rate, run-off will occur. In addition to soil texture and structure, infiltration and permeability are also governed by soil organic matter, soil compaction and the antecedent soil moisture. Wet soils have lower infiltration and permeability than dry soil. Infiltration and permeability also affect drainage below the root zone. It is the responsibility of the irrigator to manage irrigation to minimize drainage, run-off and evaporation (E) to increase stored soil water for transpiration (T). The irrigator should have knowledge of the soil so that the irrigation rate does not exceed the infiltration rate (Table 1). DI should be managed according to the soil

Table 1 Average infiltration rate for various soil textural classes

Textural class	Infiltration rate (mm hr^{-1})	Infiltration rate (in hr^{-1})
Clay	2.5–5.1	0.1–0.2
Silt loam	6.4–10.2	0.2–0.4
Sandy loam	8.9–12.7	0.4–0.5
Fine sand	12.7–17.8	0.5–0.7

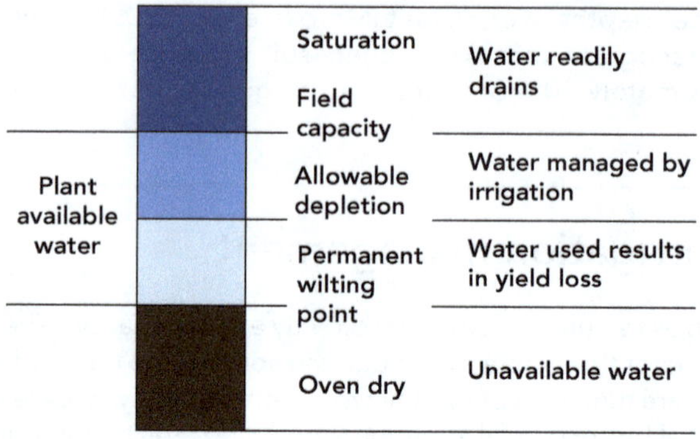

Figure 6 Categories of soil water.

texture. On fine-textured soils, the irrigation efficiency is maintained by applying irrigation at greater depths and time intervals, which minimizes soil evaporation and maximizes soil water storage. Conversely, on sandy soils where drainage is a concern, shallow, frequent irrigations are recommended to avoid drainage below the root zone.

Soil water content is generally expressed as a per cent volume, or even a 'depth of water per depth of soil'. Some important soil water content reference points are saturation, field capacity, plant available water and permanent wilting point. While a soil may have a known amount of water, only a percentage of the water is available to the plant (Fig. 6). The volume of water held by the soil between field capacity and wilting point is the plant available water. Plant available water is the fraction of the stored soil water that can be utilized by plants. This fraction varies by soil textural class due to pore size and the force (suction) with which water is held in the soil (Table 2). While plant available water can be utilized by the crop, irrigators often plan for a portion of the plant available water to be utilized by the crop (managed allowable depletion) in order to minimize plant stress and production losses that occur when plants expend energy, utilizing the lower limits of the plant available water. Sorghum yields are reduced when the allowable depletion is greater than 50% of the plant available water (Bell, 2014; Doorenbos and Pruitt (1977). Permanent wilting point is the lower limit where plants are unable to absorb water. In addition to being directly related to soils, the allowable depletion is a function of the atmospheric conditions that determine the crop water use (Tolk and Howell, 2008; Robertson and Fukai, 1994).

Table 2 Average plant available water for different soil textural classes

Textural class	Plant available water (mm m^{-1})	Plant available water (in ft^{-1})
Clay	101–126	1.2–1.5
Silty clay loam	151–168	1.8–2.0
Fine sandy loam	126–168	1.5–2.0
Fine sand	63–84	0.8–1.0
Coarse sand	21–63	0.3–0.8

6 Conclusion

Efficient water management is dependent on management decisions based on soil type, irrigation system limitations, crop species and production goals. While irrigation enhances production and reduces risk, especially in semi-arid regions or regions with temporally variable precipitation, there are also additional costs associated with irrigation including equipment, labour and energy costs. Future irrigation practices and methods of crop selection must be adopted to prolong groundwater resources. As irrigation water across Texas becomes limited by declining aquifer levels and/or restricted by groundwater district regulations, it is important that crop selection and crop water demands do not exceed the water availability. Because sorghum can withstand periods of drought, it will continue to be a valuable cropping option for prolonging limited water resources.

Globally, increasing opportunities for utilization of sorghums as livestock feed in addition to expanding food and industrial markets presents farmers in water limited regions greater production opportunities. Consequently, there are extensive research opportunities to evaluate agronomic practices to optimize sorghum quality in addition to agronomic yield while increasing crop WUE. Previous research has extensively addressed management of water resources to optimize agronomic yields; however, there are significant differences in grain quality because of drought and water stress. Research to optimize grain quality and properties is an important research area for expanding and securing future markets. Additionally, there are extensive opportunities to evaluate the deficit irrigation strategies and irrigation scheduling using emerging irrigation technologies.

7 Where to look for further information

Additional information about water management and irrigation technologies can be found through the Research and Extension Services of Texas A&M AgriLife, Kansas State University and the University of Nebraska. Specific resources are located at the Texas A&M AgriLife Water Education Network (https://water.tamu.edu/water-management-irrigation/) and Kansas State University Department of Agronomy Grain Sorghum page (https://www.agronomy.k-state.edu/extension/crop-production/grain-sorghum/). Scientific journals such as *Agronomy Journal, Crop Science, Irrigation Science, Agricultural Water Management*, and *Transactions of ASABE Irrigation* are popular journals where one can locate recent research concerning water management.

8 References

Abunyewa, A. A., R. B. Ferguson, C. S. Wortman, D. J. Lyon, S. C. Mason, S. Irmak and R. N. Klein. 2011. Grain sorghum water use with skip-row configuration in the Central Great Plains of the USA. *Afr. J. Agric. Res.* 6(23):5328–38.

Allen, R. R. and J. T. Musick. 1990. Effect of tillage and preplant irrigation on sorghum production. *Appl. Eng. Agric.* 6(5):611–18.

Assefa, Y., S. A. Staggenborg and V. P. V. Prasad. 2010. Grain sorghum water requirement and responses to drought stress: A review. *Crop Manag.* Online. doi:10.1094/CM-2010-1109-01-RV.

Baumhardt, R. L., R. C. Schwartz, O. R. Jones, B. R. Scanlon, R. C. Reedy and G. W. Marek. 2017. Long-term conventional and no-tillage effect on field hydrology and yields of a dryland crop rotation. *Soil Sci. Soc. Am. J.* 81:200–9.

Bell, J. M. 2014. Responses of grain sorghum to profile and temporal dynamics of soil water in a semi-arid environment. Ph.D. Thesis, Texas A&M University, College Station, TX.

Bordovsky, J. P., J. T. Mustian, A. M. Cranmer and C. L. Emerson. 2011. Cotton-grain sorghum rotation under extreme deficit irrigation conditions. *Appl. Eng. Agric.* 27:359–71.

Boyer, J. S. 1996. Advances in drought tolerance in plants. *Adv. Agron.* 56:187–218.

Briggs, L. J. and Shantz, H. L. (1914). Relative water requirement of plants. *J. Agric. Res. (Washington, DC)* 3:1–63.

Childs, S. W. and R. J. Hanks. 1975. Model of soil salinity effects on crop growth. *Utah Agric. Exp. Sta. Journal Paper no. 1920.*

Colaizzi, P. D., P. H. Gowdam, T. H. Marek and D. O. Porter. 2009. Irrigation in the Texas High Plains: A brief history and potential reductions in demand. *Irrig. Drain.* 58(3):257–74.

Crauford, P. Q. and J. M. Peacock. 1993. Effect of heat and drought stress on sorghum (*Sorghum bicolor*). II. Grain yield. *Exp. Agric.* 29:77–86.

Doorenbos, J. and W. H. Pruitt. 1977. Crop water requirements. *FAO Irrigation and Drain Paper 24.* FAO, Rome, Italy, 144pp.

English, M. J. 1990. Deficit irrigation. I: Analytical framework. *J. Irrig. Drain. Eng.* 116(3):399–412.

Fekade, S. G. and R. K. Daniel. 1992 Osmotic adjustment in sorghum. *Plant Physiol.* 99:577–82.

Fereres, E., E. Acevedo, D. W. Henderson and T. C. Hsiao. 1978. Seasonal changes in water potential and turgor maintenance in sorghum and maize under water stress. *Physiol. Plant.* 44:261–7.

Garces-Restrepo, C., D. Vermillion and G. Muoz. (2007). *Irrigation Management Transfer: Worldwide Efforts and Results.* Food and Agriculture Organization of the United Nations, Rome.

Gerik, T., B. Bean and R. Vanderlip. 2003. Sorghum growth and development. Texas Cooperative and Extension Publication B-6137.

Gollehon, N. and B. Winston. 2013. Groundwater irrigation and water withdrawals: The Ogallala Aquifer initiative. Economic series number 1. USDA-NRCS, Washington DC, USA. http://www.nrcs.usda.gov/Internet/FSE_DOCUMENTS/stelprdb1186440.pdf.

Grattan, S. R. Irrigation water salinity and crop production. University of California Publication 8066. FWQP Reference Sheet 9.10.

Grimes, D. W. and J. T. Musick. 1960. Effect of plant spacing, fertility, and irrigation managements on grain sorghum production. *Agron. J.* 52:647–50.

Howell, T. A. 2006. Challenges in increasing water use efficiency in irrigated agriculture. In: *International Symposium on Water and Land Management for Sustainable Irrigated Agriculture*, Adana, Turkey, 4–8 April 2006.

Howell, T. A. and E. A. Hiller. 1975. Optimization of water use efficiency under high frequency irrigation I. Evapotranspiration and yield relationship. *Trans. ASAE.* 18:873–8.

Howell, T. A., J. A. Tolk, S. R. Evett, K. S. Copeland and D. A Dusek. 2007. Evapotranspiration of deficit irrigated sorghum and winter wheat. In: Clemmens, A. J. (Ed.), *USCID Fourth International Conference on Irrigation and Drainage. The Role of Irrigation and Drainage in a Sustainable Future*, 3–6 October 2007, Sacramento, California, pp. 223–39. 2007 CDROM.

Jones, O. R. and G. L. Johnson. 1991. Row width and plant density effects on Texas High Plains sorghum. *J. Prod. Agric.* 4:613–21.

Jones, O. R., V. L. Hauser and T. W. Popham. 1994. No-tillage effects on infiltration, runoff, and water conservation on dryland. *Trans. ASABE* 37(2):473–9.

Maman, N., S. C. Mason, D. J. Lyon and P. Dhungana. 2004. Yield components of pearl millet and grain sorghum across environments in the Central Great Plains. *Crop Sci.* 44:2138–45.

Migliaccio, K. W., K. T. Morgan, G. Vellidis, L. Zotarelli, C. Fraisse, B. A. Zurweller, J. H. Andreis, J. H. Crane and D. L. Rowland. 2015. Smartphone apps for irrigation scheduling. *Trans. ASABE* 59(1):291–301.

Moroke, T. S., R. C. Schwartz, K. W. Brown and A. S. R. Juo. 2005. Soil water depletion and root distribution of three dryland crops. *Soil Sci. Soc. Am. J.* 69:197–205.

Mortlock, M. Y. and G. L. Hammer. 1999. Genotype and water limitation effects on the transpiration efficiency in sorghum. *J. Crop Prod.* 2:265–86.
Musick, J. T., F. B. Pringle and J. D. Walker. 1988. Sprinkler and furrow irrigation trends—Texas High Plains. *Appl. Eng. Agric.* 4(1):46–52.
Norwood, C. A., A. J. Schlegel, D. W. Morishita and R. E. Gwin. 1990. Cropping system and tillage effects on available soil water and yield of grain sorghum and winter wheat. *J. Prod. Agric.* 3:356–62.
Ockerby, W. E., D. J. Midmore and D. F. Yule. 2001. Leaf modification delays panicle initiation and anthesis in grain sorghum. *Aust. J. Agric. Res.* 52:127–35.
Peacock, J. M. 1982. Response and tolerance of sorghum to temperature stress. In: *Sorghum in the Eighties: Proceedings of the International Symposium on Sorghum*, 2–7 November 1981, Patancheru, Andhra Pradesh, India.
Perry, C., P. Steduto, R. G. Allen and C. M. Burt. 2009. Increasing productivity in irrigated agriculture: Agronomic constraints and hydrological realities. *Agric. Water Manag.* 96(11):1517–24.
Pierce, F. J. and T. V. Elliott. 2008. Regional and on-farm wireless sensor networks for agricultural systems in Eastern Washington. *Comput. Electron. Agric.* 61(1):32–43.
Prasad, P. V., S. R. Pisipati, R. N. Mutava and M. R. Tuinstra. 2008. Sensitivity of grain sorghum to high temperature stress during reproductive development. *Crop Sci.* 48:1911–17.
Rhoades, J. D. 1997. Sustainability of irrigation: An overview of salinity problems and control strategies, pp. 1–42. In: *CWRA 1997 Annual Conf. 'Footprints of Humanity: Reflections on Fifty Years of Water Resource Developments'*, Lethbridge, Alberta, Canada, 3–6 June 1997.
Robertson, M. J. and S. Fukai. 1994. Comparison of water extraction models for grain sorghum under continuous soil drying. *Field Crops Res.* 36:145–60.
Rogers, D. H. 2012. Introducing the web-based version of KanSched: An ET-based irrigation scheduling tool. In: *Proceedings of the 24th Annual Central Plains Irrigation Conference*, Colby, Kansas, 21–22 February 2012.
Rooney, W. L. 2004. Sorghum improvement-integrating traditional and new technology to produce improved genotypes. *Adv. Agron.* 83:37–109.
Sanchez-Daiz, M. and P. J. Kramer. 1971. Behavior of com and sorghum under water stress and during recovery. *Plant Physiol.* 48:613–16.
Schwartz, R. C., S. R. Evett and P. W. Unger. 2003. Soil hydraulic properties of cropland compared with reestablished and native grassland. *Geoderma* 116:47–60.
Staggenborg, S., B. Gordon, R. Taylor, S. Duncan and D. Fjell. 1999. Narrow-row grain sorghum production in Kansas. Kansas State University Agricultural Experiment Station and Cooperative Extension Service Publication MF-2388.
Steiner, J. L. 1986. Dryland grain sorghum water use, light interception, and growth responses to planting geometry. *Agron. J.* 78:720–6.
Tolk, J. A. and T. A. Howell. 2008. Field water supply: Yield relationships of grain sorghum grown in the USA Southern Great Plains soils. *Agric. Water Manag.* 95(12):1303–13.
Tolk, J. A., T. A. Howell and F. R. Miller. 2013. Yield component analysis of grain sorghum grown under water stress. *Field Crops Res.* 145:44–51.
Unger, P. W. and R. L. Baumhardt. 1999. Factors related to dryland grain sorghum yield increases: 1939 through 1997. *Agron. J.* 91:870–5.
Unger, P. W. and R. L. Baumhardt. 2001. Historical development of conservation tillage in the southern Great Plains. In: Stiegler, J. H. (Ed.), *Proceedings of the 24th Annual Southern Conservation Tillage Conference for Sustainable Agriculture*, Oklahoma City, July 2001. Oklahoma State University, Oklahoma City, OK.
USDA. 2012 *Census of Agriculture. Summary and State Data: Volume 1: Geographic Area Series.* National Agricultural Statistics Service, Washington DC.
USDA-NASS. 2017. Crop production: 2016 summary. ISSN: 1057–7823. http://usda.mannlib.cornell.edu/usda/current/CropProdSu/CropProdSu-01-12-2017.pdf.
Vadez, V., J. Kholova, R. S. Yadav and C. T. Hash. 2013. Small temporal differences in water uptake among varieties of pearl millet (*Pennisetum glaucum* (L.) R. Br.) are critical for grain yield under terminal drought. *Plant Soil* 371:447–62.

Vanderlip, R. L. 1979. How a sorghum plant develops (No. C040. 065). Cooperative Extension Service, Kansas State University.

Wagner, K. 2012. Status and trends of irrigated agriculture in Texas: A special report. Texas Water Resources Institute, College Station, TX.

Xin, Z., R. Aiken and J. Burke. 2009. Genetic diversity of transpiration efficiency in sorghum. *Field Crops Res.* 111:74–80

Chapter 14

Insect pests and their management in sorghum cultivation

Bonnie B. Pendleton, West Texas A&M University, USA

1 Introduction

Integrated pest management (IPM) is a practical approach that uses a combination of methods, including cultural, biological and chemical controls, to maintain pest abundance or damage without economic loss. This chapter emphasizes ways to prevent damaging infestations and estimate abundance and evaluate damage to determine when remedial action with insecticide is justified for insect pests of sorghum, *Sorghum bicolor* (L.) Moench. Insect pests, their biologies, and natures and symptoms of damage are described.

Sorghum has been domesticated to meet human needs. The crop grows well as landrace varieties with primitive cultivation in small, subsistence plots or in monocultures of hybrids produced by modern practices and equipment. Often sorghum is grown in areas too hot and too dry for other crops. Sorghum is known for production reliability, but may be damaged by insect pests.

Most insect pests are distributed widely, not host specific, and did not co-evolve with sorghum. In sorghum agro-ecosystems, one or two key pests such as greenbug, *Schizaphis graminum* (Rondani); sorghum midge, *Stenodiplosis sorghicola* (Coquillett); shoot fly, *Atherigona soccata* (Rondani); or stalk borers occur perennially and dominate control. Banks grass mite, *Oligonychus pratensis* (Banks), and corn earworm, *Helicoverpa zea* (Boddie), are secondary or induced pests injurious as a result of insecticide for a key

http://dx.doi.org/10.19103/AS.2017.0015.16

pest. Other insects are occasional pests, causing economic damage in localized areas during some years.

Cultural and biological management methods avoid insect pests, suppress pest abundance or rate of increase, delay time when pests reach damaging abundance or increase plant tolerance. Use of these non-chemical methods must be decided before problems occur, and often before the crop is planted. Chemical management uses insecticide to kill pests. Insecticides have rapid, curative action but are costly and may cause negative ecological and environmental consequences. They must be justified based on insect abundance and damage.

2 Cultural management methods to control insect pests

Cultural management methods enhance or suppress survival, abundance and severity of damage by insect pests. Practices avoid conditions that favour insect pests or activate conditions detrimental to increase in pest abundance or damage.

2.1 Crop termination and alternative host elimination

Crop termination and alternative host elimination kill insect pests and decrease damage through destruction of food and overwintering habitats. Destroying sorghum after harvest and eliminating volunteer plants that harbour insects suppress pests the following year. Herbicide can be used to kill sorghum and alternative hosts where reduced tillage is used. But, failure to destroy overwintering sites mechanically may allow certain insect pests to survive.

Destroying food sources and overwintering habitats reduce abundance of cutworms; sorghum midge; sorghum webworm, *Nola sorghiella* Riley; stalk borers; and sugarcane rootstalk weevil, *Anacentrinus deplanatus* (Casey). Johnsongrass, *Sorghum halepense* (L.) Pers., is a non-cultivated host of many sorghum insect pests, including yellow sugarcane aphid, *Sipha flava* (Forbes); greenbug; and sorghum midge.

2.2 Crop rotation

Crop rotation alternates host and non-host plants to reduce abundance and damage by insect pests, diseases and weeds. Sorghum benefits when rotated with a broadleaf or taprooted crop. Rotation is effective against insects with limited host range, long life cycle (one or fewer generations) and limited movement. For example, several species of wireworms, white grubs and some cutworms have only one generation a year, must have grass on which to develop, and because they live underground cannot during the damaging larval stage move from one field to another.

2.3 Variety selection, seedbed preparation and seed treatment

Variety selection, seedbed preparation and seed treatment are important for managing insect pests. Poor seed germination results in stand loss or weak plants vulnerable to pests. A variety adapted to the locale and least vulnerable to pests should be used. Some sorghum hybrids are more resistant to greenbug; sugarcane aphid, *Melanaphis sacchari* (Zehntner); sorghum midge; panicle bugs; and larvae of corn earworm, fall armyworm,

Spodoptera frugiperda (J. E. Smith) and sorghum webworm. Sorghum hybrids that mature early and uniformly in a locale escape infestation by sorghum midge, corn earworm, fall armyworm, sorghum webworm and stalk borers.

Insect pests stress plants, and stress combined with infection by pathogens such as maize dwarf mosaic virus transmitted by aphids increases lodging. Well-prepared seedbeds speed germination and help avoid damage by wireworms and red imported fire ant, *Solenopsis invicta* Buren. Rapidly growing plants better tolerate damage by yellow sugarcane aphid; greenbug; and chinch bug, *Blissus leucopterus leucopterus* (Say). Tillage modifies soil to manage seed placement, weeds, water infiltration and evaporation, and erosion.

Insecticide applied to seed protects against seed feeding and some seedling insect pests of sorghum. Insects controlled by systemic seed or in-furrow treatment include wireworms; red imported fire ant; cutworms; southern corn rootworm, *Diabrotica undecimpunctata howardi* Barber; aphids; and chinch bug.

2.4 Planting time

Planting time of sorghum should be as early as practical after soil is warm enough for seed germination and seedling growth. Sorghum planted early is less vulnerable when insect pests such as sorghum midge are abundant enough to cause damage. Planting sorghum early avoids large numbers of corn earworm, fall armyworm, sorghum webworm, panicle bugs and stalk borers. Early planting also takes advantage of seasonal rainfall.

2.5 Fertilizer and water

Fertilizer and water can be beneficial or detrimental to insect pests. Too much fertilizer and water can cause plants to be attractive to insect pests and extend time to maturity, increasing duration of vulnerability. Healthy, vigorously growing plants tolerate insects and other stresses. Banks grass mite and chinch bug prefer hot, dry conditions and moisture-stressed plants.

3 Biological management methods to control insect pests

Biological management methods reduce pest abundance and damage by predators, parasites and pathogens: classical control or importation, controlling insects by using introduced natural enemies; augmentation, mass culturing and periodic release of a natural enemy; and conservation or preservation, enhancement of existing natural enemies. Conservation is most applicable for suppressing sorghum insect pests.

Conservation of natural enemies enables them to suppress pests. Sorghum hosts an abundance of insect pests, primarily aphids. Predators of sorghum insect pests include lady beetles, lacewings, syrphid flies, bugs and ground beetles. Braconid wasps, chalcid wasps and tachinid flies parasitize aphids and caterpillars in sorghum.

Sorghum insect pests most affected by natural enemies are greenbug, yellow sugarcane aphid, corn earworm, fall armyworm, sorghum webworm and Banks grass mite. Predators affect greenbug and other aphids and help prevent economic damage, especially when resistant hybrids are used. Predators suppress abundance of corn earworm and fall armyworm on sorghum panicles. Parasites often terminate infestations by greenbug late in the season. Several pathogens, mostly fungi, infect pests such as chinch bug, corn earworm and fall armyworm.

Many insecticides in sorghum are broad spectrum and kill natural enemies. Negative effect on natural enemies is why insecticide must be needed before application. A pest might resurge, or a secondary pest increase in abundance, after natural enemies are destroyed.

4 Chemical management methods to control insect pests

Insecticides are chemicals that kill insects. They kill rapidly and are easy to apply. Disadvantages are cost and broad toxicity. Insecticides adversely affect non-target organisms including natural enemies. Insecticides can be timed or used to preserve natural enemies. Human health and environmental problems increase the importance of preventive management and justification for insecticides. Some pests such as spider mites are resistant to insecticides.

Insecticides should be used in proper amounts and only when necessary to prevent increasing pest abundance from causing economic loss. The economic injury level is pest abundance or damage that causes crop loss greater than cost of control. The economic threshold is pest abundance or damage when control should be applied to prevent injury. Insecticide and application costs, crop market price and yield must be considered when deciding to use insecticide. Insect pest abundance, age, and duration of attack, and stage and condition of plants attacked must be considered.

5 Insect pests of sorghum in the United States

Numerous insects attack sorghum in the United States. Insect pests are grouped according to plant parts attacked.

5.1 Pests of seed and roots

Wireworms: *Aeolus* spp.; *Eleodes* spp.; *Conoderus* spp.

Description and Biology: True wireworms are immature click beetles (Elateridae). False wireworms are immature darkling beetles (Tenebrionidae). Wireworms are cylindrical, smooth, shiny, white to yellow or brown, hard-bodied and 25-mm long. Larvae overwinter and are in soil when sorghum is planted (Fig. 1). All but adults develop in soil. A generation requires 1 year or more.

Damage and Symptoms: Wireworms feed on planted seed and less on seedling roots. Hollowed seeds do not germinate. Damage causes non-uniform stands of stunted, weak seedlings.

Monitoring: Inspect fields before planting. Assess abundance by 300 g of non-treated seed in a 12-cm hole 10 cm deep in soil. Fill the hole with soil and cover with black plastic

Figure 1 Wireworm.

to warm. Use one trap per 2.5–5 ha two weeks before planting. Examine grain and count wireworms two weeks later. Two or more larvae per trap justify insecticide, especially if germination is delayed.

Management: Damage depends on wireworms in the field at planting and at the time before seeds germinate. Sorghum should be planted where a non-grass crop was grown previously. Tillage before planting reduces non-crop plants. The seedbed should be prepared to ensure rapid germination. Insecticide applied to seed is effective.

Red imported fire ant, *Solenopsis invicta* Buren

Description and Biology: Red imported fire ants replaced native ants in infested areas. Brood (eggs, larvae and pupae) and adults are in colonies. Adults are winged males with smaller heads and darker bodies, red- or brown-winged females, queens and workers. Workers are wingless, sterile females.

Newly mated queens lay clusters of 12 eggs. The queen feeds infertile eggs, oil from her crop or protein secretions from her salivary glands to newly hatched larvae that are 6–10 days old. Older queens lay 200 eggs per day. Workers feed older larvae. Adults emerge 9–15 days after pupation. Colonies have 100 000–500 000 workers. Queens live for five or more years. Sunny pastures, meadows and fields are preferred for mounds. Size depends on soil characteristics and land disturbance, but can be 25–30 cm tall. Mounds are smaller in sandy soil.

Damage and Symptoms: Fire ants eat the embryo of planted seed and occasionally roots and leaflets of germinating seeds. Damaged seeds have ragged holes and do not germinate or produce stands. Damage is influenced by the amount of time seed is vulnerable (dry and germinating) and by abundance of ants. Fire ants are more abundant with reduced tillage.

Management: Treated seed should be planted into a prepared seedbed and the soil packed, especially when dry. Practices that prevent wireworms reduce damage by fire ants.

White grub, *Phyllophaga crinita* (Burmeister)

Description and Biology: Adults are May or June beetles. 'C'-shaped larvae are white with brown heads and food showing through the last abdominal segment (Fig. 2). Mature larvae are 20–25 mm. Larvae overwinter and are in the damaging stage when sorghum is planted. A generation requires one year or more.

Figure 2 White grub.

Damage and Symptoms: Larvae chew roots. Seeds germinate, but seedlings shorter than 15 cm die. Stand loss occurs 7–10 days after plants emerge. One white grub destroys plants in 0.3–0.5 m of row. Plants not killed are stunted and never produce grain. Larger plants lodge.

Monitoring: The last larval instar overwinters and is sampled before sorghum is planted; control may be ineffective after planting. Before planting, examine soil in 32 cm^3. White grubs may be 32 cm deep in soil in winter or 7 cm in spring. One half grub per 32 cm^3 causes economic damage.

Management: Planting after a previous non-grass crop prevents white grubs. Treated seed or insecticide applied in a band at planting is effective.

Southern corn rootworm, *Diabrotica undecimpuncta howardi* Barber

Description and Biology: The 15-mm larva has a slender, white or yellow, wrinkled body with brown head and last abdominal segment (Fig. 3). The 6-mm adult spotted cucumber beetle is yellowish-green with 11 black spots on the forewings, black head and antennae half the body length. Adults overwinter in residue or are active during winter. Females deposit eggs in soil around bases of plants in spring. Two generations develop per year.

Damage and Symptoms: Larvae eat and bore into roots or the stalk to chew the crown, destroying the apical meristem and main stem. Young plants stunt with 'deadhearts.' Tillers delay or cause non-uniform maturity. Plants lodge later.

Monitoring: Rootworms cannot be controlled after a crop is planted, and abundance cannot be determined before planting. Insecticide is based on history of rootworms in a field.

Management: Keep fields free of weeds, disk 30 days before planting, rotate with non-grass, and plant early. Treated seed or in-furrow granular or liquid insecticide at planting is effective.

5.2 Pests of leaves and stems

Cutworms (Noctuidae)

Description and Biology: Cutworms are larvae of nocturnal moths. Larvae are plump, smooth and greyish-white to greyish-black or brown depending on the species (Fig. 4).

Figure 3 Southern corn rootworm larva on roots.

Figure 4 Surface-feeding cutworm.

Mature larvae are 30–50 mm. Cutworms overwinter as larvae, adults or pupae in soil, residue or clump grass. Eggs laid on grass or in soil hatch in 2–14 days. Some species lay eggs on weeds in low areas that flood. Pupation is in soil. One to five generations are produced per year.

Damage and Symptoms: Cutworm species cut stems and kill plants near the soil surface (surface-feeding), chew and rag leaves on above-ground parts (climbing or army), or feed underground on roots of seedlings and kill small or stunt larger plants (subterranean).

Monitoring: Assess plants severed per metre by surface-feeding cutworms. Insecticide application should be based on threat to stand. Significant loss occurs when above-ground cutworms eat 30% of leaf tissue. Sample subterranean species below ground before planting.

Management: Tillage destroys residue, prepares the seedbed and controls weeds. Aerial or ground application of insecticide is effective against surface and climbing cutworms. Late-afternoon application is effective because cutworms hide in soil during the day. Baits are expensive. A 15- to 18-cm band of insecticide incorporated into 2.5–5 cm of soil at planting controls subterranean cutworms.

Yellow sugarcane aphid, *Sipha flava* (Forbes)

Description and Biology: The 2-mm aphid is lemon yellow or occasionally green, covered with short, black spines, with two double rows of dark spots on its back (Fig. 5). Cornicles are short. Winged and wingless forms live in a colony. Hosts include johnsongrass and dallisgrass, *Paspalum dilatatum* Poir. Without mating, females give birth to live young for 28 days. A female daily produces two nymphs that mature in 13–19 days.

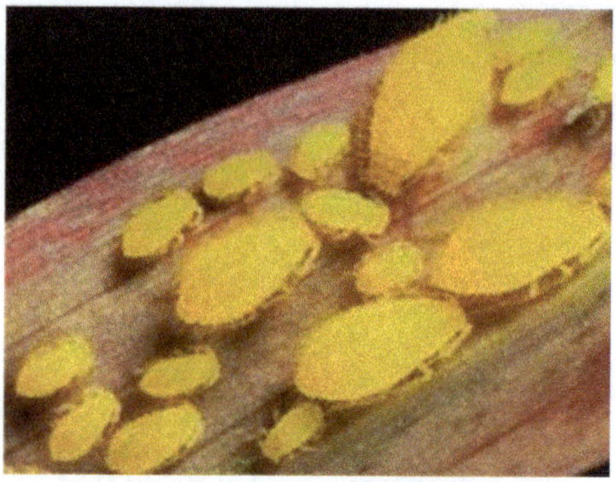

Figure 5 Yellow sugarcane aphid nymphs and adults on leaf.

Table 1 Yield loss based on symptoms of yellow sugarcane aphid damage to one to three true-leaf sorghum

Discolouration symptom	Loss per plant (%)
None	0
Localized	8
Less than one entire leaf	11
One entire leaf	31
More than one leaf	54
More than two leaves	77
Dying or dead plant	100

Damage and Symptoms: Aphids feed under lower leaves and inject toxin. Leaves of seedlings are purple in colour and older plants are yellow. Plants die or are stunted. Older plants lodge and have stalk rot.

Monitoring: Purple seedlings may indicate infestation by aphids or cold and other factors. Inspect plants the 1st week after emergence and twice weekly until five true leaves. Aphids significantly damage small seedlings (one to three leaves) after one week (Table 1). Insecticide is justified when 10–40% of seedlings are infested (Table 2). Older plants better tolerate aphids.

Management: Plant when conditions favour seedling growth. Predators such as lady beetles and lacewings eat aphids. Treated seed or foliar insecticide is effective.

Corn leaf aphid, *Rhopalosiphum maidis* (Fitch)

Description and Biology: The 2-mm aphid is dark bluish-green, with black legs, cornicles and antennae (Fig. 6). A colony has both winged and wingless forms. Females without mating give birth to live young. A generation requires seven days.

Damage and Symptoms: Aphids infest the whorl of pre-booting sorghum, leaves, stems or panicles. Nymphs and adults suck juice but do not inject toxin. Whorl leaves are mottled yellow. Aphids transmit maize dwarf mosaic virus. Moulds grow on honeydew. Honeydew on panicles hinders harvest.

Table 2 Economic injury levels based on percentage of sorghum seedlings infested by yellow sugarcane aphid

Control cost ($) per hectare	Crop market value ($) per hectare										
	250	300	350	400	450	500	550	600	650	700	750
	Infested at one true-leaf stage										
15	15	12	11	10	9	8	7	6	6	5	5
20	20	16	14	12	11	10	9	8	8	7	7
25	25	20	17	15	13	12	11	10	10	9	9
30	30	25	21	18	16	14	13	12	11	11	10
	Infested at two true-leaf stage										
15	26	21	18	16	14	13	12	11	11	10	10
20	35	28	23	20	18	17	16	15	14	13	13
25	44	35	29	25	23	21	19	18	17	16	16
30	52	43	36	31	28	25	23	21	20	19	18
	Infested at three true-leaf stage										
15	67	53	46	41	36	33	31	29	27	25	24
20	89	72	61	54	49	44	40	37	35	33	32
25	*	90	76	67	61	55	50	46	43	40	38
30	*	*	91	81	73	66	60	55	51	47	44

* Do not treat.

Figure 6 Corn leaf aphid on leaf.

Monitoring: Pull and unroll whorl leaves to sample. The aphid rarely causes economic damage and is considered beneficial because it does not inject toxin. Natural enemies increase and remain to eat greenbugs and other insects.

Management: Sorghum usually tolerates many corn leaf aphids without damage, but death of seedlings causes stand and yield loss. Aphids occasionally are abundant enough to hinder panicle exsertion and development. Insecticides are effective but usually not needed.

Greenbug, *Schizaphis graminum* (Rondani)

Description and Biology: Nymphs and adults are green, with darker green mid-dorsal stripe (Fig. 7). Adults are 1.6-mm, pear-shaped aphids. Tips of cornicles and legs are black. Colonies have both winged and wingless forms. Females without mating produce 80 live nymphs in 25 days. Nymphs mature in seven days.

Damage and Symptoms: This key pest sucks juice and injects toxin. Wheat is the winter host, with sorghum the host during summer. Grasses such as johnsongrass are intermediate hosts. Greenbugs feed in colonies under leaves and produce honeydew. Red spots indicate greenbugs. Damaged leaves turn yellow, and then brown, and eventually die. Greenbugs kill and cause stand loss of seedlings. Greenbugs on booting and older plants reduce yield due to fewer and smaller kernels and lodging. Greenbugs transmit maize dwarf mosaic virus and predispose sorghum to charcoal rot.

Monitoring: Greenbugs may be on any part including the whorl and when cool, in soil at bases of plants. Examine 40 plants per 32 ha each week. Examine more than 40 plants in fields larger than 32 ha or if decision to control is difficult. Do not inspect plants only along field borders. Determine leaf damage, greenbug abundance, predators and parasitized greenbugs, plant size, growth stage and condition of the crop.

Yield is affected by 500–1 000 greenbugs per plant. Control greenbugs when 20% of seedlings are infested or 20% of plants to boot stage have red spots or yellowing (Table 3). Greenbugs on boot to heading sorghum should be controlled when leaves have red spots or yellowing and one leaf dead. After flowering and before hard dough, control greenbugs before they kill more than two leaves on 20% of plants. A leaf is considered dead when more than 75% is red, yellow or brown. Drought-stressed plants cannot tolerate greenbugs. Insecticide is unnecessary when more than 20% of greenbugs are parasitized, or if abundance is less than the previous week.

Management: Greenbugs increase five- to sixfold per week. Rainfall and predators suppress greenbugs early, although natural enemies lag by one or two weeks. *Lysiphlebus testaceipes* (Cresson) parasitizes and decreases aphids late.

Figure 7 Greenbug nymph and adults.

Table 3 Economic threshold levels for greenbug on sorghum at different growth stages

Plant size	When to treat
Emergence to 5 cm	20% of plants visibly damaged (beginning to yellow), with greenbugs on plants
Larger plant to boot	Greenbug colonies causing red spotting or yellowing of leaves before any entire leaves on 20% of plants are killed
Boot to heading	At the death of one functional leaf on 20% of plants
Heading to hard dough	When greenbugs are sufficient to kill two normal-sized leaves on 20% of plants

Use resistant hybrids. Greenbug biotypes C, E, I and K infest sorghum, with E and I most common. Biotype I damages sorghums resistant to E, but biotype I-resistant hybrids are resistant to E. Economic threshold levels are based on damage and are the same for resistant and susceptible hybrids.

Greenbug can be controlled by some organophosphorous insecticides. Low rates spare predators and parasitoids when greenbug abundance reaches the economic threshold. Insecticide on seed, in-furrow at planting or on foliage control greenbugs.

Sugarcane aphid, *Melanaphis sacchari* (Zehntner)

Description and Biology: The sugarcane aphid occurs in Africa, Asia, Australia, tropical America and recently throughout the southern United States. Adults and nymphs resemble greenbugs but are yellow to buff with all-black cornicles and no dorsal green stripe (Fig. 8). Wingless females produce 60–100 nymphs in 13–20 days. A generation averages about six days.

Damage and Symptoms: Aphid nymphs and adults suck juice from leaves and stems. They prefer older sorghum from booting through flowering but also infest leaves of young plants. Symptoms of damage include leaf discolouration and death, reduced panicle formation and delayed or no development of grain. Pre-flowering plants may be killed. Sooty mould that interferes with harvesting (Kerns et al. 2015) grows on honeydew excreted by sugarcane aphids.

Figure 8 Sugarcane aphid nymphs and adults on leaf.

Monitoring: Sorghum through flowering should be checked twice weekly, starting on edges where wind is toward a field (Bowling et al. 2016). Count the number of aphids on a leaf from the bottom and top of each of 20 plants in at least four areas of a field. The economic threshold is 50 or more aphids per leaf when 25% of plants are infested. Yield is reduced 45–180 kg per hectare by 50–500 aphids per leaf of pre-flowering sorghum.

Management: Control johnsongrass near the sorghum field. Resistant hybrids are available. Natural enemies include several species of lady beetles, lacewings, syrphid flies and chalcid wasp. Systemic seed treatments are effective for about 30 days against early aphids. Several foliar insecticides are effective.

Chinch bug, *Blissus leucopterus leucopterus* (Say)

Description and Biology: Nymphs are yellow then red except the first two abdominal segments. Subsequent instars become darker red with a yellow band across the abdomen. The last nymphal instar is black and grey with a white spot on the back. The 4-mm adult is black, with reddish-yellow legs, and white forewings, each with a black triangle at the middle of the margin. Adults overwinter in bunch grass and move to small grains to lay eggs behind lower leaf sheaths, on roots, or in soil. First-generation nymphs move to sorghum or maize. In southern regions, adults move directly after overwintering into sorghum. A generation requires 30–40 days. A second generation develops and disperses within a field; there may be an additional generation in southern regions.

Damage and Symptoms: Adults and nymphs suck juice from stems or underground parts (Fig. 9). Older plants withstand more but become red, weak, stunted and lodged. Dry weather aids outbreaks.

Monitoring: Examine small plants and dry soil at five or more sites. Bugs live behind leaf sheaths of large plants. Apply insecticide when two or more adults are on 20% of seedlings, or when 75% of plants taller than 15 cm are infested. One bug per seedling reduces yield by 2%.

Management: Plant early and stimulate dense stands. Insecticide applied to seed or at planting for soil insects controls chinch bugs. Apply foliar insecticide with sufficient spray directed at plant bases. Control is seldom obtained on boot or older plants.

Banks grass mite, *Oligonychus pratensis* (Banks)

Description and Biology: Banks grass mite and less frequently two-spotted spider mite, *Tetranychus urticae* Koch, infest sorghum and other grasses (Fig. 10). Mites are common in arid regions. Females are 0.40–0.45 mm and larger than males. Both sexes are green

Figure 9 Chinch bug nymphs and adults on leaf.

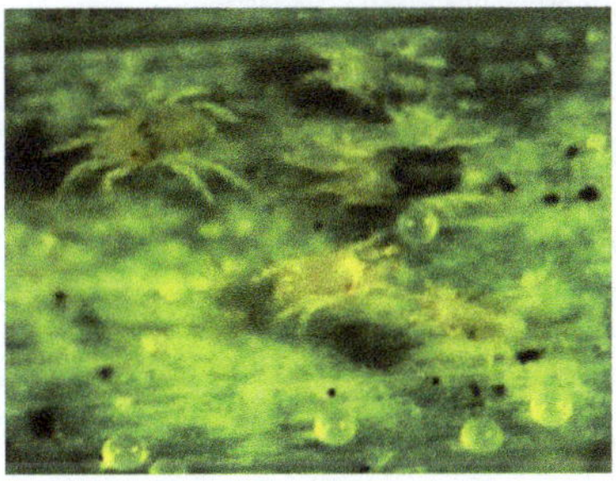

Figure 10 Banks grass mite adults, immatures and eggs.

with salmon-coloured palpi and first two pairs of legs. Two dark spots of food show through the body wall. Eggs (50 per female) are laid in webbing under leaves. The white, spherical eggs one-fourth the size of the female hatch in three or four days. Six-legged nymphs are light green; older eight-legged nymphs are darker green. A generation requires 11 days.

Damage and Symptoms: Mites suck juice and web undersides of leaves, beginning along the midrib. Infested leaves are yellow on top, later turn red and may turn brown, and eventually die. Abundant mites spread up a plant and web panicles. Mites prevent kernel filling or cause lodging and harvest loss after the hard-dough stage.

Monitoring: Inspect lower leaves for spider mites and webbing. Infestations begin along borders and spread throughout a field. Hot and dry weather favours mites. Insecticides against key pests induce mites because of tolerance to insecticide and destruction of beneficials.

Management: Not planting near small grains avoids infestation. Because mites increase rapidly on moisture-stressed plants, irrigation, where available, should be applied. Natural enemies are not always effective. Insecticide is justified when 30% of leaves of most plants are infested. Thorough coverage is required because mites often are resistant and subsequent insecticide applications are ineffective.

5.3 Pests of panicles

Sorghum midge, *Stenodiplosis sorghicola* (Coquillett)

Description and Biology: Sorghum midge is the most widely distributed sorghum insect pest and one of the most damaging in the southern United States. It occurs in most regions of the world where sorghum is grown, except Southeast Asia. The fly is 1.3 mm, orangish-red with yellow head, brown antennae and legs, and grey wings. During the single day of life, each female lays 50 yellow eggs between glumes of flowering spikelets (Fig. 11). The cylindrical eggs are 0.1–0.4 mm and hatch in two or three days. Larvae are first colourless but turn orange when grown. Larvae develop in 9–11 days and pupate between glumes of spikelets. Pupae develop in three days. The pupa moves upward to the spikelet tip where a clear pupal case remains after adult emergence. Generations require 14–16 days and multiple generations occur per year. Infestation is worse when sorghum flowering is extended by a range of planting dates or sorghum maturities. In the United States, larvae overwinter in cocoons in sorghum or johnsongrass spikelets in soil. Sorghum midges infest johnsongrass in spring before moving to sorghum where they increase until late in the season.

Damage and Symptoms: The larva feeds on the fertilized ovary, preventing kernel development. Usually not all kernels on a panicle are infested.

Monitoring: Monitor daily at mid-morning during flowering because adults live less than a day and a new brood is present each day. Visually inspect or use a clear plastic bag over panicles to detect and count adults. Inspect plants along borders, especially downwind of earlier flowering sorghum from which sorghum midges emerge. If few sorghum midges are at borders, there is no need to sample the entire field. If sorghum midges along borders equal or exceed the economic threshold of one per flowering panicle, additional panicles (avoiding plants within 45 m of borders) should be inspected and abundance calculated. Sample 20 panicles per 8 ha.

Management: Eliminating johnsongrass suppresses sorghum midge. Planting hybrids with uniform maturity early prevents late flowering and damage. Resistant hybrids are one-fifth as damaged and have economic injury levels five times higher than susceptible hybrids (Table 4). Ploughing residue kills overwintering larvae, reducing abundance the next year.

Figure 11 Sorghum midge adult laying eggs in flowering spikelet (photo by N. Yaro Diarisso).

Table 4 Economic injury levels based on number of adult sorghum midge per flowering panicle

Control cost ($) per hectare	Crop market value ($) per hectare										
	250	300	350	400	450	500	550	600	650	700	750
	Susceptible sorghum										
15	2.4	2.0	1.8	1.5	1.3	1.2	1.1	1.0	0.9	0.9	0.8
20	3.0	2.6	2.3	2.0	1.8	1.6	1.5	1.3	1.2	1.1	1.0
25	3.7	3.4	2.9	2.6	2.3	2.0	1.8	1.6	1.5	1.3	1.2
30	4.5	4.1	3.6	3.2	2.8	2.5	2.2	1.9	1.7	1.5	1.4
	Resistant sorghum										
15	12.0	10.0	9.0	7.5	6.5	6.0	5.5	5.0	4.5	4.5	4.0
20	15.0	13.0	11.5	10.0	9.0	8.0	7.5	6.5	6.0	5.5	5.0
25	18.5	17.0	14.5	13.0	11.5	10.0	9.0	8.0	7.5	6.5	6.0
30	22.5	20.5	18.0	16.0	14.0	12.5	11.0	9.5	8.5	7.5	7.0

Assess crop development, yield potential and sorghum midge abundance daily during flowering. Apply insecticide when 25–30% of panicles are flowering and again at 3- to 5-day intervals if adults still are present. Because sorghum midges lay eggs in flowering spikelets (yellow anthers), damage occurs until a panicle (7 to 9 days) or field (2 or 3 weeks) has finished flowering. Peak flowering occurs on the 8th day, when 19% of sorghum in a field is flowering.

Corn earworm, *Helicoverpa zea* (Boddie)

Description and Biology: Moths are yellowish to reddish-brown and 19 mm, with 38-mm wingspans. Females are active at evening and live for 12 days. A female deposits 350–3 000, 1.2-mm, white, ribbed eggs that darken and hatch in 3–5 days. New larvae are white, but later turn pink, green, yellow or blackish (Fig. 12). They have a dark dorsal stripe divided by a white line making the stripe appear doubled. Mature larvae are 38–50 mm. The pupa is in the soil and overwinters.

Damage and Symptoms: Larvae infest whorls and damage panicles later. Whorl leaves unfold with rows of 'shot holes' that do not affect yield. Small larvae chew florets and then developing kernels. Later instars destroy maturing kernels. The last two instars cause 80% of damage. Frass is in infested panicles and on top of upper leaves or soil. Kernels of infested panicles may mould.

Monitoring: Holes in whorl leaves as they unroll indicate corn earworms. Larvae are cannibalistic. Inspect panicles at flowering and 5-day intervals until hard dough. Shake panicles into a 19-L 'beat-bucket' to count, especially small larvae (less than 6 mm). Inspect 30 panicles per 32 ha. Insecticide may be justified if larvae reduce leaf area 30% or damage developing panicles or growing points in whorls. Two corn earworms 13 mm long or one longer than 13 mm per panicle justify insecticide (Table 5).

Management: Infestation is less in early sorghum. Use hybrids with loose or open panicles. Natural mortality, predators, parasitoids, pathogens and cannibalism suppress larvae. Insecticide is not very effective against large larvae.

Figure 12 Corn earworm on panicle.

Table 5 Economic injury levels based on number of corn earworm and fall armyworm larvae per sorghum panicle

Control cost ($) per hectare	Crop market value ($) per hectare										
	250	300	350	400	450	500	550	600	650	700	750
15	1.5	1.3	1.1	0.9	0.8	0.8	0.7	0.6	0.6	0.5	0.5
20	2.0	1.7	1.4	1.2	1.1	1.0	0.9	0.8	0.7	0.6	0.6
25	2.5	2.1	1.8	1.6	1.4	1.2	1.1	1.0	0.8	0.7	0.7
30	3.0	2.6	2.3	2.0	1.7	1.5	1.3	1.2	1.0	0.9	0.8

Fall armyworm, *Spodoptera frugiperda* (J. E. Smith)

Description and Biology: Young larvae are green with black heads. Mature larvae are green to greyish brown with a light, inverted-Y suture on the head and dorsal lines along the body (Fig. 13). Moths have variegated-grey forewings. Eggs in masses covered with scales from the moth are deposited on leaf undersides or panicles. A generation requires 1 month. Adults migrate north from tropical America each year as temperature warms.

Damage and Symptoms: Fall armyworms infest whorls and panicles and produce frass. Elongated feeding areas or rows of holes are visible when whorl leaves unfold. Larvae may feed on the boot. Larvae feed on florets and kernels. The last two larval instars cause 80% of damage. Moulds grow in infested panicles.

Monitoring: Sampling is the same as for corn earworm. Insecticide is justified with 30% leaf or panicle damage or when two larvae less than 25 mm or one larva longer than 25 mm infest panicles.

Management: Planting hybrids with loose panicles early escapes infestation. Wasps and flies parasitize, bugs and beetles prey, and pathogens infect and kill larvae. Insecticide is not very effective against large larvae.

Sorghum webworm, *Nola sorghiella* Riley

Description and Biology: The sorghum webworm is an occasional pest of late panicles in humid areas. White moths with 13-mm wingspans are active at night, laying 100 eggs singly on sorghum flowers or kernels. Eggs (0.5 mm diameter and 0.25 mm tall) are round to oval and flat. They are first white but turn yellow to brown before hatching in three or four days. Larvae are flat, yellowish or greenish brown, with four longitudinal red to black dorsal stripes (Fig. 14). Larvae 13 mm when mature are covered with spines and hair. The larval stage requires 13 days. The reddish-brown, slender, subcylindrical pupa is in a cocoon. The pupal stage lasts six days, and adult stage lasts five days. A generation requires one month, with six per year. Larvae diapause on plants.

Damage and Symptoms: Young larvae chew florets. Older larvae gnaw circular holes and eat starch in maturing kernels they partly consume. A larva can eat 12 kernels in 24 hours. Young larvae when disturbed suspend from silken threads.

Monitoring: Inspect flowering panicles every five days until kernels are hard dough. Shake panicles into a 19-L bucket. Inspect 30 plants per 32 ha.

Management: Destroy residue to kill overwintering larvae, and plant sorghum with open panicles early. Five small larvae per panicle justify insecticide.

Figure 13 Fall armyworm, whorl and panicle pest (photo by B. Drees).

Figure 14 Sorghum webworm on panicle.

Bugs: Rice stink bug, *Oebalus pugnax* (Fabricius); Southern green stink bug, *Nezara viridula* (L.); Conchuela, *Chlorochroa ligata* (Say); Leaf-footed bug, *Leptoglossus phyllopus* (L.); False chinch bug, *Nysius raphanus* Howard

Description and Biology: Bugs move to sorghum during kernel development and usually are clumped in a field.

Rice stink bug adults are straw, shield-shaped, and 12 mm (Fig. 15). Adults overwinter in grass. Females lay 10–47 green, cylindrical eggs in two rows. Eggs hatch in five days. Nymphs develop into adults in 15–28 days.

The southern green stink bug is green, shield-shaped, and 19 cm (Fig. 16). Females deposit 300–500 eggs in clusters of 30. Eggs hatch in seven days, and adults emerge in six weeks. Three to five generations per year occur at 5- to 6-week intervals. Adults live 40–60 days. Adults overwinter under litter, bark or plant material.

The conchuela is a 20-mm, dull olive to ash grey stink bug (Fig. 17). Orangish-red bands are along the lateral margins of the thorax and wings, with a spot of the same colour on the back at the base of the wings.

Figure 15 Rice stink bug.

Figure 16 Southern green stink bug.

Figure 17 Conchuela.

Figure 18 Leaf-footed bug on panicle.

The leaf-footed bug is brown, oblong and 20 mm (Fig. 18). A white band extends across the forewing. The tibiae of the hind legs are leaf-like. Eggs are in rows of 15–35. Nymphs are red.

The false chinch bug is grey to brown and 3 mm (Fig. 19). The bugs move from wild hosts into sorghum.

Damage and Symptoms: Bugs suck juice from kernels and panicles. The number of bugs that reduces yield varies by species and stage of kernel development. Bugs do more damage to early than to hard-dough kernels. Nymphs and adults reduce kernel weight, size and quality. Fungi infect, blacken and deteriorate kernels. Damaged kernels may drop during harvesting.

Monitoring: A beat-bucket can be used to estimate bugs per panicle, but adults fly. Inspect 30 plants per 32 ha.

Management: Early sorghum avoids bugs that infest when other hosts are unavailable. Some species of stink bugs prey on pests and are beneficial.

Figure 19 False chinch bugs on panicle.

Table 6 Economic injury levels based on per-panicle number of adult rice stink bugs infesting sorghum at milk or soft-dough stage of kernel development

Control cost ($) per hectare	Crop market value ($) per hectare										
	250	300	350	400	450	500	550	600	650	700	750
	Beginning at milk stage										
15	7	6	6	6	5	5	5	4	4	4	3
20	8	7	7	6	6	6	5	5	5	4	4
25	8	8	7	7	6	6	6	5	5	5	4
30	9	8	8	7	7	6	6	6	5	5	5
	Beginning at soft-dough stage										
15	9	8	8	7	7	6	6	6	5	5	5
20	10	9	9	8	8	7	7	6	6	6	5
25	12	11	10	10	9	9	8	8	7	7	6
30	13	12	11	10	10	9	9	8	8	7	7

Economic injury levels at milk- and soft-dough stages are four to nine and five to 13 rice stink bugs (Table 6) or two to six or four to 11 southern green stink bugs, conchuela or leaf-footed bugs per panicle (Table 7). The economic injury level is 140 false chinch bugs per panicle at the milk stage.

Stalk borers: Sugarcane borer, *Diatraea saccharalis* (Fabricius); Neotropical borer, *D. lineolata* (Walker); Southwestern corn borer, *D. grandiosella* Dyar

Description and Biology: Borers do not seriously damage sorghum in the United States, but are key pests in some areas of the world. Species resemble each other and have similar biologies. A mature larva is 25 mm, with brown head and thorax and white to yellow body (Fig. 20). Most segments have round, brown or black spots. Mature larvae with light or no spots overwinter inside stalks or root crowns after harvest. In spring, buff moths emerge

Table 7 Economic injury levels based on per-panicle number of adult southern green stink bugs, conchuelas and leaf-footed bugs infesting sorghum at milk or soft-dough stage of kernel development

Control cost ($) per hectare	Crop market value ($) per hectare										
	250	300	350	400	450	500	550	600	650	700	750
	Beginning at milk stage										
15	5	4	4	4	4	3	3	3	3	2	2
20	5	5	5	4	4	4	4	3	3	3	3
25	6	6	5	5	5	4	4	4	4	3	3
30	7	6	6	6	5	5	5	4	4	4	4
	Beginning at soft-dough stage										
15	8	7	7	6	6	5	5	4	4	3	3
20	9	8	8	7	7	6	6	5	5	4	4
25	10	9	9	8	8	7	7	6	6	5	5
30	11	10	9	9	8	8	7	7	6	6	5

Figure 20 Stalk borer.

from pupae and lay clusters of 12–20 oval eggs that overlap like shingles on leaves. Eggs hatch in 3–7 days. The larval stage lasts 25 days and pupal stage lasts about ten days. One to three generations develop per year.

Damage and Symptoms: Young larvae feed on leaves or leaf axes. Older larvae tunnel stalks. Infested stalks have smaller diameters and lodge. Boring causes the peduncle to break and panicle to fall. Damaged stalks are susceptible to rot.

Monitoring: Count eggs and small larvae on leaves before larvae bore into stalks. A hole near the leaf axis indicates a larva in the stalk.

Management: Plant early. Ploughing to break stubble and bury residue after harvest destroys overwintering larvae by exposure to cold. Insecticide is rarely justified.

Mexican rice borer, *Eoreuma loftini* (Dyar)

Description and Biology: The cream, 12- to 19-mm moth has a dark spot in the centre of the forewings and enlarged mouthparts protruding from the head. The cream larva has an orangish-brown head and two broken, purplish-red lines on each side along the body.

Masses of five to 100 cream, globular eggs are cemented between layers of dry leaf tissue near the ground or on upper green parts. Mature larvae, 19–25 mm, chew pupal chambers to the outer edge inside stalks. The pupa is 19–22 mm, orangish-brown, with tubercles on the abdomen. Moths emerge through 'windows' of tissue covering the pupal chambers. A generation requires 45–50 days when warm.

Damage and Symptoms: New larvae move to leaf blades or sheaths. Feeding causes the sheath to purple. Damaged leaves die. Tunnelling in the leaf sheath produces a mine. Larvae feed in leaf sheaths for several weeks and then chew the internode into the stalk. Most tunnelling is near the stalk surface. Tunnels usually are vertical, or horizontal or diagonal. Tunnels are filled with frass unlike tunnels of other borers. Larvae feeding early near the growing point cause 'deadheart'. Young plants die, but usually tiller. Moths lay eggs on deadhearts. Stalk tunnelling results in incomplete panicle filling or aborted development. Fungi enter damaged internodes and discolour the pith. Stalks lodge. Tillers and post-harvest regrowth are susceptible.

Management: Destroying debris before planting and maintaining healthy plants reduce dry leaf material for oviposition. Damage is minimal when sorghum matures at the normal time. Apply insecticide for larvae in leaves before stalk boring.

Sugarcane rootstock weevil, *Anacentrinus deplanatus* (Casey)

Description and Biology: The brown or black weevil is 9 mm long and 2.5 mm wide. Adults overwinter in residue on the ground. Weevils infest wild grasses in spring and move into sorghum. Females chew holes and deposit 16 eggs in six days. Eggs are cream, oval, 0.25 mm in diameter and 0.5 mm in length. Larvae are 5-mm, white and legless. Pupae are white and then brown. Larvae and pupae develop in 25 and ten days, respectively. A generation requires 41 days.

Damage and Symptoms: Weevils feed on young sorghum and rootstocks. Larvae tunnel into stalks near the soil surface and are at nodes and near the outer surface. Damage causes stalks to appear drought-stressed and lodge. Charcoal rot enters exit holes and feeding tunnels.

Monitoring and Management: Sugarcane rootstock weevil infests sorghum in dry years. Promote early, vigorous plant development. Insecticide is not available.

5.4 Pests of stored grain

Beetles and moths infest stored grain, with most damage by larvae. Most have short developmental times, high rates of reproduction, and long lives. Primary pests attack and develop inside whole kernels. Secondary pests eat broken kernels, grain dust or mould of stored grain. Insects directly consume grain or indirectly contaminate it with faeces, odour, webbing, cast skins or dead insects. Insect-caused heating and moisture can mould and cake grain.

Grain weevils: Rice weevil, *Sitophilus oryzae* (L.); Maize weevil, *S. zeamais* Motschulsky; Granary weevil, *S. granarius* (L.)

Weevils are primary pests of stored grain worldwide, especially in warm, humid areas. Adult rice and maize weevils are 4 mm, reddish-brown, with four red or yellow spots on the wings. Weevils infest grain in the field and storage. Females bore into kernels, deposit eggs and cover the holes with a gelatinous substance. Females lay 300–500 eggs during four or five months. Eggs hatch in three days. Larvae are legless, short and white with brown heads, and mature in 3–6 days. Five to seven generations develop per year. The granary weevil prefers temperate climates. It lacks red or yellow wing spots and cannot fly.

Grain borer: Lesser grain borer, *Rhyzopertha dominica* (Fabricius)

This primary pest is cosmopolitan, 4 mm, slender, cylindrical, shiny brown or black, with rough wing surfaces. The head is covered by a hood-shaped thorax. Adults lay 300–500 eggs on grain or grain dust. Eggs hatch in 5–11 days into grubs that bore and eat grains. The larval period is 25–50 days. Pupation is in grain or grain dust for seven or eight days. Three or four generations lasting two months develop per year.

Grain moth: Angoumois grain moth, *Sitotroga cerealella* (Olivier)

This cosmopolitan insect is a primary pest. Moths are buff or yellowish-brown with 13-mm wingspans. Hind wings have a fringe of hairs longer than the wing width. Females lay 400 eggs on kernels of panicles in the field or storage. Eggs are oval and white, but turn red and hatch in one week. Larvae develop in two or three weeks. Pupation requires one week.

Grain beetles: Flat grain beetle, *Cryptolestes pusillus* (Schnherr); Rusty grain beetle, *C. ferrugineus* (Stephens)

Flat and rusty grain beetles are widely distributed secondary pests. These smallest (1.5 mm) beetles in stored grain are flat, oblong and reddish-brown with antennae two-thirds the body length. White eggs are deposited loosely or in crevices in grain. Larvae consume dead insects or germ from kernels. Larvae form gelatinous cocoons to which food particles adhere. Egg to adult requires 5–9 weeks.

Flour beetles: Red flour beetle, *Tribolium castaneum* (Herbst); Confused flour beetle, *T. confusum* (Jacquelin du Val); Sawtoothed grain beetle, *Oryzaephilus surinamensis* (Linnaeus); Merchant grain beetle, *O. mercator* (Fauvel)

Red and confused flour beetles are cosmopolitan, secondary pests. Adults are 3 or 4 mm, flat, oblong and chestnut brown. The head and upper thorax of confused flour beetles have minute punctures. Wing covers are ridged lengthwise and have punctures between the ridges. Adults lay 450 minute, cylindrical, white eggs covered with sticky secretion covered with flour. Eggs hatch in 5–12 days. Yellowish-white, cylindrical grubs covered with hairs develop in 27–29 days. The pupa has no case. Adults emerge in 3–7 days. Four to seven generations of one month to four months develop per year. Optimal conditions are 35°C and 75% humidity.

Sawtoothed grain beetle is a secondary pest. The adult is 2.5 mm, slender, flat and brown. Six sawtooth projections are on each side of the thorax. Merchant grain beetles resemble sawtoothed grain beetles. Both have similar biologies.

Flour moths: Rice moth, *Corcyra cephalonica* (Stainton); Indian meal moth, *Plodia interpunctella* (Hubner)

Rice moth is a secondary pest. Moths are greyish brown, with 14- to 24-mm wingspans and heads with tuft of scales. Females lay 200 spherical, white eggs that hatch in 3–5 days.

Cream-coloured larvae live in galleries of webbed food and frass. Larvae develop in 20–30 days and pupae in nine or ten days.

The Indian meal moth is a widely distributed secondary pest. The wingspan is 18 mm, with red brown and copper on the outer two-thirds and grey on the inner one-third of forewings. Females lay 500 eggs. Egg to adult is 26 days.

Monitoring: Check grain every 4–6 weeks when cool or 2–4 weeks when warm. A probe should be inserted horizontally 5 cm under the surface in the centre of a bin. Collect additional samples near sides of bins. Samples should be less than 6 m apart. After sampling near the surface, insert the probe top to bottom in the grain. One primary or five secondary insects per quart of grain cause economic damage.

Monitor moths with pheromone-baited sticky traps. A grain probe that functions as a pitfall trap detects grain beetles. Trap insects in two to four grain probe traps for 48 hours at the grain surface. Ten or more insects per probe during 24 hours cause concern.

Management: Manage storage insects by sanitation, well-constructed and maintained bins, residual insecticide in bin and grain, and fumigant. Granaries should have no leaks, double walls or floors, or places for grain to accumulate. Grain should be dried to 12% moisture. Treat granaries with insecticide before storing grain. Insecticide can be applied to the grain surface, or the grain fumigated.

6 Other insect pests of sorghum in other countries

6.1 Pests of leaves

Shoot fly, *Atherigona soccata* (Rondani)

Description and Biology: The shoot fly is widespread in sorghum-growing areas of semi-arid tropics, but not Americas or Australia. It is a key pest in Asia, Africa and Mediterranean Europe. Adults are small, grey flies that deposit 2 mm, white, cigar-shaped eggs singly under leaves of one- to seven-leaf sorghum. Eggs hatch in 1–3 days and larvae move into the whorl. Mature larvae are 6 mm and yellow. The larval stage lasts 8–10 days and pupal stage last eight days. Pupation is in soil. A generation is 17–21 days.

Damage and Symptoms: Larvae cut and kill the growing point, causing 'deadheart'. The dead leaf emits an odour. Young, whitish-yellow maggots eat decaying tissue. Most damage occurs 1–4 weeks after plant emergence. Plants attacked later produce tillers that may be attacked. Late planting during the rainy season increases attack.

Management: Manage by early planting, high seeding rate, resistant varieties and insecticide application 6–12 days after plant emergence.

Shoot bug, *Peregrinus maidis* (Ashmead)

Description and Biology: The bug infests sorghum in India, West Indies and the Philippines. Yellow females are 1.5 times larger than brown males. Wings of short-winged forms extend to the sixth abdominal segment. Long-winged forms have transparent wings. Nymphs and adults live on leaves, in whorls and inside leaf sheaths. Females lay 100 eggs, one to four per slit on the upper surface of the leaf midrib. White, cylindrical, elongated and tapered eggs hatch in 7–10 days. Nymphal instars are completed in 16 days.

Damage and Symptoms: Nymphs and adults behind the leaf sheath suck stems and whorl leaves. Plants weaken, redden, stunt and lodge. Severe infestations kill top leaves, and then older leaves and plants. Infestation at boot twists top leaves and prevents panicles. Sooty mould grows on honeydew.

Management: Predators and parasitoids attack. Insecticide is effective.

Spittle bug, *Poophilus costalis* Walker

Description and Biology: Bugs occur in Africa and Asia. One to 25 nymphs rest head downward inside foam on a plant. Last instars leave the spittle. Adults are 9–11 mm, brown or grey, and able to jump.

Damage and Symptoms: Feeding on leaves and within whorls causes chlorotic spots. Plants may stunt and produce small panicles. Bugs are abundant later on.

Management: Insecticide is not required because plants recover after adult bugs leave.

Oriental armyworm, *Mythimna separata* (Walker)

Description and Biology: The oriental armyworm occurs in Asia, Australia, Pacific Islands and parts of Africa. The moths have brown forewings with dark specks, and whitish-brown hind wings. Eggs are in batches of 20–100 (total 500–900) on lower surfaces of green leaves, dry leaves and grass. Shiny, white, spherical eggs have ridges and hatch in 2–7 days. Larval development requires 14–22 days. Mature larvae are brown with three darker brown dorsal lines and lateral yellow stripes on each side. Pupation for eight or nine days is in cells in soil or inside leaf sheaths.

Damage and Symptoms: At night, larvae feed on leaves. Immature panicles are damaged. Larvae migrate in bands when food is consumed, but occurrence is sporadic.

Management: Cultivation and low plant population suppress damage. Hymenopteran and dipteran parasitoids suppress oriental armyworm. Insecticide is effective.

African armyworm, *Spodoptera exempta* (Walker)

Description and Biology: This armyworm is an occasional pest in East and West Africa. Eggs in groups of 20 hatch in three or four days under leaves. The larval period is 10–20 days. Solitary larvae are green; damaging gregarious larvae are black with green undersides. Pupation is in soil, and adults emerge in one week. Moths migrate at night over long distances. Outbreaks are associated with the rainy season. In eastern Africa, infestations begin in the south and spread northward.

Damage and Symptoms: Outbreaks cause extensive loss. Larvae attack pasture grass and other cereals.

Management: Infestation must be detected early. National or international organizations apply insecticide to control outbreaks.

Grasshoppers: *Oedaleus senegalensis* (Krauss); *Aliopus simulatrix* (Walker)

Description and Biology: These species are pests south of the Sahara. Other species are infrequent pests in other parts of the world. Adult *O. senegalensis* are 30–40 mm and green or brown. *A. simulatrix* is 25 mm and brown. Nymphs develop in two months. *O. senegalensis* survives the dry season as eggs viable for more than a year in soil. *A. simulatrix* survives the dry season as adults in soil. Eggs are laid during the rainy season. Both species breed continuously in the wet season and complete a generation in 40–50 days.

Damage and Symptoms: Most loss is from damage to seedlings or panicles. Nymphs and adults consume sorghum, other cereals and wild grass. Migrant species fly at night.

Management: Insecticide controls grasshoppers.

Migratory locusts: *Locusta migratoria* L.; *Schistocerca gregaria* L.

Description and Biology: Sometimes locusts are pests in Africa and India. Migratory locusts have solitary and swarming phases. Swarming adults are 50 mm, brown and winged. Nymphs have five instars and initially are black, becoming orange and black with age. Low-density populations are brown and grey, while solitary nymphs are green. Nymphs move together several hundred metres per day. Adults swarm and migrate by day, sometimes over long distances. Females lay 50 eggs in three or four pods. Development takes three months. Four generations occur per year.

Damage and Symptoms: Nymphs and adults eat leaves, flowers and panicles. Migratory locusts attack other grasses and occasionally other kinds of plants.

Management: National or international control prevents outbreaks. Insecticide destroys bands of nymphs.

6.2 Stalk borers

Spotted stem borer, *Chilo partellus* (Swinhoe)

Description and Biology: The borer is common in Asia and Eastern and Southern Africa. The moth is medium and straw-coloured. Females lay 500 eggs in masses of 10–18 near midribs. Eggs are oval, flat, overlap like shingles and hatch in four or five days. Larvae develop in 19–27 days. Pupation is inside the stalk. Adults emerge in 7–10 days through holes made by larvae. Last-instar larvae diapause in stalks during the dry season. Adults emerge when the rainy season starts.

Damage and Symptoms: Long, transparent windows in whorl leaves indicate borers. Larvae rag and cause shot holes in leaves. Third-instar larvae bore shoots at plant bases, damaging the growing point and causing deadheart. Larval tunnelling in stalks damages developing panicles that become chaffy or breaks peduncles.

Management: Early and uniform planting and removal of stubble and affected plants is advantageous. Parasitoids and predators suppress borers. Insecticides, including granules in the whorl, are effective.

Maize stalk borer, *Busseola fusca* (Fuller)

Description and Biology: The borer is a key pest in Africa. Eggs are in groups of 30 to150 inside leaf sheaths. The 400–500 eggs laid during five or six days hatch in 5–7 days. Larvae cluster under the leaf sheath and disperse to the whorl. They develop six or seven instars in 24–36 days. Larvae are 35–45 mm and buff to purple-brown with brown heads. The larva cuts a hole for the adult before pupating in the stalk. Adults emerge in 9–12 days. Three generations occur per year. Third-generation larvae diapause during the dry season and develop in six or seven months.

Damage and Symptoms: Young larvae eat leaves. Mature larvae bore stalks and produce deadhearts. Plant growth is retarded and flowering and grain are reduced.

Management: Early planting suppresses infestation. Plough or burn dry stalks and stubble. Parasitoids attack larvae and pupae.

Pink borer, *Sesamia inferens* Walker

Description and Biology: The borer occurs in Asia. Moths are fawn, with brown streaks on the forewings and white hind wings. Females lay 400 eggs in batches of 150 in two or three rows between the leaf sheath and stalk. Eggs are 0.7 mm, cream and spherical, with longitudinal ridges. Before hatching in 5–7 days, the colour changes to brown and then grey. Mature larvae are 25 mm and yellow with purplish-pink tinge and reddish-brown heads. The larval period is 25 days, or 75 days when cool. Pupation is in tunnels in stalks, and adults emerge in 12 days. A generation is six or seven weeks.

Damage and Symptoms: Larvae bore stalks and kill the central shoot, causing deadheart.

Management: Management is the same as for spotted stem borer.

Pink borer (Africa), *Sesamia calamistis* Hmps.

Description and Biology: The insect is present throughout Africa. Moths are beige with brown stripes. Colour differs by geographical area. Yellow, spherical eggs in groups of 40 are laid between the leaf sheath and stalk. Eggs hatch in 4–6 days. After feeding on leaves, larvae bore into stalks for two weeks. Mature larvae are 30 mm long and 3.5 mm wide, with brown heads and buff bodies with pink markings. Pupation is inside stalks for ten days. Larvae do not diapause but survive on wild grass or volunteer sorghum and millet.

Damage and Symptoms: Young larvae eat leaves, especially sheaths. Small, long windows in whorl leaves indicate larvae have eaten the upper but left the lower surfaces. Plants become ragged. Later instars bore through internodes into stalks. The stalk near the panicle is bored, causing peduncles to break, or be chaffy.

Management: Management is the same as for spotted stem borer.

African sugarcane borer, *Eldana saccharina* (Walker)

Description and Biology: The borer occurs throughout Africa south of the Sahara. Females lay 400–600 eggs in batches of two to 200 during two weeks. Eggs hatch in 5–7 days. New larvae are 2 mm and orange, becoming grey. Larvae produce much frass. The larva spins a cocoon from which it emerges and feeds. Males have six or seven and females seven or eight larval instars. The larval period is 20–60 days. Pupation occurs in a cocoon inside the stalk. The pupa is reddish-brown with dorsal ridge. Adults emerge in 8–13 days.

Damage and Symptoms: Larvae feed on leaves and bore the midrib. Mature larvae bore stalks, causing deadhearts. Larvae hang from threads and are blown onto nearby plants.

Management: Destroy volunteer and alternate plants and bury residue. Natural enemies attack the borer.

6.3 Pests of panicles

Bollworm, *Heliothis armigera* (Hubner)

Description and Biology: Bollworm is in Africa, Australia, New Zealand, southern Europe and Southeast Asia. Moths are brown or grey with specks like a 'V' on each forewing. Hind wings are dull with black borders. Females lay 700 spherical, cream eggs singly on panicles. Eggs hatch in 4–6 days. Young larvae are green. Mature larvae are 40 mm and black, brown, green, yellow or pink, with stripes. Larvae develop in three or four weeks. Pupation is in a cell in soil, with adults emerging in 2–4 weeks.

Damage and Symptoms: Larvae infest panicles and eat kernels or whorl leaves.

Management: Ploughing exposes pupae to heat and predators. Many parasitoids and predators, including birds, attack. Larvae on panicles are exposed to insects and birds. Cannibalism, bacteria and viruses reduce abundance. Insecticide controls small larvae.

Earhead webworm, *Nola analis* (Wileman and West)

Description and Biology: The caterpillar is a minor pest in Africa and Asia. Round, cream eggs are laid on spikelets and kernels. Small, white moths have two zigzag, white, vertical stripes behind two black spots in the middle near the anterior margin of the forewings. Eggs hatch in two or three days. Young larvae are grey with hair. Larvae live in webs of silken threads and frass and grow to 13.5 mm in two weeks. The conical pupa is inside webbing for eight days.

Symptoms and Damage: Larvae eat kernels.

Management: Insecticide is effective but rarely necessary.

Christmas berry webworm, *Cryptoblabes gnidiella* (Milliere)

Description and Biology: The insect damages hybrids and high-yielding cultivars in India. Adults have grey forewings and larger hind wings fringed with hairs on the anterior margin. Cream, round or conical eggs laid on spikelets and kernels hatch in three or four days. Young larvae are white with brown heads. Mature, brown, 12-mm larvae are active. Larvae develop in nine or ten days. The short pupa is covered with silk inside a silken web. Adults emerge in one week.

Damage and Symptoms: Larvae destroy kernels and produce webs of silk on panicles.

Management: Insecticide is not required.

Earhead bug, *Calocoris angustatus* (Lethiery)

Description and Biology: The bug is a serious pest in India. This and other species infest sorghum in Africa. The female is yellowish-green, 5 mm long and 1 mm wide. It lays 1.5-mm, cigar-shaped eggs between glumes or anthers of florets. Females lay 150–200 eggs that hatch in one week. In two weeks, yellow and orangish-red nymphs develop through five instars into adults. Two generations develop in the same field when panicles mature at different times.

Damage and Symptoms: Nymphs and adults suck developing kernels as panicles emerge from the boot. Young kernels shrivel and yield less. Feeding punctures reduce quality of older kernels.

Management: Open panicles are less affected. Reduviid and lygaeid bugs prey on earhead bugs.

African head bug, *Eurystylus oldi* Odhiambo

Description and Biology: Bugs are 3–5 mm, with females slightly longer. Colour is yellowish-brown to dark brown with red. Cigar-shaped eggs are inserted into kernels and hatch in

4–6 days. Females lay 24–136 eggs, one to seven per kernel. Five nymphal instars develop in 11 days. A generation develops in 18–27 days.

Damage and Symptoms: Bugs are abundant in compact panicles. Nymphs and adults suck juice from kernels that shrivel, become small and off-coloured, and yield less. Pathogens deteriorate grain quality. Older kernels have feeding punctures.

Management: Some sorghum is resistant. Open panicles are less affected. Kernel damage is less when dry. Insecticide is effective.

7 Beneficial insects

Lady beetles: Convergent lady beetle, *Hippodamia convergens* (Guerin-Meneville); *Scymnus* lady beetle, *Scymnus loewii* (Mulsant)

Convergent and *Scymnus* lady beetles are hemispherical, 1.5–6.0 mm, orange, red, tan, brown, grey, or black, with red, yellow, black or white markings. Elongated larvae are dark with bright markings and spines. *Scymnus* larvae are covered with white wax.

Adults hibernate in dry, protected places. Females lay clusters of 200–1 000 yellow eggs that hatch in three or four days. *Scymnus* lay eggs singly. Larvae feed two or three weeks. Adults and larvae eat aphids, insect eggs and small larvae, and spider mites.

Syrphid flies (Syrphidae)

Adult flower, hover or sweat flies eat nectar and pollen and pollinate plants. Many are colourful and resemble bees. Larvae are 6–19 mm, elongated and legless. Many are yellow, pink, green, or brown with black or white. Larvae use pointed jaws to grab and suck aphids. Adults lay long, white eggs among aphids.

Green lacewings: *Chrysoperla* spp.; *Chrysopa* spp.

Aphis lions use long, spindle-shaped mandibles to puncture and suck aphids, thrips, mites and insect eggs and small larvae. Eggs on 12-mm silken stalks attached to leaves or stems are protected from natural enemies, especially lacewing larvae. Adults of one species are 25 mm and green with lace-like wings and golden eyes.

Predaceous bugs: Minute pirate bug, *Orius tristicolor* (White); Common damsel bug, *Nabis americoferus* (Carayon); Large big-eyed bug, *Geocoris bullatus* (Say)

Minute pirate, damsel, and large big-eyed bugs eat insects and mites. Adult minute pirate or flower bugs are 2 mm and black with white. Long, thick, basal parts of forewings are yellow with a triangular black spot at the tip; the membranous part of the wing is white. Yellow nymphs feed on insect eggs, nymphs, and larvae, and spider mites.

Damsel bugs are 8 mm, with bodies narrowed anteriorly. The front legs have a double row of spines on the inner surface to grasp prey. Some have long- and short-winged forms. Most are yellowish-brown. Bugs hide in foliage or flowers and eat aphids and caterpillars.

Big-eyed bugs are 5 mm with large, protruding eyes, dark bodies, and white wings. Nymphs and adults capture prey.

Ground beetles (Carabidae)

Many species hide under stones, logs, or debris and prey on larval and adult insects and small animals including cutworms and worms. Adults are 10–25 mm and flat with long legs. Most are brown, but some are green, blue or copper with violet or green forewings with reddish-brown borders. Wings of most species have rows of longitudinal ridges and punctures.

Larvae are nocturnal and predaceous with sickle-shaped mandibles. Mature larvae are 10–45 mm, with flat bodies tapered to two bristly processes. Larvae are brown, black or yellow.

Parasitoids: Braconid wasps (Braconidae); Ichneumonid wasps (Ichneumonidae); Chalcid wasps (Chalcidae); Tachinid flies (Tachinidae)

A parasitoid lives on one individual that does not die until the parasitoid completes development. Several species suppress especially aphids and caterpillars.

Braconid wasps are less than 15 mm and stout-bodied. They pupate in silken cocoons outside the body of the host. Braconids parasitize aphids and larvae of moths, beetles, and flies. Parasitized aphids harden into 'mummies'. The adult wasp emerges through a hole in the mummy case. *Lysiphlebus testaceipes* (Cresson) is a shiny black wasp in aphid colonies. Aphid mummies are tan.

Ichneumonid wasps are distributed widely. Females have long ovipositors. Most are internal parasitoids of immature hosts. The parasitoid completes development in the host stage in which the egg is laid, or later. Ichneumonids parasitize most insect groups.

Chalcid wasps are 2 mm and metallic blue or green. Many live inside insect eggs or scales, aphids, caterpillars or flies.

Tachinid flies resemble house flies, are bristly, and mottled grey, brown or black. Adults feed on flowers. They glue eggs to an insect or on foliage where the host ingests the egg. Larvae are deposited on an insect where they feed. Adults emerge before or after the host dies.

8 Acknowledgements

Many colleagues, especially Dr. George L. Teetes and his graduate students who contributed for many years to the knowledge base on insect pests of sorghum, are gratefully acknowledged.

9 Where to find further information

The following references provide additional information on the description and biology, nature and symptoms of damage, and management for major insect pests of sorghum.

10 References

Bowling, R. D., Brewer, M. J., Kerns, D. L., Gordy, J., Seiter, N., Elliott, N. E., Buntin, G. D., Way, M. O., Royer, T. A., Biles, S. and Maxson, E. (2016), 'Sugarcane aphid (Hemiptera: Aphididae): a new pest on sorghum in North America', *Journal of Integrated Pest Management*, 7: 12.

Cronholm, G., Knutson, A., Parker, R. and Pendleton, B. (2007), *Managing Insect and Mite Pests of Texas Sorghum*, Texas A&M AgriLife Ext. Bull. B-1220.

ICRISAT (1989), International workshop on sorghum stem borers, International Crops Research Institute for the Semi-Arid Tropics, Patancheru, Andhra Pradesh, India.

Kerns, D. L., Brown, S., Beuzelin, J. and Guidry, K. M. (2015), 'Sugarcane aphid: a new invasive pest of sorghum', *Louisiana Agriculture*, 58: 12–14.

Kwanze, K. F. and Youm, O. (Eds) (1995), *Panicle insect pests of sorghum and pearl millet: Proceedings of an international consultative workshop*, International Crops Research Institute for the Semi-Arid Tropics, Patancheru, Andhra Pradesh, India.

Leuschner, K. and Teetes, G. L. (Eds) (1985), *Proceedings of the international sorghum entomology workshop*, International Crops Research Institute for the Semi-Arid Tropics, Patancheru, Andhra Pradesh, India.

Seshu Reddy, K. V. (1982), 'Pest management in sorghum', pp. 237–46. In L. R. House, L. K. Mughogho and J. M. Peacock (Eds), *Sorghum in the eighties*, Vol. 1, International Crops Research Institute for the Semi-Arid Tropics, Patancheru, Andhra Pradesh, India.

Sharma, H. C., Singh, F. and Nwanze, K. F. (Eds) (1997), *Plant Resistance to Insects in Sorghum*, International Crops Research Institute for the Semi-Arid Tropics, Patancheru, Andhra Pradesh, India.

Teetes, G. L. (1982) 'Sorghum insect pest management', pp. 225–35. In L. R. House, L. K. Mughogho and J. M. Peacock (eds), *Sorghum in the Eighties*, Vol. 1, International Crops Research Institute for the Semi-Arid Tropics, Patancheru, Andhra Pradesh, India.

Teetes, G. L. and Pendleton, B. B. (2000), 'Insect pests of sorghum', pp. 443–95. In C. W. Smith and R. A. Frederiksen (Eds), *Sorghum: Origin, History, Technology, and Production*, John Wiley and Sons, Inc., New York.

Teetes, G. L. and Wiseman, B. R. (1979), 'Economic threshold of *Heliothis* species in sorghum', pp. 57–62. In W. L. Sterling (Ed.), *Economic Thresholds and Sampling of Heliothis Species on Cotton, Corn, Soybeans and Other Host Plants*, USDA Southern Cooperative Series Bull. 231.

Teetes, G. L., Young, W. R. and Jotwani, M. G (1980), *Elements of Integrated Control of Sorghum Pests*, Food and Agriculture Organization of the United Nations Plant Production and Protection Paper 19, Rome.

Teetes, G. L., Seshu Reddy, K. V., Leuschner, K. and House, L. R. (1983), *Sorghum Insect Identification Handbook*, Information Bull. 12, International Crops Research Institute for the Semi-Arid Tropics, Patancheru, Andhra Pradesh, India.

10 References

Chapter 15

Sorghum diseases and their management in cultivation: seedling, seed, panicle and foliar diseases

C. R. Little and A. Y. Bandara, Kansas State University, USA; and R. Perumal, Agricultural Research Center – Hays, Kansas State University, USA

http://dx.doi.org/10.19103/AS.2017.0015.17

1 Introduction

The sorghum plant is a tropically adapted C4 crop that is photoperiod sensitive (Sage et al., 1999). However, the sorghums that are grown for grain in temperate and subtropical regions have been converted into photoperiod-insensitive plants. For high-input farming, hybrids (and their parental breeding lines) also have dwarf genes that allow for optimum combine harvest height and grain-stover ratios.

In the United States, 6.69 million acres of sorghum were planted in 2016 and 6.16 million acres were harvested. During the same year, 480 million bushels were produced at a yield of approximately 77.9 bushels/acre (USDA NASS, 2017). In 2014, 45.6 million ha of sorghum were harvested, and 71.8 million tonnes produced at a yield of 29 109 hg/ha (FAO, 2017).

Sorghum may be used for feed for poultry, cattle and pigs. The stems and foliage may be used for green chop, hay, silage and pasture grazing. In addition to the applications mentioned above, sorghum has numerous food uses. Many of the most diverse applications originated in Africa and India where the crop developed and was brought into widespread use before arriving in the New World. Examples include leavened and unleavened breads, couscous, pearled sorghum that is boiled like rice and various porridges (Hariprasana and Rakshit, 2016; Mudge et al., 2016). Sorghum is also used to make snack foods and alcoholic beverages. The use of sorghum in the dietary 'gluten-free' health food industry has opened a promising, high-value niche for food utilization and specialty sorghums. Food use sorghums that are often white-grained and white-pericarped are particularly susceptible to grain mould, grain weathering and storage fungi (see Section 5). These fungi reduce grain filling in the field, degrade endosperm tissues and impact milling and nutritional qualities, and some produce potentially harmful mycotoxins (Bhat et al., 1997; Bandyopadhyay et al., 2002).

As will be discussed, all physical structures and tissues of the sorghum plant may become diseased with one pathogen or another. Like other members of the Andropogoneae, *Sorghum bicolor* (L.) Moench produces a panicle supported by a peduncle that splits into numerous rachis branches and rachilla that support individual florets (Brown et al., 2006). Pollination, fertilization and seed development can be compromised by early infection by *Claviceps africana* Freder., Mantle & De Milliano (ergot) or grain mould fungi (Bandyopadhyay et al., 1998, 2002). In the case of ergot, if *C. africana* conidia infect an unfertilized floret prior to fertilization, the ovule and supporting tissues are converted into a mixture of fungal and plant tissue (sphacelium) that exudes sticky honeydew that may interfere with seed harvest. Grain mould fungi, such as *Fusarium thapsinum* Klittich, J.F. Leslie, P.E. Nelson & Marasas and *Curvularia lunata* (Wakker) Boedijn, can infect open florets, colonizing the pedicel and ovule. Hyphae grow into the developing seed and will cause premature deposition of the black layer, which reduces yield (Castor, 1981). Or, in the case of systemically infecting smuts, such as head smut, the entire reproductive centre is replaced by a fungal sorus filled with black teliospores.

After planting, sorghum progresses through several developmental periods including a seedling stage, a vegetative stage, a booting stage, flowering and grain development (Vanderlip and Reeves, 1972). Under cooler temperatures and high moisture, and without proper seed treatment, seeds and seedlings can become infected with numerous soilborne fungal pathogens that cause pre- and post-emergent damping-off diseases (Thakur et al., 2007) (Fig. 1). Therefore, seedling diseases may be quite destructive yet remain greatly underinvestigated.

Throughout the vegetative and booting stages, sorghum is sensitive to several foliar diseases including anthracnose, sorghum downy mildew (SDM), grey leaf spot, leaf blight and rust. Infections that occur on the upper portion of the canopy during grain-filling can have significant impacts upon yield. However, foliar infections that occur after grain maturity rarely result in significant yield losses. The majority of foliar diseases may be controlled by host resistance (if available) and cultural practices such as managing crop residues and maintaining clean fields free of pathogen reservoirs.

2 Seed rots and seedling diseases

Seed rots and seedling diseases, including pre- and post-emergent damping-off, result in poor stand establishment (Giorda, 2002). In this case, seedlings may die before emergence or at the soil surface after emergence (Swarup et al., 1962). Seedlings that do emerge will do so with reduced vigour and may appear chlorotic. In the case of seed rot, tissues within the seed will appear necrotic; embryos will acquire a darkened appearance (Leukel and Martin, 1943). Seedling roots will accumulate red pigments and become necrotic.

A wide range of soilborne and seedborne fungi cause seed rots and seedling diseases. Genera include *Aspergillus* spp., *Colletotrichum sublineola* Henn., *Fusarium* spp., *Macrophomina phaseolina* (Tassi) Goid., *Penicillium* spp., *Pythium* spp., *Rhizoctonia solani* J.G. Kühn and *Rhizopus* spp. (Christensen and Kaufmann, 1965; Pratt and Janke, 1980; McClean and Lawrence, 2001; Sumner and Bell, 1982; Leslie et al., 1990; Jardine and Leslie, 1992; Sharma et al., 2012). In general, *Aspergillus*, *Fusarium* and *Penicillium* are seedborne and will contribute to seed rot in the ground or in improper seed storage, seedling blight or damping-off when conditions favour the necrotrophic growth of these organisms (Christensen and Kaufmann, 1965). On the other hand, *Fusarium* spp., *Pythium* spp. and *Rhizoctonia solani* are common soil inhabitants and attack sorghum seeds in an external manner (Singleton et al., 1992; Little et al., 2012). Such infection may result in seed rot, damping-off and seedling blights as described above (McClean and Lawrence, 2001).

Leslie et al. (2005) and Huang and Backhouse (2005) found that *F. thapsinum*, *F. proliferatum* (Matsush.) Nirenberg, *F. andiyazi* Marasas, Rheeder, Lampr., K.A. Zeller & J.F. Leslie and *F. verticillioides* (Sacc.) Nirenberg were virulent on sorghum seedlings in artificial inoculation assays. Palmero et al. (2012) found that *F. verticillioides* was pathogenic on roots and seedlings in Spain. Seedborne *F. thapsinum* resulted in poor germination and lower fresh seedling weight when infected seed was planted (Funnell-Harris et al., 2015). *F. thapsinum* shows increased pathogenicity on sorghum seedling compared to *F. veriticillioides*. *F. graminearum* Schwabe is a common inhabitant of sorghum seedlings in wheat–sorghum rotations and was found to be a dominant seedling-associated species in southcentral Kansas.

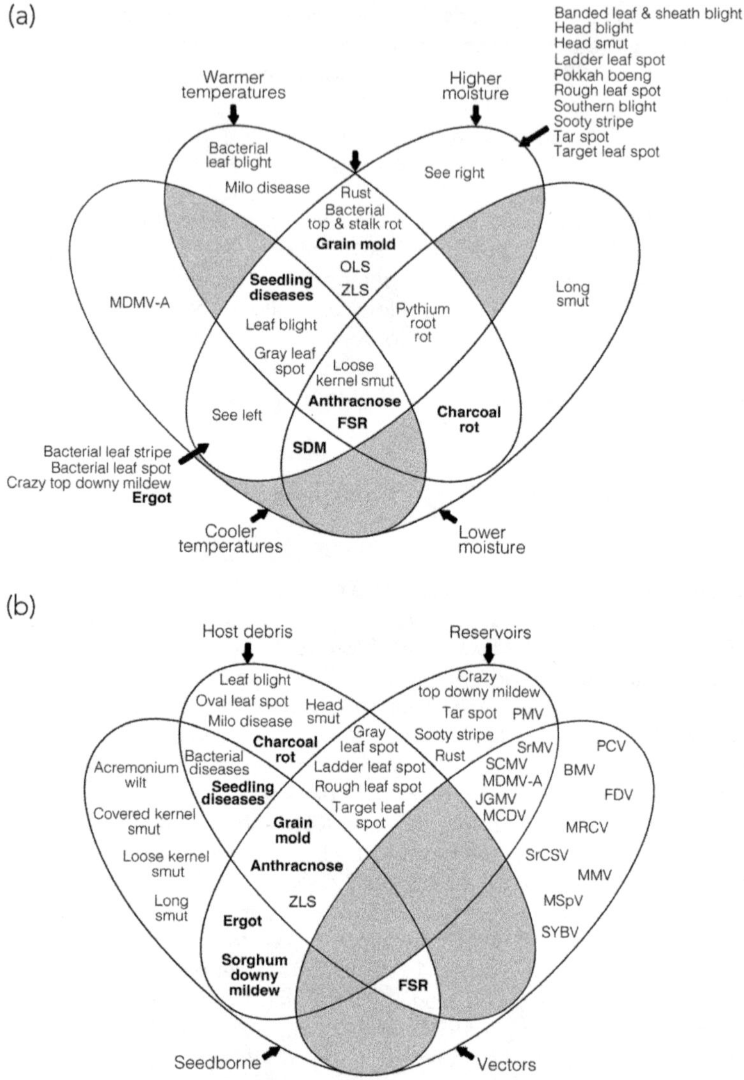

Figure 1 A working concept for multiway interactions of key environmental and disease cycle factors that influence the onset, development and persistence of sorghum diseases. (a) Temperature (cooler vs. warmer) and moisture (wetter vs. drier) conditions that may interact to exacerbate disease. Diseases such as Fusarium stalk rot and loose kernel smut, for example, have phases during the disease cycle that rely on both wet and dry periods and warm and cool temperatures. (b) Host debris, reservoirs, seedborne and vector transmission and survival are important components of some sorghum diseases. Abbreviations: BMV = Brome mosaic virus; FDV = Fiji disease virus; FSR = Fusarium stalk rot; JGMV = Johnsongrass mosaic virus; MCDV = Maize chlorotic dwarf virus; MDMV-A = Maize dwarf mosaic virus, pathotype A; MRCV = Mal de Rio Cuarto virus; MSpV = Maize stripe virus; OLS = Oval leaf spot; PCV = Peanut clump virus; SDM = Sorghum downy mildew; SCMV = Surgarcane mosaic virus; SrMV = Sorghum mosaic virus; SYBV = Sorghum yellow banding virus; ZLS = Zonate leaf spot. Images by C.R. Little.

Several *Pythium* spp. cause seed rot and damping-off in sorghum. These include *Pythium aphanidermatum* Edson (Fitzp), *Pythium arrhenomanes* Drechsler, *Pythium debaryanum* R. Hesse, *Pythium graminicola* Subraman., *Pythium myriotylum* and *Pythium ultimum* var. *ultimum* Trow (Sprague, 1950; Freeman et al., 1966; Forbes et al., 1985, 1987; Davis and Bockus, 2001; Wang et al., 2003).

Several *R. solani* anastamosis groups attack sorghum, or have been found in association with sorghum or related hosts (Jones and Belmar, 1989): AG-1A (sheath blight), AG-2 (sorghum debris) and AG-3 (johnsongrass) (Ruppel, 1985; El-Bakali et al., 2000; Pascual et al., 2000).

The same environmental conditions that favour seed germination, emergence and stand establishment also favour activity by both seedborne and soilborne pathogens. In most cases, such conditions include wet weather accompanied by lower temperatures (Leukel and Martin, 1943; Edmunds et al., 1970; Davis and Bockus, 2001; Kapanigowda et al., 2013). Heavy soils that do not drain well are particularly problematic when these environmental conditions persist during and after planting (McGee, 1992). Further, conditions such as prevailing warm temperatures at emergence and low soil pH may also stress seedlings sufficiently that many pathogens, such as *Fusarium* spp., *Pythium* and *Macrophomina phaseolina*, can have an effect (Fig. 1).

Many soilborne pathogens produce spores that allow them to persist in the soil for an extended period (Singleton et al., 1992). For example, *Pythium* spp. produce oospores that live in the soil for many years. *Fusarium* spp. produce chlamydospores that are environmentally resistant and can germinate after 1–2 years, or longer, depending on the species (Leslie and Summerell, 2006). Additionally, many isolates of *R. solani* produce sclerotia, dense masses of pseudoparenchymatous hyphal tissue that may also survive for long periods. To further complicate the matter, the pathogens mentioned above survive in host residue in the soil (Forbes and Odvody, 2000). Therefore, continuous cropping of sorghum will contribute to the accumulation of pathogen propagules over time.

Management of seedling diseases depends on three factors: (1) controlling environmental conditions (if possible); (2) understanding optimum pathogen activity and (3) ensuring rapid stand establishment (Forbes and Odvody, 2000). Under optimum temperature and moisture for seedling development, stand establishment is rapid and seedlings can 'outgrow' pathogens (Kapanigowda et al., 2013). Under dry conditions and lower than normal temperatures, seedling growth is reduced, but most pathogens are not active. Under warm and dry conditions, most pathogens are not active and, as before, seedling growth is reduced. When conditions are warm and wet, seedlings may become infected but will often outpace seedling disease. However, wet conditions coupled with lower than normal temperatures not only slow seedling development, but provide conditions that are optimum for many seedling pathogens to infect and cause disease (Forbes and Odvody, 2000).

Seed rots and seedling diseases may be managed by planting high-quality, fungicide-treated seed (Mueller et al., 2013). True fungi may be controlled using Captan, triazole, strobilurins (pyraclostrobin, azoxystrobin), fludioxonil and fluxapyroxad fungicides (PMN, 2010a,b, 2014). Metalaxyl has been used for the control of oomycete pathogens such as *Pythium* (Botha and McKenzie, 1993; Davis and Bockus, 2001). Preparation of a well-draining, low-residue seedbed, if practical or possible, is an efficient way to reduce seed rot and seedling diseases. Seed should be planted at the proper depth and at correct numbers to achieve optimum plant populations. Planting at proper depth helps seedlings escape infection or damage that may occur during emergence. Planting at the correct plant density reduces competitive stress between seedlings that may weaken seedling roots and increase their susceptibility to infection by soilborne fungi (Kapanigowda et al., 2013). Seed storage at proper moisture content, low relative humidity and cool temperatures helps to prevent the growth of seedborne fungi within seed tissues during the storage period (Christensen and Kaufman, 1965).

3 Sorghum downy mildew

Local lesions resulting from infections by conidia on leaf lamina and systemic infections resulting from oospores are two types of SDM symptoms (Prom et al., 2010). Systemic infection occurs within 2–3 weeks after planting, causing stunted plant growth with chlorotic areas from the base of the first leaves (Pande et al., 1997a; Craig, 2000). Plants with systemic infection may die prematurely at the seedling stage (Frederiksen, 1980; Pande et al., 1997a). Local lesion symptoms can occur at any crop stage.

SDM, caused by *Peronosclerospora sorghi* (Weston and Uppal) C.G. Shaw, was first observed in Africa and Asia and introduced into the Americas in the 1950s followed by an endemic infection in the southern United States (Frederiksen, 1980; Odvody and Frederiksen, 1984; Pande et al., 1997a). By the mid-1970s, Reyes et al. (1964), Futrell and Frederisken (1970) and Williams and Herron (1974) found that the pathogen was distributed throughout the sorghum-growing regions of the United States including Alabama, Georgia, Kansas, Kentucky, Louisiana, Mississippi, New Mexico, Oklahoma, Tennessee and Texas.

Among *Peronosclerospora sorghi* isolates, sorghum- and maize-infecting strains and strains that only infect maize exist. As a result, different species designations have been given to these subpopulations of *Peronosclerospora sorghi*, for example *P. maydis* (Racib.) C.G. Shaw for those that are pathogenic on maize. Further, those strains infecting sugarcane have been renamed *P. sacchari* (T. Miyake) Shirai & Hara. The primary source of inoculum responsible for initiating SDM epiphytotics are the root-infecting, soilborne oospores, which survive in the soil for several years (Pande et al., 1997a; Craig, 2000).

Both sorghum and maize are symptomatic hosts of *Peronosclerospora sorghi*. However, *Peronosclerospora sorghi* does not appear to complete sexual reproduction on maize. Therefore, no oospores are produced on this host. After infection of the germinating seed, the pathogen colonizes the foliar meristematic tissues, resulting in chlorosis of the seedling leaf (Craig, 2000). Sporangia are produced when systemically infected plants are exposed to long periods of dew or rain. The sporangia are spread to neighbouring plants where local infections occur (Bock et al., 1998; Pande et al., 1997a). *Peronosclerospora sorghi* produces conidia on erect conidiophores from the stomatal openings on the leaf surface of infected plants (Pande et al., 1997a; Craig, 2000). Although the conidia are short-lived, they play a significant role in the secondary spread of the disease (Frederiksen, 1980). In many areas, conidia serve as the primary source of inoculum (Schuh et al., 1986). As the season progresses, oospores germinate using germ tubes and invade the roots of the sorghum seedlings (Craig, 2000). Oospores survive in the soil for many years due to their thick, environmentally resistant walls. Downy mildew subsequently develops in systemically infected leaves (Fig. 2), which fall to the ground as the leaf shreds, completing

Figure 2 Examples of foliar diseases of sorghum. (a) The striped lesions of systemic sorghum downy mildew caused by *Peronosclerospora sorghi*. (b) The rectangular lesions of grey leaf spot caused by *Cercospora sorghi*. (c) The oval adaxial lesions with broad chlorotic halos as seen in sooty stripe caused by *Ramulispora sorghi*. Images by C.R. Little.

the disease cycle. In the United States, oospores are more important in the disease cycle and serve as primary inoculum for the next growing season (Schuh et al., 1986). Frederiksen (1980) noted that the initial level of SDM incidence is closely associated with the number of viable oospores in the soil.

Peronosclerospora sorghi is both a seed- and soilborne pathogen, thus rendering crop rotation less effective in controlling the disease (Perumal et al., 2008). Yield losses due to SDM of up to 11.7% have been reported in Africa (Bock et al., 1998), while 78% losses in cv. DMS 652 were reported in India (Thakur and Mathur, 2002). Numerous epidemics have occurred throughout the sorghum-growing regions of the central United States including outbreaks in Texas in the late 1960s, Kansas in 1967–8 and 1978–9 and Nebraska in 1987 (Sill et al., 1969; Futrell and Frederiksen, 1970; Lengkeek and Sim, 1979; Sim, 1980; Jensen et al., 1989). However, losses are minimal if 20–30% downy mildew incidence of systemically infected seedlings is encountered in the field under standard planting density in the United States (Frederiksen, 1980).

Cultural practices such as crop rotation, deep ploughing, early planting and the use of trap crops have also been shown to significantly reduce SDM incidence (Frederiksen, 1980; Schuh et al., 1987; Pande et al., 1997a). For a long period of time, the disease was under effective control by metalaxyl or fosetyl-A1 fungicide seed treatment (Odvody and Frederiksen, 1984). Severe epiphytotics of SDM have been more recently observed in Texas, suggesting that metalaxyl seed treatment was ineffective under certain conditions in Texas and Mexico. Plants with systemic infection were noted in 82% of the plots at Louise, Texas, 29% of the plots in Ocotlan, Mexico, and 23% of the plots at the Fucik research site, Texas. It was found that *Peronosclerospora sorghi* (pathotype P6) was responsible for the differences in disease levels (Isakeit and Odvody, 2002; Isakeit and Jaster, 2005). Pathotype P1 of *Peronosclerospora sorghi* has been documented from Nebraska and Texas (Jensen et al., 1989; Radwan et al., 2011). Pathotypes P2, P3 and P6 have been documented from Texas, although P2 is rare and the latter two are more common (Isakeit and Jaster, 2005; Radwan et al., 2011). Pathotype P6 has not been observed in the Midwest. The outbreak of SDM by P6 in Texas and Mexico warrants continuous monitoring of the disease and identification of sources for the development of resistant sorghum hybrids and the effective management of SDM (Prom et al., 2015).

4 Crazy top downy mildew

Crazy top downy mildew affects sorghum plants differently depending on age and host genotype. Generally, younger plants develop mottled and chlorotic upper leaves that resemble that of a maize dwarf mosaic virus infection (Williams et al., 1978). As plants age, leaves become rigid, deformed or curled. The texture of the leaves thicken and become rubbery (Dey et al., 1986) (Fig. 3). If plants produce panicles, phyllody of the glume tissues may be observed, which is the basis for the 'crazy top downy mildew' disease name. Often, these panicles do not produce any seed (Williams et al., 1978; Dey et al., 1986). Diseased plants also may tiller more than healthy plants (Williams et al., 1978).

Crazy top downy mildew is caused by the fungus-like oomycete *Sclerophthora macrospora* (Sacc.) Thirum., C.G. Shaw and Naras (Williams et al., 1978). This organism can infect a wide array of graminaceous hosts besides sorghum including barley, corn, crabgrass, crested wheatgrass, foxtail millet, Hungarian brome, Kentucky bluegrass, oats, reed canary grass and wheat (Whitehead, 1958; Semeniuk and Mankin, 1964). For example, in *Lolium perenne* L., perennial ryegrass, seedlings were more susceptible to *S. macrospora* than adult plants as the pathogen preferred to infect younger meristematic tissues (Dernoeden and Jackson, 1980).

Multiple sporangia germinate from stomates on any plant leaf surface (Whitehead, 1958). Like SDM, oospores are formed within the leaves, but interveinal striping and leaf shredding does not occur (Williams et al., 1978).

This disease is most problematic in areas of sorghum fields that retain standing water after heavy rains (Horne and Berry, 1980). Season-wide flooding can have a great effect if the seedbeds remain saturated. Seedling may be infected early and quiescent infections are not detected until deformed plants are apparent or after booting when panicles with phyllody symptoms are observed (Beck, 1978; Malaguti, 1978; Horne and Berry, 1980).

Like SDM, oospores of *S. macrospora* may survive in the soil for long periods (Safeeulla et al., 1976). Therefore, fields or sections of fields that perennially have a yield-impacting problem with this disease should be avoided. Low areas of fields that are prone to flooding

Figure 3 Symptoms of crazy top downy mildew (*Sclerophthora macrospora*). (a) Phyllody at the apical meristem (white arrow). (b) Chlorotic, thickened leaves. Images by C.R. Little.

may require improved drainage to ensure that water does not accumulate at planting, during seedling development, and as the plants mature. Although resistant hybrids exist, this disease has not been a specific target for breeding programmes. Seed treatment with metalaxyl may reduce seedling infection by *S. macrospora* (Narro et al., 1992). Currently, there is no evidence to suggest that *S. macrospora* is tolerant to metalaxyl.

5 Grain mould and weathering

Sorghum grain mould and weathering are characterized by grain discoloration, flecking and pigment accumulation in the pericarp, underfilled and poor-quality caryopses, and caryopses covered with characteristic mould growth (Bandyopadhyay et al., 2002; Little et al., 2012). Moulded sorghum grain is characterized by poor milling and processing characteristics (Waniska et al., 2001). Whether the problem is grain mould, grain weathering or storage moulds, fungi may penetrate the germ and endosperm and colonize these tissues (Castor, 1981; Little et al., 2012). The result is seed with low viability regarding germination and emergence and other surrogate viability assessments such as tetrazolium tests (Noll et al., 2010). Further, the vigour of moulded seeds is reduced (Little and Magill, 2003b). However, asymptomatic caryopses are typical and may represent a greater portion of moulded kernels that are visibly symptomatic. Such a range of symptoms is due to the wide array of fungi that can cause grain mould and grain weathering. The primary fungal species colonizing the caryopsis can be determined with reasonable reliability by incubating caryopses in a humidity chamber such as a glass petri dish with moistened filter paper (25°C, >95% relative humidity) for 1–3 days (Navi, 1999). After that period, fungal mycelium and sporophores diagnostic for common fungal grain mould and weathering species will become apparent on the grain surface, which can be visualized using a dissecting microscope. After periods of high humidity and warm temperatures during grain development, many of the same symptoms and signs may be observed in the field on grain-mould- or weathering-susceptible genotypes (Waniska et al., 2001). In this case, a high-powered hand lens can reveal mycelium and sporophore characteristics that are diagnostic for certain species.

Grain mould is a major disease affecting sorghum and remains a high priority for research (de Milliano, 2002) (Table 1). The disease is present to some extent wherever sorghum is grown. In environments where sorghum matures under arid conditions, slight environmental changes towards higher humidity or unusually high amounts of rainfall can result in grain mould outbreaks in locations where the disease is uncommon. Overall, grain mould and weathering remain important diseases that act as a constraint to grain production and quality worldwide (Frederiksen, 2000b).

Severe grain mould and weathering problems are encountered in areas where high humidity and rain prevail during grain development and maturation (Garud et al., 2000; Navi et al., 2005; Williams et al., 1978). Montes-Belmont et al. (2003) found that there was a significant positive relationship between temperature (i.e. 19–24°C) and grain mould disease incidence. Grain moulds are important diseases in Cameroon, Cote d'Ivoire, Central African Republic, Guinea, India, Indonesia, Malaysia, Myanmar, Niger, the Philippines, Puerto Rico, the southern United States, Tanzania, Thailand and Venezuela (Indira, 2002; Malaguti, 2002; Marley et al., 2002; Stack, 2002). Esele (2002) reported that areas of Sudan that received more than 800 mm of annual rainfall experienced the greatest problems with grain mould. However, in production regions where grain development occurs under more arid conditions, such as Australia, Iran, the Great Plains of the United States, Mexico and Pakistan, grain mould is not considered to be a major problem (Indira, 2002; Ryley et al., 2002; Stack, 2002).

Numerous fungal species are associated with sorghum caryopses (Little et al., 2012; Navi et al., 2005) (Tables 2 and 3). Many of these are capable of infecting and colonizing the seed to some extent and are generally facultative pathogens that live in the environment as saprophytes in the soil, on plant surfaces and on crop residues, but if introduced to the floret can take advantage of the developing caryopsis (Castor, 1981; Little et al., 2012).

Table 1 Proposed priorities for sorghum research based on the historical and current relative prevalence and importance[a] of the diseases covered in this chapter

Tier 1, *High priority*	Tier 2, *Medium priority*	Tier 3, *Low or no priority*
Anthracnose, foliar	Covered kernel smut	Long smut
Ergot	Crazy top downy mildew	Loose kernel smut
Grain and storage moulds	Grey leaf spot	Tar spot
Head smut	Head blight	Target leaf spot
Seedling diseases	Ladder leaf spot	
Sorghum downy mildew	Leaf blight	
	Oval leaf spot	
	Pokkah boeng	
	Rough leaf spot	
	Rust	
	Sooty stripe	
	Zonate leaf and sheath spot	

[a] References: Frederiksen (2000), Casela et al. (2002), Esele (2002), Giorda (2002), Indira (2002), Kasakova and Oleynick (2002), Malaguti (2002), Marley et al. (2002), Mtisi and McLaren (2002), Ngugi et al. (2002), Ryley et al. (2002), Stack and Pedersen (2003), Little et al. (2012) and Tesso et al. (2012).

Table 2 Panicle diseases of grain sorghum

Disease	Pathogen(s)	Symptoms and signs	Key controls	Refs.
Ergot	*Claviceps africana* *C. sorghi* *C. sorghicola*	Honeydew production Secondary mould growth Production of sclerotia	Use clean, sclerotia-free seed Rapid pollination of male-sterile plants Triazole fungicide use	Little et al. (2012)
Grain mould (see Table 4)	*Alternaria* spp. *Colletotrichum* spp. *Curvularia* spp. *Fusarium* spp. *Phoma* spp.	Grain discoloration Mould growth on grain surface Shrunken grain	Avoid grain development during periods of high relative humidity that favour mould growth	Bandyopadhyay et al. (2002), Little et al. (2012) and Prom et al. (2015)
Head blight	*Fusarium* spp. and many others	Death of rachis branches Reduced grain size	Resistance	Castor and Frederiksen (1981) and Frederiksen et al. (1982)
Smuts:				
Covered kernel smut	*Sporisorium sorghi*	Conical sori remain intact	Fungicide seed treatment	Ahmed and Reddy (1993), Esele et al. (1995) and Stoll et al. (2003)
Head smut	*S. reilianum*	Smut sori replace panicle Residual vascular strands Phyllody, dwarfing	Resistance	Little et al. (2012)
Long smut	*S. ehrenbergii*	Elongated, curved sori	Some resistance exists	Prom et al. (2015)
Loose kernel smut	*S. cruentum*	Dwarfing, thin stalks Panicle phyllody Smutted spikelets Conical sori rupture	Fungicide seed treatment	Ahmed and Reddy (1993) and Esele (1995)

Table 3 Symptoms and signs on the sorghum caryopsis associated with the dominant fungal species[a] on sorghum grain under natural or artificial conditions conducive to grain mould and weathering

Fungus	Symptom	Sign
Alternaria alternata	Discoloration, flecking	Grey mycelium, dark conidiophores; conidia in chains
Colletotrichum sublineola	Brown discoloration	Rows of brown-black acervuli with setae; conidia pill- or moon-shaped
Cuvularia lunata	Brown-black discoloration; Some seed shrivelling or shrinkage	Black mycelium, dark conidiophores; croissant-shaped conidia
Fusarium spp.	Often asymptomatic; seed shrivelling and shrinkage common, for example 'small-seed syndrome'	White, pink, salmon or orange mycelium depending upon the *Fusarium* species; conidiophores may or may not be visible under hand lens or dissecting scope; both micro- and macroconidia may be produced
Phoma sorghina	Often asymptomatic; slight necrosis at the black layer region	Pycnidia cover the pericarp; dark chlamydospores may also be present

[a] See Navi (1999) and Little et al. (2012) for exhaustive information of grain mould and grain weathering fungi.

Passive air sampling in sorghum fields showed that there was an increase of pathogen conidia from anthesis to maturity. In this study, Funnell-Harris and Pedersen (2011) found numerous fusaria including *F. andiyazi*, members of the *Fusarium incarnatum-equiseti* species complex (FIESC), *F. graminearum*, *F. proliferatum*, *F. subglutinans* (Wollenw. & Reinking) P.E. Nelson, Tousson & Marasas, *F. thapsinum* and *F. verticillioides* (Table 3).

Classically, grain mould fungi are defined as those species that infect at or near anthesis and colonize the internal tissues of the seed as it develops towards physiological maturity (Castor, 1981; Forbes, 1986; Prom and Erpelding, 2009; Little et al., 2012) (Fig. 4). Grain weathering, on the other hand, occurs when fungi colonize the exposed developing and mature grain surfaces after black layer formation. Finally, another group of fungi, the storage moulds, colonizes grain after harvest. However, many of the fungi that degrade grain in storage may have been introduced into the developing grain before storage. Poor storage conditions allow fungal inoculum to germinate, colonize and degrade grain under storage conditions (Christensen and Kaufman, 1965; Husseini et al., 2009; Sharma et al., 2010; Yassin et al., 2010). In fact, pre-existing grain mould or weathering damage may be exacerbated during transport of grain (Malaguti, 2002). An additional problem caused by some of these fungi is their potential to produce mycotoxins that may be harmful in feed or food products produced from infected grain (Waniska et al., 2001).

Traditionally, *Fusarium* and *Curvularia* spp. have been viewed as the most important grain mould fungi (Table 2). *Fusarium* spp. may infect all tissues of the panicle including the spikelets, rachis branches and peduncle (Malaguti, 2002). It is still not known if endophytic *Fusarium* infections contribute to grain mould or head blight diseases. Although the dominant fungi encountered vary from location to location, *Fusarium thapsinum* and *Curvularia lunata* (Wakker) Boedijn are the two most frequently encountered species (Little et al., 2012; Mtisi and McLaren, 2002; Waniska et al., 2001). Malaguti (2002) reported that *F. moniliforme sensu lato* (likely *F. thapsinum*) was isolated from 80% of moulded grain. Other common genera include *Alternaria alternata* (Fr.) Keissl., *Cladosporium* spp. and *Phoma sorghina* (Sacc.) Boerema, Dorenb. & Kesteren (Stack and Pedersen, 2003; Navi et al., 2005; Funnell and Pedersen, 2006) (Tables 2 and 3). Some storage moulds

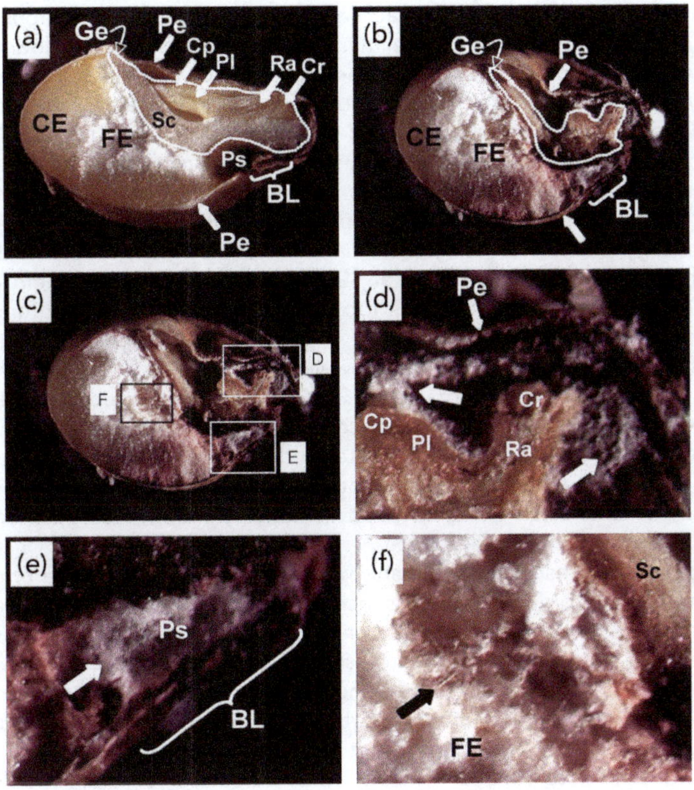

Figure 4 Cross section of healthy (a) and *Fusarium-thapsinum*-infected (b) caryopsis. Close-ups (c, insets) of mycelial mass (white arrows) in shrivelled area below the pericarp (d) and placental sac adjacent to the black layer (e). Fungal hyphae colonizing the floury endosperm adjacent to the scutellum (f). BL = Black layer; CE = corneous endosperm; FE = floury endosperm; Cp = coleoptile; Cr = coleorhiza; Ge = germ (oultined); Pe = pericarp; Pl = plumule; Ps = placental sac; Ra = radicle and Sc = scutellum. Images by C.R. Little.

include *Aspergillus*, *Penicillium* and *Rhizopus* spp. (Christensen and Kaufmann, 1965). See Little et al. (2012) for an exhaustive list of grain-associated fungi for sorghum. Also, many foliar pathogens of sorghum have a seedborne phase and may colonize caryopses. These include *Bipolaris sorghicola* (Lefebvre & Sherwin) Alcorn (target leaf spot), *Cercospora sorghi* Ellis & Everh. (grey leaf spot), *Colletotrichum sublineola* (anthracnose), *Exserohilum turcicum* (Pass.) K.J. Leonard & Suggs (leaf blight) and *Gloeocercospora sorghi* D.C. Bain & Edgerton (zonate leaf spot) (Ahmed and Ravinder Reddy, 1993) (Tables 2 and 3).

Other diseases and pest infestations influence the incidence and severity of grain mould. For example, sorghum ergot, caused by *Claviceps africana*, increases grain mould infestations by providing higher amounts of available sugar on grain and other plant surfaces (McLaren, 2002). In Africa, panicle insects are associated with grain mould (Ratnadass et al., 2003). Saprophytic fungi, including many potential grain mould pathogens, may colonize these resources and then use this as a jumping-off point, especially since honeydew accumulation occurs during grain development. In the same manner, recent infestations by the sugarcane aphid [*Melanaphis sacchari* (Zehntner)] produce copious amounts of honeydew from feeding. In a recent preliminary study in Kansas, it was found that plants with higher sugarcane aphid feeding and honeydew deposition resulted in seed that was colonized by *A. alternata*, *Fusarium thapsinum*, *F. proliferatum* and *Penicillium* spp. at a greater level than seed collected from aphid-free plants (C.R. Little, unpublished data).

In general, successful host resistance to grain mould can be divided into three areas: (1) preformed structural, (2) preformed chemical and (3) induced chemical traits (Brown et al., 2006; Little et al., 2012). The pursuit of grain mould resistance has been the most important strategy directed at long-term management of grain mould and weathering (Bandyopadhyay et al., 2002; Little et al., 2012). As early as 2001, Klein et al. (2001) found five grain mould quantitative trait loci (QTLs) associated with antifungal proteins. A_2 cytoplasm exhibited increased incidence of grain mould (Stack and Pedersen, 2003). Resistance is available in genotypes that flower during the rainy season in the 242-entry sorghum minicore collection (Sharma et al., 2010). In this case, there is no gene-for-gene resistance. Any degree of genetic resistance *per se* is polygenic and quantitative, and thus, many of those genes would contribute to the preformed structural and chemical mechanisms outlined below. Obviously, both preformed structural and chemical traits dovetail together and often must be used in concert within a sorghum genotype to provide adequate grain mould resistance.

Preformed structural resistance mechanisms include plant traits that can be specifically controlled through the breeding process including glume structure, loose panicle shape and plant height (Rooney et al., 2002; Waniska et al., 2001). Many of these structural features contribute to the plant's ability to escape or avoid infection. Caryopsis traits that form a barrier to infection by grain weathering fungi include a pigmented testa, increased endosperm hardness due to a greater proportion of corneous endosperm, a thin mesocarp and higher wax content on seeds (Rooney et al., 2002; Waniska et al., 2001).

Kernels from candidate sorghum lines may be screened for grain mould resistance using salt water float tests to check for fewer floating kernels, improved water imbibition, increased grain hardness, lower seed leachate electrical conductivity, lower ergosterol concentration in seed and direct plating approaches to check for the isolation frequency of grain mould pathogens (Indira, 2002).

Preformed chemical resistance mechanisms, in the broadest sense, include various phenolic acids, flavan-4-ols (in red pericarp sorghums), tannins and high concentrations of antifungal proteins (Indira, 2002; Rooney et al., 2002; Waniska et al., 2001). Phenolic acids, their derivatives and tannins, although effective at reducing grain mould damage in the lines that contain them, are undesirable due to their tendency to reduce food and feeding quality of grain due to dark colours, poor taste or reduced nutritional quality (Rooney et al., 2002; Waniska et al., 2001). Increased pigmentation of glumes is also associated with grain mould resistance (Indira, 2002). Funnell and Pedersen (2006) found that red and white pericarp did not matter if in a purple background. Grain from purple plant types has increased grain mould resistance compared to tan types. Funnel-Harris et al. (2015) also showed that waxy sorghums were more susceptible to grain mould than wild type.

Several antifungal proteins have been found to inhibit fungal species associated with grain mould and weathering. These include beta-glucanases, chitinases, ribosome-inactivating proteins and thaumatin-like proteins among others (Waniska et al., 2001). *In vitro*, Seetharaman et al. (1997) found that AFPs eluted from sorghum extracts using a Sephadex column could inhibit hyphal elongation of *C. lunata* and *F. moniliforme sensu lato*, inhibit spore germination of *Aspergillus flavus* and rupture *F. moniliforme sensu lato* hyphae. In the field, Rodriguez-Herrera et al. (1994) found that grain-mould-resistant sorghum lines generally had higher levels of sormatin, glucanase, chitinase and ribosome-inactivating proteins, but this was highly modified by environment. Bueso et al. (2000) found that SC719-11E amd R9025 had both high AFP content and high levels of grain mould resistance.

Induced resistance mechanisms are primarily upregulated defence response gene expression after inoculation of sorghum florets with grain mould pathogens. For example, Little and Magill (2003a) found that expression of phenylalanine ammonia lyase (PAL1-1), chalcone synthase (CHS2G), beta-glucanase (GLUC2-1) and chitinase (CHIT25-1) increased steadily after inoculation of sorghum panicles at anthesis with *F. thapsinum* or *C. lunata* conidial suspensions. And, upregulation of these genes was visibly higher on dot and northern blots compared to mock-inoculated controls. Katile et al. (2010) showed that PR-10 expression also increased after *F. thapsinum* and *C. lunata* inoculation. However, in this study, PR-10 showed greater expression in resistant lines than in susceptible ones. This response appeared to exhibit a genotype × pathogen interaction. Nithya et al. (2013) found that the upregulation of peroxidases was associated with grain-mould-resistant genotypes. Further, these authors could induce host resistance to *C. lunata* and *F. proliferatum* using several plant-growth-promoting rhizobacteria species.

Other management strategies have proven less viable. For example, treatment of grain in the field with fungicides has been less practical and not environmentally or economically sustainable (Waniska et al., 2001).

6 Head blight

Although the external symptoms of head blight and grain mould are similar, head blight differs from grain mould in that head blight infection initiates in the parts of inflorescence such as rachis branches and peduncle, whereas grain mould is a result of direct spikelet infection (McGee, 1992; Little et al., 2012). Typical head blight symptoms included necrosis of a part or the entire peduncle, rachis and rachis branches and are accompanied by reddish pith tissues in infected areas (Fig. 5). The discoloration may be seen in entire peduncle and can extend into the upper internodes of the stalk (Zummo, 1978). Head blight can result in premature death of all or parts of the seed head and open panicle with drooping rachis branches (Castor and Frederiksen, 1981; Frederiksen et al., 1982). Severe infections can cause peduncles to break over. Moreover, head blight can cause death of some to all florets in the panicle. Therefore, reduced caryopsis formation is a major

Figure 5 Typical pith discoloration of the upper stalk and peduncle (white arrow) that results from Fusarium head blight. In this case, *Fusarium fujikuroi* has infected the plant tissue. Image courtesy of J. Dahlberg.

symptom of the disease. Under severe infection, the entire panicle may be covered with heavy fungal growth, and it appears cream to pinkish tan in colour, resembling a severe grain mould infection (Zummo, 1978).

The primary causal organisms of head blight include multiple *Fusarium* spp. (including *F. thapsinum* Klittich, Leslie, Nelson and Marassas sp. nov., *F. semitectum* Berk. and Ravenel, and *F. proliferatum*) (Stack and Pedersen, 2003) (Table 2). *Colletotrichum sublineola* is also reported to be a head blight pathogen (Wall and Meckenstock, 1992). Inflorescence fragments present in soil serve as the source of primary inoculum. Prolonged wet weather conditions during anthesis are conducive for infection and disease development (Frederiksen, 2000c). Infection can occur soon after anthesis. Although the exact mechanism of infection is not clear, the waxy bloom along the outside of the stalk is believed to provide the passage for soil-inhabiting fungi to reach the head and initiate infection (Zummo, 1978). Fungal penetration may occur via wounds and cracks in the peduncle rind, rachis or panicle branches. Panicles bearing infected florets can act as a source for the secondary infection where the healthy heads are infected with airborne conidia. Although head blight can result in seed infection, evidence for the transmission is lacking (TeBeest et al., 2004).

Sorghum varieties with loose, open heads are less vulnerable to head blight infection than varieties with dense, compact heads (Zummo, 1978). Therefore, cultivation of open-headed varieties can help to control the disease. Planting dates should be adjusted to prevent overlapping of plant maturity and frequent rains to avoid the disease (TeBeest et al., 2004). Histopathological studies showed that sorghum genotypes that contain more lignin in the vascular tissue and a higher percentage of epidermal (peripheral) tissues are more resistant to head blight (Narro et al., 1992). RTx439, SC599 and GPR148 are some of the resistant sources for head blight (Rosenow and Frederiksen, 1982; Narro et al., 1992). Moreover, using SC599 derivative lines, Frederiksen et al. (1973) showed that stay-green lines are highly resistant to Fusarium head blight, and is a worthy trait to manage head blight in the field. Views on the inheritance pattern of head blight resistance are controversial. With crosses using a resistant sorghum line, Tx430 as the pollinator, Narro et al. (1992) showed that inheritance for resistance is recessive; thus, resistance needs to be present in both parents. However, Rosenow (1992) indicated that resistance is inherited in a dominant or intermediate fashion. Different conclusions may be due to the use of different genetic materials used in their experiments.

Management strategies include the cultivation of open-headed and stay-green sorghum varieties. Avoid potential overlapping of plant maturity with frequent rains by adjusting planting date and avoid delayed harvesting. Residue management can be used to reduce primary inoculum density.

7 Ergot

As far as sorghum diseases are concerned, ergot is quite distinctive. The fungus infects and invades the ovaries of unpollinated sorghum flowers. As a result, the infected ovaries convert into a mass of white fungal mycelium (sphacelium) that secretes 'honeydew' (Frederiksen et al., 1991; Frederickson et al., 1993; Tooley et al., 2000; Zummo, 1978). Honeydew is a sticky fluid that contains large quantities of macroconidia and is secreted from the infected florets. Eventually, the spore-laden honeydew will drip down to the leaves and the soil, leaving a sign of the disease (Krausz and Isakeit, 1998).

The colour of the honeydew is a good indicator of the number of conidia present. Clear honeydew contains a low concentration of conidia, whereas darker, orange honeydew contains higher levels of conidia (Alderman et al., 1999). After the honeydew darkens, it turns white as it dries. The honeydew may also serve as a substrate for other fungi including *Cerebella* spp. and *Alternaria* spp. (Bandyopadhyay et al., 1998). Such saprophytic mould growth on the panicle can be confused with grain mould and complicate disease diagnosis.

Plants with ergot suffer lower seed production, reduced quality and impeded harvestability. The higher the number of plants with ergot, and more severe the disease, the greater the potential yield loss (Krausz and Isakeit, 1998). Other direct losses include the reduction of feed quality (Blaney et al., 2000). Also, some countries do not allow entry of sorghum seeds infected by ergot pathogens (Odvody et al., 2002; Montes-Garcia et al., 2009).

In seed production fields, losses of 10–80% have been reported in India, and annual losses of 10–25% have been recorded in Zimbabwe (Bandyopadhyay et al., 1998). Grain production fields do not often experience severe outbreaks of ergot. However, research in Australia and Mexico has shown that ergot losses occur following periods of cool weather (Meinke and Ryley, 1997; Montes-Garcia et al., 2009) (Fig. 1).

However, indirect losses may be more important than the direct damage ergot causes to sorghum. Harvesting grain from ergot-infected fields can be difficult because infected florets secrete honeydew, which can plug harvesting equipment, make threshing difficult and reduce germination and seedling emergence as a result of other diseases (Krausz and Isakeit, 1998).

Sorghum ergot is caused by the fungal pathogen *Claviceps Africana* (Table 2). Although *C. africana* is the most common pathogen, two additional *Claviceps* spp. are known to infect sorghum. These include *Claviceps sorghi* B.G.P. Kulk., Seshadri & Hegde and *C. sorghicola* Tsukib. Shirman. & T. Uematsu (Frederickson et al., 1991; Isakeit et al., 1998; Pažoutova and Bogo, 2002) (Table 2). *Claviceps sorghi* was initially reported in India, whereas *C. sorghicola* was reported from Japan. *Claviceps africana* was originally reported from Zimbabwe (Pažoutová et al., 2000). Other well-known *Claviceps* spp. include *C. purpurea* (Fr.) Tul. and *C. fusiformis* Loveless, which infect rye and rice, respectively (Langham, 2004).

Claviceps sorghi is similar to *C. africana*. These two pathogens have similar conidia, but different sclerotial morphology (Bandyopadhyay et al., 1998). Unlike *C. africana*, where the sexual stage is rare or non-existent, *C. sorghi* sclerotia germinate stromata that contain perithecia. *Claviceps africana* is more widespread than *C. sorghi* or *C. sorghicola* because of its ability to produce numerous conidia as secondary inoculum (Muthusubramanian et al., 2005).

Originally, ergot was restricted to Africa and Asia but began to spread through infested seed (Aguirre et al., 1997; Pažoutová et al., 2000). The first recorded case of ergot in the Americas occurred in Brazil in 1995 (Aguirre et al., 1997). The epidemic in Brazil spread rapidly at a pace of 300 000 square miles per week. It swiftly spread throughout the continent via wind and was reported to be found in Mexico in 1996 and Texas by 1997 (Isakeit et al., 1998; Velasquez-Valle et al., 1998). Through the 1997 growing season, ergot quickly spread to the surrounding sorghum-growing areas as far north as Nebraska (Krausz and Isakeit, 1998). The disease was observed on a limited basis in Kansas in 1998 but did not pose a serious problem for producers in the state.

Claviceps africana is an ovary-infecting fungus that follows nearly the same path as the sorghum pollen tube follows during the process of fertilization (Bandyopadhyay et al., 1998;

Magill, 2013). The conidium lands on the stigma, germinates, grows down the style and penetrates the ovary (Stack, 2000). Normally, sorghum pollen only requires a few hours to fertilize the ovary, while colonization of the ovary by *C. africana* requires up to 3 days (Lansac et al., 1994; Komolong et al., 2003). Thus, florets with receptive stigmas are susceptible to infection, but fertilized ovaries render florets resistant to the fungus. Approximately 6 days after infection, *C. africana* converts the sorghum ovary into a white fungal mycelial mass known as a sphacelium (Komolong et al., 2003). Development of the sphacelium pushes the glumes apart like an oversized caryopsis. The sphacelium later begins to produce primary conidia that are immersed in the honeydew secretions. Primary conidia can survive for months in field residue, or on other hosts outside of the growing area.

As the honeydew drips down the plant panicle, the conidia of the fungus can infect additional unfertilized florets. Ergot honeydew also attracts various insects that are drawn to it for its sugar content. These insects may spread the fungus throughout a field or region (Frederickson et al., 1993; Bandyopadhyay et al., 1998; Isakeit et al., 1998; Pažoutova and Bogo, 2002).

White crusting of the honeydew, as mentioned above, indicates that primary conidia have germinated and have produced secondary conidia (Frederickson et al., 1999). Secondary conidia are airborne and rapidly spread through a field and the surrounding area (Frederickson et al., 1993; Bandyopadhyay et al., 1998; Isakeit et al., 1998; Pažoutova and Bogo, 2002). Just as urediniospores follow the '*Puccinia* Pathway' using winds moving south to north, secondary conidia of *C. africana* may move long distance in the same manner.

However, most species of ergot produce sclerotia to ensure their survival for the following season. Sclerotia are dormant structures that develop from the sphacelial tissue (Pažoutova et al., 2000; Alderman et al., 2003; Pažoutova and Frederickson, 2005). These structures are more insensitive to environmental extremes than are conidia or sphacelia. Under favourable conditions, sclerotia germinate conidia that will infect the florets of a new sorghum crop (Bandyopadhyay et al., 1998).

To determine whether there are sclerotial tissues in the field, clean sphacelia may be harvested from infected panicles, soaked in water for several hours, and then examined under a dissecting microscope (Alderman et al., 1999). Sphacelial tissues are cream or white in colour and quickly dissolve, whereas sclerotial tissues have a solid structure and are orange to brown in colour. Sclerotial development begins at the base of the sphacelium (Frederickson et al., 1999). The sclerotia of *C. africana* and *C. sorghi* can be distinguished based upon their size and morphology (Alderman et al., 1999). Generally, *C. africana* sclerotia are round and smaller than those of *C. sorghi*, whereas *C. sorghi* sclerotia are elongated and larger than those of *C. africana* (Alderman, 2003).

Losses greatly depend on environmental conditions. Climates that are conducive to ergot will contribute to greater yield loss and reduce seed quality (Meinke and Ryley, 1997). The best environment for ergot to survive and spread exhibits high humidity and cool daytime temperatures of ~70°F (21°C) and night temperatures of 50–55°F (10–13°C) (McLaren and Wehner, 1992; Miedainer and Geiger, 2015). Workneh and Rush (2006) found that, regardless of inoculum level, temperature and relative humidity had the greatest influence on infection. Daytime temperatures of 59–77°F (15–25°C) and relative humidity that approached 100% resulted in increased proportions of ergot-infected florets in sorghum panicles.

Greater numbers of unpollinated florets provide more infection sites for the pathogen, which under cool conditions results in increased epidemics (Miedainer and Geiger, 2015).

Thus, pollination timing promotes ergot infection in sorghum. Typically, pollination occurs under warm conditions. Cool temperatures before flowering delays pollination (McLaren and Wehner, 1992). Delayed pollination provides *C. africana* a window to infect ovules in a greater number of unpollinated florets throughout the field, which increases yield impact (Fig. 1). McLaren and Flett (1998) showed that viable pollen resulted in reduced ergot severity. Further, pre-flowering minimum temperatures above 18°C resulted in pollen with sufficient viability to reduce ergot severity to less than 10%.

In addition to environmental conditions, fertility rates have an impact on the effect of ergot. Male-sterile plants and their hybrids are at higher risk of infection because of a higher probability of poor pollination since they rely on pollination from plants acting as males (Krausz and Isakeit, 1998). Varieties with low fertility rates will be under-pollinated compared to typical varieties (Miedaner and Geiger, 2015).

Due to their relatedness to sorghum, johnsongrass and shattercane can also serve as a viable host plant for *C. africana* (Velasquez-Valle et al., 1998). Other host plants for sorghum ergot include many grass species, which could create a 'green bridge' for the pathogen. Because *C. africana* can infect numerous grasses and because of its ability to infest seeds, sorghum ergot can persist for years under conducive environmental conditions.

Discovering host resistance to ergot has been a challenge, although some authors have found candidate-resistant and -tolerant materials. Dahlberg et al. (2001) found that IS 8825 (Ethiopian in origin) exhibited ergot resistance in cytoplasmic male-sterile cytoplasm. Reed et al. (2002) identified wild accessions of sorghum that were resistant to spray inoculation of conidia at anthesis. Two entries, IS 14131 and IS 14257, appeared to have genes for ergot resistance when crossed with elite parents. Parh et al. (2008) identified nine QTLs linked to per cent ergot infection. In addition, they also identified nine QTL associated with pollen quantity and viability, which were related to ergot resistance in their test populations.

Pollination and cultural management strategies are still required to control the disease. Hybrids that are less sensitive to cold temperature effects on pollination may escape infection (McLaren, 1997; Bandyopadhyay et al., 1998). In this germplasm, cool temperatures do not delay pollination to the same extent as in normal germplasm. In self-fertile sorghums, resistance to *C. africana* is due to the maintenance of high self-fertility within certain environments. Often, when these genotypes are moved to environments in which they are poorly adapted, such 'tolerance' breaks down (Odvody et al., 1998).

In years that have normal summer temperatures and low humidity during flowering and pollination, losses and stresses from ergot should be minimal (Workneh and Rush, 2006). During cooler years, especially after flowering, ergot can become problematic. Cooler temperatures and high humidity during flowering delay and reduce the effectiveness of pollination (Dahlberg et al., 1998). As described above, these conditions are conducive to ergot development and spread (Bandyopadhyay et al., 1998).

In hybrid production, any factor promoting maximum pollination and fertilization of the male-sterile parent reduces the risk of ergot infection (Miedaner and Geiger, 2015). Pollen management techniques such as reduction of female-to-male ratios, multiple planting dates of male pollinators and modifying the length of female receptivity can facilitate rapid and even pollinations and reduce 'nicking' times, that is the time between flowering and pollination (Bandyopadhyay et al., 1996; Odvody et al., 1998; Cisneros-López et al., 2012). All male-sterile sorghums are susceptible to *C. africana*, but relative differences in susceptibility have been observed.

Cultural management strategies include planting at recommended dates for optimum seed production, planting clean hybrid seed with high germination rates to ensure full plant populations, providing sufficient nutrition through fertilizer application and crop rotation regimes, and grazing or cutting forage sorghums to prevent flowering until after fertilization in grain sorghum fields is complete (McLaren and Flett, 1998; Velasquez-Valle et al., 1998; Blaney et al., 2010). Forage sorghums are more susceptible to ergot due to flowering and fertilization problems as well as a high proportion of male-sterile plants. Additionally, late planting of sorghum should be avoided as these plants pollinate during periods of cooler weather, which is favourable for the disease. Preharvest use of glyphosate can kill tillers that may harbour the fungus. Velazquez-Valle et al. (1998) indicated that infected panicles could be burned as an eradicative control measure.

Various fungicides are available to control sorghum ergot. Application of fungicides during flowering can prevent the spread of the pathogen. Prom and Isakeit (2003) found that triadimefon, trifloxystrobin, tebuconazole, azoxystrobin and propiconazole could reduce ergot severity. However, from an economic point of view, fungicide applications are not feasible since three or four applications would be necessary to control the disease. At that rate, the cost on a per-acre basis would be prohibitive. In seed production fields where this is the potential for high pathogen pressure, it is more profitable to use fungicides, such as triazoles (e.g. propiconazole), to control ergot via a head spray and ground application. This approach has been employed in Brazil, the Americas and Australia since the emergence of ergot in the New World (Odvody et al., 1998). In the United States, aerial application of triazoles would be necessary to promptly treat large acreages (Odvody et al., 1998). Treating seed with a fungicide is an economical way to prevent an ergot outbreak.

8 Head smut

Head smut, caused by the facultatively biotrophic soilborne and basidiomycete fungus *Sporisorium reilianum* (Kühn) Langdon and Fullerton (syns. *Sphacelotheca reiliana* (Kühn) G.P. Clinton and *Sorosporium reilianum* (Kühn) McAlpine), is an important sorghum disease that has been reported from all parts of the world where sorghum is grown (Tables 1 and 2). Africa has been suggested as the origin of the pathogen, although different races infect sorghum, corn and sudangrass over a wide geographical area, including Africa, Europe, North and South America, Mexico, Asia, Australia, New Zealand and the West Indies (Frederiksen, 2000d; Jalali and Behdad, 2003; Little et al., 2012; Marley and Aba, 1999; Ngugi et al., 2002; Pande et al., 1993; Perumal et al., 2007; Prom et al., 2011; Reyes et al., 1964; Williams-Alanis et al., 2009). *S. reilianum* is one of the important sorghum fungal pathogens that significantly impact yield as it completely replaces the grain with smut sori (Frederiksen, 2000d; Little et al., 2012).

S. reilianum infects at the seedling stage, but obvious symptoms are not expressed until flowering. As indicated above, the disease results in the inability of infected plants to produce grain. The symptoms start with the formation of smut galls (sori) covered with a thick white to grey membrane (peridium) during the flowering stage (Williams et al., 1978; Frederiksen et al., 2000d; Little et al., 2012) (Fig. 6). Later, the peridium ruptures and releases numerous black to brown teliospores. Pathogen colonization of the vascular tissues results in broom-like filaments called 'witches' brooms'. Other symptoms such as phyllody and failure to produce a viable panicle and floret tissues have been observed in sweet sorghum and sudangrass cultivars as well as grain sorghum. Stunted plant growth

Figure 6 Characteristic symptoms of head smut caused by *Sporisorium reilianum*. (a) Panicle with smut sorus covered by white peridium (membrane) (white arrow). (b) Closer view of smut sorus showing residual vascular strands (white arrow) after teliospore liberation. (Image courtesy of Kansas State University Library/L. Claflin.)

with condensed nodal regions due to poor peduncle elongation with panicle sterility and the formation of numerous individual florets are other characteristic symptoms of *S. reilianum* (Little et al., 2012). Teliospores spread by rain and the wind through fields. Marley and Aba's (1999) survey of farmers' fields in major sorghum-growing areas within the four climatic zones of the Nigerian savanna established changing patterns in the incidence, severity and distribution of sorghum smuts.

S. reilianum requires a combination of haploid nuclei of opposite mating types to cause infection and grow through the plant as a dikaryon (Morrow and Fraser, 2009). At flowering, the fungus produces diploid teliospores in sori that replace the normal panicle tissues (Prom et al., 2011). Teliospores, which are produced from condensed masses of dikaryotic hyphae within the developing sorus, are spherical, reticulate and echiniform (Frederiksen, 2000d). Teliospores overseason in the soil and crop debris to provide inoculum for the next infection cycle (Little et al., 2012) (Fig. 1). After planting, soilborne teliospores germinate along with the seed. Germinating teliospores bear sporidia with a four-celled, branched promycelium and infect meristematic tissues in the seedling. As products of meiosis, sporidia may sprout to form yeast-like secondary sporidia or may germinate to form a germ tube to fuse with the opposite mating type to create dikaryotic hyphae that penetrate meristematic tissue in the sorghum seedling. Germination is the highest in moist soil where the temperature is 81–88°F (27–31°C) (Fig. 1). *S. reilianum* infection and symptom expression occur at two different crop stages, seedling and flowering, respectively. Therefore, it is difficult to predict infection and to implement timely control measures. Apparently, seedborne spores do not cause infection.

Head smut severity has increased in recent decades due to the cultivation of susceptible cultivars and hybrids or the appearance of more virulent races (Prom et al., 2011). In surveys of western Kenya (Ngugi et al., 2002) and the Ardestan region of Iran (Jalali and Behdad, 2003), 73–75% of plants were infected with head smut. In the United States, increased head smut incidence has been observed in sorghum fields in the coastal bend area of Texas (personal communication from Gary Odvody, Department of Plant Pathology and Microbiology, Texas A&M University, Corpus Christi, Texas). In Northern Tamaulipas, Mexico, head smut may affect susceptible genotypes up to 80% (Williams-Alanis et al., 2009). Once head smut is present in a cropping area, the pathogen is difficult to remove from the soil, even through crop rotation (Prom et al., 2011). *S. reilianum* is one of the few smuts not controllable by seed treatment, leaving host resistance as the only practical control mechanism (Frederiksen, 2000d).

Resistance to head smut is complex and can differ depending upon natural or artificial inoculation and the physiological race of the pathogen (Frederiksen and Reyes, 1980; Frederiksen, 2000b). Non-race-specific resistance through natural or artificial infection appears to suppress or slow pathogen growth in the plant towards the apical meristem. Tx430 and TAM428 are good sources that exhibit non-race-specific resistance (Frederiksen and Rosenow, 1971; Frederiksen, 1978a,b; Frederiksen and Reyes, 1980; Frederiksen, 2000d). On the other hand, race-specific resistance exists where certain genotypes are susceptible to a race of *S. reilianum*, whereas other genotypes are not. Such specific, vertical host–pathogen interactions presumably result in a hypersensitive reaction that immediately halts the pathogen infection process. SA281, SC170-6-17 and Tx414 are examples of sorghum lines that possess race-specific resistance (Frederiksen and Reyes, 1980; Frederiksen, 2000d). Other resistance classes exist and are detailed in Little et al. (2012). Beyond the germplasm already mentioned, resistance has been identified in Texas (Tx2962-Tx2978), China (Kang 4 and Kang 7) and hybrids in Mexico (Asgrow Ambar, Pioneer 82G63, RB-5×204, RB118×204 and RB-119×430; 46038×437) (Xu et al., 1997; Williams-Alanis et al., 2009; Little et al., 2012).

9 Covered kernel smut

In covered smut, the ovule is replaced by the fungus to produce a conical sorus. Sori are covered by a peridium (Esele et al., 1995; Stoll et al., 2003). Sori may be small and surrounded by the glumes, but often extend beyond the glumes and are easily distinguished from seed (Piepenbring et al., 1998). This smut may be differentiated from long smut in that most or all kernels are replaced with a sorus, which speaks to the systemic nature of the pathogen, compared to only a few spikelets replaced by sori in long smut, in which teliospores are spread by the wind (Nzioki et al., 2010). Higher numbers of infected panicles are evident in tillers and ratoon sorghum. At harvest, the peridium breaks and teliospores are spread and contaminate healthy, uninfected kernels. Although an understudied phenomenon, sori disruption at threshing may be dependent upon at least two factors: pathogen race and host genotype.

Sporisorium sorghi Ehrenb. Ex Link is a basidiomycete fungal pathogen that causes covered kernel smut (Stoll et al., 2003; Magill, 2013) (Table 2). The teliospores of *S. sorghi* may produce more than one promycelium. And, each promycelium produces more than four or more haploid sporidia (Nzioki et al., 2010).

Like *S. reilianum*, *S. sorghi* is a systemic pathogen of sorghum (Nzioki et al., 2010). Thus, only seed infested with teliospores of the pathogen will produce seedlings that are infected. It is hypothesized that infection occurs in the same manner as systemic infection occurs in head smut and *S. sorghi* mycelium grows along with the apical meristem of the seedling.

The incidence of covered kernel smut is dependent upon the location in which sorghum is being grown and varies from absent to a ubiquitous production constraint in farmers' fields (Frederiksen, 2000a). In many commercial sorghum production systems, in Australia, the United States and Zimbabwe, for example, seed treatment fungicides have been used to reduce or eliminate the incidence of the disease (Mtisi, 1997). Thus, no or reduced sori are present, and no seedborne infections take place. However, in low-input systems where farmers keep their seed for the subsequent season, and fungicides are not used, the rates of seed infection are high as is the incidence of smutted panicles (Esele, 2002). For example, in Botswana, Chad, Ethiopia, Guinea, Sudan and Tanzania, covered kernel smut has been consistently present in the past and contributes to significant losses along with the other sorghum smuts (Mansuetus, 1995; Pande et al., 1997b; Esele, 2002). In western Kenya, 42–43% of fields showed plants with covered kernel smut (Ngugi et al., 2002). In some areas of Central Africa, an endemic region for *S. sorghi*, wider use of seed dressings has reduced the incidence of covered kernel smut (Marley and Aba, 1999).

However, the overall incidence of covered kernel smut has increased in Central Africa over the past few decades (Pande et al., 1997b). At least two possibilities emerge from this phenomenon. First, it is possible that a shift in the germplasm to varieties more susceptible to the pathogen is being used by farmers. Second, a shift in the pathogen population has occurred, where *S. sorghi* has developed resistance to an active ingredient in the seed dressings that have been used. In either case, insufficient data exists to explain either hypothesis or to suggest alternative hypotheses. The use of elemental sulphur as a seed treatment is recommended to prevent pathogen evolution of resistance to other active ingredients that may have a single-site mode of action (Neergaard, 1979).

Continued efforts should focus on the use of cultural management strategies to control this smut. If seed can be collected from non-endemic areas and used for planting, this helps guarantee a disease-free crop. If seed must be collected locally, or from an infested field to be used for planting, two methods may be employed to reduce covered kernel smut incidence. First, seed should be sorted to remove smut sori. Second, seed washing removes teliospores on the pericarp and soaking the seed and drying them in the sun reduces active teliospores, but maintains seed viability (Doggett, 1988; Ahmed and Ravinder Reddy, 1993).

10 Loose kernel smut

Loose kernel smut is characterized by sori that rupture either before or after panicle exsertion. Unlike covered kernel smut, the peridium does not remain intact (Thakur et al., 2007). The wind separates the masses of teliospores from sori, leaving characteristic columellar strands (remnant vascular bundles) behind (Williams et al., 1978). In addition to the loose sori, which lend the disease its name, the plants are dwarfed, have narrow stalks and exhibit increased tillering and premature flowering (Williams et al., 1978; Esele, 1995). Tillers and ratoon crops show a higher incidence of loose smut compared to the main stem or first crop. The spikelets that do not have a sorus often show phyllody symptoms (Esele, 1995).

Loose kernel smut is caused by the fungus *Sporisorium cruentum* (J.G. Kühn) Vánky (Table 2). The teliospores are echinulate, but are less spherical and more subglobose than those of *S. reilianum*. In general, this disease is of 'minor to moderate importance' and only 'occasionally present' (Frederiksen, 2000b) (Table 1). The pathogen is seedborne and can be propagated through ratooned sorghum (Indira, 2002; Marley et al., 2002). Like covered kernel smut, loose smut incidence has been increasing in Central Africa production areas over the past two decades (Marley and Aba, 1999). The disease also exists in dual-grain- and fodder-use sorghums in Haiti (C.R. Little, personal observation).

Yield losses from this disease can be severe if it is not managed. For example, in China, yield losses range between 20% and 30% (Indira, 2002). Similar control measures, as described for covered kernel smut, have eliminated the disease in commercial production. Seed that is to be exported should be inspected. As in covered kernel smut, elemental sulphur may serve as a seed treatment that can reduce viable *S. cruentum* teliospores (Ahmed and Ravinder Reddy, 1993).

11 Long smut

Among the sorghum smuts, long smut sori are distinctive in that they are elongated, tube-like, and curved (Frederiksen, 2000b; Kollo, 2000). Sori may grow to a length of 1–2 inches, which is much longer than those produced by covered kernel smut. Like covered kernel smut sori, long smut sori are covered by a peridium (Williams et al., 1978; Marley and Aba, 1999). Although the peridium may remain intact through the season, it will dehisce as it ages. Separation of the peridium begins at the tip of the sorus and proceeds along its length to expose dark brown teliospores. Within the sorus, 8–10 columellar filaments (remnant vascular bundles) are present (Frederiksen, 2000b; Kollo, 2000). These filaments are like those observed in loose kernel smut.

Long smut is caused by the fungal pathogen *Sporisorium ehrenbergii* (J.G. Kühn) Vánky (syn. *Tolyposporium ehrenbergii*) (Mbgawa et al., 1993; Esele, 1995; Ngugi et al., 2002) (Table 2). In contrast to many of the other sorghum smuts, only a few individual spikelets of the panicle are infected, suggesting that *S. ehrenbergii* is a windborne rather than systemic smut pathogen (Prom et al., 2007). In general, this smut is only 'occasionally present' and is of 'minor importance' (Frederiksen, 2000b) (Table 1). However, in West and Central Africa, long smut is among the four sorghum smuts that have caused perennial losses (Marley et al., 2002).

Like other smuts, the use of chemical seed treatments reduced viable inoculum. Several chemicals have been suggested for the control of smuts including myclobutanil, Raxil, thiram, triadimenol and triadimefon (Kumar and Nath, 1991; Ahmed and Ravinder Reddy, 1993). Further, seed that is to be used for planting should be sorted to remove *S. ehrenbergii* sori (Ahmed and Ravinder Reddy, 1993).

12 Anthracnose

On the leaves, anthracnose produces small, round to oval lesions that are usually less than a quarter to half an inch in diameter (Knott et al., 2016; Tesso et al., 2012; Williams et al., 1978). As lesions mature, they will elongate, coalesce and become sunken in the centre (Erpelding and Prom, 2004). In susceptible genotypes, entire leaves may become necrotic, thus reducing photosynthesis capacity for the plant. Depending upon the host plant's colour genes, lesion discoloration will range from light tan to red, purple or black. In the same way, midribs may become infected and show similar symptoms (Wharton and Julian, 1996). Within the lesions, acervuli may develop. With the naked eye, these fruiting structures may look like small black spots. However, closer inspection with a hand lens will reveal hair-like setae emerging from the acervuli, which is diagnostic for anthracnose. Under humid or wet conditions, the acervuli will produce numerous falcate to lunate conidia that will appear as cream-coloured masses in and around the setae (Cardwell et al., 1989). These signs are also visible using a hand lens.

Anthracnose also occurs on and in the stalks of sorghum (Mehta et al., 2005). Stalk anthracnose is sometimes referred to as 'red rot', and is often considered to be far more damaging than the foliar phase of the disease (Erpelding and Prom, 2004). Split stalks can be easily diagnosed by the marbled discoloration pattern apparent in the pith (Williams et al., 1978). However, nodal tissue often remains normal. Like other stalk rots, anthracnose stalk rot can result in lodging and decreased yield in plants that do not lodge. Plants that have the stalk rot phase of the disease may or may not have significant foliar symptoms. Often, the peduncle, rachis branches or other panicle tissues, including inflorescences, may develop discoloured lesions that form cankers (Hess et al., 2002).

Anthracnose is caused by the hypervariable fungal pathogen *Colletotrichum sublineola* (Table 4). Moore et al. (2008) examined 87 isolates of *C. sublineola* and found 13 pathotypes. Prom et al. (2012) used amplified fragment length polymorphisms (AFLPs) to examine 232 *C. sublineola* isolated from the United States and Puerto Rico. These authors found moderate levels of genetic variability among the isolates, but high variability in terms of pathogenicity to the point where 17 new pathotypes were established. The new pathotypes were based on sets of differentials used by Cardwell et al. (1989) and Moore et al. (2008) in the United States, and Casela and Ferreira (1987) and Valèrio et al. (2005) in Brazil.

The pathogen that causes anthracnose in maize is *C. graminicola* (Ces.) G.W. Wilson. *C. sublineola* can infect numerous grass hosts including johnsongrass, wheat and wild rice. Crouch et al. (2006) found that *C. sublineola* belonged to a separate clade compared to *C. graminicola* on corn, and two clades of *C. cereale* on the pooid grasses.

Like many foliar pathogens, *C. sublineola* produces microsclerotia, especially on senescent tissues that are deposited in the soil along with host residue (Costa et al., 2003). Both acervuli and microsclerotia may overseason in host residue (Tsedaley et al., 2016). Either rehydration of acervuli or sporogenic germination of microsclerotia to produce new acervuli can initiate the disease cycle. Rain, irrigation or dew can splash conidia onto new host leaves to initiate infections (Thakur et al., 2007). If green tissue remains to be infected, the development of new conidia-laden acervuli in lesions provides secondary inoculum that can infect leaf and stalk tissues within the same season.

Colletotrichum sublineola conidia are capable of directly penetrating all plant surfaces including leaves, stalks and the pericarp of seeds. In this process, a *C. sublineola* conidium lands on the plant and germinates a germ tube. At the end of the germ tube, an appressorium swelling is produced (Lo et al., 1999). From the appressorium, a penetration peg is formed (Wharton et al., 2001). A combination of direct pressure and exoenzyme activity allows the penetration peg to push through the waxy cuticle and epidermis. At this stage, the fungus can grow intercellularly in a hemibiotrophic manner (Wharton et al., 2001). When the *Colletotrichum* fungi switches from hemibiotrophic to necrotrophic growth, lesions begin to form, which are dying cells that yield nutrients and support pathogen development including the formation of acervuli (O'Connell et al., 2011).

Table 4 Foliar diseases of grain sorghum caused by fungi

Disease	Pathogen(s)	Symptoms and signs	Key controls	Refs.
Anthracnose	*Colletotrichum sublineolum*	Small foliar lesions Acervuli within lesions Grain anthracnose Floret abortion	Resistance Rotate with non-hosts Remove weedy species hosts Residue management	Tesso et al. (2012)
Grey leaf spot	*Cercospora sorghi*	Rectangular foliar lesions Sporulation on lesions	Resistance Crop rotation Remove weedy species hosts Residue management	Odvody (2000)
Leaf blight	*Exserohilum turcicum*	Cigar-shaped lesions Sporulation on lesions Reduced grain weight	Resistance Rotate with non-hosts Foliar fungicide applications	Berquist (2000)
Oval leaf spot	*Ramulispora sorghicola*	Oval lesions with light centres Dark sclerotia may develop	Rotate with non-hosts Residue management	Girard (1980)
Rough leaf spot	*Ascochyta sorghi*	Rough pycnidia on lesion surface Large necrotic lesions	Resistance (sweet lines) Rotation with non-hosts Residue management	Zummo and Broadhead (1984)
Rust	*Puccinia purpurea*	Uredinia pustules Telia often form on abaxial side of leaves	Resistance	Frederiksen et al. (1978a,b) and Hooker (1985)
Sooty stripe	*Ramulispora sorghi*	Lesions surrounded by bright chlorotic halo Microsclerotia on underside of lesions	Resistance Crop rotation Residue management	Brady et al. (2011) and Girard (1980)
Tar spot	*Phyllachora sacchari*	Black, stromatic masses on leaves Lesions surrounded by halos	Resistance	Dalmacio (2000a)
Target leaf spot	*Bipolaris sorghicola*	Oval lesions with alternating colour	Resistance	Dalmacio (2000b)
Zonate leaf spot	*Gloeocercospora sorghi*	Bullseye lesions	Rotate with non-hosts Remove weedy species hosts	Bain and Edgerton (1943) and Girard (1980)

Sorghum anthracnose is problematic where high temperatures and high relative humidity are accompanied by intermittent periods of hot and dry weather that stress plants (Mathur et al., 2002). In addition to environment, several factors combine to result in yield losses. First, hybrid susceptibility governs the extent of infection (Perumal et al., 2017). Based on information concerning corn and sorghum anthracnose, when the infection progresses from foliar blade to midrib and then to stalk, yield impacts may be severe (Bergstrom and Nicholson, 1999; Xavier et al., 2017). Second, infection during the early vegetative stages results in a yield impact that is much higher compared to infections that occur after panicle formation and exsertion. Third, longer season duration genotypes provide more time for the pathogen to develop and the disease to progress. However, this last item requires further confirmatory research.

Host resistance exists for this disease and has been an intensive focus for sorghum breeding programmes over the last several decades. In this light, anthracnose-susceptible hybrids should not be grown in areas where anthracnose has been a perennial problem. Prom et al. (2012) observed that all 18 isolates of *C. sublineola* tested in the pathotype differential study failed to cause disease on BTx623 and QL3. Resistance in SC748-5 has remained stable across many locations and years where it has been tested. This line carries the *Cg1* resistance gene, which has been tagged with markers and could be used as a resistance source for marker-assisted breeding for durable anthracnose resistance (Perumal et al., 2009; Prom et al., 2012). Further, other anthracnose resistance loci were discovered in SC112-14 at the distal end of chromosome 5 near *Cg1* that differentially governs resistance against Arkansas, Georgia, Texas and Puerto Rico isolates (Cuevas et al., 2014).

In Arkansas, Li and Tebeest (2009) found that disease progress curves for anthracnose transitioned into the exponential phase between 40 and 50 days after inoculation for susceptible genotypes (BTx623 and Pioneer 8313). No disease was apparent on the resistant genotype (Cargill 888Y). Likewise, surrounding rows of susceptible genotypes were infected at nearly the same level as the inoculated row.

Because acervuli can overseason in the soil (Fig. 1), crop rotation may be used to reduce the viable inoculum in crop debris. From a nutritional perspective, use of 2 mM silicon in hydroponic culture significantly reduced anthracnose disease severity by 86% (Resende et al., 2012). Silicon improves host carbon assimilation, stomatal conductance to water vapour and transpiration rate in infected plants. In addition, Resende et al. (2012) indicated that silicon amendment increased antioxidant activity and reduced electrolyte leakage from infected cells.

13 Grey leaf spot

Grey leaf spot lesions begin as small oval or circular spots and expand to the parallel veins of the leaves, giving them a rectangular appearance (Odvody, 2000; Knott et al., 2016; Williams et al., 1978). Like many sorghum foliar diseases, discoloration within the lesions reflects the plant's colour genotype: tan, red or purple. Depending on the degree of host susceptibility to grey leaf spot, lesions may be quite short (<1/2 inch) to quite long and stripe-like. The latter may be confused with bacterial leaf stripe or bacterial leaf streak. Occasionally, grey leaf spot lesions may be striated, which can be confused with ladder leaf spot (Wall et al., 1989). *C. sorghi* produces the phytotoxin, cercosporin, which kills plant cells (Goodwin et al., 2001).

The key sign associated with this disease is the development of a white or grey felt-like covering on lesions during periods of high humidity or prolonged leaf wetness (Odvody, 2000; Knott et al., 2016). The conidia produced by *C. sorghi* are elongate, multicellular and borne on a fasciculate conidiophore. Germ tubes may emerge from several cells along the length of the conidium (Pascual and Raymundo, 1989).

Grey leaf spot is caused by the fungus *Cercospora sorghi* (Pande et al., 1993; Ngugi et al., 2002) (Table 4). There are over 3000 species of *Cercospora* and most have *Mycosphaerella* teleomorphs if a sexual stage is known (Groenewald et al., 2006). *C. sorghi* var. *maydis* on maize differs from *C. sorghi* on sorghum by six nucleotides in the internal transcribes spacer region, which indicates that *C. sorghi* var. *maydis* is a separate species from *C. sorghi* (Goodwin et al., 2001). Using AFLPs, *C. sorghi* was differentiated from *C. zeae-maydis* Tehon & E.Y. Daniels groups I and II by a genetic distance of 67–70% (Wang et al., 1998). The *C. zeae-maydis* clades have been recently reclassified into the pathogen name *C. zeina* Crous & U. Braun (Muller et al., 2016). In Uganda, wild sorghum (*S. bicolor* var. *verticilliflorum*) is a reservoir host for *C. sorghi* (Okori et al., 2004) (Fig. 1).

In more arid production areas, grey leaf spot appears at the end of the growing season when temperatures become cooler and the formation of overnight dew increases. When late infections take place, the disease does not impact yield because panicle and grain formation has already taken place. Ngugi et al. (2002) indicated that grey leaf spot was a prevalent disease in Kenya. In their survey, 58% and 75% of fields showed the presence of the disease in 1995 and 1996, respectively. Disease incidence was greatest in warm and humid areas with 900–1500 mm mean annual rainfall per year (Fig. 1).

Host resistance should be a central part of disease control although breeding has been limited (Okori et al., 2004) (Table 5). Sexual reproduction of *C. sorghi* may be rare in nature. Therefore, slow evolutionary responsiveness to disease control pressures such as the introduction of resistance should be the norm. For maize, crop rotation and stubble management are useful for grey leaf spot control (Ward and Nowell, 1998).

Table 5 General summary of disease evaluations for GRIN collection accessions[a]

Disease	Accessions evaluated	Resistant	Intermediate	Susceptible
Anthracnose[b]	24 556	18 588 (75.7%)	5876 (23.9%)	92 (0.4%)
Downy mildew (P1)[c]	4186	1573 (37.6%)	724 (17.3%)	1889 (45.1%)
Downy mildew (P3)[c]	2028	1247 (61.5%)	333 (16.4%)	448 (22.1%)
Downy mildew (P3, metalaxyl-resistant)	3207	175	–	3032
Ergot[d]	2022	160 (7.9%)	1697 (83.9%)	165 (8.2%)
Grey leaf spot[e]	306	4 (1.3%)	301 (98.4%)	1 (0.3%)
Ladder leaf spot[e]	1452	1302 (89.7%)	150 (10.3%)	0 (0.0%)
Leaf blight[e]	340	97 (28.5%)	241 (70.9%)	2 (0.6%)
Rust[e]	26 408	4376 (16.6%)	19 334 (73.2%)	2698 (10.2%)
Zonate leaf spot[e]	1468	1259 (85.8%)	209 (14.2%)	0 (0.0%)

[a] GRIN database (https://npgsweb.ars-grin.gov/gringlobal/descriptordetail; accessed 17 March 2017).
[b] Incidence (1.0, 1.5–4.5 and 5.0 ratings).
[c] Severity of 15 plants per plant introduction (PI) (resistant, intermediate, susceptible ratings).
[d] Percent infection (<1%, 1–50% and >50%).
[e] Incidence of plants affected (1–10%, 11–80% and 81–100%).

14 Leaf blight

Leaf blight lesions are elongated, cigar-shaped and not confined by leaf veins or the midrib (Knott et al., 2016). The margin of the leaf spot matches that of the host plant colour genotype (red, purple or tan). In general, leaf blight lesions are larger than those produced by the other foliar diseases except perhaps for sooty stripe. In susceptible genotypes, lesions may coalesce, killing entire leaf blades (Berquist, 2000; Knott et al., 2016; Williams et al., 1978). During humid conditions, lesions sporulate, giving lesions a light grey appearance. Close inspection of blighted leaves with a hand lens or dissecting microscope reveals the presence of conidiophores and conidia. This pathogen is also capable of causing seed rot and blight when the seed is planted in cool, moist soils (Fig. 1).

Leaf blight is caused by the fungus *Exserohilum turcicum* [teleomorph, *Setosphaeria turcica* (Luttr.) K.J. Leonard & Suggs.] (Bergquist, 2000) (Table 4). Population studies of *E. turcicum* have shown that there are large amounts of gene flow between populations and an important mechanism for genetic variability (Borchardt et al., 1997).

E. turcicum overseasons in infected crop residue and soil as mycelia, sclerotia or chlamydospores (Levy, 1984). The conidia are liberated from conidiophores in the morning as foliar dew dries. When released, they can be transmitted for long distances. Melanization protects the conidia from sun damage (Bergquist, 2000).

Warmer temperatures and extended periods of high relative humidity promote leaf blight (Fig. 1). Ngugi et al. (2002) found that leaf blight was one of the most prevalent foliar diseases of sorghum in western Kenya and was found in 54–78% of fields in the 2 years of their study. In their disease loss model, 1.5% of fields with ratings of ≥5 on a 1–9 disease severity scale would be sufficient to cause yield loss. The warm, humid production zones with 1300–1700 and 1600–1800 mm of average rainfall per year were conducive to severe outbreaks (Ngugi et al., 2002).

Resistance to *E. turcicum* in sorghum is well documented (Table 5). For example, in a minicore screening study, 49 accessions were resistant to leaf blight (≤3 on a 9-point scale), 141 moderately resistant (3.1–5.0), 43 susceptible (5.1–7.0) and nine highly susceptible (≥7.0) (Sharma et al., 2012). Cultivar mixtures were effective at controlling leaf blight in sorghum when either resistant/non-host maize or resistant sorghum genotypes were combined using intra-row mixtures (Ngugi et al., 2002). In this study, intra-row cultivar mixtures were more effective than inter-row mixtures.

15 Oval leaf spot

Oval leaf spot (OLS) begins as minute water-soaked spots and then develop into small circular lesions that are surrounded by a border characteristic of the plant's colour genotype (e.g. tan, red or purple) (Girard, 1980). As the lesions increase in size, they may remain circular or become oval. Black microsclerotia develop within the centres of the lesions. In the early stages, OLS resembles the foliar phase of anthracnose. However, the distinguishing sign for anthracnose is the production of acervuli with setae, which are absent in OLS.

OLS is caused by the fungal pathogen *Ramulispora sorghicola* E. Harris (Ngugi et al., 2002) (Table 4). This is a fungal pathogen that is very closely related to *R. sorghi*, which is the causal agent of sooty stripe. Both fungi are characterized by the production of dark microsclerotia that are produced within the oldest portions of the lesion. The phylogeny of this group shows that *R. sorghicola* is related to other cereal foliar pathogens such as *Cercospora*. Further, species of *Ramulispora* including *R. acuformis* (Nirenberg) Crous, *R. aestiva* (Nirenberg) E.L. Stewart & Crous, *R. anguioides* (Nirenberg) Crous and *R. herpotrichoides* (Fron) Arx share apothecial *Tapesia* teleomorphs (Goodwin et al., 2001).

OLS is considered a disease of 'minor to moderate importance' that is 'occasionally' to 'commonly present' (Fredericksen, 2000b) (Table 1). The disease occurs in Central and West Africa at relatively low levels in northern Guinea and Sudan. Ngugi et al. (2002) indicated that OLS is one of the most commonly encountered foliar diseases in western Kenya. Among foliar diseases in the region, OLS was found in 95–97% of fields, which was a greater rate of incidence than other diseases such as anthracnose or the various smuts. However, for observable losses to occur, 26.6% of the fields needed OLS severity scores of ≥5 on a 9-point scale (Ngugi et al., 2002). This rate was higher than for any other foliar disease measured, suggesting that although OLS was very common in the region, the amount of damage is generally low. Pande and Kutama (2010) indicated that OLS was a widely distributed foliar disease in Nigeria. In Central America, yield losses of 5.5% for this disease were observed in Honduras in 1985 (Wall et al., 1989). In Kenya, OLS incidence was greatest in warm, humid areas with 900–1500 mm average annual rainfall per year. However, disease severity did not differ greatly between areas with different levels of humidity (Ngugi et al., 2002) (Fig. 1).

OLS was observed in Beeville, Texas, in 2002 (Odvody et al., 2002). The pathogen can infect sorghum and johnsongrass. Grain sorghum planted near johnsongrass has greater incidence and severity of the disease (Odvody et al., 2006). In forage sorghum, Grewal et al. (1987) found that a single recessive gene conferred resistance to OLS, whereas the heterozygote was moderately susceptible. More recently, Boora et al. (2002) found randomly amplified polymorphic DNA molecular markers that co-segregated with OLS resistance.

16 Rust

Rust initially develops on resistant and susceptible host genotypes as small flecks on older leaves that may take on the natural colour of the host genotype (e.g. tan, red or purple) (White et al., 2014; Knott et al., 2016). In resistant genotypes, the infection does not progress beyond this stage. However, in susceptible genotypes, elliptical adaxial and abaxial uredinia develop but are confined between the leaf veins (Hooker, 1985; White et al., 2014) (Fig. 7). Uredinia may also occur in other tissues including leaf sheaths, the peduncle and rachis branches. The degree of host susceptibility dictates the density at which uredinia form, with the most susceptible genotypes exhibiting significant leaf coverage. For this rust, uredinia are characteristically brown. Close inspection of the uredinia reveals the production of numerous brown urediniospores (Hooker, 1985; White et al., 2014). As plants mature and tissues senesce, black telia may be visible within the uredinia. As with other closely related rusts, teliospores are two-celled and produced on an elongated stalk.

Rust is caused by the fungus *Puccinia purpurea* Cooke. *Puccinia purpurea* (Table 4) is a heteroecious, macrocyclic rust. The aecial host is *Oxalis corniculata* (Hooker, 1985). However, no evidence exists to suggest that aeciospores produced on wood sorrel

Figure 7 Adaxial pustules (uredinia) produced by the sorghum rust pathogen *Puccinia purpurea*. Image courtesy of S.R. Bean.

infect sorghum or play a significant role in the disease cycle. It is likely that repeated urediniospore infections of sorghum and reservoir hosts, such as johnsongrass, provide sufficient inoculum to perpetuate the disease (Hooker, 1985) (Fig. 1).

White et al. (2014) reported that temperatures of 16–28°C were optimum for rust development. When these temperatures coincided with 4–24 h of leaf wetness, disease severity increased. Further, normal light exposure did not influence infection. However, constant light for 16 h increased disease severity. Inoculation of plants after sowing (21–49 days post-planting) resulted in greater infection rates than those inoculated after flowering.

In many sorghum-producing areas, rust is considered a late-season disease that does not greatly impact yield since infections occur on mature leaves after grain fill has occurred (Tarr, 1962; Wang et al., 2006). However, both Frederiksen (1978a,b) and White et al. (2012) reported yield losses in late-planted sorghum. Infections are favoured by warmer temperatures and high relative humidity (Miller and Cruzado, 1969) (Fig. 1). In general, this disease is of less concern than other foliar diseases because of the late and sporadic nature of infections. However, if infections occur early, and a genotype is particularly susceptible, yield losses can occur (Hooker, 1985). Forage infections are of greater importance than infections of grain sorghum hybrids. In the southern United States, *Puccinia purpurea* can overwinter on johnsongrass (Leukel et al., 1951).

Ngugi et al. (2002) reported that rust was a ubiquitous foliar disease of sorghum in western Kenya. However, yield losses were not predicted to occur until 15.3% of fields showed disease ratings of ≥5 out of a 9-point scale. Like many other diseases in western Kenya, rust severity and incidence was greatest in warm and humid areas that received 900–1500 mm average annual rainfall.

Resistance to *Puccinia purpurea* in sorghum is complex. Rana et al. (1976) found three genes for rust resistance where susceptibility was dominant to resistance. Earlier, Miller and Cruzado (1969) found that complete resistance was controlled by one gene and the two other genes are associated with moderate susceptibility. Tao et al. (1998) identified QTLs for rust resistance in a recombinant inbred line population. After screening a minicore collection, Sharma et al. (2012) found that six accessions were resistant to rust in both field and greenhouse conditions. One of these, IS 33023, was also resistant to anthracnose and leaf blight. Also, rust susceptibility was reported to be lower in hybrids with A_2 cytoplasm compared to those with A_1 cytoplasm (Rodriguez-Herrera et al., 1994).

17 Sooty stripe

Sooty stripe is characterized by long, tan lesions surrounded by a pigmented margin (e.g. red and purple plant genotypes) and a distinct chlorotic halo. As in leaf blight, lesions may be several inches long, coalesce and kill entire leaves (Haussman et al., 2001). As the lesion matures, numerous rows of microsclerotia form on the abaxial side of the lesion, giving a 'sooty' appearance (Brady et al., 2011; Knott et al., 2016). Microsclerotia detach from leaves easily. Yield losses have been reported and can exceed 20–46% (Thomas, 1993; Thomas et al., 1993).

Sooty stripe is caused by the fungus *Ramulispora sorghi* (Ellis & Everh.) L.S. Olive & Lefebvre (Table 4). The conidia of *R. sorghi* are characteristically curved and branched (scolecosporous). They are produced in a sporogenic mass that germinates from a microsclerotium (Brady et al., 2011).

Although the nature of this pathogen's transmission is not fully understood, it is supposed that *R. sorghi* conidia are splash-dispersed onto the leaves from sporogenic microsclerotia that germinate on the soil or residue surface (Haussman et al., 2001). These authors also indicated that prevailing humid conditions during the growing season are required for successful disease development. Thus, the lower leaves of the plant are infected first. Subsequent conidiation and splash dispersal would be required to move the pathogen to upper-tier leaves. Thomas et al. (1993) found that the largest sooty stripe areas under the disease progress curve were achieved when microsclerotia were combined with soil and sown with the seed at planting and when conidial suspensions were directly applied to leaves.

Ngugi et al. (2002) reported that sooty stripe was less prevalent in western Kenya than other foliar diseases such as anthracnose, leaf blight, OLS and zonate leaf spot. Sooty stripe incidence of 30% and 20% of fields was observed in 1995 and 1996, respectively. However, sooty stripe is a common disease in West African countries including Burkina Faso, Mali, Nigeria and Senegal (Pande et al., 1993; Thomas, 1993). Pande et al. (1993) indicated that sooty stripe was a widely distributed foliar disease in Nigeria.

Resistance to sooty stripe appears to be governed by multiple genes (Haussman et al., 2001). Residue management to prevent the sporogenic germination of microsclerotia on the soil surface and spread of conidia to the lower leaves is an appropriate cultural control (Bandyopadhyay, 2000).

18 Rough leaf spot

Rough leaf spot is characterized by the development of short-lived foliar chlorosis within which numerous black pycnidia develop in and around the lesion. The bodies of the pycnidia protrude above the leaf surface, which gives the leaf surface a 'rough' feel (Knott et al., 2016). The chlorotic lesions will become necrotic and acquire an ovular or slightly elongated appearance. Often, in red and purple genotypes, the lesions will take on the appearance of the plant colour as the chlorotic lesions become necrotic. Lesions often form on the ends of leaves prior to developing on the upper portions of the leaf blades.

Rough leaf spot is caused by the fungal pathogen *Ascochyta sorghina* Sacc. Conidia spread the disease in humid areas and overseasons in crop residue. *Sorghum* spp. (grain sorghum, sweet sorghum, sudangrass, johnsongrass and wild sorghums) are all susceptible to *A. sorghina* (Zummo and Broadhead, 1984).

Rough leaf spot is of little to moderate importance, but is commonly present in most sorghum production environments (Saccas, 1954; Edmunds and Zummo, 1975; Zummo and Broadhead, 1984; Pande, 1993; Ngugi et al., 2002) (Table 1). In India, rough leaf spot is a common foliar disease. In Kenya, Ngugi et al. (2002) found rough leaf spot in 17% and 3% of fields in 1995 and 1996, respectively. Pande (1993) reported that rough leaf spot was found in Nigeria during most host development phases. Losses in the United States are minor if sorghum is produced in arid areas. Production of grain, forage and sweet sorghums in the southeast could increase the prevalence of the disease. According to Zummo and Broadhead (1984), 'By 1980, rough leaf spot had become the most prevalent sorghum disease at Meridian [Mississippi], with almost all sweet sorghum lines infected and about 90% of them showing some injury from the disease'.

19 Target leaf spot

Symptoms of target leaf spot are foliar lesions that take on an oval to elongated or blocky appearance (Knott et al., 2016; Dalmacio, 2000b). Lesions acquire the colour of the plant genotype (e.g. tan, red or purple) (Katewa, 2005; Mizuno et al., 2012; Yazawa et al., 2013).

Target leaf spot is caused by the fungus *Bipolaris sorghicola* (Lefebvre & Sherwin) Alcorn (Table 4). *B. sorghicola* is a necrotrophic fungal pathogen of sorghum (Yazawa et al., 2013). The conidia of *B. sorghicola* produce germ tubes from the apical tip and hilar ends, which lends their 'bipolar' genus name. The fungus is capable of infecting leaf tissue within 24 h of inoculation (Kawahigashi et al., 2011).

This disease is only 'occasionally present' and of 'minor importance' in the United States (Dalmaccio, 2000; Frederiksen, 2000b) (Table 1). However, in Australia, target leaf spot can cause severe infections during high rainfall years. The disease is important on local varieties throughout the humid regions of Asia (Katewa et al., 2005; Kawahigashi et al., 2011). Moreover, target leaf spot is one of the most commonly encountered leaf spot diseases in Venezuela (Borges, 1983). Target leaf spot was found on johnsongrass in Argentina (Acciaresi and Mónaco, 1999).

B. sorghicola has emerged as a model to study sorghum foliar pathogens and plant–microbe interactions in sorghum (Mizuno et al., 2012; Yazawa et al., 2013). Yazawa et al. (2013) identified 160 *B. sorghicola* transcripts expressed during infection in sorghum including those involved in fungal growth and plant-cell-wall-degrading enzymes. In sorghum, these authors found that a peroxidase (SbPRX18) and induction of Ca^{+2} signalling were important tools that resistant plants used as a defence against this pathogen. Further, work by Mizuno et al. (2012) showed that infection of BTx623 sorghum induced changes in host metabolism, particularly induction of the glycoxylate shunt of the Krebs cycle and metabolism activated phytoalexin genes. Kawahigashi et al. (2011) showed that pigmented phytoalexin compounds (e.g. 3'-deoxyanthocyanins) accumulated to a greater degree in a susceptible line, bmr-6, than in a resistant line, SIL-05.

20 Tar spot

In tar spot disease, shield-like, black fungal stromata develop on the surface of the leaves that resemble minute drips of roofing tar. The stromata are full of numerous perithecia that contain asci that will liberate ascospores for further pathogen spread and foliar infection (Thaung, 2008). These structures are often surrounded by a chlorotic or necrotic halo. The stromata are formed on both the adaxial and abaxial surfaces of the leaf and develop in rows within the leaf veins.

Tar spot is caused by the fungal pathogen *Phyllachora sacchari* Henn. (syn. *P. sorghi* Hönn) (Table 4). In addition to sorghum, *Phyllachora sacchari* also occurs on sugarcane (*Saccharum officinarum* L.) (Thaung, 2008). This disease is only 'occasionally present' and of 'moderate importance' (Frederiksen, 2000b). In Indonesia, Malaysia and the Philippines, tar spot is a common foliar disease of sorghum (Indira, 2002). Tar spot was only found in 3–4% in western Kenya (Ngugi et al., 2002).

Magill (2013) indicated that the leaf blight resistant line, CS 621, was also resistant to tar spot and grey leaf spot.

21 Zonate leaf spot

Small zonate leaf spot lesions may be difficult to distinguish from other foliar diseases. Young lesions are characterized by a water-soaked area and may have a thin, discoloured or chlorotic halo. However, zonate leaf spot is characterized by the production of mature bullseye lesions that alternate necrotic tissue with tissue discoloured to match the colour genotype of the plant (e.g. tan, red or purple) (Bain and Edgerton, 1943; Knott et al., 2016; Franklin, 2000; Williams et al., 1978). Lesions often initiate at the margins of the leaf blade and may grow to several inches in diameter. Young lesions are confined between the veins, but as the lesion increases in size, it crosses leaf veins and the midrib. On susceptible genotypes, such large lesions coalesce, turn completely necrotic and kill the entire leaf. In addition to the characteristic leaf spot, black sclerotia may be produced in the necrotic portion of the lesion as well as gelatinous masses of conidia under warm and humid conditions. In addition to foliar disease, the zonate leaf spot pathogen causes a sheath blight of sorghum (Odvody and Madden, 1984). Like foliar infections, microsclerotia may be produced on infected tissue, although these structures are often produced to a greater degree on sheaths than on leaves.

Zonate leaf spot is caused by *Gloeocercospora sorghi* D.C. Bain & Edgerton ex Deighton (Table 4). Like *R. sorghi*, the sclerotia of this fungus are soilborne and sporogenic, meaning they can germinate a mass of conidia (Dean, 1968). In this case, the conidia are long and resemble a curved needle. Conidia are produced on a sporodochium (Sprague, 1950). Conidia act as the primary inoculum and are dispersed by rain splash. Sporogenic microsclerotia that survive in soil or colonize sheaths spread infection to lower leaves in the sorghum canopy. The pathogen also causes copper spot on turfgrass species, particularly creeping bentgrass (*Agrostis stolonifera* L.) (Kaminski et al., 2005).

Interestingly, *G. sorghi* possesses and expresses a cyanide hydratase (*CHT*) gene upon infection in response to the presence of hydrogen cyanide. Wang et al. (1999) had proposed that this enzyme allowed *G. sorghi* to operate in a high HCN environment and showed that *G. sorghi* did not need to express (CHT) to infect plants.

The disease is most problematic under conditions of high humidity and warm temperatures (Odvody and Madden, 1984) (Fig. 1). Ngugi et al. (2002) reported that zonate leaf spot was one of the most commonly encountered foliar diseases in western Kenya. This disease was found in 52% and 88% of fields during a 1995 and 1996 survey, respectively. However, most fields showed that plants had < 10% leaf area affected.

Hybrids with A_2 cytoplasm were more resistant than hybrids with A_1 cytoplasm (Rodriguez-Herrera et al., 1994). Aida et al. (1996) demonstrated that the sorghum phytoalexin, apigenidin, could inhibit *G. sorghi* growth by 25% at 50–100 ppm. However, the phytoalexin had little effect on spore germination. Because of microsclerotia survival in the soil, crop rotation and residue management are the best means of cultural management (Odvody and Madden, 1984).

22 Conclusions and future trends

Numerous challenges remain for the biology and control of seed, seedling, foliar and panicle diseases. To this point, resistance to seedling diseases has been elusive. This may be due to the quantitative nature of resistance to a complex disease that is caused by such a diverse set of pathogens. It is important to integrate tolerance to abiotic stress and biotic seedling pathogens. For example, planting sorghum earlier under cooler temperatures and increased soil moisture also predisposes seedlings to infection by pathogens including *Fusarium*, *Rhizoctonia* and *Pythium* spp. (Fig. 1). The diversity of these fungi on sorghum seedlings is only partially known. Therefore, a detailed catalogue of seedling-associated fungal species is required as has been developed for the grain mould pathogens. Further, Koch's postulates should be conducted to determine which seedling-associated species are pathogens. Soil type, geography and climate are complicating environmental factors that may influence the interaction of specific species with sorghum seedlings. Genotype will interact with all of these factors, which will further complicate isolation of specific resistance to individual seedling pathogens. Further, vertical transmission, for example the ability of some pathogens to colonize seedlings and move into stalks, then peduncles and grain, has yet to be proven. Organisms such as *F. thapsinum* are associated with all of these sorghum developmental stages and may be a candidate for such a relationship with the sorghum host.

After decades of research, grain mould and weathering remain a high priority disease complex. Numerous publications have generated lists of seed-associated fungi that colonize grain before or after physiological maturity [see Navi et al. (1999) and Little et al. (2012) as examples]. Of these fungi, only a few have been explored in depth as to genetic and pathogenic species diversity. A comprehensive sorghum seed 'mycobiome' may also differ depending on the environment where the grain is produced whether or not moulding or weathering occurs. Additionally, the specific distribution of mycotoxin-producing fungal species and strains, and the toxigenic variability therein, requires additional study especially in the context of livestock feed and human foodways.

The search for ergot resistance should continue as there are no known sources of complete resistance to *C. africana*. Although materials have been found that tolerate the pathogen, or avoid infection through adaptive physiological means, mechanisms of resistance and vertical sources or resistance that may be pyramided have not been found (Table 5). Since there is an interest to expand sorghum production into cooler regions and plant earlier to take advantage of full-season moisture, tolerance to cooler temperatures in terms of pollen viability and increased fertilization efficiency is necessary. *C. africana* and *C. sorghi* isolates should be screened for resistance to the active ingredients of the commonly used, labelled fungicides. At this point, the extent of fungicide resistance that may or may not exist in these pathogen populations is unclear.

The major foliar pathogens of sorghum are complex and exhibit varying levels of diversity in the field. As it is important to understand the underlying population genetics of pathogens such as *C. sublineola*, a detailed understanding of pathotype diversity and regional diversity and the maintenance and update of host differentials remains a priority. Likewise, deeper biological investigations of the minor foliar pathogens must continue as the production of forage and biomass sorghums expands in wet and warm environments where foliar diseases predominate. Beyond basic pathogen biology, such studies should focus on how and where foliar pathogens overseason and the optimum environmental conditions where infection, reproduction and inoculum dissemination take place. In the case of an epiphytotic, resistance sources for some foliar pathogens may not exist and further screening should be pursued. Therefore, data-based recommendations for effective cultural practices including crop rotation, weed and reservoir control, and tillage for residue management are needed to update best management practices for each foliar disease.

23 Acknowledgements

This chapter is Contribution No. 18-B-409 from the Kansas Agricultural Experiment Station, Manhattan, KS, USA.

24 Where to look for further information

Frederiksen, R.A. and Odvody, G.N. 2000. *Compendium of Sorghum Diseases*, 2nd ed. APS Press, St. Paul, MN.

Leslie, J.F., ed. 2002. *Sorghum and Millets Diseases*. Iowa State Press, Ames, IA.

25 References

Acciaresi, H. and Mónaco, C. 1999. First report of *Bipolaris sorghicola* on Johnsongrass in Argentina. *Plant Disease* 83: 965.

Aguirre, J. R., Williams, H. A., Montes, N. G. and Cortinas-Escobar, H. M. 1997. First report of sorghum ergot caused by *Sphacelia sorghi* in Mexico. *Plant Disease* 81: 831.

Ahmed, K. M. and Ravinder Reddy, Ch. 1993. A pictorial guide to the identification of seedborne fungi of sorghum, pearl millet, finger millet, chickpea, pigeonpea, and groundnut. Information Bulletin No. 34. ICRISAT Press, Patnacheru, India, 192pp.

Aida, Y., Tamogami, S., Kodama, O. and Tsukiboshi, T. 1996. Synthesis of 7-methoxyapigenidin and its fungicidal activity against *Gloeocercospora sorghi*. *Bioscience, Biotechnology, and Biochemistry* 60: 1495–6.

Alderman, S. C. 2003. Diversity and speciation of *Claviceps*, pp. 195–295. In: *Clavicipitalean Fungi: Evolutionary Biology, Chemistry, Biocontrol and Cultural Impacts* (White, J. F., Bacon, C. W. and Hywel-Jones, N. L., Eds). Marcel Dekker, New York, NY.

Alderman, S., Frederickson, D., Milbrath, G., Montes, N., Narro-Sanchez, J. and Odvody, G. 1999. A laboratory guide to the identification of *Claviceps purpurea* and *Claviceps africana* in grass and sorghum seed samples. Oregon Department of Agriculture, Salem, OR, 19pp. http://oda/state.or.edu

Bain, D. C. and Edgerton, W. 1943. The zonate leaf spot, a new disease of sorghum. *Phytopathology* 33: 220–6.

Bandyopadhyay, R. 2000. Sooty stripe, pp. 14–15. In: *Compendium of Sorghum Diseases*, 2nd ed. (Frederiksen, R. A. and Odvody, G. N., Eds). APS Press, St. Paul, MN.

Bandyopadhyay, R., Frederickson, D. E., McClaren, N. W. and Odvody, G. N. 1996. Ergot – A global threat to sorghum. *International Sorghum and Millets Newsletter* 37: 1–32.

Bandyopadhyay, R., Frederickson, D. E., McLaren, N. W., Odvody, G. N. and Ryley, M. J. 1998. Ergot: A new disease threat to sorghum in the Americas and Australia. *Plant Disease* 82: 356–67.

Bandyopadhyay, R., Little, C., Waniska, R. and Butler, D. R. 2002. Sorghum grain mold: Through the 1990s into the new millennium, pp. 173–83. In: *Sorghum and Millets Diseases* (Leslie, J. F., Ed.). Iowa State Press, Ames, IA.

Beck, B. D. A. 1978. Sorghum diseases in Malawi, pp. 40–1. In: *Sorghum Diseases: A World Review* (Williams, R. J. and Frederiksen, R. A., Eds). ICRISAT, Patancheru, AP, India.

Bergquist, R. R. 2000. Leaf blight, pp. 9–10. In: *Compendium of Sorghum Diseases*, 2nd ed. (Frederiksen, R. A. and Odvody, G. N., Eds). APS Press, St. Paul, MN.

Bergstrom, G. C. and Nicholson, R. L. 1999. The biology of corn anthracnose: Knowledge to exploit for improved management. *Plant Disease* 83: 596–608.

Bhat, R. V., Shetty, P. H., Amruth, R. P. and Sudershan, R. V. 1997. A foodborne disease outbreak due to the consumption of moldy sorghum and maize containing fumonisin mycotoxins. *Clinical Toxicology* 35: 249–55.

Blaney, B. J., McKenzie, R. A., Walters, J. R., Taylor, L. F., Bewg, W. S., Ryley, H. I. and Maryam, R. 2000. Sorghum ergot (*Claviceps africana*) associated with agalactia and feed refusal in pigs and dairy cattle. *Australian Veterinary Journal* 78: 102–7.

Blaney, B. J., Ryley, M. J. and Boucher, B. D. 2010. Early harvest and ensilage of forage sorghum infected with ergot (*Claviceps africana*) reduces the risk of livestock poisoning. *Australian Veterinary Journal* 88: 311–12.

Bock, C. H., Jeger, M. J., Mughoho, L. K., Cardwell, K. F., Adeente, V., Mtisi, E., Akap, A. D., Kaula, G., Mukasambina, D. and Blair-Myers, C. 1998. Occurrence and distribution of *Peronosclerospora sorghi* [Weston and Uppal (Shaw)] in selected countries of West and Southern Africa. *Crop Protection* 17: 427–39.

Boora, K. S., Sindhu, A., Boora, P. and Magill, C. W. 2002. Identification of molecular markers for an oval leaf spot resistance gene in sorghum, pp. 239–41. In: *Sorghum and Millets Diseases* (Leslie, J. F., Ed.). Iowa State Press, Ames, IA.

Borchardt, D. S., Welz, H. G. and Geiger, H. H. 1997. Genetic structure of *Setosphaeria turcica* populations in tropical and temperate climates. *Phytopathology* 88: 322–9.

Borges, O. L. 1983. Pathogenicity of *Dreschlera sorghicola* isolates on sorghum in Venezuela. *Plant Disease* 67: 996–7.

Botha, W. J. and McKenzie, D. 1993. Glasshouse evaluation of two metalaxyl formulations against cotton and grain sorghum seedling diseases caused by *Pythium* spp. *Annals of Applied Biology* 122: 34–5.

Brady, C. R., Noll, L. W., Saleh, A. A. and Little, C. R. 2011. Disease severity and microsclerotium properties of the sorghum sooty stripe pathogen, *Ramulispora sorghi*. *Plant Disease* 95: 853–9.

Brown, P. J., Klein, P. E., Bortiri, E., Acharya, C. B., Rooney, W. L. and Kresovich, S. 2006. Inheritance of inflorescence architecture in sorghum. *Theoretical and Applied Genetics* 113: 931–42.

Bueso, F. J., Waniska, R. D., Rooney, W. L. and Bejosano, F. P. 2000. Activity of antifungal proteins against mold in sorghum caryopses in the field. *Journal of Agricultural and Food Chemistry* 48: 810–16.

Cardwell, K. F., Hepperly, P. R. and Frederiksen, R. A. 1989. Pathotypes of *Colletotrichum graminicola* and seed transmission anthracnose. *Plant Disease* 73: 255–7.

Casela, C. R. and Ferreira, A. S. 1987. A proposed system of classification of races of *Colletotrichum graminicola*, causal agent of anthracnose on sorghum (*Sorghum bicolor*). *Fitopatologia Brasileira* 12: 337–44.

Casela, C. R., Ferreira, A. S., Santos, F. G. and Guimarães, F. B. 2002. Sorghum diseases in Brazil, pp. 379–82. In: *Sorghum and Millets Diseases* (Leslie, J. F., Ed.). Iowa State Press, Ames, IA.

Castor, L. L. 1981. Grain mold histopathology, damage assessment, and resistance screening within *Sorghum bicolor* (L.) moench lines. Ph.D. dissertation, Texas A&M University, College Station, TX, 177pp.

Castor, L. L. and Frederiksen, R. A. 1981. Fusarium head blight occurrence and effects on sorghum yield and grain characteristics in Texas. *Plant Disease* 64: 1017–19.

Christensen, C. M. and Kaufmann, H. H. 1965. Deterioration of stored grains by fungi. *Annual Review of Phytopathology* 3: 69–84.

Cisneros-López, M. E., Mendoza-Enofre, L. E., González-Hernández, V. E. and Montes-Garcia, N. 2012. Male to female row ratio and fungicide application during panicle development to control sorghum ergot in the Mexican Highlands. *Australasian Plant Pathology* 41: 389–96.

Costa, R. V., Casela, C. R., Zambolim, L. and Ferreira, A. S. 2003. Antracnose de sorgo. *Fitopathologia Brasiliera* 28: 345–54.

Craig, J. 2000. Sorghum downy mildew, pp. 25–7. In: *Compendium of Sorghum Diseases*, 2nd Ed. (Frederiksen, R. A. and Odvody, G. N., Eds). APS Press, St. Paul, MN.

Crouch, J. A., Clarke, B. B. and Hillman, B. I. 2006. Unraveling evolutionary relationships among the divergent lineages of *Colletotrichum* causing anthracnose disease in turfgrass and corn. *Phytopathology* 96: 46–60.

Cuevas, H. E., Prom, L. K. and Erpelding, J. E. 2014. Inheritance and molecular mapping of anthracnose resistance genes present in sorghum line SC112-14. *Molecular Breeding* 34: 1943–53.

Dahlberg, J., Bandyopadhyay, R., Rooney, B., Odvody, G. and Frederiksen, D. 1998. Host-plant strategies within the United States. In: *Proceedings for the Status of Sorghum Ergot in North America* (Dahlberg, J. A., Ed.). 24–26 June, Corpus Christi, TX.

Dahlberg, J. A., Bandyopadhyay, R., Rooney, W. L., Odvody, G. N. and Madera-Torres, P. 2001. Evaluation of sorghum germplasm used in U.S. breeding programmes for sources of sugary disease resistance. *Plant Pathology* 50: 681–9.

Dalmacio, S. C. 2000a. Tar spot, p. 16. In: *Compendium of Sorghum Diseases* (Frederiksen, R. A. and Odvody, G. N., Eds). APS Press, St. Paul, MN.

Dalmacio, S. C. 2000b. Target leaf spot, pp. 16–17. In: *Compendium of Sorghum Diseases* (Frederiksen, R. A. and Odvody, G. N., Eds). APS Press, St. Paul, MN.

Day, J. L., Duncan, R. R., Raymer, P. L., Thomspon, D. S. and Zummo, N. 1995. Registration of 'Top 76-6' sweet sorghum. *Crop Science* 35: 1211–13.

Davis, M. A. and Bockus, W. W. 2001. Evidence for a *Pythium* sp. as a chronic yield reducer in a continuous grain sorghum field. *Plant Disease* 85: 780–4.

Dean, J. L. 1968. Germination and overwintering of sclerotia of *Gloeocercospora sorghi. Phytopathology* 58: 113–14.

de Milliano, W. A. J. 2002. Transition from the second to the third world review of sorghum and millet diseases, pp. 3–10. In: *Sorghum and Millets Diseases* (Leslie, J. F., Ed.). Iowa State Press, Ames, IA.

Dernoeden, P. H. and Jackson, N. 1980. Infection and mycelial colonization of gramineous hosts by *Sclerophthora macrospora. Phytopathology* 70: 1009–13.

Dey, S. K., Dhillon, B. S. and Malhotra, V. V. 1986. Crazy top downy mildew of maize: A new record in Punjab. *Current Science* 55: 577–8.

Doggett, O. 1988. Sorghum, 2nd ed. Tropical Agriculture Series. Longman Group Ltd., Harlow, UK, 512pp.

Edmunds, L. K. and Zummo, N. 1975. Sorghum diseases in the United States and their control. U.S. Department of Agriculture Handbook 468. U.S. Govt. Printing Office, Washington DC.

Edmunds, L. K., Futrell, M. C. and Frederiksen, R. A. 1970. Sorghum diseases, pp. 200–34. In: *Sorghum Production and Utilization* (J. S. Wall and W. M. Ross, Eds). AVI Publishing, Westport, CT.

El-Bakali, M. A., Martin, M. P., Garcia, F. F., Montón, R. C., Moret, B. A. and Nadal, P. M. 2000. First report of *Rhizoctonia solani* AG-3 on potato in Catalonia (NE Spain). *Plant Disease* 84: 806.

Erpelding, J. E. and Prom, L. K. 2004. Evaluation of Malian sorghum germplasm for resistance against anthracnose. *Plant Pathology Journal* 3: 65–71.

Esele, J. P. 1995. Foliar and head diseases of sorghum. *African Crop Science Journal* 3: 185–9.

Esele, J. P. 2002. Sorghum and pearl millet diseases in the Horn of Africa, pp. 383–7. In: *Sorghum and Millets Diseases* (Leslie, J. F., Ed.). Iowa State Press, Ames, IA.

Food & Agriculture Organization of the United Nations (FAO). 2017. Food and agricultural statistics database, FAOSTAT. Online: www.fao.org/faostat/en/#data/QC (accessed 10 November 2017).

Forbes, G. A. 1986. Characterization of grain mold resistance in sorghum (*Sorghum bicolor* L. Moench). Ph.D. dissertation, Texas A&M University, College Station, TX, 75pp.

Forbes, G. A. and Odvody, G. N. 2000. Seedling diseases, pp. 8–9. In: *Compendium of Sorghum Diseases*, 2nd ed. (Frederiksen, R. A. and Odvody, G. N., Eds). APS Press, St. Paul, MN.

Forbes, G. A., Collins, D. C., Odvody, G. N. and Frederiksen, R. A. 1985. A seedling epiphytotic of sorghum in South Texas caused by *Pythium arrhenomanes*. *Plant Disease* 69: 726.

Forbes, G. A., Ziv, O. and Frederiksen, R. A. 1987. Resistance in sorghum to seedling disease caused by *Pythium arrhenomanes*. *Plant Disease* 71: 145–8.

Franklin, C. D. 2000. Zonate leaf spot, p. 14. In: *Compendium of Sorghum Diseases*, 2nd ed. (Frederiksen, R. A. and Odvody, G. N., Eds). APS Press, St. Paul, MN.

Frederickson, D. E., Mantle, P. G. and De Milliano, W. A. J. 1991. *Claviceps africana* sp. nov.; the distinctive ergot pathogen on sorghum in Africa. *Mycological Research* 95: 1101–7.

Frederickson, D. E., Mantle, P. G. and De Milliano, W. A. J. 1993. Windborne spread of ergot disease (*Claviceps africana*) in sorghum A-lines in Zimbabwe. *Plant Pathology* 42: 368–77.

Frederickson, D. E., Odvody, G. N. and Isakeit, T. 1999. Sorghum ergot: Distinguishing sphacelia and sclerotia of *Claviceps africana* in the seed. Texas Agricultural Extension Service, Bulletin L-5315.

Frederiksen, R. A. 1978a. Sorghum rust, pp. 240–2. In: *Proceedings of the International Workshop on Sorghum Diseases*, Hyderabad, India, 11–15 December 1978. ICRISAT, Patancheru, AP, India.

Frederiksen, R. A. 1978b. Head smuts of corn and sorghum, pp. 89–105. In: *Proceedings of the 32nd Annual Corn and Sorghum Research Conference* (Loden, H. D. and Wilkinson, D., Eds). ASTA, Washington DC.

Frederiksen, R. A. 1980. Sorghum downy mildew in the United States: Overview and outlook. *Plant Disease* 64, 903–8.

Frederiksen, R. A. 2000a. Covered kernel smut, p. 22. In: *Compendium of Sorghum Diseases*, 2nd ed. (Frederiksen, R. A. and Odvody, G. N., Eds). APS Press, St. Paul, MN.

Frederiksen, R. A. 2000b. Diseases and disease management in sorghum, pp. 497–533. In: *Sorghum – Origin, History, Technology and Production* (Smith, C. and Frederiksen, R. A., Eds). John Wiley & Sons, New York, NY.

Frederiksen, R. A. 2000c. Head blight, p. 40. In: *Compendium of Sorghum Diseases*, 2nd ed. (Frederiksen, R. A. and Odvody, G. N., Eds). APS Press, St. Paul, MN.

Frederiksen, R. A. 2000d. Head smut, pp. 18–20. In: *Compendium of Sorghum Diseases*, 2nd ed. (Frederiksen, R. A. and Odvody, G. N., Eds). APS Press, St. Paul, MN.

Frederiksen, R. A. and Reyes, L. 1980. The head smut program at Texas A&M, pp. 367–72. In: *Sorghum Diseases: A World Review*. ICRISAT, Patancheru, AP, India.

Frederiksen, R. A. and Rosenow, D. T. 1971. Disease resistance in sorghum, pp. 71–82. In: *Proceedings of the 26th Annual Corn and Sorghum Research Conference*. ASTA, Washington, DC.

Frederiksen, R. A., Bockholt, A. J., Clark, L. E., Cosper, J. W., Craig, J., Johnson, J. W., Jones, B. L., Matocha, P., Miller, F. R., Reyes, L., Rosenow, D. T., Tuleen, D. and Walker, H. J. 1973. Sorghum downy mildew – a disease of maize and sorghum. Research monograph 2. Texas Agricultural Experiment Station, College Station, TX.

Frederiksen, R. A., Rosenow, D. T. and Tuleen, D. M. 1975. Resistance to *Exserohilum turcicum* in sorghum. *Plant Disease Reporter* 59: 547–8.

Frederiksen, R. A., Castor, L. L. and Rosenow, D. T. 1982. Grain mold, small seed and head blight: The *Fusarium* connection in sorghum. *Proceedings of the 37th Annual Corn and Sorghum Industry Research Conference* 37: 26–36.

Freeman, T. E., Luke, H. H. and Sechler, D. T. 1966. Pathogenicity of *Pythium aphanidermatum* on grain crops in Florida. *Plant Disease Reporter* 50: 292–4.

Funnell, D. L. and Pedersen, J. F. 2006. Association of plant color and pericarp color with colonization of grain by members of *Fusarium* and *Alternaria* in near-isogenic sorghum lines. *Plant Disease* 90: 411–18.

Funnell-Harris, D. L. and Pedersen, J. F. 2011. Presence of *Fusarium* spp. in air and soil associated with sorghum fields. *Plant Disease* 95: 648–56.

Funnell-Harris, D. L., Sattler, S. E., O'Neill, P. M., Eskridge, K. M. and Pedersen, J. F. 2015. Effect of waxy (low amylose) on fungal infection of sorghum grain. *Phytopathology* 105: 786–96.

Futrell, M. C. and Frederiksen, R. A. 1970. Distribution of sorghum downy mildew (*Sclerospora sorghi*) in the U.S.A. *Plant Disease Reporter* 54: 311–14.

Garud, T. B., Sawant, L. V. and Ismail, S. 2000. Development of grain mold in sorghum-pigeonpea intercropping systems. *International Sorghum and Millets Newsletter* 41: 55.

Giorda, L. M. 2002. Sorghum diseases in Argentina, pp. 389–91. In: *Sorghum and Millets Diseases* (Leslie, J. F., Ed.). Iowa State Press, Ames, IA.

Girard, G. C. 1980. A review of sooty stripe and rough, zonate and oval leaf spots, pp. 229–39. In: *Sorghum Diseases: A World Review* (Williams, R. J., Frederiksen, R. A. and Mughogho, L. K., Eds). ICRISAT, Patancheru, AP, India.

Goodwin, S. B., Dunkle, L. D. and Zismann, V. L. 2001. Phylogenetic analysis of *Cercospora* and *Mycosphaerella* based on the internal transcribed spacer region of ribosomal DNA. *Phytopathology* 91: 648–58.

Grewal, R. P.S., Lodhl, G. P. and Paroda, R. S. 1987. Inheritance of field resistance to sorghum oval leaf spot. *The Indian Journal of Genetics and Plant Breeding* 47: 41–5.

Groenewald, M., Groenewald, J. Z., Harrington, T. C., Abeln, E. C. A. and Crous, P. W. 2006. Mating type gene analysis in apparently asexual *Cercospora* species is suggestive of cryptic sex. *Fungal Genetics & Biology* 43: 813–25.

Hariprasanna, K. and Rakshit, S. 2016. Economic importance of sorghum, pp. 1–27. In: *The Sorghum Genome* (Rakshit, S. and Wang, Y.-H., Eds). Springer, Switzerland.

Haussman, B. I. G., Hess, D. E., Sissiko, I., Kayentao, M., Reddy, M., Reddy, B. V. S., Welz, H. G. and Geiger, H. H. 2001. Diallel analysis of sooty stripe resistance in sorghum. *Euphytica* 122: 99–104.

Hess, D. E., Bandyopadhyay, R. and Sissiko, I. 2002. Pattern analysis of sorghum genotype × environment interaction for leaf, panicle, and grain anthracnose in Mali. *Plant Disease* 86: 1374–82.

Hooker, A. L. 1985. Corn and sorghum rusts, pp. 207–36. In: *The Cereal Rusts. Vol. II. Diseases, Distribution, Epidemiology, and Control* (Roelfs, A. P. and Bushnell, W. R., Eds). Academic Press, Orlando, FL.

Horne, C. W. and Berry, R. W. 1980. Sorghum diseases atlas. Publication B-1323. Texas Agricultural Extension Service, Texas A&M University, College Station, TX, 16pp.

Huang, L.-D., and Backhouse, D. 2005. Induction of defence responses in roots and mesocotyls of sorghum seedlings by inoculation with *Fusarium thapsinum* and *F. proliferatum*, wounding and light. *Journal of Phytopathology* 153: 522–29.

Husseini, A. M., Timothy, A. G., Olufunmilayo, H. A., Ezekiel, A. S. and Godwin, H. O. 2009. Fungi and some mycotoxins found in mouldy *Sorghum* in Niger State, Nigeria. *World Journal of Agricultural Sciences* 5: 5–17.

Indira, S. 2002. Recommendations from the working group for Asia/Australia at the third global conference on sorghum and millets diseases, pp. 377–8. In: *Sorghum and Millets Diseases* (Leslie, J. F., Ed.). Iowa State Press, Ames, IA.

Isakeit, T. and Jaster, J. 2005. Texas has a new pathotype of *Peronosclerospora sorghi*, the cause of sorghum downy mildew. *Plant Disease* 89: 529.

Isakeit, T. and Odvody, G. N. 2002. Resurgence of sorghum downy mildew on the Upper Coast of Texas: Preventing a future epidemic. Texas Cooperative Extension. The Texas A&M University System, Bulletin 5/02.

Isakeit, T., Odvody, G. N. and Shelby, R. A. 1998. First report of sorghum ergot caused by *Claviceps africana* in the United States. *Plant Disease* 82: 592.

Jalali, S. and Behdad, E. 2003. Investigation on distribution and infection of *Sporisorium* species attacking sorghum in Ardestan region and comparison of some fungicides for its control. *Pajouhesh-va-Sazandegi in Agronomy & Horticulture* 59: 38–42.

Jardine, D. J. and Leslie, J. F. 1992. Aggressiveness of *Giberella fujikuroi* (*Fusarium moniliforme*) isolates to grain sorghum under greenhouse conditions. *Plant Disease* 76: 897–900.

Jensen, S. 2000. Diseases caused by viruses and viruslike organisms, pp. 43–50. In: *Compendium of Sorghum Diseases*, 2nd ed. (Frederiksen, R. A. and Odvody, G. N., Eds). American Phytopathological Society, St. Paul, MN.

Jensen, S. G., Doupnik, B., Wysong, D. and Johnson, B. 1989. An epidemic of sorghum downy mildew in Nebraska in 1987. *Plant Disease* 73: 75–8.

Jones, R. K. and Belmar, S. B. 1989. Characterization and pathogenicity of *Rhizoctonia* spp. isolated from rice, soybean, and other crops in rotation with rice in Texas. *Plant Disease* 73: 1004–10.

Kaminski, J. E., Dernoeden, P. H., O'Neill, N. R. and Wetzel III., H. C. 2005. A PCR-based method for the detection of *Ophiosphaerella agrostis* in creeping bentgrass. *Plant Disease* 89: 980–5.

Kapanigowda, M. H., Perumal, R., Aiken, R. M., Herald, T. J., Bean, S. R. and Little, C. R. 2013. Analysis of sorghum [*Sorghum bicolor* (L.) Moench] lines and hybrids in response to early-season planting and cool conditions. *Canadian Journal of Plant Science* 93: 773–84.

Kasakova, A. S. and Oleynick, A. A. 2002. The status of sorghum diseases in Russia, pp. 411–13. In: *Sorghum and Millets Diseases* (Leslie, J. F., Ed.). Iowa State Press, Ames, IA.

Katewa, R., Mathur, K. and Bunker, R. N. 2005. Variability in target leaf spot pathogen *Bipolaris sorghicola* on sorghum in Rajasthan, India. *International Sorghum and Millets Newsletter* 46: 32–5.

Katilé, S. O., Perumal, R., Rooney, W. L., Prom, L. K. and Magill, C. W. 2010. Expression of pathogenesis-related protein PR-10 in sorghum floral tissues in response to inoculation with *Fusarium thapsinum* and *Curvularia lunata*. *Molecular Plant Pathology* 11: 93–103.

Kawahigashi, H., Kasuga, S., Ando, T., Kanamori, H., Wu, J., Yonemaru, J., Sazuka, T. and Matsumoto, T. 2011. Positional cloning of ds1 the target leaf spot resistance gene against *Bipolaris sorghicola* in sorghum. *Theoretical and Applied Genetics* 123: 131–42.

Klein, R. R., Rodriguez-Herrera, R., Schlueter, J. A., Klein, P. E., Yu, Z. H. and Rooney, W. L. 2001. Identification of genomic regions that affect grain-mould incidence and other traits of agronomic importance in sorghum. *Theoretical and Applied Genetics* 102: 307–19.

Knott, C. A., Ritchey, E. L., Martin, J. R., Lee, C., Johnson, D., Bradley, C. and McNeill, S. 2016. Grain sorghum (milo) production in Kentucky. Agriculture and Natural Resources Publications 112.

Kollo, A. I. 2000. Long smut, pp. 22–3. In: *Compendium of Sorghum Diseases*, 2nd ed. (Frederiksen, R. A. and Odvody, G. N., Eds). APS Press, St. Paul, MN.

Komolong, B., Chakrabotry, S., Ryley, M. and Yates, D. 2003. Ovary colonization by *Claviceps africana* is related to ergot resistance in male-sterile sorghum lines. *Plant Pathology* 52: 620–7.

Krausz, J. and Isakeit, T. 1998. Sorghum ergot: A new disease threat to the sorghum industry. Texas Agriculture Extension Service, Publication L-5179.

Kumar, A. and Nath, V. 1991. Fungicidal control and disease rating scale of long smut of sorghum. *Indian Journal of Agricultural Science* 61: 225–7.

Kutama, A. S., Bashir, B. and James, D. 2010. Incidence of sorghum diseases in Dawakin-Kuou Local Government Area, Kano State, Nigeria. *African Journal of General Agriculture* 6: 307–13.

Langham, M. C. 2004. Ergot of rye – A plant disease impacts our health and history. *APS Education Center, News & Views*. Online: https://www.apsnet.org/edcenter/K-12/NewsViews/Pages/2004_10.aspx.

Lansac, A. R., Sullivan, C. Y., Johnson, B. E. and Lee, K. W. 1994. Viability and germination of pollen of sorghum (*Sorghum bicolor* (L.) Moench). *Annals of Botany* 94: 27–33.

Lengkeek, V. H. and Sim, T. 1979. An outbreak of sorghum downy mildew in Kansas. *Plant Disease Reporter* 63: 905–7.

Leslie, J. F. and Summerell, B. A. 2006. *The Fusarium Laboratory Manual*. Wiley-Blackwell, Ames, IA, 388pp.

Leslie, J. F., Pearson, C., Nelson, P. E. and Toussoun, T. A. 1990. *Fusarium* spp. from corn, sorghum, and soybean fields in the central and eastern United States. *Phytopathology* 80: 343–50.

Leslie, J. F., Zeller, K. A., Lamprecht, S. C., Rheeder, J. P. and Marasas, F. O. 2005. Toxicity, pathogenicity and genetic differentiation of five species of *Fusarium* from sorghum and millet. *Phytopathology* 95: 275–83.

Levy, Y. 1984. Overwintering of *Exserohilum turcicum* in Israel. *Phytoparasitica* 12: 177–82.

Leukel, R. W. and Martin, J. H. 1943. Seed rot and seedling blight of sorghum. Technical Bulletin No. 839. United States Department of Agriculture, Washington DC.

Leukel, R. W., Martin, J. H. and Lefebvre, C. L. 1951. Sorghum diseases and their control. Farmer's Bulletin 1959. United States Department of Agriculture, Washington DC.

Li, Y. H. and TeBeest, D. O. 2009. Temporal and spatial development of sorghum anthracnose in Arkansas. *Plant Disease* 93: 287–92.

Little, C. R. and Magill, C. W. 2003a. Elicitation of defense response genes in sorghum floral tissues infected by *Fusarium thapsinum* and *Curvularia lunata* at anthesis. *Physiological and Molecular Plant Pathology* 63: 271–9.

Little, C. R. and Magill, C. W. 2003b. Reduction of sorghum seedling vigor by inoculation with *Fusarium thapsinum* and *Curvularia lunata* at anthesis. *International Sorghum and Millets Newsletter* 44: 111–12.

Little, C. R., Perumal, R., Tesso, T., Prom, L. K., Odvody, G. N. and Magill, C. W. 2012. Sorghum pathology and biotechnology – A fungal disease perspective: Part I. Grain mold, head smut, and ergot. *The European Journal of Plant Science and Biotechnology* 6: 10–30.

Lo, S.-C. C., De Verdier, K. and Nicholson, R. L. 1999. Accumulation of 3-deoxyanthocyanidin phytoalexins and resistance to *Colletotrichum sublineolum* in sorghum. *Physiological and Molecular Plant Pathology* 55: 263–73.

Magill, C. W. 2013. Bridging classical and molecular genetics of sorghum disease resistance, pp. 347–66. In: *Genomics of the Saccharinae* (Patterson, A. H., Ed.). Springer, New York, NY.

Malaguti, G. 1978. Sorghum downy mildew in the Americas, pp. 184–94. In: *Sorghum Diseases: A World Review* (Williams, R. J. and Frederiksen, R. A., Eds). ICRISAT, Patancheru, AP, India.

Malaguti, G. 2002. Current status of sorghum diseases in Venezuela, pp. 411–14. In: *Sorghum and Millets Diseases* (Leslie, J. F., Ed.). Iowa State Press, Ames, IA.

Mansuetus, S. B. A. 1995. Sorghum diseases in Tanzania. *African Crop Science Journal* 3: 203–7.

Marley, P. S. and Aba, D. A. 1999. Current status of sorghum smuts. Nigerian *Journal of Sustainable Agriculture* 15: 19–132.

Marley, P. S., Diourté, M., Neya, A., Nutsugah, S. K., Sérémé, P., Katilé, S. O., Hess, D. E., Mbaye, D. F. and Ngoko, Z. 2002. Sorghum and pearl millet diseases in West and Central Africa, pp. 419–25. In: *Sorghum and Millets Disease* (Leslie, J. F., Ed.). Iowa State Press, Ames, IA.

Mathur, K., Thakur, R. P., Neya, A., Marley, P. S. and Casela, C. R. 2002. Sorghum anthracnose – problem and management strategies, pp. 211–20. In: *Sorghum and Millets Diseases* (Leslie, J. F., Ed.). Iowa State Press, Ames, IA.

Mbgawa, A. M., Pande, S., De Milliano, W. A. J. and Karunakar, R., I. 1993. Diseases and parasitic weeds of sorghum in Tanzania: Occurrence and incidence, 1986–1990. *Crop Protection* 12: 183–8.

McClean, K. S. and Lawrence, G. W. 2001. First report of *Pythium* root rot on grain sorghum in Mississippi. *Plant Disease* 85: 922.

McGee, D. C. 1992. Principles and criteria of seed transmission of sorghum pathogens, pp. 297–302. In: *Sorghum and Millets Diseases: A Second World Review*. ICRISAT, Patancheru, AP, India.

McLaren, N. W. 1987. Studies on pre- and post-emergence damping-off and seedling blight of sorghum. M.Sc. thesis, University of Natal, Pieter-Maritzburg, South Africa.

McLaren, N. W. 1993. Effect of sugary disease exudates on germination, seedling development and predisposition to seedling diseases of sorghum (*Sorghum bicolor*). *South African Journal of Plant and Soils* 10: 12–16.

McLaren, N. W. 1997. Changes in pollen viability and concomitant increase in the incidence of sorghum ergot with flowering date and implications in selection for escape resistance. *Journal of Phytopathology* 145: 261–5.

McLaren, N. W. 2002. Evaluation of sorghum hybrids for resistance to the root rot complex. *South African Journal of Plant and Soil* 19: 37–42.

McLaren, N. W. and Flett, B. C. 1998. Use of weather variables to quantify sorghum ergot potential in South Africa. *Plant Disease* 82: 26–9.

McLaren, N. W. and Wehner, F. C. 1992. Pre-flowering low temperature predisposition of sorghum to sugary disease (*Claviceps africana*). *Journal of Phytopathology* 135: 328–34.

Mehta, P. J., Wiltse, C. C., Rooney, W. L., Collins, S. D., Frederiksen, R. A., Hess, D. A., Chisi, M. and TeBeest, D. O. 2005. Classification and inheritance of genetic resistance to anthracnose in sorghum. *Field Crops Research* 93: 1–9.

Meinke, H. and Ryley, M. 1997. Effects of sorghum ergot on grain sorghum production: A preliminary climatic analysis. *Australian Journal of Agriculture Research* 48: 1241–8.

Miedainer, T. and Geiger, H. H. 2015. Biology, genetics, and management of ergot (*Claviceps* spp.) in rye, sorghum, and pearl millet. *Toxins* 7: 659–78.

Miller, F. R. and Cruzado, H. J. 1969. Allelic interactions at the *Pu* locus in *Sorghum bicolor* (L.) Moench. *Crop Science* 9: 336–8.

Mizuno, H., Kawahigashi, H., Kawahara, Y., Kanamori, H., Ogata, J., Minami, H., Itoh, T. and Matsumoto, T. 2012. Global transcriptome analysis reveals distinct expression among duplicated genes during sorghum-*Bipolaris sorghicola* interaction. *BMC Plant Biology* 12: 121.

Montes-Belmont, R., Méndez-Ramírez, I., Fores-Moctezuma, H. E. and Nava-Juárez, R. A. 2003. Impact of planting dates and climatic factors on the incidence and severity of sorghum grain mold in Morales, Mexico. *Plant Disease* 87: 1139–43.

Montes-García, N., Prom, L. K., Williams-Alanis, H. and Isakeit, T. 2009. Effect of temperature and relative humidity on sorghum ergot development in northern Mexico. *Australasian Plant Pathology* 38: 632–7.

Morrow, C. A. and Fraser, J. A. 2009. Sexual reproduction and dimorphism in the pathogenic basidiomycetes. *FEMS Yeast Research* 9: 161–77.

Moore, J. W., Ditmore, M. and TeBeest, D. O. 2008. Pathotypes of *Colletotrichum sublineolum* in Arkansas. *Plant Disease* 92: 1415–20.

Mtisi, E. 1997. Evaluation of seed treatments for the control of covered kernel smut on sorghum. *African Crop Science Journal* 3: 943–6.

Mtisi, E. and McLaren, N. W. 2002. Diseases of sorghum and pearl millet in some southern African countries, pp. 427–30. In: *Sorghum and Millets Diseases* (Leslie, J. F., Ed.). Iowa State Press, Ames, IA.

Mudge, S. R., Campbell, B. C., Mustapha, N. B. and Godwin, I. D. 2016. Genomic approaches for improving grain quality of sorghum, pp. 189–206. In: *The Sorghum Genome* (Rakshit, S. and Wang, Y.-H., Eds). Springer, New York, NY.

Mueller, D. S., Wise, K. A., Dufault, N. S., Bradley, C. A. and Chilvers, M. I. 2013. *Fungicides for Field Crops*. APS Press, St. Paul, MN, 112pp.

Muller, M. F., Barnes, I., Kunene, N. T., Crampton, B. G., Bluhm, B. H., Phillips, S. M., Olivier, N. A. and Berger, D. K. 2016. *Cercospora zeina* from maize in South Africa exhibits high genetic diversity and lack of regional population differentiation. *Phytopathology* 106: 1194–205.

Muthusubramanian, V., Bandyodadhyay, R., Reddy, D. R. and Tooley, P. W. 2005. Cultural characteristics, morphology, and variation within *C. africana* and *C. sorghi* from India. *Mycological Research* 110: 452–64

Narro, J., Betancourt, V. A. and Aguirre, J. I. 1992. Sorghum diseases in Mexico, pp. 75–84. In: *Sorghum and Millet diseases: A Second World Review*. ICRISAT, Pantancheru, AP, India.

Navi, S. S. 1999. *A Pictorial Guide for the Identification of Mold Fungi on Sorghum Grain*. ICRISAT, Patancheru, AP, India.

Navi, S. S., Bandyopadhyay, R., Reddy, R. K., Thakur, R. P. and Yang, X. B. 2005. Effects of wetness duration and grain development stages on sorghum grain mold infection. *Plant Disease* 89: 872–8.

Neergaard, P. 1979. *Seed Pathology*, Volume 1. Macmillan Press Ltd, London, 839pp.

Ngugi, H. K., King S. B., Abayo G. O. and Reddy Y. V. R. 2002. Prevalence, incidence, and severity of sorghum diseases in western Kenya. *Plant Disease* 86: 65–70.

Nithya, R., Sharma, R., Rao, V. P., Gopalokrishnan, S. and Thakur, R. P. 2013. Biochemical characterisation of grain mould resistant and susceptible genotypes and PGPR-induced resistance in the host to *Curvularia lunata* and *Fusarium proliferatum*. *Archives of Phytopathology and Plant Protection* 46: 980–9.

Nobel, M. and Richardson, M. J. 1968. Annotated list of seedborne diseases. In: *Proceedings of the International Seed Testing Association, No. 33*. ISTA, Zurich, Switzerland.

Noll, L. W., Butler, D. N. and Little, C. R. 2010. Tetrazolium violet staining of naturally and artificially moulded sorghum (*Sorghum bicolor* (L.) Moench) caryopses. *Seed Science & Technology* 38: 741–56.

Nzioki, H. S., Claflin, L. E. and Ramundo, B. A. 2010. Evaluation of screening protocols to determine genetic variability of grain sorghum germplasm to *Sporisorium sorghi* under field and greenhouse conditions. *International Journal of Pest Management* 46: 91–5.

O'Connell, R. J., Thon, M. R., Hacquard, S., Amyotte, S. G., Kleeman, J., Torres, M. F., Damm, U., Bulate, E. A., Epstein, L., Alkan, N., Altmüller, J., Alvarado-Balderrama, L., Bauser, C. A., Becker, C., Birren, B. W., Chen, Z., Choi, J., Crouch, J., Duvick, J. P., Farman, M. A., Gan, P., Heiman, D., Henrissat, B., Howard, R. J., Kabbage, M., Koch, C., Kracher, B., Kubo, Y., Law, A. D., Lebrun, M.-H., Lee, Y.-H., Miyara, I., Moore, N., Neumann, U., Nordström, K., Panaccione, D. G., Panstruga, R., Place, M., Proctor, R. H., Prusky, D., Rech, G., Reinhardt, R., Rollins, J. A., Rounsley, S., Schardl, C. L., Schwartz, D. C., Shenoy, N., Shirasu, K., Sikhakolli, U. R., Stüber, K., Sukno, S. A., Sweigard, J. A., Takano, Y., Takahara, H., Trail, F., van der Does, H. C., Voll, L. M., Will, I., Young, S., Zeng, Q., Zhang, J., Zhou, S., Dickman, M. B., Schulze-Lefert, P., van Themaat, E. V. L., Ma, L.-J. and Vaillancourt, L. J. 2011. Lifestyle transitions in plant pathogenic *Colletotrichum* fungi deciphered by genome and transcriptome analysis. *Nature Genetics* 44: 1060–5.

Odvody, G. N. 2000. Gray leaf spot, pp. 12–13. In: *Compendium of Sorghum Diseases*, 2nd ed. (Frederiksen, R. A. and Odvody, G. N., Eds). APS Press, St. Paul, MN.

Odvody, G. N. and Frederiksen, R. A. 1984. Use of systemic fungicides metalaxyl and fosetyl-Al for control of sorghum downy mildew in corn and sorghum in South Texas. 1: Seed treatment. *Plant Disease* 68: 604–7.

Odvody, G. N. and Madden, D. B. 1984. Leaf sheath blights of *Sorghum bicolor* caused by *Sclerotium rolfsii* and *Gloeocercospora sorghi* in South Texas. *Phytopathology* 74: 264–8.

Odvody, G., Bandyopadhyay, R., Frederiksen, R. A., Kaufman, H., Dahlberg, J., Velasquez, R., Torres, H. and Isakeit, T. 1998. Sorghum ergot goes global in less than three years. *APS Features*. Online

Odvody, G. N., Frederickson, D. E., Isakeit, T., Montes, N., Dahlberg, J. A. and Peterson, I. A. 2002. Quarantine issues arising from contamination of seed with ergot: An update, pp. 123–8. In: *Sorghum and Millets Diseases* (Leslie, J. F., Ed.). Iowa State Press, Ames, IA.

Odvody, G. N., Rosenow, D. T. and Black, M. C. 2006. First report of *Ramulispora sorghicola* in the United States causing oval leaf spot on Johnsongrass and sorghum in Texas. *Plant Disease* 90: 108.

Okori, P., Rubaihayo, P. R., Ekwamu, A., Fahleson, J. and Dixelius, C. 2004. Genetic characterization of *Cercospora sorghi* from cultivated and wild sorghum and its relationship to other *Cercospora* fungi. *Phytopathology* 94: 743–50.

Palmero, D., Gil-Serna, J., Gálvez, L., Curt, M. D., De Cara, M., and Tello, J. 2012. First report of Fusarium verticillioides causing stalk and root rot of sorghum in Spain. *Plant Disease* 96: 584.

Pande, S., Harikrishnan, R., Alegbejo, M. D., Mughogho, L. K., Karunakar, R. I. and Ajayi, O. 1993. Prevalence of sorghum diseases in Nigeria. *International Journal of Pest Management* 39: 297–303.

Pande, S., Bock, C. H., Bandyopadhyay, R., Narayana, Y. D., Reddy, B. V. S., Lenné, J. M. and Jeger, M. J. 1997a. Downy mildew of sorghum. Publication No. 51. ICRISAT, Patancheru, AP, India.

Pande, S., Marley, P. S. and Ajayi, O., 1997b. Increasing incidence of covered kernel (grain) smut disease of sorghum in northern Nigeria. *International Sorghum and Millets Newsletter* 38: 59–61.

Parh, D. K., Jordan, D. R., Aitken, E. A. B., Mace, E. S., Jun-ai, P., McIntyre, C. L. and Godwin, I. D. 2008. QTL analysis of ergot resistance in sorghum. *Theoretical and Applied Genetics* 117: 369–82.

Pascual, C. B. and Raymundo, A. D. 1989. Effect of resistance on *Cercospora sorghi*-sorghum-interaction. *Philippine Phytopathology* 25: 69.

Pascual, C. B., Raymundo, A. D. and Hyakumachi, M. 2000. Resistance of sorghum line CS621 to *Rhizoctonia solani* AG1-IA and other sorghum pathogens. *Journal of General Plant Pathology* 66: 23–39.

Pažoutová, S. and Bogo, A. 2002. Recovery of *Claviceps sorghi* (Ascomycotina: Clavicipitaceae) in India. *Mycopathologia* 153: 99–101.

Pažoutová, S. and Frederickson, D. E. 2005. Genetic diversity of *Claviceps africana* on sorghum and *Hyparrhenia*. *Plant Pathology* 54: 749–63.

Pažoutová, S., Bandyopadhyay, R., Frederickson, D. E., Mantle, P. G. and Frederiksen, R. A. 2000. Relations among sorghum ergot isolates from the Americas, Africa, India, and Australia. *Plant Disease* 84: 437–42.

Perumal, R., Frederiksen, R. A., Prom, L. K. and Magill, C. W. 2007. Head Smut, pp. 58–63. In: *Screening Techniques for Sorghum Diseases*. Information Bulletin # 76. International Crops Research Institute for the Semi-Arid Tropics (ICRISAT), Pantacheru, AP, India.

Perumal, R., Nimmakayala, P., Erattaimuthu, S., No, E. G., Reddy, U. K., Prom, L. K., Odvody, G. N., Luster, D. G. and Magill, C. W. 2008. Simple sequence repeat markers useful for sorghum downy mildew (*Peronosclerospora sorghi*) and related species. *BMC Genetics* 9: 77.

Perumal, R., Menz, M. A., Mehta, P. J., Katilé, S., Gutierrez-Rojas, L. A., Klein, R. R., Klein, P. E., Prom, L. K., Schlueter, J. A., Rooney, W. L. and Magill, C. W. 2009. Molecular mapping of *Cg1*, a gene for resistance to anthracnose (*Colletotrichum sublineolum*) in sorghum. *Euphytica* 165: 597–606.

Perumal, R., Rajendrakumar, P., Maulana, F., Tesso, T. and Little, C. R. 2017. Genetic changes in sorghum. In: *Sorghum: State of the Art and Future Perspective*. Agronomy Monographs No. 58. American Society of Agronomy and Crop Science Society of America, Inc., Madison, WI (doi: 10/2134/agromonogr58.2014.0053).

Piepenbring, M., Hagedorn, G. and Oberwinkler, F. 1998. Spore liberation and dispersal in smut fungi. *Botanica Acta (Plant Biology)* 111: 444–60.

Plant Management Network. 2010a. BASF announces expanded label for Stamina F3 HL fungicide seed treatment at ASTA Seed Expo 2010. Website: https://www.plantmanagementnetwork.org/pub/php/news/2010/StaminaF3HL/ (accessed 10 May 2017).

Plant Management Network. 2010b. Dynasty seed treatment fungicide now registered for use on sorghum. Website: http://www.plantmanagementnetwork.org/pub/php/news/2010/Dynasty/ (accessed 10 May 2017).

Plant Management Network. 2014. Sercadis fungicide receives EPA registration for disease control in rice. Website: http://www.plantmanagementnetwork.org/pub/php/news/2014/SercadisFungicide/ (accessed 10 May 2017).

Pratt, R. G. and Janke, G. D. 1980. Pathogenicity of three species of *Pythium* to seedlings and mature plants of grain sorghum. *Phytopathology* 70: 766–71.

Prom, L. K. and Erpelding, J. E. 2009. New sources of grain mold resistance among sorghum accessions from Sudan. *Tropical and Subtropical Agroecosystems* 10: 457–63.

Prom, L. K. and Isakeit, T. 2003. Laboratory, greenhouse, and field assessment of fourteen fungicides for activity against *Claviceps africana*, causal agent of sorghum ergot. *Plant Disease* 87: 252–8.

Prom, L. K., Cissé, N. and Ndoye, O. 2007. Assessing the vulnerability of selected sorghum lines from the United States of America to long smut (*Sporisorium ehrenbergii* Vánky) disease. *Crop Protection* 26: 1771–6.

Prom, L. K., Montes-Garcia, N., Erpelding, J. E., Perumal, R. and Medina-Ocegueda, S. 2010. Response of sorghum accessions from Chad and Uganda to natural infection by the downy mildew pathogen, *Peronosclerospora sorghi* in Mexico and the USA. *Journal of Plant Disease Protection* 117: 2–8.

Prom, L. K., Perumal, R., Erattaimuthu, S. R., Erpelding, J. E., Montes, N., Odvody, G. N., Greenwald, C., Jin, Z., Frederiksen, R. and Magill, C. W. 2011. Virulence and molecular genotyping studies of *Sporisorium reilianum* isolates in sorghum. *Plant Disease* 95: 523–9.

Prom, L. K., Perumal, R., Erattaimuthu, S. R., Little, C. R., No, E. G., Erpelding, J. E., Rooney, W. L., Odvody, G. N. and Magill, C. W. 2012. Genetic diversity and pathotype determination of *Colletotrichum sublineolum* isolates causing anthracnose in sorghum. *European Journal of Plant Pathology* 133: 671–85.

Prom, L. K., Perumal, R., Montes-Garcia, N., Isakeit, T., Odvody, G. N., Rooney, W. L., Little, C. R. and Magill, C. W. 2015. Evaluation of Gambian and Malian sorghum germplasm against the downy mildew pathogen, *Peronosclerospora sorghi*, in Mexico and the USA. *Journal of General Plant Pathology* 81: 24–81.

Radwan, G. L., Perumal, R., Isakeit, T., Magill, C. W., Prom, L. K. and Little, C. R. 2011. Screening exotic sorghum germplasm, hybrids, and elite lines for resistance to a new virulent pathotype (P6) of *Peronosclerospora sorghi* causing downy mildew. *Plant Health Progress*. Online

Rana, B. S., Tripathi, D. P. and Rao, N. G. P. 1976. Genetic analysis of some exotic × Indian crosses in sorghum. XV. Inheritance of resistance to sorghum rust. *Indian Journal of Genetics and Plant Breeding* 36: 244–9.

Ratnadass, A., Marley, P. S., Hamada, M. A., Ajayi, O., Cissé, B., Assamoi, F., Atokple, I. D. K., Beyo, J., Cisse, O., Dakouo, D., Diakite, M., Dossou-Yovo, S., Le Diambo, B., Vopeyande, M. B., Sissoko, I. and Tenkouano, A. 2003. Sorghum head-bugs and grain molds in West and Central Africa: I. Host plant resistance and bug-mold interactions on sorghum grains. *Crop Protection* 22: 837–51.

Reed, J. D., Ramundo, B. A., Claflin, L. E. and Tuinstra, M. R. 2002. Analysis of resistance to ergot in sorghum and potential alternate hosts. *Crop Science* 42: 1135–8.

Resende, R. S., Rodrigues, F. Á., Cavatte, P. C., Martins, S. C. V., Moreira, W. R., Chaves, A. R. M. and DaMatta, F. M. 2012. Leaf gas exchange and oxidative stress in sorghum plants supplied with silicon and infected by *Colletotrichum sublineolum*. *Phytopathology* 102: 892–8.

Reyes, L., Rosenow, D. T., Berry, R. W. and Futrell, M. C. 1964. Downy mildew and head smut diseases of sorghum in Texas. *Plant Disease Reporter* 48: 249–53.

Rodriguez-Herrera, R., Williams-Alanis, H., Aguirre-Rodriguez, J. and Torres-Montalvo, H. 1994. Comparative performance of isogenic sorghum in A1 and A2 cytoplasms. V. Foliar diseases. *Sorghum Newsletter* 35: 80.

Rodriguez-Herrera, R., Waniska, R. D., Rooney, W. L., Aguilar, C. N. and Contreras-Esquivel, J. C. 2006. Antifungal proteins during sorghum grain development and grain mould resistance. *Journal of Phytopathology* 154: 565–71.

Rooney, W. L., Collins, S. D., Klein, R. R., Mehta, P. J., Frederiksen, R. A. and Rodriguez-Herrera, R. 2002. Breeding sorghum for resistance to anthracnose, grain mold, downy mildew, and head smuts, pp. 273–80. In: *Sorghum and Millets Diseases* (Leslie, J. F., Ed.). Iowa State Press, Ames, IA.

Rosenow, D. T. 1992. Using germplasm from the world collection in breeding for disease resistance, pp. 319–24. In: *Sorghum and Millers Diseases: A Second World Review*. ICRISAT, Patancheru, AP, India.

Rosenow, D. T. and Frederiksen, R. A. 1982. Breeding for disease resistance in sorghum. *Sorghum in the Eighties* 1: 447–55.

Ruppel, E. G. 1985. Susceptibility of rotation crops to a root isolate of *Rhizoctonia solani* from sugar beet and survival of the pathogen in crop residues. *Plant Disease* 69: 871–3.

Ryley, M. J., Persley, D. M., Jordan, D. R. and Henzell, R. G. 2002. Status of sorghum and pearl millet diseases in Australia, pp. 441–8. In: *Sorghum and Millets Diseases* (Leslie, J. F., Ed.). Iowa State Press, Ames, IA.

Saccas, A. M. 1954. Le champignons parasites des sorghos (*Sorghum vulgare*) et des penicilliares (*Pennisetum typhoideum*) en *Afrique equitoriale* Française. *Agronomie Tropicale (Nogent-sur-Marne)* 9: 3–6.

Safeeulla, K. M. 1976. *Biology and Control of the Downy Mildews of Pearl Millet, Sorghum, and Finger Millet*. Wesley Press, Mysore, India, 304pp.
Sage, R. F., Li, M. and Monson, R. K. 1999. The taxonomic distribution of C_4 photosynthesis, pp. 551–84. In: *C_4 Plant Biology* (Sage, R. F. and Monson, R. K., Eds). Academic Press, San Diego, CA.
Schuh, W., Frederiksen, R. A. and Jeger, M. J. 1986. Analysis of spatial patterns in sorghum downy mildew with Morisita's Index of Dispersion. *Phytopathology* 76: 446–50.
Schuh, W., Jeger, M. J. and Frederiksen, R. A. 1987. The influence of soil temperature, soil moisture, soil texture, and inoculum density on the incidence of sorghum downy mildew. *Phytopathology* 77: 125–8.
Seetharaman, K., Whitehead, E., Keller, N. P., Waniska, R. D. and Rooney, L. W. 1997. *In vitro* activity of sorghum seed antifungal proteins against grain mold pathogens. *Journal of Agricultural and Food Chemistry* 45: 3666–71.
Semeniuk, G. and Mankin, C. J. 1964. Occurrence and development of *Sclerophthora macrospora* on cereals and grasses in South Dakota. *Phytopathology* 54: 409–16.
Sharma, R., Rao, V. P., Upadhyaya, H. D., Reddy, V. G. and Thakur, R. P. 2010. Resistance to grain mold and downy mildew in a mini-core collection of sorghum germplasm. *Plant Disease* 94: 439–44.
Sharma, R., Upadhyaya, H. D., Manjunatha, S. V., Rao, V. P. and Thakur, R. P. 2012. Resistance to foliar diseases in a mini-core collection of sorghum germplasm. *Plant Disease* 96: 1629–33.
Sill, W. H., Hansing, E. D., King, C. L., Dickerson, O. J., Stuteville, D. L., Schafer, J. F., Edmunds, L. K., Willis, W. G. and Paulsen, A. Q. 1969. Kansas Phytopathological Notes: 1968. *Transactions of the Kansas Academy of Science* 72:547–54.
Sim, T. 1980. Sorghum downy mildew in Kansas in 1979. *Plant Disease* 64: 499.
Singleton, L. L., Milhail, J. D. and Rush, C. M., Eds. 1992. *Methods for Research on Soilborne Phytopathogenic Fungi*. APS Press, St. Paul, MN.
Sprague, R. 1950. *Diseases of Cereals and Grasses in North America*. Ronald Press, New York, NY, 538pp.
Stack, J. 2000. EC00-1879 Sorghum ergot in the northern Great Plains. Historical Material from University of Nebraska – Lincoln Extension, Paper 1239.
Stack, J. P. 2002. Recurring and emerging sorghum diseases in North America, pp. 449–56. In: *Sorghum and Millets Disease* (Leslie, J. F., Ed.). Iowa State Press, Ames, IA.
Stack, J. P. and Pedersen, J. F. 2003. Expression of susceptibility to Fusarium head blight and grain mold in A1 and A2 cytoplasms of *Sorghum bicolor*. *Plant Disease* 87: 172–6.
Stoll, M., Piepenbring, M., Begrow, D. and Oberwinkler, F. 2003. Molecular phylogeny of *Ustilago* and *Sporisorium* species (Basidiomycota, Ustilaginales) based on internal transcribed spacer (ITS) sequences. *Canadian Journal of Botany* 81: 976–84.
Sumner, D. R. and Bell, D. K. 1982. Root diseases induced in corn by *Rhizoctonia solani* and *Rhizoctonia zeae*. *Phytopathology* 72: 86–91.
Swarup, G., Hansing, E. D. and Rogerson, C. T. 1962. Fungi associated with sorghum seed in Kansas. *Transactions of the Kansas Academy of Science* 65: 120–37.
Tao, Y. Z., Jordan, D. R., Henzell, R. G. and McIntyre, C. L. 1998. Identification of genomic regions for rust resistance in sorghum. *Euphytica* 103: 287–92.
Tarr, S. A. J. 1962. *Diseases of Sorghum, Sudan Grass, and Broom Corn*. Commonwealth Agricultural Bureaux, Oxford, UK.
TeBeest, D., Kirkpatrick, T. and Cartwright, R. 2004. Common and important diseases of grain sorghum. Grain sorghum production handbook. Univ. Ark. Coop. Ext., USA MP297.
Tesso, T., Perumal, R., Little, C. R., Adeyanju, A., Radwan, G. L., Prom, L. K. and Magill, C. W. 2012. Sorghum pathology and biotechnology – A fungal disease perspective: Part II. Anthracnose, stalk rot, and downy mildew. *The European Journal of Plant Science and Biotechnology* 6: 31–44.
Thakur, R. P. and Mathur, K. 2002. Downy mildews of India. *Crop Protection* 21: 333–45.
Thakur, R. P., Reddy, B. V. S. and Mathur, K. 2007. Screening techniques for sorghum diseases. Information Bulletin No. 76. ICRISAT, Patancheru, AP, India.

Thaung, M. M. 2008. Pathologic and taxonomic analysis of leaf spot and tar spot diseases in a tropical dry to wet monsoon ecosystem of lowland Burma. *Australasian Plant Pathology* 37: 180–97.

Thomas, M. D., Bocoum, F. and Thera, A. 1993. Field inoculations of sorghum with sclerotia and conidia of *Ramulispora sorghi* formed in vivo. *Mycologia* 85: 807–10.

Tooley, P. W., O'Neill, N. R., Goley. E. D. and Carras, M. M. 2000. Assessment of diversity of *Claviceps africana* and other *Claviceps* species by RAM and AFLP analyses. *Phytopathology* 90: 1126–30.

Tsedaley, B., Adugna, G. and Lemessa, F. 2016. Distribution and importance of sorghum anthracnose (*Colletotrichum sublineolum*) in southwestern and western Ethiopia. *Plant Pathology Journal* 15: 75–85.

Valèrio, H., Rèsende, M., Weikert-Oliviera, R. and Casela, C. 2005. Virulence and molecular diversity in *Colletotrichum graminicola* from Brazil. *Mycopathologia* 159: 449–59.

Vanderlip, R. L. and Reeves, H. E. 1972. Growth stages of sorghum [*Sorghum bicolor* (L.) Moench]. *Agronomy Journal* 64: 13–16.

Velasquez-Valle, R., Narro-Sanchez, J., Nolasco-Mora, R. and Odvody, G. N. 1998. Spread of ergot of sorghum (*Claviceps africana*) in central Mexico. *Plant Disease* 82: 447.

United States Department of Agriculture (USDA), National Agricultural Statistics Service (NASS). 2017. Online: https://www.nass.usda.gov/Statistics_by_Subject/index.php (accessed 29 March 2017).

Wall, G. C. and Meckenstock, D. H. 1992. Sorghum diseases in Central America and the Caribbean basin, p. 67. In: *Sorghum and Millets Diseases: A Second World Review*. ICRISAT, Patancheru, AP, India.

Wall, G. C., Jeger, M. J. and Frederiksen, R. A. 1989. The relationship of yield loss to foliar diseases on sorghum grown by subsistence farmers in southern Honduras. *Tropical Pet Management* 35: 57–61.

Wang, J., Levy, M. and Dunkle, L. D. 1998. Sibling species of *Cercospora* associated with gray leaf spot of maize. *Phytopathology* 88: 1269–75.

Wang, P., Sandrock, R. W. and VanEtten, H. D. 1999. Disruption of the cyanide hydratase gene in *Gloeocerecospora sorghi* increases its sensitivity to the phytoanticipin cyanide but does affect its pathogenicity on the cyanogenic plant sorghum. *Fungal Genetics & Biology* 28: 126–34.

Wang, P. H., Chung, C. Y., Lin, Y. S. and Yeh, Y. 2003. Use of polymerase chain reaction to detect the soft rot pathogen, *Pythium myriotylum*, in infected ginger rhizomes. *Letters in Applied Microbiology* 36: 116–20.

Wang, M. L., Dean, R., Erpelding, J. and Pedersen, G. 2006. Molecular genetic evaluation of sorghum germplasm differing in response to fungal diseases: Rust (*Puccinia purpurea*) and anthracnose (*Colletotrichum gramincola*). *Euphytica* 148: 319–30.

Waniska, R. D., Venkatesha, R. T., Chandrashekar, A., Krishnaveni, S., Bejosano, F. P., Jeoung, J., Jayaraj, J., Muthukrishnan, S. and Liang, G. H. 2001. Antifungal proteins and other mechanisms in the control of sorghum stalk rot and grain mold. *Journal of Agricultural and Food Chemistry* 49: 4732–42.

Ward, J. M. J. and Nowell, D. C. 1998. Integrated management practices for the control of maize gray leaf spot. *Integrated Pest Management Reviews* 3: 177–88.

Wharton, P. S. and Julian, A. M. 1996. A cytological study of compatible and incompatible interactions between *Sorghum bicolor* and *Colletotrichum sublineolum*. *New Phytologist* 134: 25–34.

Wharton, P. S., Julian, A. M. and O'Connell, R. J. 2001. Ultrastructure of the infection of *Sorghum bicolor* by *Colletotrichum sublineolum*. *Phytopathology* 91: 149–58.

White, J. A., Ryley, M. J., George, D. L., Kong, G. A. and White, S. C. 2012. Yield losses in grain sorghum due to rust infection. *Australasian Plant Pathology* 41: 85–91.

White, J. A., Ryley, M. J., George, D. L. and Kong, G. A. 2014. Optimal environmental conditions for infection and development of *Puccinia purpurea* on sorghum. *Australasian Plant Pathology* 43: 447–57.

Whitehead, M. D. 1958. Pathology and pathological histology of downy mildew, *Sclerophthora macrospora*, on six graminicolous hosts. *Phytopathology* 48: 485–93.

Williams, A. S. and Herron, J. W. 1974. Occurrence of downy mildew of sorghum in Kentucky. *Plant Disease Reporter* 40: 954–6.

Williams, R. J., Frederiksen, R. A. and Girard, J. C. 1978. Sorghum and pearl millet disease identification handbook. Information Bulletin No. 2. ICRISAT, Patancheru, AP, India.

Williams-Alanis, H., Pecina-Quintero, V., Montes-Garcia, N., Arcos-Cavazos, G., Zavala-Garcia, F. and Josue Gamez-Vazquez, A. 2009. Incidence of head smut *Sporisorium reilianum* (Kuhn) Langdon and Fullerton in grain sorghum [*Sorghum bicolor* (L.) Moench.] hybrids. *Revista Mexicana de Fitopatologia* 27: 36–44.

Workneh, F. and Rush, C. M. 2006. Weather factors associated with development of sorghum ergot in the Texas Panhandle. *Plant Disease* 90: 717–22.

Xavier, K., Pfieffer, T., Parreira, D. F., Chopra, S. and Villaincourt, L. 2017. Aggressiveness of *Colletotrichum sublineola* strains from Sorghum bicolor and *S. halapense* to sweet sorghum variety sugar drip, and their impact on yield. *Plant Disease* 101: 1578–87.

Xu, R. Y., Zhao, S. T. and Feng, W. E. 1997. Breeding Kang 4 and Kang 7 sorghum hybrids and inheritance of resistance to head smut. *Acta Agriculturae Boreali-Sinica* 12: 33–7.

Yassin, A., El-Samawaty, A., Bahkali, A., Moslem, M., Kamel, A. et al. 2010. Mycotoxin-producing fungi occurring in sorghum grains from Saudi Arabia, *Fungal Diversity*, 44: 45–52.

Yazawa, T., Kawahigashi, H., Matsumoto, T. and Mizuno, H. 2013. Simultaneous transcriptome analysis of sorghum and *Bipolaris sorghicola* by using RNAseq in combination with de novo transcriptome assembly. *PLoS One* 8: e62460.

Zummo, N. 1978. December. Fusarium disease complex of sorghum in West Africa, pp. 11–15. In: *Sorghum Diseases: A World Review*. Proc. of Int. Workshop of Sorghum Diseases, Hyderabad, India.

Zummo, N. and Broadhead, D. M. 1984. Sources of resistance to rough leaf spot disease in sweet sorghum. *Plant Disease* 68: 1048–9.

Chapter 16

Sorghum diseases and their management in cultivation: stalk, root and other diseases

C. R. Little, A. Y. Bandara and T. C. Todd, Kansas State University, USA; and R. Perumal, Agricultural Research Center – Hays, Kansas State University, USA

1 Introduction

Sorghum [*Sorghum bicolor* (L.) Moench] is an important feed and food security grain crop; it is also used for fodder and is now an advanced biofuel source. From 2012 to 2017, an average of 7.0 and 6.2 million acres were planted and harvested in the United States, respectively (USDA NASS, 2017). Productivity averaged 67.1 bu A^{-1} during this period. Throughout the world, an average of 42.8 million acres of sorghum were harvested with an average yield of 14.8 thousand hectograms ha^{-1} (FAO STAT, 2017). In 2016, the top producers of sorghum in the world were the United States, Nigeria, Sudan and Mexico.

Both abiotic and biotic stressors impact optimum yield and productivity of the sorghum crop. Among the most damaging and difficult diseases to control are the root and stalk rots (Table 1). These diseases are hidden underground or within the plant, and often do

http://dx.doi.org/10.19103/AS.2017.0015.31

Table 1 Proposed priorities for sorghum research based on the historical and current relative prevalence and importance[a] of the diseases covered in this chapter

Tier 1, *High priority*	Tier 2, *Medium priority*	Tier 3, *Low or no priority*
Charcoal rot	Bacterial leaf stripe	Acremonium wilt
Fusarium root and stalk rots	Maize dwarf mosaic virus	Bacterial leaf blight
Witchweed	Pythium root rot	Bacterial leaf spot
	Lesion nematode	Bacterial leaf streak
		Bacterial top and stalk rot
		Banded leaf and sheath blight
		Milo disease
		Southern blight
		Yellow leaf blotch
		Other virus diseases
		Other nematodes

[a] References: Frederiksen (2000), Casela et al. (2002), Esele (2002), Giorda (2002), Indira (2002), Jensen and Giorda (2002), Kasakova and Oleynick (2002), Malaguti (2002), Marley et al. (2002), Mtisi and McLaren (2002), Narayana et al. (2002), Ngugi et al. (2002), Ryley et al. (2002), Stack (2002), Seifers et al. (2012) and Tesso et al. (2012).

not present themselves in an obvious manner like the foliar and panicle diseases. Often, symptoms of the root and stalk rot 'complex' resemble abiotic stressors such as drought, high temperature or freeze damage (Jordan et al., 1984; Mughogho, 1984; Tesso et al., 2012). Further, such environmental stressors are important factors that contribute to manifestation of the disease complex. *Macrophomina phaseolina* (Tassi) Goid causes charcoal rot disease of sorghum stalks, in addition to colonizing roots, causing root rot. *Fusarium* spp., including *F. thapsinum* Klittich, J.F. Leslie, P.E. Nelson & Marasas, cause Fusarium stalk rot. But, this fungus also colonizes seedlings, roots, peduncles, panicles and developing grain (Little et al., 2012). Both of these major pathogens weaken the structural anchor and support of the plant, which can lead to lodging. Both diseases are worsened by drought and high temperatures.

To mitigate stress, balanced soil nutrition, insect management, lower plant populations and weed control are recommended. The search for resistance to these diseases is ongoing and complicated by its quantitative, environmentally modulated nature and the limited, but improving, knowledge of pathogen biology and diversity (Tesso et al., 2010; Petrovic et al., 2009; Kelly et al., 2017). The propagules of *M. phaseolina* and *F. thapsinum*, and other stalk and root pathogens may survive in soil and/or crop debris for a long time, which lessens the effectiveness of cultural control strategies such as crop rotation and residue tillage management (Mughogho and Pande, 1984; Zummo, 1984; Tesso et al., 2012).

Of the sorghum bacterial diseases, bacterial stripe caused by *Burkholderia andropogonis* Smith (syn. *Pseudomonas andropogonis* Smith) is the most widespread (Lopes-Santos et al., 2015). Unlike the fungal diseases, sorghum bacterial diseases are not yield limiting or especially problematic. However, some countries have quarantines on seed that are infested or infected by pathogenic seedborne bacteria (Ramundo and Claflin, 2005). Predominating wet conditions that maximize relative humidity, optimum leaf wetness and the potential for splash dispersal are most conducive to outbreaks of bacterial diseases in all crops (Akhtar, 1985a; Claflin et al., 1992; Claflin, 2000a; Temam, 2001). In the case of corn, epiphytotics of bacterial diseases have caused or could cause major losses. Such diseases include Stewart's wilt (*Pantoea stewartii* Smith Dye), Goss's wilt [*Clavibacter michiganensis* ssp. *nebraskensis* (Vidaver and Mandel) Davis, comb. nov.] and an emerging disease, bacterial streak and wilt (*Xanthomonas vasicola* Vauterin, sp. nov.) (Korus et al., 2017). Understanding the bacterial diseases of sorghum would provide a framework for control if new, potentially yield-limiting pathogens were to arise in the crop in the near future.

Numerous virus species occur in sorghum and cause a wide array of overlapping and often confusing symptoms (Jensen and Giorda, 2002). These include leaf banding, mosaics, mottling, necrosis, panicle abortion, spotting, sterility, streaking, striping, stunting, vein clearing and lower yields through reduced seed size and seed number (Jensen, 2000). Therefore, both serological and molecular detection methods are needed to properly diagnose virus diseases.

Of the diversity of viruses, potyviruses are the most commonly encountered in sorghum. These viruses include maize dwarf mosaic virus (MDMV), Johnsongrass mosaic virus (JgMV), sorghum mosaic virus (SrMV), sugarcane mosaic virus (SCMV) and Zea mosaic virus (ZeMV) (Seifers et al., 2012). These viruses are characterized by a +ssRNA genome and flexuous rod-type virion. Other virus groups found in sorghum include the bromoviruses, fijiviruses, furoviruses, geminiviruses, luteoviruses, mastreviruses, nucleorhabdoviruses, panicoviruses, tenuiviruses and waikaviruses (Jensen, 2000; Jensen and Giorda, 2002). Like many plant viruses, the viruses of sorghum may be persistently or non-persistently transmitted by insect vectors. Aphids, plant-, and leafhoppers transmit the bulk of the vectored sorghum viruses. However, a soilborne plasmodiophorid, *Polymyxa graminis* Ledingham, transmits soilborne wheat mosaic virus (SBWMV), which also occurs on sorghum (Jensen and Giorda, 2002; Kanyuka et al., 2003).

The parasitic plant, witchweed (*Striga* spp.), seriously impacts sorghum (Spallek et al., 2013). This is especially the case in sub-Saharan Africa. Witchweed is a hemiparasite and relies on the host for water and mineral nutrients but can undergo photosynthesis. These plants produce numerous tiny seeds that are capable of surviving in soil for many years. Germination of the host induces *Striga* seeds to germinate and produce a holdfast on the host root. The result of infection is to produce a 'xylem bridge' between the parasite and the host. Therefore, sorghum plants that are parasitized by witchweed are chlorotic and stunted and exhibit reduced yields (Doggett, 1965; Spallek et al., 2013). This is due to reduced photosynthesis and hormone imbalance in the host (Taylor et al., 1996; Frost et al., 1997; Rodenburg et al., 2008).

Like root and stalk rot diseases, the above-ground symptoms of soilborne nematode diseases often resemble abiotic stress due to damage and colonization of root tissues. Symptoms include chlorosis, stunting, wilt and yield reduction and drag (Orr and Morey, 1978; Babatola and Idowu, 1990; Sharma and McDonald, 1990; Starr, 1992; McClure et al., 1998). Inspection of underground root tissues reveals root distortions, knot-like growths and swellings and necrosis. Damaged or distorted roots are less functional and more susceptible to colonization by root and stalk rot pathogens. Together, nematode colonization and secondary fungal infections have the potential to decrease plant yields and vigour, and further predispose plants to abiotic stressors such as drought and high temperature. For example, lesion nematodes (*Pratylenchus* spp.) interact with the charcoal rot pathogen, *M. phaseolina* (Norton, 1958).

2 Charcoal rot

Typical stalk anatomical changes observed with charcoal rot disease are similar to those associated with Fusarium stalk rot (see Section 3). These include disintegration of the pith tissue at the infection site (Edmunds, 1964), which is due to the pathogen's ability to produce extracellular cell-wall-degrading enzymes (CWDE) and induce host-derived CWDE (Islam et al., 2012; Bandara et al., 2018). Infected plants often have shredded vascular tissues in both the stalk and root system that may reduce water and nutrient uptake and translocation. Severe infections can cause breakage of the stalk at the zone of infection, leading to lodging. Distinctive grey to black pigmentation in the infected area is visible to the naked eye after longitudinally splitting the stalk. Often, as the infected plants mature and reach the senescence phase, bundles of small, black sclerotia can be seen in the infected area. These are important internal signs of the disease. External symptoms include premature leaf and stalk senescence, premature ripening and impaired grain filling.

Macrophomina phaseolina (Tassi) Goid, the causal organism of charcoal rot disease, is a soilborne fungus that can cause diseases of over 500 different plant species in nearly 100 families (Mihail and Taylor, 1995) (Fig. 1; Table 2). Despite its broad host range, *Macrophomina* is a monotypic genus and contains only one species: *M. phaseolina* (Sutton, 1980). It survives in the form of microsclerotia and mycelia in crop residuals that act as the primary inocula upon overwintering. Microsclerotia can remain viable

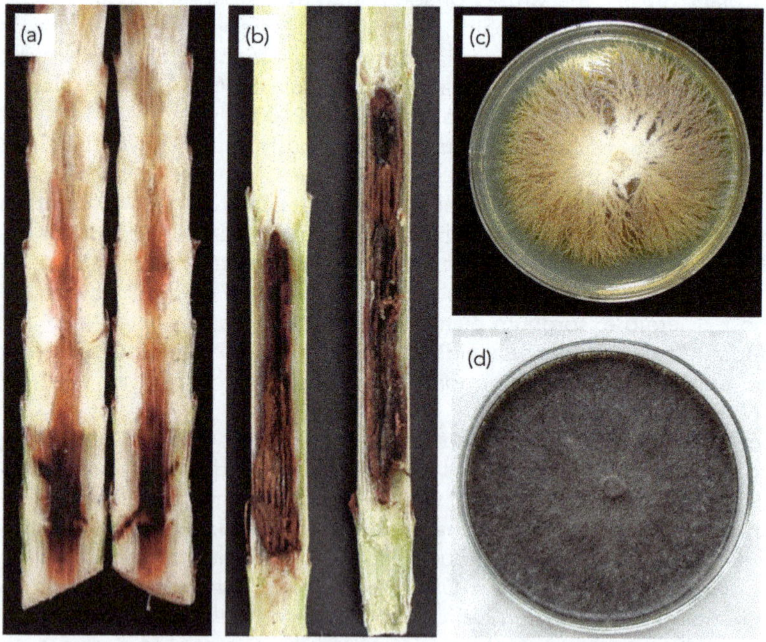

Figure 1 Characteristic pink pith lesions of Fusarium stalk rot caused by *Fusarium thapsinum* (a) and dark pith lesions produced by *Macrophomina phaseolina* (b). Example growth cultures of *F. thapsinum* (c) and *M. phaseolina* (d). Image by Y. M. A. Y. Bandara.

Table 2 Fungi and fungus-like organisms associated with sorghum roots and stalks

Genus	Species	Reference(s)
Fungi:		
Acremonium	*A. strictum*	Natural et al. (1982)
Alternaria	*Alternaria* spp.	Preston (1945) and Reed et al. (1983)
Bipolaris	*B. sorokiniana*	Preston (1945) and Sprague (1950)
Colletotrichum	*C. sublineola*	Duncan (1983) and Tesso et al. (2012)
Epicoccum	*Epicoccum* spp.	Reed et al. (1983)
Exserohilum	*E. pedicellatum*	Isakeit et al. (2007)
	E. turcicum	Preston (1945) and Sprague (1950)
Fusarium	*F. andiyazi*	Leslie et al. (2005), Tesso et al. (2010) and Kelly et al. (2017)
	F. avenaceum	Leslie et al. (1990)
	F. brevicatenulatum	Tesso et al. (2010)
	F. chlamydosporum	Leslie et al. (1990), Petrovic et al. (2009) and Kelly et al. (2017)
	F. compactum	Petrovic et al. (2009)
	F. equiseti/FIESC*	Reed et al. (1983), Leslie et al. (1990), Petrovic et al. (2009) and Kelly et al. (2017)
	F. graminearum	Reed et al. (1983), Leslie et al. (1990) and Petrovic et al. (2009)
	F. nygamai	Leslie et al. (2005), Petrovic et al. (2009) and Tesso et al. (2010)
	F. oxysporum	Sprague (1950), Mankin (1969), Reed et al. (1983), McLaren (1987), Leslie et al. (1990), Idris et al. (2007) and Petrovic et al. (2009)
	F. proliferatum	Leslie et al. (1990), Petrovic et al. (2009), Tesso et al. (2010) and Kelly et al. (2017)
	F. pseudoanthophilum	Tesso et al. (2010)
	F. scirpi	Petrovic et al. (2009)
	F. semitectum	Leslie et al. (1990) and Petrovic et al. (2009)
	F. solani	Reed et al. (1983), Petrovic et al. (2009) and Kelly et al. (2017)
	F. subglutinans	Petrovic et al. (2009)
	F. thapsinum	Leslie et al. (2005), Tesso et al. (2010), Petrovic et al. (2009) and Kelly et al. (2017)
	F. tricinctum	Reed et al. (1983)
	F. verticillioides	Petrovic et al. (2009), Tesso et al. (2010), Palmero et al. (2012) and Kelly et al. (2017)

(Continued)

Table 2 (*Continued*)

Genus	Species	Reference(s)
Macrophomina	*M. phaseolina*	Mughogho et al. (1984), Saleh et al. (2010), Tesso and Ejeta (2011) and Bandara et al. (2018)
Microdochium	*M. bolleyi*	Sprague (1950) and Mankin (1969)
Nigrospora	*N. oryzae*	Sprague (1950)
	N. sphaerica	Sprague (1950) and Reed et al. (1983)
Penicillium	*Penicillium* spp.	Luttrell (1950) and Hsi (1956)
Periconia	*P. circinata*	Leukel (1948)
	P. macrospinosa	Sprague (1950)
Phoma	*P. terrestris*	Sprague (1950)
Rhizoctonia	*R. solani*	Luttrell (1950), Stroube (1954), Hsi (1956), Bell et al. (1973) and O'Neill and Rush (1982)
Sclerotium	*S. rolfsii*	Luttrell (1950)
Trichoderma	*Trichoderma* spp.	Reed et al. (1983)
Fungus-like organisms:		
Polymyxa	*P. graminis*	Delfosse et al. (2000) and Kanyuka et al. (2003)
Pythium	*P. arrhenomanes*	Preston (1945) and Forbes et al. (1985, 1986)
	P. graminicola	Sprague (1950), Mankin (1969), Pratt and Janke (1980) and Odvody (2000)
	P. myriotylum	Pratt and Janke (1980) and Odvody (2000)
	P. periplocum	Pratt and Janke (1980) and Odvody (2000)
	*Globisporangium*** *debaryanum*	Sprague (1950) and Mankin (1969)
	*Globisporangium*** *ultimum*	Sprague (1950), Davis and Bockus (2001) and McLean and Lawrence (2001)

*FIESC = *Fusarium incarnatum-equiseti* species complex.
***Pythium* (syn.).

in soil and crop residues for more than 4 years (Short et al., 1980) and therefore are an important source of inoculum. Like Fusarium stalk rot, post-flowering stress conditions such as drought, heat and nutrient imbalance predispose *Macrophomina* infection. However, unlike Fusarium stalk rot, wet and cooler soil conditions followed by stress are dispensable to manifest charcoal rot disease (Dodd, 1980; Seetharama et al., 1987; Zummo, 1980). In fact, charcoal rot is more conspicuous when soil temperature is >32°C (Edmunds, 1964); therefore, continuous high soil temperature is conducive for charcoal rot outbreaks. Overwintering sclerotia germinate in the presence of host root exudates. Lesion nematodes contribute to enhanced charcoal rot incidences as they provide entry points for the pathogen (Norton, 1958). However, the growing hyphae can infect the roots only when the plants undergo moisture and temperature stresses (Odvody and Dunkle, 1979). Upon root invasion, the pathogen rapidly moves to above-ground basal stalk portions, attacking the lower internodes. Although disease symptoms can be observed in upper internodes (even in the peduncle) of highly susceptible genotypes, unlike *Fusarium* spp., there are no available reports that demonstrate the vertical transmission of the pathogen.

Charcoal rot is known to be a difficult-to-control pathogen due to its persistence as sclerotia in soil and plant debris (Table 3). Hence, host resistance is the key to disease control. Most of the breeding efforts for charcoal rot resistance have aimed at selecting for the stay-green trait (delayed senescence). This trait has been identified and confirmed to be associated with charcoal rot resistance (Rosenow et al., 1977; Bandara et al., 2016). Arguably, post-flowering drought tolerance and delayed senescence serve as links between stay-green and charcoal rot resistance/tolerance. Delayed stalk senescence is particularly important as necrotrophic pathogens can take advantage of rapidly senescing genotypes to fulfil their energy requirements. Stay-green genotypes SC599 and SC35 show a remarkable level of resistance to charcoal rot disease (Tesso et al., 2005; Bandara et al., 2016). However, non-co-segregation of charcoal rot resistance and stay-green traits in populations derived from the genotype SC599 indicated that stay-green does not underlie charcoal rot resistance in SC599 (Tenkouano et al., 1993), despite the positive association between the two traits. Hence, exclusion of sorghum genotypes based on non-stay-green criterion from breeding programmes aimed at developing charcoal rot resistant hybrids seems inappropriate. Most studies have revealed that charcoal rot resistance is controlled by dominant and additive genes (Thakur et al., 1996; Bramel-Cox et al., 1988; Tesso et al., 2004, 2005).

Table 3 Control measures for Fusarium stalk rot, charcoal rot and root rot diseases

Control strategy	Specific examples
Avoidance	Moisture conservation.
	Adjust planting date to avoid stressful periods.
	Avoid/minimize water deficit conditions during post-anthesis periods.
	Reduce competition by controlling plant populations and managing weeds.
	Avoid excessive use of nitrogen fertilizer; maintain the balance between nitrogen and potassium fertilizers.
	Reduce herbivory stress by controlling insects.
	Use proper rates and applications of herbicide.
Eradication	Crop rotation.
	Mix cropping systems (avoid monoculture).
	Reduce initial inoculum density through residue management in the field (fallowing, burning and tillage).
Protection	*Seed treatments and fungicide applications generally have low or no effect.*
Resistance, *physiological*	Grow stay-green hybrids with drought tolerance.
Resistance, *structural*	Grow hybrids with sturdy root systems and high stalk strength for lodging resilience.

3 Fusarium stalk rot

The typical symptoms associated with Fusarium stalk rot include degradation of the pith tissue at or near the base of the stalk, causing death of stalk pith cells (Edmunds, 1964). Infected plants often have damaged vascular and cortical tissues in both the stalk and root system that in turn contributes to reduced water, nutrient uptake and translocation and results in disrupted photosynthetic activity (Hundekar and Anahosur, 1994). Severe infections can cause stalk breakage at the zone of infection, leading to lodging (Mughogho and Pande, 1984; Maranville and Clegg, 1984). Symptoms are visible to the naked eye when an infected sorghum stalk is longitudinally split. Doing so reveals a characteristic reddish- to pink-coloured pith lesion. The lesion colour is attributed to infection-associated host anthocyanin (primarily 3-deoxyanthocyanidin) production although no published reports are available that exclusively demonstrate host anthocyanin profiles in Fusarium stalk rot lesions. Lesions may be visible in the uppermost internodes and the peduncle, particularly in highly susceptible genotypes. Other Fusarium stalk rot symptoms include premature plant death and grain ripening, and impeded grain filling. Leaves of infected plants may appear frost damaged.

At least eight *Fusarium* spp. have been described in sorghum that are associated with stalk rot (Leslie et al., 1990). However, the major *Fusarium* spp. infecting sorghum stalks include *Fusarium andiyazi* Marasas, Rheeder, Lampr., K.A. Zeller and J.F. Leslie; *F. proliferatum* (Matsush.) Nirenberg ex Gerlach and Nirenberg; *F. thapsinum*; and *F. verticillioides* (Sacc.) Nirenberg (Tesso et al., 2010). Among these, *F. thapsinum* is the most aggressive Fusarium stalk rot pathogen of sorghum (Leslie et al., 2005; Tesso et al., 2005, 2010; Tesso and Ejeta, 2011) (Fig. 1; Table 2). The pathogen survives as conidia and mycelia in crop residues and acts as the primary inoculum after over-seasoning. Mycelia or germinating conidia can enter roots and stalks via roots damaged by mechanical, biological (e.g. insect attack), or any other means (Claflin and Giorda, 2002). Initial infection occurs on root cortical tissues and subsequently enters the vascular tissues as the pathogen advances towards the base of the stalk, attacking the lower internodes (Zummo, 1980). Depending on the degree of host susceptibility, pathogens can move to the upper internodes and to the peduncle. Pathogens may eventually end up in the grains and contribute to vertical transmission (Bacon et al., 2001). Infection events are more pronounced under post-flowering stress conditions such as drought, heat, and nutrient imbalance followed by wet and cooler conditions toward physiological maturity (Dodd, 1980; Seetharama et al., 1987; Zummo, 1980).

Use of genetic resistance is the most effective approach to control Fusarium stalk rot. However, breeding for Fusarium stalk rot resistance has been challenging as the disease can be incited by a variety of *Fusarium* spp. and due to the quantitative nature of the resistance that is often confounded by the environment. Through a genome-wide association study (GWAS), Adeyanju et al. (2015) have shown a significant association between a single nucleotide polymorphic site (S9_57816733) and resistance to *F. thapsinum*. This SNP resides on the gene *sb09g029260*, which encodes chalcone synthase. However, the actual causal genes that underlie and/or contribute to Fusarium stalk rot resistance have not been cloned. Nevertheless, lodging, greenbug resistance and stay-green traits have been shown to associate with Fusarium stalk rot resistance (Giorda and Martinez, 1995). The stay-green trait (delayed senescence) associated with post-flowering drought tolerance has been identified and confirmed as a physiological means of Fusarium stalk rot resistance (Duncan, 1983; Bandara et al., 2016) (Table 3). Stay-green genotypes such as SC599, SC35, SC134 and SC1039 show excellent levels of resistance to *Fusarium* spp. that cause stalk rot (Bramel-Cox et al., 1988; Tesso et al., 2005).

4 Milo disease

Plants affected by milo disease are characterized by above-ground symptoms that are much like other diseases where a toxin metabolite acts as the key virulence factor. Leaves droop, roll, wilt, and become chlorotic and necrotic from the edges and tips inward (i.e. 'firing') (Odvody and Dunkle, 1984). Overall, plants may have a stunted appearance. Milo disease severity ranges from mild chlorosis to severe wilting and firing. When symptoms are mild, some seed loss or shrunken seeds are visible (Odvody and Dunkle, 1984). When symptoms are severe, losses are much greater with the failure to produce a full panicle of grain.

The milo disease pathogen infects roots and can destroy secondary roots, cause soft rot symptoms, watersoaking, and red discoloration in the primary root and crown (Odvody and Dunkle, 1984). Therefore, during the root rot stage of the disease, Periconia root rot, the crown becomes necrotic and discoloured. The plant produces anthocyanin pigments in response to the pathogen, and the discoloration will extend into the vascular system as those cells are stressed by the toxin produced by the pathogen.

Milo disease is caused by the necrotrophic fungal pathogen, *Periconia circinata* (L. Mangin) Sacc. & D. Sacc. (Table 2). Milo disease was first reported in Chillicothe, Texas, in 1924 and again in Garden City, Kansas, in 1926 (Odvody and Dunkle, 1984). The first report of this disease attributed it to *Pythium arrhenomanes* Drechsler (Elliot et al., 1937). However, sorghums that were resistant to *P. arrhenomanes* also exhibited symptoms of milo disease. Leukel (1948) isolated different fungi from the roots of diseased plants and through the process of elimination could reproduce the symptoms using *P. circinata*. It is thought that *P. arrhenomanes* is a secondary invader of dead root tissue.

Like other fungal pathogens such as *Alternaria alternata* (Fr.) Keissl., *A. kikuchiana* S. Tanaka, *Bipolaris sorokiniana* Shoemaker, *Cochliobolus carbonum* R.R. Nelson and *C. victoriae* R.R. Nelson, *P. circinata* produces a host-selective phytotoxin, known as peritoxin or PC-toxin (Pringle and Scheffer, 1963; Walton and Panaccione, 1993; Lamari et al., 1995; Olbe et al., 1995). Leukel (1948) was the first to suggest that *P. circinata* produced a toxin that was responsible for the symptoms observed. Scheffer and Pringle (1961) proved this and demonstrated that the pathogen produces a host-selective toxin that is responsible for disease in susceptible genotypes. The toxin causes tissue necrosis and electrolyte leakage from cells although these phenomena are not linked (Dunkle and Wolpert, 1981). Treatment of plant tissues with the pathogen toxin, including root tips, seedlings and leaves, results in many of the same symptoms and physiological responses caused by the pathogen itself including condensed heterochromatin and K^+ efflux from cells (Arias et al., 1983; Dunkle and Macko, 1995).

Now, the disease is considered only of 'minor importance' and only 'occasionally present' (Frederiksen, 2000). However, the disease was a major threat to sorghum production in the Great Plains of the United States in the 1930s. According to Heyne et al. (1944), resistant sorghum plants were observed in experimental plots that were infested with the pathogen and most other lines in the test showed symptoms. The resistant plants were selected to develop resistant varieties prior to the elucidation of the pathogen's disease-causing mechanism or the host's resistance mechanism. Through this simple selection method, milo disease has been controlled since the 1940s by a single gene (*pc*) that confers resistance to peritoxin (Dunkle and Macko, 1995). Schertz and Tai (1969) found that the gene for toxin susceptibility, *Pc*, mutates to *pc* in one of 8000 gametes, which helps to understand the ability to quickly find resistance through the selection of healthy plants in pathogen infested fields as Heyne et al. (1944) did. Because the mutation from *Pc* to *pc* is unidirectional (Nagy and Bennetzen, 2008), it is unlikely that *P. circinata* will overcome resistance. Even though milo disease is no longer a concern in sorghum production, the occurrence of a pathogen variant that could produce a modified form of peritoxin or could behave as a more aggressive root rot pathogen remains a possibility. Little work has been done to characterize isolate diversity in *P. circinata*.

Halim et al. (1973) showed that a trypsin inhibitor was found within Opaque-2 maize kernels that could retard the growth of *P. circinata* at concentrations of 200–400 mg mL^{-1}. The toxic effect of PC-toxin is inhibited by cycloheximide and cordycepin, inhibitors of protein and RNA synthesis, respectively (Wolpert and Dunkle, 1983). Further, Kiba et al. (2006) found that cycloheximide could shut down the apoptotic cell death caused by peritoxin and other host-selective toxins.

5 Acremonium wilt

Acremonium wilt causes desiccation of leaves and midrib discoloration where patches of leaf tissue become chlorotic or necrotic along the leaf axis. The apical meristem dies, leaving the basal meristem to produce abnormal tillers and the basal nodes continue to grow. However, the pathogen will infect new tillers as they are produced. Numerous genotypes are susceptible to tiller infection including 1790E, BTx398, BTx623, MR4 (4606), R5388, SC170-6-17, SC414-12, SC748-5 and Tx430. Stem cross sections reveal discoloured xylem vessels that are necrotic and plugged with pathogen structures and host exudate. Acremonium wilt can result in a 50% yield loss (Natural et al., 1982).

Acremonium wilt is caused by the fungal pathogen, *Sarocladium strictum* (W. Gams) Summerb. (syn. *Acremonium strictum* W. Gams) (Table 2). *S. strictum* is a pathogen of both monocot and dicot hosts as it has been associated with a wilt in Shasta daisy (Chase and Munnecke, 1980) and is a seedborne fungus in soybean (C.R. Little, unpublished data). The pathogen has been isolated from sorghum stalks, sheaths, and leaves. Root injury tends to increase the occurrence of the disease.

Natural et al. (1982) found that GPR-148, SC599-11E and SC630-11E were resistant to Acremonium wilt in trials conducted in College Station, Texas, in 1980 and 1981. Since that time, work on Acremonium wilt has been limited.

6 Root rot complex

The first symptoms of sorghum root rot, regardless of cause, are often observed above-ground. The aerial portions of the plant appear stunted, chlorotic, and exhibit low or no yield. Fusarium root rots often cause small patches of two or three symptomatic plants to appear. However, charcoal rot will affect a wider swath of plants surrounded by healthy plants in a field (Jim Long, *personal communication*). Root rots are often contiguous with seedling diseases, stalk rots, head blight and grain mould and weathering. Therefore, these diseases should be considered in context with their stalk rot counterparts.

Inspection of underground portions of the plant reveals root systems that may be pulled up easily due to death of primary roots, breakage of lateral roots, and lack of root hairs (Zummo, 1984; McLaren, 2002). The root cortex is discoloured and necrotic. Root discoloration depends upon the pathogen causing disease. For example, *Fusarium* spp. cause the production of circular spots and streaking on the root surface, the colour of which will depend upon plant pigmentation genes (Claflin, 2000b). *Pythium* spp. cause root necrosis with sunken lesions that appear necrotic and are less characterized by pigmentation than are *Fusarium*-induced lesions (Odvody, 2000). *Rhizoctonia* spp. produce necrotic, sunken, cankerous lesions that girdle the root cortex.

The root rot complex in sorghum is caused by numerous soilborne fungi and fungus -like organisms including *Fusarium* spp., *Macrophomina phaseolina*, *Rhizoctonia* spp., *Colletotrichum sublineola* Henn. ex Sacc. & Trotter, *Periconia circinata*, *Sarocladium strictum* and *Pythium* spp. (Sprague, 1950; Mughogho, 1984; Bandara et al., 2016) (Table 2). Hyphae, spores and sclerotia that reside and survive in soil or crop residues germinate in the presence of actively growing host roots or germinating seeds (Waniska et al., 2002; Tesso et al., 2012). Claflin (2000b) indicated that *Fusarium thapsinum* conidia could survive for six months at −16°C. Further, this author also suggested that *Fusarium* spp. could survive on maize stubble for up to 2 years. Plant roots produce exudates that have been shown to be stimulatory to pathogen germination (Idris et al., 2008). In general, pathogens enter the host through wounds and natural openings (Claflin, 2000b). Therefore, most root rot pathogens must infect or bypass cortical layers of the root and move into the vascular tissues (Zummo, 1984). Claflin (2000b) suggested that *Fusarium* was not responsible for direct pith and vascular bundle degeneration and that other fungi colonizing the stalk and root tissue were responsible. However, this is a hypothesis that requires further testing. Nevertheless, the result is a weak lower stalk, crown and root system that is prone to lodging (Trimboli, 1983). In addition to lodging, damage to the root system results in reduced water absorption, decreased nutrient uptake and decreased drought tolerance (Zummo, 1984; Claflin, 2000b).

Environmental factors that favour a pathogen during a developmental period of the host as well as abiotic stressors that predispose the host to infection will dictate the pathogen species that is most likely to infect (Mughogho and Pande, 1984). Like any soilborne disease, any environmental or human process that moves soil in any amount can move the pathogens. Such mechanisms include irrigation and rainfall, wind-blown dry soil particles, agricultural equipment and other human activity in fields. However, the pathogens associated with the root rot complex are generally considered to be ubiquitous, so dispersal and dissemination are not viewed as critical.

Fusarium species include *F. acuminatum* Ellis & Everh., *F. andiyazi*, *F. oxysporum* Schltdl., *F. thapsinum* and *F. verticillioides* (Sprague, 1950; Tesso et al., 2010; Palmero et al., 2012). Giorda et al. (1995) indicated that infection of sorghum roots by *Fusarium* spp. could occur quiescently during vegetative development. McLaren (1987) showed that *F. oxysporum* caused discoloration on sorghum seedling mesocotyls and roots and the degree of symptom development depended upon the isolate used.

Several species of *Pythium* are associated with sorghum roots, cause root necrosis and rot, and include *P. arrhenomanes*, *Globisporangium debaryanum* (R. Hesse)

Uzuhashi, Tojo & Kakish (syn. *Pythium debaryanum* R. Hesse), *P. graminicola* Subraman., *P. myriotylum* Drechsler., *P. periplocum* Drechsler. and *P. ultimum* (Sprague, 1950; Odvody, 2000; Davis and Bockus, 2001; McLean and Lawrence, 2001). *P. graminicola* was the causal agent for a root rot epidemic in the High Plains of Texas (USA) in the early 1970s (Odvody, 2000). Prior damage to sorghum roots supports colonization by *P. myriotylum* and *P. periplocum*, but these species may be more opportunistic than pathogenic.

In addition to charcoal rot, *M. phaseolina* causes root rot of sorghum (Mughogho and Pande, 1984). As in the stalk phase of the disease, charcoal rot affects roots after stress caused by high temperatures and drought (Jordan et al., 1984). Seedling infection occurs under high temperature and moisture conditions. And, as mentioned above, *M. phaseolina* may cause seedling blight and damping-off symptoms (Mughogho and Pande, 1984). Microsclerotia imbedded in host residues are the primary over-seasoning structures for this fungus. When seedlings germinate, and root exudates are produced, microsclerotia are stimulated to germinate. Multiple hyphae are produced that grow through the soil and infect root cortical layers.

The *Rhizoctonia solani* J.G. Kühn anastomosis group attacking sorghum is AG-1 (Jones and Belmar, 1989). Isakeit et al. (2007) indicated that *Exserohilum pedicellatum* (A.W. Henry) K.J. Leonard & Suggs, which caused root rot in Texas corn, is also pathogenic to sorghum.

Sorghum root rots are commonly present in all sorghum-growing regions and can be a yield-limiting constraint to production. Like stalk rots, adverse environmental factors that stress plants help to promote these diseases. McLaren (2002) found that genotype, environment and the genotype × environment interaction accounted for a 15%, 70% and 9% effect on root rot severity, respectively, showing that environment was the primary factor influencing development of root rot. Lower temperatures favour *Fusarium* infection, whereas high temperatures favour root rots caused by *M. phaseolina* (Mughogho and Pande, 1984; Mihail, 1989). Further, narrow plant row width and higher plant population result in a denser canopy, which can cool the soil and reduce charcoal rot disease severity. Plants grown under suboptimal conditions, for example experience drought stress during flowering, and experience increases in both Fusarium stalk rot and charcoal rot (Trimboli, 1983; Claflin, 2000b). Although continuous drought conditions favour charcoal rot, a period of cool and wet weather after flowering will often favour *Fusarium* if a period of stress occurred before or at flowering (Mihail, 1989). *Pythium* spp. cause decreased seed germination and emergence and reduced root growth in cool, wet soils. Under irrigation, warm and wet soils are also conducive to *Pythium* disease (Forbes et al., 1986; Odvody, 2000).

Soil tillage properties are affected by residue decomposition, crop type, residue amount, residue depth, allelopathic interactions and time (Bailey and Lazarovits, 2003). These factors also influence pathogen inoculum potential and survival. Although low tillage practices result in greater levels of residue-borne inoculum, such practices increase water uptake and alter pH and microbial communities (Mannering and Fenster, 1983). Reduced tillage also decreases soil crusting, water runoff and wind and water erosion. These factors help sorghum avoid root rots through improved stand establishment, improved stress tolerance and a deeper soil profile with improved insulative properties and water retention. Additionally, the incorporation of organic matter into soils also improves soil moisture and decreases soil temperatures (Tilander and Bonzi, 1997).

Both factors reduce stress and reduce *M. phaseolina* infection (Mughogho and Pande, 1984). Mihail (1989) found that mulching increased moisture as depth increased, which reduced drought stress and charcoal rot disease severity.

McLaren (2004) found that low pH values (i.e. <5) were associated with discoloration of the primary roots. Claflin (2000b) indicated that Fusarium root rot severity was lower when potassium and nitrogen were balanced, whereas high nitrogen to potassium ratios resulted in increased disease.

Resistance to root rots is a challenge because of the numerous pathogens involved. It is likely that a complete catalogue of root rotting fungi or fungus-like organisms has not yet been assembled. And, it is also likely that multiple pathogens and secondary root colonizers act simultaneously or synergistically. However, the extent of such interactions has not been fully explored. Thus, separate sources of resistance for *Fusarium* spp., *M. phaseolina* and *Pythium* spp. are required. Current sources of resistance to Fusarium stalk rot and charcoal rot, in addition to plants with the stay-green trait, should provide sufficient protection against these pathogens. Further efforts to develop multi-adversity resistance to sorghum root and stalk rots should be pursued.

Different defence mechanisms that may be useful for resistance to root rot pathogens have been found. Both Anahosur and Naik (1985) and Patil et al. (1985) indicated that 2 to 3× root sugars and phenols occurred in genotypes that were resistant to *M. phaseolina*. Huang and Backhouse (2006) showed that *F. thapsinum* and *F. proliferatum* induced chitinases in sorghum seedling roots. Earlier, Kumari and Chandrashekar (1994) showed that three sorghum chitinases affected growth of *Fusarium* through cell and cell wall disruption.

Several cultural management strategies are effective at controlling root rot diseases in sorghum (Table 3). Managing fertilizer rates, plant populations, weeds and insect predation can reduce plant stress, which will in turn reduce root rot severity (Claflin, 2000b). Planting date may be modulated to avoid stress and to ensure that plants avoid the most vulnerable period for root infection (approx. 40–60 days after planting) and escape the most conducive period for root rots (McLaren, 2004).

As for seedling diseases and stalk rots, crop rotation is an essential step for long-term management of root rots through the reduction of pathogen soil populations. Monocropping leads to the development of host-specific microbial communities, including plant pathogens, that lead to cumulative yield loss or 'drag' over time (Janvier et al., 2007). There is some evidence to suggest that variation exists between isolates of *M. phaseolina*, for example, which suggests that crop rotation could be an effective strategy for charcoal rot, even though the pathogen has a well-known wide host range (Saleh et al., 2010).

Fungicide seed treatments are effective for seed- and soilborne fungi but have not been historically effective for disease control in adult plant root systems (Williams and Nickel, 1984). For example, Idris et al. (2007) experimented with benomyl applications in Ethiopia for Fusarium root rot and were not successful. A year later, Idris et al. (2008) indicated that high doses of metalaxyl had low efficacy for Pythium root rot in the same region. They postulated that *Pythium* might not have been the chief cause of root rot or that local populations of that pathogen were resistant to metalaxyl. However, an additional possibility is that delivery of the chemical (i.e. soil drench) in a manner that would be effective for root rots was not achieved. This latter issue is a key constraint for management of root rot diseases using chemical controls.

7 Bacterial stripe

Bacterial stripe is characterized by lesions that run along the veins of the leaf (Akhtar, 1984). As in other foliar diseases, lesion pigmentation will reflect the colour genotype of the plant (i.e. tan, red and purple) (Akhtar, 1985b). Bacterial stripe lesions may be a foot and a half long, or longer, depending on the susceptibility of the host (Knott et al., 2016; Williams et al., 1978). The lesions do not cross veins and thus appear parallel or delimited along the long axis of the leaf. Unlike bacterial streak, lesions do not exhibit watersoaking symptoms and do not become irregular (Claflin, 2000a). Droplets of bacterial ooze accumulate within the lesions and become white and scaly when dried (Akhtar, 1984).

Bacterial stripe is caused by *Burkholderia andropogonis* (syn. *Pseudomonas andropogonis*) (Fig. 2). This bacterial pathogen has a wide host range other than sorghum including betel palm, bird-of-paradise, blueberry, *Bougainvillea*, carnation, citrus, corn and jojoba (Kobayashi et al., 1995; Cother et al., 2004; Duan et al., 2009; Tang et al., 2013; Sugiyama et al., 2017). Recent taxonomic studies of the sorghum isolates separate them from the *B. andropogonis* species complex based on 16S rRNA gene sequences and multilocus analysis and suggest placing these isolates into a new genus *Robbsia* (Lopes-Santos et al., 2017). A draft genome of this pathogen was recently published by Lopes-Santos et al. (2015).

The pathogen is seed- and residue-borne. Akhtar (1984) found that 48 of 420 infected (or infested) seeds produced bacterial stripe after planting. The disease is prevalent during cool periods with high relative humidity (Moffet et al., 1986; Claflin et al., 1992).

Figure 2 Bacterial stripe (caused by *Burkholderia andropogonis*) symptoms (a) on lower leaves in the field. Image courtesy of J. Gager. A close-up of bacterial stripe symptoms on a detached leaf (b). Image by C. R. Little.

Akhtar (1985a) showed that bacterial stripe disease incidence increased under moderate temperatures (20–26°C) and high relative humidity (>90%), whereas disease incidence decreased under high temperatures (24–36°C) and low relative humidity (16–22%). Bacterial exudate crusts are easily rehydrated. Thus, *P. andropogonis* is transmitted by rainsplash or heavy amounts of canopy dew. Ngugi et al. (2002) indicated that bacterial stripe and streak occurred in 39% and 32% of fields surveyed in western Kenya in 1995 and 1996, respectively.

Although bacterial stripe is the most commonly observed bacterial disease of sorghum, this disease has not historically caused significant yield losses (Kennedy, 1990). Murbithi and Claflin (1997) identified cultivars resistant to bacterial leaf stripe and bacterial leaf streak.

8 Bacterial streak

Bacterial streak lesions begin as watersoaked regions on the leaf. Unlike bacterial stripe, bacterial streak lesion may become more irregular and necrotic in the centre (Knott et al., 2016; Claflin, 2000a). However, the lesions will take on the colour of the plant genotype as do bacterial stripe lesions (i.e. red and purple). Like bacterial stripe, bacterial streak lesions produce exudates and form white crusts on lesions. However, this does not occur to the same extent in bacterial streak as in bacterial stripe.

Bacterial streak is caused by *Xanthomonas campestris* pv. *holcicola* (Pammel) Dowson (Schaad et al., 2000). Like *B. andropogonis*, *X.c. holcicola* is seedborne. Also, like bacterial stripe, no management is warranted. However, germplasm screening methods for bacterial leaf streak using ground disease leaf tissue application in developing whorls and inoculum spray applications have been refined in Ethiopia by Temam (2001).

9 Virus diseases of sorghum

Numerous virus species have been found in sorghum (Table 4). Not all are equally important, and most do not cause sufficient damage to warrant control.

9.1 Maize dwarf mosaic virus (and related viruses)

MDMV symptoms most often occur on the upper, youngest leaf tissue. Mosaic and mottling symptoms are characterized by interspersed regions of dark and light green tissue (Fig. 3). As with any virus mosaic disease, a piece of symptomatic leaf tissue held up to the light will reveal mosaic and mottling symptoms. In severe cases, streaking and chlorosis can occur in the youngest leaf tissue (i.e. 'Rio reaction') (Jensen, 2002). Early infections can result in stunting, delayed flowering, reduced seed set and death of the plant. Red leaf symptoms occur in some hybrids after extended periods of cool night time temperatures. MDMV-A and sugarcane mosaic virus (SCMV-MDB; syn. MDMV-B) appear to cause red leaf symptoms at different temperatures, that is <60°F and <70°F, respectively (Seifers et al., 2012). The red leaf symptoms are characterized by elongated striping, red pigment accumulation throughout and around the lesions, and necrosis in the central areas of the lesion.

Table 4 Virus species associated with *Sorghum bicolor*

Genus (family)	Virus species	Vector(s)	Genome[a]	Citation(s)
Fijivirus (Reoviridae)	Fiji disease virus (FDV)	Plant hopper	dsRNA	Jensen and Giorda (2002)
	Maize rough dwarf virus (MRDV)	Delphacid plant hoppers	dsRNA	Jensen and Giorda (2002)
	Mal de Rio Cuarto virus (MRCV)	Plant hopper	dsRNA	Jensen and Giorda (2002)
	Rice black-streaked dwarf virus (RBSDV)		dsRNA	Wu et al. (2013)
	Southern rice black-streaked dwarf virus (SRBSDV)		dsRNA	Wu et al. (2013)
Furovirus (Virgaviridae)	Sorghum chlorotic spot virus (SgCSV)	*Polymyxa graminis*	+ssRNA	Jensen and Giorda (2002)
Machlomovirus (Tombusviridae)	Maize chlorotic mottle virus (MCMV)	Beetle	+ssRNA	Huang et al. (2016) and Achon et al. (2017)
Mastrevirus (Geminiviridae)	Maize streak virus (MSV)	*Cicadulina* spp. leafhopper	ssDNA	Jensen and Giorda (2002)
Polerovirus (Luteoviridae)	Sugarcane yellow leaf virus (SYLV)	*Rhopalosiphum maidis, R. rufiabdominale, Ceratovacuna lanigera, Melanaphis sacchari*		Wei et al. (2016)
Pecluvirus (Virgaviridae)	Indian peanut clump virus (PCV)	*P. graminis* soilborne vector	+ssRNA	Reddy et al. (1998) and Delfosse et al. (2000)
Panicovirus (put.) (Tombusviridae)	Maize mild mottle virus (MMMV)	Unidentified	+ssRNA	Mandadi et al. (2014) and Pyle (2015)
Potexvirus (Alphaflexiviridae)	Foxtail mosaic virus (FoMV)	Unidentified	+ssRNA	Mandadi et al. (2014) and Pyle (2015)
Potyvirus (Potyviridae)	Johnsongrass mosaic virus (JGMV)	*Aphis* spp., *Myzus persicae, R. maidis*	+ssRNA	Seifers et al. (2012)
	Maize dwarf mosaic virus (MDMV)[b]	*Schizaphis graminum, Rhopalosiphum maidis*	+ssRNA	Seifers et al. (2012) and Shah et al. (2015)
	Sorghum mosaic virus (SrMV)	Aphid	+ssRNA	Seifers et al. (2012)
	Sugarcane mosaic virus (SCMV-MDB)[c]	Aphid	+ssRNA	Seifers et al. (2012)

(Continued)

Table 4 (*Continued*)

Genus (family)	Virus species	Vector(s)	Genome[a]	Citation(s)
Nucleorhabdovirus (Rhabdoviridae)	Maize mosaic nucleorhabdovirus (MMV)	*Peregrinus maidis* plant hopper	−ssRNA	Jensen and Giorda (2002)
	Sorghum stunt mosaic virus (SSMV)	*Graminella sonora*	−ssRNA	Creamer et al. (1997)
Waikavirus (Secoviridae)	Maize chlorotic dwarf virus (MCDV)	*G. nigrifrons, G. sonora*	+ssRNA	Stewart et al. (2014)
Tenuivirus (Phenuiviridae)	Maize stripe virus (MSpV)	*P. maidis*	−ssRNA	Creamer et al. (1997)
	Rice stripe virus (RSV)	*Leafhopper transmitted*	−ssRNA	Wu et al. (2013)
Unassigned	Sorghum yellow banding virus (SYBV)	Unidentified	Not determined	Pyle (2015)

[a] ICTV Master Species List 2016 v1.1.
[b] Previously MDMV-A.
[c] Previously MDMV-B.
put. = putative.

'Small seed' syndrome is primarily caused by MDMV (Frederiksen, 2000). However, Fusarium head blight can cause similar seed shrinking. Underdeveloped seeds are low in size and density. Like head blight, dark lesions may develop on the peduncle and rachis branches. Prior to seed development, cellular necrosis of the floret attachment point to the rachilla may occur. This is a conduit made up of transfer tissue that normally allows photosynthates to be transported into the developing seed. Therefore, premature blackening of this tissue, analogous to the black layer formation that occurs at seed maturity, reduces grain filling and results in the small seed symptom observed.

MDMV occurs throughout all areas where sorghum is grown (Giorda, 2002; Jensen et al., 2002; Narayana et al., 2002; Ryley et al., 2002; Stack, 2002; Seifers et al., 2012).

Figure 3 Mosaic and mottling symptoms produced by maize dwarf mosaic virus on sorghum. Image courtesy of Kansas State University Libraries/L. E. Claflin.

Johnsongrass serves as a reservoir host for MDMV. Aphids, such as the greenbug (*Schizaphis graminum* Rondani) and corn leaf aphid (*Rhopalosiphum maidis* Fitch), that feed on Johnsongrass can non-persistently transmit the virus to sorghum plants (Seifers et al., 2012; Shah et al., 2015).

Virus diseases can be controlled through the following three methods: (1) removal of virus sources, (2) management of vector populations and (3) host plant resistance (Shah et al., 2015). Screening of a sorghum minicore collection revealed that IS 7679 and IS 20740 were resistant to SCMV-MDB, SrMV and Johnsongrass mosaic virus-N (JGMV-N; Nigerian strain), but susceptible to JGMV-KS (Kansas strain) and JGMV-Aus (Australia strain). IS 7679 was resistant to Zea mosaic virus (ZeMV) (Seifers et al., 2012).

10 Witchweed

Plants parasitized by witchweed appear chlorotic, wilted, stunted or spindly, and will have low or no yield (Doggett, 1965; Spallek et al., 2013). Weaker infestations may result in plants that exhibit only a chlorotic leaf blotch symptom. Rodenburg et al. (2008) found that sorghum plants parasitized with *Striga hermonthica* (Delile) Benth. exhibited reduced photosynthesis. Impacts on photosynthesis are correlated with increased abscisic acid (ABA) production after *S. hermonthica* infection (Frost et al., 1997). In addition to ABA, other hormones including cytokinins and gibberellins are altered in *Striga*-infected plants (Musselman, 1980; Taylor et al., 1996).

The species of *Striga* can be differentiated by flower colour. Although there is variation in this trait, *S. hermonthica* and *S. asiatica* (L.) Kuntze produce pink to red flowers, whereas *S. densiflora* Benth. produces white flowers (Spallek et al., 2013).

Hemiparasitic plants produce their own green leaf tissue for photosynthate production but rely on the host plant for water and minerals (Spallek et al., 2013). Witchweeds (*Striga hermonthica*, *S. asiatica*, and *S. densiflora*) are hemiparasitic flowering plants that grow and reproduce on sorghum, millet, rice, corn and wheat (Musselman, 1980; Vasey et al., 2005). However, Vasey et al. (2005) reported that *Striga* developed more slowly on wheat than on sorghum. Vasudeva-Rao and Musselman (1987) asserted that *Striga* co-evolved with sorghum and should have the closest relationship with this cereal crop. *S. hermonthica* occurs in Africa; *S. asiatica* in Africa, Asia and the United States; and *S. densiflora* occurs in India (Musselman, 1980; Spallek et al., 2013). Other *Striga* species that occur on sorghum include *S. aspera*, *S. euphrasioides* and *S. brachycalyx* (Musselman, 1980). *Striga* is among the greatest constraints for sorghum production in Africa and affects 8 million ha, approximately one-third of the area used for sorghum production (Maunder, 2002). Losses are especially heavy in Eastern Africa where they have been estimated to exceed 60% (Atera and Itoh, 2011).

Striga spp. seeds can survive in the soil for several years prior to germination (Atera and Itoh, 2011). Temperatures of 25–35°C condition the seeds for germination. The term 'dust seeds' has been used to describe the seeds of *Striga* spp. since they are minute and numerous (Berner et al., 1995; Atera and Itoh, 2011; Spallek et al., 2013). On average, a *Striga* seed is only 2 to 3× the width of a human hair. The germination of sorghum seedlings and production of new roots stimulate the germination of *Striga* seeds through the production of strigolactones. Sorghum produces (+)-sorgolactone (Cardoso et al., 2011). These root exudates are capable of stimulating seed germination in a wide array of *Striga* spp. including *S. gesnerioides*, which normally parasitizes legumes and other dicots

(Parker and Reid, 1979). However, other compounds have been shown to stimulate *Striga* germination including coumarins, dihydrosorgoleone, ethylene, jasmonic acid, kinetins and various sesquiterpenes (Cordosa et al., 2011).

The production of numerous seeds dispersed evenly through the soil profile increases the chance that a germinating *Striga* seed will attach to the host root via a haustorium that connects a 'xylem bridge' between the host and the *Striga* plant (Spallek et al., 2013). Hood et al. (1998) reported that *S. asiatica* penetrated and colonized the cortex of host roots within 48 h of seed germination. It is at this stage that many non-host interactions fail. For example, non-host plants for *S. hermonthica*, such as *Lotus japonicus* L. and *Arabidopsis thaliana* (L.) Heynh., only allow infection of the root cortex in the former case, but stele infection and development to the six-leaf pair occurs in the latter case before the *Striga* seedling dies (Yoshida and Shirasu, 2009). In sorghum, by 96 h post-inoculation, the endodermis was penetrated allowing 'vascular continuity' to occur. Once the *Striga* seedling has established a viable connection between itself and the host plant, it produces leaf and shoot primoridia that will develop into stems, leaves and flowers upon emergence through the soil near the base of the host plant. At this point, the *Striga* plant produces their own photosynthates (Spallek et al., 2013). Emergence occurs up to 2 months after planting and flowering occurs after another 2–4 weeks. After flowering, *Striga* produces capsules in which up to 50 000 to half a million seeds may be produced (Berner et al., 1997). When the capsules mature, they split and release seeds into the soil. Seed maturation takes about a month after flowering has occurred (Berner et al., 1995).

An integrated pest management (IPM) strategy composed of a combination of one or more approaches including host resistance, fertilization management, herbicide use and cultural controls is required to control *Striga* (Hess et al., 2002). Due to the significant losses experienced by the sorghum farmers in Africa, growers in West and Central Africa have long participated in on-farm *Striga* IPM trials (Marley et al., 2002).

Striga resistance is the best management strategy. In Ethiopia, two varieties were developed and released with *Striga* resistance (Maunder, 2002). Although complete resistance has not been achieved, several other groups have reported resistance in sorghum to *Striga* including Arnaud et al. (1999), Maiti et al. (1984) and Olivier et al. (1991). A practical approach to sorghum resistance has been the development of low *Striga* germination lines. This trait is controlled by a single recessive gene (*lgs*, low germination stimulant) (Satish et al., 2012). Although these lines are less tolerant to *Striga* when soilborne seed populations are very high, they have been shown to have good tolerance to *S. asiatica* and *S. hermonthica* (Atera and Itoh, 2011).

Low fertility soils, which are marginal by nature or that have been depleted by repeated monocropping, are associated with increased *Striga* infestation (Cochrane and Press, 1997; Berner et al., 1997). Thus, fertilization management has shown promise for *Striga* control. For example, Babiker (2002) showed that moderate urea applications (95 kg ha^{-1}) could reduce *Striga* plant weight and capsule numbers. High urea applications (190 kg ha^{-1}) reduced *Striga* densities and reduced plant weight and capsule numbers to a greater degree than the moderate treatment. Several herbicides have been tested for *Striga* control including 2,4-D, chlorsulfuron and dicamba (Last, 1961; Babiker, 2002).

Of the cultural controls, hand-weeding remains the most effective approach to reduce *Striga* populations, even though this approach is employed after infection and symptom development (Doggett, 1965). It is thought that this simple approach greatly reduces *Striga* incidence over a period of years (Spallek et al., 2013). Intercropping sorghum or maize with non-host plants that produce high levels of strigolactones or other seed germination stimulants has also been effective. This phenomenon was first observed in cotton (Cook et al., 1966) and has been demonstrated in *Desmodium*–corn and *Lablab*–sorghum intercropping studies (Babiker, 2002; Khan et al., 2006).

Several biocontrol strategies have been tested for *Striga* control including the use of inoculative, augmentative and inundative tactics (Hess et al., 2002). Inoculative biocontrol uses introduced natural enemies for control purposes. Augmentative biocontrol increases the numbers of native, well-adapted natural enemies for control. In the case of *Striga*, several fungi could work in this manner. For example, *Fusarium* spp. are commonly associated with *Striga* plants in West Africa (Hess et al., 2002). Numerous fungi have the potential to be used in inoculative and augmentative biocontrol including *Alternaria* spp., *Cercospora* spp., *Curvularia geniculata* (Tracy & Earle) Boedijn, *Fusarium* spp. [*F. equiseti* (Corda) Sacc., *F. oxysporum*, *F. roseum* Link and *F. solani* (Mart.) Sacc.], *Phoma* spp., *Globisporangium ultimum*, *Rhizoctonia solani* and *Sclerotium rolfsii* Sacc. Insects may be used in a similar manner and include *Eulocastra argentisparsa* Hampson, *Junonia orithya* L., *Ophiomyia strigalis* Spencer and *Smicronyx* spp. (Hess et al., 2002).

11 Nematode pathogens of sorghum

Sorghum is parasitized by numerous nematode species (Table 5); yet evidence of pathogenicity continues to be limited and, in some cases, contradictory. A general lack of diagnostic symptoms combined with frequently complex biotic and abiotic interactions has likely resulted in an underestimate of nematode losses, which in turn, has contributed to a relatively insignificant research focus on nematodes as pests of sorghum. The existing evidence for nematode parasitism and pathogenicity on sorghum has been summarized by Starr (1992) and Kollo (2002). Much of the evidence for nematode pathogenicity comes from greenhouse studies, the results of which are difficult to extrapolate to field conditions. Based on the Kollo's (2002) summary of pathogenicity data for 30 West African nematode species, sorghum does appear to be relatively intolerant of nematode parasitism compared to pearl millet, although total nematode losses to US sorghum production are estimated to be less than 1% (Koenning et al., 1999). Cultural practices, principally crop rotation, have historically provided the basis for nematode management (Sharma and McDonald, 1990).

Sorghum is frequently deployed in crop rotations, and as a cover crop, and its host status, independent of its sensitivity to nematode feeding, is a principal concern for any cropping systems approach to nematode management. While sorghum is a suitable host for many nematode species, its relative host status compared to other cereal crops is important. Sorghum is often a superior host compared to millet, for example, but a relatively poor host compared to maize (Kollo, 2002; McSorley and Gallaher, 1991; Talwana et al., 2008). Crop rotation with sorghum is a particularly effective strategy for managing root-knot nematode [*Meloidogyne incognita* (Kofoid & White) Chitwood] problems in cotton and soybean in the southeastern United States (Gallaher et al., 1991, Kirkpatrick and Thomas, 2007; McSorley and Gallaher, 1993; Widmer et al., 2002). Even in this case, however, generalizations are not without risk, as the suitability of sorghum as a host for *M. incognita* varies with both the crop and nematode genotype (McClure et al., 1998; McSorley and Gallaher, 1992).

Table 5 Nematode pathogens of sorghum

Common name	Genus and species	Symptoms and signs	Key management strategies	Reference(s)
Ectoparasites				
Stunt nematodes	*Tylenchorhynchus* spp. *Quinisulcius acutus*	Stunting Root necrosis	Rotation with poor hosts (e.g. cowpea, peanut, pearl millet) *Caution:* host ranges are species specific	Kollo (2002)
Stubby-root nematodes	*Paratrichodorus minor*	Cessation of root elongation and swollen root tips (stubby root)	Rotation with peanut and pearl millet Rotation with groundnut and cowpea (unidentified West African trichodorid)	Bado et al. (2011) and Timper and Hanna (2005)
Sting nematodes	*Belonolaimus longicaudatus* *Belonolaimus* spp.	Severe stunting with root tip necrosis	Rotation with non-hosts (e.g. alfalfa, asparagus, okra, tobacco and watermelon) *Caution:* host ranges are species specific	Robbins and Barker (1973) and Todd (1991)
Needle nematodes	*Longidorus africanus*	Severe stunting with swollen, discoloured root tips	Rotation with poor hosts (e.g. *Brassica* spp.)	Kolodge et al. (1987)
Migratory endoparasites				
Root-lesion nematodes	*Pratylenchus zeae* *Pratylenchus* spp.	Stunting with discoloured or necrotic root lesions	Rotation with poor hosts (e.g. peanut, cowpea, groundnut, sunflower and pearl millet) *Caution:* host ranges are species specific	Bado et al. (2011), Bolton and De Waele (1989), Kollo (2002) and Timper and Hanna (2005)
Sedentary endoparasites				
Root-knot nematodes	*Meloidogyne incognita* *M. naasi*	Root galls on lateral roots (*M. incognita*) Elongated terminal root galls (*M. naasi*)	Resistance Rotation with non-hosts (e.g. certain tobacco, peanut and cowpea cultivars) *Caution:* host ranges are both species- and host race specific	Adegbite et al. (2005)b, Fortnum and Currin (1988) and Hartman and Sasser (1985)
	Senegalonema sorghi	Root stunting	Rotation with non-hosts (e.g. cowpea and peanut)	Baujard and Martiny (1995)

11.1 Migratory ectoparasites

11.1.1 Stunt nematodes

Stunt nematodes, primarily *Tylenchorhynchus* spp. and *Quinisulcius acutus* (Allen) Siddiqi, are among the most prevalent and abundant taxa encountered in US sorghum fields and are responsible for ~40% of nematode losses in grain sorghum production (Koenning et al., 1999). Observed frequencies of occurrence for *Q. acutus* and *Tylenchorhynchus annulatus* (Cassidy) Golden in Mississippi, USA, for example, were 78% and 30%, respectively (Cuarezma-Teran et al., 1984). This family (Telotylenchidae) is probably common in African sorghum fields as well, although it was absent in surveys of sorghum fields in South Africa (De Waele and Jordan 1988) and Uganda (Talwana et al., 2008). Its absence in Uganda is unsurprising since only root samples were collected, and these would not contain ectoparasites. As many as 11 species of *Tylenchorhynchus* and related genera have been recorded from Senegal (Kollo, 2002).

Nematode feeding often results in nonspecific above-ground symptoms that, if noticeable, mimic drought and nutrient deficiencies, and are rarely diagnostic. When stunting is present, it frequently manifests as irregular patches. Below-ground damage due to stunt nematode feeding typically is no more descriptive, and can include stunting, root pruning and root discoloration or necrosis (Anderson and Potter, 1991). Roots are significantly more susceptible to stunting than foliage (Claflin, 1984; Cuarezma-Teran and Trevathan, 1985). Accurate diagnosis requires the collection of soil samples from the root zone of affected plants to determine the types and numbers of plant-parasitic nematodes present. Since absolute thresholds are poorly documented for most species, a comparative sample from neighbouring unaffected plants can aid diagnosis.

Stunt nematodes exhibit a typical life cycle for plant-parasitic nematodes, consisting of four juvenile stages and an adult stage, each separated by moulting of the collagenous cuticle to facilitate growth. Reproduction is either sexual, characterized by the presence of males, or parthenogenetic, characterized by the absence of males, and is species dependent. Development proceeds from eggs laid in the rhizosphere and typically requires 25–35 days for completion, depending on species and temperature (Anderson and Potter, 1991).

Feeding consists of migratory browsing on epidermal cells and root hairs, with varying effects on root growth and function (Anderson and Potter, 1991). Economic yield loss in sorghum generally requires the presence of large stunt nematode population densities. In nematicide trials in Kansas, for example, *T. annulatus* suppressed sorghum yields at initial population densities exceeding 1000 nematodes per 100 cm^3 soil (Hafez and Claflin, 1982), but not at initial populations averaging 400 nematodes per 100 cm^3 soil (Todd and Claflin, 1984). Linear yield loss models estimated a ~1% loss for each 500 nematodes per 100 cm^3 across environments (Todd, unpublished).

The use of nematicides for control of stunt nematodes in sorghum is unlikely to be economically viable in most cases. Nor has resistance to these pests been developed commercially, although differences in tolerance have been observed among sorghum accessions (Claflin, 1984). Crop rotation can be an effective management strategy wherever suitable alternative crops include poor hosts or non-hosts. Unfortunately, the diversity and polyphagous habit of stunt nematodes, combined with incomplete knowledge of species-specific host ranges, currently limit this option. Nonetheless, crops such as cowpea, peanut and pearl millet have been identified as relatively poor hosts compared to sorghum for West African populations of *Tylenchorhynchus* spp. (Kollo, 2002).

11.1.2 Stubby-root nematodes

Stubby-root nematodes, *Trichodorus* and *Paratrichodorus* spp. (Trichodoridae), are widely associated with the potential for sorghum injury in both the United States (McSorley and Gallaher, 1992; Smolik, 1977) and Africa (Kollo, 2002; De Waele and Jordan, 1988). Most reports of pathogenicity involve the species *Paratrichodorus minor* (Colbran) Siddiqi. Although widely distributed, this species does exhibit a preference for sandy soils (Decraemer, 1991).

Stunting, sometimes severe, is the primary symptom of stubby-root nematode parasitism. Root development is noticeably affected by a cessation of root elongation, resulting in the characteristic stubby-root appearance (Decraemer, 1991). Although more diagnostic than most nematode damage, soil sampling for the pathogen is still necessary to confirm the diagnosis.

The life cycle of *P. minor* is similar to that of other nematodes and has been reported to be completed in 16–22 days, depending on temperature (Decraemer, 1991). Reproduction in *P. minor* is parthenogenetic.

Trichodorid feeding is usually localized at the meristematic tissues of the root tip (Decraemer, 1991). Although feeding is restricted to surface tissues, the aggregation of feeding sites at the root tip and the subsequent effect on root growth can result in substantial damage. Available evidence suggests that sorghum is much more intolerant of feeding by *P. minor* than is pearl millet (Kollo, 2002). Damage relationships, however, have not been well characterized under field conditions.

More than 100 plant species have been reported to be hosts for *P. minor* (Decraemer, 1991), but a few crops have been identified for use in rotations with sorghum to effectively manage the nematode. In the United States, population densities of *P. minor* were significantly reduced on peanut and one of two pearl millet hybrids relative to cotton and maize (Timper and Hanna, 2005), while in West Africa, rotations with legumes (groundnut and cowpea) reduced population densities of an unidentified trichodorid compared to continuous sorghum (Bado et al., 2011). Variation in host–parasite relationships among and within nematode species, as well as among crop genotypes, precludes the endorsement of general recommendations regarding crop rotation.

Stubby-root nematodes are regularly reported to be sensitive to the effects of soil disturbance, but published results are inconclusive overall. Tillage is nonetheless often recommended as a management strategy.

11.2 Other ectoparasitic nematodes associated with sorghum

Sting nematodes (*Belonolaimus* spp.) are occasional pests of sorghum in the deep sandy soils of midwestern river systems and southern coastal plains in the United States. *Belonolaimus longicaudatus* Rau is most often implicated (Koenning et al., 1999), but species delineations remain to be fully characterized. These nematodes are limited in geographic distribution to soils with greater than 80% sand content but can cause 100% losses in localized areas. Populations are polyphagous, with most grass and vegetable crops known to be hosts. Alfalfa and tobacco have been identified as non-host crops (Holdeman, 1955; Todd, 1991), but host preferences are known to be species-, and even population-, specific (Robbins and Barker, 1973).

Sorghum is among the best hosts for a southern California, USA, population of *Longidorus africanus* Merny (Kolodge et al., 1987). This species is also common in fields of various crops in the Mediterranean region (Lamberti, 1969). Another species of needle nematode, *Longidorus pisi* Edward, Misra & Singh, was frequently recovered from South African sorghum fields (De Waele and Jordan, 1988). The pathogenic potential of needle nematodes is well documented for numerous crops, but damage relationships under field conditions remain to be characterized for sorghum.

Pratylenchus spp. are often the most prevalent plant-parasitic nematodes in sorghum fields in the United States (Cuarezma-Teran et al., 1984) and Africa (Bagayoko et al., 2000; De Waele and Jordan, 1988; Talwana et al., 2008; Umar and Chubado, 2008). Based on a survey of nematologists and extension personnel in the United States, this genus is responsible for ~30% of nematode losses in grain sorghum production (Koenning et al., 1999). Internationally, *Pratylenchus zeae* Filipjev was the principal species listed as the most important nematode problem in sorghum and pearl millet production across 20 countries (Sharma and McDonald, 1990), but at least eight additional *Pratylenchus* species have been observed to reproduce on one or both crops. As noted for other nematode taxa, variability in host and parasite genotypes, as well as experimental (environmental) conditions, often results in conflicting reports of pathogenicity.

Nonspecific root and foliar stunting is typical of root-lesion nematode parasitism, with the degree of stunting related to nematode population density and the environment. Additionally, as the common name of this group suggests, discoloured or necrotic root lesions may be present, although necrotic lesion development on sorghum was observed to be sparse for *P. zeae* (Motalaote et al., 1987). Diagnosis requires isolation of nematodes from the soil or, if a sorghum crop is present, from root tissue.

Pratylenchus spp. exhibit a typical life cycle for migratory plant-parasitic nematodes, with all stages outside of the egg (second-stage juvenile through adult) infective. Parthenogenesis is a common reproductive strategy, occurring in ~40% of species (Castillo and Vovlas, 2007). Males are common in the remaining species, with sexual reproduction inferred. Duration of the life cycle varies among species, hosts and environmental conditions (especially temperature), but averages 27–28 days at 24–30°C (Castillo and Vovlas, 2007).

Pratylenchus spp. invade their host roots and migrate through the cortex, leaving a path for often necrotic cells (Castillo and Vovlas, 2007). They also migrate between the root and soil environment, feeding ectoparasitically on epidermal cells as well as endoparasitically on cortical cells. Damage relationships have been described for numerous crops but are not well characterized for sorghum. Motalaote et al. (1987) reported *P. zeae*-induced suppression of sorghum growth under greenhouse conditions but found no detectable effect in field microplots. The authors concluded that sorghum is relatively insensitive to the nematode's feeding activities. Suppression of population densities of a West African species of *Pratylenchus* through crop rotation also failed to produce any detectable effects on sorghum yields (Bado et al., 2011).

Some variation in reproduction of *P. zeae* on different sorghum cultivars has been observed (Motalaote et al., 1987), but rotations with non-host crops probably represent the best strategy for managing root-lesion nematodes in sorghum. Peanut and cowpea have been characterized as non-hosts to West African populations of *P. zeae* (Kollo, 2002), and rotation with groundnut, but not cowpea, significantly reduced root population densities of an unidentified West African *Pratylenchus* sp. in a subsequent sorghum crop (Bado et al., 2011; Bagayoko et al., 2000). Sunflower hybrids were identified as poor to non-hosts for *P. zeae* in South Africa (Bolton and De Waele, 1989).

Sorghum itself is a poor-to-moderate host for several root-lesion nematode species, including *P. brachyurus* (Godfrey) Filipv. & Schuurm.-Stekh., *P. coffeae* (Zimmerman) Filipv. & Schuurm.-Stekh. and *P. penetrans* (Cobb) Filipv. & Schuurm.-Stekh. (Motalaote et al., 1987; Thies et al., 1995). Sorghum is particularly inferior to corn in host status for *P. brachyurus* and *P. scribneri* Steiner (Gallaher et al., 1991; McSorley and Gallaher, 1991). Both sorghum and pearl millet hybrids suppressed reproduction of *P. penetrans* compared to corn, but suppression was significantly greater for the pearl millet hybrids (Jagdale et al., 2000). Pearl millet hybrids are also poor hosts for *P. brachyurus* (Timper and Hanna, 2005).

11.3 Sedentary endo- and semi-endoparasites

11.3.1 Root-knot nematodes

Host–parasite relationships vary significantly among (host) races of *Meloidogyne* spp., resulting in conflicting reports for sorghum. Reproduction and pathogenicity on sorghum has been conclusively demonstrated for the cotton root-knot nematode *M. incognita* race 3 (Babatola and Idowu, 1990; Orr and Morey, 1978) and the barley root-knot nematode *M. naasi* Franklin race 5 (Ediz and Dickerson, 1976). *M. incognita* is widely distributed in tropical and subtropical zones, with race 3 identified throughout the Americas, as well as from Africa and India (Taylor et al., 1982). *M. naasi* race 5, in contrast, is known only from Kansas, USA (Fig. 4).

The formation of root galls is a characteristic feature of root-knot nematode parasitism. On sorghum, these manifest as elongate or discrete swellings on lateral (but not brace) roots, sometimes accompanied by root proliferation for *M. incognita* (Orr and Morey,

Figure 4 Root-knot nematode (*Meloidogyne naasi*) on sorghum roots. Image by T. C. Todd.

1978), and as elongate terminal galls for *M. naasi*. Diagnosis and species identification can be confirmed by the manual extraction of embedded adult females from root galls.

Infective second-stage juveniles of *M. incognita* invade sorghum roots and establish multinucleate feeding cells (giant cells) in the cortex and stele (Orr and Morey, 1978). All subsequent juvenile stages and the adult females are immobile and embedded in root gall tissue. The adult females are swollen and pear-shaped, while adult males are vermiform and leave the root after the final moult. *M. incognita* and *M. naasi* reproduce by obligate mitotic parthenogenesis and facultative meiotic parthenogenesis, respectively. The life cycle of *M. naasi* is completed on sorghum in 34–36 days at 26–30°C (Ediz and Dickerson, 1976).

Host resistance is often an effective management strategy for root-knot nematodes. Unfortunately, little difference in the responses of commercial sorghum cultivars has been observed for individual populations of *Meloidogyne* spp., although responses can vary greatly among nematode species and even populations within species. In a comprehensive study of sorghum cultivars and *Meloidogyne* species, ten grain sorghum cultivars were all found to be resistant to *M. arenaria* (Neal) Chitwood race 2, *M. incognita* race 3 and *M. javanica* populations from Georgia and South Carolina, USA (Fortnum and Currin, 1988), while in a separate study, three sorghum cultivars that were also resistant to South Carolina populations of *M. arenaria* race 2 were susceptible (i.e. good hosts) to a South Carolina population of *M. incognita* race 3 (Ibrahim et al., 1993). Six sorghum cultivars from Nigeria likewise displayed susceptible reactions to an unidentified race of *M. incognita* (Babatola and Idowu, 1990).

Suitable rotational crops that do not support reproduction of *M. incognita* race 3 are rare, but at least certain cultivars of tobacco and peanut are, by definition, non-hosts. Resistance to a Nigerian population of *M. incognita* race 2 has also been identified in a single cowpea cultivar (Adegbite et al., 2005b). Sorghum is reported to be susceptible to *M. incognita* race 2 in Nigeria (Adegbite et al., 2005a); so a resistant cowpea cultivar would be useful for managing root-knot nematode in sorghum as well as cowpea cropping systems.

Recommendations for including sorghum as a rotation crop to manage *M. incognita* vary with location and the relative susceptibility of other crops in the rotation. Sorghum has been reported to be a better choice than corn for managing *M. incognita* in Florida cropping systems (Gallaher et al., 1991; McSorley and Gallaher, 1991, 1993). Conversely, rotation of sorghum with cotton should be avoided in the presence of this nematode (McClure et al., 1998).

11.3.2 Other sedentary nematodes associated with sorghum

Senegalonema sorghi Germani, Luc & Baldwin is a rare species described from the roots of sorghum in Senegal that is closely related to the widespread genus *Rotylenchulus* (Germani et al., 1984). Sorghum and millet are hosts, and root growth suppressions have been observed for both crops under greenhouse conditions (Baujard and Martiny, 1995). No reproduction or pathogenicity has been reported for cowpea or peanut.

Two species of cyst nematodes have been reported as parasites of sorghum: the sorghum cyst nematode *Heterodera sorghi* Jain, Sethi, Swarup & Srivastava from India and Pakistan, and the closely related species *H. gambiensis* Merny & Netscher from Gambia, Niger and India (Sharma and McDonald, 1990). Limited information on their economic impact is available.

12 Conclusions and future trends

Although progress has been made, identification of qualitative resistance to root and stalk rots has been a challenge (Bramel-Cox et al., 1988; Tenkouano et al., 1993; McLaren, 2002; Tesso et al., 2004, 2005, 2010; Adeyanju et al., 2015). This has been due to both the diversity of pathogens that cause these diseases, and the environmentally mediated and quantitative nature of resistance to these diseases. The complete diversity of fungal pathogens that cause disease on sorghum roots, and are saprophytically, endophytically and pathogenically associated with sorghum stalks, has not been fully explored, but has been recently become more clear (Petrovic et al., 2009; Kelly et al., 2017). The integration of drought tolerance through the use of the stay-green trait has been a stable source of physiological resistance to stalk rots (Tesso et al., 2012; Bandara et al., 2016). However, the impact of stay-green on root rot resistance is less understood. Non-stay-green-type resistance mechanisms to stalk and root rots require much further exploration. Resistance to *M. phaseolina*, *F. thapsinum* and other root- and stalk-associated pathogens is hypothesized to be somewhat different since these pathogens have different pathogenesis lifestyles (Tesso et al., 2012). For example, *M. phaseolina* is a necrotroph (Islam et al., 2012; Bandara et al., 2018) and *F. thapsinum* is thought to be a hemibiotroph (C.R. Little, unpublished data).

In general, little is known of the full diversity of bacterial organisms associated with sorghum and related grasses. Given the recent outbreak of bacterial leaf streak of corn caused by *Xanthomonas vasicola* in the United States (Korus et al., 2017), it is possible that new bacterial pathogens of sorghum could emerge. Therefore, the catalogue of potential bacterial pathogens should be screened for pathogenicity against a common set of sorghum germplasm. And, a smaller set of pathogenic strains should be tested against larger sets of relevant germplasm (i.e. hybrids) that are predominantly grown in regions with periods of high rainfall and relative humidity, for example the southeast United States.

Because so many virus species may co-occur in sorghum, detailed species and strain characterization is required. Although the interest to study the host range and diversity of viruses associated with sorghum has gained some traction over the last two decades (Seifers et al., 1999, 2012; Garrido and Trujillo, 2001; Stewart et al., 2014; Wei et al., 2016; Achon et al., 2017; Huang et al., 2016), the genetics of virus resistance in sorghum requires further effort. In areas of the world where viruses are problematic for the crop, understanding vectors, reservoir hosts and detection methodologies remains a priority.

Striga continues to be a major constraint for sorghum production in marginal environments where it is still responsible for major yield losses (Atera and Itoh, 2011). Efforts to discover various modes of genetic resistance and tolerance to this parasitic plant should continue (Olivier et al., 1991; Arnaud et al., 1999; Rodenburg et al., 2008; Yoshida and Shirasu, 2009; Satish et al., 2012). More work is needed for the discovery of effective, novel bioherbicides and seed treatments that can reduce colonization of host roots by *Striga* seedlings (Hess et al., 2002). Integrated management strategies that improve soil nutritional conditions, employ resistance or tolerance, reduce seed numbers or viability and protect the host plant will help control this scourge (Berner et al., 1995; Babiker, 2002; Khan et al., 2006).

As with the other pathogen groups, there are numerous nematode species that colonize sorghum (Claflin, 1984; Cuarezma-Terran and Trevathan, 1985; Fortnum and Currin, 1988; Babatola and Idowu, 1990; Sharma and McDonald, 1990; Gallaher et al., 1991;

Starr, 1992; McClure et al., 1998; Kollo et al., 2002). A priority is to further determine the regionally impactive nematode problems of sorghum and determine the secondary fungal pathogens that interact with these to cause detrimental effects from 'root nibbling', to yield 'drag', to serious yield losses. High-throughput screening of germplasm should be conducted to find resistance to lesion nematode and determine if races of this pathogen exist in the field for sorghum.

13 Acknowledgements

This chapter is Contribution No. 18-B-409 from the Kansas Agricultural Experiment Station, Manhattan, KS, USA.

14 References

Achon, M. A., Serrano, L., Clemente-Orta, G. and Sossai, S. 2017. First report of maize chlorotic mottle virus on a perennial host, *Sorghum halepense*, and maize in Spain. *Plant Disease* 101: 393.

Adegbite, A. A., Adesiyan, S. O., Agbaje, G. O. and Omoloye, A. A. 2005a. Host suitability of crops under yam intercrop to root-knot nematode (*Meloidogyne incognita* Race 2) in South-Western Nigeria. *Journal of Agriculture and Rural Development in the Tropics and Subtropics* 106: 113–18.

Adegbite, A. A., Amusa, N. A., Agbaje G. O. and Taiwo, L. B. 2005b. Screening of cowpea varieties for resistance to root-knot nematode under field conditions. *Nematropica* 35: 155–9.

Adeyanju, A., Little, C., Yu, J. and Tesso, T. 2015. Genome-wide association study on resistance to stalk rot diseases in grain sorghum. *G3: Genes, Genomes, Genetics* 5: 1165–75.

Akhtar, M. A. 1984. Bacterial stripe and bacterial leaf blight of sorghum: Effect of temperature and relative humidity on disease development, tolerance among sorghum hybrids, and overwinter survival of the incitant bacteria. M.S. thesis, Kansas State University, Manhattan, KS, 24p.

Akhtar, M. A. 1985a. Effect of temperature and relative humidity on bacterial stripe and bacterial leaf blight of sorghum. *Pakistan Journal of Agricultural Research* 6: 216–17.

Akhtar, M. A. 1985b. Genetic susceptibility of sorghum hybrids to bacterial stripe. *Pakistan Journal of Agricultural Research* 6: 122–4.

Anahosur, K. H. and Naik, S. T. 1985. Relationship of sugars and phenols on root and stalk rot of sorghum with charcoal rot. *Indian Phytopathology* 38: 131–4.

Anderson, R. V. and Potter, J. W. 1991. Stunt nematodes: *Tylenchorhynchus*, *Merlinius*, and related genera, pp. 529–86. In: *Manual of Agricultural Nematology* (Nickle, W. R., Ed.). Marcel Dekker, New York, NY.

Arias, J. A., Dunkle, L. D. and Bracker, C. E. 1983. Ultrastructural and developmental alterations induced by *Periconia circinata* toxin in the root tip of sorghum. *Canadian Journal of Botany* 61: 1491–505.

Arnaud, M. C., Veronesi, C. and Thalouarn, P. 1999. Physiology and histology of resistance to *Striga hermonthica* in *Sorghum bicolor* var. Framida. *Australian Journal of Plant Physiology* 26: 63–70.

Atera, E. and Itoh, K. 2011. Evaluation and ecologies and severity of *Striga* weed on rice in Sub-Saharan Africa. *Agriculture and Biology Journal of North America* 2: 752–60.

Babatola, O. and Idowu, K. A. 1990. Responses of some sorghum cultivars to the root-knot nematode *Meloidogyne incognita* in Nigeria. *Nematologica Mediterreneae* 18: 65–6.

Babiker, A. G. T. 2002. *Striga* control in Sudan: An integrated approach, pp. 159–63. In: *Sorghum and Millets Diseases* (Leslie, J. F., Ed.). Iowa State University Press, Ames, IA.

Bacon, C. W., Yates, I. E., Hinton, D. M. and Meredith, F. 2001. Biological control of *Fusarium moniliforme* in maize. *Environmental Health Perspectives* 109: 325–32.
Bado, V., Sawadogo, A., Thio, B., Bationo, A., Traoré, K. and Cescas, M. 2011. Nematode infestation and N-effect of legumes on soil and crop yields in legume-sorghum rotations. *Agricultural Sciences* 2: 49–55.
Bagayoko, M., Buerkert, A., Lung, G., Bationo, A. and Römheld, V. 2000. Cereal/legume rotation effects on cereal growth in Sudano-Sahelian West Africa: Soil mineral nitrogen, mycorrhizae and nematodes. *Plant and Soil* 218: 103–16.
Bailey, K. L. and Lazarovits, G. 2003. Suppressing soilborne diseases with residue management and organic amendments. *Soil & Tillage Research* 72: 169–80.
Bandara, Y. M. A. Y., Weerasooriya, D. K., Tesso, T. T. and Little, C. R. 2016. Stalk rot fungi affect leaf greenness (SPAD) of grain sorghum in a genotype- and growth stage-specific manner. *Plant Disease* 100: 2062–8.
Bandara, Y. M. A. Y., Weerasooriya, D. K., Liu, S. and Little, C. R. 2018. The necrotrophic fungus *Macrophomina phaseolina* promotes charcoal rot susceptibility through induced host cell wall-degrading enzymes. *Phytopathology* 108. doi:10.1094/PHYTO-12-17-0404-R.
Baujard, P. and Martiny, B. 1995. Ecology and pathogenicity of the Hoplolaimidae (Nemata) from the sahelian zone of West Africa: 8. *Senegalonema sorghi* Germani, Luc & Baldwin, 1984 and comparison with *Rotylenchulus reniformis* Linford & Oliveira, 1940. *Fundamental and Applied Nematology* 18: 513–22.
Bell, D. S., Harris, H. and Wells, H. D. 1973. Rhizoctonia blight of grain sorghum foliage. *Plant Disease Reporter* 57: 549–50.
Berner, D. K., Kling, J. G. and Singh, B. B. 1995. *Striga* research and control – A perspective from Africa. *Plant Disease* 79: 652–60.
Berner, D. K., Winslow, M. D., Cardwell, K. F., Raj, D. R. M. and Kim, S. K. 1997. *Striga Research Methods – A Manual*. International Institute of Tropical Agriculture, Ibadan, Nigeria.
Bolton, C. and De Waele, D. 1989. Host suitability of commercial sunflower hybrids to *Pratylenchus zeae*. *Supplemental Journal of Nematology* 21: 682–5.
Bramel-Cox, P. J., Stein, I. S., Rodgers, D. M. and Claflin, L. E. 1988. Inheritance of resistance to *Macrophomina phaseolina* (Tassi) Goid and *Fusarium monliforme* Sheldon in sorghum. *Crop Science* 28: 37–40.
Casela, C. R., Ferreira, A. S., Santos, F. G. and Guimarães, F. B. 2002. Sorghum diseases in Brazil, pp. 379–82. In: *Sorghum and Millets Diseases* (Leslie, J. F., Ed.). Iowa State Press, Ames, IA.
Castillo, P. and Vovlas, N. 2007. *Pratylenchus (Nematoda: Pratylenchidae): Diagnosis, Biology, Pathogenicity and Management. Nematology Monographs and Perspectives Volume 6* (Hunt, D. J. and Perry, R. N., Eds). Brill, Lieden, the Netherlands.
Chase, A. R. and Munnecke, D. E. 1980. Shasta daisy vascular wilt incited by *Acremonium strictum*. *Phytopathology* 70: 834–8.
Claflin, L. E. 1984. Plant-parasitic nematodes affecting sorghum, pp. 53–8. In: *Sorghum Root and Stalk Rots: A Critical Review* (Mughogho, L. K. and Rosenberg, G., Eds). International Crops Research Institute for the Semi-Arid Tropics (ICRISAT), Patancheru, AP, India.
Claflin, L. E. 2000a. Bacterial leaf streak, p. 6. In: *Compendium of Sorghum Diseases* (Frederiksen, R. A. and Odvody, G. N., Eds). APS Press, St. Paul, MN.
Claflin, L. E. 2000b. Fusarium root and stalk rot, pp. 28–30. In: *Compendium of Sorghum Diseases* (Frederiksen, R. A. and Odvody, G. N., Eds). APS Press, St. Paul, MN.
Claflin, L. E. and Giorda, L. M. 2002. Stalk rots of sorghum, pp. 185–9. In: *Sorghum and Millet Diseases* (Leslie, J. F., Ed.). Iowa State Press, Ames, IA.
Claflin, L. E., Ramundo, B. A., Leach, J. E. and Qhobela, M. 1992. Bacterial diseases of sorghum, pp. 135–51. In: *Sorghum and Millets Diseases: A Second World Review* (De Milliano, W. A. J., Frederiksen, R. A. and Bengston, A. D., Eds). ICRISAT, Patancheru, AP, India.
Cochrane, V. and Press, M. C. 1997. Geographical distribution and aspects of the ecology of the hemiparasitic angiosperm *Striga asiatica* (L) Kuntze: A herbarium study. *Journal of Tropical Ecology* 13: 371–80.

Cook, C. E., Whichard, L. P., Turner, B. and Wall, M. E. 1966. Germination of witchweed (*Striga lutea* Lour) – isolation and properties of a potent stimulator. *Science* 154: 1189–90.
Cardoso, C., Ruyter-Spira, C. and Bouwmeester, H. J. 2011. Strigolactones and root infestation by plant-parasitic *Striga, Orobanche* and *Phelipanche* spp. *Plant Science* 180: 414–20.
Cother, E. J., Noble, D., Peters, B. J., Albiston, A. and Ash, G. J. 2004. A new bacterial disease of jojoba (*Simmondsia chinensis*) caused by *Burkholderia andropogonis. Plant Pathology* 53: 129–35.
Creamer, R., He, X. and Styer, W. E. 1997. Transmission of sorghum stunt mosaic rhabdovirus by the leafhopper vector, *Graminella sonora* (Homoptera: Cicadellidae). *Plant Disease* 81: 63–5.
Cuarezma-Teran, J. A. and Trevathan, L. E. 1985. Effects of *Pratylenchus zeae* and *Quinisulcius acutus* alone and in combination on sorghum. *Journal of Nematology* 17: 169–74.
Cuarezma-Teran, J. A., Trevathan, L. E. and Bost, S. C. 1984. Nematodes associated with sorghum in Mississippi. *Plant Disease* 68: 1083–5.
Davis, M. A. and Bockus, W. W. 2001. Evidence for a *Pythium* sp. as a chronic yield reducer in a continuous grain sorghum field. *Plant Disease* 85: 780–4.
Decraemer, W. 1991. Stubby root and virus vector nematodes: *Trichodorus, Paratrichodorus, Allotrichodorus*, and *Monotrichodorus*, pp. 587–625. In: *Manual of Agricultural Nematology* (Nickle, W. R., Ed.). Marcel Dekker, New York, NY.
Delfosse, P., Reddy, A. S., Legrève, A., Thirumala Devi, K., Abdurahman, M. D., Maraite, H. and Reddy, D. V. R. 2000. Serological methods for detection of *Polymyxa graminis*, an obligate root parasite and vector of plant viruses. *Phytopathology* 90: 537–45.
De Waele, D. and Jordan, M. 1988. Plant-parasitic nematodes on field crops in South Africa. 2. Sorghum. *Revue Nématologie* 11: 203–12.
Dodd, J. L. 1980. The photosynthetic stress translocation balance concept of sorghum stalk rot, pp. 300–5. In: *Proceedings of International Workshop of Sorghum Diseases*, 11–15 December 1978. Texas A&M University/ICRISAT Press, Hyderabad, India.
Doggett, H. 1965. *Striga hermonthica* on sorghum in East Africa. *Journal of Agricultural Science* 65: 183–93.
Duan, Y. P., Sun, X., Zhou, L. J., Gabriel, D. W., Benyon, L. S. and Gottwald, T. 2009. Bacterial brown leaf spot of citrus, a new disease caused by *Burkholderia andropogonis. Plant Disease* 93: 607–14.
Duncan, R. R. 1983. Anthracnose-*Fusarium* disease complex on sorghum in southeastern USA. *Sorghum Newsletter* 26: 121–2.
Dunkle, L. D. and Macko, V. 1995. Peritoxins and their effects on sorghum. *Canadian Journal of Botany* 73: 444–52.
Dunkle, L. D. and Wolpert, T. J. 1981. Independence of milo disease symptoms and electrolyte leakage induced by the host-specific toxin from *Periconia circinata. Physiological Plant Pathology* 18: 315–23.
Ediz, S. A. and Dickerson, O. J. 1976. Life cycle, pathogenicity, histopathology, and host range of race 5 of the barley root-knot nematode. *Journal of Nematology* 8: 228–32.
Edmunds, L. K. 1964. Combined relation of plant maturity, temperature and soil moisture to charcoal stalk rot development in grain sorghum. *Phytopathology* 54: 514–17.
Elliott, C., Melchers, L. E., Lefebvre, C. L. and Wagner, F. A. 1937. Pythium root rot of milo. *Journal of Agricultural Research* 54: 797–834.
Esele, J. P. 2002. Sorghum and pearl millet diseases in the Horn of Africa, pp. 383–7. In: *Sorghum and Millets Diseases* (Leslie, J. F., Ed.). Iowa State Press, Ames, IA.
Food & Agriculture Organization of the United Nations (FAO). 2017. Food and agricultural statistics database, FAOSTAT. Online: www.fao.org/faostat/en/#data/QC (accessed 10 November 2017).
Forbes, G. A., Collins, D. C., Odvody, G. N. and Frederiksen, R. A. 1985. A seedling epiphytotic of sorghum in South Texas caused by *Pythium arrhenomanes. Plant Disease* 69: 726.
Forbes, G. A., Odvody, G. N. and Terry, J. M. 1986. Seedling diseases, p. 78. In: *Compendium of Sorghum Diseases* (Frederiksen, R. A., Ed.). American Phytopathological Society, St. Paul, MN.
Fortnum, B. A. and Currin, R. E. 1988. Host suitability of grain sorghum cultivars to *Meloidogyne* spp. *Annals of Applied Nematology* 2: 61–4.

Frederiksen, R. A. 2000. Diseases and disease management in sorghum, pp. 497–533. In: *Sorghum – Origin, History, Technology and Production* (Smith, C. and Frederiksen, R. A., Eds). John Wiley & Sons, New York, NY.

Frost, D. L., Gurney, A. L., Press, M. C. and Scholes, J. D. 1997. *Striga hermonthica* reduces photosynthesis in sorghum: The importance of stomatal limitations and a potential role for ABA? *Plant, Cell and Environment* 20: 483–92.

Gallaher, R. N., McSorley, R. and Dickson, D. W. 1991. Nematode densities associated with corn and sorghum cropping systems in Florida. *Supplemental Journal of Nematology* 23: 668–72.

Garrido, M. J. and Trujillo, G. E. 2001. Detection of sorghum yellow banding virus infecting grain sorghum in Venezuela. *Plant Disease* 85: 230.

Germani, G., Luc, M. and Baldwin, J. G. 1984. A new Rotylenculinae, *Senegulonema sorghi* n. gen., n. sp. (Nematoda: Tylenchida). *Revue Nématologia* 7: 49–56.

Giorda, L. M. 2002. Sorghum diseases in Argentina, pp. 389–91. In: *Sorghum and Millets Diseases* (Leslie, J. F., Ed.). Iowa State Press, Ames, IA.

Giorda, L. M. and Martinez, M. J. 1995. Opportunities for biotechnology research on sorghum stalk rot and virus diseases. *African Crop Science Journal* 3: 191–201.

Giorda, L. M., Martinez, M. J. and Chulze, S. 1995. Fusarium stalk rot in Argentina. Disease analysis through genetics and biotechnology, pp. 85–193. In: *Interdisciplinary Bridges to Improved Sorghum and Millet Crops* (Leslie, J. F. and Frederiksen, R. A., Eds). Iowa State University Press, Ames, IA.

Hafez, S. L. and Claflin, L. E. 1982. Control of plant parasitic nematodes on grain sorghum and yield response 1981. *Fungicide and Nematicide Tests* 37: 198.

Halim, A. H., Wassom, C. E., Mitchell, H. L. and Edmunds, L. K. 1973. Suppression of fungal growth by isolated trypsin inhibitors of corn grain. *Journal of Agricultural and Food Chemistry* 21: 1118–19.

Hess, D. E., Kroschel, J., Traoré, D., Elzein, A. E. M., Marley, P. M., Abbasher, A. A. and Diarra, C. 2002. *Striga*: Biological control strategies for a new millenium, pp. 165–70. In: *Sorghum and Millets and Diseases* (Leslie, J. F., Ed.). Iowa State University Press, Ames, IA.

Heyne, E. G., Melchers, L. E. and Lowe, A. E. 1944. Reaction of F1 sorghum plants to milo disease in the greenhouse and field. *Journal of the American Society of Agronomy* 36: 628–30.

Holdeman, Q. L. 1955. The present known distribution of the sting nematode, *Belonolaimus gracilis*, in the Coastal Plain of the southeastern United States. *Plant Disease Reporter* 39: 5–8.

Hood, M. E., Condon, J. M., Timko, M. P. and Riopel, J. L. 1998. Primary haustorial development of *Striga asiatica* on host and nonhost species. *Phytopathology* 88: 70–5.

Hsi, D. C. H. 1956. Stalk rot of sorghum in eastern New Mexico. *Plant Disease Reporter* 40: 369–71.

Huang, L. and Backhouse, D. 2006. Analysis of chitinase isoenzymes in sorghum seedlings incubated with *Fusarium thapsinum* and *F. proliferatum*. *Plant Science* 171: 539–45.

Huang, J., Wen, G. S., Li, M. J., Sun, C. C., Sun, Y., Zhao, M. F. and He, Y. Q. 2016. First report of maize chlorotic mottle virus naturally infecting sorghum and coix seed in China. *Plant Disease* 100: 1955.

Hundekar, A. R. and Anahosur, K. H. 1994. Pathogenicity of fungi associated with sorghum stalk rot. *Karnataka Journal of Agricultural Sciences* 7: 291–5.

Ibrahim, I. K. A., Lewis, S. A. and Harshman, D. C. 1993. Host suitability of graminaceous crop cultivars for isolates of *Meloidogyne arenaria* and *M. incognita*. *Supplemental Journal of Nematology* 25: 858–62.

Idris, H. A., Labuschagne, N. and Korsten, L. 2007. Screening rhizobacteria for biological control of Fusarium root and crown rot of sorghum in Ethiopia. *Biological Control* 40: 97–106.

Idris, H. A., Labuschagne, N. and Korsten, L. 2008. Suppression of *Pythium ultimum* root rot of sorghum by rhizobacterial isolates from Ethiopia and South Africa. *Biological Control* 45: 72–84.

Indira, S. 2002. Recommendations from the working group for Asia/Australia at the Third Global Conference on Sorghum and Millets Diseases, pp. 377–8. In: *Sorghum and Millets Diseases* (Leslie, J. F., Ed.). Iowa State Press, Ames, IA.

Isakeit, T., Gao, X. and Kolomiets, M. 2007. *Exserohilum pedicillatum* root rot of corn in Texas. *Plant Disease* 91: 634.

Islam, M. S., Haque, M. S., Islam, M. M., Emdad, E. M., Halim, A., Hossen, Q. M. M., Hos-sain, M. Z., Ahmed, B., Rahim, S., Rahman, M. S., Alam, M. M., Hou, S., Wan, X., Saito, J. A. and Alam, M. 2012. Tools to kill: Genome of one of the most destructive plant pathogenic fungi *Macrophomina phaseolina*. *BMC Genomics* 13: 493.

Jagdale, G. B., Ball-Coelho, B., Potter, J., Brandle, J. and Roy, R. C. 2000. Rotation crop effects on *Pratylenchus penetrans* and subsequent crop yields. *Canadian Journal of Plant Science* 80: 543–9.

Janvier, C., Villanueve, F., Alabouvette, C., Edel-Hermann, V., Matielle, T. and Steinberg, C. 2007. Soil health through disease suppression: Which strategy from descriptors to indicators? *Soil Biology & Biochemistry* 39: 1–23.

Jensen, S. 2000. Diseases caused by viruses and viruslike organisms, pp. 43–50. In: *Compendium of Sorghum Diseases*, 2nd ed. (Frederiksen, R. A. and Odvody, G. N., Eds). American Phytopathological Society, St. Paul, MN.

Jordan, W. R., Clark, R. B. and Seetharama, N. 1984. The role of edaphic factors in disease development, pp. 81–97. In: *Sorghum Root and Stalk Rots, A Critical Review: Proceedings of the Consultative Group Discussion on Research Needs and Strategies for Control of Sorghum Root and Stalk Rot Diseases*. ICRISAT, Patancheru, AP, India.

Jensen, S. G. and Giorda, L. M. 2002. Virus diseases of sorghum and millet in the Americas and Australia, pp. 403–10. In: *Sorghum and Millets Diseases* (Leslie, J. F., Ed.). Iowa State Press, Ames, IA.

Jones, R. K. and Belmar, S. B. 1989. Characterization and pathogenicity of *Rhizoctonia* spp. isolated from rice, soybean, and other crops grown in rotation with rice in Texas. *Plant Disease* 73: 1004–10.

Kanyuka, K., Ward, E. and Adams, M. J. 2003. *Polymyxa graminis* and the cereal virus it transmits: A research challenge. *Molecular Plant Pathology* 4: 393–406.

Kelly, L. A., Tan, Y. P., Ryley, M. J. and Aitken, E. A. B. 2017. *Fusarium* species associated with stalk rot and head blight of grain sorghum in Queensland and New South Wales, Australia. *Plant Pathology* 66: 1413–23.

Kennedy, B. W. 1990. Estimates of U.S. crop losses to prokaryote plant pathogens. *Plant Disease* 64: 674–6.

Khan, Z. R., Pickett, J. A., Wadhams, L. J., Hassanali, A. and Midega, C. A. O. 2006. Combined control of *Striga hermonthica* and stemborers by maize-*Desmodium* spp. intercrops. *Crop Protection* 25: 989–95.

Kiba, A., Sangawa, Y., Ohnishi, K., Yao, N., Park, P., Nakayashiki, H., Tosa, Y., Mayama, S. and Hikichi, Y. 2006. Induction of apoptotic cell death leads to the development of bacterial soft rot caused by *Pseudomonas cichorii*. *Molecular Plant-Microbe Interactions* 19: 112–22.

Kirkpatrick, T. L. and A. C. Thomas. 2007. *Crop Rotation for Management of Nematodes in Cotton and Soybean*. University of Arkansas Cooperative Extension Service Printing Services, Little Rock, AR.

Knott, C. A., Ritchey, E. L., Martin, J. R., Lee, C., Johnson, D., Bradley, C. and McNeill, S. 2016. Grain sorghum (milo) production in Kentucky. Agriculture and Natural Resources Publications 112.

Kobayashi, D. Y., Stretch, A. W. and Oudemans, P. V. 1995. A bacterial leaf spot of highbush blueberry hardwood cuttings caused by *Pseudomonas andropogonis*. *Plant Disease* 79: 839–42.

Koenning, S. R., Overstreet, C., Noling, J. W., Donald, P. A., Becker, J. O. and Fortnum, B. A. 1999. Survey of crop losses in response to phytoparasitic nematodes in the United States for 1994. *Supplemental Journal of Nematology* 31: 587–618.

Kollo, I. A. 2002. Plant-parasitic nematodes of sorghum and pearl millet: emphasis on Africa, pp. 259–66. In: *Sorghum and Millets Diseases* (Leslie, J. F., Ed.). Iowa State Press, Ames, IA.

Kolodge, C., Radewald, J. D. and Shibuya, F. 1987. Revised host range and studies on the life cycle of *Longidorus africanus*. *Journal of Nematology* 19: 77–81.

Korus, K., Lang, J. M., Adesemoye, A. O., Block, C. C., Pal, N., Leach, J. L. and Jackson-Zeims, T.A. 2017. First report of *Xanthomonas vasicola* causing bacterial leaf streak on corn in the United States. *Plant Disease* 101: 1030.

Kumari, S. R. and Chandrashekar, A. 1994. Isolation and purification of three antifungal proteins from sorghum endosperm. *Journal of the Science of Food and Agriculture* 64: 357–64.

Lamari, L., Ballance, G. M., Orolaza, N. P. and Kowatsch, R. 1995. In planta production and antibody neutralization of the *Ptr* necrosis toxin from *Pyrenophora tritici-repentis*. *Phytopathology* 85: 333–8.

Lamberti, F. C. 1969. Morphological variations and geographical distribution of *Longidorus africanus* Merny (Nematoda: Longidoridae). *Phytopathologia Mediterranea* 8: 137–41.

Last, F. T. 1961. Direct and residual effects of *Striga* control treatment on sorghum yields. *Tropical Agriculture (Trinidad)* 38: 49–56.

Leslie, J. F., Pearson, C., Nelson, P. E. and Toussoun, T. A. 1990. *Fusarium* spp. from corn, sorghum, and soybean fields in the central and eastern United States. *Phytopathology* 80: 343–50.

Leukel, R. W. 1948. *Periconia circinata* and its relation to milo disease. *Journal of Agricultural Research* 77: 201–22.

Leslie, J. F., Zeller, K. A., Lamprecht, S. C., Rheeder, J. P. and Marasas, F. O. 2005. Toxicity, pathogenicity and genetic differentiation of five species of Fusarium from sorghum and millet. *Phytopathology* 95: 275–83.

Little, C. R., Perumal, R., Tesso, T., Prom, L. K., Odvody, G. N. and Magill, C. W. 2012. Sorghum pathology and biotechnology – A fungal disease perspective: Part I. Grain mold, head smut, and ergot. *The European Journal of Plant Science and Biotechnology* 6: 10–30.

Lopes-Santos, L., Castro, D. B. A., Ottoboni, L. M. M., Park, D., Weir, B. S. and Destéfano, S. A. L. 2015. Draft genome sequence of *Burkholderia andropogonis* type strain ICMP2807 isolated from *Sorghum bicolor*. *Genome Announcements* 3: e00455–15.

Lopes-Santos, L., Castro, D. B. A., Ferreira-Tonin, M., Corrêa, D. B. A., Weir, B. S., Park, D., Ottoboni, L. M. M. and Neto, J. R. 2017. Reassessment of the taxonomic position of *Burkholderia andropogonis* and description of *Robbsia andropogonis* gen. nov. comb. nov. *Antonie Van Leeuwenhoek* 110: 727–36.

Luttrell, E. S. 1950. Grain sorghum disease in Georgia, 1949. *Plant Disease Reporter* 34: 45–52.

Maiti, R. K., Ramaiah, K. V., Bisen, S. S. and Chidley, V. L. 1984. A comparative study of the haustorial development of *Striga asiatica* (L.) Kuntze on sorghum cultivars. *Annals of Botany* 54: 447–57.

Malaguti, G. 2002. Current status of sorghum diseases in Venezuela, pp. 411–14. In: *Sorghum and Millets Diseases* (Leslie, J. F., Ed.). Iowa State Press, Ames, IA.

Mandadi, K. K., Pyle, J. D. and Scholthof, K. B. G. 2014. Comparative analysis of antiviral responses in *Brachypodium distachyon* and *Setaria viridis* reveals conserved and unique outcomes among C3 and C4 plant defenses. *Molecular Plant-Microbe Interactions* 27: 1277–90.

Mankin, C. J. 1969. Diseases of grasses and cereals in South Dakota. Agricultural Experiment Station, South Dakota State University. Technical Bulletin 35, pp. 1–27.

Mannering, J. V. and Fenster, C. R. 1983. What is conservation tillage? *Journal of Soil and Water Conservation* 38: 140–3.

Maranville, J. W. and Clegg, M. D. 1984. Morphological and physiological factors associated with stalk strength, pp. 111–18. In: *Sorghum Root and Stalk Rots, a Critical Review. Proceedings of the Consultative Group Discussion on Research Needs and Strategies for Control of Root and Stalk Rot Diseases* (Rosenberg, G., Ed.). 27 November – 2 December 1983, Bellagio, Italy. ICRISAT Press, Patancheru, AP, India.

Marley, P. S., Diourté, M., Neya, A., Nutsugah, S. K., Sérémé, P., Katilé, S. O., Hess, D. E., Mbaye, D. F. and Ngoko, Z. 2002. Sorghum and pearl millet diseases in West and Central Africa, pp. 419–25. In: *Sorghum and Millets Disease* (Leslie, J. F., Ed.). Iowa State Press, Ames, IA.

Maunder, A. B. 2002. Sorghum worldwide, pp. 11–18. In: *Sorghum and Millets Diseases* (Leslie, J. F., Ed.). Iowa State University Press, Ames, IA.

McClure, M., Husman, S. and Schmitt, M. 1998. Infection of sorghum varieties by the cotton root-knot Nematode, *Meloidogyne incognita*. In: *Cotton: A College of Agriculture Report*. College of Agriculture, The University of Arizona, Tucson, AZ.

McLaren, N. W. 1987. Studies on pre- and post-emergence damping-off and seedling blight of sorghum. M.Sc. thesis, University of Natal, Pieter-Maritzburg, South Africa.

McLaren, N. W. 2002. Evaluation of sorghum hybrids for resistance to the root rot complex. *South African Journal of Plant and Soil* 19: 37–42.

McLaren, N. W. 2004. Effect of soil nutrient status on severity of seedling diseases and root rot of sorghum (*Sorghum bicolor*). *South African Journal of Plant and Soil* 21: 263–5.

McLean, K. S. and Lawrence, G. W. 2001. First report of *Pythium* root rot on grain sorghum in Mississippi. *Plant Disease* 85: 922.

McSorley, R. and Gallaher, R. N. 1991. Nematode population changes and forage yields of six corn and sorghum cultivars. *Supplemental Journal of Nematology*. 23:673–7.

McSorley, R. and Gallaher, R. N. 1992. Comparison of nematode population densities on six summer crops at seven sites in north Florida. *Supplemental Journal of Nematology* 24: 699–706.

McSorley, R. and Gallaher, R. N. 1993. Population dynamics of plant-parasitic nematodes on cover crops of corn and sorghum. *Journal of Nematology* 25: 446–53.

Mihail, J. D. 1989. *Macrophomina phaseolina*: Spatio-temporal dynamics on inoculum and of disease in a highly susceptible crop. *Phytopathology* 79: 848–55.

Mihail, J. D. and Taylor, S. J. 1995. Interpreting variability among isolates of *Macrophomina phaseolina* in pathogenicity, pycnidium production, and chlorate utilization. *Canadian Journal of Botany* 73: 1596–603.

Mtisi, E. and McLaren, N. W. 2002. Diseases of sorghum and pearl millet in some southern African countries, pp. 427–30. In: *Sorghum and Millets Diseases* (Leslie, J. F., Ed.). Iowa State Press, Ames, IA.

Moffet, M. L., Hayward, A. C. and Fahy, P. C. 1986. Five new hosts of *Pseudomonas andropogonis* occurring in eastern Australia: Host range characterization of isolates. *Plant Pathology* 35: 34–43.

Motalaote, B., Starr, J. L., Frederiksen, R. A. and Miller, F. R. 1987. Host status and susceptibility of sorghum to *Pratylenchus* species. *Revue Nématologie* 10: 81–6.

Mughogho, L. K. 1984. Sorghum root and stalk rots: Basic disease problems, pp. 78–84. In: *Sorghum Root and Stalk Rots, a Critical Review. Proceedings of the Consultative Group Discussion on Research Needs and Strategies for Control of Root and Stalk Rot Diseases* (Rosenberg, G., Ed.). 27 November – 2 December 1983, Bellagio, Italy. ICRISAT Press, Patancheru, AP, India.

Mughogho, L. K. and Pande, S. 1984. Charcoal rot of sorghum, pp. 11–24. In: *Sorghum Root and Stalk Rots, a Critical Review. Proceedings of the Consultative Group Discussion on Research Needs and Strategies for Control of Root and Stalk Rot Diseases* (Rosenberg, G., Ed.). 27 November – 2 December 1983, Bellagio, Italy. ICRISAT Press, Patancheru, AP, India.

Murbithi, L. M. and Claflin, L. E. 1997. Genetic variation of grain sorghum germplasm for resistance to *Pseudomonas andropogonis*. *Euphytica* 98: 129–32.

Musselman, L. J. 1980. The biology of *Striga, Orobanche*, and other root parasitic weeds *Annual Review of Phytopathology* 18: 463–89.

Nagy, E. D. and Bennetzen, J. L. 2008. Pathogen corruption and site-directed recombination at a plant disease resistance gene cluster. *Genome Research* 18: 1918–23.

Narayana, Y. D., Bandyopadhyay, R., Navi, S. S. and Muniyappa, V. 2002. Sorghum viruses in Asia and Africa, pp. 431–9. In: *Sorghum and Millets Diseases* (Leslie, J. F., Ed.). Iowa State Press, Ames, IA.

Natural, M. P., Frederiksen, R. A. and Rosenow, D. T. 1982. Acremonium wilt of sorghum. *Plant Disease* 66: 863–5.

Ngugi, H. K., King S. B., Abayo G. O. and Reddy Y. V. R. 2002. Prevalence, incidence, and severity of sorghum diseases in western Kenya. *Plant Disease* 86: 65–70.

Norton, D. C. 1958. The association of *Pratylenchus hexincisus* with charcoal rot of sorghum. *Phytopathology* 48: 355–8.

Odvody, G. N. 2000. Pythium root rot, pp. 34–5. In: *Compendium of Sorghum Diseases*, 2nd ed. (Frederiksen, R. A. and Odvody, G. N., Eds). APS Press, St. Paul, MN.

Odvody, G. N. and Dunkle, L. D. 1979. Charcoal stalk rot of sorghum: Effect of environment on host parasite relations. *Phytopathology* 69: 250–4.

Odvody, G. N. and Dunkle, L. D. 1984. Periconia root rot, pp. 43–8. In: *Sorghum Root and Stalk Rots, a Critical Review: Proceedings of the Consultative Group Discussion on Research Needs and*

Strategies for Control of Sorghum Root and Stalk Diseases (Mughogho, L. K., Ed.). ICRISAT, Patancheru, AP, India.

Olbe, M., Sommarin, M., Gustafsson, M. and Lundborg, T. 1995. Effect of the fungal pathogen *Bipolaris sorokiniana* toxin prehelminthosporol on barley root plasma membrane vesicles. *Plant Pathology* 44: 625–35.

Olivier, A., Benhamou, N. and Leroux, G. D. 1991. Cell-surface interactions between sorghum roots and the parasitic weed *Striga hermonthica*-cytochemical aspects of cellulose distribution in resistant and susceptible host tissues. *Canadian Journal of Botany* 69: 1679–90.

O'Neill, N. R. and Rush, M. C. 1982. Etiology of sorghum sheath blight and pathogen virulence on rice. *Plant Disease* 66: 1115–18.

Orr, C. C. and Morey, E. D. 1978. Anatomical response of grain sorghum roots to *Meloidogyne incognita acrita*. *Journal of Nematology* 10: 48–53.

Palmero, D., Gil-Serna, J., Gálvez, L., Curt, M. D., De Cara, M. and Tello, J. 2012. First report of *Fusarium verticillioides* causing stalk and root rot of sorghum in Spain. *Plant Disease* 96: 584.

Parker, C. and Reid, D. C. 1979. Host specificity in *Striga* species – some preliminary observations, pp. 79–90. In: *Proc. Symp. Parasitic Weeds*, 2nd ed. (Musselman, L. J., Worsham, A. D. and Eplee, R. E., Eds) North Carolina State University, Raleigh, NC.

Patil, S. H., Hedge, R. K. and Anahosur, K. H. 1985. Role of sugars and phenols in charcoal rot resistance of sorghum. *Phytopathology Z.* 113: 30–35.

Petrovic, T., Walsh, J. L., Burgess, L. W. and Summerell, B. A. 2009. *Fusarium* species associated with stalk rot of grain sorghum in the northern grain belt of eastern Australia. *Australasian Plant Pathology* 38: 373–9.

Pratt, R. G. and Janke, G. D. 1980. Pathogenicity of three species of *Pythium* to seedlings and mature plants of grain sorghum. *Phytopathology* 70: 766–71.

Preston, D. A. 1945. Host index of Oklahoma plant diseases. Okla. Agric. Mech. College Agric. Exp. Stn. Tech. Bull. No. T-21.

Pringle, R. B. and Scheffer, R. P. 1963. Purification of the selective toxin of *Periconia circinata*. *Phytopathology* 53: 785–7.

Pyle, J. D. 2015. Molecular pathogenesis of viral and subviral agents in model and crop grasses. Ph.D. dissertation, Texas A&M University, 163p.

Ramundo, B. A. and Claflin, L. E. 2005. Identification of *Burkholderia andropogonis* with a repetitive sequence BOX element and PCR. *Current Microbiology* 50: 52–6.

Reddy, A. S., Hobbs, H. A., Delfosse, P., Murthy, A. K. and Reddy, D. V. R. 1998. Seed transmission of Indian peanut clump virus (IPCV) in peanut and millets. *Plant Disease* 82: 343–6.

Reed, J. E., Partridge, J. E. and Nordquist, P. T. (1983) Fungal colonization of stalks and roots of grain sorghum during the growing season. *Plant Disease* 67: 417–20.

Robbins, R. T. and Barker, K. R. 1973. Comparison of host range and reproduction among populations of *Belonolaimus longicaudatus* from North Carolina and Georgia. *Plant Disease Reporter* 57: 750–4.

Rodenburg, J., Bastiaans, L., Schapendonk, A. H. C. M., van der Putten, P. E. L., van Ast, A., Dingemanse, N. J. and Haussmann, B. I. G. 2008. Co-assimilation and chlorophyll fluorescence as indirect selection criteria for host tolerance against *Striga*. *Euphytica* 160: 75–87.

Rosenow, D. T., Johnson, J. W., Frederiksen, R. A. and Miller, F. R. 1977. Relationship of nonsenescence to lodging and charcoal rot in sorghum. Agronomy Abstracts. American Society of Agronomy, Madison, WI, p. 69.

Ryley, M. J., Persley, D. M., Jordan, D. R. and Henzell, R. G. 2002. Status of sorghum and pearl millet diseases in Australia, pp. 441–8. In: *Sorghum and Millets Diseases* (Leslie, J. F., Ed.). Iowa State Press, Ames, IA.

Saleh, A. A., Ahmed, H. U., Todd, T. C., Travers, S. E., Zeller, K. A., Leslie, J. F. and Garrett, K. A. 2010. Relatedness of *Macrophomina phaseolina* isolates from tallgrass prairie, maize, soybean, and sorghum. *Molecular Ecology* 19: 79–91.

Satish, K., Gutema, Z., Grenier, C., Rich, P. J. and Ejeta, G. 2012. Molecular tagging and validation of microsatellite markers linked to the low germination stimulant gene (*lgs*) for *Striga* resistance in sorghum (*Sorghum bicolor* (L.) Moench). *Theoretical and Applied Genetics* 124: 989–1003.

Schaad, N. W., Vidaver, A. K., Lacy, G. H., Rudolph, K. and Jones, J. B. 2000. Evaluation of proposed amended names of several pseudomonads and xanthomonads and recommendations. *Phytopathology* 90: 208–13.

Scheffer, R. P. and Pringle, R. B. 1961. A selective toxin produced by *Periconia circinata. Nature* 191: 912–13.

Schertz, K. F. and Tai, Y. P. 1969. Inheritance of reaction of *Sorghum bicolor* (L.) Moench to toxin produced by *Periconia circinata* (Mang.) sacc. *Crop Science* 9: 621–4.

Seetharama, N., Bidinger, F. R., Rao, K. N., Gill, K. S. and Mulgund, M. 1987. Effect of pattern and severity of moisture deficit stress on stalk rot incidence in sorghum. I. Use of line source irrigation technique, and the effect of time of inoculation. *Field Crops Research* 15: 289–308.

Seifers, D. L., Harvey, T. L., Haber, S., She, Y. M., Chernushevich, I., Ens, W. and Standing, K. G. 1999. Natural infection of sorghum by foxtail mosaic virus in Kansas. *Plant Disease* 83: 905–12.

Seifers, D. L., Perumal, R. and Little, C. R. 2012. New sources of resistance in sorghum (*Sorghum bicolor*) germplasm are effective against a diverse array of *Potyvirus* spp. *Plant Disease* 96: 1775–9.

Shah, M. A., Khan, A. A., Junaid, J. M., Majid, S. and Mohi-ud-din, S. 2015. Aphid vectored viral diseases and their management, pp. 511–54. In: *Insect Pests Management of Fruit Crops* (Pandey, A. K. and Mall, P., Eds). Biotech Books, New Delhi, India.

Sharma, S. B. and McDonald, D. 1990. Global status of nematode problems of groundnut, pigeonpea, chickpea, sorghum and pearl millet, and suggestions for future work. *Crop Protection* 9: 453–8.

Short, G., Wyllie, T. and Bristow, P. 1980. Survival of *Macrophomina phaseolina* in soil and in residue of soybean. *Phytopathology* 70: 13–17.

Smolik, J. D. 1977. Effects of *Trichodorus allius* and *Tylenchorhynchus nudus* on growth of sorghum. *Plant Disease Reporter* 61: 855–8.

Spallek, T., Musembu, M. and Shirasu, K. 2013. The genus *Striga*: A witch profile. *Molecular Plant Pathology* 14: 861–9.

Sprague, R. 1950. *Diseases of Cereals and Grasses in North America*. Ronald Press, New York, NY. 538 pp.

Stack, J. P. 2002. Recurring and emerging sorghum diseases in North America, pp. 449–56. In: *Sorghum and Millets Disease* (Leslie, J. F., Ed.). Iowa State Press, Ames, IA.

Starr, J. L., 1992. Nematode pathogens of sorghum, pp. 179–85. In: *Sorghum and Millets Diseases: A Second World Review* (de Milliano, W. A. J., Frederiksen, R. A. and Bengston, G. D., Eds). International Crops Research Institute for the Semi-Arid Tropics (ICRISAT), Patancheru, AP, India.

Stewart, L. R., Teplier, R., Todd, J. C., Jones, M. W., Cassone, B. J., Wijeratne, S., Wijeratne, A. and Redinbaugh, M. G. 2014. Viruses in maize and Johnsongrass in southern Ohio. *Phytopathology* 104: 1360–9.

Stroube, W. H. 1954. Host range of Rhizoctonia aerial blight fungus in Louisiana. *Plant Disease Reporter* 38: 789–90.

Sugiyama, L. S., Bushe, B. C., Heller, W. P. and Keith, L. M. 2017. First report of *Paraburkholderia andropogonis* causing bacterial leaf streak of *Strelitzia reginae* in Hawaii detected from imported plants. *Plant Disease* 101: 1030.

Sutton, B. C. 1980. *The Coelomycetes: Fungi Imperfecti with Pycnidia, Acervuli, and Stromata.* Commonwealth Mycological Institute, Kew, London, 696p.

Talwana, H. L., Butseya, M. M. and Tusiime, G. 2008. Occurrence of plant parasitic nematodes and factors that enhance population build-up in cereal-based cropping systems in Uganda. *African Crop Science Journal* 16: 119–31.

Tang, Q. H., Yu, F. Y., Zhang, S. Q., Niu, X. Q., Zhu, H., Song, W. W., Han, C. W., Wu, D. Y. and Qin, W. Q. 2013. First report of *Burkholderia andropogonis* causing bacterial leaf spot of Betel palm in Hainan Province, China. *Plant Disease* 97: 1654.

Taylor, A. L., Sasser, J. N. and Nelson, L. A. 1982. Relationship of climate and soil characteristics to geographical distribution of *Meloidogyne* species in agricultural soils. A Cooperative Publication of the Department of Plant Pathology, North Carolina State University and the United States Agency for International Development, Raleigh, NC.

Taylor, A., Martin, J. and Seel, W. E. 1996. Physiology of the parasitic association between maize and witchweed (*Striga hermonthica*): Is ABA involved? *Journal of Experimental Botany* 47: 1057–65.

Temam, H. 2001. Evaluation of inoculation methods and resistance of some sorghum genotypes to bacterial leaf streak (*Xanthomonas campestris* pv. *holcicola*). *Pest Management Journal of Ethiopia* 5: 21–8.

Tenkouano, A., Miller, F. R., Frederiksen, R. A. and Rosenow, D. T. 1993. Genetics of nonsenescence and charcoal rot resistance in sorghum. *Theoretical and Applied Genetics* 85: 644–8.

Tesso, T. and Ejeta, G. 2011. Stalk strength and reaction to infection by *Macrophomina phaseolina* of brown midrib maize (*Zea mays*) and sorghum (*Sorghum bicolor*). *Field Crops Research* 120: 271–5.

Tesso, T., Claflin, L. E. and Tuinstra, M. R. 2004. Estimation of combining ability for resistance to Fusarium stalk rot in grain sorghum. *Crop Science* 44: 1195–9.

Tesso, T. T., Claflin, L. E. and Tuinstra, M. R. 2005. Analysis of stalk rot resistance and genetic diversity among drought tolerant sorghum genotypes. *Crop Science* 45: 645–52.

Tesso, T. T., Ochanda, N., Little, C. R., Claflin, L. and Tuinstra, M. R. 2010. Analysis of host plant resistance to multiple *Fusarium* species associated with stalk rot disease in sorghum [*Sorghum bicolor* (L.) Moench]. *Field Crops Research* 118: 177–82.

Tesso, T., Perumal, R., Little, C. R., Adeyanju, A., Radwan, G. L., Prom, L. K. and Magill, C. W. 2012. Sorghum pathology and biotechnology – A fungal disease perspective: Part II. Anthracnose, stalk rot, and downy mildew. *The European Journal of Plant Science and Biotechnology* 6: 31–44.

Thakur, R. P., Fredriksen, R. A., Murty, D. S., Reddy, B. V. S., Bandyopadhyay, R., Giorda, L. M., Odvody, G. N. and Claflin, L. E. 1996. Breeding for disease resistance in sorghum, pp. 303–15. In: *Proceedings of the International Conference on Genetic Improvement of Sorghum and Pearl Millet*, 23–27 September, Lubbock, TX.

Thies, J. A., Petersen, A. D. and Barnes, D. K. 1995. Host suitability of forage grasses and legumes for root-lesion nematode *Pratylenchus penetrans*. *Crop Science* 35: 1647–51.

Tilander, Y. and Bonzi, M. 1997. Water and nutrient conservation through the use of agroforestry mulches and sorghum yield response. *Plant and Soil* 197: 219–22.

Timper, P. and Hanna, W. W. 2005. Reproduction of *Belonolaimus longicaudatus, Meloidogyne javanica, Paratrichodorus minor*, and *Pratylenchus brachyurus* on pearl millet (*Pennisetum glaucum*). *Journal of Nematology* 37. 214–19.

Todd, T. C. 1991. Effect of cropping regime on populations of *Belonolaimus* sp. and *Pratylenchus scribneri* in sandy soil. *Supplemental Journal of Nematology* 23: 646–51.

Todd, T. C. and Claflin, L. E. 1984. Evaluation of insecticides-nematicides for control of stunt nematodes in grain sorghum, 1983. *Fungicide and Nematicide Tests* 39:96.

Trimboli, D. S. 1983. Reproduction of *Fusarium moniliforme* basal stalk rot and root rot of grain sorghum in the greenhouse. *Plant Disease* 67: 891–4.

Umar, I. and Chubado, D. 2008. Survey of plant-parasitic nematodes in then Benue Valley of Adamawa State, Nigeria. *Agricultural Journal* 3: 353–6.

United States Department of Agriculture (USDA), National Agriculture Statistics Service (NASS). 2017. https://www.nass.usda.gov/Statistics_by_Subject/index.php. Accessed: 29 Mar 2017.

Vasey, R. A., Scholes, J. D. and Press, M. C. 2005. Wheat (*Triticum aestivum*) is susceptible to the parasitic angiosperm *Striga hermonthica*, a major cereal pathogen in Africa. *Phytopathology* 95: 1294–300.

Vasudeva-Rao, M. J. and Musselman, L. J. 1987. Host specificity in *Striga* spp. and physiological 'strains', pp. 13–25. In: *Parasitic Weeds in Agriculture. Volume 1, Striga* (Musselman, L. J., Ed.). CRC Press, Boca Raton, FL.

Walton, J. D. and Panaccione, D. G. 1993. Host-selective toxins and disease specificity: Perspectives and progress. *Annual Review of Phytopathology* 31: 275–303.

Waniska, R. D., Venkatesha, R. T., Chandrashekar, A., Krishnaveni, S., Bejosano, F. P., Jeoung, J., Jayaraman, J., Muthukrishnan, S. and Liang, G. H. 2002. Antifungal proteins and other mechanisms in the control of sorghum stalk rot and grain mold, pp. 287–98. In: *Sorghum and Millets Diseases* (Leslie, J. F., Ed.). Iowa State Press, Ames, IA.

Wei, C., Hincapie, M., Larsen, N. and Nuessly, G. 2016. First report of sugarcane yellow leaf virus infecting grain sorghum (*Sorghum bicolor*) in the United States. *Plant Disease* 100: 1798.

Widmer, T. L., Mitkowski, N. A. and Abawi, G. S. 2002. Soil organic matter and management of plant-parasitic nematodes. *Journal of Nematology* 34: 289–95.

Williams, R. J., Frederiksen, R. A. and Girard, J. C. 1978. Sorghum and pearl millet diseases identification handbook. Information Bulletin No. 2. ICRISAT, Patancheru, AP, India.

Williams, R. J. and Nickel, O. 1984. The role of fungicides in the control of sorghum root and stalk diseases, pp. 191–200. In: *Sorghum Root and Stalk Rots, a Critical Review: Proceedings of the Consultative Group Discussion on Research Needs and Strategies for Control of Sorghum Root and Stalk Diseases*. ICRISAT, Patancheru, AP, India.

Wolpert, T. J. and Dunkle, L. D. 1983. Alternations in gene expression in sorghum induced by the host-specific toxins from *Periconia circinata*. *Proceedings of the National Academy of Science USA* 80: 6576–80.

Wu, W.-Q., Guo, X.-G., Zhang, H.-M., Yang, J., Lv, M.-F. and Chen, J.-P. 2013. Simultaneous detection and survey of three rice viruses in China. *Plant Disease* 97: 1181–6.

Yoshida, S. and Shirasu, K. 2009. Multiple layers of incompatibility to the parasitic witchweed, *Striga hermonthica*. *New Phytologist* 183: 180–9.

Zummo, N. 1980. *Fusarium* disease complex of sorghum in West Africa, pp. 11–15. In: *Proceedings of the International Workshop on Sorghum Diseases*. ICRISAT Press, Hyderabad, India.

Zummo, N. 1984. Fusarium root and stalk disease complex, pp. 25–9. In: *Sorghum Root and Stalk Rots, a Critical Review: Proceedings of the Consultative Group Discussion on Research Needs and Strategies for Control of Sorghum Root and Stalk Diseases*. ICRISAT, Patancheru, India.

Chapter 17

Weed management in sorghum cultivation

M. Bagavathiannan, Texas A&M University, USA; W. Everman, North Carolina State University, USA; P. Govindasamy, Texas A&M University, USA; A. Dille and M. Jugulam, Kansas State University, USA; and J. Norsworthy, University of Arkansas, USA

1 Introduction

Sorghum (*Sorghum bicolor* L.), an important food and feed crop worldwide, is considered a low-input, energy-efficient, and drought-tolerant crop, capable of growing well under marginal environmental conditions. The United States is the second largest producer and the leading exporter of grain sorghum in the world (USDA-FAS, 2016). In 2017, grain sorghum was planted on about 2.5 million ha throughout the United States (USDA-NASS, 2017). The total US export of grain sorghum in 2016 was valued >$2 billion (USDA-FAS, 2016), making it an important agricultural commodity in the United States. Kansas and Texas are the top two grain-sorghum-producing states with 1.1 and 0.7 million ha, respectively (USDA-NASS, 2017). Grain sorghum is a significant crop in other states such as Colorado, Oklahoma, South Dakota and Nebraska, and to some extent in the Midsouth in Arkansas, Missouri, Mississippi and Louisiana as well as in the Southeast in North Carolina (USDA-NASS, 2017).

Weed management is an important production challenge for sorghum. Some of the dominant weeds across the US sorghum belt include, but are not limited to, barnyardgrass [*Echinochloa crus-galli* (L.) Beauv.], junglerice [*Echinochloa colona* (L.) Link], Texas panicum (*Panicum texanum* Buckl.), large crabgrass [*Digitaria sanguinalis* (L.) Scop.], Palmer amaranth (*Amaranthus palmeri* S. Wats.), waterhemp [*Amaranthus tuberculatus* (Moq.) Sauer] and kochia [*Kochia scoparia* (L.) Schrad.]. Weedy relatives such as johnsongrass [*Sorghum halepense* (L.) Pers.] and shattercane (*Sorghum bicolor* L.) are difficult to control in sorghum due to genetic similarities and the lack of selective herbicide options, except

http://dx.doi.org/10.19103/AS.2017.0015.18

for the upcoming non-transgenic acetolactate synthase (ALS)-inhibitor herbicide-resistant (Inzen™) sorghum technology. In more southern geographies, johnsongrass is the most problematic weedy relative, whereas shattercane is more common in the northern geographies (Barber et al., 2015; Thompson, 2014; Webster et al., 2012).

Weed competition typically leads to yield and grain quality losses in sorghum. Norris (1980) reported that barnyardgrass at a density of 1 plant/m of row reduced grain sorghum yield by 10%, whereas a 52% yield loss was observed when the density increased to 175 plants/m of row. Likewise, for each additional Palmer amaranth plant per 15 m of row, sorghum grain yield was reduced by 1.8–3.5% per plant (Moore et al., 2004). As Palmer amaranth density increased, grain moisture and foreign material increased, leading to additional grain quality losses. Palmer amaranth reduced grain sorghum yield from 4%, with only 1 weed/4 m of row, up to 40% if there were 6 plants/m of row (Unruh, 2013). When a naturally occurring waterhemp infestation was allowed to compete with sorghum in increasing densities throughout the season, a maximum yield loss of 80% was measured, compared to the weed-free check (Feltner et al., 1969a). Kochia at a density of 1.1 plants/m of sorghum row (spaced 51 cm apart) prevented any grain production when sorghum was grown under dryland conditions (Phillips, 1958). Yellow foxtail [*Setaria pumila* (Poir.) Roemer & J.A. Schultes] increasingly reduced sorghum yield as weed densities increased (Feltner et al., 1969b).

Though weeds are important yield-limiting factors for grain sorghum production, the options available for weed management in sorghum are limited compared to other row crops such as cotton and corn. A number of pre-emergence (PRE) and post-emergence (POST) herbicides are labelled in grain sorghum, but are not sufficient to provide effective, season-long weed control. The management of grass weeds is particularly difficult in grain sorghum due to a lack of effective POST grass control options. The anticipated ALS-inhibitor-resistant Inzen™ grain sorghum technology will allow for over-the-top application of nicosulfuron (and rimsulfuron) to provide effective grass weed control, including selective control of the weedy relatives. A number of non-chemical weed management options can be integrated with herbicides to improve management efficacy and reduce the risk of weed adaptation. More tools and technologies are currently under development to provide improved weed management options for sorghum.

In this chapter, we provide a general overview of weed management considerations and options for sorghum, with particular focus on critical period of weed control (CPWC), herbicide options available including burndown, PRE, POST and post-directed options, novel herbicide-resistant crop technologies, influence of production practices on weed adaptation and population shifts, and non-chemical weed management options suitable for integration in sorghum.

2 Critical period for weed control in sorghum

The CPWC is a time frame in the life cycle of a crop when to scout for weed populations and when to maintain a weed-free environment to prevent unacceptable yield losses. It is used for in-crop decision-making for POST weed control. The CPWC as defined by Swanton and Weise (1991) involves two separate components. First is the critical duration of weed interference, a time period that weeds can coexist with the crop and not affect yield. The second component is the critical weed-free period, a time period from planting to when later-emerging weeds will not affect crop yield. Results of the combined components determine the CPWC, also defined as the time frame during the life cycle of the crop when it must be kept weed-free to prevent unacceptable yield losses, often set at 5% loss threshold (Knezevic et al., 2002). The initiation and length of CPWC in a specific crop will be dependent on the weed species and weed density.

For grain sorghum, the CPWC time frame occurs relatively early in the crop's life cycle. Several reports suggest that first 3–4 weeks be maintained weed-free following grain sorghum planting to prevent yield losses (Burnside and Wicks, 1967; Everaarts, 1993; Fabrizius, 1998; Feltner et al., 1969a; Smith et al., 1990). This assumes that sorghum is planted into adequate soil moisture for normal seedling emergence. Redroot pigweed (*Amaranthus retroflexus* L.) that emerged when sorghum was at the 1- to 2.6-leaf stages caused significant crop yield loss at low weed densities, but if redroot pigweed did not emerge until after the 5.5-leaf stage of sorghum, no significant yield loss occurred across the weed densities (Knezevic et al., 1997). Maximum estimated yield losses varied from 3% to 46% depending on experimental location, pigweed density and time of weed emergence relative to the growth stage of sorghum. CPWC can be delayed and shortened when using grain sorghum production practices that enhance its competitive ability against weeds, such as narrow row spacing and increased seeding rates (Shaffer, 2016; Hewitt, 2015).

Grass weed management is very difficult in grain sorghum and delaying the POST control of annual grass weeds increased the extent of yield losses. Smith et al. (1990) documented that sorghum grain yield was reduced by 3.6% for each week of interference from barnyardgrass, large crabgrass or Texas panicum. The authors did observe a row width interaction as the effect of grass competition on narrow-row width (61 cm) sorghum was lesser than that on wide-row width (91 cm) sorghum. Goosegrass [*Eleusine indica* (L.) Gaertn], southern sandbur (*Cenchrus echinatus* L.) and a spurge [*Croton hirtus* (L.) Herit] grown in two different experiments did not cause grain yield loss if they emerged 30 days after planting grain sorghum (Everaarts, 1993). In the two experiments, it was concluded that 19 (goosegrass and southern sandbur) and 22 (spurge) weed-free days after planting were needed to prevent grain sorghum yield loss. The CPWC for longspine sandbur [*Cenchrus longispinus* (Hack.) Fern.] was four weeks, though there was a 10-fold difference in weed density between the two years studied (Fabrizius, 1998). If longspine sandbur was allowed to compete with grain sorghum for 4, 5 and 6 weeks after crop emergence, yields were reduced by 27%, 31% and 42%, respectively, compared to the weed-free sorghum yield (Fabrizius, 1998). Growers need to control all weeds during the CPWC, to protect sorghum yield potential. However, CPWC knowledge is scarce for the major grain sorghum production systems in the United States. Developing CPWC information will be particularly useful in guiding the timing for POST control of annual grass weeds in the upcoming Inzen™ sorghum hybrids.

3 Chemical options for weed control in sorghum

3.1 Herbicide options and considerations

Sorghum has limited herbicide options when compared to many other row crops; therefore, producers should evaluate their weed management options before planting sorghum. Excellent broadleaf weed control, including Palmer amaranth, can be achieved in sorghum where applications of POST herbicides follow a PRE herbicide, although timely application to small weeds is essential. A limited number of active ingredients are available for sorghum, but several herbicide premixes exist for both PRE and POST (Tables 1 and 2). Premixes containing atrazine will provide the greatest residual control of grass and broadleaf weeds. Atrazine use is limited to 0.91 kg in a single application and 1.13 kg in a season, with severe rotation restrictions based on rate and timing of application (e.g. after 10 June in North Carolina); therefore, it is important to plan

Table 1 Herbicide options for burndown, pre-plant or pre-emergence weed control in sorghum

Trade name	Common name	Application considerations[a]
Burndown prior to crop emergence		
Clarity	Dicamba	Apply to actively growing weeds at least 15 days prior to planting. Typically mix with glyphosate.
FirstShot SG	Thifensulfuron + tribenuron	
Glyphosate (multiple)	Glyphosate	Multiple formulations exist.
Gramoxone SL	Paraquat	
Expert[b]	*S*-metolachlor + atrazine + glyphosate	PRE from 21 days before planting through pre-emergence. Application to emerged sorghum will result in plant death.
Sequence[b]	*S*-metolachlor + glyphosate	
Pre-plant or pre-emergence		
AAtrex 4L	Atrazine	Do not exceed 2800 g atrazine/ha/season. See label for comments on rotation restrictions.
Bicep II Magnum/ Cinch ATZ[b]	*S*-metolachlor + atrazine	Caution should be used on coarse-textured soils. See label for comments on soil types.
Callisto	Mesotrione	PRE 21–7 days before planting. See label for comments on soil types.
Degree Xtra[b]	Acetochlor + atrazine	PRE after planting before sorghum emergence. See label for comments on soil types.
Dual II Magnum/ Cinch[b]	*S*-metolachlor	Apply after planting before weeds and crop emerge. See label for comments on soil types.
Intrro/Micro-Tech[b]	Alachlor	PPI within seven days before planting or PRE after planting before sorghum emergence.
Lariat[b]	Alachlor + atrazine	PPI up to 14 days before planting or PRE after planting before sorghum emergence.
Sharpen	Saflufenacil	Do not apply after emergence or when an at-planting application of organophosphate or carbamate has been used.
Warrant[b]	Acetochlor	See label for comments on soil types.
Lexar[b]	*S*-metolachlor + atrazine + mesotrione	Apply from 21 days before planting through PRE. See label for comments on soil types. Applying <7 days before sorghum planting will increase risk of crop injury.

[a] PPI – pre-plant incorporation (5 cm deep); PRE – pre-emergence application.

[b] Sorghum seed must be treated with an approved seed safener for chloroacetamide herbicides.

Table 2 Herbicide options for postemergence weed control in sorghum

Trade name	Common name	Application considerations
Over-the-top post-emergence herbicides		
2,4-D amine 3.8 SL	2,4-D amine	Apply when sorghum is 15–38 cm tall. Limited to one application. Post-directed application when sorghum is 20–38 cm tall.
AAtrex 4L[a]	Atrazine	Apply before weeds are 4 cm tall and before sorghum exceeds 30 cm height.
Aim EC	Carfentrazone-ethyl	From 10-cm tall sorghum to prior to boot stage.
Basagran	Bentazon	Excellent crop tolerance. Usually applied at the 1- to 5-leaf stage.
Buctril	Bromoxynil	Apply at the 3- to 4-leaf stage up to boot stage of sorghum.
Clarity	Dicamba	Best time to apply is the 3- to 5-leaf stage, usually within 25 days after planting. Post-directed application after sorghum is 20–38 cm tall.
Degree Xtra[a]	Acetochlor + atrazine	Apply to sorghum up to 30 cm tall.
Facet L	Quinclorac	Apply to sorghum up to 30 cm tall when weeds are 2.5–5 cm.
Huskie	Pyrasulfotole + bromoxynil	Apply from the 3-leaf stage up to 30 cm tall. Do not harvest grain within 60 days of application.
Marksman[a]	Dicamba + atrazine	Apply to sorghum up to 0 cm tall.
Peak[b]	Prosulfuron	Apply to sorghum between 13 and 50 cm tall. Post-directed application when crop is 50–76 cm tall.
Sandea[b]	Halosulfuron	Apply from 2-leaf stage but before grain head emergence.
Starane	Fluroxypyr	Apply to sorghum from the 3- to 7-leaf stage. Post-directed application when sorghum is at its 8-leaf stage to boot stage.
Warrant[c]	Acetochlor	Apply before sorghum reaches 28 cm tall.
Weedmaster	Dicamba + 2,4-D	Apply to sorghum in the 3- to 5-leaf stage. Sorghum may be most sensitive to injury if applied during stressful conditions, such as high moisture, low fertility or abnormal temperatures.
Yukon	Halosulfuron + dicamba	Apply to sorghum from the 3- to 5-leaf stage. Post-directed application when sorghum is 20–38 cm tall.

(Continued)

Table 2 (*Continued*)

Trade name	Common name	Application considerations
Zest[b]	Nicosulfuron	Apply only to grain sorghum containing the Inzen™ herbicide tolerance trait when 10–50 cm tall.
Post-directed herbicides		
Linex	Linuron	Post-directed application when sorghum is at least 38 cm tall and weeds are 5–10 cm tall. Spray no higher than the lower 8 cm of the sorghum stalk.
Gramoxone SL	Paraquat	Hooded application when sorghum is at least 38 cm tall. Keep spray off of the sorghum foliage.
Prowl H_2O[c]	Pendimethalin	Post-directed application after grain sorghum is 15 cm tall.

[a] Atrazine rates should not exceed 2.8 kg ai/ha in a single season.
[b] ALS-inhibitor-resistant biotypes will not be controlled.
[c] Provides residual control only and will not control emerged weeds.

atrazine applications accordingly. It is critical to apply a full recommended rate of a PRE grass herbicide since effective herbicide options for POST grass control once sorghum has emerged are limited to quinclorac (Facet® L) and nicosulfuron (Zest®, in the herbicide-resistant Inzen™ sorghum when the hybrids become available). Several herbicides which provide residual control of grass and broadleaf species may be applied POST or post-directed. All herbicide applications must be made following label instructions.

3.1.1 Burndown

Burndown herbicide applications are valuable to control weeds that emerge prior to crop planting both in conventional-till and no-till systems. Thus, burndown herbicide applications allow for planting into a clean seedbed. Weed control in grain sorghum begins with use of glyphosate (Roundup®), dicamba (Clarity®), 2,4-D or paraquat (Gramoxone® SL) as a burndown application several weeks prior to planting (Barber et al., 2015). Glyphosate is the most commonly applied burndown herbicide (USDA-NASS, 2012) and may be subsequently followed by pre-plant tillage in geographies where ample soil moisture is available early in the growing season. Nevertheless, several burndown herbicides are labelled for use prior to planting grain sorghum, including dicamba, glyphosate, paraquat, a premix of *S*-metolachlor and glyphosate (Sequence®), and a premix of *S*-metolachlor, atrazine and glyphosate (Expert®) (listed in Table 1). Glyphosate and paraquat provide broad-spectrum control of emerged weeds with no residual activity. The premix herbicides Expert® and Sequence® have POST and PRE activity on a wide range of grasses and small-seeded broadleaf weeds.

3.1.2 Pre-emergence

The limited number of POST herbicide options for grass control necessitates the use of a strong PRE herbicide programme with a particular focus on grasses. Many residual herbicides provide good to excellent control of grass species; however, no herbicide will provide season-long residual control. *S*-metolachlor (Dual II Magnum®, Cinch®), alachlor (Intrro®, Micro-Tech®) and acetochlor (Warrant®) are seedling shoot growth inhibitors with excellent grass activity. PRE application of chloroacetamide herbicides typically requires a safener seed treatment to prevent injury on sorghum.

For PRE broadleaf weed control in sorghum, atrazine (AAtrex®), mesotrione (Callisto®) and saflufenacil (Sharpen®) are available. Residual activity will depend on the herbicide product and rate. Atrazine use is limited to 2.8 kg ai/ha in a growing season; therefore, it is important to manage multiple atrazine applications carefully. Pendimethalin (Prowl® H_2O) and trifluralin (Treflan®) are labelled for POST application to sorghum and are commonly used at the early POST (EPOST) timing in Texas (Grichar et al., 2005). Atrazine and pendimethalin combination applied EPOST can sometimes cause severe crop injury and grain yield loss (Grichar et al., 2005). Mesotrione provides excellent residual control of broadleaf weeds, but caution should be exercised when using in sandy soils due to high potential for crop injury. Several premixes of PRE grass and broadleaf herbicides are also available for use in grain sorghum. Premix combinations of *S*-metolachlor and atrazine (Bicep II Magnum®, Cinch ATZ®) and alachlor and atrazine (Lariat®) provide broad-spectrum residual control of grasses and broadleaf weeds. The premix of mesotrione, *S*-metolachlor and atrazine (Lexar®) also has excellent residual activity on many weed species, but care must be taken when using in sandy soils. It is critical that herbicide rates for PRE applications be selected based on soil type, as recommended by specific herbicide product labels. Further, rotation restrictions to subsequent crops should also be considered when selecting residual herbicide options.

3.1.3 Post-emergence

Several effective herbicides are labelled for POST weed control in sorghum; however, most products have growth stage or plant height restrictions for use. Post-emergence herbicide programmes should be planned around grass species, which are difficult to control in grain sorghum (Thompson, 2014). Quinclorac and nicosulfuron are currently labelled for POST grass control, with additional activity on broadleaf weed species. Quinclorac is effective on small grass seedlings (up to 5 cm tall), with excellent activity on barnyardgrass, broadleaf signalgrass and large crabgrass and limited activity on fall panicum, goosegrass and Texas panicum. Nicosulfuron provides excellent control of most grass species, but the herbicide-resistant Inzen™ sorghum hybrids must be planted when they become available. Any pre-existing resistance to the ALS inhibitors will likely reduce the efficacy of nicosulfuron. Residual herbicides should be utilized at PRE and POST timings to improve weed control. Acetochlor, a premix of acetochlor and atrazine (Degree Xtra®), can be used to provide residual control of grass species in established grain sorghum.

Broadleaf weed management can be effectively accomplished with any number of POST herbicides (listed in Table 2), if timed properly. Palmer amaranth is an important weed in many sorghum fields in the United States, and timing is critical to achieve acceptable control of this species with POST herbicides; the same applies to several other weeds. Atrazine, prosulfuron (Peak®) and a premix of thiencarbazone-methyl, pyrasulfotole and bromoxynil (Huskie®) will also provide residual weed control, but atrazine provides extended residual activity compared to the other two. Though atrazine can be applied PRE or POST, most applications normally occur after crop emergence to ensure rotation flexibility if stand failure occurs. Atrazine mainly provides broadleaf weed control; hence, it is commonly tank-mixed with a chloroacetamide herbicide such as *S*-metolachlor or acetochlor for improved annual grass control through residual activity. Of these chloroacetamides, *S*-metolachlor is the most widely applied and ranks third among all herbicides applied in grain sorghum (USDA-NASS, 2012). Control of certain broadleaf species such as Palmer amaranth in grain sorghum requires tank-mixing atrazine with other herbicides such as dicamba, 2,4-D or pyrasulfotole plus bromoxynil for improved control (Barber et al., 2015; Thompson, 2014). Care should be taken with phenoxy-containing products such as 2,4-D amine, dicamba, a premix of fluroxypyr and bromoxynil (Starane® NXT), a premix of 2,4-D and dicamba (Weedmaster®) and a premix of dicamba and halosulfuron (Yukon®) to avoid off-target movement onto sensitive crops. Prosulfuron provides excellent control of broadleaf weeds that are not resistant to the ALS inhibitors. Huskie® controls a wide range of broadleaf weeds, including Palmer amaranth, along with some grass suppression, but minor crop injury (leaf discoloration) is typically expected with this herbicide. However, no yield penalties have been documented with Huskie®. Specific application considerations with these herbicides are provided in Table 2.

3.1.4 Post-directed

Three herbicide products are labelled for use as post-directed or layby application in sorghum. Pendimethalin can be applied as a directed application once sorghum has reached 15 cm height. Pendimethalin provides residual control of grasses and small-seeded broadleaf weeds, but it does not control emerged weeds. Linuron (Linex®) and paraquat can be applied post-directed after sorghum is approximately 40 cm tall. Paraquat applied alone will control emerged weeds without any residual activity, whereas linuron applied alone provides excellent residual control, with little activity on emerged weeds. A combination of these two herbicides applied post-directed controls emerged weeds, along with good residual control.

3.2 Crop rotation restrictions

A strong consideration to the subsequent crop to be planted in the field is important in selecting herbicides with soil residual activity. Several herbicides used in sorghum require long plant-back intervals (Table 3). Premixes containing atrazine have rotation restrictions based on rate and timing. When atrazine or atrazine-containing products are applied late in the growing season, only corn or sorghum may be planted in the following season. The most severe restrictions are to crops such as peanut and tobacco, especially where an atrazine-based product is applied. The product label must be consulted when planning crop rotations.

Herbicide use in the season prior to sorghum planting should be another important consideration, as several products used in cotton, peanut and wheat will negatively affect sorghum growth and yield (Table 4). With grain sorghum generally planted as a dryland crop, the risk for carryover to the crop in the Great Plains is likely greater than that for corn or cotton where irrigation may be used for the latter crops. In the Midsouth where soybean may be grown in rotation with grain sorghum, fomesafen carryover can be common (Barber et al., 2015). ALS-inhibiting herbicides such as imazapic (Cadre®), mesosulfuron-methyl (Osprey®), imazethapyr (Pursuit®), pyrithiobac (Staple®) and diclosulam (Strongarm®) will be the most restrictive products, preventing sorghum from being planted. Although visible injury may not always be observed in sorghum in the year after the ALS inhibitors are

Table 3 Recommended rotation intervals (months) fcr various herbicides used in sorghum[a]

Trade name	Common name	Soybean	Cotton	Peanut	Tobacco	Corn	Wheat
2,4-D amine 3.8 SL	2,4-D amine	FS[b]	FS	FS	FS	FS	–[b]
AAtrex	Atrazine	FS	FS	FS	18	0	3
Aim EC	Carfentrazone-ethyl	0	0	0	0	0	0
Basagran	Bentazon	0	0	0	0	0	0
Bicep II Magnum	*S*-metolachlor + atrazine	FS	FS	FS	18	0	15
Buctril	Bromoxynil	1	1	1	1	1	1
Bullet/Lariat[c]	Alachlor + atrazine	FS	FS	FS	Do not plant FS	FS	3
Callisto	Mesotrione	10	10	10	10	0	4
Clarity	Dicamba	3	3	3	3	3	1
Degree Xtra[c]	Acetochlor + atrazine	FS	FS	FS	18	0	15
Dual II Magnum	*S*-metolachlor	0	0	0	FS	0	4.5
Expert	Atrazine + *S*-metolachlor + glyphosate	FS	FS	FS	18	0	10
Facet L	Quinclorac	10	10	10	24	10	0
Gramoxone SL	Paraquat	0	0	0	0	0	0
Huskie	Pyrasulfotole + bromoxynil	4	Bioassay[d]	Bioassay[d]	Bioassay[d]	9	0.25
Intrro/Micro-Tech	Alachlor	FS	FS	FS	–[b]	FS	3
Lexar/Lumax[c]	*S*-metolachlor + atrazine + mesotrione	FS	FS	FS	Do not plant FS	0	15/4.5
Linex 4L	Linuron	4	4	4	4	4	4
Marksman[c]	Dicamba + atrazine	FS	FS	FS	Do not plant FS	0	3
Peak	Prosulfuron	10	10	10	10	1	0

(Continued)

Table 3 *(Continued)*

Trade name	Common name	Soybean	Cotton	Peanut	Tobacco	Corn	Wheat
Prowl H_2O	Pendimethalin	FS	FS	FS	FS	10	4
Sandea	Halosulfuron	9	4	6	36	1	2
Sequence	Glyphosate + *S*-metolachlor	0	0	0	FS	0	4.5
Sharpen	Saflufenacil	2	3	5	5	0	0
Starane	Fluroxypyr	FS	FS	FS	FS	3	4
Warrant	Acetochlor	0	FS	FS	FS	0	4
Weedmaster	Dicamba + 2,4-D	4	4	4	4	4	2
Yukon	Halosulfuron + dicamba	9	4	6	–[b]	1	2
Zest	Nicosulfuron	0.5	10	10	10	0	8

[a] Information provided in the table is based on specific label instructions.
[b] '–' = no plant-back interval is mentioned in the label.
[c] If only applied late in the season, can rotate to corn or sorghum the following year.
[d] Bioassay based on a small-scale field planting.
'FS' = Following spring.

Table 4 Plant-back interval for sorghum following various herbicides used in the previous season[a]

Herbicide	Interval (months)	Important considerations
Accent Q	10	18 months if pH > 7.5
Authority assist[b]	18	
Authority first	12	
Authority MTZ[b]	18	12 months if applied at 1.46 L/ha or less
Authority XL[b]	18	10 months if soil pH < 6.8
Axial XL	4	
Axiom	12	
Cadre[b]	18	
Canopy	12	9 months if applied at a reduced rate
Caparol	FS	Sorghum may be planted the following season
Capreno	10	18 months if pH > 7.5, Capreno rate >219 mL/ha or thiencarbazone-methyl >0.02 kg ai/ha
Classic	12	
Command	9	
Cotoran	9	
Devrinol	12	
Diuron/Direx	FS	Sorghum may be planted the following season
Envive	10	
Evik	11	
Finesse[b]	18	Bioassay can be carried out in the treated area to determine if injury may occur
FirstRate	9	
Flexstar	10	
Flexstar GT	10	
Harmony extra	0.5	
Laudis	10	
Lightning[b]	18	
Optill[b]	18	
Osprey	10	
PowerFlex	3	
Prefix	10	
Prowl	10	
Pursuit[b]	18	
Reflex	10	

(Continued)

Table 4 (*Continued*)

Herbicide	Interval (months)	Important considerations
Resolve Q	10	18 months if dry and <38 cm rainfall or irrigation occurs during growing season
Scepter	11	
Spartan	10	
Staple[b]	18	Do not plant sorghum the following season
Steadfast Q	10	18 months if pH > 7.5
Stout	10	18 months if pH > 7.5
Strongarm[b]	18	
Synchrony STS	9	12 months if >110 ml/ha applied and pH < 7.0
Treflan	12	
Tricor[b]	18	12 months if >50 cm rainfall
Valor XLT	10	18 months if pH ≥ 7.0

[a] Information presented here is based on specific product label instructions; herbicides with a rotation interval of ≥9 months and common wheat herbicides are listed.
[b] Sorghum is not recommended in the season following application.

applied, severe yield reductions will generally occur. The product labels must be consulted to ensure safe use of these herbicides.

4 Herbicide-resistant sorghum technology

Development of herbicide-resistant crop technology is still in its infancy in sorghum compared to other major crops such as corn, soybean or cotton. The introduction of genetically engineered (GE) crops resistant to glyphosate in 1996 gained widespread acceptance as this technology facilitated weed control throughout the cropping season. Adoption of GE crops has revolutionized agricultural production, weed management and encouraged no-till technology to conserve moisture and reduce soil erosion. Since then, many seed companies have invested significantly in the development of crop cultivars resistant to herbicides such as glufosinate, 4-hydroxyphenylpyruvate dioxygenase (HPPD) inhibitors, auxinic herbicides, among others, but only in major crops such as corn, soybean or cotton. One of the primary reasons for replacement of sorghum hectarage with corn is the availability of herbicide resistance technology in corn, which facilitates convenience and ease of weed control. To enhance the competitiveness of the grain sorghum industry, it is extremely important to develop technologies that will impact grain sorghum production and productivity. Availability of grain sorghum cultivars resistant to herbicides will help broad-spectrum weed control throughout the cropping season, thereby reducing crop loss and increasing productivity.

The discovery of resistance to ALS- and acetyl-coenzyme A carboxylase (ACCase)-inhibitor herbicides in sorghum (Tuinstra and Al-Khatib, 2008a,b) is a significant step forward for weed control options targeting both grass and broadleaf weeds infesting sorghum fields. The ALS- and ACCase-inhibitor-resistant sorghum varieties were developed via introgression of the resistance trait from weedy species (e.g. shattercane and johnsongrass) into elite sorghum lines through conventional breeding, a non-GE method. A shattercane population was found to have evolved resistance to ALS inhibitors as a result of a point mutation in the *ALS* gene, resulting in the substitution of Trp-574-Leu (Kershner, 2010). This mutation is known to bestow a high level of resistance to imidazolinones and sulfonylureas (Yu and Powles, 2014). Upon introgression, the *ALS* gene with the above mutation was transferred to elite sorghum lines (Tesso et al., 2010; Kershner, 2010). Homozygous ALS-inhibitor-resistant sorghum lines in both B- and R-line backgrounds were identified and transferred to the industry for further development. Similarly, a non-GE-based ACCase-inhibitor-resistant sorghum technology was also developed via conventional breeding by introgressing the resistance trait from johnsongrass, another wild relative of sorghum. The ACCase-inhibitor-resistant johnsongrass had a point mutation in the *ACCase* gene that resulted in Trp-2027-Cys substitution, which is known to confer a high level of resistance, predominantly to the aryloxyphenoxypropionate (APP) family herbicides (also known as 'fops'), with comparatively lower resistance to the cyclohexanedione (CHD) family herbicides (known as 'dims') (Kershner et al., 2012). Thus, upon introgression, the ACCase-inhibitor-resistant sorghum lines exhibited a high level of resistance to the APP herbicide family, but not to the CHD family. This lack of cross-resistance can ensure the longer viability of this technology (Kershner et al., 2012). Both the ALS- and ACCase-inhibitor-resistant sorghum technologies were initially developed at Kansas State University, Manhattan, USA.

The development of ALS- and ACCase-resistant sorghum technology is a great step forward towards effective weed management in sorghum. Nevertheless, these two technologies alone will not be sufficient to solve all weed problems in sorghum production. Thus, the development of new techniques and tools, particularly for grass weed control, should be a continuous exercise. HPPD inhibitors such as mesotrione or tembotrione can provide useful and flexible options for both PRE and POST weed control. These herbicides are most effective on broadleaf weeds, but have activity on a wider spectrum of weeds. Recently, considerable progress has been made in identifying HPPD-inhibitor-resistant sorghum genotypes from wide germplasm collections, yet again at Kansas State University. Genotypes possessing elevated levels of resistance to mesotrione or tembotrione were identified (Jugulam et al., unpublished). Research is in progress to characterize the HPPD-inhibitor-resistant trait in these genotypes, for subsequent introgression into elite sorghum lines. Availability of HPPD-inhibitor resistance in grain sorghum will offer additional options for weed management. The recent developments in the CRISPR/Cas9 systems may also provide additional avenues for developing novel herbicide-resistant sorghum lines through targeted gene editing.

5 Non-chemical options for weed control in sorghum

A number of non-chemical weed management practices can be implemented in sorghum, including crop rotation, weed suppressive cultivars (competition, allelopathy), narrow row spacing, high seeding rate, cover crops, inter-row cultivation, inter-row mowing, harvest weed seed control, among others (Stahlman and Wicks, 2000). However, research on the development of non-chemical weed management tools for sorghum is limited.

Crop rotation is an important cultural strategy for disrupting the dominance of weeds since the rotational crop(s) have different growth characteristics and competitive interactions with weeds, as well as offer different herbicide and other weed management tools. For example, crop rotation is an important management tactic for johnsongrass in sorghum-based production systems, wherein the weed population size is kept under control in the rotational crops such as cotton or corn through the use of glyphosate and other herbicides effective on this species (authors' personal observations).

Weed suppressive crop varieties can provide the crop an advantage over the weeds and favour improved weed control with herbicides (van der Meulen and Chauhan, 2017). Weed suppression can be due to better competitive interactions with weeds, production of allelochemicals (i.e. allelopathy) or both. Sorghum produces phenolic compounds that are shown to have phytotoxic activity on other species (e.g. Cheema et al., 2007). Panasiuk et al. (1986) illustrated the occurrence of water-soluble allelochemicals produced by germinating sorghum seeds that have an impact on weed seed germination and seedling development. Sorgoleone is an important alleochemical produced by sorghum (Dayan and Duke, 2009; Dayan et al., 2003), and its intensity is influenced by genotype (Weston and Czarnota, 2001) and environmental (Hess et al., 1992) factors. Sorghum cultivars that are more competitive and have allelopathic potential on weeds can be utilized, but more research is necessary in this regard. Additionally, planting sorghum in narrow rows and increasing sorghum seeding rates can facilitate rapid crop canopy formation and increase the competitive effects of sorghum on weeds (Thate et al., 2017).

Cover crops are becoming increasingly important in row-crop systems for weed management and other benefits. Fall-planted cover crops such as cereal rye can provide residue cover for suppressing weed emergence and establishment during spring (Teasdale, 1996); sorghum can be planted into the standing/rolled cover crop residues using a no-till or strip-till planter. Grain sorghum planted into a cereal rye stand showed high levels of weed suppression compared to a no-cover crop treatment (Samuelson et al.'s unpublished findings). Cover crop residues show the most benefit for weed suppression during early season (at crop establishment) as they gradually decompose subsequently. A living mulch (intercrop) such as a short-duration cowpea could be established between sorghum rows to suppress weeds, with an appropriate termination strategy to minimize competition with sorghum (Thate et al., 2017). Weeds present between sorghum rows can also be controlled non-chemically using cultivation or mowing. Thate et al. (2017) reported that a single inter-row cultivation reduced early-season weed infestation by about 45% compared to a weedy check. Donald (2007) showed that a combination of band application of a PRE herbicide in crop rows, followed by inter-row mowing twice, was effective in controlling weeds in sorghum.

6 Weed shifts with production practices

The principal source of weed infestation in crop fields is weed seedbank, and it is greatly manipulated by tillage and other crop production practices (Froud-Williams, 1987; Roberts, 1981). For sorghum, adoption of certain tillage practices is expected to influence a shift in the dominance of weed species associated with this crop. In conservation tillage systems (no-till, strip-till, etc.), the lack of soil disturbance and incorporation favours weed species persistence (Koskinen and McWhorter, 1986). Changes in the pattern of soil disturbance, altered soil characteristics (soil moisture, temperature, bulk density and organic matter content) and crop residue retention in conservation tillage systems exert selection towards small-seeded annuals as well as perennial weeds (Locke et al., 2002; Buhler, 1995). Small-seeded annual grasses and broadleaf weeds have the ability to germinate from shallow depths and will be more adapted to the shallow burial conditions typical in conservation systems.

In the US sorghum belt, long-term conservation tillage operations could lead to the dominance of small-seeded annual weeds. Smith et al. (1991) found that arrowleaf sida (*Sida rhombifolia* L.) has a greater ability to thrive under high soil moisture, low light, low temperature and shallow soil depth than the closely related species prickly sida (*Sida spinosa* L.), and this may make arrowleaf sida become more troublesome in conservation tillage systems. Prabhu et al. (2016) found that about 41% of the weed species were different between long-term (36 years) no-till and conventional-till systems under continuous sorghum production in College Station, Texas. Further, about 60% of the species in the no-till system were small-seeded annuals {primarily waterhemp, prostrate spurge [*Chamaesyce humistrata* (Engelm. ex Gray) Small] and common purslane [*Portulaca oleracea* L.]}, compared to only about 25% in the conventional-till system, indicating a significant weed species shift with long-term no-till systems.

Further, removal of significant tillage from the system favours the proliferation of perennial weeds that are otherwise controlled by periodic tillage operations (Tuesca et al., 2001; Triplett, 1985). In the long-term (36 years) tillage management study reported by Prabhu et al. (2016), the no-till sorghum plots had greater densities of johnsongrass, the dominant perennial weed species present in the system, compared to the conventional-till plots. Some level of strategic tillage and the use of systemic herbicides along with effective non-chemical practices are critical to manage johnsongrass and other perennial weeds in no-till sorghum production systems.

Herbicides constitute another important selection force for weed species adaptation in sorghum production. As of now, two grasses [junglerice, liverseed grass (*Urochloa panicoides* P. Beauv.)] and four broadleaf weeds (Palmer amaranth, waterhemp, redroot pigweed and kochia) have been reported to have evolved resistance to herbicides in sorghum-based production systems globally (Heap, 2017). Herbicide resistance is also confirmed in the weedy relatives shattercane and johnsongrass. The documented cases of herbicide resistance in shattercane are exclusively from the United States and for the ALS-inhibitor herbicides (Heap, 2017). For johnsongrass, resistance has been reported across the world for several herbicide sites of action, including 5-enolpyruvylshikimate-3-phosphate (EPSP) synthase-inhibitors, ALS inhibitors, ACCase inhibitors and microtubule inhibitors, primarily in corn- and soybean-based production systems. Further, multiple herbicide resistance in johnsongrass to the EPSPS inhibitor glyphosate and the ACCase inhibitor haloxyfop-methyl was reported in Argentina (Heap, 2017). Werle et al. (2017) documented the widespread occurrence of ALS-inhibitor resistance in weedy sorghum (shattercane and johnsongrass) populations in Nebraska and northern Kansas. In most cases, resistance evolution is favoured by the use of a single or few herbicide sites of action without sufficient management diversity. Moreover, the likelihood for gene flow and hybridization between cultivated sorghum and its weedy relatives could favour the transfer of herbicide resistance trait(s) from the former to the latter. The use of existing and new herbicide technologies must be combined with a range of diversified non-chemical tactics to preserve the effectiveness of all available options.

7 Conclusion

Weed management, especially grass weed management, will continue to be a major production challenge for sorghum due to the lack of effective POST herbicide options for grass control. More efforts should be invested in developing novel weed management tools and technologies for sorghum. The utilization of advanced tools such as gene editing will be particularly helpful. These new tools can provide improved options for grass control, especially selective control of weedy relatives in sorghum. However, research is vital to develop gene flow mitigation techniques and field stewardship practices to reduce the risk of gene flow between cultivated sorghum and its weedy relatives and to prolong the effectiveness of these novel tools. Sustainable weed management will require diversification of strategies that incorporate both herbicides and non-chemical tactics. More research is imperative to develop suitable non-chemical weed management options for integration with herbicide tools in sorghum production.

8 Where to look for further information

A standard introduction to the subject can be found in the following book:

Smith, C. W. and Frederiksen, R. A. (2000), *Sorghum: Origin, History, Technology, and Production*, Wiley Series in Crop Science, John Wiley & Sons Inc., New York, NY, 840p.

There are a number of other articles that cover the following general topics:

1 Critical Period of Weed Control
Knezevic, S. Z., Evans, S. P., Blankenship, E. E., Van Acker, R. C. and Lindquist, J. L. (2002), 'Critical period of weed control: The concept and data analysis', *Weed Sci.*, 50, 773–86.
2 Herbicide Resistance in Weeds
Norsworthy, J. K., Ward, S. M., Shaw, D. R., Llewellyn, R. S., Nichols, R. L., Webster, T. M., Bradley, K. W., Frisvold, G., Powles, S. B., Burgos, N. R., Witt, W. W. and Barrett, M. (2012), 'Reducing the risks of herbicide resistance: Best management practices and recommendations', *Weed Sci.*, 60(spl), 31–62.
3 Weed Population Shifts With Production Practices
Vigueira, C. C., Olsen, K. M. and Caicedo, A. L. (2013), 'The red queen in the corn: Agricultural weeds as models of rapid adaptive evolution', *Heredity*, 110, 303–11.
4 Non-Chemical Weed Management
Upadhyaya, M. K. and Blackshaw, R. E. (2007), *Non Chemical Weed Management: Principles, Concepts and Technology*, CABI Publishing, Wallingford, UK, p. 239.

9 References

Barber, T., Scott, B. and Norsworthy, J. (2015), 'Weed control in grain sorghum', in *Arkansas Grain Sorghum Production Handbook (MP297)*, https://www.uaex.edu/publications/pdf/mp297/gschapter8.pdf (accessed 22 May 2017).

Buhler, D. D. (1995), 'Influence of tillage systems on weed population dynamics and management in corn and soybean in central USA', *Crop Sci.*, 35, 1247–58.

Burnside, O. C. and Wicks, G. A. (1967), 'The effect of weed removal treatments on sorghum growth', *Weeds*, 15, 204–7.

Cheema, Z. A., Khaliq, A., Abbas, M. and Farooq, M. (2007), 'Allelopathic potential of sorghum (*Sorghum bicolor* L. Moench) cultivars for weed management', *Allelopathy J.*, 20, 167–78.

Dayan, F. E. and Duke, S. O. (2009), 'Biological activity of allelochemicals', in Osbourn, A. E. and Lanzotti, V. (Eds), *Plant-Derived Natural Products: Synthesis, Function, and Application*, Springer-Verlag, New York, NY, pp. 361–85.

Dayan, F. E., Kagan, I. A. and Rimando, A. M. (2003), 'Elucidation of the biosynthetic pathway of the allelochemical sorgoleone using retrobiosynthetic NMR analysis', *J. Biol. Chem.*, 278, 28607–11.

Donald, W. W. (2007), 'Between-row mowing systems control summer annual weeds in no-till grain sorghum', *Weed Technol.*, 21, 511–17.

Everaarts, A. P. (1993), 'Effects of competition with weeds on the growth, development and yield of sorghum', *J. Agric. Sci.*, 120, 187–96.

Fabrizius, C. H. (1998), 'Studies on the control and interference of longspine sandbur in corn and grain sorghum', MS Thesis, Kansas State University, Manhattan, KS, pp. 23–45.

Feltner, K. C., Hurst, H. R. and Anderson, L. E. (1969a), 'Tall waterhemp competition in grain sorghum', *Weed Sci.*, 17, 214–16.

Feltner, K. C., Hurst, H. R. and Anderson, L. E. (1969b), 'Yellow foxtail competition in grain sorghum', *Weed Sci.*, 17, 211–13.

Froud-Williams, R. J. (1987), 'Survival and fate of weed seed populations: Interaction with cultural practice', *In Proc. Br. Crop Prot. Conf. Weeds*, 2, 707–18.

Grichar, W. J., Besler, B. A. and Brewer, K. D. (2005), 'Weed control and grain sorghum (*Sorghum bicolor*) response to postemergence applications of atrazine, pendimethalin and trifluralin', *Weed Technol.*, 19, 999–1003.

Heap, I. (2017), 'International Survey of Herbicide Resistant Weeds', http://weedscience.org/default.aspx (accessed 20 November 2017).

Hess, D. E., Ejeta, G. and Buttler, L. G. (1992), 'Selecting sorghum genotypes expressing a quantitative biosynthetic trait that confers resistance to Striga', *Phytochem.*, 31, 493–7.

Hewitt, C. A. (2015), 'Effect of row spacing and seeding rate on grain sorghum tolerance of weeds', MS thesis, Kansas State University, Manhattan, KS, 94p.

Kershner, K. S. (2010), 'Herbicide resistance in grain sorghum', PhD thesis submitted to Kansas State University, Manhattan, KS, http://krex.k-state.edu/dspace/handle/2097/13069 (accessed 21 November 2017).

Kershner, K. S., Al-Khatib, K., Krothapalli, K. and Tuinstra, M. (2012), 'Genetic resistance to acetyl-coenzyme A carboxylase-inhibiting herbicides in grain sorghum', *Crop Sci.*, 52, 64–73.

Knezevic, S. Z., Horak, M. J. and Vanderlip, R. L. (1997), 'Relative time of redroot pigweed (*Amaranthus retroflexus* L.) emergence is critical in pigweed-sorghum [*Sorghum bicolor* (L.) Moench] competition', *Weed Sci.*, 45, 502–8.

Knezevic, S. Z., Evans, S. P., Blankenship, E. E., Van Acker, R. C. and Lindquist, J. L. (2002), 'Critical period of weed control: The concept and data analysis', *Weed Sci.*, 50, 773–86.

Koskinen, W. C. and McWhorter, C. G. (1986), 'Weed control in conservation tillage', *J. Soil Water Conserv.*, 41, 365–70.

Locke, M. A., Reddy, K. N. and Zablotowicz, R. M. (2002), 'Weed management in conservation crop production systems', *Weed Biol. Manag.*, 2, 123–32.

Moore, J. W., Murray, D. S. and Westerman, R. B. (2004), 'Palmer amaranth effects on the harvest yield of grain sorghum', *Weed Technol.*, 18, 23–9.

Norris, R. F. (1980), 'Barnyardgrass [*Echinochloa crus-galli* (L.) Beauv] competition and seed production', *Proc. Weed Sci. Soc. Am.*, 20, 5.

Panasiuk, O., Bills, D. D. and Leather, G. R. (1986), 'Allelopathic influence of *Sorghum bicolor* on weeds during germination and early development of seedlings', *J. Chem. Ecol.*, 12, 1533–43.

Phillips, W. M. (1958), 'Weed control in sorghum', Cir. 360, Fort Hays Branch, Kansas Agricultural Experiment Station.

Prabhu, G., Mowrer, J., Provin, T., Hons, F. M. and Bagavathiannan, M. (2016), 'Effect of long-term tillage practices on soil physical properties and above and below ground weed diversity in continuous sorghum production', in *70th Annual Meeting of Southern Weed Science Society*, 23–26 January 2017, Birmingham, Alabama, 30p.

Roberts, H. A. (1981), 'Seed banks in soils', *Adv. Appl. Biol.*, 6, 1–55.

Shaffer, G. E. (2016), 'Critical duration of grass weed interference in grain sorghum', MS thesis, Kansas State University, Manhattan, KS, 83p.

Smith, B. S., Murray, D. S., Green, J. D., Wanyahaya, W. M. and Weeks, D. L. (1990), 'Interference of three annual grasses with grain sorghum (*Sorghum bicolor*)', *Weed Technol.*, 4, 245–9.

Smith, C. A., Shaw, D. R. and Newsom, L. J. (1991), 'Arrowleaf sida (*Sida rhombifolia*) and prickly sida (*Sida spinosa*) germination and emergence', *Weed Res.*, 32, 103–9.

Stahlman, P. W. and Wicks, G. A. (2000). 'Weeds and their control in grain sorghum', in Smith, C. W. and Frederiksen, R. A. (Eds), *Sorghum: Origin, History, Technology and Production*, John Wiley & Sons Inc., New York, pp. 535–90.

Swanton, C. J. and Weise, S. F. (1991), 'Integrated weed management: The rationale and approach', *Weed Technol.*, 5, 648–56.

Teasdale, J. R. (1996), 'Contribution of cover crops to weed management in sustainable agricultural systems', *J. Prod. Agric.*, 9, 475–9.

Tesso, T., Kershner, K., Ochanda, N., Al-Khatib, K. and Tuinstra, M. (2010), 'Registration of 34 sorghum germplasm lines resistant to acetolactate synthase-inhibitor herbicides', *J. Plant Reg.*, 5, 215–19.

Thate, T., Samuelson, S., Bastos-Martins, M. and Bagavathiannan, M. (2017), 'Evaluating different non-chemical weed management options for grain sorghum', in *Proceedings of the Texas Plant Protection Association Annual Meeting*, 5–6 December 2017, Bryan, TX.

Thompson, C. R. (2014), 'Sorghum weed management and update 2014', http://www.agronomy.k-state.edu/extension/documents/sorghum-schools/2014-sorghum-school/Thompson-Sorghum-2014.pdf (accessed 22 May 2017).

Triplett, G. B. (1985), 'Principle of weed control for reduced -tillage corn production', in Wiese, A. F. (Ed.), *Weed Control in Limited Tillage Systems*, Weed Sci. Soc. Am., Champaign, IL, pp. 26–40.

Tuesca, D., Puricelli, E. and Papa, J. C. (2001), 'A long-term study of weed flora shifts in different tillage systems', *Weed Res.*, 41, 369–82.

Tuinstra, M. R. and Al-Khatib, K. (2008a), 'Inventors; Kansas State University Research Foundation assignee, Acetolactate synthase herbicide resistant sorghum', US Patent 2008/0216187 A1.

Tuinstra, M. R. and Al-Khatib, K. (2008b), 'Inventors; Kansas State University Research Foundation assignee, Acetyl CoA carboxylase herbicide resistant sorghum', WO Patent application 2008/089061 A1.

Unruh, B. J. (2013), 'Influence of nitrogen on weed growth and competition with grain sorghum', MS thesis, Kansas State University, Manhattan, KS, 66p.

USDA-FAS (2016), 'Global Agricultural Trade System', Foreign Agricultural Service, United States Department of Agriculture, http://apps.fas.usda.gov/gats/default.aspx (accessed 25 November 2017).

USDA-NASS (2012), 'Chemical usage', USDA National Agricultural Statistics Service, https://quickstats.nass.usda.gov/results/1DBF95A2-F6D6-31D5-987C-49033E4CA80D#B9DA685A-735E-3BE7-9500-FE4701C9700F (accessed 22 May 2017).

USDA-NASS (2017), 'Crop Acreage Report', United States Department of Agriculture, National Agricultural Statistical Service, https://usda.mannlib.cornell.edu/usda/current/Acre/Acre-06-30-2017.pdf (accessed 25 November 2017).

van der Meulen, A. and Chauhan, B. H. (2017), 'A review of weed management in wheat using crop competition', *Crop Prot.*, 95, 38–44.

Webster, T. (2012), 'Weed survey-southern states', *Proc. South Weed Sci. Soc.*, 64, 267–88, http://www.swss.ws/wp-content/uploads/docs/2012%20Proceedings-SWSS.pdf (accessed 22 May 2017).

Werle, R., Begcy, K., Yerka, M. K., Mower, J. P., Dweikat, I., Jhala, A. J. and Lindquist, J. L. (2017), 'Independent evolution of acetolactate synthase-inhibiting herbicide resistance in weedy Sorghum populations across common geographic regions', *Weed Sci.*, 65, 164–76.

Weston, L. A. and Czamota, M. A. (2001), 'Activity and persistence of sorgoleone, a long-chain hydroquinone produced by *Sorghum bicolor*', *J. Crop Prod.*, 4, 363–77.

Yu, Q. and Powles, S. B. (2014), 'Resistance to AHAS inhibitor herbicides: Current understanding', *Pest Manag. Sci.*, 70, 1340–50.

Chapter 18

Post-harvest management of sorghum

B. Tran and R. Hodges, Natural Resources Institute – University of Greenwich, UK

1 Introduction

Sorghum (*Sorghum bicolor* [L.] Moench) (Fig. 1) originated in north-eastern Africa and was domesticated there around 5000–8000 years ago. The largest diversity of cultivated and wild sorghum is also found in this part of the continent. An important secondary centre of origin of sorghum is the Indian subcontinent, with evidence for early cultivation dating back about 4500 years. There is still a high level of untapped biodiversity, with 17 of the world's 25 recognised species of sorghum being native to Australia (Dillon et al., 2004).

Sorghum has long been reported as being the world's fifth major cereal in terms of production and acreage. It is suited to hot and dry agroecologies where it is difficult to grow other food grains as these areas are frequently drought-prone and characterised by fragile environments. Consequently, it is a staple food crop for millions of the poorest and most food-insecure people in the semi-arid tropics of Africa, Asia and Latin America.

1.1 Adaptation to dry areas

Grain sorghums are generally grown in regions which are too dry or too hot for successful maize production. They are adapted to the drier climates due to several factors (Bennett et al., 1990):

- The ability to remain dormant during drought and then resume growth;
- Leaves roll up as they wilt, reducing the area of leaf exposed for transpiration;

http://dx.doi.org/10.19103/AS.2017.0015.19

Figure 1 Sorghum field. Photo by Bruce McLennan via Pixabay.

- Leaves and stalks contain an abundance of waxy coating which protects them from drying; sorghum exhibits a low transpiration ratio (kg water required to produce a kg of plant material), for example, 141 kg for sorghum, 170 kg for maize and 241 kg for wheat;
- Sorghums have a large number of fibrous roots that are particularly efficient at extracting moisture from soil (the absorption area is about twice that of maize); roots may be up to 2.5 m in length; the root absorption area is relatively large compared to the leaf area;
- Sorghums can withstand temperatures above 38°C, although dry winds coupled with hot weather during pollination reduce yields. Best yields are realised when temperatures during the season are 24–27°C;
- The water requirements for sorghum vary within the range of 350–700 mm depending on the length of the growing cycle (90 to more than 130 days). Sorghums can tolerate a wide range of soil pH and textures.

1.2 Traditional vs. industrial production

- Traditional production takes place in developing countries, primarily in Asia and Africa, where sorghum is mainly used for food. The largest group of producers are small-scale subsistence farmers with minimal access to production inputs such as fertiliser, pesticides and improved seeds (hybrids or varieties). They also lack access to good soil, water and improved credit facilities for the purchase of inputs. Yields are generally low and can vary considerably from year to year. Both yield and quality are affected by a wide array of biotic stresses (pests and diseases) and abiotic stresses such as drought and poor soils.

- In industrialised countries and also in some developing countries, sorghum production is comparatively modern, mechanised, high-input and large-scale; sorghum is primarily used for animal feed, and yields are higher.

1.3 Varieties

There are many varieties of sorghum ranging in colour from white through red to brown (Fig. 2). Traditional varieties are open-pollinated from which rural farmers retain seed for planting in the next season. Yields tend to be lower than modern hybrids. However, hybrids are only cost-effective when grown with supporting inputs of fertiliser, weed and pest control and good water management. Commercial production of hybrid seed is a problem in many developing countries, and some rural farmers do not appreciate that harvested hybrid grain cannot be retained for planting the next season. Therefore, they find sorghum production from hybrid seed expensive, even though the yields are higher than the land races.

Figure 2 Sorghum panicle. Photo by Kathryn Cox via Pixabay.

1.4 Yield

Yields from open-pollinated varieties under rain-fed conditions range from 0.3 to 1.0 tons/ha. In contrast, hybrids can yield up to 12 tons/ha under ideal inputs, soil and water conditions and higher densities of planting. Resource-poor farmers prefer varieties incorporating the characteristics of resistance to insects, disease, drought and birds and those with acceptable yields of both grain for human consumption and fodder for livestock feed. Although yields of traditional varieties are low, they are sustainable under conditions which would make maize production unfeasible or unprofitable. Commercial producers prefer dwarf varieties suitable for harvest by combine.

Sorghum is most often a dual-purpose crop: both grain and stems are highly valued outputs.

1.5 Grain structure and nutrition

Sorghum grains (Fig. 3) have a structure which is broadly similar to that of other cereals. The major components of the grain, from the outside working inwards, are the pericarp (outer covering), the testa and the endosperm (circa 84% of the grain). The germ or embryo is on the side of the endosperm and represents about 10% of the grain. The seed coat makes up the remaining 6%.

The composition of sorghum is similar to that of maize or wheat. The grain has more protein and fat than maize, but has a lower vitamin A content than yellow maize. An average sorghum grain is composed of 70% starch, 12% protein, 3% fat, 2% fibre and 1.5% ash and is relatively rich in iron, zinc, phosphorus and B-complex vitamins. However, sorghum proteins are considered to be of low quality due to a low lysine content, which

Figure 3 Sorghum grain. Photo by Vijaya Narasimha via Pixabay.

is similar to maize. A relatively high proportion of the total proteins are prolamines; these are highly cross-linked which results in a lowered rate of human digestibility. This feature is beneficial for people with diabetes; furthermore, sorghum starch is gluten-free, making sorghum a good alternative to wheat flour for individuals suffering from celiac disease.

The endosperm contains starch, and the testa, when it is present, may contain tannins; these are found in grains with a brown pericarp and pigmented testa. Tannins are characteristic of early sorghum cultivars with dark grain colour. Newer hybrids are lower in tannin and have better feed quality. Certain tannins, known as condensed tannins, form complexes with proteins and reduce their digestibility. They can also form complexes with the alimentary tract proteases, reducing the digestibility of the proteins in the grain. Despite this negative nutritional effect, high tannin varieties continue to be grown due to their bird and insect resistance, and higher malting potential than white grain varieties. In some traditional foods and beverages, the phenolics of red sorghum give a desired flavour and colour.

Tannins also contain antioxidants that are now believed to protect against cell damage, a major cause of diseases and ageing, but the negative effects of tannins on nutritional value can partially be overcome by removal of the testa by mechanical dehulling, or by alkaline treatment at the village level, traditionally by using wood ash.

Some sorghum varieties also contain nitrates and prussic acid (cyanide) in young shoots, which is mostly a problem for grazing during early stages of the plant's growth. This can be a problem with malted sorghum but generally does not affect grain that has not started germinating. When stressed by drought or heat, plants can also contain toxic levels of cyanide and/or nitrates at later stages in growth.

Other limitations of sorghum's nutritional qualities as animal feed are caused by difficult processing by milling or grinding and the grain must be crushed and cooked to improve digestibility. Its feed value is 90–95% of that of maize, which is reflected in grain price for feed.

Evidence from controlled studies is mounting and shows that aflatoxins from mould are found on sorghum grain, and this is a particular concern in the case of non-ventilated storage like pit stores. Fumonisins, another group of mycotoxins, have also frequently been found on sorghum. They are sometimes very common in India where the grain in the field becomes blackened.

1.6 World production

Sorghum, apart from being a subsistence crop, is an important commercial and export crop for the USA, Australia and Argentina. In these countries, dwarf hybrid varieties are grown and harvested mechanically, predominantly for livestock feed. The major areas of sorghum production are listed in Tables 1 and 2 (FAOSTAT, 2017).

Although not reported in 2013, Sudan is also a major producer (3.9 Mt in 2008). With less than 0.5 million tonnes, other producers of sorghum are Yemen, Venezuela, Ukraine, Italy, Uganda, Togo, France, Ghana, Myanmar, Uruguay, Mozambique, the Russian Federation, Rwanda, South Africa, El Salvador, Kenya, Pakistan, Benin, Saudi Arabia, Haiti and Paraguay. 51 other countries report production levels under 100 000 tonnes.

Production of sorghum has increased dramatically in Tanzania in recent years, to 0.8 Mt in 2013, as farmers replaced maize with the drought-resistant crop in areas where rainfall is becoming more erratic due to climate change. Following lobbying by the ICRISAT-led Hope Project, the government included improved varieties of sorghum in its seed subsidy programme and agreed to provide a fertiliser subsidy programme for sorghum. Tanzania's

Table 1 2013 Sorghum production by tonnage per country (in excess of 0.5 million t)

Country	Tonnes
United States	9 882 000
Nigeria	6 700 000
Mexico	6 308 000
India	5 280 000
Ethiopia	4 338 000
Argentina	3 636 000
China	2 895 000
Australia	2 230 000
Brazil	2 073 000
Burkina Faso	1 880 000
Niger	1 320 000
Cameroon	1 150 000
United Republic of Tanzania	832 000
Plurinational State of Bolivia	821 000
Mali	820 000
Egypt	762 000
Chad	745 000

Table 2 2013 World and regional production

Region	Tonnes
World	62 118 000
Africa	25 330 000
America	23 948 000
Asia	9 343 000
Europe	1 267 000

farmers have reported that improved sorghum varieties grow quickly, demand less labour and are more resistant to pests and diseases.

Over the past 30 years, annual world production of sorghum has not changed much, from 62.8 to 62.2 million tonnes. Africa's and Europe's shares have increased and America's and Asia's have decreased.

World average yields in 1978–80 and 2008–10 were virtually the same (1400 and 1412 kg/ha). However, these global figures mask wide variations nationally. In India, for example, between 1978 and 2010, the area planted to sorghum fell from 16.1 to 7.7 million ha and annual production fell from 11.4 to 7.0 million tonnes, but yields increased by 40% from 689 kg/ha (in 1978–80) to 965 kg/ha (in 2008–10).

1.7 World trade

The volume of trade in sorghum is small compared with that in the major grains such as wheat, maize, barley and rice. The main importers of sorghum are Japan, Mexico, the former USSR (CIS) and Venezuela, while the major producers of sorghum for domestic or foreign trade are the USA, Argentina and Australia. Most is used in livestock feed.

Within most developing countries, the sorghum crop rarely reaches the market. It is grown for home consumption unless there is a bumper crop, or if cash is needed.

The market price for sorghum is a function of its value in terms of its demand, purpose and nutritional quality. Livestock feed manufacturers procure feed materials according to a price per nutrient basis.

Since the nutritional value of sorghum is broadly 85–90% of that of maize, due to the lower digestibility of the nutrients it contains, assuming that both were equally available, sorghum would have a lower relative value. In reality, the price of grains has been influenced by political, social and agronomic factors, not always in support of sorghum, and often in support of maize, usually involving a subsidy.

2 Post-production operations

Post-production operations are listed in Fig. 4. This section is based on the INPhO sorghum factsheet produced by Natural Resources Institute of the University of Greenwich, UK, and collaborators for FAO.

2.1 Harvesting

In developing countries, almost all sorghum is harvested by hand. The panicle is cut from the standing stalk at about 16–20% moisture content, and the stalks are often left for animals to graze the best of the residual leaf material. In other communities, the stalks are cut and stored for use as dry season animal fodder, or for house thatching and fencing. Late harvesting can lead to spontaneous shedding of the grain from the panicles, resulting in significant losses and grain deterioration due to rapid changes in temperature and humidity.

In the USA, Australia and Argentina where dwarf hybrids are popular, the sorghum grain is threshed from the standing stalks by combine harvester. Grain with up to 25% moisture content can be harvested, but requires careful drying before storage.

2.2 Transport

The bulk density of sorghum grain at 520–720 kg/m^3 is similar to that of maize, and hence transport costs will be comparable for any particular type of consignment – bag or bulk.

Figure 4 Post-production operations.

2.3 Threshing

In rural Africa, threshing involves beating the dried sorghum panicles with sticks on the ground or in sacks, or using a mortar and pestle. Grain is separated from dirt and chaff by winnowing (Fig. 5). The time required for threshing depends on variety, the degree of dryness of the grain and the method of threshing. In some places in India, a common practice for threshing the grain is to place it on the road for vehicles to run over. It has been recorded that the grain from 2 to 4 ha of land (approx. 1–2 tonnes) can be threshed on the road in one day by a few people. Normally about 35 labourers are required to thresh about one half hectare (300–400 kg).

General guidelines for improved threshing efficiency and yield at rural level are:

- To reduce the amount of winnowing, thresh the panicles on mats, adobe or cement blocks, not on sand, gravel or stones;
- Thresh early to reduce field exposure to birds, rats and so forth (but ensure that the moisture content is low enough);
- Maximum moisture content of the grain before long-term storage should be 10–12%;
- Vitreous, flinty-starch type sorghums should be threshed early to reduce the number of broken grains.

The grain may be stored as un-threshed panicles or threshed before storage. Both storage methods are practised, but small-scale producers tend to store the grains un-threshed.

In many developing countries, motorised threshers (Fig. 6) have found mixed acceptability due to breakage of softer varieties of grain, problems of machine maintenance and availability of spare parts. They operate by passing the grain-bearing panicles between a

Figure 5 Winnowing sorghum in India. Photo by Bishnu Sarangi via Pixabay.

Figure 6 Mechanical threshing in India. Photo by Bishnu Sarangi via Pixabay.

moving rotor and a fixed metal plate. The loosened grains and panicle fibres are separated in a forced air current.

Where farms are sufficiently large, threshing can be achieved using mechanical combines at the time of harvest.

2.4 Drying

The moisture content of sorghum must be reduced to a safe level (ideally 10–12%) before storage.

The reasons are:

- To prevent mould growth (and thus the possibility of mycotoxin development by a range of storage fungi – especially *Aspergillus flavus* – which can, under appropriate storage and field conditions, produce carcinogenic aflatoxins);
- To reduce the likelihood of insect attack;
- To prevent grain germination.

For all stored products, the maximum 'safe' moisture content for storage is that which is in equilibrium with 70% relative humidity (RH).

During storage the moisture content of the grain will equilibrate with the surrounding air, as a function of its RH and temperature. Moulds develop if the grain's moisture content is around 18% or more, corresponding to 80% RH at usual storage temperatures.

Field drying: In many semi-arid areas, the traditional varieties tend to ripen after the rainy season and dry satisfactorily on the panicle. The most common method is to stack bundles of panicles in the field and allow them to dry in the sun. Grain on the panicles or as threshed grain should be kept off the ground on raised platforms, mats or trays whilst it is being dried. Many insects will walk away from grain spread in the sun or are killed if the temperature of the grain is high.

Figure 7 Drying sorghum in India. Photo by Bishnu Sarangi via Pixabay.

The time taken for sun drying of threshed grains depends on the ambient temperature, RH, depth of grain, bulk density and the frequency of turnover. In India, on a mud floor, sorghum grain in a layer of 20-mm depth can be dried from 16% to 9% moisture in one day (Fig. 7). Drying requires labour for spreading and turning the grain. In a study on the drying of sorghum in the shade, the moisture content of the grain was reduced from 32% to 13% in 24 h at a daytime temperature of 29°C.

Drying grain always involves a reduction of the grain weight in absolute terms. If grain was bagged prior to drying, a bag of 100 kg, subsequently dried from 32% to 13% RH would then only weigh about 80 kg. However, this situation is rare as the grain is normally dried prior to bagging and weighing and grain weight is therefore only meaningful for grain at the level of moisture content required for storage. As a consequence, appropriate grain quality control will measure moisture content at the same time as weighing.

2.4.1 Drying technologies

In developing countries, the sorghum grain is sun-dried on the panicle, and/or after threshing. Drying is traditionally achieved by spreading the grain on a flat surface such as a courtyard, a roof or a road side. Improved solar drying technologies have been developed for grains and could be used for sorghum, although the high temperatures reached would need to be tested for their effect on germination, and such drying may therefore be restricted to sorghum stored for food (Umogbai and Iorter, 2013; Osondu et al., 2015).

When the economies of scale permit, sorghum grain may be dried by modern technologies. These are based on forced air circulation, either with or without warm air. Drying temperatures and flow rates will depend upon the design of the drier and the RH of the incoming air (Table 3) (McFarlane et al., 1995).

Solar bubble dryers are variations of heated air drying technology where the sun's radiations create heat under a transparent sheet and photovoltaic panels are used to

Table 3 Three principal systems for mechanical drying of sorghum grain

Drying with unheated air forced through a drying bin	Air drying with supplementary heat when the ambient RH is greater than 75%	Heated air drying
Lowest investment. Simple to operate. Relatively ineffective in wet weather or if RH is >75%. *Recommended airflow rates:* 0.02 m^3/s per m^3 sorghum at 14% moisture content (wet basis) up to 0.04 m^3/s per m^3 sorghum at 20% moisture content to provide the most economic drying.	Heat generated from oil, gas or electric heaters during prolonged cold and wet weather. Drying temperatures should not exceed 40°C. Drying times shorter than unheated air, reducing risk of mould growth.	Batch or continuous systems using constantly heated air. Relatively sophisticated, high investment costs. Typical airflows in the range 0.5–1.5 m^3/s per m^3 of sorghum for batch drying and 1.5–2.5 m^3/s per m^3 sorghum for continuous dryers. Short drying times.

power fans (Salvatierra-Rojas et al., 2017). This technology is well suited to developing countries, but often too expensive for small-scale farmers.

2.4.2 Moisture meters

Moisture meters are required for properly assessing grain moisture content. Reliable meters are still relatively expensive and surely out of reach for widespread use at small-scale farmer's level, with prices from the US$500.00 mark for professional and quality assured meters but ranging between US$50.00 and 100.00 for cheaper online order options of less well-known origin. Really cheaper alternative methods for measuring sorghum grain moisture content include the traditional salt bottle method, which would need to be indexed for sorghum. In a similar fashion, production of a series of isotherms linking sorghum grain's moisture content to environmental RH would allow the use of much cheaper RH meters – or even papers – to become widespread. Such isotherms have to be produced for each variety of grain, and for all prevalent temperatures, which means the approach would need a concerted effort in each sorghum-growing area.

2.5 Cleaning

In traditional systems, grain cleaning is achieved by winnowing (to remove the low-density material such as leaf and stalk), while washing in water will remove most dust and stones. In mechanised systems, forced air (aspiration) is used to remove low-density material, while most stones, dust and other material are removed as the grain passes over a series of screens. Ferrous metal should be removed by a permanent magnet placed in the flow path of the grain.

2.6 Packaging

Bags used for sorghum can be made of jute, cotton, woven polypropylene or multilayer paper. Woven polypropylene bags are light in weight and of low cost and permit aeration. Their disadvantages are that hooks can irreparably damage the bags, and as they have a slippery surface, they can be difficult to stack.

Sacks are often reused, and care should be taken to prevent reinfestation of clean grain by boiling sacks in water and thorough drying.

Hermetic bags are discussed in the following section.

2.7 Storage

The goal of good storage is to be able to deliver grain from store in good quality and with no loss in quantity. This is achieved by preventing the deterioration caused by:

- Adverse climatic conditions;
- Contamination by extraneous material;
- Grain germination; and
- Pest infestation.

Ensuring that the storage environment is clean and tidy and in a good state of repair, makes a major contribution to the quality maintenance during storage, but it is insufficient to prevent losses by pests.

Karthikeyan et al. (2009; cited in Suleiman and Rugumamu, 2017) estimated that 60–70% of sorghum produced in Nigeria is stored at home level in indigenous structures ranging from bamboo baskets to mud structures, gunny bags and modern bins.

Since the introduction of high-yielding varieties, there has been a noticeable shift in the requirements to upgrade storage practices at both rural and commercial situations. Traditional varieties tend to be resistant to insect damage, whereas the hybrids are considerably more vulnerable. This difference in insect resistance has raised the profile for appropriate methods for insect control in stored sorghum.

Sorghum can be stored on the panicle or as a threshed grain. At the household level, sorghum grains are kept in a variety of stores for day to day food needs, seed and sale and as insurance against the risk of periodic grain shortages. Where containers are open to air movement, such as open-walled wooded cribs, the panicles may be put in store at 15–16% moisture content. Low RH after harvest ensures that the grain continues to dry while in store. This figure is well above the 13.5 or 14% moisture threshold for safe storage of threshed grain (Government of India recommends 14% and Codex recommends 14.5% as safe).

Storing grain on the panicle reduces the vulnerability of traditional varieties to pest and mould damage, but takes up a larger space in storage, and un-threshed grain is difficult to protect with insecticides.

Locally available materials used for storage structures include soil from termite mounds, wood, plant stalks, straw, bricks and cement. Forest products for store construction and insect control are becoming increasingly scarce as the demand for wood for fuel and construction timber increases. At the same time, skills to build granaries from local materials are disappearing as young men leave villages for work in urban areas.

Common forms of storage include jute bags, metal drums and bins, baskets, underground pits, clay pots and bins of stone or mud plaster. If grain is required for seed it is often dried on the panicle. The advantages and disadvantages of different methods of storage are summarised in Table 4.

Traditional storage systems are well suited to their environments and the varieties of grain being stored. Losses are generally low, below 5% of grain weight over a season. In India, local sorghum varieties show weight losses of 1–2% after 12 months of storage,

Table 4 The advantages and disadvantages of different methods of storage

Method of storage	Advantages and disadvantages
Jute bags	When kept on raised platforms, they allow air movement through the grain and can be repaired and reused. Disadvantages: high cost, risk of loss through theft, water or pest damage.
Woven polypropylene bags	Like jute bags, they allow air movement and can be reused. They are cheaper than jute bags. Disadvantages: they are difficult to repair, and same risk of loss through theft, water or pest damage.
Hermetic bags (e.g. PICS and others, some having the external layer impregnated with pesticide)	If used and maintained properly, they kill insect pests already present in the grain, and stop further reinfestation. Disadvantages: higher cost than woven polypropylene, must be handled carefully, susceptible to rodent damage and no air exchange for moisture control: the grain must be dry before storage.
Metal or plastic drums, bins and clay pots	Can be hermetically sealed, long life if shaded from direct sunlight. Good protection from external pest attack. Disadvantages: minimal air exchange for moisture control of damp grain so grain must be dried well before storage.
Baskets woven from stalks that can be plastered with mud/cow dung.	Large quantities stored at low cost – this is the most common storage method used by small-scale producers. Disadvantages: vulnerable to pest damage.
Underground pits. Walls usually heated with fire and lined with straw, brick or cement. Sealed with a straw/mud plug. Improved pits are lined with plastic sheeting, which improves resistance to water ingress from the walls.	Large quantities stored at low cost; relatively secure from theft. Disadvantages: grain can be seriously affected by mould growth and mycotoxin development if there is moisture ingress. Because they are cooler than the ground and air above, the grain needs to be dryer than for other stores. It is also important to air pits well before storage.
Bulk silos or granaries	Steel, aluminium or concrete structure built to heights of 30–50 m. Heat transfer from corrugated sheets is less than that from plain sheets. Plain sheets are preferred for areas with wide diurnal variations in temperature. Effective for centralised storage and distribution. Disadvantages: High operational and social costs.

compared with up to 30% for hybrid varieties. It is important to note that there has been confusion in the past between percent damaged grains and weight loss, and losses exceeding 30% were erroneously reported in some countries.

Storage treatment practices must be effective whether at the level of rural crib, underground pit or concrete silo. For effective sorghum storage, the following items should be assessed:

- Type/variety of grain;
- Post-harvest handling methods (threshing, drying, transport) and constraints;
- Advantages and disadvantages of traditional storage methods;
- Causes, extent and value of storage losses;

- What the farmer is doing to minimise the loss;
- Why the farmer is storing and how long for, and future expectations;
- Available, appropriate methods of storage loss reduction;
- The cost and benefits of existing, and alternative methods of loss reduction, taking account of cash, material, labour inputs and anticipated market prices.

For sorghum as for other grains, hermetic storage (Navarro et al., 1994) has several advantages:

- If used properly, it will control insect pests very effectively;
- It does not require the expense and health risks associated with the use of insecticides;
- It has been reported informally that it may help control aflatoxins;
- There are options available that makes it suited to small-scale farmers of larger-scale farmers or even traders, with associated cost variations.

The main disadvantages are:

- The cost is higher than traditional methods – if one does not take into account the cost of storage pesticides;
- This storage is only as good as the hermetic seal: it is crucial that this is checked carefully;
- The grain must be very dry before storing, as hermetic storage will not allow further drying.

Hermetic solutions range from lined bags such as PICS bags (Baoua et al., 2014) and several alternatives now being sold in various developing countries, sealable buckets and barrels and specially designed plastic or metal silos. A producer or trader can select the technology most appropriate to his needs and finances.

Small hermetic storage solutions like sealed buckets or small bags, if available, are ideal for seed storage and the high value of seeds makes the expense of such hermetic storage solution a better investment.

Underground pits, if well sealed, can become hermetic, and therefore protect the grain against further insect damage. However, lacking oxygen, they can then be dangerous and special care must be taken by farm workers. As for all types of hermetic storage, if farmers open them regularly to take small quantities of grain for consumption or sale, the anaerobic process must start again, and with a larger air gap above the grain, will take longer.

For an in-depth coverage and discussion of some sorghum storage cases, see the chapter on sorghum by Boxall et al. (2004), in the book *Crop Post-harvest: Science and Technology – Volume 2: Durables*. The authors present three case studies: the Ethiopian pit storage, south Indian small-scale farmers' storage management and Ethiopia's Emergency Food Security Reserve.

2.8 Post-harvest pest species

A wide range of insect pests attacks all stored sorghum grains and grain products in the arid and semi-arid tropics. Typical conditions for infestation and damage for stored sorghum grains are summarised in Table 5.

Table 5 Conditions for infestation and damage for stored sorghum grains

Insect	Optimum conditions for infestation	Damage symptoms and losses
Rhyzopertha dominica	Bores through whole grains. Is tolerant of hot dry conditions and may continue to develop at temperatures as high as 38°C and moisture contents as low as 8%.	Adults and grubs make ragged holes in grain; losses in improved cultivars often range from 17% to 66%.
Sitophilus oryzae	Preferred moisture content 12–18%. Red varieties with soft endosperm are more susceptible.	Adults and larvae feed on grain; losses range from 2% to 21% for improved cultivars in India.
Tribolium castaneum	Does not attack whole grain; feeds on damaged grains; fairly resistant to high temperatures.	Tunnelling damage; when attack is severe, grains turn greyish yellow and mouldy with pungent smell.
Sitotroga cerealella	Infestation starts in the field; develops rapidly on low moisture content sorghum stored on the panicle.	Field infestation on the upper layer of the grain; larvae bore and consume grain contents adding excreta and webbing.
Ephestia cautella, Plodia interpunctella, Corcyra cephalonica	Adult moths are active at dawn and dusk, particularly during periods of high humidity; inactive in complete darkness.	Attack the embryo of whole grains; thick webbing left on grain surface; can cause losses in excess of 10%.

2.9 Pest control during storage

Sorghum is a vital subsistence crop for rural populations in semi-arid regions, and most farmers are aware that newer high-yielding varieties are more susceptible to storage insects. Good management practices contribute to quality and reduce the extent of losses. They start with:

- Examining stores regularly for signs of insects, moulds and rodents;
- Cleanliness in and around the grain store;
- Cleaning and keeping grain dry.

Varietal resistance to insect attack, physical, mechanical and botanical pest control measures integrated with minimum use of insecticides reduce considerably the damage done by pests.

2.9.1 Synthetic insecticides

Fumigation of grain is a widely used method of controlling storage insects when large quantities are stored, such as at aggregation centres or cooperatives. Given that sorghum is mostly grown by small-scale farmers for their own consumption, it is unlikely to be routinely stored in large warehouses. Fumigation is usually restricted to commercial stores where adequate safety precautions can be taken. The only available fumigant is phosphine, and it is crucial that fumigation is undertaken by trained professionals, with all necessary equipment (fumigation sheets, weights and repair kit, joining kit, phosphine meters, appropriate personal safety equipment and warning signs). For a fumigation to be effective, it must last 5–7 days and gas concentration must be measured to ensure it stays at or above 150 ppm for at least 5 days. As phosphine gas is lethal at much lower concentrations, it is extremely important that all safety recommendations are observed.

It may well not be cost-effective to treat all grain immediately after harvest but rather only the proportion of the produce which is to be stored for three months or more. Such advice helps the farmers to decide on actions to be taken, encouraging assessment of the need to store or sell, given market prices, the quantity harvested and so on.

It has long been a standard recommendation that farmers should treat the crop to be stored with an organophosphate dust. Formulations were then improved by adding a pyrethroid to the organophosphate and more recently a neonicotinoid.

But in developing countries, good-quality pesticides are expensive and sometimes difficult to locate. Cheap pesticides are much more prevalent, but offer little to no guarantee as to their active ingredients, concentrations and even mode of application. As a consequence, farmers' health is put at risk, and often the pesticides affordable from local markets do not control the pests.

Insecticidal dusts are well adapted to the treatment of grain after harvest. Many of the same compounds are commonly applied as liquid sprays for other agricultural purposes. However, spray treatments are generally unsuitable because most farmers do not possess or cannot afford a sprayer and to prepare the spray requires dilution of an insecticide concentrate, a hazardous process especially for untrained small-holder farmers. Dilute dusts are much safer to use as they are composed of at least 95% inert material. Whilst this makes them safer, it has its own disadvantages: they are very bulky, difficult to package in quantities suitable for an individual producer's needs, and difficult to distribute to rural communities at a time when they are most needed and to where they can be conveniently obtained. Families frequently complain they have no access to insecticides when they want to apply them, and whatever little is available is expensive. Farmers who cultivate cash crops or have other sources of income may be able to afford these high prices, but most farmers in the semi-arid tropics are poor and do not have cash or credit to be able to purchase insecticides.

Much research has been undertaken on alternative methods of protecting grain against insect damage which are of low or no cost to the farmer and are therefore likely to be acceptable and utilised. In addition, these methods are more environmentally sustainable and less of a health risk than the synthetic insecticides.

2.9.2 Botanical pesticides

Throughout the developing world, farmers have traditionally used the plants and trees around them as sources of insecticides. The number and type of species used for this purpose is considerable. The literature contains many references to the effect of plants on storage insects, but this work is mostly confined to laboratory investigations. There is very little direct evidence which demonstrates that plants used by farmers are effective grain protectants. Few studies have been conducted to replicate farm methodologies, one reason being that farmers themselves are uncertain of the details of the methods and can only provide subjective assessments of effectiveness.

Plants with known insecticidal or repellent properties include: *Securidaca longipedunculata* (violet tree), *Chrysanthemum cinerariaefolium* (pyrethrum) and *Azadirachta indica* (neem). Both pyrethrum and neem are relatively unstable and may not be suitable for long-term grain storage. Although more than 130 plants have been reported as being used as storage protectants, only pyrethrum and neem have been used commercially.

In some countries, palm and coconut oils are used to protect grain, particularly those for seed. However, they are usually considered too costly for storage insect control.

2.9.3 Mechanical control

Sieving, winnowing combined with sun drying, sticky bands, water traps and baits can be used for insect control in storage. Turning grain to interrupt population development is quite feasible and effective for bulk grain in silos.

2.9.4 Physical control

In addition to hermetic storage, other physical means of controlling insect pests include high temperatures (>60°C) which inhibit development and will kill insect pests in grain if this temperature can be maintained for several hours. Dusts such as ash, sand and other mineral powders can be used to fill the interstitial spaces in grain bulk. These provide a barrier to insect movement and damage the insect cuticle, causing death by dehydration. Diatomaceous earths (naturally occurring aluminium silicates) absorb waxes from the insect cuticle and also cause death by dehydration. These are 'Generally Regarded As Safe' (GRAS) by the US Environmental Protection Agency for use with grain.

2.10 Aflatoxins and other mycotoxins

2.10.1 What are aflatoxins?

A variety of fungi attack growing crops in the field, especially when the crop is under stress and less resistant. Some of these fungi can produce toxins, such as *Aspergillus flavus* producing aflatoxins, which have been studied increasingly in recent years because of the serious threat they pose to human health.

Around 100 countries around the world have regulations governing aflatoxins in food and most include maximum permitted or recommended levels for specific commodities:

- The European Union sets limits for aflatoxin B1 and for total aflatoxins (B1, B2, G1 and G2) in nuts, dried fruits, cereals and spices. Limits vary according to the commodity, but range from 2 to 12 μg/kg for B1 and from 4 to 15 μg/kg for total aflatoxins. Sampling and analytical methods are also specified. Limits of 0.10 μg/kg for B1 and 0.025 μg/kg for M1 have been set for infant foods.
- US food safety regulations include a limit of 20 μg/kg for total aflatoxins (B1, B2, G1 and G2) in all foods except milk and a limit of 0.5 μg/kg for M1 in milk. Higher limits apply in animal feeds.

Hot and humid climates are favourable to fungus growth and toxin production by the fungus; this is therefore a particular challenge for developing countries. If grains have been infected by the fungus in the field, or if they are infected after harvest – for example, during storage in bags that have not been cleaned properly and still contain spores from the fungus – the fungus can grow on the stored grain if the RH is sufficient, typically, 70% or more.

There are many other toxins secreted by other species of fungi (mycotoxins), but aflatoxins are currently better studied. They have been found on many dried grains, maize and groundnut being the most widely reported, but many edible stored commodities are susceptible and sorghum is no exception. On sorghum, Fumonisin B1, Ochratoxin A and zearalenone have been recorded, as well as aflatoxin B1, considered the most toxic of aflatoxins (Pandey et al., 2016).

2.10.2 Presence of aflatoxins in sorghum

Although a limited number of studies have reported the presence of aflatoxins in sorghum, it appears that this is a consequence of few studies having been undertaken, not a failure to detect the toxin when a sample is taken.

Weledesemayat et al. (2016) have reported high levels of aflatoxins in sorghum in almost all samples in their study area in Ethiopia. They found higher concentrations in above-ground basket stores than underground pit stores, and concentrations increased a little with years of storage. They found that the prevalence of one type of aflatoxins, AFB1, was very high (97%) compared to a previous study in Brazil (Keller et al, 2012), which reported 39% for pre-fermented and 32% for post-fermented sorghum samples, with the latter having much higher concentrations of toxin.

Matumba et al. (2011) studied the presence of aflatoxins in Malawi. They found that all malt and beer samples, 15% and 43% of the samples of sorghum and thobwa (a local, milky, fermented drink made with maize and sorghum or millet), respectively, were contaminated with aflatoxins. The sorghum malt prepared for beer brewing had higher total aflatoxin than any other type of sample. The average aflatoxin content in the beer was 22.32 μg/L, which is higher than the permissible maximum level in ready-to-eat foods set by *Codex Alimentarius* Commission (10 μg/kg). The authors therefore concluded that the consumption of opaque sorghum-based traditional beer poses a risk of aflatoxin exposure.

In Kenya, Kange et al (2015) detected aflatoxins in almost 70% of samples taken from farmers' fields and from farmers' storage facilities in parts of the country where sorghum is more prevalent.

2.10.3 Solutions

Once present in the foodstuff, aflatoxins are in practice impossible to remove, and the only options are to destroy the grain if heavily infected (as feeding it to poultry or other farm animals would not only poison them but also customers eating their contaminated meat produce later on) or dilute the quantity of toxins by mixing the affected grain with clean grain until the quantity of toxins in the mixed grain is below the acceptable level.

In developing countries, the only currently available solution to the issue of mycotoxins in grain in general and in sorghum, in particular, is to limit the infestation of crops by fungi by avoiding crop stress through following agricultural best practice during the cropping season and then to avoid storage conditions where fungus could thrive by ensuring that the grain is stored dry enough for the corresponding RH in the store to remain below 70% (grain moisture content of 10–12%). This corresponds to post-harvest best practices, too, which include thorough cleaning of stores and equipment before the storage season, well-ventilated structures, regular grain and store quality controls and so forth [see the World Food Programme's *Training Manual for Improving Grain Postharvest Handling and Storage* (Hodges and Stathers, 2012) for in-depth coverage of grain post-harvest best practice].

Current areas of research include cheaper and easier to deploy methods for measuring aflatoxins in the field, and the use of Afla-Guard™ which aims at displacing toxin-producing strains in the field by spreading non-toxigenic strains. That approach works well in developed countries where its cost is relatively minor and large farmers' field size makes it easy to implement on a large scale. In the African context, with many small-scale farmers, it may be difficult to ensure that all farmers in an area use the non-toxigenic strains appropriately and when traders subsequently mix the harvest from several farmers, aflatoxin-contaminated stocks may be mixed with clean stocks. This may or may not be sufficient to lower the level of contamination under the threshold for health hazard. National policies will therefore be crucial in ensuring that samples are taken and aflatoxins measured on a regular basis over the whole producing area, that stocks are mixed where possible to maintain low levels of toxins and that a procedure for destroying badly affected crops is in place, probably with adequate compensation for farmers to ensure participation in the effort. As sorghum is rarely a priority crop, one can hope that it will directly benefit from such policies being put in place on other crops such as maize or groundnut.

In addition to research on aflatoxins mitigation, more research on other mycotoxins is still needed to characterise their presence and their health risks.

3 Transformation of sorghum grain

3.1 Dehulling (removal of the pericarp)

Most transformation of sorghum grain (Fig. 8) will start with dehulling and will be followed by grinding to obtain a flour.

3.1.1 Traditional methods

Traditional methods are time-consuming and arduous. In most traditional processes, the grain is dampened and dehulled by hand pounding in a mortar and pestle. There are local variations in this process, though common steps are:

- Moisten a quantity of grain in an excess of water for about 10 min;
- Drain off water, add grain to mortar and pound to abrade grains against each other;
- Winnow off separated pericarp;
- Moisten grain again with a few handfuls of water;
- Repeat pounding;
- Winnow off separated pericarp;
- Sun-dry the grain.

The time required for dehulling depends on the skill of the operator and the variety. Typical times for dehulling would be 15 kg in 1 h.

3.1.2 Machine dehulling

In the 1970s and early 1980s, the dehulling of sorghum has been successfully mechanised through the adaptation of carborundum disc barley pearling machines. These machines dehull and debran with a single pass through the machine. Bran is separated by fan

Figure 8 Sorghum grain. Photo by Vijaya Narasimha via Pixabay.

aspiration as the grain revolves between abrasive discs. The extent of dehulling depends on the size, moisture content and hardness of the grain; the dwell time in the machine and the speed and surface characteristics of the carborundum disc. Hard varieties can yield a pearled sorghum product similar to rice, whereas soft floury endosperm varieties can shatter in the machine, producing a coarse flour. Bran removed from the dehulling can be used as an animal feed.

As is often the case for mechanisation, abrasive disc dehulling has not found universal acceptance in developing countries. Problems have occurred due to limited technical and entrepreneurial skills of the operators, limited support from the manufacturers and difficulties in obtaining foreign exchange for the purchase of spare parts. Nevertheless in Botswana, where sorghum is the national grain, dehulling has been readily accepted by the small-scale private miller.

3.2 Milling

Traditional milling is by pounding sorghum grain in a mortar and pestle or stone grinding. The whole grain is thus converted to a wholemeal flour. This is the most arduous task of milling grains. To pound 15 kg of sorghum grain to a flour may take 3–5 h. At rural level, the sieving of flours to produce flour grains of specific particle size for specific culinary uses is still uncommon. In some parts of the world, for example, the Sudan, sorghum may be wet-ground by rubbing soaked grain between two stones until a fine paste is produced. This paste is used directly or left to ferment.

There are two kinds of dry flour milling methods: whole grain flour milling and decorticated grain flour milling. Making flour from whole grains can yield 90%, but digestibility is low at 18.5% with poor flavour. Making flour from peeled sorghum grain can produce high-quality flour with good digestibility, but with lower yield.

Milling is very important in China, because many sorghum foods are made of decorticated sorghum and sorghum flour. Milled sorghum is usually decorticated to remove the pericarp, followed by crushing of the decorticated grains into flour.

3.2.1 Machine grinding

3.2.1.1 Hammer mills

Power-driven hammer mills are becoming increasingly available and offer the following options for reducing grain to a flour:

- Take the cleaned grain directly to a hammer mill;
- Dehull at home and then hammer-mill;
- Dehull and debran the grain in a mechanical abrasive disc dehuller and grind the grain to a flour by hammer mill, plate mill, stone mill or roller mill.

In rural Zimbabwe, the relative charges for dehulling and milling are approximately equal. In the late 1990s, the charge for dehulling 12 kg sorghum was Z$4. The charge for milling dehulled grain to flour was Z$4. In this case, the miller kept the bran for sale as animal feed. Alternatively the charge for milling whole grain was Z$6 per 12 kg (Wood and Thomson, 1997).

3.2.1.2 Roller milling

Sorghum can be roller-milled if the variety has the desirable qualities for pericarp/bran removal, though the efficiency of bran separation from the grain is lower than that for wheat due to its different physiological structure. It is difficult to obtain a clean separation of white endosperm from the bran fraction and the final flour has a speckled appearance. Soft endosperm varieties can easily shatter in a roller mill, producing a wholemeal flour with fine bran which may be difficult to be removed by sieving (Hulse et al., 1980).

Improved milling processes have been developed, for example, roller milling after friction decortication or semi-wet milling, but these processes do not appear to have been successfully adopted in rural areas.

Mixed milling is used in China and combines the advantages of grinding milling with friction milling, with stress on the former. It is a milling method with good potential.

3.2.1.3 Wet milling for starch extraction

Starch can be extracted from sorghum by wet milling. The process is similar to that for starch extraction from maize. The pigments of sorghum may discolour the final starch, and to prevent this, low-tannin sorghums without testa are preferred.

Flour produced by wet milling is of good quality with white, fine grains and 5% higher digestibility than dry milling. The flour is easy to store. The procedure of wet flour milling is described in Fig. 9.

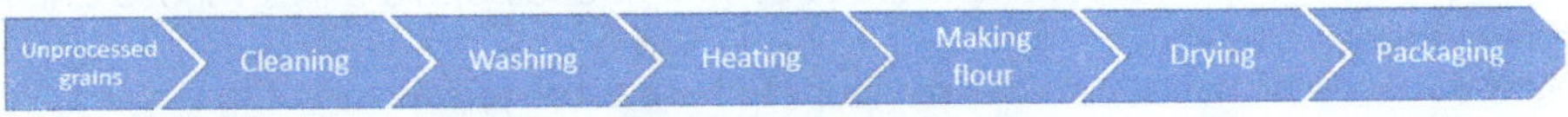

Figure 9 Processing method of wet flour milling.

3.3 Grading, sorting and packaging

For post-harvest processing of sorghum grain, machines developed for other grains like maize can generally be adapted reasonably easily. Chinese manufacturers offer machines that can cover a full, industrial value chain from threshing, dehulling, cleaning to grading and colour sorting grain, even packaging it for transport and storage.

4 Sorghum products

Sorghum's primary uses are for traditional food and beer, but it has other uses, mostly for feed and silage, but also building material, fuel, molasses and brooms, popcorn as well as wine and spirits.

4.1 Main products

In communities where sorghum is grown as a subsistence crop, the main food products prepared include thin and thick porridges, fermented and unfermented breads, couscous and snacks, lactic and alcoholic beers and beverages, malted flours for brewing, malted porridge mixes and weaning foods.

Bhakri (jolada rotti in northern Karnataka or jowar roti in many other parts of India), a variety of unleavened bread usually made from sorghum, is the staple diet in many parts of India, such as Maharashtra and the northern part of Karnataka. In eastern Karnataka and the Rayalaseema area of Andhra Pradesh, roti (jonna rotte) made with sorghum is the staple food.

In South Africa, sorghum meal is often eaten as a stiff porridge much like pap. It is called 'mabele' in Northern Sotho and 'brown porridge' in English. The porridge can be served with maswi – soured milk – or merogo – a mixture of boiled greens (much like collard greens or spinach).

In Ethiopia, sorghum is fermented to make a variety of injera flatbread, and in Sudan it is fermented to make kisra. In India, dosa flour is normally made from rice but can be diluted a little with sorghum (jowar) flour.

In the cuisine of the Southern United States, sorghum syrup was used as a sweet condiment, much as maple syrup was used in the North, usually for biscuits, maize bread, pancakes, hot cereals or baked beans. It is uncommon today.

In Arab cuisine, the unmilled grain is often cooked to make couscous, porridges, soups and cakes. Many poor use it, along with other flours or starches, to make bread.

Sorghum seeds can be popped in the same manner as popcorn (i.e. with oil or hot air, etc.), although the popped kernels are smaller than popcorn.

Sorghum sometimes is used for making tortillas (e.g. in Central America). In El Salvador, they sometimes use sorghum (maicillo) to make tortillas when there is not enough maize.

Sorghum is a staple food crop in China, especially in North China. People there can make many kinds of rich and colourful sorghum foods with varied flavours. According to Zhao (1987), there are about 40 kinds of sorghum foods. Table 6 lists various Chinese sorghum foods.

Many urban consumers consider sorghum to be a subsistence crop of low quality. This low social status for the grain constrains its desirability for inclusion in commercial products

Table 6 Sorghum foods in China

Type	Category	Name of the dish or food
Decorticated grain foods	Cooked product	gan mifan, lao mifan, shui mifan, er mifan dou mifan
	Porridge	chazi zhou, doumi zhou, ermi zhou, Jianmi zhou, naibuzi
Flour-based foods	Steamed products	YuYu, baocha, wowo, tiaotiao, jiaozi, dundun, boboye, fagao
	Boiled products	hele, miantiao, mianpian, gangsimian, miangedatang, daoxuemian, suantangzi
	Baked products	jianbing, famianbing, dabingzi
	Parched products	chao mian, youchamian
	Glutinous products	niangao, niandoubao, nianhuoshao, nianbing, suhaozi
Popping foods		penghuasu, baomihua, chaomibua

designed for urban consumers. In regions where the crop is not a staple, it may have low acceptability relative to maize due to its different organoleptic properties – unpleasant colour, aroma, mouthfeel, aftertaste and stomach-feel.

Nevertheless, since 2000, sorghum has come into increasing use in homemade and commercial breads and cereals made specifically for gluten-free diets.

Sorghum blended with wheat flour has been used over the last two decades to produce baked products, including yeast-leavened pan, hearth and flatbreads, cakes, cookies and flour tortillas. Malt drinks and malt-cocoa-based weaning food and baby foods are popular in Nigeria. However, commercial adoption remains limited. Sorghum does not contain the elastic protein, gluten, and thus the functional properties of sorghum for wheat-based bread and biscuit-type products limit its inclusion level to a practical maximum of 10–15% before changes in the structure of the product become evident. Inclusion is also dependent upon availability of sorghum, appropriate varieties and the relative price of wheat and sorghum at the mill gate. Hard endosperm sorghum is used extensively in south-east Asia for noodles.

4.2 Brewing

The steps to brewing from sorghum are:

- The grain is malted, dried, ground and mixed with water;
- An initial bacterial action on sugars forms lactic acid;
- This is followed by yeast fermentation for 4–5 days.

The resulting sorghum beer has a milky consistency with suspended starch. It also contains protein, yeast and malt. Brewing raises the nutritional value of sorghum by:

- adding vitamins;
- neutralising tannins;
- hydrolysing starch.

Earlier maturing types have superior mould resistance and are best suited for beer. Cultivars with higher tannin levels have superior malting ability and hydrolysable starch. Like barley, sorghum has some diastatic activity (capacity for enzymatic conversion of starch into sugar).

4.2.1 Lager beers

Certain varieties of red sorghums contain active amylases at concentrations suitable for certain brewing applications. For the preparation of commercial lager beers, sorghum malt is not a direct replacement for barley malt since the diastatic power of the sorghum malt is very low and variable compared to that for barley malt. Sorghum is milled for its endosperm grits as a starch source (adjunct) for hydrolysis by malt enzymes to fermentable sugars. Supplementary amylolytic and proteolytic enzymes are necessary to complete the fermentations.

In Africa, large multinational breweries have experimented with the addition of various locally available grains, legumes or roots in lager beer preparation. This includes sorghum but remains a comparatively small input compared to imported barley. In northern America, several artisanal-type breweries produce sorghum beer, which is often marketed for gluten-intolerant or celiac consumers. Tasting reviews on the Internet list well over 100 beers containing sorghum, available in the USA; so there is interest in adding sorghum to beer making, although this is not quite main stream yet.

4.2.2 Opaque beers

Africa has a tradition of making opaque beers by the use of sorghum as the source of malt and the adjunct, though for commercial brewing, maize may often be the source of the adjunct. Opaque beer is a product of a lactic and alcoholic fermentation which is sold in a microbially active state, with a shelf life of only 5–7 days. The principles of the process whether by traditional or commercial methods are illustrated in Fig. 10 (Daiber and Taylor, 1995).

Traditional sorghum beers are very important all over the African continent, and recipes vary from thick, alcoholic porridges to clear and sweet drinks. They can be brewed from sorghum only, or a mix of cereal grains. The cultural importance of sorghum beer preparation, sale and consumption cannot be overstated, and remains very much at the heart of many rural communities. These beers are also very nutritious and much richer in nutrients than European lagers. The preparation process is fast, and the beer is consumed quickly whilst it is still fermenting. But this means that the process is difficult to scale up from the local artisan model. Lyumugabe et al. (2012) reviewed several types of sorghum beers and their characteristics, and argued that pasteurisation, together with the use of starter cultures and local plants in replacement of hop, would help scaling up beer production.

4.3 Sorghum in animal feeds

Sorghum grain is one of the major ingredients in swine, poultry and cattle feed in the western hemisphere, China, Australia and South America. Sorghum is also grown for forage; in northern India, it is very common and fed to animals fresh or as silage or hay.

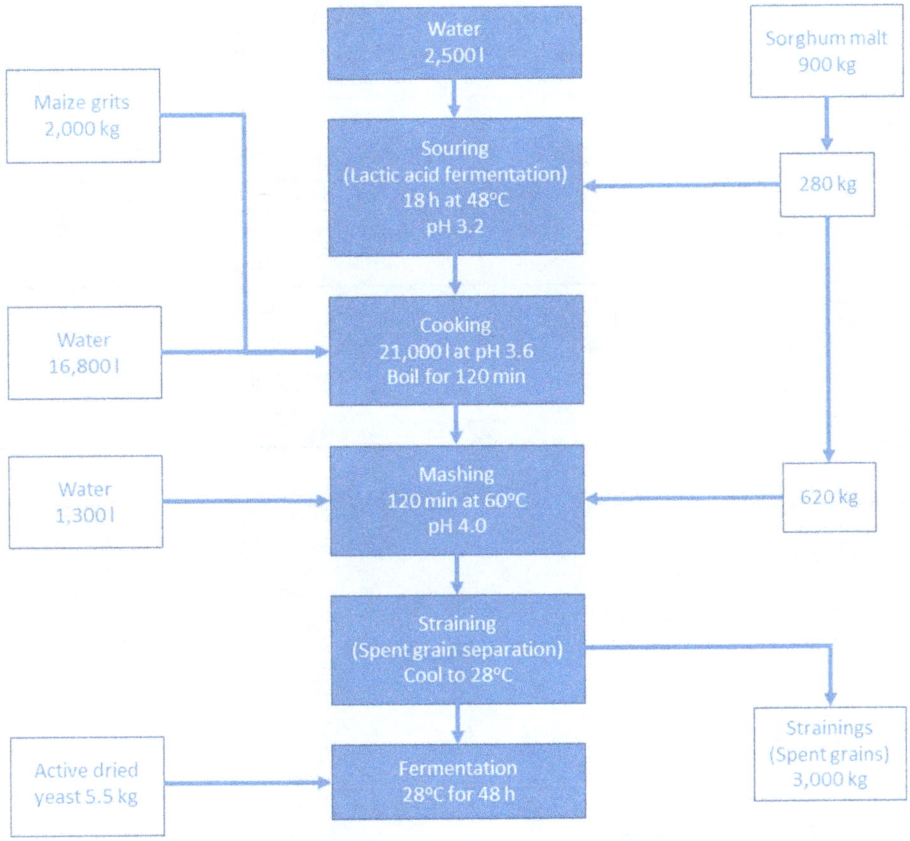

Figure 10 Principles of the production of opaque beer by the reef-type brewing process.

In the USA, sorghum grain is used primarily as a maize (corn) substitute for livestock feed because their nutritional values are very similar. Some hybrids commonly grown for feed have been developed to deter birds, and therefore contain a high concentration of tannins and phenolic compounds, which results in the need for additional processing so that cattle can digest the grain.

Livestock feed manufacturers prefer to use grains from white sorghums or low-tannin pigmented sorghums due to the effect of tannins on protein digestibility. Sorghum is therefore not a direct replacement for maize in livestock rations. Sorghum has a lower energy density and protein digestibility than maize (Table 7). This is reflected in the lower price offered for sorghum (NRI, 1988).

The typical upper inclusion limits for sorghum and maize in feeds given in Table 8. These are guideline figures since the inclusion levels will depend on the price and availability of other raw materials providing the desired protein and energy for a diet in relation to expected levels of animal performance. However, a major reason for low inclusion of sorghum in livestock feeds in developing countries is lack or inconsistency of market supply.

Table 7 Comparative data on energy and protein levels for sorghum and maize (as feed)

	Metabolisable energy for ruminants (MJ/kg)	Metabolisable energy for poultry (MJ/kg)	Protein content (%)	Lysine content (%)	Available lysine content (%)
Sorghum	12.4	13.7	11.0	0.27	0.19
Maize	12.1	14.2	9.0	0.27	0.22

Table 8 Typical upper inclusion limits for sorghum and maize in feed

	Sorghum	Maize
Poultry feeds	30%	70%
Pig feeds	30%	30%
Dairy feeds	50%	70%
Beef feeds	70%	70%

4.4 Industrial uses

Typical uses include:

- Building materials, fuel, brooms;
- Waxy sorghums for adhesives, paper, fabrics.

4.4.1 Chinese industrial uses

In China, almost every part of sorghum can be used (Table 9).

4.4.1.1 Refined sugar

Making syrup from sweet sorghum stalks has a long history in China, but the production of crystalline sugar is more recent. There are three end products in sugar refining: yellow granulated sugar, brown granulated sugar and syrup.

4.4.1.2 Vinegar

Vinegars made from sorghum grains in North China are reputed to be of very high quality. Their thickness, mellow flavour and delicate fragrance are highly valued. Shanxi Old Mature Vinegar is one of the best-known Chinese sorghum vinegars.

Table 9 Utilisation of sorghum by-products in China

Sorghum component	Utilisation
Root	Fuel
Stalk	Fuel, mat, hat, building materials, frame materials, barrier
Peduncle	Cover
Threshed spike	Broom, pot scouring brush

4.4.1.3 Monosodium glutamate

In China, the raw component to make monosodium glutamate is starch, obtained from rice, maize and sorghum.

4.4.1.4 Sorghum sweetmeat

Sorghum sweetmeat, a kind of maltose made of sorghum starch, is a popular product from the Shandong province.

4.4.1.5 Paper

The leaves and stalks of sorghum can be used for making writing paper (rough straw paper), wrapping paper and other paper products. In addition, the residue from sugar refining of sweet sorghum stalks is also used for papermaking. Compared with other papermaking materials, sorghum leaves and stalks are easy to convert into pulp, but sorghum paper is quite transparent and brittle.

4.4.1.6 Plywood from sorghum stalks

The cellulose content in sorghum stalk is very high, about 48% of dry weight. Sorghum stalks are not only light, but also pliable and hard and can be used to produce sorghum plywood.

4.4.1.7 Pigments from sorghum glume

Often, pericarps and glumes of sorghum are coloured. Depending on the variety, their colour can be dark to light, brown, red, yellow and white. Sorghum pigments belong to the isoxanthoketone galactoside family. In ancient China, people found glume pigment from sorghum to be very stable and useful to dye wool, cotton, linen and other textiles. Modern methods of extraction of pigments from sorghum glume have been developed experimentally and appear promising. They have been used in cosmetics and sweets by Chinese manufacturers.

4.4.2 Industrial alcohol

Ethanol may be produced from many high energy crops such as corn, wheat, barley, sugar cane, sugar beet, cassava, sweet potato and sorghum (Almodares and Hadi, 2009). It is produced as a biofuel either to be mixed with petrol-based fuels or used on its own in adapted engines.

Ethanol can be produced from sorghum grain or from the stalks of sweet sorghum. Sorghum grains have a high starch content and are therefore an important product for making alcohol. One tonne of sorghum grains can produce 390 L of alcohol. Sweet sorghum stalks contain high levels of sugar and are rich in fibre, and may also be used for making industrial alcohol with a theoretical yield of over 540 L/tonne of fresh stalk. In addition, the stalks left over after juice extraction from sweet sorghum, or bagasse, have good biological quality and are still rich in micronutrients. They can therefore be used as animal feed or organic manure (Almodares and Hadi, 2009).

All the characteristics of sorghum that make it a versatile grain for food and feed mean that it could also be a very good source of ethanol. These include the efficient conversion of solar energy, short growing season, the drought tolerance and generally low input requirements, the existence of varieties adapted to temperate, subtropical and tropical climates, as well as the potential uses that exist for all parts of the plant. But research on production of ethanol from sorghum is not yet fully mature, and optimised bioprocesses are still being developed. In a 1994 report on energy systems in China for the FAO, Nan et al. (1994) reported considerable levels of variation in sugar content and composition depending on the variety of sorghum, the location in the stem and the degree of maturity of the plant. They concluded that production of ethanol needed a lot more research and development before moving from the theoretical to the industrial stage. To some extent, this work is still in progress.

In Brazil, Barcelos et al. (2011) tested a process for ethanol production from sorghum grain which has the advantage of being similar to that used for corn, and could therefore be implemented relatively easily where plants for ethanol production from corn exist. In Thailand, Phukoetphim et al. (2017) tested three sugar levels and batch or fed-batch feeding regimes, and found that supplementation with nitrogen was beneficial. Optimised parameters for the production of sugar that can be used to scale up the process of ethanol production have been published by Lavudi et al. (2017).

Farrell et al. (2006) reported that producing ethanol from crops is beneficial compared to petroleum products, in terms of petrol consumption but similar in their greenhouse gas emissions. The authors studied ethanol production from corn, and it is probable that sorghum would fare even better. But they found that only cellulosic ethanol production, from grass, would be advantageous in terms of greenhouse gas emissions, although they point out that the technology is not yet available for industrial scaling up. It must of course be noted that the other usual concern for biofuels applies to sorghum, namely the diversion of agricultural land from food production, when forecasts indicate that food production is going to be under increasing stress due to climate change and population growth in years to come.

4.5 Sorghum spirits

4.5.1 Grain spirits

The traditional ingredient of alcoholic drinks in China is sorghum. 'Good liquor is always connected with sorghum' is a well-known adage. The following sorghum brews are well known in China:

- Maotai from Guizhou province, since 1704. It is the most famous spirit in China, served at state banquets. The key ingredient of Maotai is local high-quality sorghum.
- Fen Liquor from Shanxi province. It has the longest brewing history (about 1500 years).
- Wuliangye from Sichuan province. It is made with five types of crops (sorghum, glutinous rice, rice, maize and wheat), with sorghum making up 60% of Wuliangye.
- Jian Nan Chun also from Sichuan province. First produced 300 years ago during the Qing dynasty from sorghum, rice, maize, wheat and glutinous rice. Sorghum accounts for 40% of the recipe.
- Xi Feng from Shanxi province, made exclusively from sorghum.
- Luzhou tequ, also from Sichuan province, which has a 300-year history.

Sorghum and maize have very similar composition, but for making spirits, sorghum is better than maize. First, sorghum has a lower protein and lipid content than maize and this suitable protein content benefits distillation. Sorghum also has a small but useful tannin content (about 0.5–2.0%). Tannins inhibit harmful microorganisms and produce aromatics which give sorghum spirits their special flavour.

4.5.2 Sweet sorghum stalk spirits

The production method is similar to using sorghum grain. The advantages of processing sweet sorghum stalks on the spot are that the residue can be used to produce methane, feed or manure. Moreover, the waste residue of refining sugar also contains some sugar, which can be fermented for sorghum spirit.

4.5.3 Sorghum bran spirits

Sorghum bran is a by-product of milling representing more than 20% of the initial product. The composition of sorghum bran is appropriate for spirit making with 40–60% starch, 11–15% raw protein, 4–10% raw lipids and with a higher tannin content.

4.6 Promise of specialty sorghums

Various novelty varieties or uses have been developed, but have not yet found a successful market:

1. Popping sorghums;
2. Vegetable sorghum;
3. Panicle harvested at soft dough stage, roasted;
4. Vitamin-A-enriched types sorghum, with a 'yellow' grain;
5. Tannin-free sorghums; combined with bird resistance;
6. Aromatic sorghums, with an aroma of basmati rice;
7. Quality protein types from Ethiopian highlands;
8. Waxy starch types for cooking, pasting properties;
9. Sorgho, with a sweet-stem-like sugarcane: certain varieties of sorghum are characterised by the production of high levels of sugar in the stalk. These are known as sweet sorghums and attempts have been made to commercialise their production in the USA, Argentina and Brazil for the extraction of juice for the preparation of sugar syrup for alcohol production by fermentation. Sweet sorghum is used to a limited extent in producing sorghum syrup and 'jaggery' (raw sugar) in India and has recently gained importance in ethanol production;
10. Fuel and utility types, fibre uses;
11. Hybrids with Sudangrass are useful as fast-growing forage and green manure crops with heat and drought tolerance;
12. In Kenya and South Africa, there have been attempts at developing a small market for pearled sorghum as an alternative to rice;
13. In India, proposals have been made for use of dehulled sorghum within feeding regimes for infants and children (Pushpama, 1987);
14. Medieval Islamic texts list medical uses for the plant.

5 Requirements for export and quality assurance

The *Codex Alimentarius* Commission (1989, revised 1995) has established global standards for sorghum grains under Codex Standard 172-1989. However, individual producing countries have their own standards for internal procurement from farmers or by import. In commercial trading the quality standard for sorghum is agreed between buyer and seller and is usually associated with the intended use of the crop. Nevertheless, the principles of the Codex standard may be included within the specifications.

Summary of Codex Standard 172-1989 – The standard applies to sorghum for direct human consumption:

- Sorghum flour shall be safe and suitable for human consumption.
- Sorghum flour shall be free from abnormal flavours, odours and living insects.
- Sorghum flour shall be free from filth (impurities of animal origin, including dead insects) in amounts which may represent a hazard to human health.
- Moisture content 14.5% m/m max. Lower moisture limits should be required for certain destinations in relation to the climate, duration of transport and storage.
- Tannin content: (a) for whole sorghum grains, the tannin content shall not exceed 0.5% on a dry matter basis, (b) for decorticated sorghum grains, the tannin content shall not exceed 0.3% on a dry matter basis.

5.1 Hygiene

Grain should be prepared in accordance with the Recommended International Code of Practice, General Principles of Food Hygiene (CAP/RCP 1-1969, Rev. 2, 1985): 'Free from micro-organisms, substances originating from micro-organisms, or other poisonous substances in amounts which may represent a hazard to health.'

5.2 Packaging

Packed in containers (including sacks) which will safeguard the hygienic, nutritional and technological qualities of the grain.

5.3 Grain defects

A summary of general tolerances for grain defects is given in Table 10.

6 Conclusion and future trends

Sorghum is already an important cereal, worldwide, and it could become much more prevalent. There are two very different categories of sorghum production and post-harvest operations: on the one hand, western countries and Asia with their industrial, intensive farming using hybrid seeds and high levels of inputs, where the crop is mostly used for animal feed or industrial use. On the other, (mostly) African, small-scale, extensive, low input farming most often using farmers' kept seeds, and characterised by low yields.

Table 10 Tolerances for defects in sorghum grain

Defect	Limit	Definition
Blemished grains	3.0%	Insect or vermin damaged. Sprouted, diseased, frost damaged or other. Evidence of decay, mould or bacterial decomposition.
Including diseased grains	0.5%	
Broken kernels	5.0%	Pieces which pass through a screen with round holes that are 1.8 mm in diameter.
Other grains	1.0%	Non-sorghum – legumes, pulses, other edible cereals.
Foreign matter	2.0%	All organic and inorganic material which is not sorghum, broken kernels, other grains and filth. Includes loose sorghum seedcoats.
Including inorganic matter	0.55%	
Filth	0.1%	Impurities of animal origin.
Toxic or noxious seeds	Free from amounts which may be a hazard to health.	
Contaminants	Free from heavy metals in amounts hazardous to health.	

The post-harvest characteristics are equally very different: with industrial farming comes machine harvesting, threshing and winnowing. Storage is large scale, generally controlled in terms of RH and pests. Small-scale African farmers have, yet, no global access to harvest and post-harvest machines and store small quantities at farm levels and in stores where RH and pests are difficult to control. There is also, however, some larger-scale storage at trader level, which can allow for better control of the environment and of storage pests.

What are the possible developments for the crop in the future and what research and development are needed for these remain a question.

6.1 Post-harvest technologies

In developed economies, where sorghum is mostly grown for feed or industrial use, transformation technologies, including biotechnologies, can increase efficiency to the point of making some transformations more economically viable. This is the case with ethanol production, as discussed previously. The large variety of industrial uses for sorghum presented in this chapter means that other countries could learn from the Chinese experience and develop industrial value chains for sorghum. If this were to happen, governments and regional bodies should carefully weigh the impact that such industrial uses would have on food production.

In Africa, harvesters, threshers-winnowers (Fig. 11), milling machines and grinders as well as dryers would increase the efficiency of post-harvest operations and reduce the hard manual work which is currently prevalent. None of these technologies require brand new developments, merely adapting existing machines and technologies, for example, those already developed for higher value crops such as maize, so that they work on sorghum and become – or remain in some cases – affordable for small- to medium-scale farmers and therefore adequate for relatively small quantities of grain, and of course easy to maintain in rural settings.

Figure 11 Mechanical thresher. Photo by Bishnu Sarangi via Pixabay.

Improved dryers are arguably the most important technology needed in the field, because of the potential danger due to mycotoxins and all other damage caused by insufficiently dried grains. This is compounded by the expectation that climate change will make weather more unpredictable, and therefore traditional drying in the sun after harvest may no longer be a good option when the dry season is less defined.

Storage pest control is an issue that still is better dealt with using synthetic pesticides in a careful and reasoned manner. Appropriate pesticides need to be available and affordable, together with personal protective equipment and training on how to apply the pesticides. This requires a functioning extension system, whether private or public.

6.2 Improved market and trade

In the western world, sorghum could benefit from renewed interest in healthy foods. For example, sorghum was labelled the 'new quinoa' and a gluten-free health food to look out for in many forms, in *The Washington Post* at the end of 2016 (Brissette, 2016).

There are major constraints to improvements in the sorghum market. Clay (2013) listed several, and the following should be highlighted: In the developing world, a major constraint to agricultural development for sorghum and other crops is the lack of organised trade, which could offer a more stable market and reward consistent quality. Regional trade boards are being revived and will hopefully develop trade and therefore increase demand in a stable manner at a supra-national level. The other major constraint to overcome is infrastructure for transport and storage, in order for markets to function as smoothly as possible as development is, and has been for several decades, firmly rooted in the ideals of free market.

6.3 New or improved varieties

Breeding of new or improved varieties of sorghum could solve some current post-harvest problems as well as the more usual production issues of higher yield and reduced reliance on inputs and resistance to drought and pests. Post-harvest improvements for Africa could include, like hybrid varieties in the developed world, crops adapted to harvesting and threshing machinery, perhaps even to advanced dryers, and of course, resistance to fungi would reduce the risk of mycotoxins.

As for other important crops, an active field of research is that of genetic modification towards crops with added nutritional quality, most often added pro-vitamin A. Although the technology is mature and has delivered proven results from the research viewpoint, some question its validity compared to other investments in agriculture, in terms of adoption and sustainability. Botha and Viljoen (2008) have assessed the potential of genetic modification (GM) of grain sorghum to alleviate hunger in Africa. Research is already advancing in this domain, but the authors' analysis shows that 'GM sorghum is faced with the same limitations as "Golden Rice" (GM rice) in the context of combating vitamin A deficiency efficiently and sustainably. Thus, it is questionable whether the cost of developing GM sorghum can be justified when compared to the cost of investing in sustainable agricultural practice in Africa'.

6.4 Increased risk of environmental degradation

But a note of caution is required: in his 2013 book, Clay warns that habitat conversion and degradation from increased production is a direct consequence of the ability of sorghum to grow in areas with limited or erratic rains and hence fragile soils (Fig. 12). In addition, the use of agrochemicals would also increase with production. Although these are not primarily post-harvest issues, at that level in the value chain, the need for storage pest control also means reliance on pesticides.

Figure 12 Green sorghum field. Photo by Bishnu Sarangi via Pixabay.

Understanding these risks will be crucial if sorghum cultivation and in particular post-harvest practices are to be improved. Then, government and international policies will need to ensure that growing more sorghum does not come at an environmental price that cannot be afforded.

7 Where to look for further information

Natural Resources Institute's Food Security Department and Jianqiu Zou & Yuxue Shi of the Sorghum Institute of LAAS: INPhO sorghum factsheet for FAO: http://www.fao.org/in-action/inpho/crop-compendium/cereals-grains/en/. Accessed February 2017.

Oregon State University – World Food Crops course, chapter on sorghum: http://oregonstate.edu/instruct/css/330/six/index2.htm#Production. Accessed February 2017.

CGIAR factsheet on sorghum: (http://www.cgiar.org/our-strategy/crop-factsheets/sorghum/). Accessed February 2017.

FAOSTAT: http://www.fao.org/faostat/en/. Accessed February 2017.

8 References

Almodares, A. and Hadi, M. R. (2009). Production of bioethanol from sweet sorghum: A review. *African Journal of Agricultural Research* 4(9): 772–80.

Baoua, I. B., Amadou, L., Ousmane, B., Baributsa, D. and Murdock, L. L. (2014). PICS bags for Post-harvest storage of maize grain in West Africa. *Journal of Stored Products Research* 58: 20–8.

Barcelos, C. A., Maeda, R. N., Betancur, G. J. V. and Pereira Jr., N. (2011). Ethanol production from sorghum grains [*Sorghum bicolor* (L.) Moench]: Evaluation of the enzymatic hydrolysis and the hydrolysate fermentability. *Brazilian Journal of Chemical Engineering* 28(4): 597–604.

Bennett, W. F., Tucker, B. B. and Maunder, A. B. (1990). *Modern Grain Sorghum Production*. First Edition. Iowa State University Press, Ames, IA.

Botha, G. M. and Viljoen, C. D. (2008). Can GM sorghum impact Africa? *Trends in Biotechnology* 26(2): 64–9.

Boxall, R. A., Gebre-Tsadik, M., Jayaraj, K., Ramam, C. P., Patternaik, B. B. and Hodges, R. J. 2004. Chapter 5: Sorghum. In: Hodges, R. J. and Farrell, G. (Eds), *Crop Post-harvest: Science and Technology, Volume 2: Durables*. Wiley-Blackwell, Oxford, UK, pp. 128–44.

Brissette, C. (2016). Plant proteins, healthy fats and more 2017 food trends. *The Washington Post*. Accessed on-line December 2017: https://www.washingtonpost.com/lifestyle/wellness/checking-the-crystal-ball-for-2017-food-trends/2016/12/07/ead326ac-ac2a-11e6-8b45-f8e493f06fcd_story.html?utm_term=.d9d21aa1ab07.

Clay, J. (2013). *World Agriculture and the Environment: A Commodity-By-Commodity Guide to Impacts and Practices*. Island Press, Washington, Covelo and London (ISBN: 978-1-55963-370-3).

Codex Alimentarius Commission. (1989, revised 1995). Standard for sorghum (172-1989).

Daiber, K. H. and Taylor, J. R. N. (1995). Opaque beers. In: Dendy, D. (Ed.), *Sorghum and Millets: Chemistry and Technology*. American Association of Cereal Chemists Inc, St. Paul, MN, pp. 299–323.

Dillon, S. L., Lawrence, P. K., Henry, R. J., Ross, L., Price, H. J. and Johnston, J. S. (2004). *Sorghum laxiflorum* and *S. macrospermum*, the Australian native species most closely related to the cultivated *S. bicolor* based on ITS1 and ndhF sequence analysis of 25 Sorghum species. *Plant Systematics and Evolution* 249(3–4): 233–46.

Farrell, A. E., Plevin, R. J., Turner B. T., Jones, A. D., O'Hare, M. and Kammen, D. M. (2006). Ethanol can contribute to energy and environmental goals. *Science* 311(5760), 506–8.

Hodges, R. and Stathers, T. (2012). Training Manual for Improving Grain Postharvest Handling and Storage. *World Food Programme*, 248p. Available from: https://www.wfp.org/content/p4p-training-manual-improving-grain-post-harvest-handling-and-storage.

Hulse, J. H., Laing, E. M. and Pearson, O. E. (1980). *Sorghum and the Millets: Their Composition and Nutritive Value*. Academic Press, London.

Kange, A. M., Cheruiyot, E. K., Ogendo, J. O. and Arama, P. F. (2015). Effect of sorghum (*Sorghum bicolor* L. Moench) grain conditions on occurrence of mycotoxin-producing fungi. *Agriculture & Food Security* 4: 15.

Keller, L. A. M., Pereyra, C. M., Cavaglieri, L. R., Keller, K. M. and Almeida. T. X. (2012). Fungi and Aflatoxin B1 in Pre and Post fermented Sorghum trench type silos destined to Bovine intensive-Rearing in Brazil. *Julio* 2: 81–91 (cited in Weledesemayat et al. 2016).

Lavudi, S., Oberoi, H. S. and Mangamoori, L. N. (2017). Ethanol production from sweet sorghum bagasse through process optimization using response surface methodology. *3 Biotech* 7: 233.

Lyumugabe, F., Gros, J., Nzungize, J., Bajyana, E. and Thonart, P. (2012). Characteristics of African traditional beers brewed with sorghum malt: A review. *Biotechnologie, Agronomie, Société et Environnement* 16(4), 509–30.

Matumba, L., Monjerezi, M., Khonga, E. B. and Lakudzala, D. D. (2011). Aflatoxins in sorghum, sorghum malt and traditional opaque beer in southern Malawi. *Food Control* 22(2): 266–8.

McFarlane, J. A., John, A. E. and Marder, R. C. (1995). Storage of sorghum and millets: Including drying for storage with particular reference to tropical areas and the mycotoxin problem. In: Dendy, D. (Ed.), *Sorghum and Millet: Chemistry and Technology*. American Association of Cereal Chemists Inc, St. Paul, MN, pp. 169–83.

Nan, L., Best, G. and de Carvalho Neto, C. C. (1994). Integrated energy systems in China – The cold Northeastern region experience. FAO Report, FAO, Rome.

Navarro, S., Donahaye E., and Fishman S. (1994). The future of hermetic storage of dry grains in tropical and subtropical climates. In: Highley, E., Wright, E. J., Banks, H. J. and Champ, B. R. (Eds), *Proc. 6th Intl. Working Conf. Stored-Prod. Prot.* CAB International, Wallingford, UK, pp. 130–8.

NRI. (1998). Small scale manufacture of animal feed. Bulletin No 9, Natural Resources Institute (NRI), Chatham, UK.

Osondu, L. C., Ugwuoke, E. C., Chikelu, C. C., Eze, N. N. and Ukwuani, S. T. (2015). Performance evaluation of a grain (maize) solar dryer. *International Journal of Scientific & Technology Research* 4(9): 161–3

Pandey, A. K., Sain, S. K., Singh, P., Palni, U. T. and Tripathi, N. N. (2016). Mycotoxin menace in stored agricultural commodities and their management by plant volatiles: An overview. In: Kumar, P., Gupta, V. K., Tiwari, A. K. and Kamle, M. (Eds), *Current Trends in Plant Disease Diagnostics and Management Practices*. Springer International Publishing, Cham, Switzerland, p. 469 (ISBN: 978-3-319-27310-5).

Phukoetphim, N., Salakkam, A., Laopaiboon, P. and Laopaiboon, L. (2017) Improvement of ethanol production from sweet sorghum juice under batch and fed-batch fermentations: Effects of sugar levels, nitrogen supplementation, and feeding regimes. *Electronic Journal of Biotechnology* 26: 84–92.

Pushpama, P. (1987). Supplementary foods in India. IDRC monograph. No. 249e, pp. 69–77

Salvatierra-Rojas A., Nagle M., Gummert M., de Bruin T. and Müller J. (2017). Development of an inflatable solar dryer for improved post-harvest handling of paddy rice in humid climates. *International Journal of Agricultural and Biological Engineering* 10(3): 269–82.

Suleiman, M. and Rugumamu, C. P. (2017). Management of insect pests of stored sorghum using botanicals in Nigerian traditional stores. *Journal of Stored Products and Post-harvest Research* 8(9): 93–102.

Umogbai, V. I. and Iorter, H. A. (2013). Design, construction and performance evaluation of a passive solar dryer for maize cobs. *African Journal of Food Science and Technology* 4(5): 110–15 (ISSN: 2141–5455).

Weledesemayat, G. T., Gezmu, T. B., Woldegiorgis, A. Z. and Gemede, H. F. (2016). Study on Aspergillus species and Aflatoxin Levels in Sorghum (*Sorghum bicolor* L.) stored for different period and storage system in Kewet Districts, Northern Shewa, Ethiopia. *Journal of Food Sciences and Nutrition* 2: 1–8.

Wood, J.F. and Thomson, M. (1997). Natural Resources Institute (NRI). Chatham, Kent, England, UK.

Zhao, S. K. (1987). China traditional sorghum food. *Liaoning Agricultural Sciences* 2: 55–7.

Index

www.ingramcontent.com/pod-product-compliance
Lightning Source LLC
Chambersburg PA
CBHW071251180926
51613CB00001B/6